MATHEMATICAL SURVEYS AND MONOGRAPHS SERIES LIST

Volume

1 The problem of moments, J. A. Shohat and J. D. Tamarkin
2 The theory of rings, N. Jacobson
3 Geometry of polynomials, M. Marden
4 The theory of valuations, O. F. G. Schilling
5 The kernel function and conformal mapping, S. Bergman
6 Introduction to the theory of algebraic functions of one variable, C. C. Chevalley
7.1 The algebraic theory of semigroups, Volume I, A. H. Clifford and G. B. Preston
7.2 The algebraic theory of semigroups, Volume II, A. H. Clifford and G. B. Preston
8 Discontinuous groups and automorphic functions, J. Lehner
9 Linear approximation, Arthur Sard
10 An introduction to the analytic theory of numbers, R. Ayoub
11 Fixed points and topological degree in nonlinear analysis, J. Cronin
12 Uniform spaces, J. R. Isbell
13 Topics in operator theory, A. Brown, R. G. Douglas, C. Pearcy, D. Sarason, A. L. Shields; C. Pearcy, Editor
14 Geometric asymptotics, V. Guillemin and S. Sternberg
15 Vector measures, J. Diestel and J. J. Uhl, Jr.
16 Symplectic groups, O. Timothy O'Meara
17 Approximation by polynomials with integral coefficients, Le Baron O. Ferguson
18 Essentials of Brownian motion and diffusion, Frank B. Knight
19 Contributions to the theory of transcendental numbers, Gregory V. Chudnovsky
20 Partially ordered abelian groups with interpolation, Kenneth R. Goodearl
21 The Bieberbach conjecture: Proceedings of the symposium on the occasion of the proof, Albert Baernstein, David Drasin, Peter Duren, and Albert Marden, Editors
22 Noncommutative harmonic analysis, Michael E. Taylor
23 Introduction to various aspects of degree theory in Banach spaces, E. H. Rothe
24 Noetherian rings and their applications, Lance W. Small, Editor
25 Asymptotic behavior of dissipative systems, Jack K. Hale
26 Operator theory and arithmetic in H^∞, Hari Bercovici
27 Basic hypergeometric series and applications, Nathan J. Fine
28 Direct and inverse scattering on the lines, Richard Beals, Percy Deift, and Carlos Tomei
29 Amenability, Alan L. T. Paterson
30 The Markoff and Lagrange spectra, Thomas W. Cusick and Mary E. Flahive

MATHEMATICAL SURVEYS AND MONOGRAPHS SERIES LIST

Volume

31 **Representation theory and harmonic analysis on semisimple Lie groups,** Paul J. Sally, Jr. and David A. Vogan, Jr., Editors

32 **An introduction to CR structures,** Howard Jacobowitz

33 **Spectral theory and analytic geometry over non-Archimedean fields,** Vladimir G. Berkovich

34 **Inverse source problems,** Victor Isakov

35 **Algebraic geometry for scientists and engineers,** Shreeram S. Abhyankar

The Theory of Subnormal Operators

MATHEMATICAL
Surveys and Monographs

Volume 36

The Theory of Subnormal Operators

John B. Conway

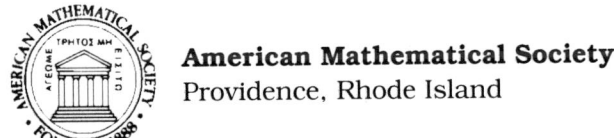

American Mathematical Society
Providence, Rhode Island

1980 *Mathematics Subject Classification* (1985 *Revision*). Primary 47B20; Secondary 32A35.

Library of Congress Cataloging-in-Publication Data

Conway, John B.
 The theory of subnormal operators/John B. Conway.
  p. cm.—(Mathematical surveys and monographs, ISSN 0076-5376; no. 36)
  Includes bibliographical references and indexes.
  ISBN 0-8218-1536-9 (acid-free)
  1. Subnormal operators. I. Title. II. Series.
QA329.2.C667   19910   90-26659
515'.7246—dc20   CIP

Copying and reprinting. Individual readers of this publication, and nonprofit libraries acting for them, are permitted to make fair use of the material, such as to copy a chapter for use in teaching or research. Permission is granted to quote brief passages from this publication in reviews, provided the customary acknowledgment of the source is given.

Republication, systematic copying, or multiple reproduction of any material in this publication (including abstracts) is permitted only under license from the American Mathematical Society. Requests for such permission should be addressed to the Manager of Editorial Services, American Mathematical Society, P.O. Box 6248, Providence, Rhode Island 02940-6248.

The owner consents to copying beyond that permitted by Sections 107 or 108 of the U.S. Copyright Law, provided that a fee of $1.00 plus $.25 per page for each copy be paid directly to the Copyright Clearance Center, Inc., 27 Congress Street, Salem, Massachusetts 01970. When paying this fee please use the code 0076-5376/91 to refer to this publication. This consent does not extend to other kinds of copying, such as copying for general distribution, for advertising or promotional purposes, for creating new collective works, or for resale.

Copyright ©1991 by the American Mathematical Society. All rights reserved.
Printed in the United States of America
The American Mathematical Society retains all rights
except those granted to the United States Government.
The paper used in this book is acid-free and falls within the guidelines
established to ensure permanence and durability. ∞
This publication was typeset using $\mathcal{A}_\mathcal{M}\mathcal{S}$-TEX,
the American Mathematical Society's TEX macro system.

10 9 8 7 6 5 4 3 2 1   95 94 93 92 91

*For Ann,
my favorite collaborator,
and
Bligh,
our best theorem.*

Contents

Preface	xiii
Chapter I. Preliminaries	1
§1 Trace class and Hilbert-Schmidt operators	1
§2 Some topologies on $\mathcal{B}(\mathcal{H})$	10
§3 Isometries	13
§4 The unilateral shift of multiplicity 1	18
§5 Direct sums	24
§6 Shifts of higher multiplicity	26
Chapter II. Subnormal Operators: The Elementary Theory	27
§1 Definition and examples	27
§2 Pure operators and the minimal normal extension	37
§3 Quasinormal operators	43
§4 Hyponormal operators	46
§5 Cyclic subnormal operators	50
§6 Weighted shifts	53
§7 Bounded point evaluations	61
§8 Bergman operators	66
§9 Spectral sets	75
§10 The commutant of a subnormal operator	79
§11 The restriction algebra and the functional calculus	85
§12 The C^*-algebra generated by a subnormal operator	88
§13 Unitary equivalence, similarity, and quasisimilarity	93
Chapter III. Function Theory On The Unit Circle	99
§1 Cesàro sums	99
§2 Convolution on the circle	103
§3 Harmonic functions on the disk	105
§4 Fatou's Theorem	108
§5 Subharmonic functions	114
§6 Hardy spaces	117
§7 The Nevanlinna class	120
§8 Factorization of functions in Nevanlinna class	125

§9 The disk algebra 132
§10 The invariant subspace lattice of the unilateral shift 135
§11 Weak-star closed ideals in H^∞ 140
§12 Szegö's Theorem 141
§13 Analytic Toeplitz operators 145

Chapter IV. Hyponormal Operators 149
§1 The real part of a hyponormal operator 150
§2 The Berger-Shaw Theorem 152
§3 The area of the spectrum of a hyponormal operator 156
§4 The self-commutator of the Bergman operator 156
§5 The decomposition of essentially normal operators 158

Chapter V. Uniform Rational Approximation 163
§1 Function algebras: examples and elementary properties 163
§2 Distributions and some results from analysis 166
§3 The Cauchy transform 173
§4 Invariant subspaces for subnormal operators 181
§5 Vitushkin localization operators 184
§6 T-invariant algebras 187
§7 The Shilov boundary 190
§8 Representing measures 193
§9 Harmonic measure 196
§10 Hardy spaces for an arbitrary region 203
§11 Peak points 211
§12 Capacity 216
§13 Some applications of analytic capacity 222
§14 Dirichlet algebras 226
§15 Gleason parts 234
§16 The Wermer Embedding Theorem 239
§17 Bands of measures 244
§18 Annihilating measures 250
§19 Mergelyan's Theorem 253
§20 The double dual of a T-invariant algebra 254
§21 The Lautzenheiser-Mlak Theorem 260
§22 Davie's Theorem 266

Chapter VI. Weak-Star Rational Approximation 277
§1 Weak-star closed subalgebras of $L^\infty(\mu)$ 277
§2 The envelope 282
§3 Chaumat's Theorem 287
§4 $H^\infty(\partial K)$ for a Dirichlet algebra 291
§5 Dirichlet algebras revisited 296
§6 The weak-star closure of a Dirichlet algebra 298
§7 The weak-star closure of the polynomials 301

Chapter VII. Some Structure Theory For Subnormal Operators	309
§1 A decomposition of subnormal operators	309
§2 The minimal normal extension problem for subnormal operators	315
§3 Spectral mapping theorems	324
§4 Spectral mapping theorems for the essential spectrum	333
§5 A factorization theorem	341
§6 An infinite factorization theorem	352
§7 Full analytic subspaces	357
§8 Reflexivity for subnormal operators	361
§9 Filling in the holes of the spectrum of a normal operator	365
§10 Quasisimilarity revisited	368
Chapter VIII. Bounded Point Evaluations	375
§1 A coloring scheme	375
§2 A sufficient condition for the existence of bounded point evaluations	384
§3 Heavy barriers exist	389
§4 Bounded point evaluations exist	397
§5 Thomson's Theorem	399
§6 Some applications of Thomson's Theorem	407
Epilogue	409
Bibliography	411
Index	431
List of symbols	435

Preface

This book is a successor to Conway [**1981a**]. In addition to reflecting the great strides in the development of subnormal operator theory since that first set of notes, the present work is oriented toward rational functions rather than polynomials. This necessitates additional function theory and a knowledge of analytic capacity, both of which are developed here. There are, however, some crucial estimates and applications of capacity to which I have only made references.

I began the preface of Conway [**1981a**] by asking the question, "What is mathematics?" I went on to explain that my definition of mathematics is that it is a collection of examples as opposed to a body of theorems. This definition remains and is still reflected in the nature of this book. I believe mathematics is a collection of examples irrespective of the particular area and that the good theorems are those that explain, classify, and interpret large classes of examples.

Paul R. Halmos introduced the concept of a subnormal operator at the same time that he defined hyponormal operators. He was led to do so by a study of the properties of the unilateral shift, the most understood non-normal operator. In a certain sense, subnormal operators were introduced too soon because the theory of function algebras and rational approximation was also in its infancy and could not be properly used to examine this class of operators. The progress in the theory of subnormal operators that has come about during the last several years grew out of applying the results of rational approximation.

This book is a research monograph but it has many of the traits of a textbook. There are exercises, some routine and some based on research papers, with most of the necessary background information contained here. The last two chapters are the objective of the book and bring us to the latest developments. Except for certain topics in subnormal operator theory that are just beginning to surface, this monograph gives the latest state of the art.

A background in function theory and functional analysis is required here. The first two entries in the bibliography are appropriate prerequisites. Indeed references are made to them in this different format, [**ACFA**] and [**FOCV**], throughout the book. In particular, the reader is assumed to have some familiarity with the Fredholm index. (See Chapter XI of [**ACFA**].) Unlike Conway [**1981a**], there is no chapter on normal operators here because this is covered in [**ACFA**]. Chapter I contains some additional background material.

Chapter II gives the basic theory of subnormal operators; that is, the results that do not depend in a significant way on results from rational approximation. Chapter III is devoted to the study of analytic functions on the unit disk and is an attempt to make the book self-contained. Not all of this material is needed for the book and the reader who is so inclined might want to steer a course through this chapter that minimizes the time spent here.

Chapter IV presents some results about hyponormal operators that are of great interest. The recent monograph of Martin and Putinar [1989] gives a thorough treatment of hyponormality.

Chapter V gives an exposition of rational approximation interspersed with applications to operator theory. In particular, §V.4 gives James Thomson's proof of Scott Brown's result that subnormal operators have nontrivial invariant subspaces. It might be taken as a sign of progress in the subject that this theorem was the culmination of Conway [1981a] and is now accessible in an early chapter.

Chapter VI studies weak-star rational approximation and proves Chaumat's Theorem characterizing the weak* closure of spaces of rational functions. Sarason's characterization of the weak-star closure of the polynomials follows as a corollary. This is a fuller explanation of the opening remark that this book is oriented toward rational functions while the first was oriented toward polynomials.

Chapter VII presents some results that can be termed structure theorems for subnormal operators. The rational functional calculus is explored, the existence of analytic subspaces is established, and this is all used to show that subnormal operators are reflexive.

Chapter VIII gives James Thomson's proof that analytic bounded point evaluations exist.

In the Epilogue we allude to several topics that are currently being worked on that are not covered in this book.

When dealing with a result on subnormal operators that is significant, I have, to the best of my knowledge, given the origins of the result. I am well aware that the probability of error in these matters is high. Let me apologize in advance to anyone who feels slighted and invite them to inform me so that a correction can be made in case a further printing of the book comes into existence. In the material not directly related to subnormal operator theory I have not been so scholarly.

I have a tendency to state exercises as I think they should be and this often results in inaccuracies. (Actually some of my favorite work started with an incorrect problem assigned to a class.) The reader, therefore, should approach the exercises with some skepticism. Something approximately like each exercise is correct. In any case it is part of the nature of research to find out what is right as well as to prove it.

Several people deserve my gratitude in varying degrees. Three people who helped a lot were John Akeroyd, Jim Dudziak, and Paul McGuire who carefully read some sections of the book and made thoughtful comments and suggestions. Paul deserves additional thanks for helping me compile a list of corrections to Conway [1981a] while he was a student. I must also thank

Jim Thomson for some very helpful correspondence during my writing of Chapter VII. Some others who have made comments on this book as well as on Conway [1981a] are Ameer Athavale, R. B. Burckel, Alp Eden, Norma Elias, Jinchuan Hou, Kyung Hee Jin, Dimitri Khavinson, John McCarthy, Bob Olin, Marc Raphael, Tavan Trent, and Derek Westwood.

Let me also thank Indiana University. Part of this work was done while I was on a sabbatical leave. The Mathematics Department there also allowed me to teach a course in which I presented much of this material.

Finally, I want to thank the two people to whom this book is dedicated. My son, Bligh, helped me with a number of tasks involved with the preparation of the final manuscript, including the index and bibliography. For this I thank him and also for being himself, a most interesting character. Ann, as always, has been Ann.

CHAPTER I

Preliminaries

In this chapter a few results are presented that are not standard in a course in functional analysis. To be sure, every definition of "standard" is a local one and is even time dependent. To compensate for this we will take the material in [ACFA] as standard, or, if you will, prerequisite, for this book. Included there and required here is some knowledge of the Fredholm index and spectral theory and a rather complete knowledge of the Spectral Theorem for bounded normal operators. For convenience sake, there will be some overlap between this book and [ACFA]; in particular, some notation will be redefined here.

§1 Trace class and Hilbert-Schmidt operators. Throughout this book, all Hilbert spaces are assumed to be separable and complex and $\mathscr{B} = \mathscr{B}(\mathscr{H})$ denotes the algebra of all bounded operators from $\mathscr{H}$ into itself. We begin with an elementary proposition.

1.1 Proposition. *If $\{e_n\}$ and $\{f_m\}$ are orthonormal bases for $\mathscr{H}$ and $A \in \mathscr{B}(H)$, then*

$$\sum_n \|Ae_n\|^2 = \sum_m \|A^* f_m\|^2 = \sum_n \sum_m |\langle Ae_n, f_m \rangle|^2.$$

(*Note*: This result is to be interpreted as stating that one of these infinite sums converges if and only if all of them do, in which case the three sums are equal.)

PROOF. It follows from Parseval's Identity that for each n, $\|Ae_n\|^2 = \sum_m |\langle Ae_n, f_m \rangle|^2$. Also for every m, $\|A^* f_m\|^2 = \sum_n |\langle e_n, A^* f_m \rangle|^2$. The result follows. □

For any operator A, define $|A|$ to be the unique positive square root of A^*A. Applying the preceding proposition to the square root of $|A|$ we obtain the following.

1.2 Corollary. *The sum $\sum_n \langle |A|e_n, e_n \rangle$ is independent of the choice of basis $\{e_n\}$.*

1.3 Definition. An operator $A: \mathscr{H} \to \mathscr{H}$ is *trace class* if there is a basis $\{e_n\}$ such that $\sum_n \langle |A|e_n, e_n \rangle < \infty$. The set of trace class operators on $\mathscr{H}$ is denoted by $\mathscr{B}_1(\mathscr{H})$, or $\mathscr{B}_1$ if the space $\mathscr{H}$ is understood.

In light of Corollary 1.2, an operator A is trace class if and only if $\sum_n \langle |A|e_n, e_n \rangle$ is finite for every choice of orthonormal basis $\{e_n\}$ and

$$\|A\|_1 \equiv \sum_n \langle |A|e_n, e_n \rangle$$

is well defined, depending only on the operator A. The number $\|A\|_1$ is called the *trace norm* of A. It will be shown below that $\mathscr{B}_1$ is a vector space, the trace norm does indeed define a norm on $\mathscr{B}_1$, and with respect to this norm $\mathscr{B}_1$ is a Banach space. But first some preliminary results must be established and another class of operators must be introduced.

1.4 DEFINITION. An operator A on $\mathscr{H}$ is called a *Hilbert-Schmidt operator* if $|A|^2$ is trace class. The set of all Hilbert-Schmidt operators is denoted by $\mathscr{B}_2(\mathscr{H})$, or $\mathscr{B}_2$ if the space $\mathscr{H}$ is understood. For any Hilbert-Schmidt operator A define

$$\|A\|_2 \equiv \left[\sum_n \langle |A|^2 e_n, e_n \rangle \right]^{1/2} = \| |A|^2 \|_1^{1/2}.$$

For any $p > 0$, the *Schatten p-class* $\mathscr{B}_p = \mathscr{B}_p(\mathscr{H})$ is defined as those operators A such that $|A|^p \in \mathscr{B}_1$. For $p = 1$ and 2, the Schatten p-class is precisely the trace class and the Hilbert-Schmidt operators, respectively. We will only be interested in these two classes, but there is a substantial literature on all of them. See Dunford and Schwartz [1963], Ringrose [1971], and Schatten [1960] for instance.

There is an analogy between the spaces $\mathscr{B}_p$ and ℓ^p. This analogy will become more dramatic as this section progresses. The example below demonstrates that there is a little more than analogy going on here and perhaps we should consider $\mathscr{B}_p$ as the "nonabelian" version of ℓ^p.

1.5 EXAMPLE. Fix an orthonormal basis $\{e_n\}$ for $\mathscr{H}$ and consider the diagonal operator $Ae_n = \alpha_n e_n$, where $\{\alpha_n\} \in \ell^\infty$. Then $A \in \mathscr{B}_p$ if and only if $\{\alpha_n\} \in \ell^p$.

It turns out that it is somewhat easier to verify certain elementary properties of $\mathscr{B}_2$ than the analogous properties for $\mathscr{B}_1$.

1.6 PROPOSITION. *Fix an operator A in $\mathscr{B}_2(\mathscr{H})$.*

(a) $\|A\|_2 = [\sum_n \|Ae_n\|^2]^{1/2}$ *for any basis $\{e_n\}$.*
(b) $\|A^*\|_2 = \|A\|_2$.
(c) $\|A\| \leq \|A\|_2$.
(d) *If $T \in \mathscr{B}(\mathscr{H})$, then AT and $TA \in \mathscr{B}_2(\mathscr{H})$ and $\|AT\|_2$ and $\|TA\|_2$ $\leq \|T\| \|A\|_2$.*
(e) $\mathscr{B}_2(\mathscr{H})$ *is an ideal of $\mathscr{B}(\mathscr{H})$ and $\|\cdot\|_2$ is a norm on $\mathscr{B}_2(\mathscr{H})$.*

PROOF. The proof of part (a) is an easy exercise. The proof of (b) follows by combining (a) and Proposition 1.1.

To prove (c), fix a unit vector e and choose an orthonormal basis $\{e_n\}$ with $e_1 = e$. From (a) it follows that $\|Ae\| \leq \|A\|_2$. Since e was arbitrary, part (c) follows.

If $\{e_n\}$ is any basis and $T \in \mathscr{B}(\mathscr{H})$, then $\|TAe_n\|^2 \leq \|T\|^2 \|Ae_n\|^2$. Thus $TA \in \mathscr{B}_2$ and $\|TA\|_2 \leq \|T\| \|A\|_2$. Using (b), $T^*A^* \in \mathscr{B}_2$ and so $AT \in \mathscr{B}_2$ with $\|AT\|_2 \leq \|T\| \|A\|_2$. This proves (d).

Now let A and $B \in \mathscr{B}_2$; so $\{\|Ae_n\|\}$ and $\{\|Be_n\|\} \in \ell^2$. Using the triangle inequality for l^2, $[\sum(\|Ae_n\| + \|Be_n\|)^2]^{1/2} \leq \|A\|_2 + \|B\|_2$. Hence $\|A + B\|_2^2 = \sum \|Ae_n + Be_n\|^2 \leq \sum(\|Ae_n\| + \|Be_n\|)^2 \leq (\|A\|_2 + \|B\|_2)^2 < \infty$. Thus $A + B \in \mathscr{B}_2$ and $\|A + B\|_2 \leq \|A\|_2 + \|B\|_2$. This essentially completes the proof of (e), the remainder of the proof being left to the reader. □

1.7 COROLLARY. *Every Hilbert-Schmidt operator is compact.*

PROOF. Since the identity operator is not in $\mathscr{B}_2$, $\mathscr{B}_2 \neq \mathscr{B}$. Since $\mathscr{B}_2$ is a proper ideal in $\mathscr{B}$, $\mathscr{B}_2 \subseteq \mathscr{B}_0$, the ideal of compact operators. (See IX.4.3 in [ACFA].) □

It is not difficult to give a direct proof of the preceding corollary (see Exercise 2). This has some merit because the proof above is only valid for separable Hilbert spaces, whereas the corollary is valid in arbitrary spaces. (How do you define Hilbert-Schmidt operators on a nonseparable space?)

We can now apply our knowledge of $\mathscr{B}_2$ to learn something of $\mathscr{B}_1$. The next result forms the bridge between the two classes. (More information on $\mathscr{B}_2$ will be obtained later.)

Recall that when we say that $A = W|A|$ is the polar decomposition of A, we mean that W is a partial isometry with initial space $\text{cl}[\text{ran }A^*] = [\ker A]^\perp$ and final space $\text{cl}[\text{ran }A]$ and that with these restrictions, W is unique (see VII.3.11 in [ACFA]). Also, it follows that if $A = W|A|$ is the polar decomposition of A, then $W^*A = |A|$.

1.8 PROPOSITION. *If $A \in \mathscr{B}(\mathscr{H})$, the following statements are equivalent.*

(a) $A \in \mathscr{B}_1(\mathscr{H})$.
(b) $|A|^{1/2} \in \mathscr{B}_2$.
(c) A *is the product of two Hilbert-Schmidt operators.*
(d) $|A|$ *is the product of two Hilbert-Schmidt operators.*

PROOF. Let $A = W|A|$ be the polar decomposition of A.

(a) *implies* (b). For any vector e, $\||A|^{1/2}e\|^2 = \langle |A|e, e \rangle$.

(b) *implies* (c). Here $A = (W|A|^{1/2})(|A|^{1/2})$ and from (b) and Proposition 1.6 both of these factors are in $\mathscr{B}_2$.

(c) *implies* (d). If $A = BC$ and B and C are in $\mathscr{B}_2$, then $|A| = (W^*B)C$. By Proposition 1.6, $W^*B \in \mathscr{B}_2$.

(d) *implies* (a). Suppose $|A| = BC$, where B and $C \in \mathscr{B}_2$. For any orthonormal basis $\{e_n\}$, $\langle |A|e_n, e_n \rangle = \langle Ce_n, B^*e_n \rangle \leq \|Ce_n\| \|B^*e_n\|$. Hence

$\sum \langle |A|e_n, e_n \rangle \leq \sum \|Ce_n\| \|B^*e_n\| \leq [\sum \|Ce_n\|^2]^{1/2} [\sum \|B^*e_n\|^2]^{1/2} = \|C\|_2 \|B\|_2$. □

1.9 PROPOSITION. *If $A \in \mathscr{B}_1$ and $\{e_n\}$ is an orthonormal basis, then $\sum |\langle Ae_n, e_n \rangle| < \infty$ and $\sum \langle Ae_n, e_n \rangle$ is independent of the choice of basis.*

PROOF. Since $A \in \mathscr{B}_1$, we can write $A = C^*B$, where B and C are Hilbert-Schmidt operators. Since $\|(B - \lambda C)e\|^2 \geq 0$ for any choice of scalar λ, it follows that $|\langle Ae, e \rangle| = |\langle Be, Ce \rangle| \leq \frac{1}{2}[\|Be\|^2 + \|Ce\|^2]$. So for an orthonormal basis $\{e_n\}$, $\sum |\langle Ae_n, e_n \rangle| \leq \frac{1}{2}[\|B\|_2^2 + \|C\|_2^2] < \infty$.

To see that $\sum \langle Ae_n, e_n \rangle$ is independent of the choice of a basis, observe that $\operatorname{Re}\langle Ae_n, e_n \rangle = \frac{1}{4}[\|(B+C)e_n\|^2 - \|(B-C)e_n\|^2]$. Hence $\operatorname{Re}\sum \langle Ae_n, e_n \rangle = \frac{1}{4}[\|B+C\|_2^2 - \|B-C\|_2^2]$. Replacing A by iA gives $\operatorname{Im}\sum\langle Ae_n, e_n\rangle = \frac{1}{4}[\|iB+C\|_2^2 - \|iB-C\|_2^2]$. The independence of $\sum\langle Ae_n, e_n\rangle$ from the choice of a basis is now immediate. □

In light of the preceding proposition we can define the trace of an operator in $\mathscr{B}_1$.

1.10 DEFINITION. If $A \in \mathscr{B}_1(\mathscr{H})$ and $\{e_n\}$ is an orthonormal basis, define

$$\operatorname{tr} A \equiv \sum_n \langle Ae_n, e_n \rangle.$$

This number, $\operatorname{tr} A$, is called the *trace* of A.

1.11 THEOREM

(a) *$\mathscr{B}_1(\mathscr{H})$ is an ideal of $\mathscr{B}(\mathscr{H})$ and $\|\cdot\|_1$ is a norm on $\mathscr{B}_1(\mathscr{H})$.*

(b) *Every trace class operator is compact. If A is a compact operator and $\alpha_1, \alpha_2, \ldots$ are the eigenvalues of $|A|$, each repeated as often as its multiplicity, then $A \in \mathscr{B}_1$ if and only if $\{\alpha_n\} \in \ell^1$ and in this case $\|A\|_1 = \sum \alpha_n$.*

(c) *$\operatorname{tr}: \mathscr{B}_1 \to \mathbb{C}$ is a positive definite linear functional. That is, if $A \in \mathscr{B}_1$, $A \geq 0$, and $A \neq 0$, then $\operatorname{tr} A > 0$.*

(d) *The normed space $(\mathscr{B}_1, \|\cdot\|_1)$ contains the finite rank operators, $\mathscr{B}_{00}$, as a dense linear manifold. Hence $\mathscr{B}_1$ is a separable normed space.*

(e) *If $A \in \mathscr{B}_1$ and $T \in \mathscr{B}(\mathscr{H})$, then $\operatorname{tr}(AT) = \operatorname{tr}(TA)$ and $|\operatorname{tr}(T|A|)| \leq \|T\| \|A\|_1$.*

(f) *$\|A\|_1 = \|A^*\|_1$ for every A in $\mathscr{B}_1(\mathscr{H})$.*

(g) *If $A \in \mathscr{B}_1$ and $T \in \mathscr{B}(\mathscr{H})$, then $\|AT\|_1$ and $\|TA\|_1 \leq \|T\| \|A\|_1$.*

PROOF. (a) Let A and $B \in \mathscr{B}_1$ and let $A = W|A|$, $B = V|B|$, and $A+B = U|A+B|$ be the respective polar decompositions. Since each operator in $\mathscr{B}_1$ is the product of two Hilbert-Schmidt operators, $\mathscr{B}_1 \subseteq \mathscr{B}_0$. Hence $|A+B|$ is a positive compact operator; let $|A+B| = \sum \gamma_n e_n \otimes e_n$ be the diagonalization of $|A+B|$, where for any vectors e and f in $\mathscr{H}$, $e \otimes f$ denotes the rank one operator defined by

(1.12) $$e \otimes f(h) = \langle h, f \rangle e,$$

for all h in $\mathscr{H}$. Thus (remember to justify the manipulation of the infinite sums that follow since it is not known initially that these sums are convergent)

$$\begin{aligned}
\sum \gamma_n &= \sum \langle |A+B|e_n, e_n \rangle \\
&= \sum \langle (A+B)e_n, Ue_n \rangle \\
&= \sum [\langle Ae_n, Ue_n \rangle + \langle Be_n, Ue_n \rangle] \\
&= \sum [\langle |A|e_n, W^*Ue_n \rangle + \langle |B|e_n, V^*Ue_n \rangle] \\
&= \sum [\langle |A|^{1/2}e_n, |A|^{1/2}W^*Ue_n \rangle + \langle |B|^{1/2}e_n, |B|^{1/2}V^*Ue_n \rangle] \\
&\leq \sum [\| |A|^{1/2}e_n \| \| |A|^{1/2}W^*Ue_n \| \\
&\quad + \| |B|^{1/2}e_n \| \| |B|^{1/2}V^*Ue_n \|] \\
&\leq \left[\sum \| |A|^{1/2}e_n \|^2\right]^{1/2} \left[\sum \| |A|^{1/2}W^*Ue_n \|^2\right]^{1/2} \\
&\quad + \left[\sum \| |B|^{1/2}e_n \|^2\right]^{1/2} \left[\sum \| |B|^{1/2}V^*Ue_n \|^2\right]^{1/2} \\
&\leq \| |A|^{1/2} \|_2^2 + \| |B|^{1/2} \|_2^2 \\
&= \|A\|_1 + \|B\|_1.
\end{aligned}$$

This shows simultaneously that $A+B \in \mathscr{B}_1$ if A and $B \in \mathscr{B}_1$ and that $\|\cdot\|_1$ satisfies the triangle inequality. The remainder of the proof that $\|\cdot\|_1$ is a norm is easy.

If $A = XY$ for X and Y in $\mathscr{B}_2$ and $T \in \mathscr{B}(\mathscr{H})$, then $TA = (TX)Y \in \mathscr{B}_1$; similarly, $AT \in \mathscr{B}_1$. This shows that $\mathscr{B}_1$ is an ideal in $\mathscr{B}(\mathscr{H})$ and completes the proof of (a).

(b) Since each trace class operator is the product of two Hilbert-Schmidt operators, $\mathscr{B}_1 \subseteq \mathscr{B}_2 \subseteq \mathscr{B}_0$.

Now assume that A is compact and let $A = W|A|$ be the polar decomposition of A. Since $|A|$ is a compact positive operator, the Spectral Theorem implies that $|A|$ can be diagonalized; let $\{e_n\}$ be an orthonormal basis $\{e_n\}$ for $\mathscr{H}$ such that $|A| = \sum \alpha_n e_n \otimes e_n$, $\alpha_n \geq 0$. If $A \in \mathscr{B}_1$, then $\infty > \operatorname{tr}(|A|) = \sum \langle |A|e_n, e_n \rangle = \sum \alpha_n$. Conversely, if $\sum \alpha_n < \infty$, then it is easily seen that $|A|$ is trace class and thus, since $\mathscr{B}_1$ is an ideal, $A = W|A| \in \mathscr{B}_1$.

(c) It is clear that the trace is a linear functional on $\mathscr{B}_1$ that is positive. Moreover, if A is a positive trace class operator with diagonalization $A = \sum \alpha_n e_n \otimes e_n$, then $\operatorname{tr} A = \sum \alpha_n$. Hence $\operatorname{tr} A = 0$ implies that $A = 0$.

(d) If $A \in \mathscr{B}_1$ and $A = W|A|$ is its polar decomposition, let $|A| = \sum \alpha_k e_k \otimes e_k$ be the diagonalization of $|A|$. If $A_n = W \sum \{\alpha_k e_k \otimes e_k : 1 \leq k \leq n\}$, then $A - A_n = W \sum \{\alpha_k e_k \otimes e_k : k \geq n+1\}$ is the polar decomposition of $A - A_n$. By part (b), $\|A - A_n\|_1 = \sum \{\alpha_k : k \geq n+1\}$ and so $A_n \to A$ in $\mathscr{B}_1$. Since each $A_n \in \mathscr{B}_{00}$, this proves (d).

(e) Let $A \in \mathscr{B}_1$ and write $A = C^*B$ with B and C in $\mathscr{B}_2$. As in the proof of Proposition 1.9, $\operatorname{Re} \operatorname{tr} A = \operatorname{Re} \operatorname{tr}(C^*B) = \frac{1}{4}[\|B + C\|_2^2 - \|B - C\|_2^2] = \frac{1}{4}[\|B^* + C^*\|_2^2 - \|B^* - C^*\|_2^2] = \operatorname{Re} \operatorname{tr}(CB^*)$. Replacing B by iB shows that $\operatorname{Im} \operatorname{tr}(C^*B) = -\operatorname{Im} \operatorname{tr}(CB^*)$. Hence

$$\operatorname{tr}(C^*B) = [\operatorname{tr}(CB^*)]^*.$$

(For a complex number α, α^* will sometimes be used to denote the complex conjugate of α.) So for any T in $\mathscr{B}(\mathscr{H})$, $\operatorname{tr}(TA) = \operatorname{tr}((TC^*)B) = [\operatorname{tr}((CT^*)B^*)]^* = [\operatorname{tr}(C(BT)^*)]^* = \operatorname{tr}(C^*(BT)) = \operatorname{tr}(AT)$.

For the second part of (e), note that $T|A|^{1/2}$ and $|A|^{1/2}T \in \mathscr{B}_2$ and $\operatorname{tr}(T|A|) = \sum \langle |A|^{1/2}e_n, |A|^{1/2}T^*e_n \rangle$. Applying the Cauchy-Schwarz Inequality gives

$$|\operatorname{tr}(T|A|)| \leq \sum |\langle |A|^{1/2}e_n, |A|^{1/2}T^*e_n \rangle|$$
$$\leq \sum \||A|^{1/2}e_n\| \||A|^{1/2}T^*e_n\|$$
$$\leq \left[\sum \||A|^{1/2}e_n\|^2\right]^{1/2} \left[\sum \||A|^{1/2}T^*e_n\|^2\right]^{1/2}$$
$$= \||A|^{1/2}\|_2 \||A|^{1/2}T^*\|_2.$$

By Proposition 1.6, $|\operatorname{tr}(T|A|)| \leq \||A|^{1/2}\|_2^2 \|T^*\| = \||A|^{1/2}\|_2^2 \|T\|$. But $\|A\|_1 = \||A|^{1/2}\|_2^2$, so the proof of (e) is complete.

(f) If $A \in \mathscr{B}_1$ and $A = W|A|$ is the polar decomposition of A, then $AA^* = W|A|^2W^*$ and so by uniqueness of the square root $|A^*| = W|A|W^*$. Hence $\|A^*\|_1 = \operatorname{tr}(|A^*|) = \operatorname{tr}(W|A|W^*)$. But by part (e) this last quantity is just $\operatorname{tr}(W^*W|A|)$ and $W^*W|A| = |A|$ since W^*W is the projection of $\mathscr{H}$ onto $\operatorname{cl}[\operatorname{ran}|A|]$.

(g) Let $A = W|A|$ and $TA = W_1|TA|$ be the respective polar decompositions; so $|TA| = S|A|$, where $S = W_1^*TW$ and hence $\|S\| \leq \|T\|$. From part (d) we get that $\|TA\|_1 = \operatorname{tr}(|TA|) = \operatorname{tr}(S|A|) \leq \|S\| \|A\|_1 \leq \|T\| \|A\|_1$. The second half of (g) can be obtained by combining the first half with part (f). □

1.13 COROLLARY. *If $A \in \mathscr{B}_1(\mathscr{H})$ and $\{e_n\}$ and $\{f_n\}$ are two bases for $\mathscr{H}$, then*

$$\sum |\langle Ae_n, f_n \rangle| \leq \|A\|_1.$$

PROOF. Let α_n be a unimodular scalar such that $\alpha_n \langle Ae_n, f_n \rangle = |\langle Ae_n, f_n \rangle|$ and let U be the diagonalizable unitary operator defined by $Ue_n = \overline{\alpha_n} f_n$. Thus $|\langle Ae_n, f_n \rangle| = \langle Ae_n, Ue_n \rangle = \langle U^*Ae_n, e_n \rangle$. Since $U^*A \in \mathscr{B}_1$, $\sum |\langle Ae_n, f_n \rangle| = \operatorname{tr}(U^*A) \leq \|U^*\| \|A\|_1 \leq \|A\|_1$. □

We now turn our attention to an investigation of the duality properties of $\mathscr{B}_1$. First, examine the rank one operator $e \otimes f$ and note that $\operatorname{ran}(e \otimes f) = \mathbb{C}e$ and $\ker(e \otimes f) = [f]^\perp$ if $e \neq 0$.

§1 TRACE CLASS AND HILBERT-SCHMIDT OPERATORS

1.14 LEMMA. *If e and $f \in \mathcal{H}$ and $T \in \mathcal{B}(\mathcal{H})$, then $T(e \otimes f) = (Te) \otimes f$ and $(e \otimes f)T = e \otimes (T^*f)$. Consequently, $\mathrm{tr}(T(e \otimes f)) = \langle Te, f \rangle = \mathrm{tr}((e \otimes f)T)$.*

PROOF. Exercise. □

For A in $\mathcal{B}_1$ define $\Phi_A: \mathcal{B}_0 \to \mathbb{C}$ by

$$\Phi_A(C) = \mathrm{tr}(CA) = \mathrm{tr}(AC)$$

for every compact operator C. Clearly Φ_A is a well-defined linear functional on $\mathcal{B}_0$.

1.15 THEOREM. *The map $A \to \Phi_A$ is an isometric isomorphism of $\mathcal{B}_1$ onto $\mathcal{B}_0^*$.*

PROOF. By Theorem 1.11(d), $\sup\{|\mathrm{tr}(AC)| : C \in \mathcal{B}_0 \text{ and } \|C\| \leq 1\} \leq \|A\|_1$. This says that Φ_A is a bounded linear functional on $\mathcal{B}_0$ and $\|\Phi_A\| \leq \|A\|_1$. Define $\rho: \mathcal{B}_1 \to \mathcal{B}_0^*$ by $\rho(A) = \Phi_A$; it is easy to check that ρ is a linear mapping and the preceding inequality shows that $\|\rho(A)\| \leq \|A\|_1$ for all A in $\mathcal{B}_1$.

It must be shown that ρ is surjective and $\|\rho(A)\| \geq \|A\|_1$ for A in $\mathcal{B}_1$. Let $\Phi \in \mathcal{B}_0^*$ and define $[g, h] = \Phi(g \otimes h)$ for all g and h in $\mathcal{H}$. It follows that $|[g, h]| \leq \|\Phi\| \|g\| \|h\|$ for all g and h and $[\cdot, \cdot]$ is sesquilinear. Hence there is a bounded operator A on $\mathcal{H}$ such that $\langle Ag, h \rangle = [g, h] = \Phi(g \otimes h)$ for all g and h in $\mathcal{H}$. It will now be shown that $A \in \mathcal{B}_1$ an $\Phi = \Phi_A$.

If $C \in \mathcal{B}_{00}$, then $C = \sum_{k=1}^n g_k \otimes h_k$ for some vectors $g_1, \ldots, g_n$, $h_1, \ldots, h_n$ in $\mathcal{H}$ (Exercise 8). Thus

$$\Phi(C) = \Phi\left(\sum_{k=1}^n g_k \otimes h_k\right) = \sum_{k=1}^n \langle Ag_k, h_k \rangle = \sum_{k=1}^n \mathrm{tr}(A(g_k \otimes h_k)) = \mathrm{tr}(AC).$$

If it can be shown that $A \in \mathcal{B}_1$, then both Φ and Φ_A are bounded linear functionals on $\mathcal{B}_0$ and the preceding equation demonstrates that they agree on the dense manifold $\mathcal{B}_{00}$. It then follows that $\Phi = \Phi_A$ and ρ is seen to be surjective.

To see that $A \in \mathcal{B}_1$, let $A = W|A|$ be the polar decomposition of A and let $\{e_n\}$ be an orthonormal basis. Thus $C_n = [\sum_{k=1}^n e_k \otimes e_k]W^*$ is a contraction in $\mathcal{B}_{00}$ and so $\|\Phi\| \geq |\Phi(C_n)| = |\Phi(\sum_{k=1}^n e_k \otimes (We_k))|$ (Why?) $= |\sum_{k=1}^n \langle Ae_k, We_k \rangle| = \sum_{k=1}^n \langle |A|e_k, e_k \rangle$. If we allow n to approach ∞, we see that $\|\Phi\| \geq \|A\|_1$ and so $A \in \mathcal{B}_1$. Thus $\Phi = \Phi_A$. But this also says that $\|\Phi_A\| \geq \|A\|_1$, so we have simultaneously shown that ρ is an isometry, completing the proof of the theorem. □

Since the dual of a Banach space is itself a Banach space, the next corollary is immediate.

1.16 COROLLARY. *$(\mathcal{B}_1, \|\cdot\|_1)$ is a Banach space.*

A direct proof of this corollary is not difficult; see Exercise 9.

What is the dual of $\mathscr{B}_1$? If $\mathscr{B}_0$ and $\mathscr{B}_1$ are the analogues of c_0 and ℓ^1, respectively, then the fact that $(\ell^1)^* = \ell^\infty$ might suggest that the dual of $\mathscr{B}_1$ is $\mathscr{B}$. This is indeed the case.

For B in $\mathscr{B}$, define $F_B: \mathscr{B}_1 \to \mathbb{C}$ by

$$F_B(A) = \mathrm{tr}(BA) = \mathrm{tr}(AB).$$

Clearly F_B is a linear functional on $\mathscr{B}_1$.

1.17 THEOREM. *The mapping $B \to F_B$ is an isometric isomorphism of $\mathscr{B}(\mathscr{H})$ onto $\mathscr{B}_1^*$.*

PROOF. The fact that $\|F_B\| \leq \|B\|$ follows from previous results of this section; thus $F_B \in \mathscr{B}_1^*$. Define $\rho: \mathscr{B} \to \mathscr{B}_1^*$ by $\rho(B) = F_B$; clearly ρ is linear. If $\varepsilon > 0$, choose a unit vector g such that $\|Bg\| > \|B\| - \varepsilon$. Now choose a unit vector h with $\langle Bg, h \rangle = \|Bg\|$. If $C = g \otimes h$, then $C \in \mathscr{B}_1$ and $\|C\|_1 = 1$. Thus $\|F_B\| \geq |\mathrm{tr}(BC)| = \langle Bg, h \rangle = \|Bg\| > \|B\| - \varepsilon$. Since ε was arbitrary, we get that $\|F_B\| = \|B\|$ and ρ is an isometry. It remains to show that ρ is surjective.

Let $F \in \mathscr{B}_1^*$; as in the proof of Theorem 1.15, there is an operator B in $\mathscr{B}$ such that $\langle Bg, h \rangle = F(g \otimes h)$ for all g and h in $\mathscr{H}$. As before, it follows that $F(T) = F_B(T)$ for every finite rank operator T. Since $\mathscr{B}_{00}$ is dense in $\mathscr{B}_1$ and both F and F_B are bounded linear functionals, $F = F_B$. □

This section concludes with a statement of a result for $\mathscr{B}_2$ analogous to Theorem 1.11. Extensive use of this result will not be made in this book and, moreover, the proof is similar to that of Theorem 1.11. So the proof will be left to the interested reader.

1.18 THEOREM. (a) *If $\langle A, B \rangle \equiv \mathrm{tr}(B^*A)$ for A and B in $\mathscr{B}_2$, then $\langle \cdot, \cdot \rangle$ is an inner product on $\mathscr{B}_2$, $\|\cdot\|_2$ is the norm defined by this inner product, and $\mathscr{B}_2$ is a Hilbert space with respect to this inner product.*

(b) *Every Hilbert-Schmidt operator is compact. If A is a compact operator and $\alpha_1, \alpha_2, \ldots$ are the eigenvalues of $|A|$, each repeated as often as its multiplicity, then $A \in \mathscr{B}_2$ if and only if $\{\alpha_n\} \in \ell^2$ and in this case $\|A\|_2^2 = \sum \alpha_n^2$.*

(c) *The Hilbert space $\mathscr{B}_2$ contains $\mathscr{B}_{00}$ as a dense linear manifold and is therefore separable.*

Exercises

1. Verify the statement in Example 1.5.
2. Give a direct proof of Corollary 1.7.
3. If $A \in \mathscr{B}(\mathscr{H})$, show that A is trace class if and only if $\sum |\langle Ae_n, e_n \rangle| < \infty$ for every orthonormal basis $\{e_n\}$.
4. Prove Lemma 1.14.
5. If $A \in \mathscr{B}$, show that $A \in \mathscr{B}_1$ if and only if the real and imaginary parts of its Cartesian decomposition are in $\mathscr{B}_1$. If $A = A^*$, then $A \in \mathscr{B}_1$ if and only if $A = A_+ - A_-$, where $A_\pm$ is a positive trace class operator. Do analogous statements hold for $\mathscr{B}_2$?

6. Compute the adjoint and polar decomposition of $g \otimes h$.
7. Show that $\|g \otimes h\| = \|g \otimes h\|_1 = \|g \otimes h\|_2 = \|g\| \|h\|$.
8. Show that every finite rank operator C can be written as $\sum_{k=1}^{n} g_k \otimes h_k$ for some vectors $g_1, \ldots, g_n, h_1, \ldots, h_n$ in $\mathcal{H}$.
9. Give a direct proof of Corollary 1.16. (That is, give a proof that does not rely on Theorem 1.15.)
10. Let $A \in \mathcal{B}(\mathcal{H})$ and show that A is compact if and only if there are orthonormal sequences $\{e_n\}$ and $\{f_n\}$ and scalars $\{\alpha_n\}$ such that $\alpha_n \to 0$ and $A = \sum \alpha_n e_n \otimes f_n$, where this infinite sum converges in the operator norm. Compute A^*, $|A|$, and $\|A\|$ in terms of this decomposition. Discuss uniqueness of the sum $\sum \alpha_n e_n \otimes f_n$.
11. Prove Theorem 1.18.
12. Fix an orthonormal basis $\{e_n\}$ for $\mathcal{H}$ and define $\psi: c_0 \to \mathcal{B}_0$ by $\psi(\{\alpha_n\}) = \sum \alpha_n e_n \otimes e_n$.
    (a) Show that ψ is an isometry whose range is the set of compact normal operators that have $\{e_n\}$ as eigenvectors.
    (b) If $T = \sum n^{-1} e_n \otimes e_n$, then $\operatorname{ran} \psi = \{T\}'' \cap \mathcal{B}_0$.
    (c) Compute $\psi^*: \mathcal{B}_1 \to \ell^1$ and $\psi^{**}: \ell^\infty \to \mathcal{B}$.
    (d) Characterize the range of ψ^{**}.
    (e) Show that there is no bounded projection of $\mathcal{B}$ onto $\mathcal{B}_0$ by using the fact that there is no bounded projection of ℓ^∞ onto c_0.
13. If $\mathcal{X}$ is any Banach space, consider the natural map $\mathcal{X} \to \mathcal{X}^{**}$. Show that the dual of this map, $\mathcal{X}^{***} \to \mathcal{X}^*$, is a projection of norm 1. (Here we are considering $\mathcal{X}^*$, in the natural way, as a subspace of its second dual, $(\mathcal{X}^*)^{**} = \mathcal{X}^{***}$.) Apply this to the space $\mathcal{X} = \mathcal{B}_0$ to show that if $L \in \mathcal{B}^*$, then there is a unique trace class operator A and a unique functional L_0 in $\mathcal{B}^*$ such that:
    (i) $L = L_0 + \Phi_A$;
    (ii) $L_0(T) = 0$ for every compact operator T;
    (iii) $\|L\| = \|L_0\| + \|A\|_1$.
Show that this operator A satisfies $\langle Ag, h \rangle = L(g \otimes h)$ for all g and h in $\mathcal{H}$.
14. (a) If $\{e_n\}$ and $\{f_m\}$ are orthonormal bases for $\mathcal{H}$, show that $\{e_n \otimes f_m : n, m \geq 1\}$ is an orthonormal basis for $\mathcal{B}_2$. For T in $\mathcal{B}(\mathcal{H})$, define the operators L_T and R_T on $\mathcal{B}_2$ by $L_T(A) = TA$ and $R_T(A) = AT$ for all A in $\mathcal{B}_2$.
    (b) Show that $L_T^* = L_{T^*}$, $R_T^* = R_{T^*}$ and $L_T R_S = R_S L_T$ for all operators S and T in $\mathcal{B}(\mathcal{H})$.
    (c) Prove that L_T is unitarily equivalent to $T \oplus T \oplus \cdots$.
15. Define (or look up) the tensor product of two Hilbert spaces. For a Hilbert space $\mathcal{H}$ define the Hilbert space $\mathcal{H}^\#$ to be the same set $\mathcal{H}$ with the same definition of vector addition but where scalar multiplication and the inner product are defined by: $\alpha \cdot^\# f = \bar{\alpha} f$ and $\langle f, g \rangle_\# = \langle g, f \rangle$.

(It is easy to see that $\mathcal{H}^\#$ is a Hilbert space.) Show that there is an isomorphism $U: \mathcal{H} \otimes \mathcal{H}^\# \to \mathcal{B}_2$ such that $U(g \otimes h) = g \otimes h$. (Sorry for the ambiguous notation, but, in fact, this exercise shows that the notation is not so ambiguous after all.) Prove (using the notation of the preceding exercise) that for any operator T on $\mathcal{H}$, $U(T \otimes S)U^* = L_T R_{S^*}$.

16. Let (X, Ω, μ) be a σ-finite measure space and $\mathcal{H} = L^2(\mu)$. For k in $L^2(\mu \times \mu)$, define $T_k: L^2(\mu) \to L^2(\mu)$ by $(T_k f)(x) = \int k(x, y) f(y) d\mu(y)$. Show that T_k is a Hilbert-Schmidt operator on $L^2(\mu)$ and the map $k \to T_k$ is an isomorphism of $L^2(\mu \times \mu)$ onto $\mathcal{B}_2(L^2(\mu))$.

§2 Some topologies on $\mathcal{B}(\mathcal{H})$. In this section a few definitions and results concerning various topologies on $\mathcal{B}(\mathcal{H})$ are collected. Some of these are recalled from [ACFA], in which case no proofs are given.

$\mathcal{B}(\mathcal{H})$ is, of course, a normed space, but this topology is not pertinent to this section. The *weak operator topology* (WOT) is the topology on $\mathcal{B}(\mathcal{H})$ defined by the collection of seminorms $\{p_{f,g} : f, g \in \mathcal{H}\}$, where $p_{f,g}(T) = |\langle Tf, g \rangle|$ for T in $\mathcal{B}(\mathcal{H})$. The *strong operator topology* (SOT) is defined by the seminorms $\{p_f : f \in \mathcal{H}\}$, where $p_f(T) = \|Tf\|$ for T in $\mathcal{B}(\mathcal{H})$. Some of the elementary properties of these two topologies can be found in [ACFA]. In particular, since $\mathcal{H}$ is separable both topologies, when restricted to the closed unit ball of $\mathcal{B}(\mathcal{H})$, ball $\mathcal{B}(\mathcal{H})$, are metrizable. We also have the following proposition.

2.1 PROPOSITION. *If $\Phi: \mathcal{B}(\mathcal{H}) \to \mathbb{C}$ is a linear functional, the following statements are equivalent.*

(a) Φ *is SOT-continuous.*
(b) Φ *is WOT-continuous.*
(c) *There are vectors $f_1, \ldots, f_n, g_1, \ldots, g_n$ in $\mathcal{H}$ such that $\Phi(T) = \sum_{k=1}^n \langle Tf_k, g_k \rangle$ for all T in $\mathcal{B}(\mathcal{H})$.*

PROOF. See IX.5.1 in [ACFA]. □

2.2 COROLLARY. $(\mathcal{B}, WOT)^* = (\mathcal{B}, SOT)^* = \mathcal{B}_{00}$.

PROOF. Note that if $C = \sum_{k=1}^n f_k \otimes g_k$, then $\text{tr}(CT) = \sum_{k=1}^n \langle Tf_k, g_k \rangle$. □

The following corollary is a consequence of this proposition and the fact that if a vector space has two locally convex topologies with identical collections of continuous linear functionals, then convex subsets have identical closures in both topologies.

2.3 COROLLARY. *If $\mathcal{S}$ is a convex subset of $\mathcal{B}(\mathcal{H})$, then the WOT closure of $\mathcal{S}$ equals the SOT closure of $\mathcal{S}$.*

2.4 PROPOSITION. *The closed unit ball of $\mathcal{B}(\mathcal{H})$ with the WOT is a compact metric space.*

PROOF. See IX.5.5 in [ACFA]. □

There is another weak topology on $\mathscr{B}(\mathscr{H})$ that must be discussed; the weak* topology that $\mathscr{B}(\mathscr{H})$ has as the dual of the trace class. Here the defining seminorms are given by $T \to |\text{tr}(TA)|$, where A can be any trace class operator.

2.5 PROPOSITION

  (a) *The closed unit ball of $\mathscr{B}(\mathscr{H})$ with the weak* topology is a compact metric space.*
  (b) *The weak* topology and the WOT agree on bounded subsets of $\mathscr{B}(\mathscr{H})$.*
  (c) *A sequence in $\mathscr{B}(\mathscr{H})$ converges weak* if and only if it converges WOT.*

PROOF. (a) In fact, $\mathscr{B}(\mathscr{H})$ is the dual of the separable Banach space $\mathscr{B}_1(\mathscr{H})$.

(b) It is easy to see that the weak* topology is bigger than the WOT, and so the identity map $i: (\mathscr{B}(\mathscr{H}), \text{weak*}) \to (\mathscr{B}(\mathscr{H}), \text{WOT})$ is continuous. If $\mathscr{S}$ is a closed bounded subset of $\mathscr{B}(\mathscr{H})$, then $i: (\mathscr{S}, \text{weak*}) \to (\mathscr{S}, \text{WOT})$ must be continuous and closed and, hence, is a homeomorphism.

(c) By the Principle of Uniform Boundedness, WOT convergent sequences must be uniformly bounded. Thus part (c) follows from part (b). □

Here is another interpretation of the weak* topology.

2.6 PROPOSITION. *A net $\{T_i\}$ in $\mathscr{B}(\mathscr{H})$ converges weak* to 0 if and only if for any two sequences $\{f_n\}$ and $\{g_n\}$ of vectors from $\mathscr{H}$ with $\sum \|f_n\|^2 < \infty$ and $\sum \|g_n\|^2 < \infty$, $\sum_n \langle T_i f_n, g_n \rangle \to 0$ with i.*

PROOF. Suppose $\{f_n\}$ and $\{g_n\}$ are sequences satisfying the condition. Since $\|f_n \otimes g_n\|_1 = \|f_n\| \|g_n\|$, it follows that $\sum_n f_n \otimes g_n$ converges in $\mathscr{B}_1$ to a trace class operator A. It is easy to check that $\sum_n \langle T f_n, g_n \rangle = \text{tr}(AT)$. The completion of the proof is left to the reader. □

The entire theory of weak* topologies on Banach spaces can be applied to $\mathscr{B}(\mathscr{H})$. One part of that theory that will be important in this book is the Krein-Smulian Theorem which states that for a Banach space $\mathscr{X}$ and a convex subset $\mathscr{S}$ in $\mathscr{X}^*$, $\mathscr{S}$ is weak* closed in $\mathscr{X}^*$ if and only if $\mathscr{S} \cap r(\text{ball } \mathscr{X}^*)$ is weak* closed for all $r > 0$. (See V.12.1 in [ACFA].) If $\mathscr{X}$ is separable, the weak* topology is metrizable on bounded sets and so the Krein-Smulian Theorem becomes the statement that a convex subset $\mathscr{S}$ in $\mathscr{X}^*$ is weak* closed if and only if it is weak* sequentially closed. Since $\mathscr{B}_1$ is separable, the following proposition is a consequence. The details are left to the reader.

2.7 PROPOSITION. (a) *If $\mathscr{S}$ is a convex subset of $\mathscr{B}(\mathscr{H})$, then $\mathscr{S}$ is weak* closed if and only if it is WOT sequentially closed.*
  (b) *If $\mathscr{X}$ is a separable Banach space and $\rho: \mathscr{X}^* \to \mathscr{B}(\mathscr{H})$ is a linear*

map, the following statements are equivalent:
  (i) $\rho: (\mathscr{X}^*, \text{weak}^*) \to (\mathscr{B}, \text{weak}^*)$ *is continuous*;
  (ii) $\rho: (\mathscr{X}^*, \text{weak}^*) \to (\mathscr{B}, \text{WOT})$ *is sequentially continuous*;
  (iii) *there is a bounded linear mapping* $\tau: \mathscr{X} \to \mathscr{B}_1$ *such that* $\rho = \tau^*$.

  (c) *If* $\mathscr{X}$ *is a separable Banach space and* $\rho: (\mathscr{X}^*, \text{weak}^*) \to (\mathscr{B}(\mathscr{H}), \text{WOT})$ *is a sequentially continuous isometry, then* $\mathscr{W} = \rho(\mathscr{X}^*)$ *is weak* closed in* $\mathscr{B}(\mathscr{H})$ *and* $\rho: \mathscr{X}^* \to \mathscr{W}$ *is a weak* homeomorphism.*

2.8 DEFINITION. A Banach algebra $\mathscr{A}$ is a *dual algebra* if there is a Banach space $\mathscr{X}$ such that $\mathscr{A}$ is isometrically isomorphic to $\mathscr{X}^*$ and multiplication in $\mathscr{A}$ is separately weak* continuous. That is, if $a \in \mathscr{A}$, then the maps $x \to ax$ and $x \to xa$ are continuous in the weak* topology.

Note that this is an abstract version of the definition of a dual algebra given in Bercovici, Foias, and Pearcy [1985] where a dual algebra is defined as a weak* closed subalgebra of $\mathscr{B}(\mathscr{H})$. Indeed the examples of dual algebras encountered here will either be of this type or weak* closed subalgebras of $L^\infty(\mu)$.

2.9 DEFINITION. If $\mathscr{A}$ and $\mathscr{B}$ are dual algebras, $\rho: \mathscr{A} \to \mathscr{B}$ is a *dual algebra homomorphism* if ρ is a Banach algebra homomorphism that is weak* continuous. The map ρ is a dual algebra isomorphism if it is an isometric isomorphism that is a weak* homeomorphism.

The proof of the next result is immediate from Proposition 2.7.

2.10 PROPOSITION. *Assume that* $\mathscr{A}$ *and* $\mathscr{B}$ *are dual algebras that are duals of separable Banach spaces. If* $\rho: \mathscr{A} \to \mathscr{B}$ *is an isometric isomorphism that is weak* sequentially continuous, then* ρ *is a dual algebra isomorphism.*

Exercises

1. Show that WOT is the weakest locally convex topology on $\mathscr{B}(\mathscr{H})$ that has $\mathscr{B}_{00}$ as its dual.
2. Show that the ball $\mathscr{B}_{00}$ is WOT (respectively, SOT) dense in ball $\mathscr{B}(\mathscr{H})$.
3. Let $\{T_i\}$ be a bounded net in $\mathscr{B}(\mathscr{H})$ and fix an orthonormal basis $\{e_n\}$.
   (a) Prove that $T_i \to 0$ (WOT) if and only if for all $n, m \geq 1$, $\langle T_i e_n, e_m \rangle \to 0$ with i.
   (b) Prove that $T_i \to 0$ (SOT) if and only if for all e_n, $\|T_i e_n\| \to 0$ with i.
4. Let $\{T_i\}$ and $\{S_i\}$ be nets in $\mathscr{B}(\mathscr{H})$ and assume that $\{T_i\}$ is uniformly bounded.
   (a) If $T_i \to 0$ (WOT) and $S_i \to 0$ (SOT), then $T_i S_i \to 0$ (WOT).
   (b) If, in addition, $T_i \to 0$ (SOT), then $T_i S_i \to 0$ (SOT).
5. Let $\mathscr{H} = \ell^2$ and define S on $\mathscr{H}$ by $Se_n = e_{n+1}$, where $\{e_n\}$ is the standard basis for ℓ^2. Examine the sequences $\{S^k\}$, $\{S^{*k}\}$, $\{S^k S^{*k}\}$, and $\{S^{*k} S^k\}$ for their convergence properties in the WOT and SOT. Compare with Exercise 4.

6. (Halmos) Fix an orthonormal basis $\{e_n\}$ for $\mathcal{H}$.
   (a) [Halmos, **1982**, Solution 28] Show that 0 belongs to the weak closure of $\{\sqrt{n} e_n : n \geq 1\}$.
   (b) Let $\{n_i\}$ be a net of integers such that $\sqrt{n_i} e_{n_i} \to 0$ weakly and define $T_i f = \sqrt{n_i} \langle f, e_{n_i} \rangle e_{n_i}$ for all f in $\mathcal{H}$. Show that $T_i \to 0$ (SOT) but $\{T_i^2\}$ does not converge to 0 (SOT).
7. Let $\mathcal{L}$ be a linear manifold in $\mathcal{B}_1$ that contains $\mathcal{B}_{00}$ and consider the topology $(\mathcal{L})$ defined on $\mathcal{B}(\mathcal{H})$ by the seminorms $\{p_A : A \in \mathcal{L}\}$, where $p_A(T) = |\text{tr}(AT)|$. Show that $(\mathcal{B}, (\mathcal{L}))^* = \mathcal{L}$ and that $(\mathcal{L})$ and WOT agree on bounded sets in $\mathcal{B}(\mathcal{H})$.

§3 Isometries.

Every unitary operator is an isometry, but the converse is well known to be false. The standard example is called the unilateral shift. This operator is, undoubtedly, the most widely studied (and understood) non-normal operator. It is a subnormal operator and is thus of interest to us.

3.1 DEFINITION. Let ℓ^2 be the set of square summable sequences $(\alpha_0, \alpha_1, \ldots)$. (There is a very good reason for starting the enumeration of the sequences from 0. The reason for this will become apparent later.) The *unilateral shift* is the operator $S: \ell^2 \to \ell^2$ defined by

$$S(\alpha_0, \alpha_1, \ldots) = (0, \alpha_0, \alpha_1, \ldots).$$

It is easy to verify that S is an isometry and clearly it fails to be surjective.

The article "the" is used in the title of this operator since there are other unilateral shifts.

3.2 DEFINITION. An operator S acting on a Hilbert space $\mathcal{H}$ is called a *unilateral shift* if there is a sequence $\mathcal{H}_0, \mathcal{H}_1, \ldots$ of pairwise orthogonal subspaces such that $\mathcal{H} = \mathcal{H}_0 \oplus \mathcal{H}_1 \oplus \cdots$ and S maps $\mathcal{H}_n$ isometrically onto $\mathcal{H}_{n+1}$.

The unilateral shift is a unilateral shift where each $\mathcal{H}_n$ is one dimensional. In fact, it is clear that in Definition 3.2 each of the spaces $\mathcal{H}_n$ has the same dimension, though each may be infinite dimensional. Call this common dimension the *multiplicity* of S. In fact, it is easy to see that if S is a unilateral shift (with the notation as above), then $\text{ran } S = \mathcal{H}_0^\perp$ and the multiplicity of S is $\dim \mathcal{H}_0 = \dim(\text{ran } S)^\perp$.

The proof of the next result is left as an exercise.

3.3 PROPOSITION. *Two unilateral shifts are unitarily equivalent if and only if they have the same multiplicity.*

3.4 COROLLARY. *If S is a unilateral shift on $\mathcal{H}$, $\mathcal{L} = \mathcal{H} \cap (S\mathcal{H})^\perp$, and T is defined on $\mathcal{L} \oplus \mathcal{L} \oplus \cdots$ by $T(f_0 \oplus f_1 \oplus \cdots) = 0 \oplus f_0 \oplus f_1 \oplus \cdots$, then S and T are unitarily equivalent.*

PROOF. In fact, T is a unilateral shift with the same multiplicity as S. □

If S is an isometry and $\mathcal{M}$ is an invariant subspace, then clearly $S \mid \mathcal{M}$ is also an isometry. If $S\mathcal{M} = \mathcal{M}$, then the restriction of S to $\mathcal{M}$ is a surjective isometry, and hence is unitary. More can be said.

3.5 Proposition. *If S is an isometry and $\mathcal{M}$ is an invariant subspace such that $S\mathcal{M} = \mathcal{M}$, then $\mathcal{M}$ reduces S, $S \mid \mathcal{M}$ is unitary, and $\mathcal{M} \subseteq \bigcap_n S^n \mathcal{H}$.*

PROOF. Recall that an operator S is an isometry if and only if $S^*S = 1$. Thus $S\mathcal{M} = \mathcal{M}$ implies that $S^*\mathcal{M} = S^*S\mathcal{M} = \mathcal{M}$; so $\mathcal{M}$ reduces S. Clearly $S \mid \mathcal{M}$ is unitary. Also $\mathcal{M} = S^n \mathcal{M} \subseteq S^n \mathcal{H}$ for every $n \geq 1$. □

If S is an isometry, it fails to be unitary if it is not surjective. So if S is an isometry we can consider the closed subspaces of $\mathcal{H}, S\mathcal{H}, S^2\mathcal{H}, S^3\mathcal{H}, \ldots$; either they are all equal (in which case S is unitary) or none are equal (Why?). Also $S\mathcal{H} \supseteq S^2\mathcal{H} \supseteq \cdots$; so consider $\mathcal{L}_\infty \equiv \bigcap_n S^n \mathcal{H}$. Note that since S is one-to-one, $S\mathcal{L}_\infty = \bigcap_n S^{n+1} \mathcal{H} = \mathcal{L}_\infty$, since $S^{n+1}\mathcal{H} \subseteq S^n \mathcal{H}$. So by the preceding proposition, $S \mid \mathcal{L}_\infty$ is unitary and $\mathcal{L}_\infty$ reduces S. Also, every invariant subspace $\mathcal{M}$ with $S\mathcal{M} = \mathcal{M}$ is contained in $\mathcal{L}_\infty$; that is, $\mathcal{L}_\infty$ is the "unitary part" of S. All these matters are collected in the next result.

3.6 The von Neumann-Wold decomposition. *If S is an isometry on $\mathcal{H}$ and $\mathcal{L}_\infty = \bigcap_n S^n \mathcal{H}$, then $\mathcal{L}_\infty$ reduces S, $S \mid \mathcal{L}_\infty$ is unitary, and $S \mid \mathcal{L}_\infty^\perp$ is a unilateral shift. Thus every isometry on a Hilbert space is the direct sum of a unilateral shift and a unitary.*

PROOF. Put $\mathcal{L}_n = S^n \mathcal{H}$ for $n \geq 0$ (so $\mathcal{L}_0 = \mathcal{H}$); clearly $\mathcal{L}_n \supseteq \mathcal{L}_{n+1}$. Put $\mathcal{H}_n = \mathcal{L}_n \cap (\mathcal{L}_{n+1})^\perp$. Note that the fact S is an isometry implies that each $\mathcal{L}_n$, and hence each $\mathcal{H}_n$, is closed.

CLAIM 1. $\mathcal{H}_n \perp \mathcal{H}_m$ for $n \neq m$.

In fact, if $n < m$, then $\mathcal{H}_m \subseteq \mathcal{L}_m \subseteq \mathcal{L}_{n+1} \subseteq \mathcal{H}_n^\perp$.

CLAIM 2. $\mathcal{L}_\infty^\perp = \mathcal{H}_0 \oplus \mathcal{H}_1 \oplus \cdots$.

If $m > n$, then $\mathcal{H}_n^\perp \supseteq \mathcal{L}_m$. Hence $\mathcal{H}_n^\perp \supseteq \bigcap_{m \geq n+1} \mathcal{L}_m = \mathcal{L}_\infty$. Thus $\mathcal{H}_0 \oplus \mathcal{H}_1 \oplus \cdots \subseteq \mathcal{L}_\infty^\perp$. On the other hand, if $f \in (\mathcal{H}_0 \oplus \mathcal{H}_1 \oplus \cdots)^\perp$, then for every $n \geq 1$, $f \in (\mathcal{H}_0 \oplus \cdots \mathcal{H}_n)^\perp = \mathcal{L}_{n+1}$. Thus $f \in \mathcal{L}_\infty$.

CLAIM 3. $\mathcal{L}_\infty$ reduces S, $S\mathcal{L}_\infty = \mathcal{L}_\infty$, and $S \mid \mathcal{L}_\infty$ is unitary.

This is a consequence of the preceding proposition, as discussed prior to the statement of this theorem.

CLAIM 4. $S\mathcal{H}_n = \mathcal{H}_{n+1}$.

It is easy to see that $S\mathcal{H}_n \subseteq S^{n+1}\mathcal{H}$. Also, if $f \in \mathcal{H}_n$ and $h \in \mathcal{H}$, then $\langle Sf, S^{n+2}h \rangle = \langle f, S^{n+1}h \rangle = 0$. Hence $S\mathcal{H}_n \subseteq \mathcal{H}_{n+1}$. Conversely, if $f \in \mathcal{H}_{n+1} = \mathcal{L}_{n+1} \cap (\mathcal{L}_{n+2})^\perp$, then there is a vector g in $\mathcal{L}_n$ with $Sg = f$. If $h \in \mathcal{L}_{n+1}$, then $Sh \in \mathcal{L}_{n+2}$ and so $0 = \langle f, Sh \rangle = \langle Sg, Sh \rangle = \langle g, h \rangle$. Hence $g \in \mathcal{L}_n \cap (\mathcal{L}_{n+1})^\perp = \mathcal{H}_n$ and $f = Sg \in S\mathcal{H}_n$.

Combining Claims 2 and 4, we see that $S \mid \mathcal{L}_\infty^\perp$ is a unilateral shift of multiplicity $\dim \mathcal{H}_0$. □

An isometry is said to be *pure* if there is no reducing subspace on which it is unitary. (In light of Proposition 3.5, this is equivalent to saying that there is no invariant subspace $\mathcal{M}$ such that $S \mid \mathcal{M}$ is unitary.)

3.7 COROLLARY. *An isometry is pure if and only if it is a unilateral shift.*

PROOF. If S is pure, the preceding theorem implies that S must be a unilateral shift. Conversely, if S is a unilateral shift and $\mathcal{M}$ is an invariant subspace for S such that $S\mathcal{M} = \mathcal{M}$, then Proposition 3.5 implies $\mathcal{M} \subseteq \bigcap_n S^n \mathcal{H} = (0)$. □

3.8 DEFINITION. An operator U on a Hilbert space $\mathcal{H}$ is a *bilateral shift* if there are pairwise orthogonal subspaces $\{\mathcal{H}_n : n \in \mathbb{Z}\}$ such that $\mathcal{H} = \oplus_n \mathcal{H}_n$ and $U\mathcal{H}_n = \mathcal{H}_{n+1}$ for all n in $\mathbb{Z}$. The *multiplicity* of U is the dimension of $\mathcal{H}_0$ ($= \dim \mathcal{H}_n$ for all n).

It is easy to see that a bilateral shift is unitary and its multiplicity is a complete unitary invariant. Define *the bilateral shift* to be the bilateral shift of multiplicity 1.

There is a relation between unilateral and bilateral shifts, as you might expect. For example, if U is a bilateral shift relative to the spaces $\{\mathcal{H}_n\}$ and $\mathcal{H} = \mathcal{H}_0 \oplus \mathcal{H}_1 \oplus \cdots$, then $\mathcal{H}$ is invariant for U and $S = U \mid \mathcal{H}$ is a unilateral shift. Note that the multiplicity of S is the same as the multiplicity of U. A converse of this is also true and the astute reader may see it now. But before formally presenting this, some groundwork will be done.

Suppose that S is a unilateral shift on $\mathcal{H} = \mathcal{H}_0 \oplus \mathcal{H}_1 \oplus \cdots$, $S\mathcal{H}_n = \mathcal{H}_{n+1}$. Let e_0 be a unit vector in $\mathcal{H}_0$ and put $e_n = S^n e_0$ for $n \geq 1$. It is easy to see that $\{e_n\}$ is an orthonormal sequence. Using this technique, we get the following result, the details of whose proof are left to the reader.

3.9 PROPOSITION. *If S is a unilateral shift on $\mathcal{H}$ of multiplicity α and $\{e_i : 1 \leq i \leq \alpha\}$ is a basis for $\mathcal{H} \cap (S\mathcal{H})^\perp$, then $\{S^n e_i : n \geq 0 \text{ and } 1 \leq i \leq \alpha\}$ is a basis for $\mathcal{H}$.*

3.10 COROLLARY. *If S is a unilateral shift of multiplicity α and $1 \leq \beta < \alpha$, then there is an invariant subspace $\mathcal{M}$ for S such that $S \mid \mathcal{M}$ is a unilateral shift of multiplicity β.*

3.11 THEOREM. *If S is an isometry on $\mathcal{H}$, then there is a Hilbert space $\hat{\mathcal{H}}$ containing $\mathcal{H}$ and a unitary $U: \hat{\mathcal{H}} \to \hat{\mathcal{H}}$ such that $U\mathcal{H} \subseteq \mathcal{H}$ and $U\mid \mathcal{H} = S$. Moreover, $\hat{\mathcal{H}}$ can be chosen to be the smallest reducing subspace for U that contains $\mathcal{H}$; in this case, U and $\hat{\mathcal{H}}$ are unique up to unitary equivalence.*

PROOF. First assume that S is a unilateral shift. By Corollary 3.4 it may be assumed that $\mathcal{H} = \mathcal{L} \oplus \mathcal{L} \oplus \cdots$ and $S(f_0 \oplus f_1 \oplus \cdots) = 0 \oplus f_0 \oplus f_1 \oplus \cdots$. For convenience write this as $\mathcal{H} = \mathcal{L}_0 \oplus \mathcal{L}_1 \oplus \cdots$, where $\mathcal{L}_n = \mathcal{L}$ for all $n \geq 1$. Let $\hat{\mathcal{H}} = \bigoplus_{n=-\infty}^{\infty} \mathcal{L}_n$, where $\mathcal{L}_n = \mathcal{L}$ for all n. Define U on $\hat{\mathcal{H}}$ by $U(\cdots \oplus f_{-2} \oplus f_{-1} \oplus \hat{f}_0 \oplus f_1 \oplus f_2 \oplus \cdots) = \cdots \oplus f_{-2} \oplus \hat{f}_{-1} \oplus f_0 \oplus f_1 \oplus f_2 \oplus \cdots$, where the $\,\hat{}\,$ denotes the location of the zero-th coordinate. Clearly U is a unitary; in fact, U is a bilateral shift of the same multiplicity as S.

It is also clear that $U\mathscr{H} \subseteq \mathscr{K}$ and $U \mid \mathscr{H} = S$. Suppose $\mathscr{R}$ is a reducing subspace for U and $\mathscr{H} \subseteq \mathscr{R}$. So $\mathscr{R} \supseteq U^{*n}\mathscr{L}_0$ for every $n \geq 1$. But $\mathscr{L}_{-n} = U^{*n}\mathscr{L}_0$ and so it follows that $\mathscr{R} = \mathscr{K}$. It is left as an exercise to show that U and $\mathscr{K}$ are unique up to unitary equivalence.

Now for the arbitrary isometry. Using the von Neumann-Wold decomposition, $S = S_1 \oplus S_0$ on $\mathscr{H} = \mathscr{H}_1 \oplus \mathscr{H}_0$, where S_1 is a unilateral shift and S_0 is unitary. Let U_1 be a bilateral shift on $\mathscr{K}_1 \supseteq \mathscr{H}_1$ such that $U_1 \mid \mathscr{H}_1 = S_1$. Put $\mathscr{K} = \mathscr{K}_1 \oplus \mathscr{H}_0$ and $U = U_1 \oplus S_0$. The verification that U has the appropriate properties is left to the reader. $\square$

If U is the unitary extension of the isometry S for which $\mathscr{K}$ is the smallest reducing subspace of U that contains $\mathscr{H}$, then U is called the *minimal unitary extension* of S. This terminology is justified by the uniqueness part of the theorem.

The next corollary is actually a scholium.

3.12 COROLLARY. *The minimal unitary extension of a unilateral shift of multiplicity α is a bilateral shift of multiplicity α.*

3.13 COROLLARY. *Let S be a unilateral shift of multiplicity α and let U be its minimal unitary extension acting on $\mathscr{K}$. If $\{e_i : 1 \leq i \leq \alpha\}$ is a basis for $\mathscr{H} \cap (S\mathscr{H})^\perp$, then $\{U^n e_i : n \in \mathbb{Z} \text{ and } 1 \leq i \leq \alpha\}$ is a basis for $\mathscr{K}$.*

The next result will be of importance later.

3.14 PROPOSITION. *If S is a unilateral shift of multiplicity α acting on $\mathscr{H}$, and $\mathscr{M}$ is an invariant subspace for S, then $S \mid \mathscr{M}$ is a unilateral shift of multiplicity $\beta \leq \alpha$.*

PROOF. It is clear that $S \mid \mathscr{M}$ is an isometry and since S is pure it must be that $S \mid \mathscr{M}$ is pure. By Corollary 3.7, S is a unilateral shift; let β be the multiplicity of $S \mid \mathscr{M}$. It remains to show that $\beta \leq \alpha$. To do this it suffices to assume that $\alpha < \infty$.

Let U, acting on $\mathscr{K}$, be the minimal unitary extension of S and let $\{e_i : 1 \leq i \leq \alpha\}$ be an orthonormal basis for $\mathscr{H} \cap (S\mathscr{H})^\perp$. By the preceding corollary, $\{U^n e_i : n \in \mathbb{Z} \text{ and } 1 \leq i \leq \alpha\}$ is a basis for $\mathscr{K}$. Let $f_1, \ldots, f_\beta$ be a basis for $\mathscr{M} \cap (S\mathscr{M})^\perp$. Hence

$$\beta = \sum_{j=1}^{\beta} \|f_j\|^2$$

$$= \sum_{j=1}^{\beta} \sum_{i=1}^{\alpha} \sum_{n=0}^{\infty} |\langle f_j, U^n e_i \rangle|^2$$

$$= \sum_{j=1}^{\beta} \sum_{i=1}^{\alpha} \sum_{n=0}^{\infty} |\langle U^{*n} f_j, e_i \rangle|^2$$

$$= \sum_{i=1}^{\alpha} \sum_{j=1}^{\beta} \sum_{n=0}^{\infty} |\langle U^{*n} f_j, e_i \rangle|^2$$

$$\leq \sum_{i=1}^{\alpha} \|e_i\|^2 \text{ (by Bessel's Inequality)}$$

$$= \alpha. \quad \square$$

The preceding result is from Halmos [**1961**], though the proof is due to I. Halperin; see Halmos [**1982**], Solution 15. Also see Robertson [**1965**].

Finally, we consider the spectral properties of unilateral shifts. Recall (see [**ACFA**]) that for any operator T in $\mathscr{B}(\mathscr{H})$, the *approximate spectrum* of T, $\sigma_{\text{ap}}(T)$, is defined to be the set of complex numbers λ satisfying $\inf\{\|(T-\lambda)f\| : \|f\| = 1\} = 0$. Equivalently, $\lambda \in \sigma_{\text{ap}}(T)$ if and only if either $\ker(\lambda - T) \neq (0)$ or $\operatorname{ran}(\lambda - T)$ is not closed [**ACFA**], page 213. Thus $\lambda \in \sigma_{\text{ap}}(T)$ if and only if $\lambda \in \sigma_\ell(T)$, the *left spectrum* of T (see [**ACFA**], page 200). Also $\sigma_{\ell e}(T)$ is the *left essential spectrum* of T; that is, the left spectrum of the coset of T in the Calkin algebra $\mathscr{B}(\mathscr{H})/\mathscr{B}_0(\mathscr{H})$. The *right essential spectrum* and the *essential spectrum* of T are denoted by $\sigma_{\text{re}}(T)$ and $\sigma_e(T)$, respectively. The reader should read §XI.4 of [**ACFA**] for the properties of these subsets of $\sigma(T)$ as well as §XI.6 for some additional information. This book will assume familiarity with these topics as well as properties of the Fredholm index as found in [**ACFA**].

3.15 PROPOSITION. *If S is the unilateral shift of multiplicity 1, then the following statements are valid.*
  (a) $\sigma(S) = \operatorname{cl}\mathbb{D}$, $\sigma_{\text{ap}}(S) = \partial\mathbb{D}$, $\sigma_p(S) = \varnothing$.
  (b) *For $|\lambda| < 1$, $\dim \ker(S^* - \lambda) = 1$ and $(1, \lambda, \lambda^2, \ldots) \in \ker(S^* - \lambda)$.*
  (c) $\sigma_e(S) = \sigma_{\ell e}(S) = \sigma_{\text{re}}(S) = \partial\mathbb{D}$.

A proof of this can be found on page 209 of [**ACFA**], but you are urged to prove it on your own.

Because a shift of multiplicity α is the direct sum of α copies of the shift of multiplicity 1, the proof of the next result can be obtained from the preceding one. (See Proposition 5.1 below.)

3.16 PROPOSITION. *If S is a unilateral shift of multiplicity α, then the following statements are valid.*
  (a) $\sigma(S) = \operatorname{cl}\mathbb{D}$, $\sigma_{\text{ap}}(S) = \partial\mathbb{D}$, $\sigma_p(S) = \varnothing$.
  (b) *For $|\lambda| < 1$, $\dim \ker(S^* - \lambda) = \alpha$.*
  (c) $\sigma_{\ell e}(S) = \partial\mathbb{D}$.
  (d) *If $\alpha < \infty$, $\sigma_{\text{re}}(S) = \sigma_e(S) = \partial\mathbb{D}$.*
  (e) *If $\alpha = \infty$, $\sigma_{\text{re}}(S) = \sigma_e(S) = \operatorname{cl}\mathbb{D}$.*

Exercises

1. Prove Proposition 3.3.

2. Show that a normal isometry is unitary.
3. If S is a unilateral shift of multiplicity α and n is a positive integer, show that S^n is a unilateral shift of multiplicity $n\alpha$.
4. Show that a bilateral shift is unitary and two bilateral shifts are unitarily equivalent if and only if they have the same multiplicity.
5. Define $S: L^2(0, \infty) \to L^2(0, \infty)$ by $(Sf)(x) = f(x-1)$ if $x \geq 1$ and $(Sf)(x) = 0$ for $x \leq 1$. Show that S is a unilateral shift and calculate its multiplicity and its minimal unitary extension.
6. Compute the adjoint of a unilateral and a bilateral shift.
7. Prove that for any bilateral shift U, U is unitarily equivalent to U^*.
8. Give an example of a unitary operator U that is not a bilateral shift and satisfies $U \cong U^*$.
9. Prove Proposition 3.15.
10. Prove Proposition 3.16.
11. Let S_1, S_2 be unilateral shifts on $\mathcal{H}_1$, $\mathcal{H}_2$ with minimal unitary extensions U_1, U_2 acting on $\mathcal{K}_1$, $\mathcal{K}_2$. Show that if $X: \mathcal{H}_1 \to \mathcal{H}_2$ is an operator such that $XS_1 = S_2 X$, then there is a unique operator $Y: \mathcal{K}_1 \to \mathcal{K}_2$ such that:
    (i) $Y \mid \mathcal{H}_1 = X$;
    (ii) $YU_1 = U_2 Y$;
    (iii) $\|Y\| = \|X\|$.
12. Let S be the unilateral shift and put $T = S \oplus S^*$; prove the following statements.
    (a) $\sigma(T) = \sigma_\ell(T) = \sigma_r(T) = \operatorname{cl} \mathbb{D}$ and $\sigma_p(T) = \mathbb{D}$.
    (b) $\sigma_e(T) = \partial \mathbb{D}$ and if $|\lambda| < 1$, then $\operatorname{ind}(T - \lambda) = 0$.
    (c) There is a rank 1 operator F such that $T + F$ is the bilateral shift.
    (d) There is a sequence of rank 1 operators $\{F_n\}$ such that $F_n \to 0$ and $\sigma(T + F_n) = \partial \mathbb{D}$ for all n. What does this say about the continuity of the spectrum?

§4 The unilateral shift of multiplicity one. In this section we will make a more detailed study of the unilateral shift of multiplicity 1.

Let U be the bilateral shift on $\ell^2(\mathbb{Z}): Ue_n = e_{n+1}$, where $\{e_n\}$ is the usual basis for $\ell^2(\mathbb{Z})$. U is a star-cyclic unitary operator and is thus unitarily equivalent to multiplication by z, the independent variable, on $L^2(\mu)$ for some measure μ supported on $\partial \mathbb{D}$ [ACFA, IX.3.4]. We begin by showing that the isomorphism that implements this unitary equivalence is the Fourier transform.

Normalized Lebesgue measure on $\partial \mathbb{D}$ will be denoted by m and the Lebesgue spaces of this measure will be denoted by $L^p(\partial \mathbb{D})$.

4.1 DEFINITION. If $f \in L^1(\partial \mathbb{D})$, then the *Fourier transform* of f is the function $\hat{f}: \mathbb{Z} \to \mathbb{C}$ defined by

$$\hat{f}(n) = \int_{\partial \mathbb{D}} f \bar{z}^n \, dm.$$

Note that since m is a finite measure, $L^p(\partial\mathbb{D}) \subseteq L^1(\partial\mathbb{D})$ for $1 \le p \le \infty$. Hence $\hat{f}$ is defined for f in $L^p(\partial\mathbb{D})$ and all p.

A *trigonometric polynomial* is a function in $C(\partial\mathbb{D})$ of the form $\sum_{k=-m}^{n} a_k z^k$.

4.2 LEMMA. *The trigonometric polynomials are uniformly dense in $C(\partial\mathbb{D})$ and hence dense in $L^p(\partial\mathbb{D})$ for $1 \le p < \infty$; they are weak* dense in $L^\infty(\partial\mathbb{D})$. Thus $\{z^n : n \in \mathbb{Z}\}$ is an orthonormal basis for $L^2(\partial\mathbb{D})$.*

PROOF. The first part is an easy consequence of the Stone-Weierstrass Theorem. The last statement only needs the calculation necessary to show that the functions z^n are orthonormal. □

4.3 THEOREM. *If $f \in L^2(\partial\mathbb{D})$, then $\hat{f} \in \ell^2(\mathbb{Z})$. If $V: L^2(\partial\mathbb{D}) \to \ell^2(\mathbb{Z})$ is defined by $Vf = \hat{f}$, then V is an isomorphism and if W denotes multiplication by z on $L^2(\partial\mathbb{D})$, then VWV^{-1} is the bilateral shift.*

PROOF. The first part, that $\hat{f} \in \ell^2(\mathbb{Z})$, as well as the statement that V is an isometry, is a direct consequence of Parseval's Identity and the fact that $\{z^n\}$ is a basis for $L^2(\partial\mathbb{D})$. If $f = z^m$, then it is straightforward to check that $\hat{f}(n) = 0$ if $n \ne m$ and $\hat{f}(m) = 1$. Thus ran V is dense and so V must be an isomorphism.

If $\{e_n\}$ is the usual basis in $\ell^2(\mathbb{Z})$, we have just seen that $V(z^n) = e_n$. Thus $VW(z^n) = V(z^{n+1}) = e_{n+1} = UV(z^n)$. This completes the proof. □

4.4 COROLLARY. *If U is a bilateral shift of any multiplicity, then $\sigma(U) = \partial\mathbb{D}$.*

Recall that the spectral properties of the unilateral shift S of multiplicity α were determined in Proposition 3.16. In particular, if $|\lambda| < 1$, then $S - \lambda \in \mathscr{SF}$, the set of semi-Fredholm operators, and $\text{ind}(S - \lambda) = -\alpha$.

4.5 DEFINITION. For $1 \le p \le \infty$, $H^p = $ all functions f in $L^p(\partial\mathbb{D})$ such that $\hat{f}(n) = 0$ for $n < 0$. (*Note*: The letter H here stands for "Hardy." G. H. Hardy was a pioneer in the study of these spaces and they are universally called Hardy spaces.)

Note that H^p is the annihilator of $\{z^{-n} : n \ge 1\}$. Since $\{z^{-n} : n \ge 1\} \subseteq L^q(\partial\mathbb{D})$ for the index q that is dual to p, H^p is a closed subspace of $L^p(\partial\mathbb{D})$. (H^∞ is a weak* closed subspace of $L^\infty(\partial\mathbb{D})$.) For consistency, let $L^p = L^p(\partial\mathbb{D})$.

Now return to the notation in Theorem 4.3. If $\ell^2 \equiv \ell^2(\mathbb{N} \cup \{0\})$, then $V^{-1}\ell^2 = H^2$. Since $V^{-1}e_n = z^n$, it follows that the analytic polynomials are dense in H^2. Also, the unilateral shift, $U \mid \ell^2$, is unitarily equivalent to $W \mid H^2$. This information is gathered together in the next theorem.

4.6 THEOREM. *The analytic polynomials are dense in H^2 and the unilateral shift is unitarily equivalent to multiplication by z on H^2. Moreover, the* *isomorphism that implements this unitary equivalence is the restriction of the inverse Fourier transform, $\ell^2 \to H^2$.*

Thus the unilateral shift will be identified with multiplication by z on H^2 and both will be denoted by S.

Since two unilateral shifts are unitarily equivalent if they have the same multiplicity, we could have observed that multiplication by z on H^2 is a unilateral shift since it shifts the basis $\{z^n : n \geq 0\}$ and concluded that it is unitarily equivalent to the shift on ℓ^2. However, it is important to realize that this unitary equivalence is related to the classical Fourier transform. This allows the application of function theory to the study of the shift.

For the unilateral shift it is helpful to know the eigenvectors for $S^* - \bar{\lambda}$, $|\lambda| < 1$. Since $\ker(S - \lambda)^*$ is one dimensional for $|\lambda| < 1$, we need only find one such eigenvector.

4.7 PROPOSITION. *If $|\lambda| < 1$ and $k_\lambda(x) = (1 - \bar{\lambda}z)^{-1}$, then $k_\lambda \in H^2$, $\|k_\lambda\| = (1 - |\lambda|^2)^{-1/2}$, $\langle p, k_\lambda \rangle = p(\lambda)$ for all polynomials, and $S^* k_\lambda = \bar{\lambda} k_\lambda$.*

PROOF. Since $|\lambda| < 1$ it follows that k_λ is an analytic function on a neighborhood of $\mathrm{cl}\,\mathbb{D}$ and $k_\lambda(z) = \sum_{n=0}^\infty \bar{\lambda}^n z^n$ for z in $\mathrm{cl}\,\mathbb{D}$, with this series converging absolutely and uniformly on $\partial \mathbb{D}$. Since $z^n \in H^2$, $k_\lambda \in H^2$. Also, Parseval's Identity implies that $\|k_\lambda\|^2 = \sum |\lambda|^{2n} = (1 - |\lambda|^2)^{-1}$. Moreover, if $j \geq 0$, then $\langle z^j, k_\lambda \rangle = \lambda^j$. Hence $\langle p, k_\lambda \rangle = p(\lambda)$ for every polynomial. Finally, if p is any polynomial and $|\lambda| < 1$, then $\langle p, S^* k_\lambda \rangle = \langle Sp, k_\lambda \rangle = \langle zp, k_\lambda \rangle = \lambda p(\lambda) = \lambda \langle p, k_\lambda \rangle = \langle p, \bar{\lambda} k_\lambda \rangle$. Since the polynomials are dense in H^2, $S^* k_\lambda = \bar{\lambda} k_\lambda$. □

The reason that the eigenvector k_λ is distinguished from its scalar multiples is the fact that $\langle p, k_\lambda \rangle = p(\lambda)$ for all polynomials p. This "reproducing" property will be seen to be important, not only for the study of the shift, but also for general subnormal operators.

We have seen that H^2 is a closed subspace of L^2 and $\{z^n : n \geq 0\}$ is an orthonormal basis for H^2. So if $f \in H^2$, $f = \sum_{n \geq 0} \hat{f}(n) z^n$, where $\hat{f}(n)$ is the nth Fourier coefficient and $\sum_n |\hat{f}(n)|^2 = \|f\|^2 < \infty$. It follows that the radius of convergence of the power series $\sum_{n \geq 0} \hat{f}(n) z^n$ is at least 1. Thus to each f in H^2 we can associate a unique analytic function defined on the open unit disk, $\mathbb{D}$. This connection with analytic functions will be exploited in Chapter III and, as a result, deep information will be obtained about H^2. In fact, there is a way to associate an analytic function on $\mathbb{D}$ to each element in any Hardy space H^p. These matters, however, must be postponed, and we must temporarily content ourselves with less ambitious projects.

The next result will only be stated. The proof is postponed until §III.1 because only a weakened version of the result is required at present and the full theorem is more appropriately proved there. The reader can, if he or she wishes, turn to §III.1 for the proof; no other prerequisites are needed.

It will be quite beneficial to consider H^∞ with its relative weak* topology.

For any formal Fourier series $\sum_{n=-\infty}^{\infty} c_n e^{in\theta} = \sum_{n=-\infty}^{\infty} c_n z^n$, let $s_n(z)$ be the nth partial sum, $s_n(z) = \sum_{k=-n}^{n} c_n z^n =$. The nth Cesàro mean of the series is defied for z on $\partial \mathbb{D}$ by

$$\sigma_n(z) = \frac{1}{n} \sum_{k=0}^{n-1} s_k(z).$$

Now if $f \in L^1$, f has a formal Fourier series $\sum_{n=-\infty}^{\infty} \hat{f}(n) z^n$. By the nth partial sum and the nth Cesàro mean for the function f, we mean the corresponding quantity for this formal series associated with f. To indicate the dependence on f, these sums will be denoted by $s_n(f, z)$ and $\sigma_n(f, z)$. Note that each Cesàro mean is a trigonometric polynomial.

4.8 Proposition. *If $\phi \in L^\infty$, $\|\sigma_n(\phi)\|_\infty \leq \|\phi\|_\infty$ and $\sigma_n(\phi) \to \phi$ weak* in L^∞.*

This appears as part (b) of Corollary III.1.7.

Now assume that $\phi \in H^\infty$. Since $\hat{\phi}(n) = 0$ for $n \leq -1$, $s_n(\phi)$, and hence $\sigma_n(\phi)$, are analytic polynomials. That is, $\sigma_n(\phi)$ is a polynomial in $1, z, \ldots, z^n$. This gives the following.

4.9 Corollary. *H^∞ is the weak* sequential closure of the analytic polynomials.*

Let $\mathcal{M}$ be the closed linear span of $\{z^{-n} : n \geq 1\}$ in L^1. So H^∞ is the annihilator of $\mathcal{M}$ in L^∞: $H^\infty = \mathcal{M}^\perp$. By general Banach space theory, H^∞ is naturally equal (isometrically isomorphic) to $(L^1/\mathcal{M})^*$. In fact, if $L^1 \to L^1/\mathcal{M}$ is the quotient map, the range of the adjoint $(L^1/\mathcal{M})^* \to L^\infty$ is the space H^∞. Thus H^∞ is the dual of a separable Banach space and has its own weak* topology. Moreover, this weak* topology is the same as the relative weak* topology that it inherits as a subspace of L^∞. See §V.2 in [ACFA] for these details.

This will enable us to apply duality theory to the study of H^∞ and the unilateral shift. But first, we give another characterization of H^∞ as a subspace of L^∞.

4.10 Proposition. *If $\phi \in H^\infty$ and $f \in H^2$, then $\phi f \in H^2$. Conversely, if $\phi \in L^\infty$ and $\phi f \in H^2$ for all f in H^2, then $\phi \in H^\infty$.*

PROOF. Let $\phi \in L^\infty$ and $k \in \mathbb{Z}$. For any n in $\mathbb{Z}$ it follows that $\widehat{\phi z^k}(n) = \int \phi z^k \bar{z}^n \, dm = \int \phi \bar{z}^{(n-k)} \, dm = \hat{\phi}(n-k)$. If $\phi \in H^\infty$, $k \geq 0$ and $n \leq -1$, then $\widehat{\phi z^k}(n) = \hat{\phi}(n-k) = 0$ since $n - k \leq -1$. Thus $\phi z^k \in H^2$ for all $k \geq 0$. Hence $\phi p \in H^2$ for all polynomials p. Since ϕ is bounded, $\phi H^2 \subseteq H^2$ by Proposition 4.6.

Conversely, assume that $\phi \in L^\infty$ and $\phi f \in H^2$ for all f in H^2. In particular, $\phi = \phi 1 \in H^2$. Hence the negative Fourier coefficients of ϕ vanish and so $\phi \in H^\infty$. □

We therefore have that for each ϕ in H^∞ the operator $\phi(S)$ can be defined on H^2 by $\phi(S)f = \phi f$. The more traditional notation is $\phi(S) = T_\phi$, the analytic Toeplitz operator with symbol ϕ.

4.11 THEOREM. *If S is the unilateral shift on H^2, then*
$$\{S\}' = \{S\}'' = P^\infty(S) = \{\phi(S) : \phi \in H^\infty\},$$
where $P^\infty(S)$ is the weak closed algebra generated by S and the identity. Moreover, the map $\phi \to \phi(S)$ of H^∞ onto $P^\infty(S)$ is an isometric algebraic isomorphism and a weak* homeomorphism.*

PROOF. If p is a polynomial, then it is easy to see that $p(S) \in \{S\}''$. So by general operator theory, $P^\infty(S) \subseteq \{S\}'' \subseteq \{S\}'$.

Define $\rho : H^\infty \to \mathscr{B}(H^2)$ by $\rho(\phi) = \phi(S)$. It is easy to check that ρ is an algebraic homomorphism. Also for ϕ in H^∞ and f in H^2, $\|\phi f\|^2 = \int |\phi f|^2 \, dm \leq \|\phi\|_\infty^2 \|f\|^2$ and so $\|\phi(S)\| \leq \|\phi\|_\infty$.

Suppose $\{\phi_i\}$ is a net in H^∞ and $\phi_i \to 0$ weak*. If $f, g \in H^2$, then $\langle \phi_i(S)f, g \rangle = \int \phi_i f \bar{g} \, dm$. But $f\bar{g} \in L^1$ and so $\phi_i(S) \to 0$ (WOT). That is, the map $\rho : (H^\infty, \text{weak*}) \to (\mathscr{B}(H^2), \text{WOT})$ is continuous. Since the polynomials are weak* dense in H^∞, this shows that $\rho(H^\infty) \subseteq P^\infty(S) \subseteq \{S\}'' \subseteq \{S\}'$. It must be shown that $\{S\}' \subseteq \rho(H^\infty)$.

Let $T \in \{S\}'$ and put $\phi = T(1)$; so $\phi \in H^2$. It is easy to check that for any polynomial p, $T(p) = \phi p$. If $f \in H^2$, let $\{p_n\}$ be a sequence of polynomials such that $p_n \to f$ in H^2. By passing to a subsequence if necessary, it may be assumed that $p_n(z) \to f(z)$ a.e. $[m]$. Thus $\phi p_n = T(p_n) \to T(f)$ in H^2 and $\phi p_n \to \phi f$ a.e. $[m]$. Thus $Tf = \phi f$ for all f in H^2.

Without loss of generality we may assume that $\|T\| = 1$. Algebraic manipulation shows that $T^k f = \phi^k f$ for f in H^2 and $k \geq 1$. Hence $\|\phi^k f\|_2 \leq \|f\|_2$ for all $k \geq 1$. Taking $f = 1$ shows that $\int |\phi|^{2k} \, dm \leq 1$ for all $k \geq 1$. If $\Delta = \{z \in \partial \mathbb{D} : |\phi(z)| > 1\}$, then $1 \geq \int_\Delta |\phi|^{2k} \, dm$ for all $k \geq 1$. If $m(\Delta) \neq 0$, then this sequence of integrals converges monotonically to ∞, giving a contradiction. Thus $m(\Delta) = 0$, and so $\phi \in L^\infty$ and $\|\phi\|_\infty \leq 1$. This shows two things. First $\phi H^2 = TH^2 \subseteq H^2$, so, in fact, $\phi \in H^\infty$ and $T = \phi(S)$. This completes the proof that $\{S\}' = \{S\}'' = P^\infty(S) = \{\phi(S) : \phi \in H^\infty\}$.

But this argument also shows that for any ϕ in H^∞, $\|\phi\|_\infty \leq \|\phi(S)\|$ and hence ρ is an isometry of H^∞ onto $P^\infty(S)$. A consideration of the weak* continuity properties of ρ that have already been established, shows, by Proposition 2.7, that $\rho(H^\infty) = P^\infty(S)$ is a weak* closed algebra in $\mathscr{B}(H^2)$ and ρ is a weak* homeomorphism. □

We can now prove one of the most celebrated theorems in operator theory. Let $\phi \in H^\infty$ and consider $\mathscr{M} = \phi H^2$. If S is the unilateral shift, it is clear that $S\mathscr{M} \subseteq \mathscr{M}$. So if $\mathscr{M}$ is closed, $\mathscr{M} \in \operatorname{Lat} S$, the lattice of invariant subspaces of S. When is $\mathscr{M}$ closed? If there is an $\varepsilon > 0$ such that $|\phi| \geq \varepsilon$ a.e. $[m]$ on $\partial \mathbb{D}$, then $\mathscr{M} = \phi H^2$ is closed. Note that it is quite likely that $\phi H^2 \neq H^2$. For example, take $\phi = z$. The next theorem gives a converse of this observation.

4.12 BEURLING'S THEOREM (Beurling [1949]). *If S is the unilateral shift, then*

$$\operatorname{Lat} S = \{\phi H^2 : \phi \in H^\infty \text{ and } |\phi(z)| = 1 \text{ a.e. } [m]\}.$$

PROOF. In the paragraph prior to the statement of the theorem, it was shown that $\phi H^2 \in \operatorname{Lat} S$ if $|\phi| = 1$. For the converse, let $\mathscr{M} \in \operatorname{Lat} S$ and let U be the bilateral shift on L^2; so $S = U \mid H^2$. By Proposition 3.14, $S \mid \mathscr{M} = U \mid \mathscr{M}$ is a unilateral shift of multiplicity 1 and, hence, is unitarily equivalent to S. Let $W: H^2 \to \mathscr{M}$ be an isomorphism such that $WSW^{-1} = S \mid \mathscr{M}$. Note that U is a unitary extension of $S \mid \mathscr{M}$, so if $\mathscr{K}$ is the smallest reducing subspace for U that contains $\mathscr{M}$, $U \mid \mathscr{K}$ is the minimal unitary extension of $S \mid \mathscr{M}$. But U is a cyclic unitary, so there is a Borel subset Δ of $\partial \mathbb{D}$ such that $\mathscr{K} = \mathscr{X}_\Delta L^2$ (see IX.6.9 in [ACFA]). But $U \cong U \mid \mathscr{K}$ since both are bilateral shifts of multiplicity 1. Hence $m(\partial \mathbb{D} \backslash \Delta) = 0$; that is, $\mathscr{K} = L^2$ and U is the minimal unitary extension of $S \mid \mathscr{M}$.

By Exercise 3.11, there is an operator $V: L^2 \to L^2$ such that $V \mid H^2 = W$ and $VU = UV$. Hence there is a ϕ in L^∞ such that $V = M_\phi$ (See IX.6.9 in [ACFA]). But the fact that W is an isometry implies that V is an isometry. (You must examine the proof of Exercise 3.11 to see this, but it is easy.) Since V is an isometry, $|\phi| = 1$. Finally, if $f \in H^2$, then $\phi f = Vf = Wf \in \mathscr{M} \subseteq H^2$ so that by Proposition 4.10, $\phi \in H^\infty$. $\square$

There is a catch here. The preceding result is really not the full Beurling Theorem. The full strength of the theorem will only come to the fore when we can fully understand the nature of the functions ϕ in H^∞ with $|\phi| = 1$. For example, we observed before the statement of the theorem that for any ψ in H^2, $\operatorname{cl}[\psi H^2]$ is a closed invariant subspace for the shift. According to Beurling's Theorem, for any such ψ there is a ϕ in H^∞ with $|\phi| = 1$ and $\phi H^2 = \operatorname{cl}[\psi H^2]$. How do we get ϕ from ψ? If ϕ_1 and $\phi_2 \in H^\infty$ and $|\phi_1| = |\phi_2| = 1$, then $\phi_1 H^2 \cap \phi_2 H^2$ and $\phi_1 H^2 \vee \phi_2 H^2 \in \operatorname{Lat} S$. Thus there are functions ψ and $\mathscr{X}$ in H^∞ with $|\psi| = |\mathscr{X}| = 1$ and $\phi_1 H^2 \cap \phi_2 H^2 = \psi H^2$ and $\phi_1 H^2 \vee \phi_2 H^2 = \mathscr{X} H^2$. What is the relation of ψ and $\mathscr{X}$ to ϕ_1 and ϕ_2? Can it ever be that $\phi_1 H^2 \cap \phi_2 H^2$ is (0)? All these are questions that we should be able to answer before we can state with full justification that we "know" the lattice of invariant subspaces for the shift. All these are questions that we will answer, but only after we devote considerable effort to the study of the function theory associated with the Hardy spaces, which will be taken up in Chapter III.

Exercises

1. For $f = \sum_{n \geq 0} c_n z^n$ in H^2, extend f to an analytic function on $\mathbb{D}$ by letting $f(z) = \sum_{n \geq 0} c_n z^n$ for $|z| < 1$. Show that $\langle f, k_\lambda \rangle = f(\lambda)$.
2. If $\phi \in H^\infty$ and $|\lambda| < 1$, show that $\phi(S)^* k_\lambda = \overline{\phi(\lambda)} k_\lambda$. ($H^\infty \subseteq H^2$ and so $\phi(\lambda)$ is defined for λ in $\mathbb{D}$ as in Exercise 1.)
3. Show that for every ϕ in H^∞, $\sigma(\phi(S)) = \text{cl}[\phi(\mathbb{D})]$.
4. Let U be the bilateral shift on L^2. If $\mathcal{M} \in \text{Lat } U$ such that $\mathcal{M}$ does not reduce U, then there is a ϕ in L^∞ with $|\phi| = 1$ such that $\mathcal{M} = \phi H^2$. Discuss the uniqueness of ϕ. (This exercise is continued as Exercise III.10.11.)
5. Let p be a polynomial that does not vanish on $\partial \mathbb{D}$ and show that $p(S)$ is a Fredholm operator and
$$\text{ind } p(S) = -\frac{1}{2\pi i} \int_{|z|=1} \frac{p'(z)}{p(z)} dz.$$

§5 Direct sums. If $\mathcal{H}$ is the direct sum of the Hilbert spaces $\mathcal{H}_1, \mathcal{H}_2, \ldots$ and $T \in \mathcal{B}(\mathcal{H})$, then T can be represented as a matrix $T = [T_{ij}]$, where $T_{ij}: \mathcal{H}_j \to \mathcal{H}_i$ is a bounded operator. If this direct sum is finite, the converse is also true. If, however, the direct sum is infinite, then there are no known conditions on an infinite matrix of bounded operators $[T_{ij}]$ so that it defines a bounded operator of $\mathcal{H}_1 \oplus \mathcal{H}_2 \oplus \cdots$ into itself.

If $T_1, T_2, \ldots$ are bounded operators on $\mathcal{H}_1, \mathcal{H}_2, \ldots$, then it is easy to see that $T = T_1 \oplus T_2 \oplus \cdots$ is a bounded operator on $\mathcal{H}_1 \oplus \mathcal{H}_2 \oplus \cdots$ if and only if $\sup_n \|T_n\| < \infty$, in which case $\|T\| = \sup_n \|T_n\|$. This operator T corresponds to the diagonal matrix with entries $T_1, T_2, \ldots$. The proof of the next proposition is left to the reader.

5.1 PROPOSITION. *Let $T_k \in \mathcal{B}(\mathcal{H}_k)$ for $k \geq 1$.*
   (a) $\sigma(T_1 \oplus \cdots \oplus T_n) = \bigcup_{k=1}^n \sigma(T_k)$.
   (b) *If* $T = T_1 \oplus T_2 \oplus \cdots$ *and for each* $k \geq 1$, $\|(\lambda - T_k)^{-1}\| = [\text{dist}(\lambda, \sigma(T_k))]^{-1}$ *whenever* $\lambda \notin \sigma(T_k)$, *then* $\sigma(T) = \text{cl}[\bigcup_k \sigma(T_k)]$.

Without some extra condition such as that given in part (b), the conclusion of (b) is not valid. See, for example, Solution 98 in Halmos [1982].

5.2 DEFINITION. If $\mathcal{H}$ is a Hilbert space and $1 \leq n \leq \infty$, let $\mathcal{H}^{(n)}$ denote the direct sum of $\mathcal{H}$ with itself n times. If $T \in \mathcal{B}(\mathcal{H})$, $T^{(n)}$ is the direct sum of T with itself n times, acting on $\mathcal{H}^{(n)}$. Thus $T^{(n)}$ corresponds to a diagonal matrix with T constantly along the diagonal.

The proof of the next result is conceptually easy; the only difficulty (if there is any) lies in establishing good notation to facilitate good bookkeeping. This is left to the reader.

5.3 PROPOSITION. (a) *If* $T = T_1 \oplus T_2 \oplus \cdots$ *is a bounded operator on* $\mathcal{H} = \mathcal{H}_1 \oplus \mathcal{H}_2 \oplus \cdots$ *and* $A = [A_{ij}] \in \mathcal{B}(\mathcal{H})$, *then* $AT = TA$ *if and only if* $A_{ij}T_j = T_i A_{ij}$ *for all i and j.*

(b) *If $\mathcal{H}$ is any Hilbert space and* $T \in \mathcal{B}(\mathcal{H})$, *then* $A = [A_{ij}]$ *in* $\mathcal{B}(\mathcal{H}^{(n)})$ *commutes with* $T^{(n)}$ *if and only if* $A_{ij}T = TA_{ij}$ *for all i and j.*

Recall that if $\mathcal{S} \subseteq \mathcal{B}(\mathcal{H})$, the *commutant* of $\mathcal{S}$, $\mathcal{S}'$, is defined to be the set of all those operators A in $\mathcal{B}(\mathcal{H})$ such that $AT = TA$ for all T in $\mathcal{S}$. The *double commutant* of $\mathcal{S}$, $\mathcal{S}''$, is the commutant of $\mathcal{S}'$. That is, $\mathcal{S}'' = \{\mathcal{S}'\}'$. Once again the next proposition can be obtained by the reader on his own or he can see IX.6.2 in [ACFA].

5.4 PROPOSITION. *If $T \in \mathcal{B}(\mathcal{H})$, then* $\{T^{(n)}\}'' = \{A^{(n)} : A \in \{T\}''\} = \{\{T\}''\}^{(n)}$.

Exercises

1. Supply the proofs for the results in this section.
2. (a) If S is a unilateral shift of multiplicity α and $\mathcal{L}$ is a Hilbert space of dimension α, show that S is unitarily equivalent to the operator on $\mathcal{L}^{(\infty)}$ defined by the matrix

$$(5.5) \quad \begin{bmatrix} 0 & 0 & 0 & \cdots \\ 1 & 0 & 0 & \cdots \\ 0 & 1 & 0 & \cdots \\ \cdot & \cdot & \cdot & \end{bmatrix}.$$

(b) If S is the matrix shown in (5.5), show that $A = [A_{ij}] \in \{S\}'$ if and only if $A_{ij} = 0$ for $j > i$ and $A_{ij} = A_{i+1, j+1}$ for $i \geq j$.

(c) Prove that $A = [A_{ij}] \in \{S\}''$ if and only if $A_{ij} = 0$ for $j > i$ and $A_{ij} = A_{i+1, j+1} = $ a multiple of the identity for $i \geq j$.

3. Let $\{e_n\}$ be an orthonormal basis for $\mathcal{H}$ and let $\{\lambda_n\} \subseteq \mathbb{C}$ with $\sup_n |\lambda_n| < \infty$. If T is defined by $Te_n = \lambda_n e_n$ for all $n \geq 1$, determine the commutant of T. Give necessary and sufficient conditions on $\{\lambda_n\}$ such that $\{T\}' = \{T\}''$.
4. (Conway and Wu [1977]) For any operator T let $P(T)$ denote the norm closed algebra generated by T and 1; that is, $P(T)$ is the norm closure of $\{p(T) : p$ is an analytic polynomial$\}$. Show that $P(T_1 \oplus T_2) = P(T_1) \oplus P(T_2)$ if and only if the polynomially convex hulls of $\sigma(T_1)$ and $\sigma(T_2)$ are disjoint. If this happens, then every invariant subspace of $T_1 \oplus T_2$ is of the form $\mathcal{M}_1 \oplus \mathcal{M}_2$, where $\mathcal{M}_1 \in \text{Lat } T_1$ and $\mathcal{M}_2 \in \text{Lat } T_2$.

5. (Conway and Wu [**1977**]) For an operator T let $R(T)$ be the norm closure of $\{r(T) : r \text{ is a rational function with poles off } \sigma(T)\}$. Show that $R(T_1 \oplus T_2) = R(T_1) \oplus R(T_2)$ if and only if $\sigma(T_1) \cap \sigma(T_2) = \varnothing$.
6. Let $A \in \mathscr{B}(\mathscr{H})$ and put $T = A^{(\infty)}$. Prove the following statements.
   (a) $\sigma(T) = \sigma_e(T) = \sigma(A)$.
   (b) $\sigma_p(T) = \sigma_p(A)$.
   (c) $\sigma_{\ell e}(T) = \sigma_\ell(T) = \sigma_l(A)$ $(= \sigma_{ap}(A))$ and $\sigma_{re}(T) = \sigma_r(T) = \sigma_r(A)$.
   (d) If $\lambda \in \sigma(T) \backslash \sigma_{le}(T) \cap \sigma_{re}(T)$, give a necessary and sufficient condition that $\operatorname{ind}(T - \lambda) = -\infty$.

§6 Shifts of higher multiplicity. Since a unilateral shift of multiplicity α is the direct sum of α shifts of multiplicity 1, the results of the last two sections can be combined to yield information about shifts of higher multiplicity. The treatment here is not exhaustive. For example, the analogue of Beurling's Theorem is not given. The reader interested in the extension of this result to shifts of higher multiplicity can consult Halmos [**1961**], Helson [**1964**], or Sz-Nagy and Foias [**1970**].

If S is the unilateral shift, then (5.3) and (5.4) characterize $\{S^{(n)}\}'$ and $\{S^{(n)}\}''$ in terms of $\{S\}' = \{S\}'' = P^\infty(S) = \{\phi(S) : \phi \in H^\infty\}$ (4.11). Piecing together these results, the following theorems can be derived. The details are left to the reader.

6.1 THEOREM. *If S is the unilateral shift on H^2 and $1 \le n \le \infty$, then*
$$\{S^{(n)}\}' = \{[\phi_{ij}(S)] \in \mathscr{B}((H^2)^{(n)}) : \phi_{ij} \in H^\infty\}.$$

6.2 THEOREM. *If S is the unilateral shift on H^2 and $1 \le n \le \infty$, then*
$$P^\infty(S^{(n)}) = \{S^{(n)}\}'' = \{\phi(S)^{(n)} : \phi \in H^\infty\}.$$

Moreover, the map $\phi \to \phi(S)^{(n)}$ is an isometric algebraic isomorphism of H^∞ onto $P^\infty(S^{(n)})$ that is a weak* homeomorphism.

CHAPTER II

Subnormal Operators: The Element Theory

In this chapter most of the elementary theory of subnormal operators is presented. This is the part of the theory that does not rely heavily on a knowledge of function algebras. In addition, several examples will be explored at length. As the reader will see after the definition, all unilateral shifts are subnormal operators, so that §1.4 and §1.6 are instances of the kind of in-depth information we would like to have about all subnormal operators. Unfortunately, our desires remain as yet unfulfilled.

§1 Definition and examples. In this section the definition of a subnormal operator is given and several equivalent formulations are presented.

1.1 DEFINITION. An operator S on a Hilbert space $\mathscr{H}$ is *subnormal* if there is a Hilbert space $\mathscr{K}$ containing $\mathscr{H}$ and a normal operator N on $\mathscr{K}$ such that $N\mathscr{H} \subseteq \mathscr{H}$ and $S = N|\mathscr{H}$.

In other words, an operator is subnormal if it has a normal extension. The study of normal operators has been distinctly successful. The Spectral Theorem gives a concrete structure to these operators and, as a consequence, necessary and sufficient conditions for two normal operators to be unitarily equivalent can be stated in terms of objects that can usually be calculated for any concretely defined normal operator. In fact, it is fair to say that almost any question concerning normal operators can be answered. (This last statement is, of course, not one hundred percent true. There are questions about normal operators that continue to defy solution. See Nikolskii [**1969**] and Makarov [**1987**], for example.)

We would like to understand the structure of as many operators as possible. The concept of "subnormality" is sufficiently close to "normal" that there is a reasonable expectation that an understanding of subnormal operators is within grasp. After all, success tends to be contagious. Indeed, many of the questions and conjectures concerning subnormal operators, especially in the early development of the area, have been inspired by the theory of normal operators. For the elementary properties, this analogy is valid and revealing. It will become clear as this book progresses, however, that there is an essential difference between subnormal and normal operators: the theory of normal operators rests on measure theory while subnormal operator theory calls for great reliance on analyticfunction theory. Indeed much of the progress in subnormal operators in the last twenty years is linked to and relies on the prior success in function theory—particularly in the study of function algebras.

There is a philosophical point in the definition of a subnormal operator that bears a little emphasis now so that it can be laid to rest and not concern us as we progress through the book. When, in the definition, we say that $\mathscr{K}$ "contains" $\mathscr{H}$, we actually mean that there is an isometry $U : \mathscr{H} \to \mathscr{K}$. With this, the remainder of the definition translates to the existence of a normal operator N on $\mathscr{K}$ such that $NU\mathscr{H} \subseteq U\mathscr{H}$ and $N|U\mathscr{H} = USU^*|U\mathscr{H}$. In the definition above, and in the rest of the book, this technicality will be ignored. The reader, however, might avoid some confusion if this technically precise formulation is kept in mind. This will be especially true, for example, when we consider the uniqueness of the normal extension.

By Theorem I.3.11, every isometry is subnormal. Here are additional examples that are important not only because they serve as illustrations and sources of intuition.

1.2 EXAMPLE. Let G be a bounded open subset of $\mathbb{C}$ and let $L^2(G)$ denote the L^2-space of Lebesgue area measure restricted to G. The *Bergman space* is the subspace $L_a^2(G)$ of $L^2(G)$ consisting of those functions that are equal a.e. to an analytic function. $L_a^2(G)$ will be shown to be complete in §8. The *Bergman operator* for G is the operator S in $\mathscr{B}(L_a^2(G))$ defined by $(Sf)(z) = zf(z)$ for all f in $L_a^2(G)$. If $N : L^2(G) \to L^2(G)$ is defined by $(Nf)(z) = zf(z)$ for f in $L^2(G)$, then N is normal and N is an extension of S. Thus S is subnormal. These operators are studied more closely in §8.

1.3 EXAMPLE. Let μ be a compactly supported measure on $\mathbb{C}$ and let $P^2(\mu)$ denote the closure in $P^2(\mu)$ of the analytic polynomials. Define $S_\mu : P^2(\mu) \to P^2(\mu)$ by $(S_\mu f)(z) = zf(z)$ for all f in $P^2(\mu)$. If N_μ is defined as multiplication by z on $L^2(\mu)$, then N_μ is normal. In fact, N_μ is the general star-cyclic normal operator (see IX.3.4 in [ACFA]). If p is an analytic polynomial, then $N_\mu p = zp$ is also a polynomial. By passing to limits we see that N_μ leaves $P^2(\mu)$ invariant and thus S_μ is a subnormal operator with N_μ as a normal extension. It will be shown in §5 that S_μ is the general cyclic subnormal operator.

By Theorem I.4.6, the unilateral shift is the operator S_m, where m is normalized Lebesgue measure on $\partial \mathbb{D}$. A generalization of the preceding result is the following.

1.4 EXAMPLE. Let μ be as in the preceding example and let K be an arbitrary compact subset of $\mathbb{C}$ which contains the support of μ. Let $R^2(K, \mu)$ be the closure in $L^2(\mu)$ of $\text{Rat}(K)$, the rational functions with poles off K. If S is defined on $R^2(K, \mu)$ as multiplication by z, then S is subnormal with N_μ as a normal extension.

If the set K in the preceding example is polynomially convex, then Runge's Theorem implies that $R^2(K, \mu) = P^2(\mu)$.

The concept of subnormality was introduced in Halmos [1950], where these operators were called "completely subnormal". The term "subnormal" as it is used here was first introduced in Halmos [1952].

The next class of operators was introduced in Brown [1953].

1.5 DEFINITION. An operator S is *quasinormal* if S and S^*S commute.

1.6 PROPOSITION. *If $S = UA$ is the polar decomposition of S, then S is quasinormal if and only if A and U commute.*

PROOF. If $UA = AU$, then $SA^2 = A^2 S$ and so S is quasinormal. Conversely, if S is quasinormal, then S commutes with A^2 by definition. Since A can be approximated by polynomials in A^2, $SA = AS$. Hence $(UA - AU)A = SA - AS = 0$. Thus $(UA - AU) = 0$ on $\operatorname{ran} A$. But if $f \in (\operatorname{ran} A)^\perp = \ker A$, then, by the definition of U, $Uf = 0$. Therefore, $UA - AU = 0$. □

The main reason for introducing quasinormal operators is the following.

1.7 PROPOSITION. *Every quasinormal operator is subnormal.*

PROOF. Consider two cases. First assume that $\ker S = (0)$. If $S = UA$ is the polar decomposition of S, then U must be an isometry. If $E = UU^*$, then E is the projection onto the final space of U; thus $E^\perp U = U^* E^\perp = 0$. (Here $E^\perp = 1 - E$.) Define operators V and B on $\mathscr{K} = \mathscr{H} \oplus \mathscr{H}$ by

$$V = \begin{bmatrix} U & E^\perp \\ 0 & U^* \end{bmatrix}, \quad B = \begin{bmatrix} A & 0 \\ 0 & A \end{bmatrix},$$

and let $N = VB$. Since $UA = AU$ and $U^*A = AU^*$, it is easy to check that N is normal. Since

$$N = \begin{bmatrix} S & E^\perp A \\ 0 & U^* A \end{bmatrix} = \begin{bmatrix} S & E^\perp A \\ 0 & S^* \end{bmatrix},$$

it follows that N leaves $\mathscr{H} = \mathscr{H} \oplus (0)$ invariant and $N|\mathscr{H} = S$.

Now suppose that $\ker S \neq (0)$. Here $\ker S = \mathscr{L} \subseteq \ker S^*$, since $S^* = AU^* = U^*A$. Let $S_1 = S|\mathscr{L}^\perp$; so $S = S_1 \oplus 0$ on $\mathscr{L}^\perp \oplus \mathscr{L} = \mathscr{H}$. Now $S^*S = S_1^*S_1 \oplus 0$ and it is easy to see that S_1 is quasinormal. By the first part, S_1 is subnormal. Clearly S is subnormal. □

1.8 EXAMPLE. If A is a positive operator on a Hilbert space $\mathscr{L}$, define S on $\mathscr{H} = \mathscr{L}^{(\infty)}$ by the matrix

$$S = \begin{bmatrix} 0 & 0 & 0 & \dots \\ A & 0 & 0 & \dots \\ 0 & A & 0 & \dots \\ . & . & A & \dots \end{bmatrix}.$$

It is left as an easy exercise for the reader to show that S is a bounded quasinormal operator. (What is the normal extension of such an operator?) Later, in §3, it will be shown that every quasinormal operator is the direct sum of a normal operator and an operator of this form.

We now turn our attention to giving several equivalent formulations of subnormality. Each of these formulations has something to commend it, even if this is only the fact that some contemplated generalization of subnormality is not a generalization but only a reformulation of the same concept.

If X is a locally compact space, a *positive operator-valued measure* (POM) on X is a function Q that assigns to each Borel set Δ contained in X a positive operator $Q(\Delta)$ in $\mathscr{B}(\mathscr{H})$ such that $Q(X) = 1$ and for each f in $\mathscr{H}$, $\langle Q(\cdot)f, f\rangle$ is a regular Borel measure on X. Every spectral measure on X is a POM, but the converse is false. For example, let E be a spectral measure on X with values in $\mathscr{B}(\mathscr{K})$ and let $\mathscr{H}$ be any subspace of $\mathscr{K}$; define $Q(\Delta) = PE(\Delta)|\mathscr{H}$, where P is the orthogonal projection of $\mathscr{K}$ onto $\mathscr{H}$. It can be checked that Q is a POM with $\|Q(\Delta)\| \leq 1$ for all Δ, but Q is a spectral measure if and only if P commutes with $E(\Delta)$ for every Borel set Δ. It is a result of Naimark that the converse of this construction is valid; that is, if Q is a POM with $\|Q(\Delta)\| \leq 1$ for all Borel sets Δ, then there is a spectral measure E with values in $\mathscr{B}(\mathscr{K})$ for some larger Hilbert space $\mathscr{K}$ such that $Q(\cdot) = PE(\cdot)|\mathscr{H}$. See Berberian [**1966**] and Fillmore [**1970**] for this fact and more concerning POMs.

If Q is a POM and ϕ is a bounded Borel function on X, then $\int \phi\, dQ$ denotes the unique operator T defined by the bounded quadratic form

$$\langle Tf, f\rangle = \int \phi(x)\, d\langle Q(x)f, f\rangle.$$

1.9 THEOREM. *If $S \in \mathscr{B}(\mathscr{H})$, then the following statements are equivalent.*

(a) S *is subnormal.*

(b) S *has a quasinormal extension.*

(c) (*Halmos* [**1950**]) *If $f_0, \ldots, f_n \in \mathscr{H}$, then*

(1.10) $$\sum_{j,k} \langle S^j f_k, S^k f_j\rangle \geq 0;$$

and there is a constant $c > 0$ such that for any $f_0, \ldots, f_n$ in $\mathscr{H}$,

(1.11) $$\sum_{j,k} \langle S^{j+1} f_k, S^{k+1} f_j\rangle \leq c \sum_{j,k} \langle S^j f_k, S^k f_j\rangle.$$

(d) (*Syzmanski* [**preprint**]) *There is a constant $c > 0$ such that condition (1.11) holds for any $f_0, \ldots, f_n$ in $\mathscr{H}$.*

(e) (*Bram* [**1955**]) *Condition (1.10) holds for any $f_0, \ldots, f_n$ in $\mathscr{H}$.*

(f) (*Embry* [**1973**]) *For any $f_0, \ldots, f_n$ in $\mathscr{H}$*

(1.12) $$\sum_{j,k} \langle S^{j+k} f_j, S^{j+k} f_k\rangle \geq 0.$$

(g) (*Bunce and Deddens* [**1977**]) *If $B_0, \ldots, B_n \in C^*(S)$, the C^* algebra generated by S, then*

$$\sum_{j,k} B_j^* S^{*k} S^j B_k \geq 0.$$

(h) (*Embry* [**1973**]) *There is a positive operator-valued measure Q on some interval $[0, a]$ in $\mathbb{R}$ such that for every $n \geq 0$,*

$$S^{*n} S^n = \int t^{2n} \, dQ(t).$$

(i) (*Bram* [**1955**]) *There is a POM Q supported on a compact subset of $\mathbb{C}$ such that for all $m, n \geq 0$,*

(1.13)
$$S^{*n} S^m = \int \bar{z}^n z^m \, dQ(z).$$

PROOF. Clearly (a) implies (b). Conversely, if (b) holds, then S is subnormal by Proposition 1.7. Thus (a) and (b) are equivalent.

(a) *implies* (c). Let N be a normal operator on $\mathscr{K}$ with $\mathscr{H} \subseteq \mathscr{K}$ and $N|\mathscr{H} = S$ and let P be the projection of $\mathscr{K}$ onto $\mathscr{H}$. It is easy to see that for f in $\mathscr{H}$, $S^{*n} f = P N^{*n} f$. If $f_0, \ldots, f_n \in \mathscr{H}$, then

$$\sum_{j,k} \langle S^j f_k, S^k f_j \rangle = \sum_{j,k} \langle N^j f_k, N^k f_j \rangle = \sum_{j,k} \langle N^{*k} N^j f_k, f_j \rangle = \sum_{j,k} \langle N^j N^{*k} f_k, f_j \rangle$$

$$= \sum_{j,k} \langle N^{*k} f_k, N^{*j} f_j \rangle = \left\| \sum_k N^{*k} f_k \right\|^2.$$

Thus (1.10) holds. If $g_k = S f_k$ and each f_k in the preceding equations is replaced by g_k, then we get that

$$\sum_{j,k} \langle S^{j+1} f_k, S^{k+1} f_j \rangle = \sum_{j,k} \langle S^j g_k, S^k g_j \rangle = \left\| \sum_k N^{*k} N f_k \right\|^2$$

$$= \left\| N \sum_k N^{*k} f_k \right\|^2 \leq \|N\|^2 \sum_{j,k} \langle S^j f_k, S^k f_j \rangle.$$

Therefore (1.11) is valid with $c = \|N\|^2$.

(c) *implies* (d). Clear.

(d) *implies* (e). First observe that it suffices to assume that $\|S\| < 1$. In fact for an arbitrary constant $\rho > 0$ and $S_1 = \rho^{-1} S$, $\sum_{j,k} \langle S_1^j f_k, S_1^k f_j \rangle = \sum_{j,k} \langle S^j h_k, S^k h_j \rangle$ and $\sum_{j,k} \langle S_1^{j+1} f_k, S_1^{k+1} f_j \rangle = \rho^{-2} \sum_{j,k} \langle S^{j+1} h_k, S^{k+1} h_j \rangle$, where $h_k = \rho^{-k} f_k$. Thus (1.11) holds if and only if the same condition holds for S_1. Similarly, (1.10) holds for S if and only if it holds for S_1. Thus we may assume that $\|S\| < 1$.

Next take just 2 vectors in (1.11), $f_0 = 0$ and f_1 arbitrary, and see that the constant c in question must satisfy $c \geq \|S\|^2$. Since $\|S\| < 1$ and (1.11) remains valid for any larger value of c, it is clear that it may be assumed that $c \geq 1$.

Let $\mathscr{K} = \mathscr{H}^{(\infty)}$ and let $\mathscr{K}_0$ be the finitely nonzero sequences in $\mathscr{K}$. Let M be the infinite operator matrix whose (j, k) entry is $S^{*k}S^j$ and let M act on $\mathscr{K}_0$. If $f = (f_0, \ldots, f_n, 0, \ldots) \in \mathscr{K}_0$, then

$$\sum_j \|(Mf)_j\|^2 = \sum_j \left\| \sum_k S^{*k} S^j f_k \right\|^2$$

$$\leq \sum_j \left[\sum_k \|S\|^{k+j} \|f_k\| \right]^2$$

$$\leq \sum_j \left[\sum_k \|S\|^{2k+2j} \right] \left[\sum_k \|f_k\|^2 \right]$$

$$\leq (1 - \|S\|^2)^{-2} \|f\|^2;$$

so that since $\|S\| < 1$, $Mf \in \mathscr{K}$ and M extends to a bounded operator on $\mathscr{K}$. Clearly M is hermitian.

Note that $\langle Mf, f \rangle_{\mathscr{K}} = \sum_{j,k} \langle S^j f_k, S^k f_j \rangle$; so to prove that (1.10) holds, we must prove that M is positive. (We will drop the subscript $\mathscr{K}$ in the notation for the inner product for $\mathscr{K}$.) Note that $\langle S^{(\infty)*} M S^{(\infty)} f, f \rangle = \sum_{j,k} \langle S^{j+1} f_k, S^{k+1} f_j \rangle$, so that $S^{(\infty)*} M S^{(\infty)} \leq cM$ on $\mathscr{K}$ by (1.11). Iterating this inequality $\langle S^{(\infty)*2} M S^{(\infty)2} f, f \rangle = \langle (S^{(\infty)*} M S^{(\infty)}) S^{(\infty)} f, S^{(\infty)} f \rangle \leq c \langle M S^{(\infty)} f, S^{(\infty)} f \rangle = c \langle S^{(\infty)*} M S^{(\infty)} f, f \rangle \leq c^2 \langle Mf, f \rangle$. Continuing we get that $\langle Mf, f \rangle \geq c^{-n} \langle S^{(\infty)*n} M S^{(\infty)n} f, f \rangle$ for all f in $\mathscr{K}$. But $c \geq 1$ implies that $\{c^{-n}\}$ is a bounded sequence. Also $\|S^{(\infty)*n} M S^{(\infty)n}\| \leq \|S\|^{2n} \|M\|$ and this converges to 0 since $\|S\| < 1$. Thus $\langle Mf, f \rangle \geq \lim_n c^{-n} \langle S^{(\infty)*n} M S^{(\infty)n} f, f \rangle = 0$ and (1.10) is established.

(e) *implies* (f). If $g_k = S^k f_k$, then (1.10) implies that $\sum_{j,k} \langle S^j g_k, S^k g_j \rangle = \sum_{j,k} \langle S^{j+k} f_k, S^{j+k} f_j \rangle$, so (1.12) holds.

(e) *implies* (g). If $B_0, \ldots, B_n \in C^*(S)$ and $f \in \mathscr{H}$, let $f_k = B_k f$ and write out (1.10) to obtain that (g) is valid.

(g) *implies* (e). A routine application of Zorn's Lemma shows that any operator is the direct sum of operators that have a star-cyclic vector. Thus it suffices to prove that (1.10) holds when S is assumed to have a star-cyclic vector e_0; that is, assume that $\mathscr{H} = \text{cl}[C^*(S) e_0]$. If $B_0, \ldots, B_n \in C^*(S)$, then (g) implies that (1.10) holds for $f_k = B_k e_0$. But since (1.10) holds for a dense set of vectors, it holds for all vectors.

(a) *implies* (i). Let $N = \int z \, dE(z)$ be the spectral decomposition of N, a normal extension of S acting on $\mathscr{\tilde{K}} \supseteq \mathscr{H}$. Let P be the orthogonal projection of $\mathscr{\tilde{K}}$ onto $\mathscr{H}$ and define $Q(\Delta) = PE(\Delta)|\mathscr{H}$ for every Borel subset Δ of $\mathbb{C}$. Clearly Q is a POM and is supported on $\sigma(N)$. Also for any h in $\mathscr{H}$, $\langle S^{*n} S^m h, h \rangle = \langle N^{*n} N^m h, h \rangle = \int \bar{z}^n z^m \, d\langle E(z) h, h \rangle = \int \bar{z}^n z^m \, d\langle E(z) h, Ph \rangle = \int \bar{z}^n z^m \, d\langle PE(z) h, h \rangle = \int \bar{z}^n z^m \, d\langle Q(z), h, h \rangle$. This establishes (i).

§1 DEFINITION AND EXAMPLES   33

(i) *implies* (h). Let Q be the POM hypothesized in (i) and let K be the support of Q. For a Borel set Δ contained in $[0, \infty)$, define $Q_+(\Delta) = Q(\{z \in \mathbb{C} : |z| \in \Delta\})$. In fact, $Q_+(\Delta) = Q(\tau^{-1}(\Delta))$, where $\tau(z) = |z|$, and so Q_+ is a POM and is supported on the interval $[0, a]$, where $a = \max\{|z| : z \in K\}$. For any f in $\mathscr{H}$, $\int t^{2n} d\langle Q_+(t)f, f\rangle = \int |z|^{2n} d\langle Q(z)f, f\rangle$ by the change of variables formula for integrals. This proves (h).

(h) *implies* (f). Fix $f_0, \ldots, f_n$ in $\mathscr{H}$ and define scalar-valued measures μ_{jk} by $\mu_{jk}(\Delta) = \langle Q(\Delta)f_j, f_k\rangle$. Let μ be a positive measure on $[0, a]$ such that $\mu_{jk} \ll \mu$ for all j, k; let $h_{jk} = d\mu_{jk}/d\mu$, the Radon-Nikodym derivative. Now for each u in $C([0, a])$, $\rho(u) \equiv \int u \, dQ$ defines a bounded operator and $\rho : C([0, a]) \to \mathscr{B}(\mathscr{H})$ is a positive linear map. Note that for any u, $\langle \rho(u)f_j, f_k\rangle = \int u \, d\mu_{jk} = \int u h_{jk} d\mu$. Moreover, if $\lambda_0, \ldots, \lambda_n \in \mathbb{C}$ and $u \geq 0$, then $\sum_{j,k}(\int u h_{jk} d\mu)\lambda_j \bar{\lambda}_k = \|\sum_j \rho(u)^{1/2}\lambda_j f_j\|^2 \geq 0$. It follows (Exercise 8) that for $[\mu]$ almost every t, $(h_{jk}(t))_{j,k}$ is a positive $(n+1) \times (n+1)$ matrix. This implies that $\sum_{j,k} h_{jk}(t) t^{2j} t^{2k} \geq 0$ a.e. $[\mu]$. Therefore

$$0 \leq \int \sum_{j,k} h_{jk}(t) t^{2(j+k)} d\mu(t)$$

$$= \sum_{j,k} \int t^{2(j+k)} d\mu_{jk}(t)$$

$$= \sum_{j,k} \langle S^{j+k} f_j, S^{j+k} f_k\rangle.$$

and (1.12) holds.

(f) *implies* (b). Parts of this proof are similar to the proof that (d) implies (e) and so some details will be omitted. First it suffices to assume that $\|S\| < 1$. Let $\mathscr{K} = \mathscr{H}^{(\infty)}$. There is a bounded operator T defined on $\mathscr{K}$ whose operator matrix relative to this decomposition has entries $T_{jk} = S^{*j+k} S^{j+k}$. (Verify!) In fact, $\|T\| \leq (1 - \|S\|^4)^{-1}$. If $f = (f_0, \ldots, f_n, 0, \ldots) \in \mathscr{K}$, then $\langle Tf, f\rangle = \sum_{j,k} \langle S^{*j+k} S^{j+k} f_j, f_k\rangle$. By hypothesis, $T \geq 0$.

Now define $A = S^{(\infty)}$ on $\mathscr{K}$. For f in $\mathscr{K}$, $(A^*TAf)_j = S^*(TAf)_j = S^* \sum_k S^{*j+k} S^{j+k} (Af)_k = \sum_k S^{*j+k+1} S^{j+k+1} f_k$. Thus

$$\|A^*TAf\|^2 = \sum_{j=1}^{\infty} \left\|\sum_{k=0}^{\infty} S^{*j+k} S^{j+k} f_k\right\|^2$$

$$\leq \sum_{j=0}^{\infty} \left\|\sum_{k=0}^{\infty} S^{*j+k} S^{j+k} f_k\right\|^2$$

$$= \|Tf\|^2.$$

This implies that $(A^*TA)^2 \leq T^2$. Since both A^*TA and T are positive operators, a standard result from operator theory gives that $A^*TA \leq T$.

(See Exercise VIII.3.12 in [ACFA].) Hence if $f = (f_0, \ldots, f_n, 0, \ldots) \in \widetilde{\mathscr{H}}$ and we write out the inequality $\langle A^* T A f, f \rangle \leq \langle T f, f \rangle$, we get

$$\text{(1.14)} \qquad \sum_{j,k} \langle S^{j+k+1} f_j, S^{j+k+1} f_k \rangle \leq \sum_{j,k} \langle S^{j+k} f_j, S^{j+k} f_k \rangle.$$

Now we construct a quasinormal extension of S. Let $\widetilde{\mathscr{L}}$ be all finitely nonzero sequences of elements of $\mathscr{H}$ and define a sesquilinear form on $\widetilde{\mathscr{L}}$ by

$$[f, g] = \sum_{j,k} \langle S^{j+k} f_j, S^{j+k} g_k \rangle,$$

for all $f = (f_0, f_1, \ldots)$ and $g = (g_0, g_1, \ldots)$ in $\widetilde{\mathscr{L}}$. By (1.12), $[f, f] \geq 0$. Let $\mathscr{L}_0 \equiv \{f \in \widetilde{\mathscr{L}} : [f, f] = 0\}$. It is left as an exercise for the reader to show that $\mathscr{L}_0$ is a linear manifold in $\widetilde{\mathscr{L}}$ and $[f + \mathscr{L}_0, g + \mathscr{L}_0] \equiv [f, g]$ defines an inner product on $\widetilde{\mathscr{L}}/\mathscr{L}_0$. Let $\mathscr{L}$ be the completion of $\widetilde{\mathscr{L}}/\mathscr{L}_0$ with respect to the norm defined by this inner product.

Define a function R_0 on $\widetilde{\mathscr{L}}$ by $(R_0 f)_k = S f_k$. It is easy to see that R_0 is linear and $[R_0 f, R_0 f] = \sum_{j,k} \langle S^{j+k+1} f_j, S^{j+k+1} f_k \rangle \leq [f, f]$ by (1.14). It follows that $R_0 \mathscr{L}_0 \subseteq \mathscr{L}_0$ and R_0 induces a contraction R on $\mathscr{L}$ defined by $R(f + \mathscr{L}_0) = R_0 f + \mathscr{L}_0$.

Now define B_0 on $\widetilde{\mathscr{L}}$ by $(B_0 f)_0 = 0$ and $(B_0 f)_k = f_{k-1}$ for $k \geq 1$. Using (1.14), it follows that $B_0 \mathscr{L}_0 \subseteq \mathscr{L}_0$ and $B(f + \mathscr{L}_0) = B_0 f + \mathscr{L}_0$ extends to a well-defined bounded operator on $\mathscr{L}$. An easy computation shows that R and B commute. Furthermore, $[R_0 f, R_0 f] = [f, B_0 f]$, so that $B = R^* R$. Therefore R is quasinormal.

It is left to the reader to show that the map $f \to (f, 0, 0, \ldots) + \mathscr{L}_0$ is an isometry of $\mathscr{H}$ into $\mathscr{L}$ and if $\mathscr{H}$ is identified with its image under this embedding, then R leaves $\mathscr{H}$ invariant and $R|\mathscr{H} = S$. Thus S has a quasinormal extension. □

Each of the conditions in Theorem 1.9 has its virtues. Conditions (c) through (f) give an internal characterization of subnormality; that is, a characterization solely in terms of S and its action on $\mathscr{H}$ with no reference to a normal operator acting on another space. Conditions (h) and (i) make a connection between subnormal operators and moment theorems; this connection will surface several times again. Condition (i) has not seen wide application, but it probably deserves more study.

Condition (g) gives a "spaceless" characterization of subnormality since the concept of positivity is a C^*-algebra concept. Indeed, (g) permits the definition of a subnormal element of any C^*-algebra. Let $\mathscr{A}$ be a C^*-algebra and define s in $\mathscr{A}$ to be *subnormal* if $\sum_{j,k} a_j^* s^{*k} s^j a_k \geq 0$ for any choice of $a_0, \ldots, a_n$ in $C^*(s)$. (See Exercise 2.) The proof of the next result is left to the reader.

1.15 PROPOSITION (Bram [**1955**]). *If $\mathscr{A}$ and $\mathscr{B}$ are C^*-algebras and $\phi : \mathscr{A} \to \mathscr{B}$ is a $*$ homomorphism, then $\phi(s)$ is subnormal whenever s is subnormal.*

REMARKS. 1. Another characterization of subnormal operators can be found in Theorem 2 of Bram [**1955**]. Others are Stochel and Szafraniec [**1984**] and Trent [**1981**]. The reader should also see Szymanski [**1987**]. In connection with subnormal operators and C^*-algebras also see Szafraniec [**1982**]. In Athavale and Pedersen [**preprint**] the connection between subnormal operators and moment problems is looked at.

2. Agler [**1985a**] provides another characterization of subnormal operators that has other interesting aspects. He proves the following.

1.16 THEOREM. *Let S be a contraction and for each $n \geq 1$ define*

$$(1 - S^*S)^{[n]} = \sum_{k=0}^{n} C_{n,k} S^{*k} S^k,$$

where $C_{n,k}$ is the binomial coefficient. Then S is subnormal if and only if $(1 - S^*S)^{[n]} \geq 0$ for all $n \geq 1$.

Also look at Agler [**1988**] for an abstract approach to these matters as well as to subnormal operators and other classes of operators.

3. There is a theory of subnormal semigroups that will not be covered in this book. See Embry and Lambert [**1977**], Frankfurt [**1977b**], Ito [**1958**], Lubin [**1977**] and [**1978**], Masani [**1978**], and Nussbaum [**1976**].

4. Another example of a subnormal operator is the *Cesàro operator*. This is the operator $C : l^2 \to l^2$ defined by

$$C(a_1, a_2, \dots) = \left(a_1, \frac{a_1 + a_2}{2}, \frac{a_1 + a_2 + a_3}{3}, \dots \right).$$

This was shown to be subnormal in Kriete and Trutt [**1971**]. Another proof of this can be found in Cowen [**1984**]. The difficulty with showing that C is subnormal is that there is no clear normal extension. Additional references on the Cesàro operator and related operators are Ghosh, Rhoades, and Trutt [**1977**], Kay, Soul, and Trutt [**1976**], and Kriete and Trutt [**1974**].

5. There are weighted translation operators that are subnormal. See Bastian [**1976**].

6. Some miscellaneous results on subnormal operators are in Conway [**1985a**] and [**1985b**].

This section concludes with a topological characterization of subnormality, but first an observation. If S is a subnormal operator on $\mathscr{H}$ and N is a normal extension on $\mathscr{K}$, then

$$\mathscr{K}_0 = \bigvee \{ N^{*k} h : h \in \mathscr{H} \text{ and } k \geq 0 \}$$

is a reducing subspace for N that contains $\mathscr{H}$. Hence $N_0 \equiv N | \mathscr{K}_0$ is also a normal extension of S. This fact will be more closely examined in the next section when we define the minimal normal extension of a subnormal operator. Here we need to know that the normal extension of a subnormal operator can be assumed to act on a separable Hilbert space, since the separability of $\mathscr{H}$ implies that of $\mathscr{H}_0$.

1.17 THEOREM (Bishop [**1957**]). *If $S \in \mathscr{B}(\mathscr{H})$, then the following statements are equivalent.*

(a) *S is subnormal.*
(b) *S is the SOT limit of a sequence of normal operators.*
(c) *S belongs to the SOT closure of the set of normal operators.*

The proof requires Exercise I.2.4 as well as an additional lemma and is a simplification of the original proof of Bishop that is due to Conway and Hadwin [**1983**].

1.18 LEMMA. *If $\{N_i\}$ is a net of normal operators with $N_i = \int z\, dE_i(z)$ and $N_i \to S$ (SOT), then for any open set G containing $\sigma(S)$, $E_i(G) \to 1$ (SOT).*

PROOF. For each i, let $Q_i = E_i(\mathbb{C} \setminus G)$; it must be shown that $Q_i \to 0$ (SOT). Equivalently it must be shown that $Q_i \to 0$ (WOT). Since $\{Q_i\} \subseteq$ ball $\mathscr{B}(\mathscr{H})$, this net has a WOT cluster point A in ball $\mathscr{B}(\mathscr{H})$. Let $\{Q_j : j \in J\}$ be a subnet such that $Q_j \to A$ (WOT). Define $u_j : G \to \mathscr{B}(\mathscr{H})$ by

$$u_j(z) = \int_{\mathbb{C} \setminus G} (w - z)^{-1} dE_j(w).$$

Hence $\|u_j(z)\| \leq [\mathrm{dist}(z, \mathbb{C} \setminus G)]^{-1}$ for all z in G and $\{u_j\}$ is a locally bounded net of analytic functions on G. By a vector-valued version of Montel's Theorem (see Exercise 4), there is an analytic function $u : G \to \mathscr{B}(\mathscr{H})$ and a subnet $\{u_k : k \in K\}$ such that $u_k(z) \to u(z)$ (WOT) for each z in G.

Fix z in G. Since $N_k - z \to S - z$ (SOT), Exercise I.2.4 implies that $u_k(z)(N_k - z) \to u(z)(S - z)$ (WOT). But $u_k(z)(N_k - z) = Q_k \to A$ (WOT). Thus $u(z)(S - z) = A$. But z was arbitrary. Therefore, if $f : \mathbb{C} \to \mathscr{B}(\mathscr{H})$ is defined by $f(z) = u(z)$ for z in G and $f(z) = A(S - z)^{-1}$ for $z \notin \sigma(S)$, f is well defined. It is easy to check that f is analytic and thus an entire function. But $f(z) = A(S - z)^{-1} \to 0$ as $z \to \infty$, so $f \equiv 0$ by Liouville's Theorem. Therefore $A = 0$. This argument shows that 0 is the unique (WOT) limit point of the original net $\{Q_i\}$. Hence $Q_i \to 0$ (WOT). □

PROOF OF THEOREM 1.17. (a) *implies* (b). Let N be a normal extension on a separable space $\mathscr{K}$ and let $\{e_n\}$ be an orthonormal basis for $\mathscr{H}$. For each $n \geq 1$, let $\mathscr{H}_n = \vee\{e_1, \ldots, e_n, Se_1, \ldots, Se_n\}$ and let $U_n : \mathscr{H} \to \mathscr{K}$ be an isomorphism such that U_n is the identity of $\mathscr{H}_n$. Put $N_n = U_n^{-1}NU_n$; so each N_n is normal. The claim is that $N_n \to S$ (SOT).

In fact, note that for $n \geq k$, $N_n e_k = U_n^{-1}NU_n e_k = U_n^{-1}Ne_k = U_n^{-1}Se_k = Se_k$. Thus $N_n e_k \to Se_k$ as $n \to \infty$. By Exercise I.2.3, $N_n \to S$ (SOT).

(b) *implies* (c). Obvious.

(c) *implies* (a). Let $\{N_i\}$ be a net of normal operators such that $N_i \to S$ (SOT). Let $N_i = \int z \, dE_i(z)$ be the spectral decomposition of N_i and let $G = \{z : |z| < \|S\| + 1\}$. According to the preceding lemma, $E_i(G) \to 1$ (SOT). If $M_i = E_i(G)N_i$, then M_i is normal and $M_i \to S$ (SOT). Moreover, $\sigma(M_i) \subseteq \operatorname{cl} G$ for all i and so $\|M_i\| \leq \|S\| + 1$ for all i. Again Exercise I.2.4 implies that $M_i^k \to S^k$ (SOT) for every $k \geq 1$. It now follows from Theorem 1.9(e) that S is subnormal. □

Exercises

1. Find the polar decomposition of the operator in Example 1.8 as well as a normal extension.
2. If $\mathscr{A}$ is a C^*-algebra and $s \in \mathscr{A}$, show that s is subnormal if and only if $\sum_{j,k} a_j^* s^{*j+k} s^{j+k} a_k \geq 0$ for any $a_0, \ldots, a_n$ in $\mathscr{A}$. (One approach to this problem is to take a faithful representation of $\mathscr{A}$. Is there a C^*-algebra proof; that is, one that is internal to $\mathscr{A}$ and does not rely on representation theory?)
3. Find the POM Q satisfying (1.9h) for the operator in Example 1.8.
4. Fix a Banach space $\mathscr{X}$ and an open set G in $\mathbb{C}$. Define a function $u: G \to \mathscr{X}^*$ to be analytic if $z \to \langle u(z), x \rangle$ is analytic for every choice of x in $\mathscr{X}$. (Is this equivalent to assuming that $\lim_{h \to 0} \{h^{-1}[u(z+h) - u(z)]\}$ exists in norm for all z in G?) Suppose that a net of analytic functions $\{u_i\}$ from G into $\mathscr{X}^*$ is locally bounded. That is, if $\overline{B}(a; r) \subseteq G$, then there is a constant $M > 0$ such that $\|u_i(z)\| \leq M$ for $|z - a| \leq r$ and for all i. Show that there is an analytic function $u: G \to \mathscr{X}^*$ and a subnet $\{u_j\}$ of $\{u_i\}$ such that $u_j(z) \to u(z)$ weak* in $\mathscr{X}^*$ for each z in G.
5. Let $A \in \mathscr{B}(\mathscr{K})$ and let P be the projection onto a subspace $\mathscr{H}$ of $\mathscr{K}$. Let $B = PA|\mathscr{H} \in \mathscr{B}(\mathscr{H})$ and show that $\mathscr{H} \in \operatorname{Lat} A$ if and only if $B^*B = PA^*A|\mathscr{H}$.
6. Show that the tensor product of two subnormal operators is subnormal.
7. Show that Condition (1.10) is equivalent to the condition that for each $n \geq 1$, the operator defined on $\mathscr{H}^{(n)}$ by the matrix whose (j, k) entry is $S^{*k}S^j$, $0 \leq j, k \leq n-1$, is positive.
8. Let μ be a positive measure on the compact space X and assume that for $1 \leq j, k \leq n$, $a_{jk} : X \to \mathbb{C}$ is a bounded Borel function. If for every continuous positive function $u : X \to \mathbb{R}$ and for all complex scalars $\lambda_1, \ldots, \lambda_n$, $\int [\sum_{j,k} a_{jk}(x) \lambda_j \overline{\lambda}_m] d\mu(x) \geq 0$, then the $n \times n$ matrix $[a_{jk}(x)]$ is a non-negative matrix a.e. $[\mu]$.

§2 Pure operators and the minimal normal extension. If S is a subnormal operator and N is a normal operator, then $S \oplus N$ is also subnormal. It is often helpful to eliminate subnormal operators of this form from our consideration or, at least, to only consider them after we have understood the property under discussion for those subnormal operators that are not of this form. This comment, being highly nonmathematical, is perhaps confusing. To eliminate the confusion and rejoin the world of mathematics, we need to define precisely the idea of a pure subnormal operator.

2.1 PROPOSITION. *If $A \in \mathscr{B}(\mathscr{H})$, then there is a reducing subspace $\mathscr{H}_0$ for A such that*:

(a) $A_0 = A|\mathscr{H}_0$ *is a normal operator*;
(b) $A_1 = A|\mathscr{H}_0^\perp$ *has no reducing subspace on which it is normal.*

PROOF. Let $\mathscr{H}_0 = \bigvee \{\mathscr{M} : \mathscr{M}$ reduces A and $A|\mathscr{M}$ is normal$\}$. It is easy to see that $\mathscr{H}_0$ is invariant for A as well as A^*, and so $\mathscr{H}_0$ reduces A. If $\mathscr{M}$ reduces A and $A|\mathscr{M}$ is normal, then $\mathscr{M} \subseteq \ker(A^*A - AA^*)$. It follows that $\mathscr{H}_0$ is contained in this kernel and hence $A|\mathscr{H}_0$ is normal. Part (b) is clear. □

2.2 DEFINITION. An operator A is *pure* if it has no reducing subspaces $\mathscr{R}$ such that $A|\mathscr{R}$ is normal.

Thus A is pure if the subspace $\mathscr{H}_0$ of Proposition 2.1 is (0). Note that for an isometry U, the space $\mathscr{H}_0$ is $\cap_n U^n \mathscr{H}$ and the definition of a pure operator is consistent with the definition of a pure isometry (§I.3).

The normal extension of a subnormal operator is never unique. If N is a normal extension of S and M is any normal operator, then $N \oplus M$ is also a normal extension of S. How can this "overkill" be eliminated?

2.3 DEFINITION. If S is a subnormal operator on $\mathscr{H}$ and N is a normal extension of S acting on $\mathscr{K}$, say that N is a *minimal normal extension* (mne) of S if $\mathscr{K}$ has no proper subspace that reduces N and contains $\mathscr{H}$.

Notice that if N is a normal extension of S and M is a normal operator, then $N \oplus M$ cannot be a minimal normal extension of S. The minimal normal extension can be characterized in a very easy but useful way.

2.4 PROPOSITION. *If S is a subnormal operator on $\mathscr{H}$ and N is a normal extension of S acting on $\mathscr{K}$, then N is a minimal normal extension of S if and only if*

$$\mathscr{K} = \bigvee \{N^{*k} h : h \in \mathscr{H}, k \geq 0\}.$$

PROOF. Let $\mathscr{L} = \vee \{N^{*k} h : h \in \mathscr{H}, k \geq 0\}$. It is clear that $\mathscr{L}$ contains $\mathscr{H}$ and reduces N since $\mathscr{H}$ is invariant for N. A moment's reflection will also reveal that $\mathscr{L}$ cannot properly contain any reducing subspace of N that contains $\mathscr{H}$; $\mathscr{L}$ is the minimal reducing subspace for N that contains $\mathscr{H}$. The proposition is now immediate. □

The next proposition shows that minimal normal extensions are unique.

2.5 PROPOSITION. *For $k = 1, 2$, let S_k be a subnormal operator on $\mathscr{H}_k$ and let N_k be a minimal normal extension of S_k acting on the Hilbert space*

$\mathscr{K}_k$. If $U: \mathscr{H}_1 \to \mathscr{H}_2$ is an isomorphism such that $US_1 = S_2 U$, then there is an isomorphism $V: \mathscr{K}_1 \to \mathscr{K}_2$ such that $V|\mathscr{H}_1 = U$ and $VN_1 V^* = N_2$.

PROOF. In light of the preceding proposition and the properties desired for V, we want to define V on $\mathscr{K}_1$ by the formula $V(N_1^{*n} h_1) = N_2^{*n} U h_1$ for h_1 in $\mathscr{H}_1$ and $n \geq 0$. It must be shown that this indeed defines an operator that is an isomorphism.

If $h_1, \ldots, h_m \in \mathscr{H}_1$ and $n_1, \ldots, n_m \geq 0$, then

$$\left\| \sum_k N_2^{*n_k} U h_k \right\|^2 = \left\langle \sum_k N_2^{*n_k} U h_k, \sum_j N_2^{*n_j} U h_j \right\rangle = \sum_{j,k} \langle N_2^{n_j} U h_k, N_2^{n_k} U h_j \rangle$$

$$= \sum_{j,k} \langle U N_1^{n_j} h_k, U N_1^{n_k} h_j \rangle \sum_{j,k} \langle N_1^{n_j} h_k, N_1^{n_k} h_j \rangle = \left\| \sum_k N_1^{*n_k} h_k \right\|^2.$$

This shows simultaneously that

(2.6) $$V \left[\sum_k N_1^{*n_k} h_k \right] = \sum_k N_2^{*n_k} U h_k$$

is a well-defined linear operator from some linear manifold in $\mathscr{K}_1$ into a linear manifold in $\mathscr{K}_2$. By Proposition 2.4, the fact that N_1 and N_2 are the minimal normal extensions of S_1 and S_2 implies that the domain and range of V are dense in their host spaces. Therefore V extends to an isomorphism of $\mathscr{K}_1$ onto $\mathscr{K}_2$. It is routine to show that V has the desired properties. □

2.7 COROLLARY. *If S is a subnormal operator and N_1 and N_2 are minimal normal extensions of S, then N_1 and N_2 are unitarily equivalent.*

Because of this corollary, we can now legitimately speak of *the* minimal normal extension of a subnormal operator S; this will be denoted by $N = \operatorname{mne} S$. It is efficient and useful to fix some notation.

2.8 CONVENTION. S will always denote a subnormal operator acting on a Hilbert space $\mathscr{H}$ and N will be its minimal normal extension acting on $\mathscr{K} \supseteq \mathscr{H}$.

If we write $\mathscr{K} = \mathscr{H} \oplus \mathscr{H}^\perp$, then N has the 2×2 matrix representation

(2.9) $$N = \begin{bmatrix} S & X \\ 0 & T^* \end{bmatrix},$$

the 0 appearing since $N\mathscr{H} \subseteq \mathscr{H}$. If the matrix for N^* is written relative to this same decomposition, $\mathscr{K} = \mathscr{H} \oplus \mathscr{H}^\perp$, it is

$$N^* = \begin{bmatrix} S^* & 0 \\ X^* & T \end{bmatrix}.$$

However, if we write the matrix of N^* with respect to the decomposition $\mathscr{K} = \mathscr{H}^\perp \oplus \mathscr{H}$, it becomes

$$N^* = \begin{bmatrix} T & X^* \\ 0 & S^* \end{bmatrix}.$$

From here it follows that the operator T appearing in (2.9) is subnormal and N^* is a normal extension of T. Is N^* the minimal normal extension of T? The next proposition shows that purity and minimality are related concepts. (This relation exists only in operator theory and not in social behavior.)

2.10 PROPOSITION. *The following statements are equivalent.*

(a) *S is pure.*
(b) *N^* is the minimal normal extension of T.*
(c) *The smallest subspace of $\mathscr{H}$ that reduces S and contains $X(\mathscr{H}^\perp)$ is $\mathscr{H}$.*
(d) *There is no nonzero projection P on $\mathscr{H}$ such that $PS = SP$ and $PX = 0$.*

PROOF. It will only be shown that (a) and (b) are equivalent; the remainder of the proof is left as an exercise.

Let $\mathscr{L} = \bigvee \{N^n g : g \in \mathscr{H}^\perp \text{ and } n \geq 0\}$. By Proposition 2.4, $N^* = \text{mne } T$ if and only if $\mathscr{L} = \mathscr{K}$. Now it is easy to see that $\mathscr{L}$ reduces N and $\mathscr{L}^\perp \subseteq \mathscr{H}$ since $\mathscr{H}^\perp \subseteq \mathscr{L}$. Hence $\mathscr{L}^\perp$ reduces N and $N|\mathscr{L}^\perp = S|\mathscr{L}^\perp$ is normal. If S is pure, this implies that $(0) = \mathscr{H} \cap \mathscr{L}^\perp$, so $\mathscr{L} = \mathscr{K}$ and $N^* = \text{mne } T$.

Now assume that S is not pure; so there is a proper subspace $\mathscr{M}$ of $\mathscr{H}$ such that $\mathscr{M}$ reduces N. Thus $\mathscr{M}^\perp$ reduces N^* and $\mathscr{H}^\perp \subseteq \mathscr{M}^\perp$. Thus $N^*|\mathscr{M}^\perp$ is a normal extension of $N^*|\mathscr{H}^\perp = T$. □

REMARKS. 1. Morrel [1973] gives a formula for the space $\mathscr{H}_0$ in Proposition 2.1.

2. Andô [1963a] gives a matrix representation of a normal extension of a subnormal operator, but this matrix is not always the minimal normal extension. Stampfli [1966] gives a matrix representation of the minimal normal extension of S.

3. Morrel [1973] gives a matrix representation for an arbitrary operator and examines this further in the case of a subnormal operator.

4. Bunce [1978] characterizes the minimal normal extension in terms of a universal diagram property. This involves the concept of a completely positive map and is related to the theorem of Naimark on dilation of a POM to a spectral measure.

5. Conway [1981b] calls the operator T in the matrix representation (2.9) the *dual* of S when S is pure and investigates the relation between a pure subnormal operator and its dual. The dual of a subnormal operator was also studied in Murphy [1982] and Yan [1985].

Now that we know that the minimal normal extension of a subnormal operator is unique up to unitary equivalence, we can define the *normal spectrum* of a subnormal operator S to be the spectrum of its minimal normal extension; denote this by $\sigma_n(S)$. Thus, if $N = \text{mne } S$, $\sigma_n(S) = \sigma(N)$. The next result gives the basic properties of the normal spectrum.

2.11 THEOREM. *If S is a subnormal operator, the following statements hold.*

(a) *(Halmos [1952])* $\sigma_n(S) \subseteq \sigma(S)$.
(b) $\sigma_{ap}(S) \subseteq \sigma_n(S)$ *and* $\partial\sigma(S) \subseteq \partial\sigma_n(S)$.
(c) *(Bram [1955]) If U is a bounded component of $\mathbb{C}\setminus\sigma_n(S)$, then either U and $\sigma(S)$ are disjoint or $U \subseteq \sigma(S)$.*

PROOF. (a) It suffices to show that if S is invertible, then N is invertible. (Why?) If $N = \int z\, dE(z)$ is the spectral decomposition of N, $\varepsilon > 0$, and $\mathscr{M} = E(B(0;\varepsilon))\mathscr{K}$, then for f in $\mathscr{M}$, $\|N^k f\| \leq \varepsilon^k \|f\|$ for all $k \geq 1$. So if $f \in \mathscr{M}$ and $h \in \mathscr{H}$, $|\langle f, h\rangle| = |\langle f, S^k S^{-k} h\rangle| = |\langle f, N^k S^{-k} h\rangle| = |\langle N^{*k} f, S^{-k} h\rangle| \leq \|N^{*k} f\| \|S^{-k} h\| \leq \varepsilon^k \|f\| \|S^{-k} h\| \leq \varepsilon^k \|S^{-1}\|^k \|f\| \|h\|$. Letting $k \to \infty$ shows that if $\varepsilon < \|S^{-1}\|^{-1}$, then $\langle f, h\rangle = 0$. Thus $\mathscr{M} \perp \mathscr{H}$ if $\varepsilon < \|S^{-1}\|^{-1}$ and so $\mathscr{H} \subseteq \mathscr{M}^\perp$. But this implies that $N|\mathscr{M}^\perp$ is a normal extension of S. By the minimality of N, $\mathscr{M} = (0)$ and so N is invertible.

(b) This is easy. If $\lambda \in \sigma_{ap}(S)$, there is a sequence of unit vectors $\{h_n\}$ in $\mathscr{H}$ such that $\|(\lambda - S)h_n\| \to 0$. But $(\lambda - S)h_n = (\lambda - N)h_n$. Hence $\sigma_{ap}(S) \subseteq \sigma_{ap}(N) = \sigma_n(S)$. Also, if $\lambda \in \partial\sigma(S)$, then $\lambda \in \sigma_{ap}(S)$ and so $\lambda \in \sigma_n(S)$. But part (a) implies that $\lambda \notin \text{int}\,\sigma_n(S)$ and so $\lambda \in \partial\sigma_n(S)$.

(c) (Due to S. K. Parrott) Let U be a bounded component of $\mathbb{C}\setminus\sigma_n(S)$ and put $U_+ = U\setminus\sigma(S)$ and $U_- = U\cap\sigma(S)$. So $U = U_- \cup U_+$, $U_+ \cap U_- = \varnothing$, and U_+ is open. If it can be shown that U_- is also open, then the connectedness of U implies that either U_+ or U_- is empty, completing the proof. But part (b) implies $U_- = U \cap [\text{int}\,\sigma(S)]$. □

Recall that for any compact subset K of $\mathbb{C}$, $\hat{K}$ denotes the *polynomially convex hull* of K. (See [**FOCV**].) Since $\hat{K}$ is the union of K and the bounded components of its complement, $\hat{K} = \widehat{\partial K}$. Since $\partial\sigma(S) \subseteq \sigma_{ap}(S) \subseteq \sigma_n(S)$, we get the following.

2.12 COROLLARY. $\widehat{\sigma(S)} = \widehat{\sigma_n(S)}$ *and the norm of S equals its spectral radius.*

PROOF. The first part of the corollary is an immediate geometric consequence of Theorem 2.11. The second part follows from the fact that N and S have the same spectral radius and $\|S\| \leq \|N\|$. □

2.13 EXAMPLE. If μ is a compactly supported measure on $\mathbb{C}$ and S_μ is multiplication by z on $P^2(\mu)$, then the minimal normal extension of S_μ is the normal operator N_μ, multiplication by z on $L^2(\mu)$. In fact, the linear span of $\{\bar{z}^n p : p \text{ is a polynomial and } n \geq 0\}$ contains all polynomials in z and $\bar{z}$. By the Stone-Weierstrass Theorem, this is dense in $L^2(\mu)$ and so by (2.4) N_μ is the minimal normal extension of S_μ. Hence $\sigma_n(S_\mu) = \text{support}\,\mu$. The exact determination of $\sigma(S_\mu)$ is unknown and very difficult.

2.14 EXAMPLE. Let G be a bounded open subset of $\mathbb{C}$ and let S be the Bergman operator on $L_a^2(G)$. The minimal normal extension of S is multiplication by z on $L^2(G)$.

It will be shown later (see §8) that $\partial[\operatorname{cl} G] \subseteq \sigma_{ap}(S) \subseteq \partial G$. So if G is an annulus, $\sigma_{ap}(S) = \partial G$. Here, of course, $\sigma_n(S) = \operatorname{cl} G$. So it is not necessarily true for an arbitrary subnormal operator that the normal spectrum and the approximate point spectrum are equal.

By similar reasoning, the minimal normal extension of the unilateral shift is the bilateral shift.

If U is a bounded component of $\mathbb{C} \setminus \sigma_n(S)$, how can you tell whether U is contained in or disjoint from $\sigma(S)$? The answer isn't always easy, but the following result is often useful.

2.15 PROPOSITION. *If $N = \operatorname{mne} S$ and U is a bounded component of the complement of $\sigma(N)$, then the following statements are equivalent.*
  (a) *$U \cap \sigma(S) = \varnothing$.*
  (b) *For each λ in U, $(N - \lambda)\mathscr{H} = \mathscr{H}$.*
  (c) *There is a λ in U with $(N - \lambda)\mathscr{H} = \mathscr{H}$.*
  (d) *For each λ in U, $\mathscr{H} \in \operatorname{Lat}(N - \lambda)^{-1}$.*
  (e) *There is a λ in U for which $\mathscr{H} \in \operatorname{Lat}(N - \lambda)^{-1}$.*

PROOF. (a) *implies* (b). If $\lambda \in U$, then $S - \lambda$ is invertible (by (a)) and so $\mathscr{H} = (S - \lambda)\mathscr{H} = (N - \lambda)\mathscr{H}$.

(b) *implies* (d). Because $(N - \lambda)\mathscr{H} = \mathscr{H}$ and $\lambda \notin \sigma(N)$, $\mathscr{H} = (N - \lambda)^{-1}\mathscr{H}$ and so $\mathscr{H} \in \operatorname{Lat}(N - \lambda)^{-1}$.

(d) *implies* (e). Clear.

(e) *implies* (c). Since $(N - \lambda)^{-1}\mathscr{H} \subseteq \mathscr{H}$, $\mathscr{H} \subseteq (N - \lambda)\mathscr{H} \subseteq \mathscr{H}$.

(c) *implies* (a). By hypothesis, $S - \lambda$ is surjective. But $\ker(S - \lambda) \subseteq \ker(N - \lambda)$ and since $N - \lambda$ is invertible, $S - \lambda$ is also one-to-one. Hence $\lambda \notin \sigma(S)$ and so by Theorem 2.11(c), $U \cap \sigma(S) = \varnothing$. □

The section closes with a result that is somewhat technical, but seems appropriate to give here. It will have many important uses later.

Since N is normal, $W^*(N)$, the von Neumann algebra generated by N, is abelian. Thus if $A \in W^*(N)$, A is normal. If, in addition, $A\mathscr{H} \subseteq \mathscr{H}$, then $A|\mathscr{H}$ is a subnormal operator. However, A need not be the minimal normal extension of $A|\mathscr{H}$. A trivial example of this phenomenon is to take $A = 1$, but more subtle examples can be found. Nevertheless, there are relations between A and the minimal normal extension of $A|\mathscr{H}$, as the next proposition indicates. Recall that $W^*(N) = \{f(N) : f \text{ is a bounded Borel function on } \sigma(N)\}$.

2.16 PROPOSITION (Olin [1976]). *Let $N = \operatorname{mne} S$ and suppose f is a bounded Borel function on $\sigma(N)$ such that $f(N)\mathscr{H} \subseteq \mathscr{H}$. If $f(S) \equiv f(N)|\mathscr{H}$, then $\sigma(f(N))$ equals the normal spectrum of $f(S)$, $\sigma_n(f(S))$.*

PROOF. Let $\mathscr{M} = \bigvee\{f(N)^{*k}h : h \in \mathscr{H}$ and $k \geq 0\}$ and put $M = f(N)|\mathscr{M}$; so M is the minimal normal extension of $f(S)$. We want to show that $\sigma(M) = \sigma(f(N))$. Because $\mathscr{M}$ reduces $f(N)$, $\sigma(M) \subseteq \sigma(f(N))$. Suppose $\lambda \notin \sigma(M)$. Choose ε such that $0 < 2\varepsilon < \text{dist}(\lambda, \sigma(M))$ and put $G = B(\lambda; \varepsilon) \cap \sigma(f(N))$. If $f(N) = \int z\, dE(z)$ is the spectral decomposition of $f(N)$ and P is the projection of $\mathscr{H}$ onto $\mathscr{M}$, then $F(\Delta) = PE(\Delta) = E(\Delta)P$ is the spectral measure for M. But G is disjoint from $\sigma(M)$, so $F(G) = 0$. Thus $\mathscr{H} \leq \mathscr{M} \leq E(G)^\perp \mathscr{H}$. Thus $E(G)^\perp \mathscr{H} \neq (0)$. But N commutes with $E(G)$ (Why?), so $E(G)^\perp \mathscr{H}$ reduces N and contains $\mathscr{H}$. Since N is the minimal normal extension of S, this space must be $\mathscr{H}$. Thus $E(G) = 0$ and so $\lambda \notin \sigma(f(N))$. □

2.17 COROLLARY. *If S, N, and f are as in the preceding proposition, then $\|f(S)\| = \|f(N)\|$.*

For additional spectral information about subnormal operators see Hastings [**1981**] and McGuire [**1988b**].

Exercises

1. (*Important*) Show that if N is normal and $\mathscr{L} \in \text{Lat } N$ such that $N|\mathscr{L}$ is normal, then $\mathscr{L}$ reduces N. (Also see Exercise 4.1 below.)
2. Show that the minimal normal extension of the unilateral shift is the bilateral shift.
3. If the minimal normal extension of S has the representation (2.9), show that the subnormal operator T is always pure.
4. Using the notation (2.9), show that $\sigma(T) = \sigma(S)^*$ provided S is pure.
5. If S is the unilateral shift, find the entries in the matrix (2.9).
6. With the notation of (2.9), show:
   (a) $XX^* = S^*S - SS^*$;
   (b) $X^*X = T^*T - TT^*$;
   (c) $XT = S^*X$.
7. Show that $\ker(S^*S - SS^*)$ is an invariant subspace for S. What is this subspace for the unilateral shift?
8. Verify the statements in Example 2.14.
9. Let γ be a smooth Jordan curve and let m = normalized arc length measure. Use the Poisson kernel to show that each f in $P^2(m)$ has a natural analytic extension to the inside of γ.

§3 Quasinormal operators. Quasinormal operators have received a lot of attention since their introduction in Brown [**1953**]; probably too much attention. They constitute a very well understood class of subnormal operators, undoubtedly the most understood class. Because they are so well understood, it is convenient to use them as a first test of any question that arises.

The following lemma appeared in Campbell [**1975**], but the proof given is from Conway and Olin [**1977**].

3.1 LEMMA. *If $N = \operatorname{mne} S$, then S is quasinormal if and only if $\mathscr{H}$ is invariant for N^*N.*

PROOF. If S is quasinormal, P is the projection of $\mathscr{K}$ onto $\mathscr{H}$, and $h \in \mathscr{H}$, then $\|PN^*Nh\|^2 = \|S^*Sh\|^2 = \langle S^*Sh, S^*Sh \rangle = \langle Sh, SS^*Sh \rangle = \langle Sh, S^*S^2h \rangle = \|S^*h\|^2 = \|N^2h\|^2 = \|N^*Nh\|^2$. Hence $N^*Nh \in \mathscr{H}$. The converse is clear. □

3.2 THEOREM (Brown [1953]). *If S is any operator on $\mathscr{H}$, then S is a pure quasinormal operator if and only if there is a positive operator A on a Hilbert space $\mathscr{L}$ with $\ker A = (0)$ such that S is unitarily equivalent to*

$$\begin{bmatrix} 0 & 0 & 0 & \cdots \\ A & 0 & 0 & \cdots \\ 0 & A & 0 & \cdots \\ \cdot & \cdot & A & \cdots \end{bmatrix}$$

acting on $\mathscr{L}^{(\infty)}$.

PROOF. First assume that S is a pure quasinormal operator and let $S = UB$ be its polar decomposition. Now $\ker S = \ker B$, so if $h \in \ker S$, then $S^*h = BU^*h = U^*Bh = 0$. So $\ker S \subseteq \ker S^*$ and $\ker S$ reduces S. Since S is assumed pure, it must be that $\ker S = (0)$. Hence S^*S is injective and has dense range and U is an isometry.

Let $N = \operatorname{mne} S$. If $f \in \ker N$ and $h \in \mathscr{H}$, then $\langle f, SS^*h \rangle = \langle f, NN^*h \rangle$ (Lemma 3.1) $= \langle Nf, Nh \rangle = 0$. Since $\operatorname{ran}(SS^*)$ is dense in $\mathscr{H}$, $\ker N \perp \mathscr{H}$. That is, $(\ker N)^\perp \supseteq \mathscr{H}$ and $(\ker N)^\perp$ reduces N. By the minimality of N, $\ker N = (0)$.

Let $N = VC$, $C = |N|$, be the polar decomposition of N. Since $\ker N = (0)$, V is unitary. Also, it follows that $U = V|\mathscr{H}$ and $B = C|\mathscr{H}$. (Why?)

We want to show that the isometry U is a unilateral shift; to do this, we consider the von Neumann-Wold decomposition of U. Let $\mathscr{H}_0 = \cap_n U^n \mathscr{H}$; we want to show that $\mathscr{H}_0 = (0)$. If $n \geq 1$, then $NU^n \mathscr{H} = VCV^n \mathscr{H} = V^{n+1}C\mathscr{H} \subseteq U^{n+1}\mathscr{H}$; thus, $\mathscr{H}_0$ is invariant for N. Also, $N^*U^n \mathscr{H} = V^*CV^n \mathscr{H} = V^{n-1}B\mathscr{H} \subseteq U^{n-1}\mathscr{H}$. Thus $\mathscr{H}_0$ reduces N, and hence S. But $S|\mathscr{H}_0 = N|\mathscr{H}_0$ (Why?). Since S was assumed pure, $\mathscr{H}_0 = (0)$ and U must be a unilateral shift.

Let $\mathscr{L} = \mathscr{H} \cap (U\mathscr{H})^\perp$ and $\mathscr{L}_n = U^{n-1}\mathscr{H} \cap (U^n \mathscr{H})^\perp = U^n \mathscr{L}$ for $n \geq 0$. It follows that $B\mathscr{L}_n \subseteq \mathscr{L}_n$, so put $A_n = B|\mathscr{L}_n$. Thus $B = A_0 \oplus A_1 \oplus \cdots$. It is left to the reader to check that S is unitarily equivalent to

$$\begin{bmatrix} 0 & 0 & 0 & \cdots \\ A_0 & 0 & 0 & \cdots \\ 0 & A_1 & 0 & \cdots \\ \cdot & \cdot & A_2 & \cdots \end{bmatrix}$$

acting on $\mathscr{L}^{(\infty)}$. If $U_n : \mathscr{L}_0 \to \mathscr{L}_n$ is defined by $U_n = U^n|\mathscr{L}_0$, then $U_n A_0 U_n^{-1} = A_n$. It is now clear that S has the asserted form.

The converse is an exercise. □

Another way to state Theorem 3.2 is to state that every pure quasinormal operator is unitarily equivalent to $U \otimes A$, where U is the unilateral shift and A is a positive operator with trivial kernel. Without worrying too much about the definition on the tensor product of two operators, let's agree to denote by $U \otimes A$ the quasinormal operator that appears in Example 1.8. (Anyone who wishes may worry about this and everyone should at some time worry about it, but no one should make it a fetish.)

Notice that if $A_1, A_2, \ldots$ are positive operators with $\sup_n \|A_n\| < \infty$, then $U \otimes [\bigoplus_n A_n] = \oplus_n (U \otimes A_n)$. Since every positive operator can be written as a direct sum of cyclic positive operators, this says that every pure quasinormal operator can be written as a direct sum of operators of the form $U \otimes A$, for a cyclic positive operator A with no kernel.

But if A is a cyclic positive operator, then there is a positive measure ν supported on some bounded interval $[0, a]$ in $\mathbb{R}$ such that $A \cong N_\nu$. If $\ker A = (0)$, $\nu(\{0\}) = 0$. With these preliminaries, the next is not too surprising, though a proof is required. Nevertheless the proof will be left for the interested reader to discover or to look up (depending on the level of interest).

3.3 THEOREM (Conway and Wu [**1982**]). *Let ν be a positive measure on $[0, a]$ with $\nu(\{0\}) = 0$ and let $A = N_\nu$ on $L^2(\nu)$. Define a compactly supported measure on $\mathbb{C}$ by*

$$(3.4) \qquad d\mu(re^{i\theta}) = (2\pi)^{-1} d\theta \, d\nu(r),$$

and let $\mathscr{L} =$ the closed linear span of $\{|z|^k z^n : k, n \geq 0\}$ in $L^2(\mu)$. Then $U \otimes A$ is unitarily equivalent to the restriction of N_μ to $\mathscr{L}$.

It is worthwhile to be certain that the meaning of (3.4) is understood. The measure is actually defined in terms of the corresponding linear functional on $C_c(\mathbb{C})$, the space of continuous functions from $\mathbb{C}$ into $\mathbb{C}$ that have compact support, by $f \to \iint f(re^{i\theta}) \, d\theta \, d\nu(r)$. Equation (3.4) can also be interpreted using Fubini's Theorem.

For further information on quasinormal operators, the reader can see Brown [**1953**], Conway and Wu [**1982**], Embry-Wardrop [**1981**], Whitley [**1978**], Wogen [**1979**], and Yoshino [**1973**].

Exercises

1. Let $S = U \otimes A$ as in Example 1.8. Without using Theorem 3.3, show that A is cyclic if and only if there is a vector e in $\mathscr{H} = \mathscr{L}^{(\infty)}$ such that $\mathscr{H}$ is the closed linear span of $\{|S|^k S^n e : k, n \geq 0\}$.

2. Show that $U \otimes A_1 \cong U \otimes A_2$ if and only if $A_1 \cong A_2$. Note that this enables us to use the complete unitary invariants for positive operators (see §IX.10 in [**ACFA**]). To give a complete set of unitary invariants for quasinormal operators.

3. If S is the unilateral shift and A is an arbitrary operator, give a necessary and sufficient condition on A for $S \otimes A$ to be quasinormal.

4. If S is a pure quasinormal operator with $\|S\| = 1$ and $\ker(1 - S^*S) = (0)$, show that $S^n \to 0$ (SOT).

5. If S is a pure quasinormal operator with $\|S\| = 1$, show that $\ker(1 - S^*S) \neq (0)$ if and only if there is a reducing subspace $\mathscr{R}$ for S such that $S|\mathscr{R}$ is the unilateral shift.

6. Show that if S is a pure quasinormal operator with $\|S\| = 1$, then $S^{*n} \to 0$ (SOT).

7. (G. E. Keough) For a positive invertible operator A on a Hilbert space $\mathscr{L}$, define H_A^2 to be the collection of all formal power series $F(z) = \sum_n z^n f_n$, where $f_n \in \mathscr{L}$ and $\sum_n \|A^n f_n\|^2 < \infty$. If also $G(z) = \sum_n z^n g_n \in H_A^2$, define
$$\langle F, G \rangle = \sum_{n=0}^{\infty} \langle A^n f_n, A^n g_n \rangle.$$

(a) Prove that H_A^2 is a Hilbert space.

(b) If $r = \|A^{-1}\|^{-1}$, show that every F in H_A^2 defines an analytic function from the disk $B(0; r)$ into $\mathscr{L}$.

(c) Define $S : H_A^2 \to H_A^2$ by $(SF)(z) = zF(z)$. Show that S is quasinormal and S^*S is invertible.

(d) If S is a quasinormal operator and S^*S is invertible, show that there is a positive invertible operator A such that S is unitarily equivalent to multiplication by z on H_A^2.

8. If S is a pure quasinormal operator represented as in (3.2), show that $C = [C_{ij}] \in \{S\}'$ if and only if $C_{ij} = 0$ for $j > i$ and $AC_{ij} = C_{i+1,j+1}A$ for $j \leq i$. Show that $C \in \{S\}''$ if and only if $C_{ij} = 0$ for $j > 1$ and $C_{ij} = C_{i+1,j+1} \in \{A\}''$ for $j \leq i$.

9. (Conway and Wu [1982]) Suppose ν is a measure on $[0, 1]$ and $A = N_\nu$. Define $V : L^2(\nu) \to L^2(\nu)$ by
$$(Vf)(x) = \int_{[x, 1]} f(t)\, d\nu(t).$$
Show that if $C_{ij} = 0$ for $i \neq j$ and $C_{jj} = A^j V A^{-j}$, then $C = [C_{ij}]$ is a bounded operator in $\{U \otimes A\}'$.

§4 Hyponormal operators. When he introduced the concept of subnormality, Halmos [1950] also introduced a larger class of operators that are today called "hyponormal" (the term first appeared in Berberian [1961]).

4.1 DEFINITION. An operator A is *hyponormal* if $A^*A \geq AA^*$.

4.2 PROPOSITION. *Every subnormal operator is hyponormal.*

PROOF. Let S be subnormal and write its minimal normal extension as in (2.9). Perform the required matrix multiplications to calculate $0 =$

$N^*N - NN^*$. From the $(1,1)$ entry we get $0 = S^*S - SS^* - XX^*$, so $S^*S - SS^* = XX^* \geq 0$, and S is hyponormal. □

There are hyponormal operators that are not subnormal, and the reader will undoubtedly notice many as we progress through this book. One easy example which also illustrates a difficulty in the study of hyponormal operators is $A = U^* + 2U$; this operator A is hyponormal but A^2 is not hyponormal (Exercise 3). Of course, the square of a subnormal operator is subnormal, so this shows that A is not subnormal.

The verification of the following equivalent formulation of hyponormality is left to the reader.

4.3 PROPOSITION. *An operator A is hyponormal if and only if $\|Ah\| \geq \|A^*h\|$ for all vectors h.*

If $A^*A \leq AA^*$ or, equivalently, $\|A^*h\| \geq \|Ah\|$ for all vectors h, the A is said to be a *cohyponormal* operator. Operators that are either hyponormal or cohyponormal are called *seminormal*.

4.4 PROPOSITION. *Let A be a hyponormal operator.*

(a) *If A is invertible, then A^{-1} is hyponormal.*
(b) *If $\lambda \in \mathbb{C}$, then $A - \lambda$ is hyponormal.*
(c) *If $\lambda \in \sigma_p(A)$ and $f \in \mathscr{H}$ such that $Af = \lambda f$, then $A^*f = \bar{\lambda}f$.*
(d) *If f and g are eigenvectors corresponding to distinct eigenvalues of A, then $f \perp g$.*
(e) *If $\mathscr{M} \in \operatorname{Lat} A$, then $A|\mathscr{M}$ is hyponormal.*

PROOF. (a) This proof uses the fact that if T is a positive invertible operator and $T \geq 1$, then $T^{-1} \leq 1$. Since $A^*A \geq AA^*$ and A is invertible, $A^{-1}(A^*A)A^{*-1} \geq A^{-1}(AA^*)A^{*-1} = 1$; taking inverses gives $A^*A^{-1}A^{*-1}A \leq 1$. So $A^{-1}A^{*-1} = A^{*-1}(A^*A^{-1}A^{*-1}A)A^{-1} \leq A^{*-1}A^{-1}$ and A^{-1} is hyponormal.

(b) $(A - \lambda)(A^* - \bar{\lambda}) = AA^* - \lambda A^* - \bar{\lambda}A + |\lambda|^2 \leq A^*A - \lambda A^* - \bar{\lambda}A + |\lambda|^2 = (A^* - \bar{\lambda})(A - \lambda)$.

(c) Using part (b) and the preceding proposition, if $f \in \ker(A - \lambda)$, $\|(A^* - \bar{\lambda})f\| \leq \|(A - \lambda)f\| = 0$.

(d) If $Af = \lambda f$ and $Ag = \mu g$, then $\lambda \langle f, g \rangle = \langle Af, g \rangle = \langle f, A^*g \rangle = \langle f, \bar{\mu}g \rangle = \mu \langle f, g \rangle$.

(e) If P is the projection of $\mathscr{H}$ onto $\mathscr{M}$ and $B = A|\mathscr{M}$, then $B^* = PA^*|\mathscr{M}$. So for f in $\mathscr{M}$, $\|B^*f\| = \|PA^*f\| \leq \|A^*f\| \leq \|Af\| = \|Bf\|$. □

4.5 COROLLARY. *If A is hyponormal and $\lambda \in \sigma_p(A)$, then $\ker(A - \lambda)$ reduces A. Hence, if A is a pure hyponormal operator, then $\sigma_p(A) = \emptyset$.*

4.6 PROPOSITION (Stampfli [1962]). *If A is hyponormal, then $\|A^n\| = \|A\|^n$ and so $\|A\| = r(A)$, the spectral radius of A.*

PROOF. If $f \in \mathscr{H}$ and $n \geq 1$, then $\|A^n f\|^2 = \langle A^n f, A^n f \rangle = \langle A^* A^n f, A^{n-1} f \rangle \leq \|A^* A^n f\| \|A^{n-1} f\| \leq \|A^{n+1} f\| \|A^{n-1} f\|$. Hence $\|A^n\|^2 \leq \|A^{n+1}\| \|A^{n-1}\|$. We now proceed to prove the equality by induction. Clearly it is true for $n = 1$; so suppose $\|A^k\| = \|A\|^k$ for $1 \leq k \leq n$. Then $\|A\|^{2n} = \|A^n\|^2 \leq \|A^{n+1}\| \|A^{n-1}\| = \|A^{n+1}\| \|A\|^{n-1}$. Cancelling gives that $\|A\|^{n+1} \leq \|A^{n+1}\|$. The reverse inequality holds for all operators.

Finally, $r(A) = \lim \|A^n\|^{1/n} = \|A\|$. □

4.7 COROLLARY. *If A is hyponormal and $\lambda \notin \sigma(A)$, then $\|(\lambda - A)^{-1}\| = [\operatorname{dist}(\lambda, \sigma(A))]^{-1}$.*

4.8 PROPOSITION (Stampfli [**1962**]). *If A is hyponormal and λ is an isolated point of $\sigma(A)$, then $\lambda \in \sigma_p(A)$. Thus pure hyponormal operators have no isolated points in their spectrum.*

PROOF. Replacing A by $A - \lambda$, we may assume that $\lambda = 0$. Let $\rho > 0$ such that $|z| > \rho$ for all z in $\sigma(A)$ different from 0, let γ be the circle $|z| = \rho$, and put $E = (2\pi i)^{-1} \int_\gamma (z - A)^{-1} dz$, the Riesz idempotent corresponding to the isolated point 0. So $\mathscr{M} = E\mathscr{H}$ is a closed invariant subspace for A and $\sigma(A|\mathscr{M}) = \{0\}$. But $A|\mathscr{M}$ is hyponormal and so Proposition 4.6 implies $\|A|\mathscr{M}\| = 0$. That is, $\mathscr{M} = \ker A$. □

Since every nonzero point of the spectrum of a compact operator is isolated, the next corollary follows from the preceding proposition plus a small argument left to the reader.

4.9 COROLLARY. *If A is hyponormal compact operator, then A is normal.*

This last corollary says that the theory of hyponormal operators (and hence subnormal operators) is strictly an infinite dimensional theory.

4.10 PROPOSITION. *Let A be a hyponormal operator.*

(a) *A is invertible if and only if A is right invertible.*
(b) *A is a Fredholm operator if and only if A is a right Fredholm operator.*
(c) *$\sigma(A) = \sigma_r(A)$ and $\sigma_e(A) = \sigma_{re}(A)$.*
(d) *If A is pure and $\lambda \in \sigma(A) \setminus \sigma_e(A)$, then $\operatorname{ind}(A - \lambda) \leq 1$.*

PROOF. To prove (a), suppose A is right invertible so that there is a bounded operator B such that $AB = 1$. Therefore A is surjective and $\ker A^* = (\operatorname{ran} A)^\perp = (0)$. By Corollary 4.4(c), $\ker A = (0)$ and A is invertible.

The proof of (b) is similar to that of (a) and part (c) is immediate from (a) and (b). It remains to prove (d).

Because A is a pure hyponormal operator, $A - \lambda$ is a pure hyponormal operator; hence $\ker(A - \lambda) = (0)$ by Corollary 4.5. Since $\lambda \in \sigma(A)$, $A - \lambda$ is not surjective. Thus $\operatorname{ind}(A - \lambda) = \dim[\ker(A - \lambda)] - \dim[\operatorname{ran}(A - \lambda)]^\perp = -\dim[\operatorname{ran}(A - \lambda)]^\perp \leq -1$. □

Notice that the unilateral shift is a left invertible hyponormal operator that is not invertible.

Recall that the *Weyl spectrum* of an operator A, $\sigma_w(A)$, is defined by $\sigma_w(A) = \bigcap\{\sigma(A+K) : K \in \mathscr{B}_0\}$, and, by Schechter's Theorem, $\sigma_w(A) = \sigma_e(A) \cup \{\lambda \in \mathbb{C} : A - \lambda \in \mathscr{SF} \text{ and } \operatorname{ind}(A - \lambda) \neq 0\}$. (See XI.6.12 in [**ACFA**].) The next result is an extension of Weyl's Theorem on normal operators.

4.11 Proposition (Coburn [**1966**]). *If A is hyponormal, then $\sigma_w(A) = \{\lambda \in \sigma(A) : \lambda \text{ is not an isolated eigenvalue of finite multiplicity}\}$.*

Proof. Suppose that λ is an isolated eigenvalue of A having finite multiplicity. So $\ker(A - \lambda)$ reduces A and so by Corollary 4.5 and Proposition 4.8, $A = \lambda I \oplus B$, where I is the identity on a finite dimensional space, B is hyponormal, and $\lambda \notin \sigma(B)$. Thus $\lambda \notin \sigma_w(A)$.

By Schechter's Theorem (see XI.6.10 in [**ACFA**]), for the converse it suffices to assume that $A - \lambda$ is not invertible and $A - \lambda$ is Fredholm with $\operatorname{ind}(A - \lambda) = 0$, and prove that λ must be an isolated eigenvalue of finite multiplicity. It suffices to assume that $\lambda = 0$. Since $A \in \mathscr{F}$ and $\operatorname{ind} A = 0$, 0 is an eigenvalue of finite multiplicity; it remains to show that 0 is isolated. But $\ker A \subseteq \ker A^*$ (4.4(c)) $= (\operatorname{ran} A)^\perp$ and $0 = \operatorname{ind} A = \dim[\ker A] - \dim[(\operatorname{ran} A)^\perp]$, and therefore $\ker A = (\operatorname{ran} A)^\perp$. Thus $A = 0 \oplus B$ for an invertible operator B. Since $\sigma(A) = \{0\} \cup \sigma(B)$, 0 must be an isolated point of $\sigma(A)$. □

4.12 Corollary (Coburn [**1966**]). *If A is a pure hyponormal operator, then $\|A\| \leq \|A + K\|$ for every compact operator K.*

Proof. Since A is pure, $\sigma_p(A) = \emptyset$ and so $\sigma(A) = \sigma_w(A) = \bigcap\{\sigma(A+K) : K \in \mathscr{B}_0\}$. So for every compact operator K, $\sigma(A) \subseteq \sigma(A+K)$. Thus $\|A\| = r(A) \leq r(A+K) \leq \|A+K\|$. □

For a further extension of Weyl's Theorem, see Berberian [**1969**].

The theory of hyponormal operators is an extensive and highly developed area. The recent book by Martin and Putinar [**1989**] gives a current view of this subject. An older and less encyclopedic treatment is Clancey [**1979**], where the exposition is very good. Some further topics in hyponormal operators will be given in this book (see Chapter IV), but only if they are relevant to subnormal operators. One result that will not be covered is the recent theorem proved by Scott Brown [**1987**] that a hyponormal operator whose spectrum is "thick" has an invariant subspace. (A compact set that has interior is "thick" in this sense, but this is not the hypothesis of Brown's theorem.) This will not be done here since an easier proof of the existence of an invariant subspace for an arbitrary subnormal operator can be established directly.

Some additional results on hyponormal operators that will not be covered in this book can be found in Andô [**1963b**], Berberian [**1962**], Berger [**1978**], Carey and Pincus [**1975, 1977, and 1979**], Helton and Howe [**1973**], Herrero [**preprint**], Howe [**1974**], Martin and Putinar [**1987**], Mlak [**1967**], Sz-Nagy and Foias [**1975**], Putnam [**1967, 1974a, and 1974c**], Saito [**1972**], Stampfli [**1962, 1965, and 1969**]. This list is by no means exhaustive, only indicative.

Exercises

1. If A is hyponormal and $\mathscr{M} \in \text{Lat}\, A$ such that $A|\mathscr{M}$ is normal, then $\mathscr{M}$ reduces A. (See Exercise 3.1.)
2. If E is an idempotent and hyponormal, then E is a projection.
3. (Ito and Wong [1972]) If U is the unilateral shift and $A = U^* + 2U$, then U is hyponormal but U^2 is not.
4. (Douglas [1966]) If A and B are bounded operators on $\mathscr{H}$, then the following statements are equivalent:
   (a) $\operatorname{ran} A \subseteq \operatorname{ran} B$;
   (b) there is a $\lambda \geq 0$ such that $AA^* \leq \lambda^2 BB^*$;
   (c) there is a bounded operator C such that $A = BC$.

   If, moreover, these conditions hold, then the operator C in part (c) can be chosen so that
   (i) $\|C\|^2 = \inf\{\mu : AA^* \leq \mu BB^*\}$;
   (ii) $\ker A = \ker C$;
   (iii) $\operatorname{ran} C \subseteq (\ker B)^{\perp}$.

   Under these restrictions, C is unique. (According to Douglas, the equivalence of (a) and (c) is due to Halmos.)
5. If A is hyponormal, there is an operator B such that $A = A^*B$, $\ker A = \ker B$, and $\operatorname{ran} B \subseteq \text{cl}[\operatorname{ran} A]$.
6. (Stampfli and Wadhwa [1977]) An operator A is *dominant* if $\operatorname{ran}(A-\lambda) \subseteq \operatorname{ran}(A-\lambda)^*$ for all λ in $\sigma(A)$.
   (a) Every hyponormal operator is dominant.
   (b) If A is dominant and $\lambda \in \sigma_p(A)$, then $\ker(A - \lambda)$ reduces A.
7. (Berberian [1978]) Use the notation of Exercise I.1.14.
   (a) L_A is hyponormal if and only if A is hyponormal.
   (b) R_B is hyponormal if and only if B^* is hyponormal.
   (c) $L_A - R_B$ is hyponormal if and only if A and B^* are hyponormal.
   (d) If A and B^* are hyponormal and X is a Hilbert-Schmidt operator such that $AX = XB$, then $A^*X = XB^*$. If X is one-to-one and has dense range, then A and B are diagonalizable normal operators that are unitarily equivalent.

   (Note that (d) is a partial extension of the Fuglede-Putnam Theorem. An investigation of extensions of this result for subnormal operators will be seen in §10.)
8. If A is a hyponormal operator and $\sigma(A) = K \cup L$, where K and L are disjoint compact sets, then the Riesz idempotent corresponding to K is a projection (that is, it has norm 1).

§5 Cyclic subnormal operators.

We know (see IX.3.4 in [ACFA]) that if N is a *-cyclic normal operator on a Hilbert space $\mathscr{H}$, then there is a compactly supported measure μ on $\mathbb{C}$ such that N is unitarily equivalent to N_μ, the operator defined as multiplication by z on $L^2(\mu)$. In fact, if e is the star-cyclic vector for N, then the unitary $U : \mathscr{K} \to L^2(\mu)$ can be chosen so that $Ue = 1$, the constantly 1 function. In this section we will see some analogous results for subnormal operators. In particular, the operator S_μ will be characterized (see Corollary 5.3 below).

5.1 DEFINITION. If $A \in \mathscr{B}(\mathscr{H})$, $e_0 \in \mathscr{H}$, and K is a compact subset of $\mathbb{C}$ containing $\sigma(A)$, then e_0 is a $\mathrm{Rat}(K)$ *cyclic vector* for A if $\{u(A)e_0 : u \in \mathrm{Rat}(K)\}$ is dense in $\mathscr{H}$. An operator is $\mathrm{Rat}(K)$ *cyclic* if it has a $\mathrm{Rat}(K)$ cyclic vector. If e_0 is a $\mathrm{Rat}(\sigma(A))$ cyclic vector, then it is called an *i-cyclic vector* and A is a *rationally cyclic operator*. (The letter i here stands for "invertible".)

Recall that e_0 is a *cyclic vector* for A (A is a *cyclic operator*) if $\{p(A)e_0 : p$ is a polynomial$\}$ is dense in $\mathscr{H}$. By Runge's Theorem, this is equivalent to the statement that e_0 is a $\mathrm{Rat}(\widehat{\sigma(A)})$ cyclic vector for A.

Note that if S is subnormal and $N = \mathrm{mne}\, S$, then the fact that $\sigma(N) \subseteq \sigma(S)$ implies that $u(N)$ is well defined for any u in $\mathrm{Rat}(K)$ if K contains $\sigma(S)$.

5.2 THEOREM. *If S is subnormal and has a $\mathrm{Rat}(K)$ cyclic vector e_0, then there is a compactly supported measure μ on $\mathbb{C}$ and an isomorphism $U : \mathscr{K} \to L^2(\mu)$ such that*:

(a) $U\mathscr{H} = R^2(K, \mu)$;
(b) $Ue_0 = 1$;
(c) $UNU^{-1} = N_\mu$;
(d) *if $V = U|\mathscr{H}$, V is an isomorphism of $\mathscr{H}$ onto $R^2(K, \mu)$ and $VSV^{-1} = N_\mu | R^2(K, \mu)$.*

PROOF. Since N is the minimal normal extension of S, $\mathscr{K} = \bigvee\{N^{*n}u(N)e_0 : n \geq 0$ and $u \in \mathrm{Rat}(K)\}$. It is claimed that e_0 is a *-cyclic vector for N. Indeed, let $\mathscr{L} = \bigvee\{N^{*n}N^k e_0 : n, k \geq 0\}$; clearly $\mathscr{L}$ is a reducing subspace for N. On the other hand, the Stone-Weierstrass Theorem implies that $C(K)$ is the uniform closed linear span of $\{\bar{z}^n z^k : n, k \geq 0\}$ and, since $\mathrm{Rat}(K) \subseteq C(K)$, $u(N)e_0 \in \mathscr{L}$ for every u in $\mathrm{Rat}(K)$. Thus $\mathscr{K} \subseteq \mathscr{L}$, and since $N = \mathrm{mne}\, S$, $\mathscr{L} = \mathscr{K}$. Hence e_0 is a *-cyclic vector for N.

Therefore there is a compactly supported measure μ and an isomorphism $U : \mathscr{K} \to L^2(\mu)$ such that $Ue_0 = 1$ and $UNU^{-1} = N_\mu$. That is, (b) and (c) hold. By a standard argument, $U\phi(N) = \phi(N_\mu)U$ for every bounded Borel function ϕ. In particular, $Uu(S)e_0 = Uu(N)e_0 = u$; by taking limits, (a) follows. Part (d) is an immediate consequence. □

The next corollary was obtained independently by Bram [1955] and I. M. Singer (see Wermer [1955]).

5.3 COROLLARY. *S is a cyclic subnormal operator if and only if S is unitarily equivalent to S_μ for some compactly supported measure μ on the plane.*

Here is a small application of measure theory that will be used frequently in this book. If $f \in R^2(K, \mu)$, there is a sequence $\{f_n\} \subseteq \text{Rat}(K)$ such that $\int |f - f_n|^2 d\mu \to 0$. By passing to a subsequence if necessary, it may always be assumed that $f_n(z) \to f(z)$ a.e. $[\mu]$. The proof and statement of the next result should be compared to Theorem I.4.11.

5.4 THEOREM (Yoshino [1969]). *Let μ be a measure with support contained in the compact subset K of $\mathbb{C}$. If $S = N_\mu | R^2(K, \mu)$, then*

$$\{S\}' = \{M_\phi : \phi \in R^2(K, \mu) \cap L^\infty(\mu)\},$$

where $M_\phi f = \phi f$ for all f in $R^2(K, \mu)$.

PROOF. Clearly if $\phi \in R^2(K, \mu) \cap L^\infty(\mu)$, then M_ϕ is a well-defined operator in $\{S\}'$. Conversely, fix an operator A in $\{S\}'$ and put $\phi = A(1)$; so $\phi \in R^2(K, \mu)$. It is easy to see that if $f \in \text{Rat}(K)$, then $AM_f = M_f A$. Hence for any f in $\text{Rat}(K)$, $Af = AM_f 1 = M_f A 1 = f\phi$. If $f \in R^2(K, \mu)$, there is a sequence $\{f_n\} \subseteq \text{Rat}(K)$ such that $\int |f - f_n|^2 d\mu \to 0$ and $f_n(z) \to f(z)$ a.e. $[\mu]$. Hence $0 = \lim \|Af - Af_n\| = \lim \|Af - \phi f_n\|$. But $\phi f_n \to \phi f$ a.e. $[\mu]$. Thus, $Af = \phi f$ for all f in $R^2(K, \mu)$.

It remains to show that $\phi \in L^\infty(\mu)$. To do this we may suppose that $\|A\| = 1$. So if $\Delta = \{z : |\phi(z)| > 1\}$, then for every $n \geq 1$, $\|1\|^2 \geq \|A^n(1)\|^2 = \|\phi^n\|^2 = \int |\phi|^{2n} d\mu \geq \int_\Delta |\phi|^{2n} d\mu$. But $|\phi(z)|^{2n} \to \infty$ monotonically on Δ, so it must be that $\mu(\Delta) = 0$ and $\|\phi\|_\infty \leq 1$. □

5.5 COROLLARY. *If μ is a compactly supported measure on $\mathbb{C}$, then*

$$\{S_\mu\}' = \{M_\phi : \phi \in P^2(\mu) \cap L^\infty(\mu)\}.$$

5.6 COROLLARY. *If $S = N_\mu | R^2(K, \mu)$ and T is a bounded operator such that $TS = ST$, then $\ker T$ is a reducing subspace for S.*

PROOF. In fact, there is a function ϕ in $R^2(K, \mu) \cap L^\infty(\mu)$ such that $Tf = \phi f$ for all f. So $\ker T = \{f \in R^2(K, \mu) : f = 0$ a.e. $[\mu]$ on $\{z : \phi(z) = 0\}\}$. Clearly $\ker T$ reduces T. □

An understanding of the spaces $R^2(K, \mu)$ and $P^2(\mu)$ is one of the most important undertakings in the theory of subnormal operators and one which is far from complete. Such an understanding would undoubtedly lead to a greater understanding of the general subnormal operator, not just in the sense that it will give a larger class of examples, but also because it will reveal much about the structure of these operators. Moreover, the techniques needed to solve the $P^2(\mu)$ riddle will be useful in studying the general operator.

A giant step forward in this quest was recently taken by James E. Thomson [**preprint**]. This paper solves the long open bounded point evaluation problem (introduced in §7 below). This work is presented in detail in Chapter VIII of this book, where more comments on its importance to subnormal operator theory are made.

An additional reference on rationally cyclic subnormal operators is McCarthy [**1990**]. Amongst other things he shows that if a subnormal operator S has the property that each of its invariant subspaces must be invariant for all the operators in $\{S\}'$, then S must have a star-cyclic vector.

§6 Weighted shifts. If $A \in \mathscr{B}(\mathscr{H})$ and $A = S|A|$ is the polar decomposition of A, then S is an isometry if and only if $\ker A = (0)$. It may be that S is the unilateral shift. If this is the case and $|A|$ is a diagonalizable positive operator, then A is called a weighted shift.

6.1 DEFINITION. An operator A is called a *unilateral weighted shift* if there is an orthonormal basis $\{e_n : n \geq 0\}$ and a sequence of scalars $\{\alpha_n\}$ such that $Ae_n = \alpha_n e_{n+1}$ for all $n \geq 0$. Say that A is a weighted shift with weight sequence $\{\alpha_n\}$.

It is not hard to see that if A is as in the definition, then $A = SD$, where S is the unilateral shift and D is the diagonal operator with $De_n = \alpha_n e_n$ for all n. Thus $|A| = |D|$ and $|A|e_n = |\alpha_n|e_n$ for all n. Also note that a necessary and sufficient condition for A to be bounded is that the weight sequence is bounded.

Bilateral weighted shifts are defined analogously. An operator A is a *bilateral weighted shift* with weight sequence $\{\alpha_n : n \in \mathbb{Z}\}$ if there is an orthonormal basis $\{e_n : n \in \mathbb{Z}\}$ such that $Ae_n = \alpha_n e_{n+1}$ for all n.

Throughout this section and beyond, results occur in pairs; one for unilateral shifts and one for bilateral shifts. If there is no dramatic difference in the proofs, only one of the proofs will be given and the other will be left to the reader without comment.

The first result establishes that it will always suffice to assume that the weight sequence consists of non-negative numbers.

6.2 PROPOSITION. *If A is a unilateral (bilateral) weighted shift with weight sequence $\{\alpha_n\}$, then A is unitarily equivalent to a unilateral (bilateral) weighted shift with weight sequence $\{|\alpha_n|\}$.*

PROOF. If $\{e_n\}$ is the basis for $\mathscr{H}$ that is shifted by A, define $U : \mathscr{H} \to \mathscr{H}$ by $Ue_n = \lambda_n e_n$, where $|\lambda_n| = 1$ for all n. It follows that $UAU^* e_n = \bar{\lambda}_n \lambda_{n+1} \alpha_n e_{n+1}$. If $\lambda_0 = 1$, then UAU^* is a weighted shift with weight sequence $\{\lambda_1 \alpha_0, \bar{\lambda}_1 \lambda_2 \alpha_1, \bar{\lambda}_2 \lambda_3 \alpha_2, \ldots\}$. It is easy to verify that $\{\lambda_n\}$ can be chosen so that this sequence becomes $\{|\alpha_0|, |\alpha_1|, |\alpha_2|, \ldots\}$. □

What happens if one of the weights is 0? If $\alpha_n = 0$, then it is easy to see that $\mathscr{L} = \bigvee\{e_k : 0 \leq k \leq n\}$ and $\mathscr{M} = \bigvee\{e_k : k \geq n+1\}$ are both invariant subspaces for the weighted shift A. Since $\mathscr{L} \perp \mathscr{M}$, these spaces reduce A.

Thus $A = B \oplus C$, where $B = A|\mathscr{L}$ and $C = A|\mathscr{M}$. Moreover, C is a weighted shift with weight sequence $\{\alpha_{n+1}, \alpha_{n+2}, \ldots\}$ and B is a nilpotent operator on a finite dimensional space. In fact, if α_n is the first nonzero weight, B is a cyclic nilpotent operator. This leads to two possibilities.

If the weight sequence contains an infinite number of 0's, A is unitarily equivalent to the direct sum of cyclic nilpotent operators on finite dimensional spaces. Such operators as this are not readily understood; indeed, they can be quite complicated. For example, their spectrum can be any size disk, though it must be a disk since such an operator is unitarily equivalent to $e^{i\theta}$ times itself for any choice of θ.

If there are a finite number of 0's in the weight sequence, then A is unitarily equivalent to the direct sum of a finite number of cyclic nilpotent operators on finite dimensional spaces and a weighted shift with nonzero weights. The direct sum of a finite number of cyclic nilpotent operators on finite dimensional spaces is well understood. This leaves the weighted shift with no zero weights.

In light of these comments and the preceding proposition, we are therefore led to make the following binding agreement.

Henceforward, all weight sequences will be strictly positive.

Also, to avoid repetitive statements about notation, in this section A will always denote a weighted shift with weight sequence $\{\alpha_n\}$, $\alpha_n > 0$ for $n \geq 0$, and A will shift the basis $\{e_n\}$. Similar notational conventions apply to bilateral shifts.

The proof of the next proposition is an exercise.

6.3 PROPOSITION. (a) *If A is a unilateral shift, then $A^* e_0 = 0$ and $A^* e_n = \alpha_{n-1} e_{n-1}$ for $n \geq 1$. Consequently, $(A^*A - AA^*)e_0 = \alpha_0^2 e_0$ and $(A^*A - AA^*)e_n = (\alpha_n^2 - \alpha_{n-1}^2)e_n$ for $n \geq 1$.*

(b) *If A is a bilateral shift, then $A^* e_n = \alpha_{n-1} e_{n-1}$ for all n. Consequently, $(A^*A - AA^*)e_n = (\alpha_n^2 - \alpha_{n-1}^2)e_n$ for all n.*

6.4 COROLLARY. *If A is a unilateral (bilateral) weighted shift, S (respectively, U) is the corresponding unilateral (bilateral) unweighted shift, and $D = \mathrm{diag}\{\alpha_n\}$, then $A = SD$ ($A = UD$) is the polar decomposition of A.*

6.5 PROPOSITION. *If A is a weighted shift and $\omega \in \partial \mathbb{D}$, then $A \cong \omega A$.*

PROOF. If $Ue_n = \omega^n e_n$ for all n, then a computation shows that $UAU^* = \omega A$. □

As a consequence of the preceding proposition, $\sigma(A) = \sigma(\omega A) = \omega \sigma(A)$; that is, the spectrum of a weighted shift must be circularly symmetric.

We here leave the general theory of weighted shifts. The interested reader can continue this topic with an excellent survey article by Shields [**1974**], which has become the starting point for all work on weightedshifts. In the remainder of this section we will concentrate on weighted shifts that are more germane to the principal aims of this book.

The first result with this restricted focus is an easy consequence of Proposition 6.3.

6.6 PROPOSITION. *A weighted shift is hyponormal if and only if its weight sequence is increasing.*

To begin the study of hyponormal weighted shifts, let us decipher their spectral properties. It is easy to see that $\|A\| = \sup_n \alpha_n$ (for any weighted shift A). So if A is hyponormal, let $a_+ = \lim \alpha_n = \|A\|$ and let $a_- = \lim_{n \to -\infty} \alpha_n$. The next theorem, some of whose parts are actually only a rephrasing and combination of other parts, done for convenience and ready reference, contains the most significant spectral information about hyponormal weighted shifts.

6.7 THEOREM. *Let A be a hyponormal weighted shift.*
  (a) *If A is a unilateral weighted shift, then:*
    (i) $\sigma(A) = \{\lambda : |\lambda| \leq \|A\|\}$;
    (ii) $\sigma_p(A) = \emptyset$ and $\sigma_p(A^*) = \{\lambda : |\lambda| < \|A\|\}$;
    (iii) *if $|\lambda| < \|A\|$, then $\operatorname{ran}(A - \lambda)$ is closed and $\ker(A - \lambda)^*$ is one dimensional;*
    (iv) $\sigma_e(A) = \sigma_{le}(A) = \sigma_{re}(A) = \{\lambda : |\lambda| = \|A\|\}$;
    (v) *if $|\lambda| < \|A\|$, then $A - \lambda$ is Fredholm and $\operatorname{ind}(A - \lambda) = -1$.*
  (b) *If A is a bilateral hyponormal weighted shift, then*
    (i) $\sigma(A) = \{\lambda : a_- \leq |\lambda| \leq \|A\|\}$;
    (ii) $\sigma_p(A) = \emptyset$ and $\sigma_p(A^*) = \{\lambda : a_- < |\lambda| < \|A\|\}$;
    (iii) *if $a_- < |\lambda| < \|A\|$, then $\operatorname{ran}(A - \lambda)$ is closed and $\ker(A^* - \lambda)$ is one dimensional;*
    (iv) $\sigma_e(A) = \sigma_{le}(A) = \sigma_{re}(A) = \{\lambda : |\lambda| = a_- \text{ or } \|A\|\}$;
    (v) *if $a_- < |\lambda| < \|A\|$, then $A - \lambda$ is Fredholm and $\operatorname{ind}(A - \lambda) = -1$.*

This theorem will quickly follow from the next slightly more general result.

6.8 PROPOSITION. (a) *If A is a unilateral weighted shift, $\sigma_p(A) = \emptyset$. If $\alpha_n \to \alpha_+$, then $\sigma_e(A) = \{\lambda : |\lambda| = \alpha_+\}$, $\sigma(A) = \{\lambda : |\lambda| \leq \alpha_+\}$, and for $|\lambda| < \alpha_+$, $A - \lambda$ is Fredholm and $\operatorname{ind}(\lambda - A) = -1$.*

(b) *If A is a bilateral weighted shift and $\alpha_n \to \alpha_\pm$ as $n \to \pm\infty$, then $\sigma_e(A) = \{\lambda : |\lambda| = \alpha_+ \text{ or } \alpha_-\}$. Also $\sigma_p(A) \subseteq \{\lambda : \alpha_+ \leq |\lambda| \leq \alpha_-\}$. In particular, if $\alpha_- < \alpha_+$, $\sigma_p(A) = \emptyset$, $\sigma(A) = \{\lambda : \alpha_- \leq |\lambda| \leq \alpha_+\}$, and for $\alpha_- < |\lambda| < \alpha_+$, $A - \lambda$ is Fredholm with $\operatorname{ind}(A - \lambda) = -1$.*

PROOF. (a) The fact that $\sigma_p(A) = \emptyset$ should be left as an exercise, but the computation needed to prove it is also needed for the bilateral case. If $f = \sum_n \hat{f}(n) e_n \in \mathscr{H}$, then $(A - \lambda)f = \sum_n \alpha_n \hat{f}(n) e_{n+1} - \sum_n \lambda \hat{f}(n) e_n$. So if $(A - \lambda)f = 0$, $\lambda \hat{f}(0) = 0$ and $\lambda \hat{f}(n) = \alpha_{n-1} \hat{f}(n-1)$ for $n \geq 1$. It is easy to see that these equations are valid only if $f = 0$.

Now assume that $\lim \alpha_n = \alpha_+$. If $\alpha_+ = 0$, then A is compact since $|A|$ is. If $\alpha_+ > 0$ and S is the unilateral shift relative to the basis $\{e_n\}$, then it follows that $A - \alpha_+ S$ is a unilateral weighted shift whose weight sequence converges to 0. Hence $A - \alpha_+ S$ is compact and so $\sigma_e(A) = \sigma_e(\alpha_+ S) = \{\lambda : |\lambda| = \alpha_+\}$. Also, for $|\lambda| < \alpha_+$, $A - \lambda \in \mathscr{F}$ and $\mathrm{ind}(A - \lambda) = \mathrm{ind}(\alpha_+ S - \lambda) = -1$. In particular, this implies that $\{\lambda : |\lambda| \leq \alpha_+\} \subseteq \sigma(A)$.

It remains to show that if $|\lambda| > \alpha_+$, then $A - \lambda$ is invertible. But such a λ is not in $\sigma_e(A)$ so that $A - \lambda \in \mathscr{F}$ and so $\mathrm{ind}(A - \lambda) = \mathrm{ind}(\alpha_+ S - \lambda) = 0$. Since $\ker(A - \lambda) = (0)$, it follows that $\mathrm{ran}(A - \lambda)^\perp = (0)$ and so $A - \lambda$ is invertible.

(b) Let $\mathscr{H}_+ = \bigvee\{e_n : n \geq 0\}$ and $\mathscr{H}_- = \bigvee\{e_{-n} : n \geq 1\}$; so $\mathscr{H} = \mathscr{H}_- \oplus \mathscr{H}_+$. Define unilateral weighted shifts $A_\pm$ on $\mathscr{H}_\pm$ by letting $A_+ = A | \mathscr{H}_+$ and $A_- e_{-n} = \alpha_{-n} e_{-n-1}$ for $n \geq 1$. Also define the rank one operator F on $\mathscr{H}$ by letting $Fe_{-1} = \alpha_{-1} e_0$ and $Fe_n = 0$ for $n \neq -1$. It is left to the reader to check that $A = (A_-^* \oplus A_+) + F$. Since F has finite rank, part (a) implies that $\sigma_e(A) = \sigma_e(A_-^* \oplus A_+) = \{\lambda : |\lambda| = \alpha_+ \text{ or } \alpha_-\}$.

To prove the statement about $\sigma_p(A)$, note that if $(A - \lambda)f = 0$, then, as in part (a), $\lambda \hat{f}(n) = \alpha_{n-1} \hat{f}(n-1)$ for all n in $\mathbb{Z}$. Without loss of generality we may assume that $\hat{f}(0) = 1$. Hence we can solve these equations to obtain $\hat{f}(n) = (\alpha_0 \cdots \alpha_{n-1}) \lambda^{-n}$ and $\hat{f}(-n) = \lambda^n (\alpha_{-1} \cdots \alpha_{-n})^{-1}$ for $n \geq 1$. Thus

$$\|f\|^2 = \sum_{n=1}^\infty |\lambda|^{2n} (\alpha_{-1} \cdots \alpha_{-n})^{-2} + 1 + \sum_{n=1}^\infty |\lambda|^{-2n} (\alpha_0 \cdots \alpha_{n-1})^2$$

So if $|\lambda| < \alpha_+$, let $|\lambda| < \rho < \alpha_+$ and choose m such that $\rho < \alpha_n$ for $n \geq m$. So $|\lambda|^{-2n} (\alpha_0 \cdots \alpha_{n-1})^2 = [|\lambda|^{-2m} (\alpha_0 \cdots \alpha_{m-1})^2][|\lambda|^{-2n+2m} (\alpha_m \cdots \alpha_{n-1})^2] > [|\lambda|^{-2m} (\alpha_0 \cdots \alpha_{m-1})^2] > 0$, contradicting the fact that the above series converges. Thus it must be that $|\lambda| \geq \alpha_+$. In a similar fashion, $|\lambda| \leq \alpha_-$.

Now assume that $\alpha_- < \alpha_+$. Clearly $\sigma_p(A) = \varnothing$. Also let $A_\pm$ and F be defined as above. If $\alpha_- < |\lambda| < \alpha_+$, then $A - \lambda$ is Fredholm and $\mathrm{ind}(A - \lambda) = \mathrm{ind}(A_-^* \oplus A_+ - \lambda) = \mathrm{ind}(A_-^* - \lambda) + \mathrm{ind}(A_+ - \lambda)$. But from part (a), $A_-^* - \lambda$ is invertible and $\mathrm{ind}(A_+ - \lambda) = -1$. Thus $\mathrm{ind}(A - \lambda) = -1$.

If $|\lambda| < \alpha_-$, then part (a) implies that $\mathrm{ind}(A - \lambda) = \mathrm{ind}(A_-^* - \lambda) + \mathrm{ind}(A_+ - \lambda) = 1 + (-1) = 0$. But $\ker(A - \lambda) = (0)$, so $A - \lambda$ is invertible. Similarly, $A - \lambda$ is invertible for $|\lambda| > \alpha_+$ and so $\sigma(A) = \{\lambda : \alpha_- \leq |\lambda| \leq \alpha_+\}$. $\square$

PROOF OF THEOREM 6.7. (a) Part (i) follows quickly from (6.8). Since $A - \lambda$ is Fredholm for $|\lambda| < \|A\|$ and $\mathrm{ind}(A - \lambda) = -1$, the fact that $\ker(A - \lambda) = (0)$ implies that $\ker(A - \lambda)^* = [\mathrm{ran}(A - \lambda)]^\perp$ is 1 dimensional. Since Fredholm operators have closed range, the remainder of (a) follows.

The proof of (b) is left to the reader. $\square$

6.9 COROLLARY. *If A is a hyponormal unilateral weighted shift, then A does not have an nth root.*

PROOF. If $A = B^n$ for some B in $\mathscr{B}(\mathscr{H})$, then B must be Fredholm by the Spectral Mapping Theorem, and $-1 = \operatorname{ind} A = \operatorname{ind} B^n = n(\operatorname{ind} B)$, a contradiction. □

Further information about the spectral theory of arbitrary weighted shifts can be found in the survey article by Shields [1974]. The approximate point spectrum of a general weighted shift is rather difficult to determine; it can be found in Ridge [1970].

Shields [1974], as mentioned before, is encyclopedic. You can also find there a characterization of the commutant of a general weighted shift. This will not be done here even for a hyponormal weighted shift, in which case the answer is appealing (viz, the commutant of a hyponormal weighted shift A is the algebra of bounded analytic functions on the disk $\{\lambda : |\lambda| < \|A\|\}$). Instead we focus our attention on subnormal weighted shifts.

6.10 THEOREM. *Let $\{e_0, e_1, \ldots\}$ be an orthonormal basis for $\mathscr{H}$ and let S be a weighted shift relative to this basis with weight sequence $\{\alpha_n\}$, where $\sup_n \alpha_n = 1$. The following statements are equivalent.*

(a) *S is subnormal.*

(b) *There is a probability measure ν on $[0, 1]$ containing 1 in its support such that for $n \geq 1$*

$$(\alpha_0 \cdots \alpha_{n-1})^2 = \int r^{2n} d\nu(r).$$

(c) *There is a probability measure ν on $[0, 1]$ containing 1 in its support such that if μ is the measure defined on $\mathbb{C}$ by*

$$d\mu(re^{i\theta}) = (2\pi)^{-1} d\theta \, d\nu(r),$$

then S is unitarily equivalent to S_μ.

The correspondence between subnormal weighted shifts of norm 1 and probability measures on $[0, 1]$ with 1 in their support, as described in (b), is bijective.

PROOF. (a) *implies* (b). Since S is a cyclic subnormal operator with cyclic vector e_0, Theorem 5.2 implies there is a compactly supported measure μ on $\operatorname{cl} \mathbb{D}$ and an isomorphism $U : \mathscr{H} \to P^2(\mu)$ such that $Ue_0 = 1$ and $USU^{-1} = S_\mu$. If $\beta_n = (\alpha_0 \cdots \alpha_{n-1})$ for $n \geq 1$ and $\beta_0 = 1$, then $S^n e_0 = \beta_n e_n$ for all n. Therefore, $U(S^n e_0) = S_\mu^n 1 = z^n$. So $\int |z|^{2n} d\mu = \beta_n^2 \|Ue_n\|^2 = \beta_n^2$. If ν is defined on $[0, 1]$ by $\nu(\Delta) = \mu(\{z : |z| \in \Delta\})$, then ν is a probability measure and $\beta_n^2 = \int r^{2n} d\nu(r)$. Since $\|S\| = 1$ and $S \cong \omega S$ for all scalars ω with $|\omega| = 1$, $\partial \mathbb{D} \subseteq \operatorname{support} \mu$. Hence $1 \in \operatorname{support} \nu$.

(b) *implies* (c). Let μ be the measure defined in (c), and let β_n be as defined in the first part of this proof. Condition (b) can be restated as $\beta_n^2 = \|z^n\|^2$, the square of the norm of z^n in $P^2(\mu)$. Moreover, a routine calculation reveals that in $P^2(\mu)$, $z^n \perp z^m$ for $n \neq m$. Therefore,

$\{z^n \beta_n^{-1} : n \geq 0\}$ is an orthonormal basis for $P^2(\mu)$. If $U : \mathscr{H} \to P^2(\mu)$ is defined by $Ue_n = z^n \beta_n^{-1}$, then U is an isomorphism and $USU^{-1} = S_\mu$.

Since it is clear that (c) implies (a), this completes the proof of the equivalence of (a) through (c). The uniqueness statement in the theorem is an immediate consequence of the Stone-Weierstrass Theorem. □

According to Halmos [**1982**], the equivalence of (a) and (b) is due to C. A. Berger. This equivalence was independently established and first published in Gellar and Wallen [**1970**].

If $d\nu(r) = 2r\,dr$ on $[0, 1]$, the corresponding subnormal weighted shift is the Bergman operator for $\mathbb{D}$. Let δ_0 and δ_1 be the unit point masses at 0 and 1 for any $\alpha > 0$ put $\nu = (\alpha \delta_0 + \delta_1)/(1 + \alpha)$. So ν is a probability measure containing 1 in its support. The corresponding subnormal shift has weight sequence $\{(1 + \alpha)^{-1/2}, 1, 1, \ldots\}$. It turns out that this is the only subnormal weighted shift whose weight sequence is eventually constant.

In fact, Stampfli [**1966**] gives another characterization of subnormal weighted shifts and using this he proved the next result. The proof presented here will use the preceding theorem and not rely on his criterion.

6.11 PROPOSITION (Stampfli [**1966**]). *If S is a subnormal weighted shift and $\alpha_n = \alpha_{n+1}$ for some $n \geq 1$, then $\alpha_k = \alpha_1$ for all $k \geq 1$.*

PROOF. Without loss of generality it may be assumed that $\|S\| = 1$. Let ν be the probability measure on $[0, 1]$ such that $\beta_n^2 = (\alpha_0 \cdots \alpha_{n-1})^2 = \int r^{2n} d\nu(r)$ for $n \geq 1$. By hypothesis, there is an $n \geq 1$ such that $\alpha_n = \alpha_{n+1}$, or $\beta_{n+1}/\beta_n = \beta_{n+2}/\beta_{n+1}$; equivalently, $\beta_{n+1}^2 = \beta_n \beta_{n+2}$. This implies that

$$\int r^{2n+2} d\nu(r) = \left(\int r^{2n} d\nu(r)\right)^{1/2} \left(\int r^{2n+4} d\nu(r)\right)^{1/2}.$$

But this says that equality holds in the Cauchy-Schwarz Inequality. Hence, there is a constant c such that $r^n = cr^{n+2}$ a.e. $[\nu]$. But $r^n - cr^{n+2}$ is continuous and so $r^n - cr^{n+2} = 0$ everywhere on the support of ν. Since $1 \in \operatorname{support} \nu$, $c = 1$ and $r^n - r^{n+2} = 0$ on $\operatorname{support} \nu$. This implies that $\operatorname{support} \nu$ is contained in the two point set $\{0, 1\}$. That is $\nu = a\delta_0 + b\delta_1$, where $a, b \geq 0$ and $a + b = 1$. The result follows by computing the moments β_n^2. □

Stampfli [**1966**] also shows that if $0 < \alpha_0 < \alpha_1 < \alpha_2$ are arbitrarily given, then there is a subnormal weighted shift with these three weights in the initial spots, but that the first four weights cannot be arbitrarily assigned.

Now for a bilateral version of Theorem 6.10, which was first done in Herrero [**1972**].

6.12 THEOREM. *Let $\{e_n : n \in \mathbb{Z}\}$ be an orthonormal basis for $\mathscr{H}$ and let S be a bilateral weighted shift relative to the basis with weight sequence $\{\alpha_n\}$, where $\sup_n \alpha_n = 1$. The following statements are equivalent.*

(a) S is subnormal.

(b) *There is a probability measure ν on $[0, 1]$ with 1 in its support such that $r^n \in L^1(\nu)$ for all n in $\mathbb{Z}$ and for $n \geq 1$,*

$$(\alpha_0 \cdots \alpha_{n-1})^2 = \int r^{2n} \, d\nu(r),$$

$$(\alpha_{-1} \cdots \alpha_{-n})^2 = \int r^{-2n} \, d\nu(r).$$

(c) *There is a probability measure ν on $[0, 1]$ with 1 in its support such that $r^n \in L^1(\nu)$ for all n in $\mathbb{Z}$ and if $d\mu(re^{i\theta}) = (2\pi)^{-1} d\theta \, d\nu(r)$ and $R^2(\nu)$ is the closed linear span in $L^2(\mu)$ of $\{z^n : n \in \mathbb{Z}\}$, then S is unitarily equivalent to multiplication by z on $R^2(\mu)$.*

The correspondence given in (b) *is bijective.*

PROOF. Only the proof that (a) implies (b) will be given; the proofs of the remaining implications are similar to the corresponding proofs in Theorem 6.10. The details are left to the reader.

For $n \leq 0$, let $\mathscr{H}_n = \bigvee \{e_k : k \geq n\}$; so $S | \mathscr{H}_n$ is a subnormal unilateral weighted shift with weight sequence $\{\alpha_n, \alpha_{n+1}, \ldots\}$. By Theorem 6.10 there is a probability measure ν_n on $[0, 1]$ with $1 \in \text{support } \nu$ such that for $k \geq 1$, $(\alpha_n \cdots \alpha_{n+k-1})^2 = \int r^{2k} d\nu_n(r)$. In particular, $\alpha_n^2 = \int r^2 d\nu_n(r)$ so that $\alpha_n^{-2} r^2 d\nu_n(r)$ is a probability measure. Moreover, $\int r^{2k} (\alpha_n^{-2} r^2) d\nu_n(r) = \alpha_n^{-2} \int r^{2(k+1)} d\nu_n(r) = (\alpha_{n+1} \cdots \alpha_{n+k})^2$. By the uniqueness of the measures, $\alpha_n^{-2} r^2 d\nu_n(r) = d\nu_{n+1}(r)$.

Put $\nu = \nu_0$; so $(\alpha_0 \cdots \alpha_{n-1})^2 = \int r^{2n} d\nu(r)$ if $n \geq 1$. Since $d\nu(r) = \alpha_{-1}^2 r^2 d\nu_{-1}(r)$, $r^{-2} \in L^1(\nu)$. By induction, $r^{-2n} \in L^1(\nu)$ for all $n \geq 1$ and $r^{-2n} d\nu(r) = (\alpha_{-1} \cdots \alpha_{-n})^{-2} d\nu_{-n}(r)$. Since ν_n is a probability measure, this implies that $\int r^{-2n} d\nu(r) = (\alpha_{-1} \cdots \alpha_{-n})^{-2}$. □

REMARKS. 1. Frankfurt [1975, 1976, and 1977a] contains a lot of information about subnormal unilateral weighted shifts and the related function spaces $P^2(\mu)$, where μ is the measure appearing in Theorem 6.10. In particular, Frankfurt [1975] characterizes the functions in $P^2(\mu)$.

2. If S is a unilateral weighted shift and if $U_\omega : \mathscr{H} \to \mathscr{H}$ is defined for $|\omega| = 1$ by $U_\omega e_n = \omega^n e_n$, then $U_\omega S U_\omega^* = \omega S$ for all ω in $\partial \mathbb{D}$. Note that $\omega \to U_\omega$ is a (SOT) continuous unitary representation of the circle group. Gellar [1977] has shown that this characterizes the subnormal unilateral weighted shifts amongst all the pure cyclic subnormal operators. That is, if S is a pure cyclic subnormal operator and $\omega \to U_\omega$ is a (SOT) continuous unitary representation of the circle group such that $U_\omega S U_\omega^* = \omega S$ for all ω in $\partial \mathbb{D}$, then S is a unilateral weighted shift.

3. The preceding remark raises the question whether a bilateral shift can be cyclic. Note that the bilateral shift of multiplicity 1 is cyclic. However it has been independently shown by Herrero [1972] and Frankfurt [1975] that a bilateral hyponormal weighted shift cannot be cyclic unless it is normal.

A good paper with some general information on cyclic operators is Herrero [**1978**]. Also, Herrero [**preprint**] shows that most hyponormal bilateral operator valued weighted shifts cannot be cyclic.

4. Ghatage [**1976**] and Lambert [**1976**] both prove the extension of Theorem 6.10 to invertible operator valued weighted shifts.

5. Kulkarni [**1970**] extended the Stampfli [**1966**] criterion for the subnormality of a unilateral weighted shift to bilateral weighted shifts and characterized the invertible bilateral subnormal weighted shifts by moment sequences.

6. Ivanovski [**1973**] generalized the criterion of Stampfli [**1966**] for subnormality to unilateral operator valued weighted shifts. He also characterized bilateral invertible operator valued weighted shifts in terms of moment sequences for positive operator valued measures.

7. Lambert [**1976**] shows how to derive Theorem 6.10 directly from Theorem 1.9(h).

8. Lambert [**1976**] also proves the following interesting characterization of subnormality using weighted shifts.

PROPOSITION (Lambert [**1976**]). *Let* $S \in \mathscr{B}(H)$ *with* $\ker S = (0)$ *and* $\|S\| = 1$. *For each nonzero vector* f *in* $\mathscr{H}$, *define* A_f *to be the weighted shift with weight sequence* $\{\|S^{n+1}f\|/\|S^n f\|\}$. *Then* S *is subnormal if and only if* A_f *is subnormal for every nonzero vector* f.

9. Lubin [**1977**] studies weighted shifts "in several variables". These are families of commuting weighted shifts.

10. Applications of subnormal weighted shifts to operator theory can be found in Bastian and Harrison [**1974**] and Deddens and Stampfli [**1973**].

Exercises

1. Give necessary and sufficient conditions that two weighted shifts be unitarily equivalent.
2. Give necessary and sufficient conditions that two weighted shifts be similar.
3. Show that a weighted shift with nonzero weights is irreducible.
4. Show that if A is a unilateral weighted shift with $\|A\| \leq 1$, then $A^{*n} \to 0$ (SOT). Is the condition that A be a contraction needed?
5. Which weighted shifts are normal?
6. If T is an arbitrary operator in $\mathscr{B}(\mathscr{H})$, define $m(T) \equiv \inf\{\|Tf\| : \|f\| = 1\}$.
   (a) Show that if T is invertible, then $m(T) = \|T^{-1}\|^{-1}$.
   (b) If $T \in \mathscr{B}(\mathscr{H})$ and $\lambda \in \sigma_{\mathrm{ap}}(T)$, then $|\lambda| \geq m(T)$.
   (c) If A is a hyponormal weighted shift, then $\|A\| = \lim m(A^n)^{1/n}$.
   (*Hint*: find a formula for $m(A^n)$ in terms of the weights.)
7. Let S be a subnormal unilateral weighted shift with $\|S\| \leq 1$ and let ν be the probability measure on $[0, 1]$ as in Theorem 6.10.

(a) Show that S is similar to the unilateral shift if and only if $\nu(\{0\}) > 0$.
(b) Show that $S^n \to 0$ (SOT) if and only if $\nu(\{0\}) = 0$.

§7 **Bounded point evaluations.** Recall the notation from Example 1.4. Let K be a compact subset of the complex plane, let μ be a positive measure supported on K, and let $R^2(K,\mu)$ be the closure of Rat(K) in $L^2(\mu)$. Define $S: R^2(K,\mu) \to R^2(K,\mu)$ as multiplication by z.

7.1 DEFINITION. A point λ in $\mathbb{C}$ is called a *bounded point evaluation* (bpe) for $R^2(K,\mu)$ if there is a constant $C > 0$ such that

$$|f(\lambda)| \le C\|f\|$$

for every f in Rat(K). Let bpe(K,μ) denote the set on bounded point evaluations for $R^2(K,\mu)$.

If K is polynomially convex, then $R^2(K,\mu) = P^2(\mu)$. Let bpe(μ) be the set of bounded point evaluations for $P^2(\mu)$.

Notice that if $\lambda \in$ bpe(K,μ), then the linear functional $f \to f(\lambda)$ on Rat(K) has a unique bounded extension to $R^2(K,\mu)$. Thus there is a unique function k_λ in $R^2(K,\mu)$ such that $\langle f, k_\lambda \rangle = f(\lambda)$ for all f in Rat(K). This is an important fact about bounded point evaluations that is equivalent to the definition.

7.2 PROPOSITION. *If $\lambda \in \mathbb{C}$, the following statements are equivalent.*

(a) *λ is a bounded point evaluation for $R^2(K,\mu)$.*
(b) *There is a k_λ in $R^2(K,\mu)$ such that $f(\lambda) = \langle f, k_\lambda \rangle$ for every f in* Rat(K).
(c) $\ker(S-\lambda)^*$ *is one dimensional.*
(d) $\bar\lambda \in \sigma_p(S^*)$.

PROOF. It is clear from the remarks preceding the proposition that (a) and (b) are equivalent.

(d) *implies* (b). Let e be a nonzero vector in $[\operatorname{ran}(S-\lambda)]^\perp = \ker(S-\lambda)^*$. Note that $\operatorname{ran}(S-\lambda) \supseteq \{f \in \operatorname{Rat}(K) : f(\lambda) = 0\}$, so that $\langle f, e \rangle = 0$ if $f \in \operatorname{Rat}(K)$ and $f(\lambda) = 0$. But if f is an arbitrary rational function with poles off K, then $f = [f - f(\lambda)] + f(\lambda)$, where $f(\lambda)$ is the constant function. Thus $\langle f, e \rangle = f(\lambda)\langle 1, e \rangle$. Since $e \ne 0$, it follows that $\langle 1, e \rangle \ne 0$. Put $k_\lambda = ce$ where $\bar c^{-1} = \langle 1, e \rangle$. The preceding equations show that (b) holds.

(b) *implies* (c). It is left to the reader to verify that k_λ belongs to $\ker(S-\lambda)^*$; note that k_λ is unique. In the preceding paragraph an argument was given that shows that if $e \in \ker(S-\lambda)^*$ and $\alpha = \langle 1, e \rangle \ne 0$, then $\langle f, \alpha^{-1}e \rangle = f(\lambda)$ for every f in Rat(K). By the uniqueness of k_λ, $e = \alpha k_\lambda$ and so $\ker(S-\lambda)^*$ is the one dimensional space spanned by k_λ.

Since it is clear that (c) implies (d), this finishes the proof. □

For many choices of K and μ, $R^2(K, \mu) = L^2(\mu)$. (For example, take $K = \partial \mathbb{D}$ and let $\mu = m =$ Lebesque measure on $\partial \mathbb{D}$.) If this is the case, then $\mathrm{bpe}(K, \mu)$ is the set of atoms of μ and for each bounded point evaluation λ, $k_\lambda = \mu(\{\lambda\})^{-1}\chi_{\{\lambda\}}$. (Verify!)

Note that $\bar{\lambda} \in \sigma_p(S^*)$ if and only if λ belongs to the compression spectrum of S (see Halmos, [1982]). So the set of bounded point evaluations is precisely the compression spectrum of S. In particular, this implies that $\mathrm{bpe}(K, \mu)$ is a subset of K.

If $\lambda \in \mathrm{bpe}(K, \mu)$, then k_λ will always denote the unique element in $R^2(K, \mu)$ such that $\langle f, k_\lambda \rangle = f(\lambda)$ for every f in $\mathrm{Rat}(K)$. This function k_λ associated with the bounded point evaluation λ is called the *reproducing kernel* for $R^2(K, \mu)$ at λ. Often the function $k : \mathrm{bpe}(K, \mu) \to R^2(K, \mu)$ defined by $k(\lambda) = k_\lambda$ is also called the reproducing kernel. If f is an arbitrary function in $R^2(K, \mu)$, let $\hat{f}(\lambda) = \langle f, k_\lambda \rangle$ for λ in $\mathrm{bpe}(K, \mu)$.

7.3 PROPOSITION. (a) *If* $f \in R^2(K, \mu)$, *then* $f(\lambda) = \hat{f}(\lambda)$ *a.e.* $[\mu]$ *on* $\mathrm{bpe}(K, \mu)$.

(b) *If* $f \in R^2(K, \mu)$ *and* $\phi \in R^2(K, \mu) \cap L^\infty(\mu)$, *then* $\widehat{\phi f}(\lambda) = \hat{\phi}(\lambda)\hat{f}(\lambda)$ *for all* λ *in* $\mathrm{bpe}(K, \mu)$.

PROOF. (a) Let $f \in R^2(K, \mu)$ and let $\{f_n\}$ be a sequence in $\mathrm{Rat}(K)$ such that $\|f_n - f\| \to 0$. By passing to a subsequence if necessary, we may assume that $f_n(\lambda) \to f(\lambda)$ a.e. $[\mu]$. But then for λ in $\mathrm{bpe}(K, \mu)$, $\langle f_n, k_\lambda \rangle \to \langle f, k_\lambda \rangle$. The result follows.

(b) First note that if λ is a bounded point evaluation, then $\widehat{z\phi}(\lambda) = \langle z\phi, k_\lambda \rangle = \langle \phi, S_\mu^* k_\lambda \rangle = \langle \phi, \bar{\lambda} k_\lambda \rangle = \lambda \hat{\phi}(\lambda)$; a little algebra shows that for every polynomial p, $\widehat{\phi p}(\lambda) = \hat{\phi}(\lambda) p(\lambda)$. More algebra shows that $\widehat{\phi f}(\lambda) = \hat{\phi}(\lambda)\hat{f}(\lambda)$ for every rational function f with poles off K. If f is any function in $R^2(K, \mu)$, let $\{f_n\}$ be a sequence in $\mathrm{Rat}(K)$ such that $\|f_n - f\| \to 0$. Since ϕ is bounded, $\|\phi f_n - \phi f\| \to 0$. Thus $\widehat{\phi f_n}(\lambda) \to \widehat{\phi f}(\lambda)$. But $\widehat{\phi f_n}(\lambda) = \hat{\phi}(\lambda)\hat{f_n}(\lambda) \to \hat{\phi}(\lambda)\hat{f}(\lambda)$. This proves part (b). □

As a consequence of this lemma we can assume that $f(\lambda) = \hat{f}(\lambda)$ for every f in $R^2(K, \mu)$ and for every λ in $\mathrm{bpe}(K, \mu)$; this assumption will remain in force throughout this book.

Also note that Proposition 7.2 implies that $\mathrm{bpe}(K, \mu) \subseteq \sigma(S)$, but nothing can be said about the relation between $\mathrm{bpe}(K, \mu)$ and $\sigma_n(S)$, the spectrum of the minimal normal extension of S. In fact, if S is the unilateral shift, multiplication by z on $P^2(m) = H^2$, then Proposition I.3.15 implies that $\mathrm{bpe}(m) = \mathbb{D}$ which is disjoint from $\mathrm{support}(m) = \sigma(N)$. Note that by Proposition 6.7, if $S = S_\mu$ is a subnormal unilateral weighted shift with $\|S\| = 1$, then $\mathrm{bpe}(\mu) = \mathbb{D}$. However, we can always find a measure μ as in (6.10(c)) such that its support is any compact circularly symmetric subset of $\mathrm{cl}\,\mathbb{D}$ that includes $\partial \mathbb{D}$. If S is an invertible subnormal bilateral weighted shift and $K = \{z : \|S^{-1}\|^{-1} \leq |z| \leq \|S\|\}$, then $\mathrm{bpe}(K, \mu) = \mathrm{int}\,K$.

§7 BOUNDED POINT EVALUATIONS

In fact there is a more refined concept that we need; one that gives an analytic structure to the study of rationally cyclic subnormal operators.

7.4 DEFINITION. A point λ in $\text{bpe}(K, \mu)$ is called an *analytic bounded point evaluation* (abpe) for $R^2(K, \mu)$ if $\lambda \in \text{int}\{\text{bpe}(K, \mu)\}$ and for every f in $R^2(K, \mu)$, the function $z \to f(z)$ is analytic in a neighborhood of λ. Let $\text{abpe}(K, \mu)$ be the collection of analytic bounded point evaluations. In case K is polynomially convex, $\text{abpe}(\mu) = \text{abpe}(K, \mu)$.

Technically, the condition in this definition should read that $z \to f(z)$ agrees a.e. $[\mu]$ with an analytic function in a neighborhood of λ. But in light of Proposition 7.3 the above phrasing is not only acceptable but preferred.

The first question that arises about analytic bounded point evaluations is whether a certain uniformity exists. Look at the definition closely. As stated the neighborhood of λ on which the function $z \to f(z)$ is analytic depends on the function f in $R^2(K, \mu)$. Is $\text{abpe}(K, \mu)$ an open set?

7.5 LEMMA. *If $\lambda \in \text{abpe}(K, \mu)$, then there is an $r > 0$ such that $B(\lambda; r) \subseteq \text{abpe}(K, \mu)$; in particular, $\text{abpe}(K, \mu)$ is an open subset of $\mathbb{C}$.*

PROOF. By definition, $\lambda \in \text{int}\{\text{bpe}(K, \mu)\}$; so let $\delta > 0$ such that $B(\lambda; \delta) \subseteq \text{bpe}(K, \mu)$. For $m, n \geq 1$, let $F_{m,n} \equiv \{f \in R^2(K, \mu) : f \text{ is analytic on } B(\lambda; \delta/n) \text{ and } |f(z)| \leq m\|f\| \text{ for } |z - \lambda| < \delta/n\}$. First note that $\bigcup_{m,n} F_{m,n} = R^2(K, \mu)$. In fact, if $f \in R^2(K, \mu)$, there is an $r < \delta$ such that f is analytic on $B(\lambda; r)$. Choose n with $\delta/n < r$. Since $|f|$ is uniformly bounded on $\overline{B}(\lambda; \delta/n)$, m can be found with f in $F_{m,n}$.

If it can be shown that $F_{m,n}$ is closed in $R^2(K, \mu)$, then the lemma will follow by the Baire Category Theorem. Indeed, suppose it has been established that $F_{m,n}$ is closed; so $\text{int } F_{m,n} \neq \varnothing$ for some choice of m and n. Thus there is an f_0 in $F_{m,n}$ and an $\varepsilon > 0$ such that $\|f - f_0\| < \varepsilon$ implies $f \in F_{m,n}$. This says that if g is an arbitrary element in $R^2(K, \mu)$, then $f = f_0 + \frac{\varepsilon}{2\|g\|} g \in F_{m,n}$. Hence $g = \frac{2\|g\|}{\varepsilon}(f - f_0)$ is analytic on $B(\lambda; \delta/n)$.

To show that $F_{m,n}$ is closed, let $\{f_k\} \subseteq F_{m,n}$ and assume that $f_k \to f$ in $R^2(K, \mu)$. But by hypothesis $\{f_k\}$ is uniformly bounded on $B(\lambda; \delta/n)$, so that Montel's Theorem implies that there is a subsequence $\{f_{k_j}\}$ such that $g(z) \equiv \lim_{k_j} f_{k_j}(z)$ is analytic on $B(\lambda; \delta/n)$. But $f_{k_j}(z) = \langle f_{k_j}, k_z \rangle \to \langle f, k_z \rangle$ for all z in $B(\lambda; \delta/n)$. So $f(z) = g(z)$ on $B(\lambda; \delta/n)$ and is thus analytic there. Also, $|f(z)| \leq |f(z) - f_{k_j}(z)| + |f_{k_j}(z)| \leq |f(z) - f_{k_j}(z)| + m\|f_{k_j}\| \to m\|f\|$. Therefore $f \in F_{m,n}$ and $F_{m,n}$ is closed. □

7.6 PROPOSITION. *If $\lambda \in \text{bpe}(K, \mu)$, then λ is an analytic bounded point evaluation if and only if there is a number $r > 0$ such that $B(\lambda; r) \subseteq \text{bpe}(K, \mu)$ and*

$$\sup\{\|k_z\| : |z - \lambda| < r\} < \infty.$$

PROOF. If $\lambda \in \text{abpe}(K, \mu)$, the preceding lemma guarantees the existence of a number $r > 0$ such that each f in $R^2(K, \mu)$ is analytic on a neighborhood of $\overline{B}(\lambda; r)$. Thus for each f in $R^2(K, \mu)$, $\sup\{|\langle f, k_z\rangle| : |z - \lambda| < r\} < \infty$. But this implies that $\{k_z : |z - \lambda| < r\}$ is a weakly bounded subset of $R^2(K, \mu)$. By the Principle of Uniform Boundedness, $\sup\{\|k_z\| : |z - \lambda| < r\} < \infty$.

Conversely, assume that $r > 0$ is given such that $B(\lambda; r) \subseteq \text{bpe}(K, \mu)$ and $\sup\{\|k_z\| : |z-\lambda| < r\} = M < \infty$. Fix f in $R^2(K, \mu)$ and let $\{f_n\} \subseteq \text{Rat}(K)$ such that $\|f_n - f\| \to 0$. But the assumption implies that $f_n(z) \to f(z)$ uniformly on $B(\lambda; r)$ and so f must be analytic there. □

7.7 THEOREM (Trent [1979a]). *If S is multiplication by z on $R^2(K, \mu)$, then*

$$\text{abpe}(K, \mu) = \sigma(S) \setminus \sigma_{\text{ap}}(S) = \{\text{int } \sigma(S)\} \setminus \sigma_{\text{ap}}(S).$$

Before proving this theorem, we need the following lemma. This is a special case of a much more general result from Subin [1967] and Cowen and Douglas [1978].

7.8 LEMMA. *If $A \in \mathcal{B}(\mathcal{H})$ and $0 \in \sigma(A) \setminus \sigma_{\text{ap}}(A)$, then there is an open neighborhood U of 0 and a function $h : U \to \mathcal{H}$ such that:*

(i) $h(\lambda) \neq 0$ *and* $h(\lambda) \in \ker(A - \lambda)^*$ *for all λ in U;*
(ii) *for each f in $\mathcal{H}$, $\lambda \to \langle f, h(\lambda)\rangle$ is analytic on U.*

PROOF. Since $\sigma_l(A) = \sigma_{\text{ap}}(A)$, the hypothesis implies that A is left invertible; let $B \in \mathcal{B}(\mathcal{H})$ such that $BA = 1$. Now if $|\lambda| < \|B\|^{-1}$, then $(1 - \lambda B)$ is invertible and $(1 - \lambda B)^{-1}(A - \lambda) = (1 + \lambda B + \lambda^2 B^2 + \cdots)(A - \lambda) = (A + \lambda BA + \lambda^2 B^2 A + \cdots) - (\lambda + \lambda^2 B + \lambda^3 B^2 + \cdots) = A$. Hence $A^* = (A^* - \overline{\lambda})(1 - \overline{\lambda}B^*)^{-1}$ for $|\lambda| < \|B\|^{-1}$. Put $U = \{\lambda : |\lambda| < \|B\|^{-1}\}$ and let h_0 be any nonzero vector in $\ker A^*$. Define $h : U \to \mathcal{H}$ by $h(\lambda) = (1 - \overline{\lambda}B^*)^{-1}h_0$. Clearly $h(\lambda) \neq 0$ and $(A - \lambda)^* h(\lambda) = A^* h_0 = 0$, thus proving (i). But if $f \in \mathcal{H}$, then $\langle f, h(\lambda)\rangle = \langle f, (1 - \overline{\lambda}B^*)^{-1}h_0\rangle = \langle (1 - \lambda B)^{-1}f, h_0\rangle$ and so (ii) also holds. □

PROOF OF THEOREM 7.7. Since $\partial \sigma(S) \subseteq \sigma_{\text{ap}}(S)$, the second equality holds for all operators. Suppose $\lambda \in \sigma(S) \setminus \sigma_{\text{ap}}(S)$; without loss of generality we assume that $\lambda = 0$. By the preceding lemma, there is a neighborhood U of 0 and a nonvanishing function $h : U \to \mathcal{H}$ such that $(S^* - \overline{\lambda})h(\lambda) \equiv 0$ and $\lambda \to \langle f, h(\lambda)\rangle$ is analytic on U for every f in $R^2(K, \mu)$. Thus $\langle 1, h(\lambda)\rangle$ is analytic and never vanishes on U; put $k_\lambda = \langle h(\lambda), 1\rangle^{-1}h(\lambda)$ for λ in U, so that $\langle 1, k_\lambda\rangle = 1$. If $g \in \text{Rat}(K)$ and $f = (z - \lambda)g$, then $\langle f, k_\lambda\rangle = \langle 1, h(\lambda)\rangle^{-1}\langle (S - \lambda)g, h(\lambda)\rangle = 0$. If f is an arbitrary element of $\text{Rat}(K)$, then $f = (z - \lambda)g + f(\lambda)$ for some g in $\text{Rat}(K)$ and so it follows that $\langle f, k_\lambda\rangle = f(\lambda)$. This implies that $U \subseteq \text{bpe}(K, \mu)$. If $f \in R^2(K, \mu)$, then for λ in U, $f(\lambda) = \langle f, k_\lambda \rangle = \langle 1, h(\lambda) \rangle^{-1} \langle f, h(\lambda) \rangle$; by the preceding lemma, $\lambda \to f(\lambda)$ is analytic on U.

Now suppose that $0 \in \text{abpe}(K, \mu)$; so $0 \in \sigma(S)$. Let $r > 0$ such that $B = \{\lambda : |\lambda| \leq r\} \subseteq \text{abpe}(K, \mu)$ and let M be a constant such that $\|k_\lambda\| \leq M$ for all λ in B. If $f \in R^2(K, \mu)$, f is analytic on B and for $|\lambda| = r$, $|f(\lambda)| = |\lambda^{-1} \langle zf, k_\lambda \rangle| \leq (M/r)\|Sf\|$. By the Maximum Principle, $|f(\lambda)| \leq (M/r)\|Sf\|$ for all λ in B. Hence

$$\|f\|^2 = \int_B |f|^2 \, d\mu + \int_{\mathbb{C} \setminus B} |(z/z)f|^2 \, d\mu$$
$$\leq (M^2/r^2)\mu(B)\|Sf\|^2 + r^{-2}\|Sf\|^2$$
$$\leq C^2 \|Sf\|^2,$$

where $C^2 = r^{-2}(M^2 \mu(B) + 1)$. That is, $\|Sf\| \geq C^{-1}\|f\|$ for every f in $R^2(K, \mu)$ and so $0 \notin \sigma_{\text{ap}}(S)$. □

7.9 COROLLARY. *If S is pure, then*

$$\text{abpe}(K, \mu) = \sigma(S) \setminus \sigma_{\text{ap}}(S) = \{\lambda : S - \lambda \in \mathscr{F} \text{ and } \text{ind}(S - \lambda) = -1\}.$$

7.10 PROPOSITION (Trent [**1979a**]). *For any compact set K and for any measure μ supported on K, $\text{abpe}(K, \mu)$ is dense in $\text{int}\{\text{bpe}(K, \mu)\}$.*

PROOF. If $\lambda \in \text{int}\{\text{bpe}(K, \mu)\}$, let $r > 0$ such that $B = \overline{B}(\lambda; r) \subseteq \text{int}\{\text{bpe}(K, \mu)\}$. Put $B_n = \{z \in B : \|k_\lambda\| \leq n\}$; so $B = \cup_n B_n$. But each B_n is closed (Why?) and so the Baire Category Theorem implies that some B_n has interior. If $\lambda_1 \in \text{int } B_n$, then $\lambda_1 \in \text{abpe}(K, \mu)$ by Proposition 7.6. □

In the case of a subnormal unilateral weighted shift, $\text{abpe}(\mu) = \text{bpe}(\mu) = \text{int } \sigma(S_\mu)$. This is far from the truth in general. Olin [**1977**] gives an example of a compact set K and a measure μ living on K such that there is a point λ in ∂K that belongs to $\text{bpe}(K, \mu)$; thus $\lambda \notin \text{abpe}(K, \mu)$ and $\text{bpe}(K, \mu)$ is not always open. Also see Radjabalipour [**1975**].

The case of polynomials, however, is typified by what happens for unilateral shifts. In fact, $\text{bpe}(\mu) = \text{abpe}(\mu)$ whenever S_μ is pure. See Theorem VIII.4.4. Moreover in Chapter VIII the recent result of James E. Thomson will be demonstrated showing that if $P^2(\mu) \neq L^2(\mu)$, then $\text{abpe}(\mu) \neq \emptyset$.

There are several unanswered questions about bounded point evaluations.

7.11 OPEN QUESTION. Is it always the case that $\text{int}\{\text{bpe}(K, \mu)\} = \text{abpe}(K, \mu)$?

If K is polynomially convex, the answer to this question is "yes"; in fact, $\text{abpe}(\mu) = \text{bpe}(\mu)$ (VIII.4.4).

If $R^2(K, \mu) \neq L^2(\mu)$, must $\text{bpe}(K, \mu)$ be nonempty? The reason you might be led to believe that this question has an affirmative answer is that this is the case when K is polynomially convex. Also, rational functions are analytic and often limits of analytic functions are analytic. That this is not the case, however, was shown in Brennan [**1971a**] and Fernstrom [**1976**].

This section will close with a result on $P^2(\mu)$ and its bounded point evaluations.

7.12 PROPOSITION. *If S is a cyclic subnormal operator, then every component of $\sigma(S) \setminus \sigma_{\mathrm{ap}}(S)$ is simply connected.*

PROOF. By Theorem 7.7, $\sigma(S) \setminus \sigma_{\mathrm{ap}}(S) = \mathrm{abpe}(K, \mu)$. Let $S = S_\mu$ on $P^2(\mu)$ and fix a simple closed curve γ in $\mathrm{abpe}(\mu)$; it must be shown that the inside of γ, $\mathrm{ins}\,\gamma$, is contained in $\mathrm{abpe}(\mu)$. But $z \to k_z$ is a weakly continuous function from $\gamma \to P^2(\mu)$. Hence $\{k_z : z \in \gamma\}$ is weakly bounded, and thus norm bounded. Let $M > 0$ such that $\|k_z\| \leq M$ for all z on γ. So if p is any polynomial and $z \in \gamma$, $|p(z)| \leq M\|p\|$. By the Maximum Modulus Theorem, $|p(z)| \leq M\|p\|$ for all z in $\mathrm{ins}\,\gamma$. But this implies that $\mathrm{ins}\,\gamma \subseteq \mathrm{bpe}(\mu)$ and $\sup\{\|k_z\| : z \in \mathrm{ins}\,\gamma\} \leq M$. It follows from Proposition 7.6 that $\mathrm{ins}\,\gamma \subseteq \mathrm{abpe}(\mu)$. □

A version of this result is true for arbitrary cyclic operators. Herrero [1978] shows that if T is a cyclic operator, then the components of $\{\lambda : T - \lambda \in \mathscr{F}$ and $\mathrm{ind}(T - \lambda) = -1\}$ are simply connected. (Actually a more general result than this is proved.)

Additional results on bounded point evaluations can be found in Brennan [**1971a, 1971b, 1973, 1979a,** and **1979b**], Kriete [**1979**], Kriete and Trent [**1977**], Trent [**1979a, 1979b, 1984,** and **1985**], and Trent and Wang [**1984**].

Exercises

1. Show that if S is a pure subnormal operator, then $\mathrm{abpe}(K, \mu) = \{\lambda \in \mathbb{C} : S - \lambda \in \mathscr{F}$ and $\mathrm{ind}(S - \lambda) = -1\}$.
2. Show that the map $z \to k_z$ of $\mathrm{abpe}(K, \mu) \to R^2(K, \mu)$ is continuous if $R^2(K, \mu)$ has its norm topology. (*Hint*: Use Schwarz's Lemma.)
3. Let μ be a compactly supported measure on $\mathbb{C}$ and show that the following statements are equivalent.
   (a) $\mathrm{bpe}(\mu)$ is nonempty.
   (b) S_μ has an invariant subspace $\mathscr{M}$ such that $\dim \mathscr{M}^\perp = 1$.
   (c) S_μ has an invariant subspace $\mathscr{M}$ such that $\dim \mathscr{M}^\perp < \infty$.

§8 Bergman operators. For a bounded open subset G of $\mathbb{C}$, let $L^p(G)$ denote the Lebesgue space for area measure restricted to G. Area measure will be denoted by Area, $d\mathscr{A}$, or $\mathscr{A}$.

Let $L_a^p(G)$ denote the *Bergman space* of those functions in $L^p(G)$ that are analytic; technically, those elements in $L^p(G)$ that are equal almost everywhere to a function that is analytic on G.

8.1 DEFINITION. The *Bergman operator* for G is the operator S defined on $L_a^2(G)$ by $(Sf)(z) = zf(z)$.

Note that since G is bounded, S is a bounded operator on the normed space $L_a^2(G)$, but we do not know yet that $L_a^p(G)$ is a Banach space and

$L_a^2(G)$ is a Hilbert space. The first order of business is to rectify this. First we need a standard result, the area mean value theorem for analytic functions.

8.2 LEMMA. *If f is an analytic function in a neighborhood of the closed disk $\overline{B}(a;r)$, then*
$$f(a) = (\pi r^2)^{-1} \int_{B(a;r)} f \, d\mathscr{A} .$$

PROOF. By the mean value property, $f(a) = (2\pi)^{-1} \int_{-\pi}^{\pi} f(a+te^{i\theta}) d\theta$ for $0 < t < r$. Hence
$$(\pi r^2)^{-1} \int_{B(a;r)} f \, d\mathscr{A} = (\pi r^2)^{-1} \int_0^r t \left[\int_{-\pi}^{\pi} f(a+te^{i\theta}) \, d\theta \right] dt$$
$$= 2r^{-2} \int_0^r f(a) t \, dt$$
$$= f(a). \quad \square$$

8.3 PROPOSITION. *If $f \in L_a^p(G)$, $1 \le p < \infty$, $a \in G$, and $\mathrm{dist}(a, \partial G) > r > 0$, then*
$$|f(a)| \le (\pi^{-1} r^2)^{1/p} \|f\|_p .$$

PROOF. By the preceding lemma,
$$|f(a)| = (\pi r^2)^{-1} \left| \int_{B(a;r)} f \, d\mathscr{A} \right| \le (\pi r^2)^{-1} \left(\int_{B(a;r)} |f|^p \, d\mathscr{A} \right)^{1/p} \left(\int_{B(a;r)} 1^q \, d\mathscr{A} \right)^{1/q}$$
$$\le (\pi r^2)^{-1} \|f\|_p (\pi r^2)^{1/q} = (\pi^{-1} r^2)^{1/p} \|f\|_p . \quad \square$$

8.4 PROPOSITION. *For $1 \le p < \infty$, $L_a^p(G)$ is a Banach space and $L_a^2(G)$ is a Hilbert space.*

PROOF. It must be shown that $L_a^p(G)$ is complete; equivalently, $L_a^p(G)$ is closed in $L^p(G)$. Let $\{f_n\} \subseteq L_a^p(G)$ and suppose $f_n \to f$ in $L^p(G)$; without loss of generality we can assume that $f_n(z) \to f(z)$ a.e. Let K be a compact subset of G and let $0 < r < \mathrm{dist}(K, \partial G)$. By Proposition 8.3 there is a constant C such that $|h(z)| \le C\|h\|_p$ for every h in $L_a^p(G)$ and every z in K. In particular, $|f_n(z) - f_m(z)| \le C\|f_n - f_m\|_p$ for all m, n. Thus $\{f_n\}$ is a uniformly Cauchy sequence of analytic functions on K. Therefore there is an analytic function g on G such that $f_n(z) \to g(z)$ uniformly on compact subsets of G. It must be that $f(z) = g(z)$ a.e. and so $f \in L_a^p(G)$. $\quad \square$

8.5 THEOREM. *If S is the Bergman operator for the bounded open set G, then S is a bounded subnormal operator and the following statements hold.*
  (a) *$\sigma(S) = \mathrm{cl}\, G$ and $\sigma_p(S) = \varnothing$.*
  (b) *For λ in G there is a k_λ in $L_a^2(G)$ such that $\langle f, k_\lambda \rangle = f(\lambda)$ for all f in $L_a^2(G)$.*
  (c) *For λ in G, $\mathrm{ran}(S - \lambda) = \{f \in L_a^2(G) : f(\lambda) = 0\}$ is closed, $S - \lambda$ is Fredholm, and $\mathrm{ind}(S - \lambda) = -1$.*

PROOF. The proof that S is a bounded subnormal operator is an exercise.

(a) If $\lambda \notin \operatorname{cl} G$, then $(z - \lambda)^{-1}$ is a bounded analytic function on G and multiplication by this function is the inverse of $S - \lambda$. Thus $\sigma(S) \subseteq \operatorname{cl} G$. On the other hand, if $\lambda \in G$, then $1 \notin (z - \lambda) L_a^2(G) = (S - \lambda) L_a^2(G)$, and so $G \subseteq \sigma(S)$. Since the spectrum is closed, $\operatorname{cl} G \subseteq \sigma(S)$. Note that if $0 = (S - \lambda) f = (z - \lambda) f$, then $f(z) = 0$ for $z \neq \lambda$; hence $\ker(S - \lambda) = (0)$ for λ in G.

(b) Proposition 8.3 implies that for λ in G, the linear functional $f \to f(\lambda)$ is bounded on $L_a^2(G)$. Hence there is a k_λ in $L_a^2(G)$ such that $f(\lambda) = \langle f, k_\lambda \rangle$ if $f \in L_a^2(G)$. This proves (b).

(c) Note that $\{f \in L_a^2(G) : f(\lambda) = 0\}$ is the kernel of this linear functional and so this linear manifold is closed. Clearly $\operatorname{ran}(S - \lambda) \subseteq \{f \in L_a^2(G) : f(\lambda) = 0\}$. Conversely, if $f \in L_a^2(G)$ and $f(\lambda) = 0$, then there is an analytic function $g : G \to \mathbb{C}$ such that $f = (z - \lambda) g$. We will now see that $g \in L_a^2(G)$. From here it follows that $f = (S - \lambda) g$, thus establishing that $\operatorname{ran}(S - \lambda) = \{f \in L_a^2(G) : f(\lambda) = 0\}$. To show that $g \in L_a^2(G)$, let $\delta > 0$ such that $B = \{z : |z - \lambda| \leq \delta\} \subseteq G$. By Proposition 8.3, there is a constant $C > 0$ such that $|f(z)| \leq C \|f\|$ for every f in $L_a^2(G)$ and z in B. If $|z - \lambda| = \delta$, then $|g(z)| = |z - \lambda|^{-1} |f(z)| \leq \delta^{-1} C \|f\|$; by the Maximum Principle, this inequality holds in B. Also $\int_{G \setminus B} |g|^2 d\mathscr{A} = \int_{G \setminus B} |z - \lambda|^{-2} |f|^2 d\mathscr{A} \leq \delta^{-2} \|f\|^2$. Combining these facts shows that $g \in L_a^2(G)$.

Finally, if $f \in L_a^2(G)$, then $f = (f - f(\lambda)) + f(\lambda)$, so that $\operatorname{ran}(S - \lambda) + \mathbb{C} = L_a^2(G)$ and $\dim[\operatorname{ran}(S - \lambda)]^\perp = 1$. Since $\ker(S - \lambda) = (0)$, this shows that $\operatorname{ind}(S - \lambda) = -1$. □

The noncomatose reader will have noticed the connection between this theorem and the results of the preceding section on bounded point evaluations. From the way that we have defined a bounded point evaluation, the Bergman operator for a bounded open set G does not have bounded point evaluations since $L_a^2(G)$ may not equal $R^2(K, \mu)$ for any choice of K and μ. Undoubtedly there exists a general theory that will cover both cases. The details of this theory have not been worked out, though something like it can be found in Seddighi [1983] and Shields and Wallen [1971].

The exact determination of the essential spectrum of the Bergman operator is not trivial. This is done in Axler, Conway, and McDonald [1982]; the result will be stated here without proof.

If $\lambda \in \partial G$, say that λ is a *removable boundary point* of G if for every f in $L_a^2(G)$ there is a neighborhood U of λ such that f has an analytic continuation to $G \cup U$. Call the points in ∂G that are not removable the *essential boundary points*.

8.6 THEOREM (Axler, Conway, and McDonald [1982]). *If G is a bounded open subset of $\mathbb{C}$ and S is the Bergman operator for G, then*:

(a) $\sigma_e(S) =$ the essential boundary points of G;
(b) if $\lambda \in \operatorname{cl} G$ and $\lambda \notin \sigma_e(S)$, then $\operatorname{ind}(S - \lambda) = -1$.

It is not difficult to see that $\partial[\operatorname{cl} G] \subseteq \sigma_e(S)$ and it is rather easy to show that isolated points of ∂G are removable boundary points. Also $\operatorname{Area}(\partial G) = \operatorname{Area}(\sigma_e(S))$. (See Exercises 3 and 5.)

8.7 EXAMPLE. Let K be a compact subset of $\mathbb{D}$ such that $\operatorname{Area} K > 0$ and $\operatorname{int} K = \varnothing$. Also assume that the support of the restriction of area measure to K is the set K itself; that is, if U is an open set and $\varnothing \neq U \cap K$, then $\operatorname{Area}(U \cap K) \neq 0$. (Give an example of such a set.) If $G = \mathbb{D} \setminus K$, then the essential spectrum of the Bergman operator for G is $\partial \mathbb{D} \cup K = \partial G$. In fact, if $f(z) = \int_K (z - \zeta)^{-1} d\mathscr{A}(\zeta)$, then f is a bounded analytic function on G that does not analytically continue to any point of K.

It is shown in Axler, Conway, and McDonald [1982] that $\lambda \in \sigma_e(S)$ if and only if for every neighborhood U of λ, the set $U \cap \partial G$ has positive logarithmic capacity.

We will now focus our attention on the question of the cyclicity of Bergman operators.

8.8 DEFINITION. For a bounded open set G, let $P^2(G)$ be the closure of the polynomials in $L^2(G)$. $R^2(G)$ is the closure of the set of rational functions that belong to $L^2(G)$.

It follows that $P^2(G) \subseteq R^2(G) \subseteq L_a^2(G)$ and $P^2(G)$ and $R^2(G)$ are invariant subspaces for the Bergman operator. Note that if r is a rational function with poles off $\operatorname{cl} G$, then $r \in L^2(G)$. However, in the definition of $R^2(G)$, the rational functions are allowed to have poles on ∂G as long as the functions belong to $L^2(G)$. If $G =$ the punctured unit disk, then z^{-1} has its poles off G but does not belong to $L^2(G)$. If, on the other hand, $G = \{z = x + iy : 0 < x < 1 \text{ and } |y| < \exp(-x^2)\}$, then $z^{-1} \in L^2(G)$.

It is not difficult to construct an example of a set G for which $P^2(G) \neq L_a^2(G)$. (Like what?) Finding a G with $R^2(G) \neq L_a^2(G)$ is a little more difficult.

If K is a compact subset of $\mathbb{C}$, then the open set $\mathbb{C} \setminus K$ has at most a countable number of components, exactly one of which is unbounded. Call the boundary of this unique unbounded component of $\mathbb{C} \setminus K$ the *outer boundary* of K. Note that the outer boundary of K is a subset of ∂K. In fact, the outer boundary of K is precisely $\partial \widehat{K}$, the boundary of the polynomially convex hull of K.

8.9 DEFINITION. A *Carathéodory region* is an open connected subset of $\mathbb{C}$ whose boundary equals its outer boundary.

8.10 PROPOSITION. *A Carathéodory region is a component of* $\operatorname{int}\{\widehat{G}\}$ *and hence is simply connected.*

PROOF. Let $K = \widehat{G}$ and let H be the component of $\operatorname{int} K$ that contains G; it must be shown that $H = G$. Suppose there is a point z_1 in $H \setminus G$ and fix a point z_0 in G. Let $\gamma : [0, 1] \to H$ be a path such that $\gamma(0) = z_0$ and $\gamma(1) = z_1$. Put $\alpha = \inf\{t : \gamma(t) \in H \setminus G\}$. Thus $0 < \alpha \le 1$ and $\gamma(t) \in G$ for $0 \le t < \alpha$. Since $H \setminus G$ is relatively closed in H, $w = \gamma(\alpha) \notin G$. Thus $w \in \partial G$. But since G is a Carathéodory region, $\partial G = \partial K$. Hence $w \in \partial K$. But $w \in H \subseteq \operatorname{int} K$, a contradiction.

It is left as an exercise for the reader to show that the components of the interior of any polynomially convex subset of $\mathbb{C}$ are simply connected. □

There are simply connected regions that are not Carathéodory regions. For example, let G be the slit disk, $\mathbb{D} \setminus (-1, 0]$. Carathéodory regions tend to be well-behaved simply connected regions, however there can be some rather bizarre ones.

8.11 Example. A *cornucopia* is an open ribbon G that winds about the unit circle so that each point of $\partial \mathbb{D}$ belongs to ∂G. (See Figure 1.)

If G is the cornucopia, then $\operatorname{cl} G$ consists of the closed ribbon together with $\partial \mathbb{D}$. Hence $\mathbb{C} \setminus \operatorname{cl} G$ has 2 components: the unbounded component and $\mathbb{D}$. Nevertheless, G is a Carathéodory region.

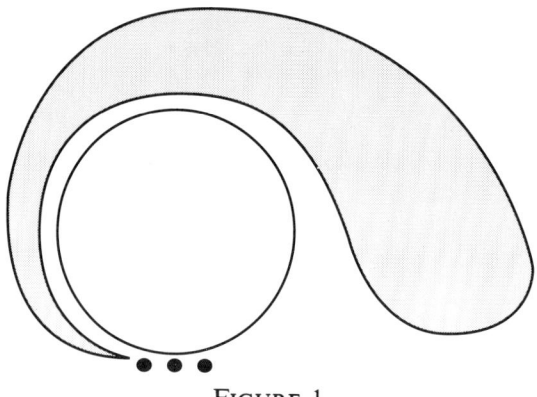

FIGURE 1

8.12 Proposition. *If G is a Carathéodory region, then $G = \operatorname{int}\{\operatorname{cl} G\}$. If G is a simply connected region such that $G = \operatorname{int}\{\operatorname{cl} G\}$ and $\mathbb{C} \setminus \operatorname{cl} G$ is connected, then G is a Carathéodory region.*

PROOF. Exercise.

8.13 Lemma. *If $\{f_n\}$ is a sequence in $L_a^2(G)$, then $\{f_n\}$ converges weakly to f if and only if $\sup_n \|f_n\| < \infty$ and $f_n(z) \to f(z)$ for all z in G.*

PROOF. If $f_n \to f$ weakly, then $\sup_n \|f_n\| < \infty$ by the Principle of Uniform Boundedness. Also $f_n(z) = \langle f_n, k_z \rangle \to \langle f, k_z \rangle = f(z)$. If τ is the topology of pointwise convergence on $L_a^2(G)$, then we have just seen that the identity map $i : (L_a^2(G), \text{weak}) \to (L_a^2(G), \tau)$ is continuous. Since (ball $L_a^2(G)$, weak) is compact and τ is a Hausdorff topology, i must be a homeomorphism. □

8.14 Proposition. *Let G be a bounded Carathéodory region and suppose $\{G_n\}$ is a sequence of simply connected regions such that $\operatorname{cl} G_{n+1} \subseteq G_n$ and $\bigcap \{G_n : n \geq 1\} = \hat{G}$. If a is a fixed point in G and, for each n, $\phi_n : G \to G_n$ is the Riemann map such that $\phi_n(a) = a$ and $\phi_n'(a) > 0$, then $\phi_n(z) \to z$ uniformly on compact subsets of G.*

Proof. Let $K = \hat{G}$ and let $\psi_n : G \to G$ be the restriction of ϕ_n^{-1} to G. Now $\{\phi_n\}$ and $\{\psi_n\}$ are uniformly bounded sequences of analytic functions on G. Let ϕ and ψ be cluster points of $\{\phi_n\}$ and $\{\psi_n\}$ in the topology of uniform convergence on compact sets. It will be shown that $\phi(z) = z$ and so z is the only cluster point of $\{\phi_n\}$; this will imply the proposition. By passing to a subsequence if necessary, it may be assumed that $\phi_n(z) \to \phi(z)$ and $\psi_n(z) \to \psi(z)$ uniformly on compact subsets of G.

Hurwitz's Theorem implies either ϕ (respectively, ψ) is one-to-one or it is constant. It will be shown that neither function is constant. To do this, let $\overline{B}(a;r) \subseteq G$ and let $M \geq |\phi_n'(z)|$ for all $n \geq 1$ and all z in $\overline{B}(a;r)$. Thus if $|w - a| \leq r/M$, then $|\phi_n(w) - a| = |\phi_n(w) - \phi_n(a)| \leq M|w - a| \leq r$. So $\phi_n(\overline{B}(a;r/M)) \subseteq \overline{B}(a;r)$. Fix w with $|w - a| \leq r/M$ and put $z_n = \phi_n(w)$. There is a point z in $\overline{B}(a;r)$ and a subsequence $\{z_{n_k}\}$ such that $z_{n_k} \to z$. Hence $w = \psi_{n_k}(z_{n_k}) \to \psi(z)$. That is, $\psi(z) = w$. Therefore $\overline{B}(a;r/M) \subseteq \psi(G)$ and ψ is not constant. Hence ψ is one-to-one. Also $z_{n_k} = \phi_{n_k}(w) \to \phi(w)$, so that $\phi(w) = z$. Therefore

$$\psi(\phi(w)) = w$$

for $|w - a| \leq r/M$. This implies that ϕ is not constant.

Since ϕ is not constant, $\phi(G)$ is a connected open set. From the conditions imposed on $\{G_n\}$, $\phi(G) \subseteq K$. But $a = \phi(a) \in \phi(G)$. Because G is a Carathéodory region, the lemma implies that $\phi(G) \subseteq G$. But $\psi(\phi(w)) = w$ on $\overline{B}(a;r/M)$ and hence $\psi(\phi(w)) = w$ for all w in G. This implies that $\psi : G \to G$ is a Riemann map. Clearly $\psi(a) = a$ since $\psi_n(a) = a$ for each $n \geq 1$; by similar reasoning $\psi'(a) > 0$. By the uniqueness of Riemann maps, $\psi(z) = z$ for all z in G. Thus $\phi(z) = z$ on G. □

8.15 Theorem (Farrell [**1934**] and Markusevic [**1934**]). *If G is a bounded Carathéodory region, then $P^2(G) = L_a^2(G)$.*

Proof. Let $K = \hat{G}$ and let $\tau : \mathbb{D} \to \mathbb{C}_\infty \setminus K$ be a Riemann map with $\tau(0) = \infty$. Put $G_n = \mathbb{C} \setminus \tau(\{z : |z| \leq 1 - 1/n\})$. The sequence $\{G_n\}$ satisfies the hypothesis of the preceding proposition. So fix a in G and let ϕ_n be the Riemann map of G onto G_n with $\phi_n(a) = a$ and $\phi_n'(a) > 0$. By Proposition 8.14, $\phi_n(z) \to z$ uniformly on compact subsets of G. Let $\psi_n = \phi_n^{-1} : G_n \to G$. Fix f in $L_a^2(G)$ and put $f_n = (f \circ \psi_n)\psi_n'$. Thus f_n is analytic in a neighborhood of K and so, by Runge's Theorem, f_n can be approximated uniformly on K by polynomials. Thus $f_n | G \in P^2(G)$.

Also $\|f_n\|^2 \le \int_{G_n} |f_n|^2 \, d\mathscr{A} = \int_{G_n} |f_n \circ \psi_n|^2 |\psi_n'|^2 \, d\mathscr{A} = \int_G |f|^2 \, d\mathscr{A} = \|f\|^2$ by the change of variables formula for area integrals. If $z \in G$, $\psi_n(z) \to z$ and $\psi_n'(z) \to 1$. Therefore, $f_n(z) \to f(z)$ as $n \to \infty$. By Lemma 8.13, $f_n \to f$ weakly and so $f \in P^2(G)$. □

Rubel and Shields [**1964**] prove that if G is a bounded open set whose boundary coincides with the boundary of its polynomially convex hull and if $f \in H^\infty(G)$, then there exists a sequence of polynomials $\{p_n\}$ such that $\|p_n\|_G \le \|f\|_G$ and $p_n(z) \to f(z)$ uniformly on compact subsets of G. (Note that this condition on G is the same as the condition for a region to be a Carathéodory region, but G is not assumed to be connected here.) A corollary of this result is that the algebra $\{M_p : p \text{ is a polynomial}\}$ is SOT dense in $\{M_\phi : \phi \in H^\infty(G)\}$ acting on the Bergman space $L_a^2(G)$. In particular, one can approximate with polynomials the H^∞ function that is 1 on the open unit disk and 0 on the cornucopia.

Some hypothesis is needed in Theorem 8.15 besides the simple connectedness of G. For example, if $G = \mathbb{D} \setminus (-1, 0]$, then $z^{1/2} \in L_a^2(G)$ but $z^{1/2} \notin P^2(G)$. In fact, it is not difficult to see that the functions in $P^2(G)$ are precisely those functions in $L_a^2(G)$ that have an analytic continuation to $\mathbb{D}$.

An exact description of the functions in $P^2(G)$ is difficult, though many properties of these functions can be given. Exercise 12 shows that if G is an annulus, then every f in $P^2(G)$ has an analytic extension to the open disk. In general, if U is a bounded component of $\mathbb{C} \setminus [\text{cl } G]$ such that ∂U is disjoint from the outer boundary of G, then every function in $P^2(G)$ has an analytic extension to $G \cup [\text{cl } U]$ that belongs to $P^2(G \cup [\text{cl } U])$, though the norm of the extension is larger.

What happens if U is a bounded component of $\mathbb{C} \setminus [\text{cl } G]$ and ∂U meets the outer boundary of G? The answer to this question is quite complex and the continuing subject of research. See Mergeljan [**1953**], Cima and Matheson [**1985**], and Brennan [**1977**].

The next theorem can be proved by reasoning similar to that used to prove Theorem 8.15, or Theorem 8.15 can be used to prove it. See Mergeljan [**1953**] for details.

8.16 THEOREM. *Let G be a bounded region in $\mathbb{C}$ such that $\mathbb{C} \setminus [\text{cl } G]$ has components $U_1, \ldots, U_m$. Let $K_j = \partial U_j$ and let K_0 be the outer boundary; assume $K_i \cap K_j = \emptyset$ for $i \ne j$ and fix a point z_j in U_j, $1 \le j \le m$. If $f \in L_a^2(G)$, then there is a sequence $\{f_n\}$ of rational functions with poles in $\{\infty, z_1, \ldots, z_m\}$ such that $f_n \to f$ in $L_a^2(G)$. In particular, $R^2(G) = L_a^2(G)$.*

REMARKS. 1. The space $L_a^2(G)$ is called the Bergman space because of the work of Bergman [**1947, 1950**].

2. To the best of my knowledge, the first place the operator multiplication by z on $L_a^2(G)$ is studied is Halmos, Lumer, and Schäffer [**1953**].

3. The question of when $P^2(G)$ or $R^2(G)$ equals $L_a^2(G)$ has attracted considerable attention. There are several related questions. For example, if w is a positive function in $L^1(G)$, let $L_a^2(w\,d\mathscr{A})$ denote the space of analytic functions in $L^2(w\,d\mathscr{A})$; this is a weighted Bergman space. When are the polynomials or rational functions dense in $L_a^2(w\,d\mathscr{A})$? For that matter, when is the weighted Bergman space complete? For some results and further references, see Brennan [**1979a** and **1977**], Hedberg [**1972**], and Mergeljan [**1953**].

4. The Bergman operator on a weighted Bergman space was studied in Elias [**1988**] where amongst other things the spectral properties are established for a large class of weights.

5. This question of the density of the polynomials in $L_a^2(G)$ is the same as the question of when the Bergman operator for G has the function 1 as a cyclic vector. The question arises as to whether the Bergman operator can be cyclic even if 1 is not a cyclic vector. The answer is yes. See Akeroyd, Khavinson, and Shapiro [**preprint**].

6. Kyung Hee Jin [**1989**] studies unbounded Bergman operators.

If S is the Bergman operator for G and $\phi \in H^\infty(G)$, the bounded analytic functions on G, then $\phi(S)$ can be defined as multiplication by ϕ on $L_a^2(G)$. It is easy to see that $\phi(S)$ is a bounded operator and $\|\phi(S)\| = \|\phi(N)\| = \|\phi\|_\infty$. (Use Corollary 2.17.) It is also easy to see that $\phi(S) \in \{S\}'$. The proof of the following theorem is from Shields and Wallen [**1971**].

8.17 THEOREM. *If S is the Bergman operator for G, then $\{S\}' = \{\phi(S) : \phi \in H^\infty(G)\}$.*

PROOF. For each z in G let $k_z \in L_a^2(G)$ such that $\langle f, k_z \rangle = f(z)$ for all f in $L_a^2(G)$; it is easy to check that $S^* k_z = \bar{z} k_z$ and $\ker(S^* - \bar{z}) = \mathbb{C} k_z$ for every z in G. Now if $AS = SA$, then $A^* S^* = S^* A^*$ and so $\ker(S^* - \bar{z})$ is invariant for A^*. Thus for each z in G there is a complex number $\phi(z)$ such that $A^* k_z = \overline{\phi(z)} k_z$. Since $\phi(z) \in \sigma(A)$, $|\phi(z)| \leq \|A\|$; that is, $\phi: G \to \mathbb{C}$ is a bounded function. On the other hand, $\phi(z) = \langle 1, A^* k_z \rangle = \langle A(1), k_z \rangle$ for each z in G, so that $\phi = A(1)$ and hence ϕ is an analytic function. That is, $\phi \in H^\infty(G)$ and $\|\phi\|_\infty \leq \|A\|$. Moreover, for every f in $L_a^2(G)$, $(Af)(z) = \langle Af, k_z \rangle = \langle f, A^* k_z \rangle = \langle f, \overline{\phi(z)} k_z \rangle = \phi(z) f(z)$; so $A = \phi(S)$. □

Even though the commutant of the Bergman operator was found with minimal pain, finding $P^\infty(S)$, the weak* closed algebra generated by S and 1, is far more complicated. If the open set G is rather nice, then $P^\infty(S)$ is not complicated and can be easily found. (See Exercises 15, 16, and 17.) But consider the following example. Let C be the usual Cantor ternary set in $[0, 1]$ and let $G = B(1/2; 1) \setminus C$. If S is the Bergman operator for G, then it follows that $\sigma_e(S) = C \cup \partial B(1/2; 1)$. However, $P^\infty(S) = H^\infty(B(1/2;1))$ because any bounded analytic function on G has an analytic continuation to $B(1/2;1)$. This is due to the fact that though C has positive logarithmic capacity, it has zero analytic capacity (§V.12).

Many problems concerning Bergman operators remain unsolved. For example, what are the invariant subspaces? We know that even in the case where $G = \mathbb{D}$, the structure of $\operatorname{Lat} S$ is fantastically complicated. In fact, there are spaces $\mathcal{M}$ and $\mathcal{N}$ in $\operatorname{Lat} S$ with $\mathcal{N} \subseteq \mathcal{M}$, such that $\dim(\mathcal{M} \cap \mathcal{N}^\perp) = \infty$ and $\mathcal{L} \in \operatorname{Lat} S$ whenever $\mathcal{N} \subseteq \mathcal{L} \subseteq \mathcal{M}$. (Apostol, Bercovici, Foias, and Pearcy [**1985**]. Also see Bercovici, Foias, and Pearcy [**1985**], page 101.) That is, $\operatorname{Lat} S$ contains copies of the lattice of all subspaces of an infinite dimensional Hilbert space.

Additional references: Brennan [**1971a, 1971b, 1973**], Hastings [**1975**], Horowitz [**1974**], Janas [**1983b**], Shelburne [**1982**].

Exercises

For these exercises, G will always be a bounded open subset of $\mathbb{C}$ and S is the Bergman operator for G.

1. If f is analytic in the punctured disk $G = \{z : 0 < |z| < 1\}$, for which values of p does the condition $\int_G |f|^p d\mathcal{A} < \infty$ imply that f has a removable singularity at 0?
2. Let G be a bounded open set and let $G_1, G_2, \ldots$ be its components. If S_n is the Bergman operator for G_n, show that S is unitarily equivalent to $\bigoplus_n S_n$.
3. Show that $\partial[\operatorname{cl} G] \subseteq \sigma_e(S)$.
4. Find a formula for the evaluation function k_z in $L_a^2(\mathbb{D})$ when $|z| < 1$.
5. Verify the statements in Example 8.7. Use the same technique to show that $\operatorname{Area}(\partial G) = \operatorname{Area}(\sigma_e(S))$ for any open set G.
6. If $G = \mathbb{D} \setminus \{0, 1/2, 1/3, \ldots\}$, find $\sigma_e(S)$.
7. If $G = \mathbb{D} \setminus [-1/2, 1/2]$, find $\sigma_e(S)$.
8. Find an example of a set G such that $0 \in \operatorname{int}\{\operatorname{cl} G\}$ and $z^{-n} \in L_a^2(G)$ for all $n \geq 1$.
9. Show that if K is a polynomially convex subset of $\mathbb{C}$, then the components of $\operatorname{int} K$ are simply connected.
10. If G is a Carathéodory region, then $\partial G = \partial[\operatorname{cl} G]$.
11. Give an example of a simply connected region G that is not a Carathéodory region but satisfies $G = \operatorname{int}\{\operatorname{cl} G\}$.
12. If G is a bounded open set in $\mathbb{C}$ and K is a compact subset of G, then every function f in $P^2(G \setminus K)$ has an analytic continuation to G that belongs to $P^2(G)$. Show that if G is connected, then the restriction map $f \to f|(G \setminus K)$ is a bijection of $P^2(G)$ onto $P^2(G \setminus K)$.
13. Let $\{a_n\}$ be an increasing sequence of positive numbers such that $1 = \lim_n a_n$. Choose $r_1, r_2, \ldots$ such that the closed balls $B_n = \overline{B}(a_n; r_n)$ are

pairwise disjoint and contained in $\mathbb{D}$; put $G = \mathbb{D} \setminus \cup_n B_n$. Show that each f in $P^2(G)$ has an analytic continuation to $\mathbb{D}$. Must this continuation belong to $L_a^2(\mathbb{D})$?

14. Prove that the Bergman operator for G is irreducible if and only if G is connected.
15. Show that if S is the Bergman operator for $\mathbb{D}$, then $P^\infty(S) = \{S\}' = H^\infty(\mathbb{D})$.
16. If S is the Bergman operator for $G = \mathbb{D} \setminus [-1/2, 1/2]$, then $P^\infty(S) = H^\infty(\mathbb{D})$ and $\{S\}' = H^\infty(G)$. So $P^\infty(S) \neq \{S\}'$.
17. If S is the Bergman operator for $G = \{z : 1/2 < |z| < 1\}$, then $\{S\}' = H^\infty(G)$ and $P^\infty(S) = H^\infty(\mathbb{D})$. However, $H^\infty(G) = R^\infty(S)$, the weak* closure of the rational functions in S.
18. Let G_1 and G_2 be bounded open sets with $\text{cl } G_1 \subseteq \text{cl } G_2$ and define $R : L_a^2(G_1) \to L_a^2(G_2)$ by $Rf = f|G_2$. Prove that R is a compact operator. In fact, R^*R is a Hilbert-Schmidt operator on $L_a^2(G_2)$.

§9 **Spectral sets.** For any set E and any function $f : E \to \mathbb{C}$, let
$$\|f\|_E \equiv \sup\{|f(x)| : x \in E\}.$$

9.1 DEFINITION. If $T = \mathscr{B}(\mathscr{H})$ and K is a compact subset of $\mathbb{C}$, then K is a *spectral set* for T if $\sigma(T) \subseteq K$ and $\|f(T)\| \leq \|f\|_K$ for every rational function f with poles off K. If $\sigma(T)$ is a spectral set, say that T is a *von Neumann operator*.

The concept of a spectral set was introduced in von Neumann [**1951**], where several properties are obtained. In particular, it is shown there that if T is a contraction, then $\text{cl } \mathbb{D}$ is a spectral set for T. The next result establishes the pertinence of the concept to this book.

9.2 PROPOSITION. *Every subnormal operator is a von Neumann operator.*

PROOF. If $N = \text{mne } S$, then $\sigma(N) \subseteq \sigma(S)$; so if $f \in \text{Rat}(\sigma(S))$, $f(N)$ is well defined. In fact, $f(S) = f(N)|\mathscr{H}$ and so $f(S)$ is a subnormal operator. Since $\sigma(f(S)) = f(\sigma(S))$, $\|f(S)\| = r(f(S)) = \sup\{|f(z)| : z \in \sigma(S)\}$. □

The preceding proposition does not extend to hyponormal operators, as seen by an example in Wadhwa [**1973**]. The difficulty here is that even though the norm of a hyponormal operator equals its spectral radius, $f(T)$ may not be hyponormal when T is hyponormal and f is a rational function. (See Exercise 4.3.) Indeed, if T is a hyponormal operator and $f(T)$ is hyponormal for all f in $\text{Rat}(\sigma(T))$, then T is a von Neumann operator. This leads to the following question.

9.3 OPEN QUESTION. If $T \in \mathscr{B}(\mathscr{H})$ and $f(T)$ is hyponormal for every f in $\text{Rat}(\sigma(T))$, must T be subnormal? More generally, if $p(T)$ is hyponormal for every polynomial p, must T be subnormal?

Some recent progress on this problem was made in McCullough and Paulsen [**1989**] where it is shown that the polynomial question is equivalent to the same question for weighted shifts.

9.4 Definition. For a compact subset K of $\mathbb{C}$, let $R(K)$ be the uniform closure of $\text{Rat}(K)$ in $C(K)$.

So $R(K)$ is a Banach algebra with identity. This algebra is the subject of Chapter V. For now we limit ourselves to some uses of $R(K)$ in operator theory.

9.5 Proposition. *If K is a spectral set for T, then there is an algebraic homomorphism $\rho : R(K) \to \mathscr{B}(\mathscr{H})$ such that $\|\rho(f)\| \leq \|f\|_K$ and the following hold.*

(a) *If $f \in \text{Rat}(K)$, then $\rho(f) = f(T)$.*
(b) $f(\sigma(T)) \subseteq \sigma(\rho(f)) \subseteq f(K)$ *for all f in $R(K)$.*

Proof. If $f \in \text{Rat}(K)$, then the definition of a spectral set implies that $\|f(T)\| \leq \|f\|_K$. Thus the map $f \to f(T)$ extends to a contraction $\rho : R(K) \to \mathscr{B}(\mathscr{H})$; it is elementary to check that ρ is a homomorphism. It remains to verify (b).

Fix f in $R(K)$ and let $\{f_n\} \subseteq \text{Rat}(K)$ such that $\|f_n - f\|_K \to 0$. If $\lambda \notin f(K)$, then there is an n_0 such that for $n \geq n_0$, $\lambda \notin f_n(K)$. It follows that $(f_n - \lambda)^{-1} \to (f - \lambda)^{-1}$ uniformly on K; thus $(f - \lambda)^{-1} \in R(K)$. It is easy to verify that $\rho((f - \lambda)^{-1}) = (\rho(f) - \lambda)^{-1}$ and so $\lambda \notin \sigma(\rho(f))$. Therefore, $\sigma(\rho(f)) \subseteq f(K)$.

Let $\varepsilon > 0$ and put $\Lambda(\varepsilon) = \{\lambda : \text{dist}(\lambda, \sigma(\rho(f))) < \varepsilon\}$. Maintaining our notation, it follows from the fact that ρ is continuous that $\|f_n(T) - \rho(f)\| \to 0$. By standard spectral theory (see Exercise 1), there is an n_0 such that $\sigma(f_n(T)) \subseteq \Lambda(\varepsilon)$ for $n \geq n_0$. Also, choose n_0 such that $\|f_n - f\|_K < \varepsilon$ for $n \geq n_0$. It follows that $f(\sigma(T)) \subseteq \Lambda(2\varepsilon)$; since ε was arbitrary, $f(\sigma(T)) \subseteq \sigma(\rho(f))$. □

Notice that property (a) says that the map ρ extends the Riesz-Dunford functional calculus. We formalize this extension as a new functional calculus.

9.6 Definition. If T is an operator, K is a spectral set for T, and ρ is the homomorphism described in Proposition 9.5, then $f(T) \equiv \rho(f)$ for every f in $R(K)$.

In the case of a von Neumann operator this functional calculus enjoys more attributes.

9.7 Theorem. *If T is a von Neumann operator, then the following hold.*

(a) *For f in $R(\sigma(T))$, $\|f(T)\| = \|f\|_{\sigma(T)}$.*
(b) *(Spectral Mapping Theorem) $\sigma(\rho(f)) = f(\sigma(T))$ for all f in $R(\sigma(T))$.*
(c) $\rho(R(\sigma(T))) = R(T) \equiv$ *the norm closure of $\{f(T) : f \in \text{Rat}(\sigma(T))\}$.*

Proof. By Proposition 9.5, $\|f(T)\| \leq \|f\|_{\sigma(T)} \leq r(f(T))$, the spectral radius of $f(T)$. But $r(f(T)) \leq \|f(T)\|$, so (a) is satisfied. Part (b) is a consequence of Proposition 9.5(b) and part (c) is immediate from the fact that $f \to f(T)$ is an isometry. □

Therefore for a von Neumann operator, and hence for a subnormal operator, we have extended the Riesz-Dunford functional calculus so that the Spectral Mapping Theorem remains in effect. If S is a subnormal operator and N is its minimal normal extension, what is the relation between $f(S)$ and $f(N)$ for f in $R(\sigma(S))$?

9.8 Proposition. *If $f \in R(\sigma(S))$, then $f(S) = f(N)|\mathscr{H}$.*

PROOF. This is immediate for any f in $\mathrm{Rat}(\sigma(S))$ and the general result follows by taking limits. Equivalently, this can be proven by appealing to the uniqueness of the functional calculus. □

In many ways the next result is a justification for studying $R(K)$ in connection with operator theory. It indicates a deep connection between the nature of $R(\sigma(T))$ and the nature of T.

9.9 Theorem (von Neumann [1951]). *If K is a spectral set for T and $R(K) = C(K)$, then T is normal.*

Two preliminary lemmas are needed.

9.10 Lemma. *If K is a spectral set for T and $K \subseteq \mathbb{R}$, then T is selfadjoint.*

PROOF. If $f(z) = (z+i)(z-i)^{-1}$ for z in $\mathbb{C}$, then $f(\infty) = 1$ and $f(\mathbb{R}) = \partial\mathbb{D} \setminus \{1\}$. Since $K \subseteq \mathbb{R}$, $\|f\|_K = 1$; hence, by hypothesis, $\|f(T)\| \leq 1$. But also $\|1/f\|_K = 1$ and so $\|f(T)^{-1}\| \leq 1$. This implies that $f(T)$ is unitary (see Exercise 2). Thus for h in $\mathscr{H}$,

$$(9.11) \qquad \|(T+i)(T-i)^{-1}h\| = \|h\|.$$

Now $\|(T\pm i)h\|^2 = \|Th\|^2 \pm 2\operatorname{Im}\langle Th, h\rangle + \|h\|^2$. Letting $h = (T-i)g$, so that $g = (T-i)^{-1}h$, and substituting this in (9.11) we get $\|(T+i)g\| = \|(T-i)g\|$ for every g in $\mathscr{H}$. Squaring both sides of this equation and performing the necessary arithmetic we arrive at $\operatorname{Im}\langle Tg, g\rangle = -\operatorname{Im}\langle Tg, g\rangle$ for every g in $\mathscr{H}$. Thus T must be selfadjoint. □

9.12 Lemma. *If K is a spectral set for T, $f \in R(K)$, and $A = f(T)$, then $f(K)$ is a spectral set for A.*

PROOF. Let $K_1 = f(K)$; so if $r \in \mathrm{Rat}(K_1)$, then $r \circ f \in R(K)$ and $r \circ f(T) = r(A)$. Hence $\|r(A)\| = \|r \circ f(T)\| \leq \|r \circ f\|_K = \|r\|_{K_1}$. □

PROOF OF THEOREM 9.9. Let $f(z) = \bar{z}$; by hypothesis, $f \in R(K)$. Let $g = [z + f]/2$ and $h = [z - f]/2i$. So $g, h \in R(K)$ and hence $g(T)$ and $h(T)$ are well-defined operators. But by Lemma 9.12, $g(K)$ and $h(K)$ are spectral sets for $g(T)$ and $h(T)$. But $g(K)$ and $h(K) \subseteq \mathbb{R}$ and so Lemma 9.10 implies $g(T)$ and $h(T)$ are selfadjoint. But $z = g + ih$ and so $T = g(T) + ih(T)$ and it follows by the uniqueness of the Cartesian decomposition of an operator that $g(T) = \text{Re}\, T$ and $h(T) = \text{Im}\, T$. Since $g(T)$ and $h(T)$ clearly commute, T must be normal. □

There is a substantial literature on spectral sets. For example, see Foias [**1959**], Lautzenheiser [**1973**], Lebow [**1963**], Mlak [**1972b**], Sz-Nagy [**1960**], Sarason [**1965b**], and J. P. Williams [**1967**]. It is also shown in Agler [**1979** and **1980**] that a von Neumann operator has a nontrivial invariant subspace. In Conway and Dudziak [**1990**], it is shown that von Neumann operators have a very large collection of invariant subspaces so that they are reflexive (see VII.8.1 for the definition). Each of these results followed the proofs that subnormal operators enjoy the same properties.

What are some examples of von Neumann operators, besides the subnormal ones? Note that the adjoint of a subnormal operator is a von Neumann operator and direct sums of von Neumann operators are von Neumann operators. Here is another way to construct an operator with a given spectral set.

9.13 EXAMPLE. Fix a compact subset K of $\mathbb{C}$ and let N be a normal operator on $\mathscr{K}$ with $\sigma(N) \subseteq \partial K$. Let $\mathscr{M}$ and $\mathscr{N}$ be subspaces of $\mathscr{K}$ that are invariant for $f(N)$ for every f in $R(K)$. Assume that $\mathscr{N} \subseteq \mathscr{M}$ and put $\mathscr{H} = \mathscr{M} \cap \mathscr{N}^\perp$; let P be the projection of $\mathscr{K}$ onto $\mathscr{H}$. If $T \equiv PN|\mathscr{H}$, then $f(T) = Pf(N)|\mathscr{H}$ for every f in $\text{Rat}(K)$ and K is a spectral set for T.

Which of the operators in the preceding example are von Neumann operators? I don't know.

The most important question in the study of spectral sets is whether every operator having K as a spectral set arises as in Example 9.13. If K is the closed unit disk, the answer is yes and this is the celebrated dilation theorem of B. Sz-Nagy (see Sz-Nagy and Foias [**1970**], page 13). If K is an annulus the answer is also in the affirmative and can be found in Agler [**1985b**]. The result is not known to hold for any other sets. See Arveson [**1969**], Douglas and Paulsen [**1986**], and Paulsen [**1988**] for additional information.

Exercises

1. Suppose $T_n \to T$ in $\mathscr{B}(\mathscr{H})$ and show that for every $\varepsilon > 0$ there is an n_0 such that $\sigma(T_n) \subseteq \{\lambda : \text{dist}(\lambda, \sigma(T)) < \varepsilon\}$ for $n \geq n_0$. If each T_n is hyponormal, show that for λ in $\sigma(T)$ there is a sequence $\{\lambda_n\}$ with each λ_n in $\sigma(T_n)$ such that $\lambda_n \to \lambda$.

2. Show that if T is an invertible operator with $\|T\|$ and $\|T^{-1}\| \leq 1$, then T is unitary.

3. The direct sum of a uniformly bounded sequence of von Neumann operators is a von Neumann operator.

4. Verify the statements in Example 9.13.

5. (Sarason [**1965b**]) Let $\mathscr{S}$ be a semigroup of operators in $\mathscr{B}(\mathscr{K})$, let $\mathscr{H}$ be a closed linear subspace of $\mathscr{K}$, and let P be the projection of

$\mathcal{K}$ onto $\mathcal{H}$. Show that $PAPBP = PABP$ for all A and B in $\mathcal{S}$ if and only if there are invariant subspaces $\mathcal{M}$ and $\mathcal{N}$ for $\mathcal{S}$ such that $\mathcal{N} \subseteq \mathcal{M}$ and $\mathcal{H} = \mathcal{M} \cap \mathcal{N}^\perp$. Compare this with Example 9.13.

§10 The commutant of a subnormal operator. Some results on the commutant of a subnormal operator have already been encountered. In Theorem 5.4 the commutant of a rationally cyclic subnormal operator was characterized as $\{M_\phi : \phi \in R^2(K, \mu) \cap L^\infty(\mu)\}$. This also says that every operator in the commutant of a rationally cyclic subnormal operator "lifts" to the commutant of its minimal normal extension.

10.1 DEFINITION. If S is a subnormal operator and N is its minimal normal extension, then an operator A in $\{S\}'$ *lifts* to $\{N\}'$ if there is an operator B in $\{N\}'$ such that $B\mathcal{H} \subseteq \mathcal{H}$ and $A = B|\mathcal{H}$.

Not every operator in $\{S\}'$ lifts to $\{N\}'$. An easy example of this phenomenon is furnished in Deddens [**1971**]. If T is the unilateral shift and $E = 1 - TT^*$, then $\begin{bmatrix} 0 & 0 \\ E & 0 \end{bmatrix} \in \{\begin{bmatrix} T & 0 \\ 0 & 0 \end{bmatrix}\}'$ but does not lift (Exercise 1). A more complicated example that reveals a bit more of the complexity of the problem is the following.

10.2 EXAMPLE (Abrahamse [**preprint**]). Let T_β be the subnormal weighted shift with weight sequence $\beta, 1, 1, \ldots$. Define $D : \ell^2 \to \ell^2$ by $De_0 = 2e_0$ and $De_n = e_n$ for $n \geq 1$. Thus D is a positive invertible operator on ℓ^2 and it is easy to check that $DT_1 = T_{1/2}D$. Let N_1 and $N_{1/2}$ be the minimal normal extensions of T_1 and $T_{1/2}$ acting on $\mathcal{K}_1$ and $\mathcal{K}_{1/2}$.

CLAIM. There is no operator $X : \mathcal{K}_1 \to \mathcal{K}_{1/2}$ such that $XN_1 = N_{1/2}X$, $X\ell^2 \subseteq \ell^2$, and $X|\ell^2 = D$.

In fact, if there were, then $\ell^2 = \operatorname{ran} D \subseteq \operatorname{ran} X$. Since $N_{1/2} = \operatorname{mne} T_{1/2}$, it must be that $\operatorname{ran} X$ is dense in $\mathcal{K}_{1/2}$ (Why?) By Proposition IX.6.10 of [**ACFA**], $N_1 | (\ker X)^\perp$ is unitarily equivalent to $N_{1/2}$. But $N_{1/2}$ has a kernel and N_1 has none since it is the bilateral shift. Thus no such operator X exists and this proves the claim.

But this implies that $\begin{bmatrix} 0 & 0 \\ D & 0 \end{bmatrix} \in \{T_1 \oplus T_{1/2}\}'$ and it does not lift.

There is a possible philosophical objection to this example: the operator $\begin{bmatrix} 0 & 0 \\ D & 0 \end{bmatrix}$ is not subnormal. So let's complicate the example somewhat. Let S be the quasinormal operator as in Theorem 3.2 with D in place of A; that is, S is an infinite matrix with D constantly along the subdiagonal and 0's elsewhere. Let $T = T_1 \oplus T_{1/2} \oplus T_{1/4} \oplus \cdots$; so T is also subnormal. Moreover, $TS = ST$. (Verify!) Let S and T operate on $\mathcal{H} = \mathcal{H}_0 \oplus \mathcal{H}_1 \oplus \mathcal{H}_2 \oplus \cdots$, where each $\mathcal{H}_n$ is ℓ^2. Let N be the minimal normal extension of T acting on $\mathcal{K}$; so $\mathcal{K} = \mathcal{K}_0 \oplus \mathcal{K}_1 \oplus \cdots$, each $\mathcal{K}_n$ reduces N, and $N|\mathcal{K}_n$ is the minimal normal extension of $T_{1/2^n}$. Suppose S lifts to $\{N\}'$; so there is an operator A in $\{N\}'$ such that $A\mathcal{H} \subseteq \mathcal{H}$ and $A|\mathcal{H} = S$. Now $A\mathcal{H}_0 = \mathcal{H}_1$ and so if $f \in \mathcal{H}_0$ and $n \geq 0$, then $AN^{*n}f = N^{*n}Af \in \mathcal{H}_1$. Since $N|\mathcal{K}_n$

is the minimal normal extension of $T_{1/2}n$, $A\mathcal{K}_0 \subseteq \mathcal{K}_1$ and $A|\mathcal{K}_0$ lifts $D : \mathcal{K}_0 \to \mathcal{K}_1$, contradicting the preceding paragraph. Thus S is a subnormal element of $\{T\}'$ that does not lift.

There are other examples of this same phenomenon. For example, Lubin [1977] produces two commuting subnormal operators S_1 and S_2 such that $S_1 + S_2$ and $S_1 S_2$ are not hyponormal. This implies that S_1 and S_2 cannot have commuting normal extensions N_1 and N_2, for then $N_1 + N_2$ and $N_1 N_2$ would be normal extensions of $S_1 + S_2$ and $S_1 S_2$.

These examples are far from being cyclic. Of course if S is a cyclic subnormal operator, Yoshino's Theorem implies that every operator in $\{S\}'$ lifts. The following result, whose proof is left as an exercise, complements Yoshino's Theorem.

10.3 PROPOSITION (Embry-Wardrop, private communication). *If S is a subnormal operator, $N = \mathrm{mne}\, S$, and if T is a cyclic subnormal operator in $\{S\}'$, then T lifts to $\{N\}'$.*

There is an internal criterion for lifting.

10.4 THEOREM (Bram [1955]). *Let S be subnormal with minimal normal extension N. If $T_0 \in \{S\}'$, then T_0 lifts to $\{N\}'$ if and only if there is a constant $c > 0$ such that for every finite subset $\{f_0, f_1, \ldots, f_n\}$ of $\mathcal{H}$,*

$$\sum_{j,k=0}^{n} \langle S^j T_0 f_k, S^k T_0 f_j \rangle \leq c \sum_{j,k=0}^{n} \langle S^j f_k, S^k f_j \rangle.$$

If a lifting exists, it is unique.

PROOF. The proof of necessity is left to the reader. For the proof of sufficiency, assume that $T_0 \in \{S\}'$ and T_0 satisfies the given inequality for any choice of $f_0, \ldots, f_n$ in $\mathcal{H}$. It is easy to see that the condition implies that $\|\sum_k N^{*k} T_0 f_k\|^2 \leq c \|\sum_k N^{*k} f_k\|^2$. So if we set

$$T\left(\sum_k N^{*k} f_k\right) = \sum_k N^{*k} T_0 f_k,$$

T is a well-defined bounded operator on a dense linear manifold in $\mathcal{K}$; extend T to a bounded operator $T : \mathcal{K} \to \mathcal{K}$. It is left as an exercise for the reader to verify that T is a lift of T_0 to the commutant of N.

To show that T is unique it suffices to show that if $T \in \{N\}'$ and $T|\mathcal{H} = 0$, then $T = 0$. But if $f \in \mathcal{H}$ and $n \geq 1$, then Fuglede's Theorem implies that $TN^{*n}f = N^{*n}Tf = 0$ and so $T = 0$. □

The preceding theorem does not completely settle the question of lifting elements of the commutant of S because the criterion is often impossible to apply. The situation is similar to the various equivalent formulations of subnormality (see Theorem 1.9). There is much work to be done here. For example, what is the obstruction to lifting every operator in $\{S\}'$? Every operator in $R^\infty(S)$, the weak* closure of the rational functions in S, clearly is a liftable element of $\{S\}'$. For which subnormal operators is it true that an operator in $\{S\}'$ lifts if and only if it belongs to $R^\infty(S)$? When is $R^\infty(S) = \{S\}'$?

Yoshino [**1976**] and Embry-Wardrop (*private communication*) have pointed out that if $T_0 \in \mathscr{B}(\mathscr{H})$ and T_0 satisfies the inequality in Theorem 10.3 for all $f_0, \ldots, f_n$ in $\mathscr{H}$, then it follows that $T_0 \in \{S\}'$.

If B commutes with both S and S^*, then the situation is completely different. In fact, note that B commutes with both S and S^* if and only if $B \in W^*(S)'$. Thus the next result shows that in this case B always lifts.

10.5 THEOREM (Bram [**1955**]). *Let S be a subnormal operator, $N = \text{mne } S$, and let P be the projection of $\mathscr{K}$ onto $\mathscr{H}$. If $\mathscr{A} \equiv W^*(N)' \cap \{P\}'$ and $\rho : \mathscr{A} \to W^*(S)'$ is defined by $\rho(A) = A|\mathscr{H}$, then ρ is a *-isomorphism and a weak* homeomorphism.*

PROOF. First note that if $A \in \mathscr{A}$, then A is reduced by $\mathscr{H}$ and $\rho(A) \in W^*(S)'$. Thus ρ is well defined and has its range in $W^*(S)'$. Since N is the minimal normal extension of S, it is easy to see that ρ is a *-monomorphism. But every *-monomorphism is an isometry (see VIII.4.8 in [**ACFA**]). It must be shown that ρ is surjective.

Let Q be a projection in $W^*(S)'$; so $Q\mathscr{H}$ reduces S. Let $\widetilde{Q}$ be the projection of $\mathscr{K}$ onto $\bigvee\{N^{*k}f : k \geq 0 \text{ and } f \in Q\mathscr{H}\}$. Clearly $\widetilde{Q}\mathscr{K}$ reduces N and so $\widetilde{Q} \in W^*(N)'$. Let $E = P - Q$ and define $\widetilde{E}$ analogously; since $EQ = 0$, it is not hard to show that $\widetilde{E}\widetilde{Q} = 0$. Now $N|\widetilde{E}\mathscr{K}$ and $N|\widetilde{Q}\mathscr{K}$ are minimal normal extensions of $S|E\mathscr{H}$ and $S|Q\mathscr{H}$, so $N|(\widetilde{Q}\mathscr{K} \oplus \widetilde{E}\mathscr{K}) = \text{mne}[S|(Q\mathscr{H} \oplus E\mathscr{H})] = \text{mne } S = N$. Therefore, $\widetilde{E} = \widetilde{Q}^\perp$. It follows that $\widetilde{Q}\mathscr{K} \cap \mathscr{H} = Q\mathscr{H}$. (Why?) Thus $\widetilde{Q}P = P\widetilde{Q} = Q$ and so $\widetilde{Q} \in \mathscr{A}$. Moreover, $\rho(\widetilde{Q}) = Q$.

What we have just shown is that each projection in $W^*(S)'$ is the image of a projection in $\mathscr{A}$. If $A \in W^*(S)'$ and $A = A^*$, then for every $\varepsilon > 0$ there are pairwise orthogonal projections $Q_1, \ldots, Q_n$ in $W^*(S)'$ and real scalars $\alpha_1, \ldots, \alpha_n$ such that $\|A - \sum_k \alpha_k Q_k\| < \varepsilon$. Now $\sum_k \alpha_k \widetilde{Q}_k \in \mathscr{A}$ and $\rho(\sum_k \alpha_k \widetilde{Q}_k) = \sum_k \alpha_k Q_k$. A straightforward argument shows that there is an $\widetilde{A}$ in $\mathscr{A}$ such that $\rho(\widetilde{A}) = A$. The details are left to the reader. But since ρ is a *-monomorphism, the fact that every hermitian element of $W^*(S)'$ belongs to the range of ρ implies that ρ is surjective.

It is trivial to show that ρ is WOT sequentially continuous. But then Proposition I.2.7 implies that ρ is a weak* homeomorphism, and the proof is complete. □

Can the Fuglede-Putnam Theorem [**ACFA**] be generalized to subnormal operators? No! If this did hold for subnormal operators, then $\{S\}' = W^*(S)'$ and this is clearly false when S is the unilateral shift. Example 10.2 gives another example of this. There is, however, an "anti-Fuglede-Putnam Theorem."

10.6 PROPOSITION (Radjavi and Rosenthal [1971]). *If S_1 and S_2 are subnormal operators on $\mathscr{H}$ and if $A: \mathscr{H} \to \mathscr{H}$ is injective, has dense range, and $S_1 A = A S_2^*$, then S_1 and S_2 are normal operators and $S_1 \cong S_2^*$.*

PROOF. Let N_1 and N_2 be the normal extensions of S_1 and S_2 acting on $\mathscr{K}$. (Note that N_1 and N_2 are not assumed to be minimal. To get N_1 and N_2 acting on the same space, minimality must be sacrificed. How is it done?) Let N_1 and N_2 have the matrices relative to the decomposition $\mathscr{K} = \mathscr{H} \oplus \mathscr{H}^\perp$

$$N_1 = \begin{bmatrix} S_1 & X_1 \\ 0 & T^* \end{bmatrix}, \quad N_2 = \begin{bmatrix} S_2 & X_2 \\ 0 & T^* \end{bmatrix}.$$

If $A_0 = \begin{bmatrix} A & 0 \\ 0 & 0 \end{bmatrix}$, then it can be checked that $N_1 A_0 = A_0 N_2^*$. It follows (see IX.6.10 in [ACFA]) that $\ker A_0$ reduces N_2, $\mathrm{cl}[\operatorname{ran} A_0]$ reduces N_1, and $N_2^*|(\ker A_0)^\perp \cong N_1|\mathrm{cl}[\operatorname{ran} A_0]$. But $(\ker A_0)^\perp = \mathrm{cl}[\operatorname{ran} A_0] = \mathscr{H}$ by the definition of A_0 and the hypothesis concerning A. That is, $\mathscr{H}$ reduces both N_1 and N_2 and so S_1 and S_2 are normal. □

The preceding result has been extended to hyponormal operators, but a different proof is needed. In fact, Stampfli and Wadhwa [1976] have proved a further extension to a class of operators that includes the hyponormal operators.

If μ and ν are equivalent (mutually absolutely continuous) measures on the compact set X, then $L^\infty(\mu) = L^\infty(\nu)$ and the norms are identical. The spaces of functions $L^1(\mu)$ and $L^1(\nu)$ are not equal, but there is a way of getting around this. Namely, using the Radon-Nikodym Theorem, $L^1(\mu)$ can be identified with the subspace of $M(X)$, $\{\eta \in M(X) : \eta \ll \mu\}$, and this identification is isometric if $L^1(\mu)$ has its usual norm and the subspace of $M(X)$ has the total variation norm from $M(X)$. So $L^1(\mu)$ and $L^1(\nu)$ are identified via this same subspace of $M(X)$ whenever $[\mu] = [\nu]$.

The remaining L^p spaces can be radically different for mutually absolutely continuous measures. For example, there are equivalent measures μ and ν such that $L^2(\mu)$ and $L^2(\nu)$ are incomparable. In fact, a result from Bram [1955] will be proved below (V.14.21) which shows that for any compactly supported measure on $\mathbb{C}$, there is a mutually absolutely continuous measure ν such that $P^2(\nu) = L^2(\nu)$. One of the tools used in that proof will be a construction of Hastings [1979b], which we now examine.

If K is a compact subset of $\mathbb{C}$ and E is a compact subset of K, let $R(E, K)$ be the closure in $C(E)$ of $\mathrm{Rat}(K)$.

10.7 LEMMA. *If f is a Borel function on K, $\mu \in M(K)$, if there exist compact sets $E_1 \subseteq E_2 \subseteq \cdots \subseteq K$ such that $\mu(E_n) \to \|\mu\|$, and if for each $n \geq 1$ there is a function g_n in $R(E_n, K)$ with $f|E_n = g_n$ a.e. $[\mu]$, then there is a measure ν and positive constants c_n such that the following properties are satisfied.*

(a) $[\nu] = [\mu]$ *and* $d\nu/d\mu$ *is bounded.*

(b) $\nu|E_n \geq c_n\mu|E_n$ for all n.
(c) $f \in R^2(K,\nu)$.

PROOF. Let $f_n \in \text{Rat}(K)$ such that $|f(z) - f_n(z)| < 1/n$ a.e. $[\mu]$ on E_n and put $M_n = \max\{|f_n(z)| : z \in E_n\}$. Choose positive constants c_n having the following properties:

$$1 \geq c_1 \geq c_2 \geq \ldots ;$$
$$2M_n^2 c_n \leq 1/n^2 ;$$
$$c_1 \int_{E_1} |f|^2 d\mu + \sum_{n=1}^{\infty} c_{n+1} \int_{E_{n+1}\setminus E_n} |f|^2 d\mu < \infty .$$

(How can you find these constants?) Now define the measure ν by letting $\nu|E_1 = \mu|E_1$ and $\nu|(E_{n+1}\setminus E_n) = c_{n+1}\mu|(E_{n+1}\setminus E_n)$ for $n \geq 1$.
Clearly (a) and (b) hold with $d\nu/d\mu \leq 1$. Also,

$$\int |f - f_n|^2 d\nu = \int_{E_n} |f - f_n|^2 d\nu + \int_{\mathbb{C}\setminus E_n} |f - f_n|^2 d\nu$$
$$\leq n^{-2}\mu(E_n) + 2c_n \int_{\mathbb{C}\setminus E_n} |f_n|^2 d\mu + 2\int_{\mathbb{C}\setminus E_n} |f|^2 d\nu$$
$$\leq n^{-2}\|\mu\| + 2\sum_{k=n}^{\infty} c_{k+1} \int_{E_{k+1}\setminus E_k} |f|^2 d\mu$$
$$< \varepsilon$$

for sufficiently large n. Hence $f \in R^2(K,\nu)$. □

10.8 PROPOSITION (Hastings [1979b]). *If U is an open subset of $\mathbb{C}$ such that $U \subseteq K$, $\mu(K\setminus U) = 0$, and each component of $\mathbb{C}\setminus U$ meets $\mathbb{C}\setminus K$, then for any Borel function f on K that is analytic on U, there is a measure ν and a sequence of positive constants $\{c_n\}$ such that:*

(a) $[\nu] = [\mu]$;
(b) $\nu|E_n \geq c_n\mu|E_n$, where $E_n = \{z \in U : \text{dist}(z, \mathbb{C}\setminus U) \geq n^{-1}\}$;
(c) $f \in R^2(K,\nu)$.

PROOF. If E_n is defined as in the statement of the proposition, then each component of $\mathbb{C}\setminus E_n$ contains a component of $\mathbb{C}\setminus U$. Hence each component of $\mathbb{C}\setminus E_n$ meets $\mathbb{C}\setminus K$. Thus, $f|E_n \in R(E_n, K)$ by Runge's Theorem. The result now follows by the preceding lemma. □

The virtue of the preceding proposition is its ability to produce several interesting and unusual examples.

10.9 EXAMPLE. If μ is the restriction of Area measure to $\mathbb{D}$, then there is a measure ν equivalent to μ such that $z^{1/2} \in P^2(\nu)$. Indeed, let $U = \mathbb{D}\setminus(-1,0]$, $K = \text{cl}\,\mathbb{D}$, and $f(z) = z^{1/2}$ in Proposition 10.8.

10.10 EXAMPLE. If μ is the restriction of Area measure to $\mathbb{D}$, then there is a measure ν equivalent to μ such that $z^{-1} \in P^2(\nu)$. For this, apply

Proposition 10.8 with $U = \mathbb{D} \setminus \{0\}$ and K as in the preceding example and with $f(z) = z^{-1}$.

REMARKS. 1. Olin and Thomson [**1979**] give an example of a subnormal operator and an operator A in $\{S\}''$ that does not lift to $\{N\}'$.

2. Embry-Wardrop [**1981**] gives a necessary and sufficient condition for lifting an operator that intertwines two subnormal operators to one that intertwines their minimal normal extensions. If S_j is a subnormal operator on $\mathscr{H}_j$, $j = 1, 2$, with minimal normal extension N_j on $\mathscr{K}_j$ and $T_0 : \mathscr{H}_1 \to \mathscr{H}_2$ such that $T_0 S_1 = S_2 T_0$, then $\begin{bmatrix} 0 & 0 \\ T_0 & 0 \end{bmatrix} \in \{S_1 \oplus S_2\}'$. There is an operator $T : \mathscr{K}_1 \to \mathscr{K}_2$ such that $TN_1 = N_2 T$ and $T|\mathscr{H}_1 = T_0$ if and only if the operator $\begin{bmatrix} 0 & 0 \\ T_0 & 0 \end{bmatrix}$ lifts to the commutant of $N_1 \oplus N_2$. The interested reader can now apply Theorem 10.4 to obtain conditions for T_0 to lift to such an operator T.

3. Some additional papers on the lifting of the commutant of a subnormal operator are Abrahamse [**1978**], Ito [**1958**], Lubin [**1977, 1978, 1979**, and **1980**], Mlak [**1971** and **1972a**], Slocinski [**1975**], and Yoshino [**1973**].

4. If $\phi \in H^\infty$ and T_ϕ is the corresponding Toeplitz operator on H^2 (T_ϕ = multiplication by ϕ), then T_ϕ is a subnormal operator. There is a whole separate literature devoted to the commutant of one of these analytic Toeplitz operators. See Abrahamse [**1975, 1976**], Abrahamse and Ball [**1976**], Clancey and Morrel [**1974**], Cowen [**1978, 1980a**, and **1980b**], Deddens [**1971**], Deddens and Wong [**1973**], Nordgren [**1967**], Rudol [**preprint**], Solomyak [**1986**], Thomson [**1975, 1976a, 1976b**, and **1977**].

5. Example 10.9 was used in Clancey, Conway, and Raphael [**1983**].

Exercises

1. Show that if T is the unilateral shift and $E = 1 - TT^*$, then $\begin{bmatrix} 0 & 0 \\ E & 0 \end{bmatrix} \in \{T \oplus 0\}'$ but does not lift.
2. Prove Proposition 10.6 by applying Theorem 10.4.
3. Show that if S_1 and S_2 are subnormal operators and R is an invertible operator such that $S_1 R = R S_2$ and $S_1^* R = R S_2^*$, then $S_1 \cong S_2$.
4. If S_1 and S_2 are subnormal operators and R is a bounded operator such that $S_1 R = R S_2$ and $S_1^* R = R S_2^*$, what can be said?
5. (Radjavi and Rosenthal [**1971**]) If T is any operator, then T^2 is normal if and only if
$$T = N \oplus \begin{bmatrix} M & A \\ 0 & -M \end{bmatrix},$$
where N and M are normal, $A \geq 0$, $\ker A = (0)$, and $AM = MA$.
6. Prove that if S is subnormal and S^2 is normal, then S is normal.
7. (Hastings [**1979b**]) If ν and K are as in Example 10.9, determine $\text{bpe}(K, \mu)$ and $\text{abpe}(K, \mu)$.
8. (Hastings [**1979b**]) If ν and K are as in Example 10.10, determine $\text{bpe}(K, \mu)$ and $\text{abpe}(K, \mu)$.

9. (Hastings [**1979b**]) Let $\mu_1 = \text{Area}|\mathbb{D}$, $\mu_2 = $ arc length measure on $\partial \mathbb{D}$. Show that there is a measure ν with $[\nu] = [\mu_1 + \mu_2]$ such that $\text{abpe}(\text{cl}\,\mathbb{D}, \nu) = \mathbb{D}$ and $P^2(\nu) = P^2(\nu|\mathbb{D}) \oplus P^2(\nu|\partial\mathbb{D})$.

10. (Wogen [**1985**]) Show that there is a pure subnormal operator S such that there is an operator T that is not subnormal but $T^2 = S$.

§11 The restriction algebra and the functional calculus. If S is a subnormal operator with $N = \text{mne}\,S$, then $\sigma(N) \subseteq \sigma(S)$ and so $f(S)$ and $f(N)$ are both defined for any rational function with poles off $\sigma(S)$. As we have already seen, $f(S) = f(N)|\mathscr{H}$. In fact, S is a von Neumann operator and for any f in $R(\sigma(S))$, $f(S)$ is defined and $f(S) = f(N)|\mathscr{H}$. In this section we extend the functional calculus for a subnormal operator as far as is reasonable.

11.1 DEFINITION. If S is a subnormal operator with $N = \text{mne}\,S$ and μ is a scalar-valued spectral measure for N (see IX.8.4 in [ACFA]), let

$$\mathscr{R}(N, H) \equiv \{\phi \in L^\infty(\mu) : \phi(N)\mathscr{H} \subseteq \mathscr{H}\}.$$

$\mathscr{R}(N, \mathscr{H})$ is called the *restriction algebra* for S. The scalar-valued spectral measure μ for N will also be called the *scalar-valued spectral measure* for S.

We know from Proposition 9.7 that $R(\sigma(S)) \subseteq \mathscr{R}(N, \mathscr{H})$. It is also easy to see that $\mathscr{R}(N, \mathscr{H})$ is an algebra, thereby justifying the terminology.

If $\phi \in \mathscr{R}(N, \mathscr{H})$, define

$$\phi(S) \equiv \phi(N)|\mathscr{H}.$$

Clearly $\phi(S)$ is subnormal since $\phi(N)$ is a normal extension of $\phi(S)$. However, $\phi(N)$ need not be the minimal normal extension of $\phi(S)$. An easy example is found by taking $\phi \equiv 1$. A variation on this example is found by letting S_1 and S_2 be two copies of the unilateral shift and putting $S = S_1 \oplus (S_2 + 3)$. If N_1 and N_2 are the minimal normal extensions of S_1 and S_2, then $N = N_1 \oplus (N_2 + 3)$ is the minimal normal extension of S. Let Δ_1 be the closed unit disk and $\Delta_2 = \Delta_1 + 3$. So $\sigma(S) = \Delta_1 \cup \Delta_2$ and $\Delta_1 \cap \Delta_2 = \emptyset$. Also, $\sigma(N) = \partial\Delta_1 \cup \partial\Delta_2$. If $f(z) = z\chi_{\Delta_1}(z)$, then $f(N) = N_1 \oplus 0$ and so $f \in \mathscr{R}(N, \mathscr{H})$. Also, $f(S) = S_1 \oplus 0$. However, $f(N)$ is not the minimal normal extension of $f(S)$; that honor goes to $N_1 \oplus 0$, where this time the zero operator acts on $\mathscr{K}_2$, not $\mathscr{H}_2$.

The following is the basic result on the functional calculus for a subnormal operator.

11.2 THEOREM. *If S is a subnormal operator with $N = \text{mne}\,S$ and μ is a scalar-valued spectral measure for N, then the following statements hold.*

(a) $\mathscr{R}(N, \mathscr{H})$ *is a weak* closed subalgebra of* $L^\infty(\mu)$.

(b) *If $\rho : \mathscr{R}(N, \mathscr{H}) \to \mathscr{B}(\mathscr{H})$ is defined by $\rho(\phi) = \phi(S)$, then ρ is a dual algebra isomorphism onto its image.*

PROOF. (a) It will only be shown that $\mathscr{R}(N, \mathscr{H})$ is weak* closed in $L^\infty(\mu)$. Let $\{\phi_i\}$ be a net in $\mathscr{R}(N, \mathscr{H})$ and suppose that $\phi_i \to \phi$ weak* in $L^\infty(\mu)$. By the functional calculus for normal operators (see IX.8.10 in [ACFA]), $\phi_i(N) \to \phi(N)$ (WOT). (Actually, the convergence is weak* here even though the reference only establishes WOT convergence.) Hence for each h in $\mathscr{H}$, $\phi_i(N)h \to \phi(N)h$ weakly. Since $\phi_i(N)h \in \mathscr{H}$ for all i, $\phi(N)h \in \mathscr{H}$. Thus $\phi \in \mathscr{R}(N, \mathscr{H})$.

(b) Corollary 2.17 implies that $\|\phi(S)\| = \|\phi(N)\|$; since $\|\phi(N)\| = \|\phi\|_\infty$, this establishes that ρ is an isometry. It is easy to check that ρ is a homomorphism. It remains to check that the range of ρ is weak* closed and that ρ is a weak* homeomorphism. By Proposition I.2.7 and I.2.10 it suffices to show that if $\{\phi_k\}$ is a sequence in $\mathscr{R}(N, \mathscr{H})$ such that $\phi_k \to 0$ weak* in $L^\infty(\mu)$, then $\phi_k(S) \to 0$ (WOT) in $\mathscr{B}(\mathscr{H})$. But $\phi_k(N) \to 0$ (WOT) in $\mathscr{B}(\mathscr{K})$, so if $g, h \in \mathscr{H}$, then $\langle \phi_k(S)g, h \rangle = \langle \phi_k(N)g, h \rangle \to 0$. □

For any compact subset K of $\mathbb{C}$ and any positive measure μ on K, let $R^\infty(K, \mu)$ be the weak* closure of $\text{Rat}(K)$ in $L^\infty(\mu)$. $P^\infty(\mu)$ is the weak* closure of the polynomials in $L^\infty(\mu)$.

As we saw before, $R(\sigma(S)) \subseteq \mathscr{R}(N, \mathscr{H})$ and so $P^\infty(\mu) \subseteq R^\infty(\sigma(S), \mu) \subseteq \mathscr{R}(N, \mathscr{H})$.

11.3 COROLLARY (Conway and Olin [1977]). *If $\rho: R^\infty(\sigma(S), \mu) \to \mathscr{B}(\mathscr{H})$ is defined by $\rho(\phi) = \phi(S)$, then $\rho(R^\infty(\sigma(S), \mu)) = R^\infty(S)$ and $\rho: R^\infty(\sigma(S), \mu) \to R^\infty(S)$ is a dual algebra isomorphism.*

11.4 COROLLARY. *If $\rho: P^\infty(\mu) \to \mathscr{B}(\mathscr{H})$ is defined by $\rho(\phi) = \phi(S)$, then $\rho(P^\infty(\mu)) = P^\infty(S)$ and $\rho: P^\infty(\mu) \to P^\infty(S)$ is a dual algebra isomorphism.*

Later (Chapter VI) the algebra $R^\infty(K, \mu)$ will be studied more fully and the results of this study will be applied to obtain a deeper understanding of this functional calculus for subnormal operators.

Notice that if $\phi \in \mathscr{R}(N, \mathscr{H})$, then $\phi(S) \in \{S\}'$. In particular, if S is multiplication by z on $R^2(K, \mu)$, then $\mathscr{R}(N, \mathscr{H}) \subseteq R^2(K, \mu) \cap L^\infty(\mu)$. But clearly $R^\infty(K, \mu) \cap L^\infty(\mu) \subseteq \mathscr{R}(N, \mathscr{H})$, and so we get the following result.

11.5 PROPOSITION. *If S is multiplication by z on $R^2(K, \mu)$, then the restriction algebra is $R^2(K, \mu) \cap L^\infty(\mu)$.*

The restriction algebra is the largest algebra of functions for which one can hope to have a reasonable functional calculus; that is, a functional calculus which is multiplicative and enjoys other agreeable attributes. In general, however, the restriction algebra remains shrouded in fog. It is never equal to $L^\infty(\mu)$ unless S is normal. Other than this statement, not much else has been said and no one has ever studied this algebra.

The difficulty with $\mathscr{R}(N, \mathscr{H})$ is that it depends very much on the subnormal operator S and is not something that can be only associated with N. The algebra $P^\infty(\mu)$, for example, only depends on the measure class $[\mu]$ (or

N) and does not take into account any of the subtleties of the subspace $\mathscr{H}$ and the action of N on it. The algebra $P^\infty(\mu)$ is well understood (Sarason [1972] and §VI.7). But because of its loose connection with S it might be suspected that this understanding would not contribute to a better comprehension of subnormality. This turns out to be false, as we will see later in this book.

This section closes with a rather unusual result about $\mathscr{M}^\infty(\mu)$, the weak* closure of a linear manifold $\mathscr{M}$ in $L^\infty(\mu)$. Observe that if μ and ν are mutually absolutely continuous measures, their L^∞ spaces are equal. If $L^1(\mu)$ is identified via the Radon-Nikodym Theorem with $\{\eta : \eta \ll \mu\}$, then we see that the weak* topology on the spaces $L^\infty(\mu)$ and $L^\infty(\nu)$ are identical when μ and ν are mutually absolutely continuous. It is also rather easy to see that whenever $[\nu] = [\mu]$, $\mathscr{M}^\infty(\nu) \cap \mathscr{M}^2(\nu)$ is a weak* closed subalgebra of $L^\infty(\mu)$ that contains $\mathscr{M}^\infty(\mu)$. The next proposition has proved quite useful when $\mathscr{M}$ is the set of polynomials and μ is a compactly supported measure on the plane. Note, however, that it applies to any finite measure space. (The result will be proved for any value of p, $1 \leq p < \infty$.)

11.6 PROPOSITION. *If μ is a finite measure such that $L^1(\mu)$ is separable and $\mathscr{M}$ is a linear manifold in $L^\infty(\mu)$, then there exists a measure ν such that μ and ν are mutually absolutely continuous and $\mathscr{M}^\infty(\mu) = \mathscr{M}^p(\nu) \cap L^\infty(\nu)$. (Here the obvious notation holds: $\mathscr{M}^p(\nu)$ denotes the closure of $\mathscr{M}$ in $L^p(\mu)$.)*

PROOF. Let $\mathscr{M}^\perp = \{h \in L^q(\mu) : \int fh \, d\mu = 0 \text{ for all } f \text{ in } \mathscr{M}\}$ and let $\{h_n\}$ be a dense sequence in the unit ball of $\mathscr{M}^\perp$ (Exercise 3). Put $\nu = (1 + \sum_n 2^{-n}|h_n|)\mu$; note that $\mu \leq \nu$ and ν is absolutely continuous with respect to μ. If $f \in \mathscr{M}^p(\nu) \cap L^\infty(\nu)$, then there is a sequence $\{f_k\}$ in $\mathscr{M}$ such that $\int |f_k - f|^p \, d\nu \to 0$. It follows that $\int |f_k - f|^p |h_n| \, d\mu \to 0$. Write h_n as the product of two functions h_n^1 and h_n^2 such that $|h_n^1|^p = |h_n^2|^q = |h_n|$. So $|\int (f_k - f) h_n \, d\mu| \leq [\int |f_k - f|^p |h_n| \, d\mu]^{1/p} [\int |h_n| \, d\mu]^{1/q} \to 0$ as $k \to \infty$. Hence $\int fh_n \, d\mu = 0$ for all $n \geq 1$. By the Hahn-Banach Theorem, $f \in \mathscr{M}^\infty(\mu)$. The other inclusion follows from the fact that $\mathscr{M}^2(\nu) \cap L^\infty(\nu)$ is weak* closed. □

The preceding proposition has a checkered past. Ball, Olin, and Thomson [1978] proved that for any compactly supported measure on $\mathbb{C}$, $P^\infty(\mu) = \bigcap \{P^2(\nu) \cap L^\infty(\mu) : [\nu] = [\mu]\}$. In 1989 John Murphy was the first to prove (11.6) but James Thomson and T. W. Gamelin, independently, quickly followed this with simple proofs. The above proof is from Thomson (*private communication*).

An interesting question still remains.

11.7 OPEN QUESTION. Does the collection $\{P^2(\nu) \cap L^\infty(\mu) : [\nu] = [\mu]\}$ of weak* closed subalgebras of $L^\infty(\mu)$ form a lattice?

Exercises

1. Find the restriction algebra of a quasinormal operator.
2. How are the restriction algebras of S and $S \oplus S$ related?
3. Prove that if $L^1(\mu)$ is a separable Banach space, then so is $L^p(\mu)$ for $1 \leq p < \infty$.

§12 The C^*-algebra generated by a subnormal operator.

For any operator T, $C^*(T)$ denotes the C^*-algebra generated by the operator T and the identity. Thus, $C^*(T)$ is the closed linear span in $\mathscr{B}(\mathscr{H})$ of all words in T and T^*.

First we encounter a result about the C^*-algebra generated by a hyponormal operator. Recall that if N is normal, there is a natural *-isomorphism from $C^*(N)$ onto $C(\sigma(N))$ that takes N to the function z. Thus for each λ in $\sigma(N)$ there is a homomorphism $\phi : C^*(N) \to \mathbb{C}$ such that $\phi(N) = \lambda$. For a normal operator $\sigma(N) = \sigma_{\text{ap}}(N)$ and $C^*(N)$ is abelian. This is not the case for a hyponormal operator, but we can get a reasonable facsimile to the result for normal operators.

12.1 PROPOSITION (Bunce [**1970**]). *If A is a hyponormal operator, then $\lambda \in \sigma_{\text{ap}}(A)$ if and only if there exists a nonzero *-homomorphism ϕ on $C^*(A)$ such that $\phi(A) = \lambda$.*

PROOF. There is no loss in generality in assuming that $\lambda = 0$. So let $\phi : C^*(A) \to \mathbb{C}$ be a homomorphism and assume that $0 \notin \sigma_{\text{ap}}(A)$. Thus there is a constant $c > 0$ with $\|Ah\| \geq c\|h\|$ for all h in $\mathscr{H}$. This implies that $A^*A - c^2$ is a positive operator. Hence $0 \leq \phi(A^*A - c^2) = |\phi(A)|^2 - c^2$. Thus $|\phi(A)|^2 \geq c^2 > 0$ and so $\phi(A) \neq 0$.

Conversely, assume that $0 \in \sigma_{\text{ap}}(A)$ and let $\{f_n\}$ be a sequence of unit vectors in $\mathscr{H}$ such that $\|Af_n\| \to 0$. Let LIM denote a Banach limit on ℓ^∞ (see §III.7 in [**ACFA**]) and define $\phi : \mathscr{B}(\mathscr{H}) \to \mathbb{C}$ by $\phi(T) = \text{LIM}\langle Tf_n, f_n \rangle$. Note that ϕ is a positive linear functional and $\phi(TA) = 0$ for all T since $\|TAf_n\| \to 0$. But A is hyponormal and so $\|A^*f_n\| \leq \|Af_n\|$. Thus $\|A^*f_n\| \to 0$ and so $\phi(TA^*) = 0$ for all T in $\mathscr{B}(\mathscr{H})$. On the other hand, $\phi(1) = 1$. So if $p(A^*, A)$ is any noncommuting polynomial in A and A^* that has no constant term, $\phi(p(A^*, A) + \alpha) = \alpha$ for all scalars α. But this implies that ϕ is multiplicative on a dense subalgebra of $C^*(A)$. Therefore, $\phi|C^*(A)$ is a homomorphism and $\phi(A) = 0$. □

It is not uncommon for a nonabelian Banach algebra to entertain very few homomorphisms. The preceding proposition says that the (nonabelian) C^*-algebra generated by a hyponormal operator possesses quite a rich supply of these homomorphisms. What is going on?

Let $\mathscr{A}$ be a Banach algebra with identity and let $\phi : \mathscr{A} \to \mathbb{C}$ be a nonzero homomorphism. If a and $b \in \mathscr{A}$, then $\phi(ab) = \phi(a)\phi(b) = \phi(ba)$. Thus $\phi(ab - ba) = 0$ and so ϕ annihilates the commutator ideal of $\mathscr{A}$.

§12 THE C*-ALGEBRA

12.2 LEMMA. *If $\mathscr{A}$ is a C^*-algebra with identity, Φ is the set of nonzero homomorphisms of $\mathscr{A}$ into $\mathbb{C}$, and J is the commutator ideal of $\mathscr{A}$ (that is, J is the norm closed ideal generated by the set of all elements of $\mathscr{A}$ of the form $ab - ba$), then*
$$J = \bigcap \{\phi^{-1}(0) : \phi \in \Phi\}.$$

PROOF. Let $L = \bigcap\{\phi^{-1}(0) : \phi \in \Phi\}$; clearly L is a closed ideal of $\mathscr{A}$ and, as we saw before the statement of this lemma, $L \supseteq J$. Now suppose $A \in \mathscr{A} \setminus J$. So $A + J$ is a nonzero element of the abelian C^*-algebra $\mathscr{A}/J$. Hence there is a homomorphism $\phi_1 : \mathscr{A}/J \to \mathbb{C}$ such that $\phi_1(A+J) \neq 0$. Thus $\phi(B) = \phi_1(B+J)$ defines a homomorphism on $\mathscr{A}$ and $\phi(A) \neq 0$. □

12.3 PROPOSITION. *With the notation of the preceding lemma, Φ is the maximal ideal space of $\mathscr{A}/J$. Hence $\mathscr{A}/J \cong C(\Phi)$ under the Gelfand transform, $A + J \to \widehat{A}$, where $\widehat{A}(\phi) = \phi(A)$ for A in $\mathscr{A}$ and ϕ in Φ.*

PROOF. Exercise.

12.4 COROLLARY (Bunce [1970]). *If A is a hyponormal operator, there is an isometric *-isomorphism of $C^*(A)/J$ onto $C(\sigma_{ap}(A))$, where $A+J$ is mapped to the function z.*

PROOF. The proof has much in common with the proof of the corresponding result for normal operators (see VIII.2.3 in [ACFA]). In fact the proof of the present proposition will follow along the lines of that of the normal result once it is established that $\phi \to \phi(A)$ is a homeomorphism of Φ onto $\sigma_{ap}(A)$. But Proposition 12.1 implies that this map is surjective. On the other hand, if ϕ and $\psi \in \Phi$ and $\phi(A) = \psi(A)$, then the fact that A is a generator of the C^*-algebra implies that $\phi = \psi$. So the map $\phi \to \phi(A)$ is injective as well. It remains to show that the map is continuous—an easy exercise. □

Let's return to the case of a subnormal operator. The idea here is to relate the algebras $C^*(S)$ and $C^*(N)$. If μ is the scalar-valued spectral measure for N and $u \in L^\infty(\mu)$, define the operator $T_u : \mathscr{K} \to \mathscr{H}$ by
$$T_u h = P(u(N)h),$$
where P is the orthogonal projection of $\mathscr{K}$ onto $\mathscr{H}$. The operator T_u is called the *Toeplitz operator* (associated with S) with *symbol* u. If S is the unilateral shift, then these Toeplitz operators are the classical ones (Douglas [1972]). Note that if $u \in \mathscr{R}(N, \mathscr{H})$, then $T_u = u(S)$.

It is not difficult to see that $u \to T_u$ is a linear mapping from $L^\infty(\mu)$ into $\mathscr{B}(\mathscr{H})$, $\|T_u\| \leq \|u\|_\infty$, and that $T_u \geq 0$ whenever $u \geq 0$. This map is not, however, multiplicative. There is a partial multiplicative property that it enjoys. If $f, h \in \mathscr{H}$ and n and m are positive integers, then $\langle S^{*n}S^m f, h \rangle = \langle S^m f, S^n h \rangle = \langle N^m f, N^n h \rangle = \langle PN^{*n}N^m f, h \rangle$. This implies

that
$$T_{\bar{z}^n}T_{z^m} = T_{\bar{z}^n z^m}$$
for all $n, m \geq 0$. This leads almost immediately to the following.

12.5 LEMMA. $\{T_u : u \in C(\sigma(N))\} \subseteq C^*(S)$.

Define $\Theta : C^*(N) \to C^*(S)$ by $\Theta(u) = T_u$. As was already pointed out, Θ is a positive linear map. Let $\rho_N : C^*(N) \to C(\sigma(N))$ be the *-isomorphism such that $\rho_N(N) = z$ and let $\rho_S : C^*(S)/J \to C\{\sigma_{\mathrm{ap}}(S)\}$ be the *-isomorphism obtained in Corollary 12.4 with $\rho_S(S+J) = z$. Define $\pi : C^*(S) \to C^*(S)/J$ to be the natural map.

Because $\sigma(N) \subseteq \sigma(S)$, the restriction map $\tau : C(\sigma(N)) \to C(\sigma_{\mathrm{ap}}(S))$ is a well-defined surjective homomorphism.

12.6 THEOREM (Olin and Thomson [1982]). *If* $\tau : C(\sigma(N)) \to C(\sigma_{\mathrm{ap}}(S))$ *is the restriction map, then the diagram*

$$\begin{array}{ccc} C^*(N) & \xrightarrow{\rho_N} & C(\sigma(N)) \\ \Theta \downarrow & & \downarrow \tau \\ C^*(S) & \xrightarrow{\pi} C^*(S)/J \xrightarrow{\rho_S} & C(\sigma_{\mathrm{ap}}(S)) \end{array}$$

commutes.

PROOF. Since polynomials in z and $\bar{z}$ are dense in $C(\sigma(N))$, it suffices to show that for $n, m \geq 0$, $\rho_S(T_{\bar{z}^n z^m} + J)(\lambda) = \bar{\lambda}^n \lambda^m$. But $\rho_S(T_{\bar{z}^n z^m} + J) = \rho_S \pi(S^{*n} S^m) = \bar{z}^n z^m$, since $\rho_S \pi(S) = z$. □

Bunce [1978] discusses the map Θ above and uses it to characterize the minimal normal extension.

12.7 COROLLARY. *If* $u \in C(\sigma(N))$, *then* $T_u \in J$ *if and only if* $u(\lambda) = 0$ *for all* λ *in* $\sigma_{\mathrm{ap}}(S)$.

12.8 PROPOSITION (Olin and Thomson [1982]). *If* $T \in C^*(S)$ *and* J *is the commutator ideal, then the following statements are equivalent.*

(a) $T \in J$.
(b) *If* $\lambda \in \sigma_{\mathrm{ap}}(S)$ *and* $\{h_n\}$ *is any sequence of unit vectors in* $\mathscr{H}$ *such that* $\|(S - \lambda)h_n\| \to 0$, *then* $\|Th_n\| \to 0$.
(c) *If* $\lambda \in \sigma_{\mathrm{ap}}(S))$, *then there exists a sequence of unit vectors* $\{h_n\}$ *in* $\mathscr{H}$ *such that* $\|(S - \lambda)h_n\| \to 0$ *and* $\|Th_n\| \to 0$.

PROOF. It will only be shown here that (a) implies (b); the remaining implications are left to the reader. If p is a polynomial in 2 noncommuting variables, then $\|[p(S^*, S) - p(\bar{\lambda}, \lambda)]h_n\| \to 0$. If $\phi \in \Phi$ such that $\phi(S) = \lambda$, it follows that $\|[T - \phi(T)]h_n\| \to 0$ for any T in $C^*(S)$. But for T in the commutator ideal, $\phi(T) = 0$. □

It sometimes happens that a subnormal operator has the additional property that the self-commutator of S, $[S^*, S] = S^*S - SS^*$, is compact. Indeed, it will be shown (Theorem IV.2.1) that if A is a hyponormal operator with the property that there are a finite number of vectors $h_1, \ldots, h_n$ such that $\{u(A)h_j : u \in \text{Rat}(\sigma(A)) \text{ and } 1 \leq j \leq n\}$ has dense linear span in $\mathcal{H}$, then $A^*A - AA^*$ is trace class. Thus the additional assumption that the self-commutator is compact is one that is often satisfied.

Recall that an operator A is *irreducible* if the only reducing subspaces for A are the trivial ones, (0) and $\mathcal{H}$. Equivalently, A is irreducible if and only if the only projections that commute with A are 0 and 1. If A is an irreducible operator, then $C^*(A)$ is an irreducible C^*-algebra. It then follows from von Neumann's Double Commutant Theorem (see IX.6.4 in [ACFA]) that the weak closure of $C^*(A)$ ($= W^*(A)$) is all of $\mathcal{B}(\mathcal{H})$. In the same vein, it follows that if $C^*(A)$ contains a nonzero compact operator, then $C^*(A) \supseteq \mathcal{B}_0(\mathcal{H})$. (See page 18 in Arveson [1976].)

12.9 LEMMA. *If A is an irreducible operator with compact self-commutator, then the commutator ideal of $C^*(A)$ is $\mathcal{B}_0(\mathcal{H})$.*

PROOF. $C^*(A)$ is an irreducible C^*-algebra and it contains a nonzero compact operator (viz, $A^*A - AA^*$). Hence, as mentioned before, $\mathcal{B}_0(\mathcal{H}) \subseteq C^*(A)$. Now $\mathcal{B}_0(\mathcal{H})$ must be an ideal of $C^*(A)$ and $A + \mathcal{B}_0(\mathcal{H})$ is a normal generator of $C^*(A)/\mathcal{B}_0(\mathcal{H})$. Hence $C^*(A)/\mathcal{B}_0(\mathcal{H})$ is abelian. So if B and $C \in C^*(A)$, $BC - CB \in \mathcal{B}_0(\mathcal{H})$; that is, $J \subseteq \mathcal{B}_0(\mathcal{H})$.

On the other hand, if $\phi : C^*(A) \to \mathbb{C}$ is a homomorphism, $\phi | \mathcal{B}_0(\mathcal{H})$ is also a homomorphism. But $\mathcal{B}_0(\mathcal{H})$ has no nonzero homomorphisms, so it must be that $\phi | \mathcal{B}_0(\mathcal{H}) \equiv 0$. By Lemma 12.2, $\mathcal{B}_0(\mathcal{H}) \subseteq J$. □

12.10 THEOREM (Olin and Thomson [1982]). *If S is an irreducible subnormal operator such that $S^*S - SS^*$ is compact, then $\sigma_{\text{ap}}(S) = \sigma_e(S)$ and the following hold for all u and v in $C(\sigma(N))$.*

(a) T_u *is compact if and only if $u(\sigma_e(S)) = 0$.*
(b) $\|T_u + \mathcal{B}_0\| = \|u\|_{\sigma(S)}$.
(c) $T_{uv} - T_u T_v \in \mathcal{B}_0$.
(d) $\sigma_e(T_u) = u(\sigma_e(S))$.

PROOF. By the preceding lemma, we know that $\mathcal{B}_0$ is the commutator ideal of $C^*(S)$. By Corollary 12.4, $\sigma_{\text{ap}}(S) = \sigma_e(S)$. Parts (b) and (d) are now immediate from Corollary 12.4. Part (a) follows from (b) and part (c) follows from Theorem 12.6. □

If S is the unilateral shift, Theorem 12.10 is well-known and classical (see Douglas [1972]). If S is the Bergman operator on a bounded open set in $\mathbb{C}$, then S satisfies the hypothesis of the preceding theorem; the results for this case were obtained in Axler, Conway, and McDonald [1982]. In connection with this, the reader might also consult Axler [1982] and Elias [1988].

For the unilateral shift, a classical result is that spectral inclusion holds: $u(\sigma(N)) \subseteq \sigma(T_u)$ for all u in $L^\infty(\mu)$. This does not hold in general. The following result will be stated without proof.

12.11 THEOREM (Keough [1981]). *If S is a subnormal operator with $N = $ mne S, P is the projection of $\mathcal{K}$ onto $\mathcal{H}$, and $N = \int z\, dE(z)$ is the spectral resolution of N, then the following statements are equivalent.*

(a) $u(\sigma(N)) \subseteq \sigma(T_u)$ *for all u in $C(\sigma(N))$.*
(b) $\|T_u\| = \|u\|_{\sigma(N)}$ *for all u in $C(\sigma(N))$.*
(c) $\sigma(N) = \sigma_{\mathrm{ap}}(S)$.
(d) *For every relatively open subset Δ of $\sigma(N)$, $\|PE(\Delta)P\| = 1$.*
(e) *For every relatively open subset Δ of $\sigma(N)$, the angle between the spaces $E(\Delta)\mathcal{K}$ and $\mathcal{H}$ is 0.*
(f) *If $u \in C(\sigma(N))$ and $\lambda \in u(\sigma(N))$, then there is a sequence $\{h_n\}$ of unit vectors in $\mathcal{H}$ such that $\|[u(N) - \lambda]h_n\| \to 0$.*

REMARKS. 1. Keough [1981] also gives a necessary and sufficient condition for spectral inclusion to hold for all u in $L^\infty(\mu)$. (See Exercise 5.)

2. For which operators A is there an irreducible subnormal operator S such that $C^*(A) = C^*(S)$? It is not difficult to see that A must be irreducible. Additional restrictions must be placed on the spectral properties of A. If S is desired to be the unilateral shift, this was solved in Conway and McGuire [1984]. For almost all the remaining cases this is done in McGuire [1988a].

3. Behncke [1970] and Wogen [1971] have discussed the von Neumann algebra generated by a subnormal operator.

Exercises

1. Which unilateral shifts A have $A^*A - AA^*$ compact? Show that every hyponormal weighted shift A has $A^*A - AA^*$ in $\mathcal{B}_1$.
2. If A is a hyponormal bilateral weighted shift, is $A^*A - AA^*$ compact?
3. (Keough [1981]) Show that $\{\phi \in L^\infty(\mu) : \sigma(\phi(N)) \subseteq \sigma(T_\phi)\}$ is norm closed in $L^\infty(\mu)$.
4. (Keough [1981]) The only subnormal unilateral weighted shift S with $\|S\| = 1$ that satisfies the conditions of Theorem 12.11 is the unilateral shift.
5. (Keough [1981]) With the notation of Theorem 12.11, the following statements are equivalent.
   (a) $\sigma(\phi(N)) \subseteq \sigma(T_\phi)$ for every ϕ in $L^\infty(\mu)$.
   (b) $\|T_\phi\| = \|\phi\|_\infty$.
   (c) For every Borel set Δ with $0 < E(\Delta) < 1$, the angle between $E(\Delta)\mathcal{K}$ and $\mathcal{H}$ is 0.
   (d) If $\phi \in L^\infty(\mu)$ and $\lambda \in \sigma(\phi(N))$, there is a sequence of unit vectors $\{h_n\}$ in $\mathcal{H}$ such that $\|[\phi(N) - \lambda]h_n\| \to 0$.

6. (Keough [1981]) Suppose S satisfies the conditions of Theorem 12.11. If S is irreducible and $S^*S - SS^*$ is compact, then $C^*(S) = \{T_u + C : u \in C(\sigma(N)) \text{ and } C \in \mathscr{B}_0(\mathscr{H})\}$.

§13 Unitary equivalence, similarity, and quasisimilarity. We have already seen that unitarily equivalent subnormal operators have unitarily equivalent minimal normal extensions. This furnishes a necessary condition for unitary equivalence, but it is far from sufficient. For example, if μ is the restriction of area measure to $\mathbb{D}$, let $\mathscr{H}_1$ be the closed linear span of $\{z^n|z|^m : n, m \geq 0\}$ in $L^2(\mu)$ and let $\mathscr{H}_2 = P^2(\mu)$. If $S_j = N_\mu|\mathscr{H}_j$, then S_1 and S_2 have the same minimal normal extensions but they are not unitarily equivalent. In fact, one is quasinormal and the other is not.

The following result for rationally cyclic subnormal operators is the extent of our knowledge of unitary equivalence for subnormal operators.

13.1 PROPOSITION. *For $i = 1, 2$, let S_i be multiplication by z on $R^2(K, \mu_i)$. The operators S_1 and S_2 are unitarily equivalent if and only if the following conditions are satisfied.*

(a) $[\mu_1] = [\mu_2]$.
(b) *There is a function f in $R^2(K, \mu_2)$ such that f is a rationally cyclic vector for S_2, $|f|^2 = d\mu_1/d\mu_2$, $1/f$ belongs to $R^2(K, \mu_1)$, and $1/f$ is a rationally cyclic vector for S_1.*

PROOF. Suppose there is an isomorphism $U : R^2(K, \mu_1) \to R^2(K, \mu_2)$ such that $US_1U^{-1} = S_2$. By Proposition 2.5 there is an isomorphism $V : L^2(\mu_1) \to L^2(\mu_2)$ such that $V|R^2(K, \mu_1) = U$ and $VN_{\mu_1}V^{-1} = N_{\mu_2}$. This proves that $[\mu_1] = [\mu_2]$; let $u = d\mu_1/d\mu_2$ and define $W : L^2(\mu_1) \to L^2(\mu_2)$ by $Wf = u^{1/2}f$. It is easy to see that W is also an isomorphism and $WN_{\mu_1}W^{-1} = N_{\mu_2}$. Hence $W^{-1}V$ is a unitary operator in $\{N_{\mu_1}\}'$. Therefore, there is a function v in $L^\infty(\mu_1)$ with $|v| \equiv 1$ and $W^{-1}V = v(N_{\mu_1})$, multiplication by v on $L^2(\mu_1)$. But this implies that $Vh = vu^{1/2}h$ for all h in $L^2(\mu_1)$. In particular, $f \equiv U(1) = V(1) = vu^{1/2} \in R^2(K, \mu_2)$ and $|f|^2 = u$. Since 1 is a Rat(K) cyclic vector for S_1, it follows that f is a Rat(K) cyclic vector for S_2.

It is easy to see that $Ur = rf$ for all rational functions r with poles off K. If $h \in R^2(K, \mu_1)$, let $\{r_n\}$ be a sequence of such rational functions such that $\int |h - r_n|^2 d\mu_1 \to 0$. By passing to a subsequence if necessary, it may be assumed that $r_n(z) \to h(z)$ a.e. $[\mu_1]$. Hence, $Ur_n = r_nf \to hf$ a.e. $[\mu_1]$. But we also have that $Ur_n \to Uh$ in norm. Hence, $Uh = fh$ for all h in $R^2(K, \mu_1)$. Since the function 1 belongs to the range of U, it follows that $1/f \in R^2(K, \mu_1)$. It is easy to see that $1/f$ is a rationally cyclic vector for S_1.

The proof of the converse is left to the reader. □

Perhaps by changing the concept of equivalence to something weaker than unitary equivalence more can be said. Granted, if we know that two subnormal operators are equivalent in some way that is weaker than unitary equivalence, this information is less valuable. Therefore, when the equivalence is weakened, some effort must be devoted to deriving the consequences of this equivalence for subnormal operators.

13.2 DEFINITION. If A_1, A_2 are operators on Hilbert spaces $\mathscr{H}_1$, $\mathscr{H}_2$, then A_1 is *similar* to A_2 (in symbols, $A_1 \approx A_2$) if there is an invertible operator $X : \mathscr{H}_1 \to \mathscr{H}_2$ such that $XA_1 = A_2 X$. The operators A_1 and A_2 are *quasisimilar* (in symbols, $A_1 \sim A_2$) if there are operators $X_{12} : \mathscr{H}_2 \to \mathscr{H}_2$ and $X_{21} : \mathscr{H}_1 \to \mathscr{H}_2$ that satisfy the following conditions for $i, j = 1, 2$:

(a) $\ker X_{ij} = 0$ and $\operatorname{ran} X_{ij}$ is dense;
(b) $X_{ij} A_j = A_i X_{ij}$.

(Take X_{11} and X_{22} to be the identity operators.)

An operator that satisfies condition (a) is called *quasi-invertible* or a *quasiaffinity*.

It is not difficult to show that similarity and quasisimilarity are equivalence relations and both are weaker than unitary equivalence. Moreover, similarity implies quasisimilarity. In a certain sense these relations are unnatural for the study of subnormal operators. It is possible for a subnormal operator, or even a normal operator, to be similar to an operator that is not subnormal. Also, even though it is easy to see that similarity preserves the spectrum and the various parts of the spectrum, the same cannot be said about quasisimilarity. In spite of all these obstacles, we persist.

One thing was already established before quasisimilarity was even introduced. If two normal operators are quasisimilar, then they are unitarily equivalent. In fact, if N_1 and N_2 are normal operators and X is any quasi-invertible operator such that $N_1 X = X N_2$, then Proposition IX.6.10 of [ACFA] implies that $N_1 \cong N_2$.

Criteria for two subnormal operators to be quasisimilar are still sparse, but there is a growing body of literature about the consequences of quasisimilarity for subnormal operators. We begin with the fact that two quasisimilar hyponormal operators have the same spectrum.

13.3 LEMMA. *If A is hyponormal and $\{h_n\}$ is a sequence of vectors in $\mathscr{H}$ such that $Ah_{n+1} = h_n$, then either $\|h_0\| \geq \|h_1\| \geq \cdots$ or $\|h_n\| \to \infty$.*

PROOF. If $h \in \mathscr{H}$, then $\|Ah\|^2 = \langle Ah, Ah \rangle = \langle A^*Ah, h \rangle \leq \|A^*Ah\| \|h\| \leq \|A^2 h\| \|h\| \leq \frac{1}{4}[\|A^2 h\| + \|h\|]^2$. Letting $h = h_{n+2}$ we get $\|h_{n+1}\| \leq \frac{1}{2}[\|h_n\| + \|h_{n+2}\|]$. The completion of the proof is now an exercise. □

13.4 THEOREM (Clary [1975]). *If A is a hyponormal operator, B is any operator, and X is a quasi-invertible operator such that $XB = AX$, then $\sigma(A) \subseteq \sigma(B)$.*

PROOF. It suffices to show that if B is invertible, then so is A. Let $c = \|B^{-1}\|$ and for $n \geq 1$, let $h_n = c^{-n} X B^{-n} h$. Thus, $cAh_{n+1} = c^{-n} AXB^{-n-1} h = c^{-n} XBB^{-n-1} h = c^{-n} XB^{-n} h = h_n$. Hence, the preceding lemma applies (with cA replacing A) to the sequence $\{h_n\}$. But $\|h_n\| \leq c^{-n} \|X\| \|B^{-n}\| \|h\| \leq \|X\| \|h\|$ and so $\|h_n\|$ cannot converge to ∞. Thus, the preceding lemma implies that $\|Xh\| = \|h_0\| \geq \|h_1\| = c^{-1} \|XB^{-1}h\|$ for any vector h in $\mathscr{H}$. Therefore if $h \in \mathscr{H}$, $\|AXB^{-1}h\| = \|Xh\| \geq c^{-1} \|XB^{-1}h\|$. But this says that A is bounded below on ran X. Since ran X is dense, A is bounded below. But ran $A = A\mathscr{H} \supseteq AX\mathscr{H} = XB\mathscr{H} = \text{ran } X$, so ran A is also dense. Therefore, A is invertible. □

13.5 COROLLARY. *Quasisimilar hyponormal operators have equal spectra.*

If two subnormal operators are quasisimilar, what is the relation between their normal and pure parts, if any? Before answering this, we need a lemma which is an operator theoretic version of the Cantor-Bernstein Theorem. The proof is left as an exercise.

13.6 LEMMA (Kadison and Singer [1957]). *If A_1, A_2 are operators on $\mathscr{H}_1$, $\mathscr{H}_2$ and $\mathscr{L}_1$, $\mathscr{L}_2$ are reducing subspaces for these operators such that $A_1 \cong A_2 | \mathscr{L}_2$ and $A_2 \cong A_1 | \mathscr{L}_1$, then $A_1 \cong A_2$.*

13.7 PROPOSITION (Conway [1980]). *For $j = 1, 2$, let S_j be a subnormal operator on $\mathscr{H}_j$ and suppose $\mathscr{H}_j = \mathscr{M}_j \oplus \mathscr{N}_j$, where $\mathscr{M}_j$ and $\mathscr{N}_j$ reduce S_j, $M_j = S_j | \mathscr{M}_j$ is normal, and $S_j | \mathscr{N}_j$ is pure. If $S_1 \sim S_2$, then $M_1 \cong M_2$.*

PROOF. Let X_{ij} be quasi-invertible operators such that $X_{ij} S_j = S_i X_{ij}$. It follows that $\mathscr{L}_2 \equiv \text{cl}[X_{21} \mathscr{M}_1] \in \text{Lat } S_2$, $X \equiv X_{21} | \mathscr{M}_1 : \mathscr{M}_1 \to \mathscr{L}_2$ is quasi-invertible, and $XM_1 = (S_2 | \mathscr{L}_2) X$. By Proposition 10.6, $S_2 | \mathscr{L}_2$ is normal and $M_1 \cong S_2 | \mathscr{L}_2$. Hence, $\mathscr{L}_2 \subseteq \mathscr{M}_2$ and $M_1 \cong M_2 | \mathscr{L}_2$. Similarly, $\mathscr{L}_1 \equiv \text{cl}[X_{12} \mathscr{M}_2] \subseteq \mathscr{M}_1$ and $M_2 \cong M_1 | \mathscr{L}_1$. The result now follows by the preceding lemma. □

Note that the proof of the preceding proposition shows that $X_{ij} \mathscr{M}_j \subseteq \mathscr{M}_i$. We will use this fact later.

What about the pure parts of the quasisimilar subnormal operators? Unfortunately, they need not be quasisimilar.

13.8 EXAMPLE. Define $\mathscr{F} = H^2 \oplus H^2 \oplus \cdots$ and $\mathscr{G} = L^2(0, \tfrac{1}{3}) \oplus L^2(0, \tfrac{1}{3}) \oplus \cdots$, let $\Delta = \{z \in \mathbb{C} : |z| < \tfrac{1}{2}\}$, and put

$$\mathscr{H}_1 = \mathscr{F} \oplus \mathscr{G} \oplus L_a^2(\Delta), \quad \mathscr{H}_2 = \mathscr{F} \oplus \mathscr{G}.$$

The function on $(0, 1/3)$ defined by $t \to t$ is denoted by x. If $f = (f_1, f_2, \ldots) \in \mathscr{F}$, $g = (g_1, g_2, \ldots) \in \mathscr{G}$, and $h \in L_a^2(\Delta)$, define $S_j : \mathscr{H}_j \to \mathscr{H}_j$ by

$$S_1(f \oplus g \oplus h) = zf \oplus xg \oplus zh,$$
$$S_2(f \oplus g) = zf \oplus xg.$$

Define $X_{ij}: \mathscr{H}_j \to \mathscr{H}_i$ by

$$X_{21}(f \oplus g \oplus h) = f \oplus \left[h \bigg| \left(0, \frac{1}{3}\right) \oplus g_1 \oplus g_2 \oplus \cdots \right],$$

$$X_{12}(f \oplus g) = (f_2, f_3, \ldots) \oplus g \oplus f_1 | \Delta.$$

(Here $f_1|\Delta$ means the restriction of the function f_1 from H^2 to $\Delta \subseteq \mathbb{D}$ by means of analytic bounded point evaluations. See Proposition I.4.7 and §7.) Each of the above operators is well defined. It is easy to check using direct observations and the Closed Graph Theorem that they are also bounded operators. It is left to the reader to verify that X_{ij} is quasi-invertible and $X_{ij}S_j = S_i X_{ij}$ ($i, j = 1, 2$).

Let T_j be the pure part of S_j acting on $\mathscr{L}_j$. So

$$\mathscr{L}_1 = \mathscr{F} \oplus L_a^2(\Delta) \text{ and } \mathscr{L}_2 = \mathscr{F}.$$

It is claimed that T_1 and T_2 are not quasisimilar. In fact, suppose there is a quasi-invertible operator $Z: \mathscr{L}_1 \to \mathscr{L}_2$ with $ZT_1 = T_2 Z$. Put $\mathscr{M} = \text{cl}[Z((0) \oplus L_a^2(\Delta))] \subseteq \mathscr{L}_2 = \mathscr{F}$. Thus, $\mathscr{M} \in \text{Lat } T_2$ and $A_2 = T_2|\mathscr{M}$ is an isometry. If $W: L_a^2(\Delta) \to \mathscr{M}$ is defined by $Wh = Z(0 \oplus h)$, W is quasi-invertible and $WA_1 = A_2 W$, where A_1 is the Bergman operator for Δ. But this contradicts Theorem 13.4.

This example is from Conway [**1980**]; another example can be found in Hastings [**1979a**].

The fact that infinite direct sums of subnormal operators were taken to construct the above example does not seem to be an accident. At least in the case of cyclic subnormal operators the situation can be improved.

13.9 LEMMA. *Let S_1, S_2 be subnormal operators on $\mathscr{H}_1$, $\mathscr{H}_2$ and let $X_{ij}: \mathscr{H}_j \to \mathscr{H}_i$ be bounded operators with dense range such that $X_{ij}S_j = S_i X_{ij}$. If S_1 is cyclic, then S_2 is cyclic and each X_{ij} is injective. In particular, $S_1 \sim S_2$.*

PROOF. If p is a polynomial, then $X_{21}p(S_1) = p(S_2)X_{21}$. So if e_1 is a cyclic vector for S_1, then $e_2 = X_{21}e_1$ is a cyclic vector for S_2. Thus, we may assume that S_j is multiplication by z on $P^2(\mu_j)$, where μ_j is a compactly supported measure on $\mathbb{C}$.

An easy algebraic manipulation shows that $X_{12}X_{21} \in \{S_1\}'$. Thus, Yoshino's Theorem implies there is a function ϕ_1 in $P^2(\mu_1) \cap L^\infty(\mu_1)$ such that $X_{12}X_{21} = \phi_1(S_1)$. Since each X_{ij} has dense range, so does $\phi_1(S_1)$. Therefore, ϕ_1 cannot vanish on a set of positive μ_1 measure. Indeed, if it did vanish on Δ and $\mu_1(\Delta) > 0$, then since $\phi_1 P^2(\mu_1)$ is dense in $P^2(\mu_1)$ it follows that $P^2(\mu_1) \subseteq L^2(\mu_1|(\mathbb{C} \setminus \Delta))$, contradicting the fact that N_{μ_1} is the minimal normal extension of S_{μ_1}. But since $|\phi_1| > 0$ a.e. $[\mu_1]$, $\phi_1(S_1)$ must be injective. But $\ker X_{21} \subseteq \ker \phi_1(S_1) = (0)$, so X_{21} is injective. Similarly, X_{12} is injective. □

§13 EQUIVALENCE

13.10 PROPOSITION. *If S_1, S_2 are cyclic subnormal operators on $\mathcal{H}_1, \mathcal{H}_2, S_j = M_j \oplus T_j$ on $\mathcal{H}_j = \mathcal{M}_j \oplus \mathcal{N}_j$, where M_j is normal and T_j is pure, then $S_1 \sim S_2$ if and only if $M_1 \cong M_2$ and $T_1 \sim T_2$.*

PROOF. Half of the proposition is clear. So assume that $S_1 \sim S_2$ and let $X_{ij} : \mathcal{H}_j \to \mathcal{H}_i$ be quasi-invertible operators such that $X_{ij}S_j = S_i X_{ij}$. Let Q_j be the projection of $\mathcal{H}_j$ onto $\mathcal{N}_j$, and define $Y_{ij} : \mathcal{N}_j \to \mathcal{N}_i$ by $Y_{ij} = Q_i X_{ij} | \mathcal{N}_j$.

If $f \in \mathcal{N}_i$, then $f = \lim X_{ij} h_n$ for some sequence $\{h_n\}$ in $\mathcal{H}_j$. Let $h_n = m_n + g_n$, where $m_n \in \mathcal{M}_j$ and $g_n \in \mathcal{N}_j$. Now the proof of Proposition 13.7 shows that $X_{ij} \mathcal{M}_j \subseteq \mathcal{M}_i$; so $f = Q_i f = \lim Q_i X_{ij} h_n = \lim Q_i X_{ij} g_n = \lim Y_{ij} g_n$ and hence Y_{ij} has dense range. Since $Q_i \in \{S_i\}'$, it is easy to check that $Y_{ij} T_j = T_i Y_{ij}$. But S_j is cyclic, so the preceding lemma implies that Y_{ij} is injective and $T_1 \sim T_2$. □

The next result says that if you want to study quasisimilarity for cyclic subnormal operators, it is possible to represent the relevant operators in a rather pleasant and simple way.

13.11 PROPOSITION (Conway [1980]). *If S_1 and S_2 are cyclic subnormal operators, then $S_1 \sim S_2$ if and only if there are compactly supported measures μ_1 and μ_2 such that $S_1 \cong S_{\mu_1}$, $S_2 \cong S_{\mu_2}$, and there are constants c_1 and c_2 and a function ϕ in $P^2(\mu_1) \cap L^\infty(\mu_1)$, such that the following hold.*

(a) *$\{\phi p : p \text{ is a polynomial}\}$ is dense in $P^2(\mu_1)$.*
(b) *For every polynomial p,*

$$c_2 \int |\phi|^2 |p|^2 \, d\mu_1 \le \int |p|^2 \, d\mu_2 \le c_1 \int |p|^2 \, d\mu_1.$$

PROOF. Suppose $S_1 \sim S_2$ and let $Y_{ij} : \mathcal{H}_j \to \mathcal{H}_i$ be quasi-invertible operators such that $Y_{ij} S_j = S_i Y_{ij}$. If e_1 is a cyclic vector for S_1, then it follows that $e_2 = Y_{21} e_1$ is a cyclic vector for S_2. Choose compactly supported measures μ_1 and μ_2 such that there are isomorphisms $U_j : \mathcal{H}_j \to P^2(\mu_j)$ with $U_j e_j = 1$ and $U_j S_j U_j^{-1} = S_{\mu_j}$. Put $X_{ij} = U_i Y_{ij} U_j^{-1}$. Thus $X_{ij} : P^2(\mu_j) \to P^2(\mu_i)$ is quasi-invertible and it is straightforward to verify that $X_{ij} S_{\mu_j} = S_{\mu_i} X_{ij}$.

If p is a polynomial, then $X_{21} p = X_{21} p(S_{\mu_1})(1) = p(S_{\mu_2}) X_{21}(1) = p(S_{\mu_2}) U_2 Y_{21} e_1 = p(S_{\mu_2})(1) = p$. If $c_1 = \|X_{21}\|^2$, this shows that $\int |p|^2 d\mu_2 \le c_1 \int |p|^2 d\mu_1$.

To find the constant c_2, notice that $X_{12} X_{21}$ commutes with S_{μ_1}. By Yoshino's Theorem there is a ϕ in $P^2(\mu_1) \cap L^\infty(\mu_1)$ such that $X_{12} X_{21} f = \phi f$ for all f in $P^2(\mu_1)$. If $c_2 = \|X_{12}\|^{-2}$ it follows that the other inequality is valid.

The converse is left to the reader. □

The study of quasisimilarity is still quite young. Raphael [1982] showed that quasisimilar cyclic subnormal operators have equal essential spectra, while L. R. Williams [1980b] showed that quasisimilar quasinormal operators have the same essential spectrum. Recently these results were generalized to the following, whose proof will not be given though a proof for the cyclic case will be presented in §VIII.6 after we have proved Thomson's Theorem.

13.12 THEOREM (Yang [1990]). *Two quasisimilar subnormal operators have the same essential spectrum.*

It remains unknown whether quasisimilar hyponormal operators have equal essential spectra.

Notice that Example 13.8 shows that two quasisimilar subnormal operators need not have the same approximate point spectrum. Clary [1973] contains an easier example illustrating this.

Another open question about quasisimilar subnormal operators is as follows.

13.13 OPEN QUESTION. Do quasisimilar subnormal operators have naturally isomorphic commutants?

In the case of rationally cyclic subnormal operators, this was answered in the affirmative by Raphael [1986]. Recently McCarthy [**preprint**] gave another proof. For the cyclic case this is proved in §VIII.6 as a consequence of Thomson's Theorem on the existence of a plentiful supply of analytic bounded point evaluations.

Some additional papers on quasisimilarity for subnormal operators are Conway [**1982**], Hastings [**1975, 1978,** and **1979a**], Raphael [**1985b**], and L. R. Williams [**1980a**].

Exercises

1. Prove Lemma 13.6.
2. With the same notation as in Proposition 13.7, show that $S_1 \approx S_2$ if and only if $M_1 \cong M_2$ and $S_1|\mathscr{N}_1 \approx S_2|\mathscr{N}_2$.
3. (Allan Lambert) Two unilateral shifts are quasisimilar if and only if they are similar.
4. Let S be a subnormal unilateral weighted shift with $\|S\| = 1$, and let ν be the corresponding probability measure on $[0, 1]$ (see Theorem 6.10). Show that S is similar to the unilateral shift if and only if $\nu(\{1\}) > 0$.
5. (Hoover [1972]) Quasisimilar isometries are unitarily equivalent.
6. (Hoover [1972]) If $\{S_i\}$ and $\{T_i\}$ are collections of operators such that $S_i \sim T_i$ for all i, then $\oplus_i S_i \sim \oplus_i T_i$.
7. (Hoover [1972]) Give an example of two quasisimilar operators with unequal spectra.
8. Give an example of a non-hyponormal operator that is quasisimilar to a normal operator.
9. State and prove a result for similar rationally cyclic subnormal operators analogous to Proposition 13.11.

CHAPTER III

Function Theory on the Unit Circle

In this chapter we will focus on the Hardy spaces H^p. Our initial aim is to better understand the unilateral shift. As was said after Beurling's Theorem (I.4.12), even though we formally have characterized the invariant subspaces of the shift as the subspaces of H^2 having the form ϕH^2 for some ϕ in H^∞ with $|\phi| \equiv 1$, until we fully understand the structure of these functions in H^∞ we cannot truthfully say that we understand Beurling's Theorem.

As it turns out, understanding H^∞ will also help us to get more information about the general subnormal operator.

In this chapter there won't be an attempt to attribute the results. For a fuller account of this area together with historical attributions, the reader can consult Duren [**1970**], Hoffman [**1962**], and Koosis [**1980**].

§1 Cesàro sums. Recall from §I.4 that for any formal Fourier series $\sum_{n=-\infty}^{\infty} c_n z^n$, z in $\partial \mathbb{D}$, $s_n(z)$ is the nth partial sum of the series, $s_n(z) = \sum_{k=-n}^{n} c_n z^n$. The nth *Cesàro means* of the series is defined for z in $\partial \mathbb{D}$ by

$$\sigma_n(z) = \frac{1}{n} \sum_{k=0}^{n-1} s_k(z).$$

If μ is a measure (that is, a complex valued regular Borel measure) on $\partial \mathbb{D}$, a formal Fourier series can also be associated with μ. Namely, let

$$\hat{\mu}(n) = \int \bar{z}^n \, d\mu(z)$$

and call $\{\hat{\mu}(n)\}$ the Fourier coefficients of μ. This gives rise to the Fourier series $\sum_{n=-\infty}^{\infty} \hat{\mu}(n) z^n$.

Now if $f \in L^1$, and $d\mu = f \, dm$ (m is normalized Lebesgue measure on $\partial \mathbb{D}$), then $\hat{\mu}(n) = \hat{f}(n)$ for all n and so this notion of the Fourier series associated with a measure generalizes that of a Fourier series associated with a function.

By the nth partial sum and the nth Cesàro means for the measure μ, we mean the corresponding quantity for the associated series. To indicate the dependence on μ, these sums are denoted by $s_n(\mu, z)$ and $\sigma_n(\mu, z)$. Note that each Cesàro sum is a trigonometric polynomial.

For consistency, $M = M(\partial\mathbb{D})$, the Banach space of all complex valued measure on $\partial\mathbb{D}$, and $C = C(\partial\mathbb{D})$, the space of continuous functions on $\partial\mathbb{D}$. Recall that $M = C^*$, and hence M has a natural weak* topology.

Here is the main result of this section.

1.1 THEOREM

(a) If $f \in L^p$, $1 \leq p < \infty$, the Cesàro sums for f converge to f in the L^p norm.
(b) If $f \in C$, the Cesàro sums converge to f uniformly on $\partial\mathbb{D}$.
(c) If $f \in L^\infty$, the Cesàro sums converge to f in the weak* topology of L^∞.
(d) If $\mu \in M$, the Cesàro sums converge to μ in the weak* topology of M.

The proof will be obtained by a recourse to operator theory on a Banach space. If $\mathscr{X}$ is one of the Banach spaces under consideration (that is, $\mathscr{X} = L^p, C$, or M), define $\sigma_n : \mathscr{X} \to \mathscr{X}$ by letting $\sigma_n(x) =$ the nth Cesàro sum of x. (If $\mathscr{X} = M$ and $\mu \in M$, then $\sigma_n(\mu)$ is the measure that is absolutely continuous with respect to m whose Radon-Nikodym derivative is the trigonometric polynomial $\sigma_n(\mu)$.) To prove the theorem, it must be shown that if $\mathscr{X} = L^p$, $1 \leq p < \infty$, or C, then $\|\sigma_n(x) - x\| \to 0$ and if $\mathscr{X} = L^\infty$ or M, then $\sigma_n(x) \to x$ weak* for every x in $\mathscr{X}$. Actually, we will see that the last part follows from the first part and a duality argument. But first we will see that σ_n is actually an integral operator.

If $f \in L^1$ with Fourier series $\sum_{n=-\infty}^{\infty} c_n z^n$, then

$$s_n(f, z) = \sum_{k=-n}^{n} \left[\int f(w)\overline{w}^k\, dm(w)\right] z^k$$

$$= \int f(w) \left[\sum_{k=-n}^{n} (\overline{w}z)^k\right] dm(w).$$

Therefore

(1.2(i)) $$\sigma_n(f, z) = \int f(w) K_n(\overline{w}z)\, dm(w),$$

where K_n is the nth Cesàro mean of the series $\sum_{n=-\infty}^{\infty} \zeta^n$. The same type of formula holds for a measure μ:

(1.2(ii)) $$\sigma_n(\mu, z) = \int K_n(\overline{w}z)\, d\mu(w).$$

To properly study σ_n, we need to get a better hold on the kernel K_n. To do this, let's first look at the nth partial sum of the series $\sum_{n=-\infty}^{\infty} \zeta^n$. So if

$\zeta = e^{i\theta}$,

$$s_n(\zeta) = \sum_{k=-n}^{n} \zeta^k$$

$$= \sum_{k=0}^{n} \zeta^k + \sum_{k=1}^{n} \overline{\zeta}^k$$

$$= \frac{1 - \zeta^{n+1}}{1 - \zeta} + \frac{1 - \overline{\zeta}^{n+1}}{1 - \overline{\zeta}} - 1$$

$$= \frac{\operatorname{Re}\zeta^n - \operatorname{Re}\zeta^{n+1}}{1 - \operatorname{Re}\zeta}$$

$$= \frac{\cos n\theta - \cos(n+1)\theta}{1 - \cos\theta}.$$

From here it follows that

(1.3) $$K_n(\zeta) = \frac{1}{n}\left[\frac{1 - \operatorname{Re}\zeta^n}{1 - \operatorname{Re}\zeta}\right] = \frac{1}{n}\left[\frac{1 - \cos n\theta}{1 - \cos\theta}\right].$$

1.4 LEMMA. *For each $n \geq 1$, $K_n \geq 0$, $K_n(\zeta) = K_n(\overline{\zeta})$, and $\int K_n\, dm = 1$.*

PROOF. Applying the half angle formulas from trigonometry to (1.3), it follows that $K_n(\zeta) = n^{-1}[(\sin(n\theta/2))/\sin(\theta/2)]^2 \geq 0$. It is also clear from (1.3) that the second part is valid. For the last part, use (1.2) with $f \equiv 1$. □

Before stating the next lemma, you should prick your consciousness and be aware that you are dealing with Banach space duality below and so there is an absence of a complex conjugate in the various expressions for the inner product $\langle \cdot, \cdot \rangle$. (Maybe we should call this an outer product.)

1.5 LEMMA.

(a) *If $1 \leq p < \infty$ and q is the index dual to p, then the adjoint of the operator $\sigma_n: L^p \to L^p$ is the operator $\sigma_n: L^q \to L^q$.*
(b) *The adjoint of the operator $\sigma_n: C \to C$ is the operator $\sigma_n: M \to M$.*

PROOF. Only part (a) will be proved. So let $f \in L^p$ and $g \in L^q$. By interchanging the order of integration and using the preceding lemma, it follows that

$$\langle \sigma_n(f), g \rangle = \int g(z)\left[\int f(w) K_n(\overline{w}z)\, dm(w)\right] dm(z)$$

$$= \int f(w)\left[\int g(z) K_n(\overline{w}z)\, dm(z)\right] dm(w)$$

$$= \int f(w)\left[\int g(z) K_n(\overline{z}w)\, dm(z)\right] dm(w)$$

$$= \langle f, \sigma_n(g) \rangle. \quad \square$$

We can now state a general Banach space result that, when combined with the preceding lemma, will show how parts (c) and (d) of Theorem 1.1 follow from parts (a) and (b).

1.6 PROPOSITION. *Let $\mathscr{X}$ be a Banach space and let $\{T_n\}$ be a sequence of bounded operators from $\mathscr{X}$ into $\mathscr{X}$.*

(a) *If $\sup_n \|T_n\| < \infty$, D is a dense subset of $\mathscr{X}$, and $\|T_n x - x\| \to 0$ for all x in D, then $\|T_n x - x\| \to 0$ for all x in $\mathscr{X}$.*

(b) *If $\|T_n(x) - x\| \to 0$ for all x in $\mathscr{X}$, then for every x^* in $\mathscr{X}^*$, $T_n^* x^* \to x^*$ weak* in $\mathscr{X}^*$.*

PROOF. (a) If $x \in \mathscr{X}$ and $\varepsilon > 0$, let $y \in D$ such that $\|x - y\| < \min(\varepsilon/2c, \varepsilon/2)$, where $c = \sup_n \|T_n\|$. Then $\|T_n x - x\| \le \|T_n(x - y)\| + \|T_n y - y\| + \|y - x\| \le 2\varepsilon/3 + \|T_n y - y\|$. This can be made less than ε for all sufficiently large n.

(b) This is easy: $|\langle x, T_n^* x^* - x^* \rangle| = |\langle T_n x - x, x^* \rangle| \le \|T_n x - x\| \|x^*\| \to 0$. $\square$

We can now prove the main theorem.

PROOF OF THEOREM 1.1. (a) Let $f \in L^p$ and let $g \in L^q$, where q is dual to p. First note that by a change of variables, $\sigma_n(f, z) = \int f(w) K_n(\overline{w}z) \, dm(w) = \int f(\overline{\zeta}z) K_n(\zeta) \, dm(\zeta)$. Hence

$$|\langle \sigma_n(f), g \rangle| = \left| \int g(z) \left[\int f(w) K_n(\overline{w}z) \, dm(w) \right] dm(z) \right|$$

$$= \left| \int g(z) \left[\int f(\overline{\zeta}z) K_n(\zeta) \, dm(\zeta) \right] dm(z) \right|$$

$$= \left| \int K_n(\zeta) \left[\int g(z) f(\overline{\zeta}z) \, dm(z) \right] dm(\zeta) \right|$$

$$\le \int K_n(\zeta) \int |g(z) f(\overline{\zeta}z)| \, dm(z) \, dm(\zeta).$$

Applying Hölder's inequality and using the fact that $\int f(\overline{\zeta}z) \, dm(z) = \int f \, dm$, we get

$$|\langle \sigma_n(f), g \rangle| \le \int K_n(\zeta) \|g\|_q \|f\|_p \, dm(\zeta)$$

$$\le \|g\|_q \|f\|_p$$

since $\int K_n \, dm = 1$.

Hence $\|\sigma_n\| \le 1$ for all n. It is easy to check that for any integer k, $\sigma_n(z^k) \to z^k$ as $n \to \infty$. Thus for a trigonometric polynomial f, $\sigma_n(f) \to f$ as $n \to \infty$. Since the trigonometric polynomials are dense in L^p for all finite p, part (a) follows from part (a) of the preceding proposition.

(b) This is easier than part (a). If $f \in C$, then $|\sigma_n(f, z)| \le \int |f(w)| K_n(\overline{w}z) \, dm(w) \le \|f\|_\infty$ and so $\|\sigma_n\| \le 1$.

As mentioned before, parts (c) and (d) follow from parts (a) and (b) via the second part of the preceding proposition. $\square$

1.7 COROLLARY. (a) *If $1 \le p < \infty$, the analytic polynomials are dense in H^p. In fact, if $f \in H^p$, there is a sequence of polynomials $\{p_n\}$ such that $\|p_n\|_p \le \|f\|_p$ and $p_n \to f$.*

(b) *The analytic polynomials are weak* * *sequentially dense in* H^∞. *In fact, if* $f \in H^\infty$, *there is a sequence of polynomials* $\{p_n\}$ *such that* $\|p_n\|_\infty \leq \|f\|_\infty$ *and* $p_n \to f$ *weak* *.

Exercises

1. For any function f and w in $\partial \mathbb{D}$, let $f_w: \partial \mathbb{D} \to \mathbb{C}$ be defined by $f_w(z) = f(\overline{w}z)$.
   (a) If $1 \leq p < \infty$, show that the map $w \to f_w$ is a continuous function from $\partial \mathbb{D} \to L^p$.
   (b) Show that $w \to f_w$ is a continuous map from $\partial \mathbb{D} \to C$.
   (c) Show that $w \to f_w$ is a continuous map from $\partial \mathbb{D}$ into L^∞ with the weak* topology.
   (d) If $\mu \in M$ and $\mu_w(\Delta) = \mu(\overline{w}\Delta)$, then $\mu_w \in M$ and the map $w \to \mu_w$ is a continuous map from $\partial \mathbb{D}$ into M with its weak* topology.
2. Compute the Cesàro means of δ_1, the unit point mass at 1.
3. If $\mu \in M(\partial \mathbb{D})$ and $\hat{\mu}(n) = 0$ for all n in $\mathbb{Z}$, then $\mu = 0$.

§2 Convolution on the circle. If μ and ν are two measures on $\partial \mathbb{D}$, define $L: C(\partial \mathbb{D}) \to \mathbb{C}$ by

$$(2.1) \qquad L(f) = \int_{\partial \mathbb{D}} \int_{\partial \mathbb{D}} f(zw) \, d\mu(z) \, d\nu(w)$$

for all f in C. The proof of the next proposition is straightforward.

2.2 PROPOSITION. *If* $L: C \to \mathbb{C}$ *is defined as in* (2.1), *then* L *is a bounded linear functional and* $\|L\| \leq \|\mu\| \|\nu\|$.

Since this linear functional L is bounded, the Riesz Representation Theorem implies that there is a unique measure on $\partial \mathbb{D}$ corresponding to it.

2.3 DEFINITION. *If* μ *and* $\nu \in M$, *then* $\mu * \nu$ *is the unique measure such that*

$$\int f \, d\mu * \nu = \int_{\partial \mathbb{D}} \int_{\partial \mathbb{D}} f(zw) \, d\mu(z) \, d\nu(w)$$

for all f in C. The measure $\mu * \nu$ is called the *convolution* of μ and ν.

Here are some properties of the convolution; their verification is left to the reader.

2.4 PROPOSITION. *If* $\mu, \nu,$ *and* $\eta \in M$, *the following hold.*
   (a) $\mu * \nu = \nu * \mu$ *and* $\|\mu * \nu\| \leq \|\mu\| \|\nu\|$.
   (b) *If* μ *and* ν *are positive, then* $\mu * \nu \geq 0$.
   (c) $(\mu * \nu) * \eta = \mu * (\nu * \eta)$.
   (d) $\mu * (\nu + \eta) = \mu * \nu + \mu * \eta$.
   (e) *If* δ_1 *is the unit point mass at* 1, *then* $\delta_1 * \mu = \mu = \mu * \delta_1$.

There is an equivalent way to define $\mu * \nu$ as a function defined on the Borel subsets of $\partial \mathbb{D}$. See Exercise 3.

2.5 PROPOSITION. *If μ and $\nu \in M$, then $\widehat{\mu * \nu}(n) = \hat{\mu}(n)\hat{\nu}(n)$ for every n in $\mathbb{Z}$.*

PROOF. In fact, $\widehat{\mu * \nu}(n) = \int \bar{z}^n \, d\mu * \nu = \iint \bar{z}^n \bar{w}^n \, d\mu(z) \, d\nu(w) = (\int \bar{z}^n \, d\mu(z))(\int \bar{w}^n \, d\nu(w)) = \hat{\mu}(n)\hat{\nu}(n)$. □

2.6 PROPOSITION. *If f and $g \in L^1$, $\mu = fm$, $\nu = gm$, and $h \colon \partial \mathbb{D} \to \mathbb{C}$ is defined by*
$$h(z) = \int f(z\bar{w})g(w) \, dm(w)$$
for z in $\partial \mathbb{D}$, then $h \in L^1$ and $\mu * \nu = hm$.

PROOF. The proof that h is an L^1 function is left to the reader. If $u \in C$, then
$$\int u \, d\mu * \nu = \iint u(zw)f(z)g(w) \, dm(z) \, dm(w)$$
$$= \int g(w) \left[\int u(zw)f(z) \, dm(z) \right] dm(w).$$

Fix w in $\partial \mathbb{D}$ and substitute $z\bar{w}$ for z in the integral $\int u(zw)f(z) \, dm(z)$. Noting that $m(E) = m(\bar{w}E)$ for any Borel set and any w in $\partial \mathbb{D}$, we get $\int u(zw)f(z) \, dm(z) = \int u(z)f(z\bar{w}) \, dm(z)$. Thus
$$\int u \, d\mu * \nu = \int g(w)[\int u(z)f(z\bar{w}) \, dm(z)] \, dm(w)$$
$$= \int u(z)[\int f(z\bar{w})g(w) \, dm(w)] \, dm(z) = \int u(z)h(z) \, dm(z) \quad □.$$

2.7 DEFINITION. *If f and $g \in L^1$, then the* convolution *of f and g is the function*
$$f * g(z) = \int f(z\bar{w})g(w) \, dm(w).$$

Note that the preceding proposition shows that the definitions of convolution for measures and functions are consistent. Also, the basic algebraic properties for the convolution of two functions can be read off from Proposition 2.4. In particular, it follows from part (a) of Proposition 2.4 that
$$\|f * g\|_1 \leq \|f\|_1 \|g\|_1.$$

Exercises

1. Compute $\hat{m}$ and $\hat{\delta}_1$.
2. If $f \in L^1(\mu * \nu)$, show that $\int f \, d\mu * \nu = \iint f(zw) \, d\mu(z) \, d\nu(w)$.
3. If E is a Borel subset of $\partial \mathbb{D}$, show that $\mu * \nu(E) = (\mu \times \nu)(\{(z, w) \in \partial \mathbb{D} \times \partial \mathbb{D} : zw \in E\})$.
4. If μ and $\nu \in M$ and $\nu \ll m$, then $\mu * \nu \ll m$.

§3 Harmonic functions on the disk. For w in $\partial \mathbb{D}$ and $|z| < 1$, define

$$P_z(w) = \operatorname{Re}\left(\frac{w+z}{w-z}\right)$$
$$= \operatorname{Re}\left(\frac{1+z\overline{w}}{1-z\overline{w}}\right)$$
$$= \sum_{n=0}^{\infty}(z\overline{w})^n + \sum_{n=1}^{\infty}(\overline{z}w)^n.$$

If $z = re^{i\theta}$ and $w = e^{it}$, then this is precisely the *Poisson kernel* (see page 256 in [**FOCV**]) $P_r(\theta - t)$. Here are the properties of this kernel. The reader can consult [**FOCV**] for the proof.

3.1 PROPOSITION. *For $|z| < 1$ and w in $\partial \mathbb{D}$, the following hold.*
(a) $P_z(w) > 0$ and $\int P_z(w)\, dm(w) = 1$.
(b) $P_{\overline{z}}(\overline{w}) = P_z(w)$.
(c) *If* $\zeta \in \partial \mathbb{D}$ *and* $0 < r < 1$, *then* $P_{r\zeta}(w) = P_r(w\overline{\zeta}) = P_{r\zeta\overline{w}}(1)$.
(d) *If* $n \in \mathbb{Z}$, *then* $\int P_z(w) w^n\, dm(w) = z^n$.
(e) *If* $z = re^{i\theta}$, *then* $P_z(e^{it}) = P_r(\theta - t) = \frac{1-r^2}{1-2r\cos(\theta-t)+r^2}$ *and for any* $\delta > 0$

$$\lim_{r \to 1-}[\sup\{P_r(\theta): |\theta| \geq \delta\}] = 0.$$

If $\mu \in M$ and $|z| < 1$, define $\tilde{\mu}(z) = \int P_z(w)\, d\mu(w)$. Similarly, if $f \in L^1$, define $\tilde{f}(z) = \int f P_z\, dm$. These definitions are consistent since $\widetilde{fm} = \tilde{f}$. It is not difficult to prove the following (see [**FOCV**]).

3.2 PROPOSITION. *If* $\mu \in M(\partial \mathbb{D})$, *then $\tilde{\mu}$ is a harmonic function in $\mathbb{D}$.*

Note that we are dealing with complex valued harmonic functions here. If $f \in C(\partial \mathbb{D})$, then we know that $\tilde{f}$ is the solution of the Dirichlet problem with boundary values f (see [**FOCV**]). Indeed, this follows from Theorem 3.5(a) below.

If $u: \mathbb{D} \to \mathbb{C}$ and $0 < r < 1$, define $u_r: \partial \mathbb{D} \to \mathbb{C}$ by $u_r(w) = u(rw)$ for w in $\partial \mathbb{D}$. We will sometimes want to consider the function u_r as defined on $\mathbb{D}$ or $\operatorname{cl}\mathbb{D}$ by the same formula, but no separate notation will be employed.

If $\mu \in M$ and $0 < r < 1$, then $\tilde{\mu}_r$ is an element of $C(\partial \mathbb{D})$ by the preceding proposition. Thus if $1 \leq p \leq \infty$ and $f \in L^p$, $\tilde{f}_r \in L^p$ and so we can define the operator $T_r: L^p \to L^p$ by $T_r f = \tilde{f}_r$; similarly we can define $T_r: C \to C$ by $T_r f = \tilde{f}_r$ and $T_r: M \to M$ by $T_r \mu = \tilde{\mu}_r \cdot m$.

3.3 PROPOSITION
(a) *For $\mathscr{X} = L^p$, C, or M, $T_r: \mathscr{X} \to \mathscr{X}$ is a bounded linear operator with $\|T_r\| \leq 1$ for all r.*
(b) *If $1 \leq p < \infty$ and q is conjugate to p, the dual of the map $T_r: L^p \to L^p$ is the map $T_r: L^q \to L^q$.*

(c) *The dual of the map $T_r: C \to C$ is the map $T_r: M \to M$.*

PROOF. (a) This will only be proved for $\mathscr{X} = L^p$. Let $f \in L^p$ and let $h \in L^q$, where q is the index that is conjugate to p. Thus

$$\begin{aligned}(3.4)\quad \langle T_r f, h \rangle &= \int \tilde{f}_r(\zeta) h(\zeta) \, dm(\zeta) \\ &= \int \left[\int f(w) P_{r\zeta}(w) \, dm(w) \right] h(\zeta) \, dm(\zeta).\end{aligned}$$

Substitute $w\zeta$ for w in this equation and use the following two facts: this change of variables does not change the value of the integral and $P_{r\zeta}(w\zeta) = P_r(w)$ (3.1(c)). This gives

$$|\langle T_r f, h \rangle| = \left| \int \left[\int f(w\zeta) P_r(w) \, dm(w) \right] h\zeta \, dm(\zeta) \right|$$
$$\leq \int P_r(w) \left[\int |f(w\zeta)| |h(\zeta)| \, dm(\zeta) \right] dm(w)$$
$$\leq \|f\|_p \|h\|_q,$$

since $\int P_r(w) \, dm(w) = 1$. This shows that $\|T_r\| \leq 1$ for all r. It is easy to see that T_r is linear.

(b) If $f \in L^p$ and $h \in L^q$, then in (3.4) use the fact that $P_{r\zeta}(w) = P_{r\overline{w}}(\overline{\zeta}) = P_{rw}(\zeta)$ by (3.1(c)) to obtain

$$\langle T_r f, h \rangle = \int \left[\int f(w) P_{rw}(\zeta) \, dm(w) \right] h(\zeta) \, dm(\zeta)$$
$$= \int \left[\int h(\zeta) P_{rw}(\zeta) \, dm(\zeta) \right] f(w) \, dm(w)$$
$$= \langle f, T_r h \rangle.$$

The proof of (c) is similar. □

3.5 THEOREM

(a) *If $f \in C(\partial \mathbb{D})$, then $\|f - \tilde{f}_r\|_{\partial \mathbb{D}} \to 0$ as $r \to 1-$.*
(b) *If $1 \leq p < \infty$ and $f \in L^p(\partial \mathbb{D})$, then $\|f - \tilde{f}_r\|_p \to 0$.*
(c) *If $\mu \in M(\partial \mathbb{D})$ and if μ_r is the element of $M(\partial \mathbb{D})$ defined by $\mu_r = (\tilde{\mu})_r \cdot m$, then $\mu_r \to \mu$ weak* in $M(\partial \mathbb{D})$.*
(d) *If $f \in L^\infty(\partial \mathbb{D})$, then $\tilde{f}_r \to f$ weak* in $L^\infty(\partial \mathbb{D})$.*

PROOF. From (3.1(d)) we have that $T_r z^n = r^n z^n$ for all n in $\mathbb{Z}$. Hence for a trigonometric polynomial f, $T_r f \to f$ uniformly on $\partial \mathbb{D}$ as $r \to 1$. Since $\|T_r\| \leq 1$ for all r, Proposition 1.6(a) implies that (a) and (b) hold. Parts (c) and (d) hold by applying Proposition 1.6(b) and Proposition 3.3(c). □

As was said before, part (a) of the preceding theorem shows that for f in $C(\partial \mathbb{D})$, $\tilde{f}$ has a continuous extension to $\text{cl } \mathbb{D}$, thus solving the Dirichlet problem with boundary values f. We can legitimately consider $\tilde{f}$ for

f in $L^p(\partial \mathbb{D})$ as a solution of the Dirichlet problem with the noncontinuous boundary values f. Indeed, such a perspective is justified by the last theorem.

Now suppose that $u \colon \mathbb{D} \to \mathbb{C}$ is a harmonic function. What are necessary and sufficient conditions that $u = \tilde{\mu}$ for some measure μ? Before providing an answer to this question it is helpful to observe the consequences of a few elementary manipulations with the basic properties of a harmonic function.

If u is harmonic and real valued, there is an analytic function $f \colon \mathbb{D} \to \mathbb{C}$ with $u = \operatorname{Re} f$. Let $f(z) = \sum_n a_n z^n$ be the power series expansion of f in $\mathbb{D}$. So for w in $\partial \mathbb{D}$ and $0 < r < 1$ the series $\sum_n a_n r^n w^n$ converges absolutely and thus

$$u_r(w) = \frac{1}{2}[f(rw) + f(rw)^*]$$
$$= \frac{1}{2}\left[\sum_{n=0}^{\infty} a_n r^n w^n + \sum_{n=0}^{\infty} \overline{a_n} r^n \overline{w}^n\right]$$
$$= \sum_{n=-\infty}^{\infty} c_n r^{|n|} w^n,$$

where $c_0 = \operatorname{Re} a_0$, $c_n = 1/2(a_n)$ for $n > 0$, and $c_n = 1/2(\overline{a_n})$ for $n < 0$. For a complex-valued harmonic function $u \colon \mathbb{D} \to \mathbb{C}$ a consideration of its real and imaginary parts shows that for $|w| = 1$ and $r < 1$,

$$(3.6) \qquad u_r(w) = \sum_{n=-\infty}^{\infty} c_n r^{|n|} w^n$$

for some choice of constants c_n. Moreover, the series (3.6) converges uniformly and absolutely for w in $\partial \mathbb{D}$.

On the other hand, for any harmonic function $u \colon \mathbb{D} \to \mathbb{C}$ and $0 < r < 1$, u_r is a continuous function on $\partial \mathbb{D}$ and hence has a Fourier series. Using (3.6) we can calculate the Fourier coefficients of u_r. Namely, $\hat{u}_r(n) = \int u_r(w) \overline{w}^n \, dm(w) = \sum_k c_k r^{|k|} \int w^k \overline{w}^n \, dm(w) = c_n r^{|n|}$. We have thus discovered half of the following lemma; the other half can be obtained by a consideration of the expansion of $P_z(w)$ in powers of $z\overline{w}$ and $\overline{z}w$.

3.7 LEMMA
 (a) *If $u \colon \mathbb{D} \to \mathbb{C}$ is a harmonic function and $0 < r < 1$, the Fourier series for u_r is given by the formula* (3.6).
 (b) *If $\mu \in M(\partial \mathbb{D})$, then the Fourier series of the function $\tilde{\mu}_r$ is given by*

$$\sum_n r^{|n|} \hat{\mu}(n) w^n$$

and the convergence is uniform and absolute for w on $\partial \mathbb{D}$.

3.8 THEOREM. *Suppose $u \colon \mathbb{D} \to \mathbb{C}$ is a harmonic function.*
 (a) *There is a μ in $M(\partial \mathbb{D})$ with $u = \tilde{\mu}$ if and only if $\sup_r \|u_r\|_1 < \infty$.*

(b) *If $1 < p \leq \infty$, there is a function f in $L^p(\partial \mathbb{D})$ with $u = \tilde{f}$ if and only if $\sup_r \|u_r\|_p < \infty$.*

(c) *There is a function f in $L^1(\partial \mathbb{D})$ with $u = \tilde{f}$ if and only if $\{u_r\}$ is L^1 convergent.*

(d) *There is a function f in $C(\partial \mathbb{D})$ with $u = \tilde{f}$ if and only if $\{u_r\}$ is uniformly convergent.*

PROOF. (a) If $u = \tilde{\mu}$, then $\|u_r\|_1 = \int |u_r| \, dm \leq \int |\int P_{rz}(w) \, d\mu(w)| \, dm(z) \leq \iint P_{rz}(w) \, d|\mu|(w) \, dm(z) = \int [\int P_{rz}(w) \, dm(z)] \, d|\mu|(w) = \|\mu\|$.

Now assume that u is a harmonic function on $\mathbb{D}$ and L is a constant such that $\|u_r\|_1 \leq L$ for all $r < 1$. Put $\nu_r =$ the measure $u_r \cdot m$ in $M(\partial \mathbb{D})$. So $\{\nu_r\}$ is a uniformly bounded net of measures on $\partial \mathbb{D}$. By Alaoglu's Theorem there exists a measure μ in $M(\partial \mathbb{D})$ that is a weak* cluster point of this net. Hence

$$\hat{\nu}_r(n) = \int \overline{w}^n \, d\nu_r = \int u_r \overline{w}^n \, dm \to_{cl} \hat{\mu}(n).$$

But Lemma 3.7 implies that $\hat{\nu}_r(n) = \hat{u}_r(n) = r^{|n|} c_n \to c_n$ as $r \to 1$. Hence $\hat{\mu}(n) = c_n$. By Exercise 1.3, this implies that the weak* cluster point of $\{\nu_r\}$ is unique. Hence $\nu_r \to \mu$ weak* in $M(\partial \mathbb{D})$. An examination of the series in Lemma 3.7 shows that $u = \tilde{\mu}$.

(b) This proof is like that of part (a). For $1 < p < \infty$, the weak compactness of bounded sets in $L^p(\partial \mathbb{D})$ is used instead of weak* compactness. The weak* topology on $L^\infty(\partial \mathbb{D})$ is used when $p = \infty$.

The proofs of (c) and (d) are left as exercises. □

There is actually part of the preceding proof that should be recorded.

3.9 COROLLARY. *If $u: \mathbb{D} \to \mathbb{C}$ is a harmonic function and $\sup_r \|u_r\|_1 < \infty$, then the measures $u_r \cdot m \to \mu$ weak* in $M(\partial \mathbb{D})$, where μ is the measure such that $u = \tilde{\mu}$.*

Of course a similar result holds for the L^p portion of the preceding theorem, but we won't record this.

This section concludes with an interesting result that in some ways does not fit in with the preceding theorem and in other ways is directly connected with it.

3.10 HERGLOTZ'S THEOREM. *If u is a harmonic function on $\mathbb{D}$, then $u = \tilde{\mu}$ for a positive measure μ on $\partial \mathbb{D}$ if and only if $u \geq 0$ on $\mathbb{D}$.*

PROOF. It is easy to see that since the Poisson kernel is positive, for any positive measure μ, $u = \tilde{\mu} \geq 0$ on $\mathbb{D}$. Conversely, assume that $u(z) \geq 0$ for $|z| < 1$. Then $\|u_r\|_1 = \int u_r(w) \, dm(w) = u(0)$ by the Mean Value Theorem. By Corollary 3.9, there is a μ in $M(\partial \mathbb{D})$ such that $u = \tilde{\mu}$ and $u_r \cdot m \to \mu$ weak*. Since $u_r \geq 0$, it must be that $\mu \geq 0$. The theorem now follows. □

§4 Fatou's Theorem. As was mentioned before, if $\mu \in M(\partial \mathbb{D})$, we can consider $\tilde{\mu}$ as the solution of the Dirichlet problem with boundary values μ.

§4 FATOU'S THEOREM

Theorem 3.5 tells us that $\tilde{\mu}_r \to \mu$ weak* and so we have that the boundary value, μ, is the limit of the harmonic extension, $\tilde{\mu}$. But can we say more? For f in $C(\partial \mathbb{D})$ we know that for any ζ in $\partial \mathbb{D}$, $\tilde{f}(z) \to f(\zeta)$ as $z \to \zeta$. If $f \in L^p(\partial \mathbb{D})$, can we conclude that $\tilde{f}(z) \to f(\zeta)$ as $z \to \zeta$? By taking f to be a discontinuous function, such as the characteristic function of the upper half circle, we can see that this simple extension of the result for continuous functions will fail. It turns out that for f in $L^1(\partial \mathbb{D})$ and $|\zeta| = 1$, $\tilde{f}(r\zeta) \to f(\zeta)$ as $r \to 1$ for a.e. $[m]$. This is part of a theorem of Fatou that is discussed here.

One reason you might believe this result is an application of standard measure theory to Theorem 3.5. If $f \in L^p$, then we know that $\tilde{f}_r \to f$ in L^p norm. Thus there is a sequence $\{r_n\}$ with $r_n \to 1$ and $\tilde{f}_{r_n}(\zeta) \to f(\zeta)$ a.e.. Since $\tilde{f}_r$ is harmonic and thus quite well behaved, it is reasonable (or believable) that we actually get that $\tilde{f}_r(\zeta) \to f(\zeta)$ a.e.. It is somewhat sad that this simple approach to the theorem doesn't seem to give way to a proof. The proof is more involved.

We begin by recalling some measure theory. If $\mu \in M(\partial \mathbb{D})$, there is a corresponding measure on $[0, 2\pi]$, which will also be denoted by μ, such that $\int f d\mu = \int f(e^{it}) d\mu(t)$ for every f in $C(\partial \mathbb{D})$. The corresponding measure on $[0, 2\pi]$ is not unique. For example, if $\mu = \delta_1$ in $M(\partial \mathbb{D})$, then either δ_0 or $\delta_{2\pi}$ can be chosen as the corresponding measure on $[0, 2\pi]$. This is, however, essentially the only way in which uniqueness fails. (What does this mean?)

For a measure μ on $[0, 2\pi]$ there is a function of bounded variation u on $[0, 2\pi]$ such that $\int f d\mu = \int f(t) du(t)$ for every continuous function f, where this second integral is a Lebesgue-Stieltjes integral. It might be worthwhile to recall how this correspondence is established, though no proofs will be given here. The proofs can be found in many of the standard treatments of integration theory.

If μ is a positive measure on $[0, 2\pi]$, define a function $u: [0, 2\pi] \to \mathbb{R}$ by letting $u(0) = 0$ and $u(t) = \mu([0, t))$ for $t > 0$. The function u is left continuous, increasing, and $\int f d\mu = \int f(t) du(t)$ for all continuous functions f on $[0, 2\pi]$. If μ is an arbitrary complex-valued Borel measure on $[0, 2\pi]$, let $\mu = \mu_1 - \mu_2 + i(\mu_3 - \mu_4)$ be the Jordan decomposition and let $u = u_1 - u_2 + i(u_3 - u_4)$, where u_j is the increasing function corresponding to the positive measure μ_j. This establishes a bijective correspondence between complex-valued measures μ on $[0, 2\pi]$ and left continuous functions u of bounded variation that are normalized by requiring that $u(0) = 0$.

The next proposition gives the basic properties of this correspondence between measures and functions of bounded variation.

4.1 PROPOSITION. *Let $\mu \in M[0, 2\pi]$ and let u be the corresponding normalized function of bounded variation.*

(a) *The function u is continuous at t_0 if and only if $\mu(\{t_0\}) = 0$.*

(b) *The measure μ is absolutely continuous with respect to Lebesgue measure if and only if u is an absolutely continuous function, in which case $\int f \, d\mu = \int f(t) u'(t) \, dt$ for every continuous function f.*

(c) *If $E = \{t : u'(t) \text{ exists and is not } 0\}$, then E is measurable, $\mu | E$ is absolutely continuous with respect to Lebesgue measure, and $\mu | ([0, 2\pi] \backslash E)$ is singular with respect to Lebesgue measure.*

Remember that functions of bounded variation have derivatives a.e.. If u and μ correspond as above, then part (c) shows that μ is singular with respect to Lebesgue measure if and only if $u' = 0$ a.e..

Now fix θ_0, $0 \leq \theta_0 \leq 2\pi$, and consider the portion of the open unit disk $\mathbb{D}$ contained in an angle with vertex $e^{i\theta_0} = w_0$, symmetric about the radius $z = r w_0$, $0 \leq r \leq 1$, and having opening 2α, where $0 < \alpha < \frac{\pi}{2}$. See Figure 1.

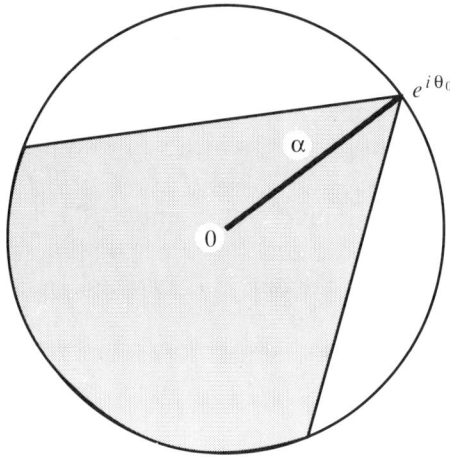

Figure 1

Call such a region a *Stoltz angle* with vertex w_0 and opening α. The variable z is said to *approach w_0 nontangentially* if $z \to w_0$ through some Stoltz angle. This will be abbreviated $z \to w_0$ (n.t.).

4.2 LEMMA. *Given a Stoltz angle with vertex $w_0 = e^{i\theta_0}$ and opening α, there is a constant C and a $\delta > 0$ such that if $z = re^{i\theta}$ belongs to the Stoltz angle and $|z - w_0| < \delta$, then $|\theta - \theta_0| \leq C(1 - r)$.*

PROOF. It suffices to assume that $\theta_0 = 0$ so that $w_0 = 1$. If L is the straight line that forms an edge of the Stoltz angle, then a reference to Figure 2 will show that for $z = re^{i\theta}$ on L, $\sin(\alpha + \theta) = \sin \alpha / r$. Hence

$$\frac{1-r}{\theta} = \frac{1}{\sin(\alpha + \theta)} \left[\frac{\sin(\alpha + \theta) - \sin \alpha}{\theta} \right] \xrightarrow[\theta \to 0+]{} \cot \alpha.$$

Thus $\theta / (1 - r) \to \tan \alpha$ as $\theta \to 0+$. Since the tangent function is increasing, the lemma now follows. □

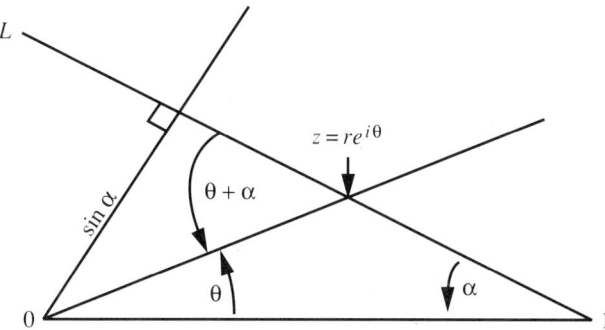

Figure 2

4.3 Fatou's Theorem. *Let $\mu \in M[0, 2\pi]$ and let u be the corresponding function of bounded variation; extend u to be defined on $\mathbb{R}$ by making u periodic with period 2π. If u is differentiable at θ_0, then $\tilde{\mu}(z) \to 2\pi u'(\theta_0)$ as $z \to e^{i\theta_0}$ (n.t.).*

(*Note*: We are identifying $M(\partial \mathbb{D})$ and $M[0, 2\pi]$. Also, the only reason for extending u to be defined on $\mathbb{R}$ is to facilitate the discussion at 0 and 2π.)

PROOF. It suffices to only consider the case where $\theta_0 = 0$; so we are assuming that $u'(0)$ exists. We may also assume that $u'(0) = 0$. In fact, if $u'(0) \neq 0$, let $\nu = \mu - 2\pi u'(0)m$. The function of bounded variation corresponding to ν is $v(\theta) = u(\theta) - u'(0)\theta$, since m is normalized Lebesgue measure. So $v'(0)$ exists and $v'(0) = 0$. If we know that $\tilde{\nu}(z) \to 0$ as $z \to 1$ (n.t.), then $\tilde{\mu}(z) = \tilde{\nu}(z) + 2\pi u'(0) \to 2\pi u'(0)$ as $z \to 1$ (n.t.).

So assume that $u'(0) = 0$. We want to show that

(4.4) $$\int_{-\pi}^{\pi} \frac{1-r^2}{1-2r\cos(t-\theta)+r^2} \, d\mu(t) \to 0$$

as $z = re^{i\theta} \to 1$ (n.t.). Using the preceding lemma, it suffices to show that (4.4) holds if, for some fixed positive constant C, $\theta \to 0$ and $r \to 1$ while satisfying

(4.5) $$|\theta| \leq C(1-r).$$

Let Γ be the set of z satisfying (4.5).

Recall that if the Poisson kernel is considered as a function of θ with r fixed, then $P_r(\theta) = (1-r^2)/(1-2r\cos\theta+r^2)$ and so differentiation with respect to θ gives that

$$P_r'(\theta) = \frac{2r(1-r^2)\sin\theta}{(1-2r\cos\theta+r^2)^2}.$$

Fix $\varepsilon > 0$. Since $u'(0)$ exists and equals 0, there is a $\delta > 0$ such that

$|u(t)| \leq \varepsilon|t|$ for $|t| < \delta$. Thus if $z = re^{i\theta} \in \Gamma$,

$$\tilde{\mu}(z) = \int_{-\pi}^{\pi} P_r(t - \theta)\, d\mu(t) = \left(\int_{-\delta}^{\delta} + \int_{\pi \geq |t| \geq \delta}\right) P_r(t - \theta)\, d\mu(t).$$

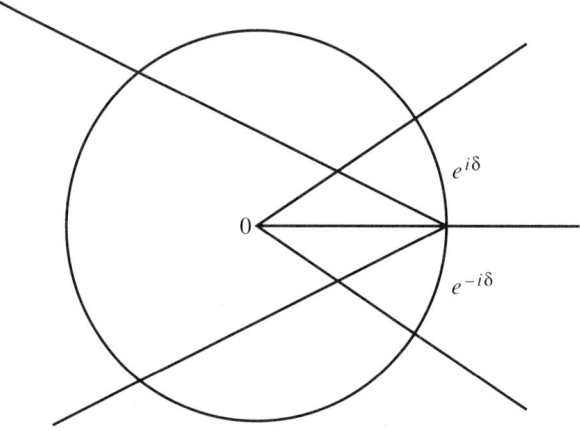

Figure 3

Examining Figure 3 we see that if $0 < \delta_1 < \delta$, there is a neighborhood U_1 of 1 such that if $z \in \Gamma \cap U_1$ and $|t| \geq \delta$, then $|t - \theta| \geq \delta_1$. Thus (3.1e) implies δ_1 and U_1 can be chosen so that $P_r(t - \theta) < \varepsilon$ for $|t| \geq \delta$ and $z \in \Gamma \cap U_1$. Therefore

$$(4.6) \qquad |\tilde{\mu}(z)| \leq \varepsilon \|\mu\| + \left|\int_{-\delta}^{\delta} P_r(t - \theta)\, d\mu(t)\right|.$$

Using integration by parts, for $z = re^{i\theta}$ in $\Gamma \cap U_1$,

$$\left|\int_{-\delta}^{\delta} P_r(t - \theta)\, d\mu(t)\right| \leq |u(t) P_r(t - \theta)|_{-\delta}^{\delta} + \left|\int_{-\delta}^{\delta} u(t) P_r'(t - \theta)\, dt\right|$$

$$= |u(\delta) P_r(\delta - \theta) - u(-\delta) P_r(-\delta - \theta)| +$$

$$+ \left|\int_{-\delta}^{\delta} u(t) P_r'(t - \theta)\, dt\right|$$

$$\leq 2\delta \varepsilon^2 + \left|\int_{-\delta}^{\delta} u(t) P_r'(t - \theta)\, dt\right|.$$

From (4.6) we infer that

$$(4.7) \qquad |\tilde{\mu}(z)| \leq \varepsilon \|\mu\| + 2\varepsilon^2 + \left|\int_{-\delta}^{\delta} u(t) P_r'(t - \theta)\, dt\right|.$$

Now fix z and assume that $\theta \geq 0$. The case where $\theta \leq 0$ is treated similarly and will be left to the reader. Also, assume that U_1 is sufficiently small that $\theta < \delta/2$ for $z = re^{i\theta}$ in U_1. Hence

(4.8)
$$\int_{-\delta}^{\delta} u(t) P'_r(t-\theta)\,dt = \left(\int_{-\delta}^{0} + \int_{0}^{2\theta} + \int_{2\theta}^{\delta}\right) P'_r(t-\theta) u(t)\,dt$$
$$= X + Y + Z.$$

Now since $|u(t)| \leq \varepsilon |t|$ for $|t| \leq \delta$,

$$|Y| = \left| \int_{0}^{2\theta} \frac{2r(1-r^2)\sin(t-\theta) u(t)}{(1-2r\cos(t-\theta)+r^2)^2}\,dt \right|$$
$$\leq 2r(1-r^2) \int_{0}^{2\theta} \frac{|\sin(t-\theta)| \varepsilon t}{(1-2r\cos(t-\theta)+r^2)^2}\,dt.$$

But $(1-2r\cos(t-\theta)+r^2) \geq 1-2r+r^2 = (1-r)^2$ and $|\sin(t-\theta)| \leq |t-\theta| \leq \theta$ for $0 \leq t \leq 2\theta$. Hence

$$|Y| \leq \frac{2\varepsilon r(1-r^2)}{(1-r)^4} \theta \int_{0}^{2\theta} t\,dt$$
$$= \frac{\varepsilon r(1+r)\theta(4\theta^2)}{(1-r)^3}$$
$$\leq \frac{8\varepsilon \theta^3}{(1-r)^3}.$$

By (4.5) we get
$$|Y| \leq 8\varepsilon C^3.$$

Now for the term Z in (4.8). If $2\theta \leq t \leq \delta$, then $0 \leq t - 2\theta = 2(t-\theta) - t$ and so $t \leq 2(t-\theta)$. Hence $|u(t)| \leq 2\varepsilon(t-\theta)$. Thus

$$|Z| = \left| \int_{2\theta}^{\delta} P'_r(t-\theta) u(t)\,dt \right|$$
$$\leq 2\varepsilon \int_{2\theta}^{\delta} (-P'_r(t-\theta))(t-\theta)\,dt$$
$$= 2\varepsilon \int_{\theta}^{\delta-\theta} (-P'_r(t))t\,dt$$
$$\leq 2\varepsilon \int_{0}^{\pi} (-P'_r(t))t\,dt$$
$$= 2\varepsilon(-tP_r(t))\big|_0^\pi + 2\varepsilon \int_0^\pi P_r(t)\,dt$$
$$\leq 2\varepsilon\pi \left(\frac{1-r^2}{(1+r)^2}\right) + 2\varepsilon\pi$$
$$< 4\pi\varepsilon,$$

provide U_1 is chosen sufficiently small. (That is, we force r to be very close to 1.)

For the term X in (4.8), observe that
$$\int_{-\delta}^{0} P_r'(t-\theta)u(t)\,dt = \int_{\theta}^{\theta+\delta} P_r'(t)u(\theta-t)\,dt.$$
Now for $\theta \leq t \leq \theta+\delta$, $0 \leq t-\theta \leq \delta$ and so
$$|X| \leq \varepsilon \int_{\theta}^{\theta+\delta} (-P_r'(t))(t-\theta)\,dt.$$
Using the preceding methods we obtain the fact that for some constant M, $|X| \leq M\varepsilon$.

Referring to (4.7) and (4.8) we get that there is a constant C' that is independent of ε such that for all z in Γ and in a sufficiently small neighborhood U_1 of 1, $|\tilde{\mu}(z)| \leq C'\varepsilon$. □

4.9 COROLLARY. *If $\mu \in M(\partial \mathbb{D})$, then $\tilde{\mu}$ has nontangential limits a.e. $[m]$ on $\partial \mathbb{D}$.*

PROOF. Functions of bounded variation have finite derivatives a.e.. □

4.10 COROLLARY. *If u is a non-negative harmonic function on $\mathbb{D}$, then $\lim_{r \to 1-} u(re^{i\theta})$ exists and is finite a.e. on $[0, 2\pi]$.*

The next result is also a corollary of Fatou's Theorem but it is sufficiently important to merit a more proclamatory label.

4.11 THEOREM. *If $1 \leq p \leq \infty$ and $u \colon \mathbb{D} \to \mathbb{C}$ is a harmonic function such that $\sup_{r<1} \|u_r\|_p < \infty$, then*
$$f(w) \equiv \lim_{r \to 1-} u(rw)$$
exists and is finite a.e. $[m]$ on $\partial \mathbb{D}$. If $1 < p \leq \infty$, then $f \in L^p(m)$ and $u = \tilde{f}$. If $p = 1$, then $u = \tilde{\mu}$ for some measure μ in $M[0, 2\pi]$ and f is the Radon-Nikodym derivative of the absolutely continuous part of μ.

PROOF. This proof is actually a collage of several preceding results. First assume that $1 < p \leq \infty$. By Theorem 3.8 there is a function g in L^p such that $u = \tilde{g}$. By Fatou's Theorem, $g = f$ a.e. $[m]$. Now suppose $p = 1$. Again Theorem 3.8 implies that $u = \tilde{\mu}$ for some μ in $M[0, 2\pi]$. Let $\mu = \mu_a + \mu_s$ be the Lebesgue decomposition of μ with respect to m. Let g be the Radon-Nikodym derivative of μ_a with respect to m. Thus if w is the function of bounded variation on $[0, 2\pi]$ corresponding to μ, then $w' = g$ a.e.. It follows by Fatou's Theorem that $g = f$ a.e.. □

4.12 EXAMPLE. If $\mu = \delta_1$, the unit point mass at 1 on $\partial \mathbb{D}$, $\tilde{\mu}(z) = \int P_z\,d\mu = P_z(1) = \text{Re}(1+z/1-z)$. Here the conclusion of Fatou's Theorem can be directly verified. (Why doesn't this contradict the Maximum Principle for harmonic functions?)

§5 **Subharmonic functions.** We begin by modifying the definition of a subharmonic function from that given in [**FOCV**].

§5 SUBHARMONIC FUNCTIONS

5.1 Definition. If G is an open subset of $\mathbb{C}$, a function $u: G \to [-\infty, +\infty)$ is *subharmonic* if u is upper semicontinuous and for every closed disk $\overline{B}(a; r)$ contained in G we have

$$u(a) \leq \frac{1}{2\pi} \int_0^{2\pi} u(a + re^{i\theta}) \, d\theta.$$

A function $u: G \to (-\infty, +\infty]$ is *superharmonic* if $-u$ is subharmonic.

Some remarks are in order here. In [**FOCV**] only continuous finite valued functions were allowed to be subharmonic. This was done only because the Lebesgue integral was not assumed there as a prerequisite. Also there is another slight difference between this definition of a subharmonic function and that given by many authors in that the function that is identically equal to $-\infty$ is allowed to be subharmonic here. In fact, since G is not assumed to be connected, u may be constantly equal to $-\infty$ on some components and finite valued on others. This is usually excluded as a possibility in the definition of a subharmonic function.

Next it should be emphasized that the integral in (5.1) is not assumed to be finite. On the other hand, u is upper semicontinuous (abbreviated by usc) and $u(z) < +\infty$ for all z so that if $\overline{B}(a; r) \subseteq G$, then u is bounded above on $\overline{B}(a; r)$. Therefore, $\int_0^{2\pi} u(a + re^{i\theta}) \, d\theta < +\infty$. From the definition it may occur, however, that $\int_0^{2\pi} u(a + re^{i\theta}) \, d\theta = -\infty$ (in which case it follows, of course, that $u(a) = -\infty$). It can be shown that this does not occur unless u is constantly $-\infty$ on the component that contains $\overline{B}(a; r)$.

All results here will be phrased in terms of subharmonic functions. For each of these there is a corresponding result for superharmonic functions whose phrasing is left to the reader. If such a result for superharmonic functions is needed in the course of this book, the reference will be made to the subharmonic result.

Our main concern with subharmonic functions at present is to know that functions such as $|f|^p$ are subharmonic when f is an analytic function. This will follow from the following more general result. This next proposition is sometimes called Jensen's Inequality.

5.2 Proposition. *If (X, Ω, μ) is a probability measure space, $h \in L^1_{\mathbb{R}}(\mu)$, $-\infty \leq a < h(x) < b \leq +\infty$ a.e. $[\mu]$, and $\phi: (a, b) \to \mathbb{R}$ is a convex function, then*

$$\phi\left(\int h \, d\mu\right) \leq \int \phi \circ h \, d\mu.$$

A few remarks about this proposition might be appropriate before launching into the proof. First, even though it is assumed that h is integrable, it does not follow that $\phi \circ h$ is integrable. (Example?) If this is indeed the case, then the proposition says that the worst that can happen is that $\int \phi \circ h \, d\mu$ exists but is equal to $+\infty$. Second, a convex function is automatically continuous (see Exercise VI.3.2 in [**FOCV**]).

Also the definition of a convex function is that $\phi(tx+(1-t)y) \leq t\phi(x) + (1-t)\phi(y)$ for all x and y and $0 \leq t \leq 1$. This is equivalent to the requirement that $(\phi(u)-\phi(x))/(u-x) \leq (\phi(y)-\phi(x))/(y-x)$ for $a < x < u < y < b$. (See Exercise VI.3.3 in [**FOCV**].) This last fact will be used in the proof of the proposition, to which we now turn our attention.

PROOF OF PROPOSITION 5.2. It may be assumed that h is not constant. Since μ has total mass 1, $a < I = \int f \, d\mu < b$. Put

$$S = \sup\{(\phi(I) - \phi(t))/(I-t): a < t < I\}.$$

By the remarks preceding the proof, $S < \infty$. So for $a < t < I$,

(5.3) $$\phi(I) + S(t-I) \leq \phi(t).$$

If $I < t < b$, then we also have that $S \leq (\phi(t) - \phi(I))/(t-I)$; equivalently, (5.3) holds for $I < t < b$ and hence for all t in (a, b). In particular, letting $t = h(x)$ implies that $0 \leq \phi(h(x)) - \phi(I) - S[h(x) - I]$ a.e. $[\mu]$. Since μ is a probability measure, $0 \leq \int \phi \circ h \, d\mu - \phi(I) - S[\int h \, d\mu - I] = \int \phi \circ f \, d\mu - \phi(I)$. □

5.4 THEOREM. *If u is a subharmonic function on G with $-\infty \leq a \leq u(z) \leq b \leq \infty$ for all z in G and $\phi: [a, b] \to [-\infty, \infty)$ is an increasing convex function, then $\phi \circ u$ is subharmonic.*

PROOF. Since ϕ is continuous, $\phi \circ u$ is upper semicontinuous. On the other hand, if $\overline{B}(a; r) \subseteq G$, then Proposition 5.2 and the fact that ϕ is increasing imply that

$$\phi(u(a)) \leq \frac{1}{2\pi} \int_0^{2\pi} \phi \circ u(a + re^{i\theta}) \, d\theta.$$

Hence $\phi \circ u$ is subharmonic. □

5.5 EXAMPLE. If $f: G \to \mathbb{C}$ is an analytic function, then $\log|f|$ is subharmonic. In fact, if $\overline{B}(a; r) \subseteq G$, then $f(a) = \frac{1}{2\pi} \int_0^{2\pi} f(a + re^{i\theta}) \, d\theta$ and so $|f(a)| \leq \frac{1}{2\pi} \int_0^{2\pi} |f(a + re^{i\theta})| \, d\theta$. That is, $|f|$ is subharmonic. Now the logarithm is an increasing function but it is concave and and not convex so we cannot use the preceding proposition to conclude that $\log|f|$ is subharmonic. This conclusion follows, however, by an appeal to Jensen's Formula (see XI.1.2 in [**FOCV**]).

5.6 EXAMPLE. If $f: G \to \mathbb{C}$ is an analytic function, then $|z|^p$ is a subharmonic function on G for $0 < p < \infty$. From Example 5.5 we know that $\log|f|$ is subharmonic. If $\phi(t) = e^{pt}$, then ϕ is increasing and convex. By Proposition 5.4, $|f(z)|^p = \phi(\log|f(z)|)$ is subharmonic.

5.7 PROPOSITION. *If $u: \mathbb{D} \to \mathbb{R}$ is subharmonic and continuous, then, as a function of r, $\frac{1}{2\pi} \int_0^{2\pi} u(re^{i\theta}) \, d\theta$ is increasing.*

PROOF. Let $0 \leq r_1 < r_2 < 1$. Solving the Dirichlet Problem for the boundary values $u|\partial B(0; r_2)$ we obtain a continuous function $f: \overline{B}(0; r_2) \to$

§6 HARDY SPACES

$\mathbb{R}$ such that f is harmonic on $B(0; r_2)$ and $f(z) = u(z)$ for $|z| = r_2$. By the Maximum Principle, $f \geq u$ on $\overline{B}(0; r_2)$. Hence

$$\frac{1}{2\pi} \int_0^{2\pi} u(r_1 e^{i\theta}) \, d\theta \leq \frac{1}{2\pi} \int_0^{2\pi} f(r_1 e^{i\theta}) \, d\theta$$
$$= f(0)$$
$$= \frac{1}{2\pi} \int_0^{2\pi} f(r_2 e^{i\theta}) \, d\theta$$
$$= \frac{1}{2\pi} \int_0^{2\pi} u(r_2 e^{i\theta}) \, d\theta. \quad \square$$

Exercises

1. Let $\log^+ x = \log x$ if $x \geq 1$ and $\log^+ x = 0$ for $0 < x \leq 1$. If $f: G \to \mathbb{C}$ is an analytic function, show that $\log^+ |f|$ is subharmonic.
2. If $u: G \to \mathbb{R}$ is a harmonic function and $1 \leq p < \infty$, show that $|u|^p$ is subharmonic.

§6 Hardy spaces. Here we introduce the Hardy spaces H^p of analytic functions on the open unit disk. The definition given here differs from that given in (I.4.5). These two definitions are equivalent, however. This is proved in Theorem 6.5 for $1 < p \leq \infty$. The proof of the equivalence for $p = 1$ must be postponed until Theorem 8.13.

6.1 DEFINITION. If $f: \mathbb{D} \to \mathbb{C}$ is a measurable function and $1 \leq p < \infty$, define

$$M_p(r, f) \equiv \left[\frac{1}{2\pi} \int_0^{2\pi} |f(re^{i\theta})|^p \, d\theta \right]^{1/p} ;$$

also define

$$M_\infty(r, f) \equiv \sup\{|f(re^{i\theta})|: 0 \leq \theta \leq 2\pi\}.$$

For any value of p, $1 \leq p \leq \infty$, let H^p denote the space of all analytic functions on $\mathbb{D}$ for which $\|f\|_p \equiv \sup_{r<1} M_p(r, f) < \infty$.

If $f: \mathbb{D} \to \mathbb{C}$ and $0 < r < 1$, denote by f_r the function defined on $\partial \mathbb{D}$ by $f_r(z) = f(rz)$. Thus for any such r, $M_p(r, f)$ is the L^p norm of f_r. From this observation it follows that $\|\cdot\|_p$ is a norm on H^p. In fact, we will see that H^p is a Banach space (6.5).

Also, from standard L^p space theory, $H^p \subseteq H^q \subseteq H^1$ if $1 \leq q \leq p$. In particular H^∞ is the space of bounded analytic functions on $\mathbb{D}$ and $H^\infty \subseteq H^p$ for all p.

It is possible to define H^p spaces for $0 < p < 1$, but this will not be done here. The interested reader can consult Duren [**1970**], Hoffman [**1962**], or Koosis [**1980**] for this as well as a great deal more information on these spaces.

6.2 PROPOSITION. *If $f: \mathbb{D} \to \mathbb{C}$ is an analytic function, then $\|f\|_p = \lim_{r \to 1^-} M_p(r, f)$.*

PROOF. Assume that p is finite. From Example 5.6 we know that $z \to |f(z)|^p$ is a continuous subharmonic function and so, by Proposition 5.7, $M_p(r, f)$ is an increasing function of r. Therefore the supremum must equal the limit.

By the Maximum Modulus Theorem $M_\infty(r, f)$ is also an increasing function of r. Thus the proposition also holds for $p = \infty$. □

Note that by Corollary 4.11 if $f \in H^1$, then there is a measure μ on $\partial \mathbb{D}$ such that $f = \tilde{\mu}$. Also if $1 \leq p < \infty$ and $f \in H^p$, then $f = \tilde{g}$ for some g in $L^p(\partial \mathbb{D})$. Moreover, in the case that $p > 1$, $f_r \to g$ a.e. $[m]$ as $r \to 1$. In particular, each function in H^p, $1 \leq p \leq \infty$, has nontangential limits at almost every point of $\partial \mathbb{D}$.

There are two questions that occur here. First, when $p > 1$, which functions g in L^p can arise in this way? Note that when we answer this question we will have identified H^p with a certain subspace of L^p and thus have the possibility of combining measure theory with the theory of analytic functions.

The second question concerns the case when $p = 1$. Here the theory becomes more subtle and difficult. It will turn out that if $f \in H^1$ and $\mu \in M$ such that $f = \tilde{\mu}$, then μ is absolutely continuous with respect to Lebesgue measure and the nontangential limit of f is the Radon-Nikodym derivative of μ with respect to m. This is the F. and M. Riesz Theorem proved in §8 below. For now we content ourselves with complete information when $p > 1$ and partial information when $p = 1$. Recall that for any function f (respectively, measure μ), $\hat{f}$ (respectively, $\hat{\mu}$) denotes the Fourier transform of f (respectively, μ).

6.3 THEOREM. *If $1 \leq p \leq \infty$ and $f \in L^p$ such that $\hat{f}(n) = 0$ for $n < 0$, then $\tilde{f}$, the Poisson integral of f, belongs to H^p. Moreover:*

(a) $\|f\|_p = \|\tilde{f}\|_p$;
(b) *if $1 \leq p < \infty$, $\|\tilde{f}_r - f\|_p \to 0$ as $r \to 1-$;*
(c) *if $p = \infty$, $\tilde{f}_r \to f$ weak* in L^∞ as $r \to 1-$.*

PROOF. If $z = re^{i\theta} \in \mathbb{D}$ and $|w| = 1$, then $P_z(w) = \sum_{n=-\infty}^{\infty} r^{|n|} e^{in\theta} \overline{w}^n$. Hence

$$\tilde{f}(z) = \int f(w) P_z(w) \, dm(w)$$

$$= \int f(w) \sum_{n=-\infty}^{\infty} r^{|n|} e^{in\theta} \overline{w}^n \, dm(w)$$

$$= \sum_{n=-\infty}^{\infty} r^{|n|} e^{in\theta} \int f(w) \overline{w}^n \, dm(w),$$

since the series converges uniformly in w. Therefore

$$\tilde{f}(z) = \sum_{n=-\infty}^{\infty} r^{|n|} e^{i\theta} \hat{f}(n) = \sum_{n=0}^{\infty} \hat{f}(n) z^n$$

since, by hypothesis, $\hat{f}(n) = 0$ when $n < 0$. Thus $\tilde{f}$ has a power series expansion in $\mathbb{D}$ and so $\tilde{f}$ is analytic on $\mathbb{D}$.

Also, for $1 \leq p < \infty$, $\|\tilde{f}_r - f\|_p \to 0$ as $r \to 1-$ (3.5). If $p = \infty$, then $\|\tilde{f}_r\|_\infty \leq \|f\|_\infty$. Hence, in either case, $\sup_r M_p(r, f) = \sup_r \|\tilde{f}_r\|_p < \infty$ and so $\tilde{f} \in H^p$. The remaining details are easily deduced from Theorem 3.5. □

What about the converse of the preceding theorem? Here is where we must assume that $p > 1$ and postpone consideration of the case where $p = 1$ until later. Suppose that $f \in H^p$, $1 \leq p \leq \infty$, and let g be the nontangential limit function of f. That is, $g(w) = \lim_{r \to 1-} f(rw) = \lim_{r \to 1-} f_r(w)$ a.e. $[m]$ on $\partial \mathbb{D}$. By Corollary 4.11, $g \in L^p$. Now if $1 < p < \infty$, $\|f_r - g\|_p \to 0$. Also

$$\hat{g}(n) = \int g(w) \overline{w}^n \, dm(w)$$
$$= \lim_{r \to 1-} \int f_r(w) \overline{w}^n \, dm(w)$$
$$= \lim_{r \to 1-} \hat{f}_r(n).$$

But $f(z) = \sum_0^\infty a_n z^n$ for $|z| < 1$, with convergence uniform on proper subdisks of $\mathbb{D}$. It follows (how?) that $\hat{f}_r(n) = a_n r^n$ if $n \geq 0$ and $\hat{f}_r(n) = 0$ if $n < 0$. Therefore $\hat{g}(n) = a_n$ if $n \geq 0$ and $\hat{g}(n) = 0$ if $n < 0$. By Theorem 6.3, $\tilde{g} \in H^p$. But also (see the proof of Theorem 6.3) for $|z| < 1$,

$$\tilde{g}(z) = \sum_0^\infty \hat{g}(n) z^n = \sum_0^\infty a_n z^n.$$

Hence $\tilde{g} = f$ and the desired converse is obtained.

If $p = \infty$, then it is not necessarily true that $f_r \to g$ in L^∞ norm, but it is true that $f_r \to g$ weak* as $r \to 1-$. Since $\overline{w}^n \in L^1$ for every n, the argument of the preceding paragraph shows that $\hat{g}(n) = 0$ for $n < 0$ and $\tilde{g} = f$. This discussion can be summarized as follows.

6.4 THEOREM. *If $1 < p \leq \infty$ and $f \in H^p$, then $g(w) = \lim_{r \to 1-} f(rw)$ defines a function g in L^p with $\hat{g}(n) = 0$ for $n < 0$ and $\tilde{g} = f$.*

These last two theorems establish a correspondence between functions in H^p (defined on $\mathbb{D}$) and the functions in

$$\mathcal{H}^p \equiv \{f \in L^p : \hat{f}(n) = 0 \text{ for } n < 0\},$$

but only when $p > 1$. Now it is easy to see that $\mathcal{H}^p$ is a closed subspace of L^p and we are led to the following theorem.

6.5 THEOREM. *If $1 < p \leq \infty$, then the map that takes f in H^p to its boundary values establishes an isometry of H^p onto the closed subspace $\mathscr{H}^p$ of L^p. Thus H^p is a Banach space. For $p = 2$, H^2 is a Hilbert space with $\{1, z, z^2, \ldots\}$ as an orthonormal basis. For $p = \infty$, H^∞ is the dual of a Banach space as $\mathscr{H}^\infty$ is a weak* closed subspace of L^∞.*

Because of this result we can identify the functions in H^p, $1 < p \leq \infty$, with their radial or nontangential limits. In §8 we will prove this assertion for $p = 1$, but first, in the next section, we will investigate another class of analytic functions on $\mathbb{D}$ and use this information to derive this correspondence between H^1 and $\mathscr{H}^1$.

Exercises

1. Supply the details in the proof of Corollary 6.5.
2. Let $\mathscr{A} = \{f \in C(\partial\mathbb{D}): \hat{f}(n) = 0 \text{ for } n < 0\}$. If $f \in \mathscr{A}$, show that its Poisson integral, $\tilde{f}$, is an analytic function on $\mathbb{D}$. Also if $g: \text{cl}\,\mathbb{D} \to \mathbb{C}$ is defined by $g(z) = \tilde{f}(z)$ for $|z| < 1$ and $g(z) = f(z)$ for $|z| = 1$, then g is continuous. Conversely, if g is a continuous function on $\text{cl}\,\mathbb{D}$ that is analytic on $\mathbb{D}$ and $f = g|\partial\mathbb{D}$, then $g(z) = \tilde{f}(z)$ for $|z| < 1$.
3. Give a direct proof that H^1 is a Banach space.
4. If $1 \leq p < \infty$ and $|a| < 1$, define $L_a: H^p \to \mathbb{C}$ by $L_a(f) = f(a)$ for all f in H^p. Show that $L_a \in (H^p)^*$ and $\|L_a\| = (1 - |a|^2)^{-1/p}$.

§7 The Nevanlinna Class. In this section we will study another collection of analytic functions which is not a Banach space but includes all the Hardy spaces H^p.

7.1 DEFINITION. A function f is of *Nevanlinna class* (or of *bounded characteristic*) if f is an analytic function on $\mathbb{D}$ and

$$\sup_{r<1} \frac{1}{2\pi} \int_0^{2\pi} \log^+ |f(re^{i\theta})|\, d\theta < \infty.$$

The Nevanlinna class is denoted by N.

Note that since $\log^+ |f(z)|$ is a subharmonic function (Exercise 5.1) and is also continuous, $\frac{1}{2\pi} \int_0^{2\pi} \log^+ |f(re^{i\theta})|\, d\theta$ is an increasing function of r (5.7). Thus the definition of a function in Nevanlinna class can be weakened by only stipulating the finiteness of the supremum over a sequence $\{r_n\}$ with $r_n \to 1$. Also

$$\sup_{r<1} \frac{1}{2\pi} \int_0^{2\pi} \log^+ |f(re^{i\theta})|\, d\theta = \lim_{r \to 1-} \frac{1}{2\pi} \int_0^{2\pi} \log^+ |f(re^{i\theta})|\, d\theta.$$

Since $\log x \leq x^p$ for $x \geq 1$, we have the following.

7.2 PROPOSITION. *If $1 \leq p \leq \infty$, then $H^p \subseteq N$.*

Thus every result for Nevanlinna class is a result about functions belonging to all the Hardy classes. Our immediate goal is to give a characterization of functions in N (Theorem 7.10 below) and to study the zeros of these functions. In the next section we will obtain a factorization theorem for this class. We begin with an elementary but important result.

7.3 LEMMA. *If $\{a_n\}$ is a sequence in $\mathbb{D}$, the following statements are equivalent.*

(a) $\sum_{n=1}^{\infty}(1 - |a_n|) < \infty$.
(b) $\prod_{n=1}^{\infty} |a_n|$ *converges.*
(c) $\sum_{n=1}^{\infty} \log|a_n|^{-1} < \infty$.
(d) $\prod_{n=1}^{\infty} \frac{|a_n|}{a_n}\left(\frac{a_n - z}{1 - \overline{a_n}z}\right)$ *converges uniformly and absolutely on compact subsets of $\mathbb{D}$.*

PROOF. The proof that (a), (b), and (c) are equivalent can be found in §VII.5 of [**FOCV**]. The fact that (a) implies (d) is Exercise VII.5.4 in [**FOCV**]. By evaluating the infinite product in (d) at $z = 0$, (b) can be deduced from (d). □

7.4 DEFINITION. A sequence $\{a_n\}$ satisfying one of the equivalent conditions in the preceding lemma is called a *Blaschke sequence*. If $\{a_n\}$ is a Blaschke sequence and m is an integer, $m \geq 0$, then the function

$$(7.5) \qquad b(z) = z^m \prod_{n=1}^{\infty} \frac{|a_n|}{a_n} \left(\frac{a_n - z}{1 - \overline{a_n}z} \right)$$

is called a *Blaschke product*.

The factor z^m in the definition of a Blaschke product is there to allow b to have a zero at the origin.

7.6 PROPOSITION. *If b is the Blaschke product defined by (7.5), then $b \in H^\infty$ and $|b(w)| = 1$ a.e. $[m]$ on $\partial \mathbb{D}$. The zeros of b are precisely the points $a_1, a_2, \ldots$ and, provided $m > 0$, the origin.*

PROOF. Let b_n denote the product of z^m with the first n factors in (7.5). It is easy to see that $|b_n(z)| \leq 1$ on cl$\mathbb{D}$ and $|b_n(w)| = 1$ for w in ∂D. By (7.3) $|b(z)| \leq 1$ on $\mathbb{D}$ and so $b \in H^\infty$. Also for $n > k$,

$$\int |b_n - b_k|^2 \, dm = 2\left[1 - \operatorname{Re} \int b_n \overline{b_k} \, dm\right]$$
$$= 2\left[1 - \operatorname{Re} \int \frac{b_n}{b_k} \, dm\right].$$

Because $n > k$, b_n/b_k is analytic on $\mathbb{D}$. Thus the mean value property implies that

$$\int \frac{b_n}{b_k} \, dm = \frac{b_n}{b_k}(0) = \prod_{j=k+1}^{n} |a_j|.$$

Hence

$$\int |b_n - b_k|^2 \, dm = 2\left[1 - \prod_{j=k+1}^{n} |a_j|\right].$$

But Lemma 7.3 implies that the right-hand side of this last equation can be made arbitrarily small for sufficiently large n and k. Therefore $\{b_n\}$ is a Cauchy sequence in H^2 and must converge to some function f. If $|z| < 1$ and P_z is the Poisson kernel, then $P_z \in L^2$ and so $b_n(z) = \int P_z b_n \, dm \to \int P_z f \, dm$. Hence it must be that $f = b$. That is, $b_n \to b$ in H^2.

Because we have convergence in the L^2 norm, there is a subsequence $\{b_{n_k}\}$ such that $b_{n_k}(w) \to b(w)$ a.e. $[m]$. Since $|b_{n_k}| = 1$ a.e. on $\partial \mathbb{D}$, it follows that $|b| = 1$ a.e. on $\partial \mathbb{D}$. Finally, the statement about the zeros of the Blaschke product follows from (VII.5.9) of [**FOCV**]. □

We obtained a useful fact in the course of the preceding proof that is worth recording.

7.7 COROLLARY. *If $\{a_n\}$ is a Blaschke sequence, b is the corresponding Blaschke product, and b_n is the finite Blaschke product with zeros $a_1, \ldots, a_n$, then there is a sequence of integers $\{n_k\}$ such that $b_{n_k} \to b$ a.e. on $\partial \mathbb{D}$.*

The next result is the reason for our concern with Blaschke sequences.

7.8 THEOREM. *If f is in the Nevanlinna class and f is not identically 0, then the zeros of f form a Blaschke sequence. Moreover, if b is the Blaschke product with the same zeros as f, then $f/b \in N$. If $f \in H^p$, then $f/b \in H^p$.*

PROOF. It suffices to assume that $f(0) \neq 0$. By Jensen's Formula (see page 280 in [**FOCV**])

$$(7.9) \qquad \frac{1}{2\pi} \int_0^{2\pi} \log|f(re^{i\theta})| \, d\theta = \log|f(0)| + \sum_{|a_k| < r} \log\left(\frac{r}{|a_k|}\right),$$

where $a_1, a_2, \ldots$ are the zeros of f, repeated as often as their multiplicity, and r is chosen so that $|a_k| \neq r$ for any k. But $\log x \leq \log^+ x$ and so, since $f \in N$, there is a finite constant $M > 0$ such that $\frac{1}{2\pi} \int_0^{2\pi} \log|f(re^{i\theta})| \, d\theta \leq M$ for all $r < 1$. This implies, with a small argument, that $\sum_{k=1}^\infty \log(1/|a_k|) < \infty$. By Lemma 7.3, $\{a_k\}$ is a Blaschke sequence.

Now to show that $f/b \in N$, when b is the Blaschke product with the same zeros as f. It is left as an exercise to show that if f has a zero at $z = 0$ of order m, then $f/z^m \in N$. Thus we may assume that $f(0) \neq 0$ and hence $b(0) \neq 0$. If $g = f/b$, then g is an analytic function on $\mathbb{D}$ and never vanishes. If $z \in \mathbb{D}$ with $|g(z)| \leq 1$, then $|f(z)| \leq |b(z)| \leq 1$ and so $\log^+ |g(z)| = 0 < -\log|b(z)| = \log^+ |f(z)| - \log|b(z)|$. If $|g(z)| > 1$, then $\log^+ |g(z)| = \log|g(z)| = \log|f(z)| - \log|b(z)| \leq \log^+ |f(z)| - \log|b(z)|$. Thus we have that for all z in $\mathbb{D}$

$$\log^+ |g(z)| \leq \log^+ |f(z)| - \log|b(z)|.$$

But $\log|b|$ is a subharmonic function and so for $0 < r < 1$,

$$\frac{1}{2\pi}\int_0^{2\pi} \log^+|g(re^{i\theta})|\,d\theta \le \frac{1}{2\pi}\int_0^{2\pi}\log^+|f(re^{i\theta})|\,d\theta - \frac{1}{2\pi}\int_0^{2\pi}\log|b(re^{i\theta})|\,d\theta$$

$$\le \frac{1}{2\pi}\int_0^{2\pi}\log^+|f(re^{i\theta})|\,d\theta - \log|b(0)|.$$

Since $b(0) \ne 0$, this implies that $g \in N$.

Now to show that $f/b \in H^p$ when $f \in H^p$. Let $a_1, a_2, \ldots$ be the zeros of f, repeated as often as their multiplicities and let b_n be the Blaschke product with zeros $a_1, \ldots, a_n$. Put $g_n = f/b_n$ and let $M \ge \int |f(rw)|^p\,dm(w)$ for all $r < 1$. If $\varepsilon > 0$, there is an r_0 such that for $r_0 \le |z| \le 1$, $|b_n(z)| > 1 - \varepsilon$. Therefore

$$\int |g_n(rw)|^p\,dm(w) \le \frac{1}{(1-\varepsilon)^p}\int |f(rw)|^p\,dm(w)$$

$$\le \frac{M}{(1-\varepsilon)^p}$$

for $r_0 < r < 1$. Letting $r \to 1$, we get that g_n belongs to H^p. Since ε was arbitrary, we can let $\varepsilon \to 0$ to see that $\|g_n\|_p \le M$ for all n. By Corollary 7.7 there is a subsequence $\{g_{n_k}\}$ such that $g_{n_k} \to g \equiv f/b$ a.e. on $\partial \mathbb{D}$. Therefore $g_{n_k} \to g$ weakly in L^p. Since each g_n belongs to H^p (Why?), $g \in H^p$. □

Here is the promised characterization of functions in Nevanlinna class.

7.10 THEOREM (F. and R. Nevanlinna). *If f is an analytic function on $\mathbb{D}$, then f belongs to the Nevanlinna class if and only if $f = g_1/g_2$ for two bounded analytic functions g_1 and g_2.*

PROOF. First assume that f is the quotient of the two bounded analytic functions g_1 and g_2. It can further be assumed that $|g_i(z)| \le 1$ for z in $\mathbb{D}$, $i = 1, 2$, and, since f must be analytic, it can also be assumed that g_2 has no zeros in $\mathbb{D}$. It follows that $\log|g_i(z)| \le 0$ and so $\log|f| = \log|g_1| - \log|g_2| \le -\log|g_2|$; hence $\log^+|f| \le -\log|g_2|$. This implies that

$$\frac{1}{2\pi}\int_0^{2\pi}\log^+|f(re^{i\theta})|\,d\theta \le -\frac{1}{2\pi}\int_0^{2\pi}\log|g_2(re^{i\theta})|\,d\theta$$
$$= -\log|g_2(0)|$$

since $\log|g_2|$ is harmonic. Thus $f \in N$.

Now assume that $f \in N$. By Theorem 7.8 we may assume that f does not vanish on $\mathbb{D}$. Hence $u = \log|f|$ is a harmonic function. Thus

$$u(0) = \int \log|f_r|\,dm$$
$$= \int \log^+|f_r|\,dm - \int \log^-|f_r|\,dm.$$

(Here $\log^- x = -\min\{\log x, 0\}$.) Since $f \in N$ and the left-hand side of this equation is independent of r, it follows that $\sup_{r<1} \int \log^- |f_r| \, dm < \infty$. Therefore

$$\sup_{r<1} \|u_r\|_1 = \sup_{r<1} \int |\log|f_r|| \, dm$$
$$= \sup_{r<1} \left\{ \int \log^+ |f_r| \, dm + \int \log^- |f_r| \, dm \right\}$$
$$< \infty.$$

By Theorem 3.8 there is a measure μ on $\partial \mathbb{D}$ such that $u(z) = \tilde{\mu}(z) = \int P_z \, dm$. Since u is real-valued, μ is a real-valued measure. Let $\mu = \mu_+ - \mu_-$ be the Hahn decomposition of μ and put $u_\pm = \tilde{\mu}_\pm$; so u_+ and u_- are non-negative harmonic functions on $\mathbb{D}$.

Now $\mathbb{D}$ is simply connected and f is a nonvanishing analytic function, so there is an analytic function h on $\mathbb{D}$ such that $f = e^h$; stipulate that $h(0) = \log|f(0)|$ so that h is unique. Thus $u = \operatorname{Re} h$ and it follows (by uniqueness) that

$$h(z) = \int \frac{w+z}{w-z} \, d\mu(w).$$

Also let

$$h_\pm(z) = \int \frac{w+z}{w-z} \, d\mu_\pm(w).$$

Thus $h = h_+ - h_-$ and $\operatorname{Re} h_\pm = u_\pm \geq 0$. Put $g_1 = e^{-h_-}$ and $g_2 = e^{-h_+}$; so g_1 and g_2 are analytic functions on $\mathbb{D}$ with no zeros. Also $|g_i(z)| = \exp(-\operatorname{Re} h_\pm(z)) = \exp(-u_\pm(z)) \leq 1$. That is, g_1 and g_2 belong to H^∞. Finally, $g_1/g_2 = \exp(-h_-)/\exp(-h_+) = \exp(h_+ - h_-) = f$. □

7.11 COROLLARY. *If $f \in N$ and f is not constantly 0, then f has a nontangential limit a.e. on $\partial \mathbb{D}$ and $\log|f(e^{i\theta})| \in L^1$.*

PROOF. We begin by proving the corollary if $f \in H^\infty$. So assume that $\|f\|_\infty \leq 1$; we may also assume that $f(0) \neq 0$. By Fatou's Theorem, f has nontangential limits a.e. on $\partial \mathbb{D}$; thus $f_r(w) \to f(w)$ a.e. on $\partial \mathbb{D}$. By Fatou's Lemma (from real variables)

$$\int_{\partial \mathbb{D}} |\log|f(w)|| \, dm(w) \leq \liminf_{r \to 1^-} \int |\log|f_r(w)|| \, dm(w)$$
$$= \liminf_{r \to 1^-} \left[-\int \log|f_r(w)| \, dm(w) \right].$$

But $\log|f|$ is subharmonic so that $\log|f(0)| \leq \int \log|f_r| \, dm$. Hence

$$\int_{\partial \mathbb{D}} |\log|f(w)|| \, dm(w) \leq -\log|f(0)| < \infty.$$

Thus $\log|f| \in L^1$.

Now let $f \in N$. By the preceding theorem, $f = g_1/g_2$ for two bounded analytic functions g_1 and g_2. Since $\log|g_i| \in L^1$, neither g_1 nor g_2 can vanish on a subset of $\partial \mathbb{D}$ with positive measure. Thus the fact that both g_1 and g_2 have nonzero nontangential limits a.e. on $\partial \mathbb{D}$ implies that the same is true of f. Also, because $\log|g_i| \in L^1$, it follows that $\log|f| = \log|g_1| - \log|g_2| \in L^1$. $\square$

The condition that $\log|f(e^{i\theta})|$ is integrable is phrased by saying that f is *log integrable*.

7.12 COROLLARY. *If $f \in N$ and $f(w) = 0$ on a subset of $\partial \mathbb{D}$ having positive measure, then f is identically 0.*

7.13 COROLLARY. *If $f \in H^p$ and $f(w) = 0$ on a subset of $\partial \mathbb{D}$ having positive measure, then f is identically 0.*

§8 Factorization of functions in the Nevanlinna Class. In this section we will give a canonical factorization of functions in the class N. This will also factor functions in the Hardy spaces H^p. Actually, the strategy is to factor bounded analytic functions and then use Theorem 7.10 to factor functions in N.

8.1 DEFINITION. An *inner function* is a bounded analytic function ϕ on $\mathbb{D}$ such that $|\phi(w)| = 1$ a.e. $[m]$.

It follows that every Blaschke product is an inner function, but there are some additional ones.

8.2 PROPOSITION. *If μ is a positive singular measure on $\partial \mathbb{D}$ and*

$$(8.3) \qquad \phi(z) = \exp\left(-\int \frac{w+z}{w-z} d\mu(w)\right),$$

then ϕ is an inner function.

PROOF. It is easy to see that ϕ is well defined and analytic on $\mathbb{D}$. Let $u(z) = -\tilde{\mu}(z) = -\int P_z(w) d\mu(w) = -\text{Re} \int (w+z)/(w-z) d\mu(w)$; so $|\phi(z)| = e^{u(z)}$. Since μ is a positive measure, $u(z) \leq 0$ for all z in $\mathbb{D}$. Also the fact that μ is a singular measure implies, by Fatou's Theorem, that $u(rw) \to 0$ as $r \to 1-$ a.e. $[m]$. Hence if $|w| = 1$ and both ϕ and u have a nontangential limit at w, then $|\phi(w)| = \lim_{r \to 1-} |\phi(rw)| = \lim_{r \to 1-} e^{-u(rw)} = 1$ a.e. $[m]$. Therefore, ϕ is inner. $\square$

An inner function ϕ as defined in (8.3) is called a *singular inner function*. It will turn out that singular functions are the inner functions with no zeros in $\mathbb{D}$, while the Blaschke products are the singular functions with zeros. At this point the reader might be advised to work Exercise 11 to see that the correspondence between singular inner functions and positive singular measures μ as described by (8.3) is bijective.

We come now to another class of analytic functions that are, in a certain sense, complementary to the inner functions. The idea here is to use formula (8.3) but with an absolutely continuous measure. It is also not required that the measure be positive.

8.4 Definition. An analytic function $f: \mathbb{D} \to \mathbb{C}$ is an *outer function* if there is a real-valued function h on $\mathbb{D}$ that is integrable with respect to Lebesgue measure and such that

$$(8.5) \qquad f(z) = \exp\left(\int \frac{w+z}{w-z} h(w)\, dm(w)\right)$$

for all z in $\mathbb{D}$.

It is clear that (8.5) defines an analytic function on $\mathbb{D}$ which has no zeros. Also the fact that h is real-valued implies that the harmonic function $\log|f|$ is precisely the Poisson transform of h.

8.6 Proposition. *If f is the outer function defined by (8.5), then f is in the Nevanlinna class and $h = \log|f|$ a.e. $[m]$ on $\partial \mathbb{D}$. Moreover, $f \in H^p$ if and only if $e^h \in L^p$.*

Proof. As we already observed, $\log|f| = \tilde{h}$ on $\mathbb{D}$. Thus

$$\frac{1}{2\pi}\int_0^{2\pi} \log^+ |f(re^{i\theta})|\, d\theta = \frac{1}{2\pi}\int_0^{2\pi} \max\{\tilde{h}_r(e^{i\theta}), 0\}\, d\theta$$
$$\leq \|\tilde{h}_r\|_1.$$

But Theorem 3.8 implies that this last term is uniformly bounded in r. Therefore $f \in N$. Also, Fatou's Theorem implies that f has nontangential limits a.e. on $\partial \mathbb{D}$ and so, when the limit exists, $h(w) = \lim_{r \to 1_-} \tilde{h}(rw) = \lim_{r \to 1_-} \log|f(rw)|$. This proves the first part of the proposition.

Since $|f|^p = (e^{\tilde{h}})^p = e^{p\tilde{h}}$,

$$\int |f(rw)|^p\, dm(w) = \int e^{p\tilde{h}_r(w)}\, dm(w)$$
$$= \int \exp\left(\int ph(z) P_{rw}(z)\, dm(z)\right) dm(w).$$

Since the exponential function is convex and $P_{rw}(z)\, dm(z)$ is a probability measure, Proposition 5.2 implies that

$$\int |f(rw)|^p\, dm(w) \leq \iint \exp(ph(z)) P_{rw}(z)\, dm(z)\, dm(w)$$
$$= \int \exp(ph(z))\left(\int P_{rw}(z)\, dm(w)\right) dm(z)$$
$$= \int \exp(ph(z))\, dm(z)$$
$$= \int (e^h)^p\, dm,$$

since $\int P_{rw}(z)\, dm(w) = \int P_{rz}(w)\, dm(w) = 1$. So if $e^h \in L^p$, then $f \in H^p$.

Conversely, if $f \in H^p$, then $f(e^{i\theta}) \in L^p$ and, since $e^h = |f|$, $e^h \in L^p$. □

We are now in a position to prove one of the main results of this section.

§8 FACTORIZATION

8.7 THEOREM. *If $f \in H^\infty$, then*
$$f(z) = cb(z)\phi(z)F(z),$$
where c is a constant with $|c| = 1$, b is a Blaschke product, ϕ is a singular inner function, and F is an outer function in H^∞. Conversely, any function having this form belongs to H^∞.

PROOF. Assume that f is a bounded analytic function with $\|f\|_\infty \leq 1$. By Theorem 7.8, $f = cbg$, where b is a Blaschke product, g is a bounded analytic function on $\mathbb{D}$ with $g(0) > 0$, and c is a constant with $|c| = 1$. It also follows that $\|g\|_\infty \leq 1$. Let $g = e^{-k}$, for a unique analytic function $k \colon \mathbb{D} \to \mathbb{C}$ with $k(0) = \log|g(0)|$. Thus $u \equiv \operatorname{Re} k = -\log|g| \geq 0$. That is, u is a non-negative harmonic function on $\mathbb{D}$. By Herglotz's Theorem $u(z) = \int P_z(w) \, d\mu(w)$, for some positive measure μ on $\partial \mathbb{D}$. Therefore,
$$k(z) = \int \frac{w+z}{w-z} \, d\mu(w).$$
(Why?) Let $\mu = \mu_a + \mu_s$ be the Lebesgue decomposition of μ with respect to m, with $\mu_a \ll m$ and $\mu_s \perp m$ and let h be the Radon-Nikodym derivative of μ_a with respect to m; so $\mu_a = hm$. It follows that $h \geq 0$ a.e. $[m]$ since μ is a positive measure. Define
$$F(z) = \exp\left(\int \frac{w+z}{w-z}(-h(w)) \, dm(w)\right)$$
and let ϕ be the singular inner function corresponding to the measure μ_s. It is easy to see that $g = e^{-k} = \phi F$. By (8.6), $-h = \log|F|$ a.e. $[m]$. Also $\log|g| = -\operatorname{Re} k = -u$ and $u(rw) \to h(w)$ a.e. $[m]$ by Corollary 4.11. Thus $-h = \log|g|$ on $\partial \mathbb{D}$ and this implies that $|g| = |F|$ on $\partial \mathbb{D}$. Therefore, $F \in H^\infty$.

The converse is clear. □

In light of Theorem 7.10 we can now factor functions in N.

8.8 COROLLARY. *If f is a function in the Nevanlinna class, then*
$$f(z) = cb(z) \left(\frac{\phi_1(z)}{\phi_2(z)}\right) F(z),$$
where c is a constant with $|c| = 1$, b is a Blaschke product, ϕ_1 and ϕ_2 are singular inner functions, and F is an outer function in N.

Now concentrate on H^p for finite p. It will be shown that Theorem 8.7 extends to this situation with the factor F an outer function in H^p. To do this we will first prove a result which has some independent interest.

8.9 PROPOSITION. *If $f \in H^1$, then $f = hk$ for two functions h and k in H^2.*

PROOF. Let b be the Blaschke product with the same zeros as f. By Theorem 7.8, $f = bg$, where $g \in H^1$ and g has no zeros in $\mathbb{D}$. Since $\mathbb{D}$ is simply connected, there is an analytic function h on $\mathbb{D}$ such that $h^2 = g$. Since $g \in H^1$, $h \in H^2$. Also, $k = bh \in H^2$. Clearly, $f = hk$. □

8.10 THEOREM. *If $1 \leq p \leq \infty$ and $f \in H^p$, then*
$$f(z) = cb(z)\phi(z)F(z),$$
where c is a constant with $|c| = 1$, b is a Blaschke product, ϕ is a singular inner function, and F is an outer function in H^p. Conversely, any function having this form belongs to H^p.

PROOF. In light of Theorem 8.7 it remains to consider the case where $p < \infty$. Before getting into the proper part of the proof, the reader is asked to establish the inequality
$$|\log^+ a - \log^+ b| \leq |a - b|,$$
valid for all positive numbers a and b. (Just consider various cases.) Hence
$$\int |\log^+|f(rw)| - \log^+|f(w)|| \, dm(w) \leq \int |f(rw) - f(w)| \, dm(w)$$
$$\leq \left[\int |f(rw) - f(w)|^p \, dm(w) \right]^{1/p}.$$
Since this last quantity converges to 0 as $r \to 1-$, we obtain that

(8.11) $$\lim_{r \to 1-} \int |\log^+|f(rw)| - \log^+|f(w)|| \, dm(w) = 0.$$

(8.12) Claim. If $f \in H^p$, then $\log|f(z)| \leq \int P_z(w) \log|f(w)| \, dm(w)$ for $|z| < 1$.

To see this, fix z in $\mathbb{D}$. By Theorem 7.8, $f = bg$, where b is a Blaschke product and g is a function in H^p with no zeros. Hence, $\log|f| = \log|b| + \log|g| \leq \log|g|$. Since $|f(w)| = |g(w)|$ on $\partial \mathbb{D}$, it suffices to prove the claim for the function g.

Because g does not vanish on $\mathbb{D}$, $\log|g|$ is a harmonic function on $\mathbb{D}$ and so for $0 < r < 1$, $\log|g_r|$ is harmonic in a neighborhood of $\operatorname{cl}\mathbb{D}$. Thus $\log|g_r(z)| = \int P_z(w) \log|g_r(w)| \, dm(w)$. But for $|z| < 1$, $P_z \in L^\infty$. Therefore (8.11) implies that
$$\lim_{r \to 1-} \int P_z(w) \log^+|g_r(w)| \, dm(w) = \int P_z(w) \log^+|g(w)| \, dm(w).$$
Now Fatou's Lemma implies that $\int P_z(w) \log^-|g(w)| \, dm(w) \leq \liminf_{r \to 1-} \int P_z(w) \log^-|g_r(w)| \, dm(w)$. Hence
$$\log|g(w)| = \lim_{r \to 1-} \log|g_r(w)|$$
$$= \lim_{r \to 1-} \int P_z(w)[\log^+|g_r(w)| - \log^-|g_r(w)|] \, dm(w)$$
$$\leq \int P_z \log^+|g| \, dm - \int P_z \log^-|g| \, dm$$
$$= \int P_z \log|g| \, dm,$$
thus proving (8.12).

Now to complete the proof of this theorem. Let $f = cb(\phi_1/\phi_2)F$ as in Corollary 8.8 and put $\phi = \phi_1/\phi_2$. Since $|f(w)| = |F(w)|$ a.e. $[m]$ on $\partial \mathbb{D}$, $F \in H^p$. Also $|\phi| = 1$ a.e. $[m]$ on $\partial \mathbb{D}$. So if it can be shown that $|b(z)\phi(z)| \leq 1$ on $\mathbb{D}$, this will show that $b\phi$ is an inner function and complete the proof. But for $|z| < 1$,

$$\log|F(z)| = \int P_z(w)\log|F(w)|\,dm(w)$$
$$= \int P_z(w)\log|f(w)|\,dm(w)$$
$$\geq \log|f(z)|$$

by (8.12). Therefore, $1 \geq |f(z)/F(z)| = |b(z)\phi(z)|$. □

It is now possible to obtain the promised extension to H^1 of Theorems 6.4 and 6.5. The result is stated for all p, though it is only necessary to offer a proof for the case that $p = 1$.

8.13 THEOREM. *If $f \in H^p$, $1 \leq p \leq \infty$, then $g(w) = \lim_{r\to 1_-} f(rw)$ defines a function in L^p such that $\hat{g}(n) = 0$ for $n < 0$, $\tilde{g} = f$, and $\|g\|_p = \lim_{r\to 1_-} \|f_r\|_p$. Thus, the map that takes a function f in H^p onto its boundary values g is an isometric isomorphism of H^p onto the subspace $\mathscr{H}^p$ of L^p. For $p = \infty$, this map is also a weak* homeomorphism.*

PROOF. We can assume that $p = 1$. If $f \in H^1$, then $f = hk$ for functions h and k in H^2 (8.9); also let h and k denote the respective boundary functions in $\mathscr{H}^2$. By Theorem 6.4, $\tilde{h} = h$, $\tilde{k} = k$, $\|h_r - h\|_2 \to 0$, and $\|k_r - k\|_2 \to 0$. If $0 < r, s < 1$, then

$$\int |f(rw) - f(sw)|\,dm(w) \leq \int |h(rw)[k(rw) - k(sw)]|\,dm(w)$$
$$+ \int |k(sw)[h(rw) - h(sw)]|\,dm(w)$$
$$\leq \|h_r\|_2\|k_r - k_s\|_2 + \|k_s\|_2\|h_r - h_s\|_2.$$

If $M = \max\{\|h\|_2, \|k\|_2\}$ and $\varepsilon > 0$, then there is an $r_0 < 1$ such that $\|k_r - k_s\|_2 < \varepsilon/2M$ and $\|h_r - h_s\|_2 < \varepsilon/2M$ for $r, s > r_0$. Hence $\|f_r - f_s\|_1 < \varepsilon$ whenever $r, s > r_0$. That is, $\{f_r\}$ is a Cauchy net in L^1. Let $F \in L^1$ such that $\|f_r - F\|_1 \to 0$ as $r \to 1_-$. By Theorem 3.8, $f = \tilde{F}$. By Fatou's Theorem, $F(w) = \lim_{r\to 1_-} f(rw) = g(w)$ a.e. $[m]$ on $\partial\mathbb{D}$. But since $f_r \to g$ in L^1 norm, $\hat{f}_r(n) \to \hat{g}(n)$ for all n. Thus $\hat{g}(n) = 0$ for $n < 0$. It is also clear that $\|f_r\|_1 \to \|g\|_1$. □

We have seen that if $f \in L^p$, $1 \leq p \leq \infty$, and $\hat{f}(n) = 0$ for all $n < 0$, then f is the boundary function of a function in H^p. We are therefore justified in calling functions in L^p whose negative Fourier coefficients are 0 analytic functions. What are the "analytic measures?" That is, if μ is a regular Borel measure on $\partial\mathbb{D}$ and $\hat{\mu}(n) = 0$ for $n < 0$, what can we conclude, if anything, about μ and $\tilde{\mu}$? The answer, contained in the F. and M. Riesz Theorem below, is that we get nothing new since such measures must be absolutely continuous with respect to Lebesgue measure and therefore have Radon-Nikodym derivative equal to a function in the Hardy space H^1.

8.14 THE F. AND M. RIESZ THEOREM. *If $\mu \in M(\partial \mathbb{D})$ and $\hat{\mu}(n) = 0$ for $n < 0$, then $\mu \ll m$ and $d\mu/dm \in H^1$.*

PROOF. By familiar arguments (see, for example, the proof of Theorem 6.3), if $f = \tilde{\mu}$, then $f(z) = \sum_{n=0}^{\infty} \hat{\mu}(n) z^n$ for $|z| < 1$. Thus f is an analytic function. Also, $\|f_r\|_1 \leq \|\mu\|$ and so $f \in H^1$. If $g(w) = \lim_{r \to 1-} f(rw)$, then Theorem 8.13 implies that $\tilde{g} = f$. Hence $\widetilde{\mu - g} = 0$. But now an easy computation shows that for every integer n, $\widehat{\mu - g}(n) = 0$ and so $\hat{\mu}(n) = \hat{g}(n)$ for all n. Since the trigonometric polynomials are dense in $C(\partial \mathbb{D})$, it follows that $\mu = gm$. □

We close this section with a discussion of weak convergence in the H^p spaces. Part of this discussion (the part for $p > 1$) could have been presented earlier, while the consideration of the case where $p = 1$ is dependent on the F. and M. Riesz Theorem. We begin with the case $p > 1$.

8.15 PROPOSITION. *If $1 < p \leq \infty$, $f \in H^p$, and $\{f_n\}$ is a sequence in H^p, then the following statements are equivalent.*

(a) $f_n \to f$ weakly in L^p (weak* in L^∞ if $p = \infty$).
(b) $\sup_n \|f_n\|_p < \infty$ and $f_n(z) \to f(z)$ uniformly on compact subsets of $\mathbb{D}$.
(c) $\sup_n \|f_n\|_p < \infty$ and $f_n(z) \to f(z)$ for all z in $\mathbb{D}$.
(d) $\sup_n \|f_n\|_p < \infty$ and $f_n^{(k)}(0) \to f^{(k)}(0)$ for all $k \geq 0$.

PROOF. (a) *implies* (b). By the Principle of Uniform Boundedness, $\sup_n \|f_n\|_p < \infty$. If K is a compact subset of $\mathbb{D}$, let $s = \sup\{|z| : z \in K\}$; so $s < 1$. Let $s < r < 1$; by Cauchy's Theorem for any analytic function h on $\mathbb{D}$ and z in K,

$$h(z) = \frac{1}{2\pi i} \int_{|\zeta|=r} \frac{h(\zeta)}{\zeta - z} d\zeta = \frac{1}{2\pi} \int_0^{2\pi} \frac{h(re^{i\theta})}{re^{i\theta} - z} d\theta.$$

Now if $h \in H^p$, $h_r \to h$ in L^p norm. So letting $r \to 1$ in the preceding equation gives that

$$h(z) = \frac{1}{2\pi} \int_0^{2\pi} \frac{h(e^{i\theta})}{e^{i\theta} - z} d\theta$$

for all z in K. By Exercise 5, $\{(e^{i\theta} - z)^{-1} : z \in K\}$ is compact in L^q, where q is the index dual to p. By Exercise 6, $f_n(z) \to f(z)$ uniformly on K.

(c) *implies* (a). Assume that $1 < p < \infty$. The fact that $\{f_n\}$ is norm bounded implies, by the reflexivity of L^p, that there is a function g in

L^p such that $f_n \to_{\text{cl}} g$ weakly in L^p. Since L^q is separable, there is a subsequence $\{f_{n_k}\}$ such that $f_{n_k} \to g$ weakly. If $p = \infty$, then the fact that L^1 is separable implies there is a g in L^∞ and a subsequence $\{f_{n_k}\}$ such that $f_{n_k} \to g$ weak*. In either case, for all integers m, $\hat{g}(m) = \lim_{n_k \to \infty} \hat{f}_{n_k}(m)$. Hence $g \in H^p$. Also, the fact that $(e^{i\theta} - z)^{-1} \in L^q$ implies that $\tilde{g}(z) = \langle g, (e^{i\theta} - z)^{-1}\rangle = \lim_{n_k \to \infty}\langle f_{n_k}, (e^{i\theta} - z)^{-1}\rangle = \lim_{n_k \to \infty} f_{n_k}(z) = f(z)$. Hence f is the unique weak (respectively, weak*) cluster point of $\{f_n\}$ and so $f_n \to f$ weakly (respectively, weak*).

It is clear that (b) implies (c) and the proof that (d) is equivalent to the remaining conditions is left to the reader. □

What happens if $p = 1$? By the F. and M. Riesz Theorem, H^1 can be identified, isometrically and isomorphically, with $\mathscr{L} \equiv \{\mu \in M(\partial\mathbb{D}): \hat{\mu}(n) = 0 \text{ for } n < 0\}$. Now it is easy to see that this is a weak* closed subspace of $M = M(\partial\mathbb{D})$. So if $\mathscr{L}_\perp = \{f \in C(\partial\mathbb{D}): \int fh\,dm = 0 \text{ for all } h \text{ in } H^1\}$, then $H^1 \cong (C\partial\mathbb{D})/\mathscr{L}_\perp)^*$. That is, H^1 is the dual of a Banach space and therefore has a weak* topology. In fact, this is precisely the relative weak* topology it inherits via its identification with the subspace $\mathscr{L}$ of M. Thus a sequence $\{f_n\}$ in H^1 converges weak* to f in H^1 if and only if $\int gf_n\,dm \to \int gf\,dm$ for all g in $C(\partial\mathbb{D})$. We therefore have the following analogue of the preceding proposition.

8.16 PROPOSITION. *If $f \in H^1$ and $\{f_n\}$ is a sequence in H^1, then the following statements are equivalent.*

(a) $f_n \to f$ weak* in H^1.
(b) $\sup_n \|f_n\|_1 < \infty$ and $f_n(z) \to f(z)$ uniformly on compact subsets of $\mathbb{D}$.
(c) $\sup_n \|f_n\|_1 < \infty$ and $f_n(z) \to f(z)$ for all z in $\mathbb{D}$.
(d) $\sup_n \|f_n\|_1 < \infty$ and $f_n^{(k)}(0) \to f^{(k)}(0)$ for all $k \geq 0$.

The proof is left to the reader.

Exercises

1. Let ϕ_1 and ϕ_2 be singular inner functions corresponding to the singular measures μ_1 and μ_2. If ϕ is the function ϕ_1/ϕ_2 in the Nevanlinna class, show that ϕ is an inner function if and only if $\mu_1 \geq \mu_2$.
2. Let $\mathscr{F}$ be the collection of all inner functions and observe that $\mathscr{F}$ is a semigroup under multiplication. If $\phi \in \mathscr{F}$, characterize all the divisors of ϕ. Apply this to the singular inner function ϕ corresponding to the measure $\alpha\delta_1$, where δ_1 is the unit point mass at 1 and $\alpha > 0$.
3. If ϕ is an inner function, is it possible for $1/\phi$ to belong to some H^p space? Can $1/\phi$ belong to some L^p space?
4. If Δ is a measurable subset of $\partial\mathbb{D}$ having positive Lebesgue measure and a and b are two positive numbers, show that there is an outer function

f in H^∞ with $|f(w)| = a$ a.e. on Δ and $|f(w)| = b$ a.e. on $\partial \mathbb{D} \setminus \Delta$. Show that if $h \in L^p$, $1 \leq p \leq \infty$, and $h \geq 0$, then there is a function f in H^p such that $h = |f|$ a.e. on $\partial \mathbb{D}$ if and only if $\log h \in L^1$ and the function f can always be chosen to be an outer function.

5. If $1 \leq q < \infty$, the function $z \to (e^{i\theta} - z)^{-1}$ is a continuous function from $\mathbb{D}$ into L^q. Thus for any compact subset K of $\mathbb{D}$, $\{(e^{i\theta} - z)^{-1} : z \in K\}$ is a compact subset of L^q.

6. If $\mathscr{X}$ is a Banach space, $\{x_n\}$ is a sequence in $\mathscr{X}$ such that $x_n \to 0$ weakly, and K is a norm compact subset of $\mathscr{X}^*$, then $\sup\{|x^*(x_n)| : x^* \in K\} \to 0$ as $n \to \infty$.

7. Prove that condition (d) in Proposition 8.15 is equivalent to the remaining ones.

8. Prove Proposition 8.16.

9. Show that for f in L^1, $\int f\phi\, dm = 0$ for all ϕ in H^∞ if and only if $f \in H^1$ and $f(0) = 0$. Denote this subspace of H^1 by H_0^1. Show that H^∞ is isometrically isomorphic to $(L^1/H_0^1)^*$.

10. Let b_n be the Blaschke product with a zero of multiplicity n at $1 - 1/n$. Show that $\{b_n\}$ converges weak* in H^∞. What is its limit?

11. (a) Show that if μ is a complex-valued measure on $\partial \mathbb{D}$, (8.3) defines an analytic function ϕ on $\mathbb{D}$ with no zeros.
    (b) Show that $f(z) = -\int (w+z)/(w-z)\, d\mu(w)$ is the unique analytic function on $\mathbb{D}$ such that $\phi = e^f$ and $f(0) = -\mu(\partial \mathbb{D})$.
    (c) Show that $f^{(n)}(0) = -2n! \int \overline{w}^n\, d\mu(w)$ for $n \geq 0$.
    (d) Show that if μ is a singular measure and $\phi \equiv 1$, then $\mu = 0$.
    (e) Show that if μ is a real-valued measure and $\phi \equiv 1$, then $\mu = 0$.
    (f) Now assume that μ_1 and μ_2 are two positive singular measures that represent the same singular inner function ϕ. That is, assume that $\exp(-\int (w+z)/(w-z)\, d\mu_1(w)) = \exp(-\int (w+z)/(w-z)\, d\mu_2(w)) = \phi(z)$. Show that $\mu_1 = \mu_2$.

§9 The disk algebra.

Here we study an algebra of continuous functions on the closed disk (or the unit circle) that is related to the Hardy spaces. Strictly speaking, we do not need this algebra for the study of subnormal operators, but it does furnish intuition for the study of the algebras $R(K)$ which is crucial for subnormal operator theory.

9.1 DEFINITION. The *disk algebra* is the algebra A of all continuous functions f on cl $\mathbb{D}$ that are analytic on $\mathbb{D}$.

It is easy to see that A is a Banach algebra with the supremum norm. In fact, it is a closed subalgebra of H^∞. In light of the recent sections of this book, the proof of the next result should offer little difficulty to the reader.

9.2 THEOREM. *The map $f \to f|\partial \mathbb{D}$ is an isometric isomorphism of A onto the subalgebra $\mathscr{A} = \{g \in C(\partial \mathbb{D}) : \hat{g}(n) = 0 \text{ for } n < 0\}$ of $C(\partial \mathbb{D})$.*

Furthermore, if $g \in \mathscr{A}$ and $f = \tilde{g}$, then $f|\partial \mathbb{D} = g$ and $f_r \to g$ uniformly on $\partial \mathbb{D}$.

In fact the last part of the preceding theorem is the statement that f is the solution of the Dirichlet problem for the disk with boundary values g. We will no longer make a distinction between the disk algebra A and the algebra $\mathscr{A}$ consisting of its boundary values. That is, we will often think of A as a subalgebra of $C(\partial \mathbb{D})$.

9.3 PROPOSITION. *The analytic polynomials are uniformly dense in the disk algebra.*

PROOF. This is easy to prove if you first observe that the Cesàro means of a function in the disk algebra is an analytic polynomial and then apply Theorem 1.1. □

9.4 THEOREM. *If $\rho \colon A \to \mathbb{C}$ is a nonzero homomorphism, then there is a point a in $\operatorname{cl} \mathbb{D}$ such that $\rho(f) = f(a)$ for all f in A. Thus the maximal ideal space of A is homeomorphic to $\operatorname{cl} \mathbb{D}$ and under this identification the Gelfand transform is the identity map.*

PROOF. If $\rho \colon A \to \mathbb{C}$ is a nonzero homomorphism, let $a = \rho(z)$. Since $\|\rho\| = 1$, $|a| \leq 1$. It follows by algebraic manipulation that $\rho(p) = p(a)$ for all polynomials in z. In light of the preceding proposition, $\rho(f) = f(a)$ for all f in A. Conversely, if $|a| \leq 1$ and $\rho(f) = f(a)$ for all f in A, then ρ is a homomorphism. Thus $\rho \to \rho(z)$ is a one-to-one correspondence between the maximal ideal space and $\operatorname{cl} \mathbb{D}$. The proof of the fact that this correspondence is a homeomorphism and the concomitant fact about the Gelfand transform is left to the reader. □

9.5 PROPOSITOIN. *$\{\operatorname{Re} p|\partial \mathbb{D} \colon p \text{ is an analytic polynomial}\}$ is uniformly dense in $C_{\mathbb{R}}(\partial \mathbb{D})$.*

PROOF. If $g(w) = \sum_{k=-n}^{n} C_k w^k$, where $C_{-k} = \overline{C}_k$, then $g = \operatorname{Re} p$ for some analytic polynomial. On the other hand, the Cesàro means of any function in $C_{\mathbb{R}}(\partial \mathbb{D})$ is such a trigonometric polynomial and the means converge uniformly on $\partial \mathbb{D}$ (Theorem 1.1). □

9.6 COROLLARY. *If μ is a real-valued measure on $\partial \mathbb{D}$ such that $\int p \, d\mu = 0$ for every analytic polynomial p, then $\mu = 0$.*

9.7 COROLLARY. *If μ is a real-valued measure on $\partial \mathbb{D}$ such that $\int p \, d\mu = 0$ for every analytic polynomial p with $p(0) = 0$, then $\mu = cm$ for some real constant c.*

PROOF. Let $c = \int 1 \, dm = \mu(\partial \mathbb{D})$. It is easy to check that if $\nu = \mu - cm$, then $\int p \, d\nu = 0$ for all polynomials p. The result now follows from the preceding corollary. □

This essentially completes the information we will see here about the disk algebra. There is more information available in the references. For example, what are the closed ideals of A?

Now we turn our attention to some related matters that will be of use later. We will henceforth make no distinction between functions in the Hardy spaces H^p and their boundary values. That is, with no warning we will consider functions in H^p as functions on $\mathbb{D}$ or on $\partial \mathbb{D}$ unless there is a distinct expository advantage in making a distinction.

9.8 THEOREM. *If $f \in H^1$, the following statements are equivalent.*

(a) *The function $\theta \to f(e^{i\theta})$ is of bounded variation on $[0, 2\pi]$.*
(b) *The function f belongs to the disk algebra and $\theta \to f(e^{i\theta})$ is absolutely continuous.*
(c) *The derivative of f belongs to H^1.*

PROOF. (a) *implies* (b). Let $u(\theta) = f(e^{i\theta})$; we are assuming that u is a function of bounded variation. Since $f \in H^1$, $(1/2\pi) \int_0^{2\pi} u(\theta) e^{in\theta} d\theta = \int f(w) w^n dm(w) = 0$ for $n \geq 1$. Using integration by parts, this implies that for $n \geq 1$,

$$0 = \left. \frac{u(\theta)}{in} e^{in\theta} \right|_0^{2\pi} - \frac{1}{in} \int_0^{2\pi} e^{in\theta} du(\theta)$$

$$= -\frac{1}{in} \int_0^{2\pi} e^{in\theta} du(\theta).$$

The F. and M. Riesz Theorem now implies that u is an absolutely continuous function whose negative Fourier coefficients vanish. In particular, u is continuous and since $f = \tilde{u}$, $f \in A$.

(b) *implies* (c). Since $u(\theta) = f(e^{i\theta})$ is absolutely continuous, for $0 < r < 1$ and for all θ,

$$f(re^{i\theta}) = \frac{1}{2\pi} \int_0^{2\pi} P_r(\theta - t) f(e^{it}) dt.$$

Differentiating both sides with respect to θ gives

$$ire^{i\theta} f'(re^{i\theta}) = \frac{1}{2\pi} \int_0^{2\pi} \frac{\partial}{\partial \theta} [P_r(\theta - t)] f(e^{it}) dt.$$

Since P_r is an even function, this implies

$$ire^{i\theta} f'(re^{i\theta}) = \frac{1}{2\pi} \int_0^{2\pi} \frac{\partial}{\partial t} [P_r(\theta - t)] u(t) dt.$$

Since u is absolutely continuous, integration by parts yields

$$ire^{i\theta} f'(re^{i\theta}) = \frac{1}{2\pi} \int_0^{2\pi} P_r(\theta - t) u'(t) dt.$$

This implies that $izf'(z)$ is an analytic function on $\mathbb{D}$ that is the Poisson integral of the L^1 function u'. Hence $izf'(z)$ belongs to H^1. But this implies that $f' \in H^1$.

(c) *implies* (b). Let h denote the boundary values of f'. So $h \in L^1$, $\hat{h}(-n) = 0$ for $n > 0$, and f' is the Poisson integral of h. Let $g(\theta) = \int_0^\theta ie^{it}h(t)\,dt$; so $g(0) = 0$. Also, $g(2\pi) = \int_0^{2\pi} ie^{it}h(t)\,dt = i\int_0^{2\pi} e^{-i(-t)}\,dt = i\hat{h}(-1) = 0$. Now g is absolutely continuous and $g'(\theta) = ie^{i\theta}h(\theta)$ a.e.. For $n < 0$,

$$\hat{g}(n) = \frac{1}{2\pi}\int_0^{2\pi} g(\theta)e^{-in\theta}\,d\theta$$

$$= -\frac{1}{in}\frac{1}{2\pi}\int_0^{2\pi} g(\theta)\,d(e^{-in\theta})$$

$$= -\frac{1}{2\pi in}e^{-in\theta}g(\theta)\Big|_0^{2\pi} + \frac{1}{2\pi in}\int_0^{2\pi} e^{-in\theta}g'(\theta)\,d\theta$$

$$= \frac{1}{2\pi n}\int_0^{2\pi} h(\theta)e^{-i(n-1)\theta}\,d\theta$$

$$= \frac{1}{2\pi n}\hat{h}(n-1)$$

$$= 0.$$

Thus $\tilde{g}$ belongs to the disk algebra and $\tilde{g}|\partial\mathbb{D} = g$ is absolutely continuous. Since we have already shown that (b) implies (c), we have that $\tilde{g}' \in H^1$. Moreover, an examination of the proof that (b) implies (c) reveals that

$$\frac{d}{d\theta}g(e^{i\theta}) = \lim_{r\to 1-} ie^{i\theta}\tilde{g}(re^{i\theta}).$$

But $\frac{d}{d\theta}g(e^{i\theta}) = ie^{i\theta}h(\theta) = ie^{i\theta}f'(e^{i\theta})$. Thus

$$\lim_{r\to 1-}[f'(re^{i\theta}) - \tilde{g}'(re^{i\theta})] = 0$$

a.e.. Since $f' - \tilde{g}' \in H^1$, $f' = \tilde{g}'$. Therefore, there is a constant C such that $f = \tilde{g} + C$. This implies that $f \in A$ and $\theta \to f(e^{i\theta})$ is absolutely continuous.

Since it is clear that (b) implies (a), this completes the proof. □

It is worth recording the following fact that surfaced in the preceding proof.

9.9 COROLLARY. *If f belongs to the disk algebra and $\theta \to f(e^{i\theta})$ is absolutely continuous, then*

$$\frac{d}{d\theta}f(e^{i\theta}) = ie^{i\theta}\lim_{r\to 1-}f'(re^{i\theta}) \text{ a.e.}.$$

9.10 COROLLARY. *If f belongs to the disk algebra and $\theta \to f(e^{i\theta})$ is absolutely continuous, then the length of the curve $\theta \to f(e^{i\theta})$ is $2\pi\|f'\|_1$.*

§10 The invariant subspace lattice of the unilateral shift. Beurling's Theorem (I.4.12) characterizes the invariant subspaces of the unilateral shift of multiplicity 1 as those subspaces of H^2 of the form ϕH^2 for some inner function ϕ in H^∞. From §8 we know that such functions in H^∞ can be factored as a Blaschke product times a singular inner function. In the present section this knowledge will be applied to determine the structure of the lattice of invariant subspaces of the shift.

To begin we note that if ϕ is an inner function, α is a scalar with $|\alpha| = 1$, and $\psi = \alpha\phi$, then $\phi H^2 = \psi H^2$. The converse of this is also true.

10.1 PROPOSITION. *If ϕ and ψ are inner functions, then $\phi H^2 = \psi H^2$ if and only if there is a scalar α with $|\alpha| = 1$ and $\psi = \alpha\phi$.*

PROOF. Exercise.

Now that we have determined the ambiguity in representing the invariant subspaces of the shift, we restrict our attention to certain specific inner functions.

10.2 DEFINITION. Let **I** denote the set of all inner functions ϕ of the form

$$\phi = z^k b(z) \psi(z),$$

where $k \geq 0$, b is a Blaschke product with $b(0) > 0$ (possibly $b \equiv 1$), and ψ is a singular inner function (possibly $\psi \equiv 1$). Call this the *canonical factorization* of a function in **I**.

Another way to define **I** is to say that it is the set of all inner functions ϕ such that if n is the first non-negative integer with $\phi^{(n)}(0) \neq 0$, then $\phi^{(n)}(0) > 0$. The proof of the next proposition is immediate.

10.3 PROPOSITION. *If S is the unilateral shift of multiplicity 1, then the map $\phi \to \phi H^2$ is a bijection from **I** onto Lat S.*

If $\mathscr{M}_1$ and $\mathscr{M}_2$ are subspaces of a Hilbert space $\mathscr{H}$, then define $\mathscr{M}_1 \vee \mathscr{M}_2 \equiv \text{cl}[\mathscr{M}_1 + \mathscr{M}_2]$ and $\mathscr{M}_1 \wedge \mathscr{M}_2 \equiv \mathscr{M}_1 \cap \mathscr{M}_2$. This definition of join and meet makes the set of all subspaces of $\mathscr{H}$ into a lattice with a largest element and a smallest element. It is also a complete lattice. That is, every nonempty collection of subspaces has a supremum and an infimum. If $T \in \mathscr{B}(\mathscr{H})$, then it is easy to see that Lat T is a complete sublattice of the lattice of all subspaces of $\mathscr{H}$.

Now consider the shift S. In light of the preceding proposition, the lattice structure of Lat S induces a lattice structure on **I**. What is this structure? That is, if ϕ_1 and ϕ_2 are functions in **I**, then there are functions ϕ and ψ in **I** such that $\phi_1 H^2 \wedge \phi_2 H^2 = \phi H^2$ and $\phi_1 H^2 \vee \phi_2 H^2 = \psi H^2$. How can ϕ and ψ be obtained from ϕ_1 and ϕ_2?

Note that **I** is a semigroup under multiplication and that 1 is the identity on **I**. No element of **I** has an inverse other than 1. In fact, this semigroup has no zero divisors; that is, if ϕ and $\psi \in \mathbf{I}$ and $\phi\psi = 1$, then $\phi = \psi = 1$. (Why?) This makes possible the definition of a greatest common divisor and least common multiple of two functions in **I**.

If ϕ_1 and $\phi_2 \in \mathbf{I}$, then $\phi = \gcd(\phi_1, \phi_2)$ (the *greatest common divisor* of ϕ_1 and ϕ_2) if ϕ is a divisor of both ϕ_1 and ϕ_2 and for any ψ in **I** such that $\psi|\phi_1$ and $\psi|\phi_2$, it follows that $\psi|\phi$. Similarly, $\phi = \text{lcm}(\phi_1, \phi_2)$ (the *least common multiple* of ϕ_1 and ϕ_2) if ϕ is a multiple of both ϕ_1 and ϕ_2 and if $\psi \in \mathbf{I}$ and ψ is a multiple of both ϕ_1 and ϕ_2, then ψ is a multiple of ϕ. Because of the lack of zero divisors, if either the greatest common divisor or the least common multiple of two functions in $\mathbf{I}$ exists, it is unique.

The existence of the greatest common divisor and least common multiple of two functions in $\mathbf{I}$ is not clear, though it is true.

10.4 PROPOSITION. *If ϕ_1 and $\phi_2 \in \mathbf{I}$, then:*

(a) $\phi_1 H^2 \wedge \phi_2 H^2 = \phi H^2$, where $\phi = \text{lcm}(\phi_1, \phi_2)$;
(b) $\phi_1 H^2 \vee \phi_2 H^2 = \phi H^2$, where $\phi = \gcd(\phi_1, \phi_2)$.

PROOF. First observe that if $\phi, \psi \in \mathbf{I}$, then $\phi H^2 \leq \psi H^2$ if and only if $\phi = \psi g$ for some g in $\mathbf{I}$. Now Beurling's Theorem implies that $\phi_1 H^2 \wedge \phi_2 H^2 = \phi H^2$ for a unique function ϕ in $\mathbf{I}$. Hence $\phi H^2 \subseteq \phi_k H^2$ for both k. This implies that ϕ is a multiple of ϕ_k for $k = 1, 2$. On the other hand, if $\psi \in \mathbf{I}$ such that ψ is a multiple of both ϕ_1 and ϕ_2, then $\psi H^2 \leq \phi_k H^2$ for both k. Hence $\psi H^2 \leq \phi H^2$ and so ψ is a multiple of ϕ. Therefore, ϕ is the least common multiple of ϕ_1 and ϕ_2.

The proof of (b) is similar. □

For ϕ_1 and ϕ_2 in $\mathbf{I}$, say that $\phi_1 \geq \phi_2$ if $\phi_2|\phi_1$; that is, if ϕ_1 is a multiple of ϕ_2. Henceforth it will always be assumed that $\mathbf{I}$ has this ordering. It is customary to define $\phi_1 \vee \phi_2 \equiv \text{lcm}(\phi_1, \phi_2)$ and $\phi_1 \wedge \phi_2 \equiv \gcd(\phi_1, \phi_2)$; with these definitions $\mathbf{I}$ becomes a lattice. These operations don't coincide with the corresponding lattice operations in Lat S, but, nevertheless, the following is a direct consequence of the preceding proposition.

10.5 COROLLARY. *The map $\phi \to \phi H^2$ is a lattice anti-isomorphism form $\mathbf{I}$ onto the invariant subspace lattice of the unilateral shift.*

We are still not finished. If ϕ_1 and ϕ_2 are given functions in $\mathbf{I}$, how can we compute $\phi_1 \wedge \phi_2$ and $\phi_1 \vee \phi_2$? We start by giving a more explicit formulation of the inequality $\phi_1 \leq \phi_2$ in $\mathbf{I}$.

10.6 PROPOSITION. *Let $\phi_1, \phi_2 \in \mathbf{I}$ and let $\phi_j = z^{k_j} b_j \psi_j$ be the canonical factorization of ϕ_j, $j = 1, 2$. Furthermore let μ_j be the positive singular measure on $\partial \mathbb{D}$ associated with the singular function ψ_j. Then $\phi_1 \geq \phi_2$ if and only if*

(i) $k_1 \geq k_2$;
(ii) *the zeros of b_1 contain the zeros of b_2, counting multiplicities;*
(iii) $\mu_1 \geq \mu_2$.

PROOF. Let $\phi \in \mathbf{I}$ such that $\phi_1 = \phi \phi_2$ and let $\phi = z^k b \psi$ be the canonical factorization of ϕ. Thus $z^{k_1} b_1 \psi_1 = (z^k b \psi)(z^{k_2} b_2 \psi_2) = z^{k+k_2} b b_2 \psi \psi_2$ and so $z^{k_1} = z^{k+k_2}$, $b_1 = bb_2$, and $\psi_1 = \psi\psi_2$. Thus (i) is immediate and (ii) is an easy exercise. If μ is the singular measure corresponding to ψ, then Exercise 8.11 implies that $\mu_1 = \mu + \mu_2$ and so (iii) holds.

The converse is left as an exercise. □

This gives a better idea of the ordering on **I**. Now to make the meet and join operations more precise. For this we introduce the meet and join of two positive measures, a topic not often seen in standard courses in measure theory though it is rather easy. This will furnish an easily applicable recipe for finding the join and meet of two singular inner functions, from which the join and meet of any two inner functions is readily obtained.

10.7 LEMMA. *Let μ_1 and μ_2 be two positive measures on a compact set X and set $\mu = \mu_1 + \mu_2$; set $f_j =$ the Radon-Nikodym derivative of μ_j with respect to μ, $g = \max(f_1, f_2)$, and $h = \min(f_1, f_2)$. Put $\nu = g\mu$ and $\eta = h\mu$.*

(a) *$\mu_1, \mu_2 \leq \nu$ and if σ is any positive measure on X such that $\mu_1, \mu_2 \leq \sigma$, then $\nu \leq \sigma$.*

(b) *$\mu_1, \mu_2 \geq \eta$ and if σ is any positive measure on X such that $\mu_1, \mu_2 \geq \sigma$, then $\eta \geq \sigma$.*

PROOF. It is clear that $\mu_1, \mu_2 \leq \nu$. Let σ be as described in (a) and let $\Delta = \{x \in X : f_1(x) \geq f_2(x)\}$. For any Borel set E, $\nu(E) = \int_E g\, d\mu = \int_{E \cap \Delta} f_1\, d\mu + \int_{E \setminus \Delta} f_2\, d\mu = \mu_1(E \cap \Delta) + \mu_2(E \setminus \Delta) \leq \sigma(E \cap \Delta) + \sigma(E \setminus \Delta) = \sigma(E)$. The proof of (b) is similar. □

For two positive measures μ_1 and μ_2, the measures ν and η in the preceding lemma will be denoted by $\mu_1 \vee \mu_2$ and $\mu_1 \wedge \mu_2$, respectively.

The proof of the next result consists in piecing together the various parts of this section that have preceded.

10.8 THEOREM. *Let ϕ_1 and ϕ_2 be functions in **I** with canonical factorizations $\phi_j = z^{k_j} b_j \psi_j$, where b_j is a Blaschke product with zeros Z_j, each repeated as often as its multiplicity, and ψ_j is a singular inner function with corresponding singular measure μ_j. Let $k_1 \vee k_2 = \max(k_1, k_2)$ and $k_1 \wedge k_2 = \min(k_1, k_2)$; $b_1 \vee b_2 =$ the Blaschke product with zeros $Z_1 \cup Z_2$ and $b_1 \wedge b_2 =$ the Blaschke product with zeros $Z_1 \cap Z_2$; $\psi_1 \vee \psi_2 =$ the singular inner function with measure $\mu_1 \vee \mu_2$ and $\psi_1 \wedge \psi_2 =$ the singular inner function with measure $\mu_1 \wedge \mu_2$. Then*

(a) $\phi_1 \vee \phi_2 = z^{k_1 \vee k_2}(b_1 \vee b_2)(\psi_1 \vee \psi_2)$ *and* $\phi_1 H^2 \wedge \phi_2 H^2 = (\phi_1 \vee \phi_2)H^2$.

(b) $\phi_1 \wedge \phi_2 = z^{k_1 \wedge k_2}(b_1 \wedge b_2)(\psi_1 \wedge \psi_2)$ *and* $\phi_1 H^2 \vee \phi_2 H^2 = (\phi_1 \wedge \phi_2)H^2$.

The next corollaries illustrate the power of this theorem. Also see the exercises.

10.9 COROLLARY. *If S is the unilateral shift of multiplicity 1 and $\mathcal{M}_1$ and $\mathcal{M}_2$ are nonzero invariant subspaces for S, then $\mathcal{M}_1 \cap \mathcal{M}_2 \neq (0)$.*

10.10 COROLLARY. *With the notation of the preceding theorem,* $\phi_1 H^2 \vee \phi_2 H^2 = H^2$ *if and only if* $k_1 = k_2 = 0$, $Z_1 \cap Z_2 = \emptyset$, *and* $\mu_1 \perp \mu_2$.

Exercises

1. Show that for an inner function ϕ, $\phi^{(n)}(0) > 0$ for the first positive integer n with $\phi^{(n)}(0) \neq 0$ if and only if $\phi \in \mathbf{I}$.
2. Suppose ϕ_1 and ϕ_2 are inner functions and $g \in H^2$ such that $\phi_1 = g\phi_2$. Show that g is an inner function. If ϕ_1 and $\phi_2 \in \mathbf{I}$, show that $g \in \mathbf{I}$.
3. If ϕ_1 and $\phi_2 \in \mathbf{I}$, show that $\phi_1 H^2 + \phi_2 H^2 = H^2$ if and only if there are functions u_1 and u_2 in H^∞ such that $\phi_1 u_1 + \phi_2 u_2 = 1$.
    Remark. The condition in Exercise 3 is equivalent to the requirement that $\inf\{|\phi_1(z)| + |\phi_2(z)|: |z| < 1\} > 0$. This is a consequence of the Corona Theorem (Carleson [**1962**]). Is it possible to give an easier proof in this special case? I don't know.
4. Using the notation of Theorem 10.8, show that if $k_1 = k_2 = 0$, Z_1 and Z_2 have disjoint closures, and μ_1 and μ_2 have disjoint closed supports, then $\phi_1 H^2 + \phi_2 H^2 = H^2$.
5. For ϕ in $\mathbf{I}$, define $\mathscr{H}_\phi = H^2 \cap (\phi H^2)^\perp$ and let P_ϕ be the projection of H^2 onto $\mathscr{H}_\phi$. Let $S_\phi : \mathscr{H}_\phi \to \mathscr{H}_\phi$ be the compression of S to $\mathscr{H}_\phi$; that is, $S_\phi f = P_\phi(zf)$ for all f in $\mathscr{H}_\phi$. Show that $\|S_\phi\| = 1$. Assume that $\phi = b\psi$, where b is a Blaschke product with zeros Z and ψ is a singular inner function with corresponding measure μ. Show that $\sigma(S_\phi) = \operatorname{cl} Z \cup (\operatorname{support} \mu)$. What is $\operatorname{Lat} S_\phi$?
6. If μ and ν are positive measures on a compact space X, show that the following statements are equivalent
    (a) $\mu \wedge \nu = 0$.
    (b) $\mu \vee \nu = \mu + \nu$.
    (c) $\mu \perp \nu$.
7. If $\{\phi_i\} \subseteq \mathbf{I}$, describe the inner functions ϕ and ψ such that $\bigvee \phi_i H^2 = \phi H^2$ and $\bigwedge \phi_i H^2 = \psi H^2$. Give necessary and sufficient conditions that $\bigvee \phi_i H^2 = H^2$. That $\bigwedge \phi_i H^2 = (0)$.
8. Let ϕ be the singular inner function corresponding to the measure $\alpha \delta_1$, $\alpha > 0$. If $\mathscr{M}_1$ and $\mathscr{M}_2$ are invariant subspaces for the shift and both contain ϕH^2, show that either $\mathscr{M}_1 \subseteq \mathscr{M}_2$ or $\mathscr{M}_2 \subseteq \mathscr{M}_1$. (See Sarason [**1965c**] for more on this situation.)
9. Let $\mathscr{M}_1$ and $\mathscr{M}_2$ be invariant subspaces for the shift such that $\mathscr{M}_2 \subseteq \mathscr{M}_1$. If ϕ_1 and ϕ_2 are the corresponding functions in $\mathbf{I}$, give a necessary and sufficient condition in terms of ϕ_1 and ϕ_2 that $\dim(\mathscr{M}_1 \cap \mathscr{M}_2^\perp) < \infty$. If $\dim(\mathscr{M}_1 \cap \mathscr{M}_2^\perp) > 1$, show that there is an invariant subspace $\mathscr{M}$ for the shift such that $\mathscr{M}_2 \subseteq \mathscr{M} \subseteq \mathscr{M}_1$ and $\mathscr{M} \neq \mathscr{M}_1$ or $\mathscr{M}_2$.
10. Show that the inner functions are weak* dense in the unit ball of H^∞.

11. (Conway [1973]) Endow **I** with the relative weak* topology from H^∞ and for each ϕ in **I** let P_ϕ be the projection of H^2 onto ϕH^2. Show that if a sequence $\{\phi_n\}$ in **I** converges weak* to ϕ in **I**, then $P_{\phi_n} \to P_\phi$ (SOT).

§11 Weak-star closed ideals in H^∞. A discussion of the Banach Algebra H^∞ is story by itself. (For example, see Garnett [1981].) There are a few properties of this algebra that will be of interest to us and are easily accessible from the material that has been developed so far in this chapter.

11.1 PROPOSITION. *A homomorphism $\rho \colon H^\infty \to \mathbb{C}$ is weak* continuous if and only if there is a point λ in $\mathbb{D}$ such that $\rho(\phi) = \phi(\lambda)$ for all ϕ in H^∞.*

PROOF. Assume that ρ is weak* continuous and put $\lambda = \rho(z)$. Since ρ has norm 1, $|\lambda| \leq 1$. Also, $\rho(p) = p(\lambda)$ for all polynomials p. Now if $\phi \in H^\infty$, there is a sequence $\{p_n\}$ of polynomials such that $\|p_n\|_\infty \leq \|\phi\|_\infty$ and $p_n \to \phi$ weak* in H^∞ (1.7). But ρ is weak* continuous and so $p_n(\lambda) = \rho(p_n) \to \rho(\phi)$. If it can be shown that $|\lambda| < 1$, then it also follows that $p_n(\lambda) \to \phi(\lambda)$ (8.15) and so $\rho(\phi) = \phi(\lambda)$, completing the proof.

To show that $|\lambda| < 1$, it suffices to show that if $|\lambda| = 1$, then $\{p \colon p$ is a polynomial and $p(\lambda) = 0\}$ is weak* dense in H^∞. So fix λ in $\partial \mathbb{D}$ and let B be the weak* closure of $\{(z - \lambda)p \colon p$ is a polynomial$\}$. If it can be shown that $1 \in B$, then $B = H^\infty$ and we are done. To this end, note that if, for $a > 1$, $f_a(z) = (z - \lambda)/(z - a\lambda)$, then $f_a \in B$ (Why?). It is left to the reader to show that $\|f_a\|_\infty \leq 1$ for all $a > 1$. But for $|z| < 1$, $f_a(z) \to 1$ as $a \to 1+$. By (8.15), $f_a \to 1$ weak* as $a \to 1+$. Thus $1 \in B$.

The converse is an easy consequence of (8.15). □

A reference for the next result is Rubel and Shields [1966].

11.2 THEOREM. *If ϕ is an inner function, then ϕH^∞ is a weak* closed ideal of H^∞. Conversely, every weak* closed ideal of H^∞ has this form.*

PROOF. It is straightforward to show that ϕH^∞ is a weak* closed ideal of H^∞ whenever ϕ is an inner function. Now assume that I is a weak* closed ideal in H^∞ and let $\mathcal{M}$ be the closure of I in H^2. It is immediate that $\mathcal{M}$ is an invariant subspace for the shift. By Beurling's Theorem, there is an inner function ϕ such that $\mathcal{M} = \phi H^2$. It is claimed that $I = \phi H^\infty$. Acutally, it is easy to see that $I \subseteq \phi H^\infty$ since $I \subseteq \mathcal{M} \cap H^\infty$.

For the other inclusion, let $\{f_n\} \subseteq I$ such that $\|f_n - \phi\|_2 \to 0$. By passing to a subsequence if necessary, we may assume that $f_n \to \phi$ a.e. $[m]$ on $\partial \mathbb{D}$. Since $I \subseteq \phi H^\infty$, for each n there is a function h_n in H^∞ such that $f_n = \phi h_n$. Define v_n on $\partial \mathbb{D}$ by $v_n(w) = 1$ when $|h_n(w)| \leq 1$ and $v_n(w) = |h_n(w)|$ otherwise. Now $v_n^{-1} \in L^1$ and $\log(v_n^{-1}) \in L^1$. Thus there is an outer function g_n in H^∞ such that $|g_n| = v_n$ on $\partial \mathbb{D}$ (Proposition 8.6). But $f_n \to \phi$ a.e. on $\partial \mathbb{D}$ and so $h_n = \bar{\phi} f_n \to \bar{\phi}\phi = 1$ a.e. on $\partial \mathbb{D}$. This in turn implies that $g_n \to 1$ a.e. on $\partial \mathbb{D}$. But $\|g_n\|_\infty \leq 1$ for all n and so $g_n \to 1$ weak* in L^∞. By Proposition 8.15, $g_n(z) \to 1$ for all z in $\mathbb{D}$. Thus $g_n(z)f_n(z) \to \phi(z)$ for z in $\mathbb{D}$. But $\|g_n f_n\|_\infty = \|v_n^{-1} h_n\|_\infty \le 1$. By Proposition 8.15, $g_n f_n \to \phi$ weak* in H^∞. Since $g_n f_n \in I$, $\phi \in I$. Therefore, $\phi H^\infty \subseteq I$. □

§12 Szegö's Theorem. The reader is aware that the spaces $P^2(\mu)$, the closure of the polynomials in $L^2(\mu)$ for a compactly supported measure μ, are of importance in the pursuit of subnormal operators. In this section we will study these spaces when μ is supported on the unit circle. In this situation, multiplication by z on $P^2(\mu)$ is an isometry and isometries are a well understood class of subnormal operators. However, the information obtained here will be of use in the study of other subnormal operators as well as other parts of analysis.

Let A denote that disk algebra (§9) and let $A_0 = \{f \in A : f(0) = 0\}$; so A_0 is a maximal ideal in A. If μ is a positive measure on $\partial \mathbb{D}$, let $P_0^2(\mu)$ be the closure of A_0 in $L^2(\mu)$. Since $P^2(\mu)$ is the closure of A in $L^2(\mu)$, we have that $\dim[P^2(\mu) \cap P_0^2(\mu)^\perp] \le 1$. If $\mu = m$, Lebesgue measure on $\partial \mathbb{D}$, then this dimension equals 1. Can it be 0? The answer is easily seen to be yes by taking μ to be the unit point mass at 1. However, the general answer is the key to this section.

12.1 THEOREM. *If μ is a positive measure on $\partial \mathbb{D}$ such that $1 \notin P_0^2(\mu)$ and F is the orthogonal projection of 1 onto $P_0^2(\mu)$, then the following statements hold.*

(a) *The measure $|1 - F|^2 \mu$ is a constant multiple of m.*
(b) *The function $(1 - F)^{-1}$ belongs to H^2.*
(c) *If $\mu = hm + \mu_s$ is the Lebesgue decomposition of μ with respect to m, then $(1 - F)h \in L^2(m)$.*
(d) *The function $F = 1$ a.e. $[\mu_s]$.*

PROOF. (a) Let $\{f_n\} \subseteq A_0$ such that $f_n \to F$ in $P_0^2(\mu)$. If $f \in A_0$, then $(1 - f_n)f \in A_0$ and $(1 - f_n)f \to (1 - F)f$ in $P_0^2(\mu)$. Thus $(1 - F)f \in P_0^2(\mu)$ for all f in A_0. On the other hand, by the definition of F, $1 - F \perp P_0^2(\mu)$. Hence for every f in A_0, $0 = \langle (1 - F)f, 1 - F \rangle = \int |1 - F|^2 f \, d\mu$. By Corollary 9.7, $|1 - F|^2 \mu = cm$ for some real constant c. Since $1 - F \ne 0$, $c \ne 0$.

(d) Let $\mu = \mu_a + \mu_s$ be the Lebesgue decomposition of μ with respect to m. Since $cm = |1 - F|^2 \mu = |1 - F|^2 \mu_a + |1 - F|^2 \mu_s$, it must be that $|1 - F|^2 \mu_s = 0$.

(b) Maintaining the notation of the preceding paragraph, we have that $\mu_a = c|1 - F|^{-2} m$. That is, $h \equiv d\mu_a/dm = c|1 - F|^{-2}$. Since $h \in L^1(m)$, $|1 - F|^{-1} \in L^2(m)$. Also if $n \ge 1$, then $\int (1 - F)^{-1} z^n \, dm =$

$\int (1-\overline{F})z^n |1-F|^{-2}\, dm = c^{-1}\int(1-\overline{F})z^n\, d\mu_a = c^{-1}\int(1-\overline{F})z^n\, d\mu = c^{-1}\langle z^n, 1-F\rangle = 0$. Hence (b).

(c) Since $h = c|1-F|^{-2}$, $|1-F|h = c|1-F|^{-1} \in L^2(m)$. □

12.2 COROLLARY. *If μ is a positive measure on $\partial \mathbb{D}$ with absolutely continuous part μ_a, then*

$$\inf_{f \in A_0} \int |1-f|^2\, d\mu = \inf_{f \in A_0} \int |1-f|^2\, d\mu_a.$$

Note that if μ is a measure and $\mu = \nu + \eta$, where ν and η are mutually singular measures, then $L^2(\mu) = L^2(\nu) \oplus L^2(\eta)$ in a natural way (using the Hahn decomposition). The next proposition examines this relative to the Lebesgue decomposition of μ and at the level of the P^2 spaces.

12.3 PROPOSITION. *If μ is a positive measure on $\partial \mathbb{D}$ with $\mu = \mu_a + \mu_s$ the Lebesgue decomposition of μ with respect to m, then*

$$P_0^2(\mu) = P_0^2(\mu_a) \oplus L^2(\mu_s).$$

PROOF. Clearly $P_0^2(\mu) \subseteq P_0^2(\mu_a) \oplus L^2(\mu_s)$. It must be shown that $[P_0^2(\mu)]^\perp \subseteq [P_0^2(\mu_a) \oplus L^2(\mu_s)]^\perp$. That is, it must be shown that if $g \in L^2(\mu)$ and $g \perp P_0^2(\mu)$, then $g = 0$ a.e. $[\mu_s]$. In fact if g is such a function and $n > 0$, then $0 = \langle z^n, g \rangle = \int z^n \overline{g}\, d\mu = \widehat{\overline{g}\mu}(-n)$. That is, the negative Fourier coefficients of the measure $\overline{g}\mu$ are all zero. By the F. and M. Riesz Theorem, $\overline{g}\mu = hm$ for some h in $L^1(m)$. Hence $hm = \overline{g}\mu_a + \overline{g}\mu_s$, and so $\overline{g}\mu_s = 0$. It follows that $g = 0$ a.e. $[\mu_s]$. □

We have seen that if h is a non-negative function in $L^1(m)$, then $\log h$ may fail to be integrable. In fact, $\log h \in L^1$ if and only if there is a function f in H^1 with $h = |f|$ (Proposition 8.6). Because $\log x \leq x - 1$ for $x > 0$, the only way that $\log h$ can fail to be integrable is for $\int \log h\, dm = -\infty$; that is, the approximating sums for the integral must diverge to $-\infty$. If this is indeed the case, then the expression $\exp[\int \log h\, dm]$ that appears in the subsequent text is to be interpreted as 0 (What else?).

12.4 PROPOSITION. *If h is a non-negative function in $L^1(m)$, then*

$$\exp\left[\int \log h\, dm\right] = \inf\left\{\int h e^g\, dm : g \in L_{\mathbb{R}}^1(m) \text{ and } \int g\, dm = 0\right\}.$$

PROOF. If $g \in L_{\mathbb{R}}^1(m)$ and $\int g\, dm = 0$, then $\int \log(he^g)\, dm = \int (\log h + g)\, dm = \int \log h\, dm$. Thus letting $\mu = m$ in Proposition 5.2 and replacing h by he^g, we get

$$\exp\left[\int \log h\, dm\right] \leq \inf\left\{\int h e^g\, dm : g \in L_{\mathbb{R}}^1(m) \text{ and } \int g\, dm = 0\right\}.$$

If $\varepsilon > 0$, let $c_\varepsilon = \int \log(h+\varepsilon)\, dm$ and put $g_\varepsilon = c_\varepsilon - \log(h+\varepsilon)$. Thus $g_\varepsilon \in L_{\mathbb{R}}^1(m)$ and $\int g_\varepsilon\, dm = 0$. Also, $\int h e^{g_\varepsilon}\, dm = \int h e^{c_\varepsilon}(h+\varepsilon)^{-1}\, dm =$

$e^{c_\varepsilon} \int h(h+\varepsilon)^{-1} dm$. Now $c_\varepsilon \to \int \log h\, dm$ by monotone convergence. On the other hand, $\int h(h+\varepsilon)^{-1} dm \to 1$. Thus $\int h e^{g_\varepsilon} dm \to \exp[\int \log h\, dm]$, proving the proposition. □

12.5 LEMMA. *If h is a non-negative function in $L^1(m)$ and $g \in L^1_{\mathbb{R}}(m)$ with $\int g\, dm = 0$, then there exists a sequence of functions $\{g_n\}$ in $L^\infty_{\mathbb{R}}(m)$ such that $\int g_n\, dm = 0$ for all $n \geq 1$ and $\int h e^{g_n} dm \to \int h e^g dm$ as $n \to \infty$.*

PROOF. Let $f_n = g$ if $|g| \leq n$ and 0 otherwise. Then

$$\int h e^{f_n} dm = \int_{|g| \leq n} h e^g dm + \int_{|g| > n} h\, dm.$$

Since h and $g \in L^1(m)$, $\int_{|g|>n} h\, dm \to 0$ as $n \to \infty$. By the Monotone Convergence Theorem, $\int_{|g| \leq n} h e^g dm \to \int h e^g dm$. Therefore, $\int h e^{f_n} dm \to \int h e^g dm$. Let $g_n = f_n - \int f_n$. Then $\int h e^{g_n} dm = \exp(\int f_n) \int h e^{f_n} dm$. Since $\int f_n \to \int g = 0$, $\exp(\int f_n) \to 1$ and so $\int h e^{g_n} dm \to \int h e^g dm$. □

2.6 PROPOSITION. *If $h \in L^1(m)$ and $h \geq 0$, then*

$$\exp\left[\int \log h\, dm\right] = \inf_{f \in A_0}\left[\int h e^{\mathrm{Re}\, f} dm\right].$$

PROOF. Let $\alpha \equiv \inf_{f \in A_0}[\int h e^{\mathrm{Re}\, f} dm]$ and let $\beta \equiv \inf\{\int h e^g dm : g \in L^1_{\mathbb{R}}(m)$ and $\int g\, dm = 0\}$. By Proposition 12.4, it must be shown that $\alpha = \beta$. But if $f \in A_0$, then $\int \mathrm{Re}\, f\, dm = 0$ and so $\alpha \geq \beta$. To obtain the other inequality, we first use the preceding lemma to see that $\beta = \inf\{\int h e^g dm : g \in L^\infty_{\mathbb{R}}(m)$ and $\int g\, dm = 0\}$. Since for any polynomial p, $\int \mathrm{Re}\, p\, dm = \mathrm{Re}\, p(0)$, it follows from Proposition 9.5 that $\{\mathrm{Re}\, f : f \in A_0\}$ is uniformly dense in $\{g \in C_{\mathbb{R}}(\partial \mathbb{D}) : \int g\, dm = 0\}$. Thus if $g \in L^\infty_{\mathbb{R}}(m)$ with $\int g\, dm = 0$, then there is a sequence of functions $\{f_n\}$ from A_0 such that $\{\mathrm{Re}\, f_n\}$ is uniformly bounded and $\mathrm{Re}\, f_n \to g$ a.e. on $\partial \mathbb{D}$. By the Lebesgue Dominated Convergence Theorem, $\int h e^{\mathrm{Re}\, f_n} dm \to \int h e^g dm$. Hence $\alpha \leq \int h e^g dm$. But g was arbitrary, so $\alpha \leq \beta$. □

12.7 SZEGÖ'S THEOREM. *If μ is a positive measure on $\partial \mathbb{D}$, $\mu = \mu_a + \mu_s$ is the Lebesgue decomposition of μ with respect to m, and h is the Radon-Nikodym derivative of μ_a with respect to m, then*

$$\inf_{f \in A_0} \int |1 - f|^2 d\mu = \exp\left[\int \log h\, dm\right].$$

PROOF. Using the preceding proposition,

$$\exp\left[\int \log h\, dm\right] = \inf_{g \in A_0} \int h e^{2\mathrm{Re}\, g} dm.$$

But $e^{2\operatorname{Re} g} = |e^g|^2$ and if $g \in A_0$, then $f = 1 - e^g \in A_0$. Hence $e^{2\operatorname{Re} g} = |1 - f|^2$ and

(12.8) $$\exp\left[\int \log h \, dm\right] \geq \inf_{f \in A_0} \int |1 - f|^2 h \, dm.$$

Now let $g \in A_0$ and apply (12.8) to the function $|1 - g|^2$. This yields

$$\exp\left[\int \log|1 - g|^2 \, dm\right] \geq \inf_{f \in A_0} \int |1 - f - g + fg|^2 \, dm$$
$$\geq 1,$$

since $|1 - f - g + fg|^2$ is subharmonic. This says that $\log|1 - g|^2 \in L^1$ and $a \equiv \int \log|1-g|^2 \, dm \geq 0$. Put $k = \log|1-g|^2 - a$ and $c = e^a$. Thus $\int k \, dm = 0$, $c \geq 1$, and $ce^k = |1 - g|^2$. By Proposition 12.4, applied to the original function h, $\exp[\int \log h \, dm] \leq \int h e^k \, dm \leq c \int h e^k \, dm = \int |1 - g|^2 h \, dm$. Combining this with (12.8) we get that

$$\exp\left[\int \log h \, dm\right] = \inf_{f \in A_0} \int |1 - f|^2 h \, dm.$$

The theorem now follows from Corollary 12.2. □

Of course the left-hand side of the equation in Szegö's Theorem is precisely the distance in $L^2(\mu)$ from the constant function 1 to the space $P_0^2(\mu)$. If this distance is zero, then $1 \in P_0^2(\mu)$. But this implies that for $n \geq 1$, $\bar{z}z^n = z^{n-1} \in P_0^2(\mu)$. Thus $P_0^2(\mu)$ is invariant for multiplication by $\bar{z}$ as well as z. Therefore, $P_0^2(\mu)$ contains all polynomials in z and $\bar{z}$ and must equal $L^2(\mu)$. This proves the next corollary.

12.9 COROLLARY. *If μ is a positive measure on $\partial \mathbb{D}$, $\mu = \mu_a + \mu_s$ is the Lebesgue decomposition of μ with respect to m, then $P_0^2(\mu) = L^2(\mu)$ if and only if*

$$\int \log\left(\frac{d\mu_a}{dm}\right) dm = -\infty.$$

Exercises

1. Without using any of the results of this section, show that if $\alpha > 0$ and $\Gamma = \{e^{i\theta} : \pi \geq |\theta| \geq \alpha\}$, then every continuous function on Γ is the uniform limit of analytic polynomials.
2. If $\tau: \mathbb{D} \to \mathbb{D}$ is an analytic function and $f \in H^2$, is $f \circ \tau$ in H^2?
3. Give an example of a measure μ on $\partial \mathbb{D}$ such that μ and m are mutually absolutely continuous and $P_0^2(\mu) = L^2(\mu)$. Note that $P^\infty(\mu) \neq L^\infty(\mu)$.
4. Use the F. and M. Riesz Theorem to show that for a measure μ on $\partial \mathbb{D}$, $P^\infty(\mu) \neq L^\infty(\mu)$ if and only if $m \ll \mu$.
5. Let $N = N_m$ on $L^2(m)$ and let S be the unilateral shift of multiplicity 1. Show that if $A: L^2 \to H^2$ is a bounded operator such that $AN = SA$,

then $A = 0$. Show that there is a bounded operator $B: H^2 \to L^2$ such that $\ker B = (0)$, ran B is dense, and $BS = NB$.

§13 Analytic Toeplitz operators. Here we will collect a few facts about the unilateral shift S and its associated operators, the analytic Toeplitz operators. Represent S as multiplication by z on H^2. If $\phi \in H^\infty$, $T_\phi \equiv \phi(S) =$ multiplication by ϕ on H^2. From Theorem I.4.11 we know that $\{S\}' = \{S\}'' = P^\infty(S) = \{T_\phi : \phi \in H^\infty\}$. Moreover the map $\phi \to T_\phi$ is a dual algebra isomorphism of H^∞ onto $P^\infty(S)$.

13.1 PROPOSITION. *If $\phi \in H^\infty$, then $\sigma(T_\phi) = \operatorname{cl} \phi(\mathbb{D})$.*

PROOF. If $\lambda \notin \operatorname{cl} \phi(\mathbb{D})$, then $\inf\{|\phi(z) - \lambda| : z \in \mathbb{D}\} > 0$ so that $(\phi - \lambda)^{-1} \in H^\infty$. It follows that if $\psi = (\phi - \lambda)^{-1}$, then $T_\psi = (T_\phi - \lambda)^{-1}$ and so $\sigma(T_\phi) \subseteq \operatorname{cl} \mathbb{D}$. If $\lambda \notin \sigma(T_\phi)$, let $A = (T_\phi - \lambda)^{-1}$. It is easy to see that since T_ϕ commutes with S, so does A. Therefore there is an analytic function ψ in H^∞ such that $A = T_\psi$. But $1 = A(T_\phi - \lambda) = T_{\psi(\phi - \lambda)}$, and so $\phi - \lambda$ is invertible in the Banach algebra H^∞. Hence $\phi - \lambda$ must be bounded below on $\mathbb{D}$. □

Each analytic Toeplitz operator is subnormal and M_ϕ acting on L^2 is a normal extension of T_ϕ. Is this the minimal normal extension? Clearly the answer is no when ϕ is constant. For a nonconstant function ϕ, however, it turns out that this is indeed the case. The first step in the proof is the following lemma.

13.2 LEMMA. *If $\rho: L^\infty \to \mathbb{C}$ is a positive linear functional such that $\rho(\phi) = \phi(0)$ for all functions ϕ in H^∞, then $\rho(f) = \int f \, dm$ for all f in L^∞.*

PROOF. Define $\rho_1 : L^\infty \to \mathbb{C}$ by $\rho_1(f) = \int f \, dm$ for all f in L^∞. If $u \in L^\infty$ and u is real-valued, then there is an invertible function ϕ in H^∞ such that $|\phi| = e^u$ (Why?). Therefore

$$|\phi(0)| = |\rho_1(\phi)| \le \rho_1(|\phi|) = \rho_1(e^u),$$
$$|\phi(0)|^{-1} = |\rho(\phi^{-1})| \le \rho(|\phi|^{-1}) = \rho(e^{-u}),$$

the last inequality due to the positivity of ρ. Thus $1 = |\phi(0)| |\phi(0)|^{-1} \le \rho_1(e^u) \rho(e^{-u})$. This implies that the function $g: \mathbb{R} \to \mathbb{R}$ defined by $g(t) = \rho_1(e^{tu}) \rho(e^{-tu})$ has a minimum value at $t = 0$. Hence $0 = g'(0) = \rho_1(u)\rho(1) - \rho_1(1)\rho(u) = \rho_1(u) - \rho(u)$. Therefore $\rho(u) = \rho_1(u) = \int u \, dm$ for every real-valued bounded function u. This proves the lemma. □

The preceding lemma is a special case of the Gleason-Whitney Theorem. (See page 164 in Douglas [1972].)

13.3 THEOREM. *If B is a weak* closed subalgebra of L^∞ that contains H^∞, then either $B = H^\infty$ or $B = L^\infty$.*

PROOF. The proof is divided into two cases.

CASE 1. There is a homomorphism $\rho\colon B \to \mathbb{C}$ such that $\rho(\phi) = \phi(0)$ for all ϕ in H^∞.

Let $\tilde{\rho}\colon L^\infty \to \mathbb{C}$ be a norm preserving extension of ρ. Thus $1 = \|\rho\| = \|\tilde{\rho}\| = \tilde{\rho}(1)$ and so $\tilde{\rho}$ is positive. By the preceding lemma, $\tilde{\rho}(f) = \int f\,dm$ for all f in L^∞. Hence if $f \in B$ and $n > 0$, $\hat{f}(-n) = \int fz^n\,dm = \rho(fz^n) = \rho(f)\rho(z^n) = 0$. Therefore in Case 1, $B = H^\infty$.

CASE 2. There is no homomorphism $\rho\colon B \to \mathbb{C}$ such that $\rho(\phi) = \phi(0)$ for all ϕ in H^∞.

Under these circumstances, $\rho(z) \neq 0$ for all nonzero homomorphisms ρ on B. Indeed, if there were a homomorphism $\rho\colon B \to \mathbb{C}$ such that $\rho(z) = 0$ and $\phi \in H^\infty$, then there is a ψ in H^∞ such that $\phi - \phi(0) = z\psi$. Thus $\rho(\phi) - \phi(0) = \rho(\psi)\rho(z) = 0$, contradicting the assumption of Case 2.

But this says that the Gelfand transform of z on the maximal ideal space of B never vanishes. Hence z is invertible in B; that is, $z^{-1} = \bar{z} \in B$. Since B is a weak* closed algebra, $B = L^\infty$. □

13.4 COROLLARY. *If $\phi \in H^\infty$ and ϕ is not constant, then* $\mathrm{mne}(T_\phi) = M_\phi$ *on L^2.*

PROOF. Let $\mathscr{K}$ be the smallest reducing subspace for M_ϕ that contains H^2. If $B = \{f \in L^\infty : \mathscr{K}$ is invariant for $M_f\}$, then it is easy to see that B is a weak* closed subalgebra of L^∞ that contains H^∞ and $\bar{\phi}$. Since ϕ is not constant, the preceding theorem implies that $B = L^\infty$. Thus $\mathscr{K}$ reduces the bilateral shift and contains H^2. Therefore $\mathscr{K} = L^2$. □

There is an extensive literature on analytic Toeplitz operators. In addition to the references cited at the end of §II.10, the reader can also look at Douglas [1972] and the references there.

Several open questions about analytic Toeplitz opertors remain.

13.5 OPEN QUESTION. Which analytic Toeplitz opertors are cyclic?

If T_ϕ is cyclic, then it is not difficult to see that ϕ must be one-to-one on $\mathbb{D}$ (see Exercise 3 below). However, this condition is not sufficient. If ϕ is a one-to-one function in H^∞ and $\Omega = \phi(\mathbb{D})$, then T_ϕ is cyclic if and only if the polynomials are dense in the Hardy space $H^2(\Omega)$ (see §V.10 for the definition). If Ω is a Carathéodory region (II.8.9), then this is the case and T_ϕ is cyclic (Exercise 4). If Ω is a crescent (that is, $\partial\Omega$ consists of two Jordan curves that meet at precisely one point), then the answer depends on the geometry of Ω. Akeroyd [1987, 1989, and **preprint**] has given necessary conditions and sufficient conditions that T_ϕ be cyclic and these conditions are essentially the same. Also, Solomyak and Volberg [1989] give a necessary and sufficient condition for T_ϕ to be cyclic if ϕ is analytic in a neighborhood of $\mathrm{cl}\,\mathbb{D}$. They actually prove much more than this in this quite interesting paper.

13.6 OPEN QUESTION. What are necessary and sufficient conditions for two analytic Toeplitz operators to be unitarily equivalent?

13.7 OPEN QUESTION. If two analytic Toeplitz operators are quasisimilar (respectively, similar), must they be unitarily equivalent?

Exercises

1. If $\phi \in H^\infty$ and T_ϕ is either compact or normal, show that ϕ must be constant.
2. If $\phi \in H^\infty$, show that T_ϕ is an isometry if and only if ϕ is inner. What is its multiplicity?
3. If $\phi \in H^\infty$ and T_ϕ is cyclic, show that ϕ is one-to-one on $\mathbb{D}$.
4. If $\phi \in H^\infty$, ϕ is one-to-one on $\mathbb{D}$, and $\Omega = \phi(\mathbb{D})$ is a Carathéodory region, show that T_ϕ is cyclic.

CHAPTER IV

Hyponormal Operators

Hyponormal operators were introduced in §II.4 and some of their elementary properties were derived. In this chapter deeper results are established for these operators, though this is by no means an exhaustive treatment. The new book of Martin and Putinar [**1989**] provides such a treatment. Clancey [**1979**] is a good introduction to some of the topics not covered here; in particular this book treats the principal function. Also Brown [**1987**] shows that if T is a hyponormal operator and $R(\sigma(T)) \neq C(\sigma(T))$, then T has a nontrivial invariant subspace. (This hypothesis is fulfilled when $\sigma(T)$ has nonempty interior. See the next chapter for a more thorough discussion of the algebra $R(K)$.) This theorem of Brown will not be proved here. In addition, we will not discuss the unitary invariants of Carey and Pincus [**1975, 1977, 1979,** and **1981**] or the recent work of Martin and Putinar [**1987**].

Instead attention is concentrated on two results that are pertinent to the study of subnormal operators. The first section looks at the real part of a hyponormal operator and the second focuses on the Berger-Shaw Theorem. At one point it seemed that the assumption of subnormality rather than hyponormality might simplify the proofs of these results. This is probably not the case. The proof of Theorem 1.1 below has always involved only elementary aspects of operator theory, even though the use of these ideas is rather involved. And recently Hadwin and Nordgren have simplified the proof of the Berger-Shaw Theorem to a point where it is hard to imagine a further simplification.

Nevertheless, special proofs for subnormal operators would have a value. To begin with, such a specialized proof might reveal additional information. This has happened in the case of Putnam's Inequality (3.1). This theorem follows as a corollary of the Berger-Shaw Theorem, though it predates that result. A direct proof in the subnormal case has been found by Axler and Shapiro [**1985**]. This will be postponed until §V.3 because a piece of additional analysis is needed which, for purposes of expository flow, is best left until then.

Therefore, the reader who is so inclined is encouraged to try to find new proofs of the results in the next two sections that are special to subnormal operators. I'd be interested.

§1 The real part of a hyponormal operator. If A is a selfadjoint operator, then A is said to be *absolutely continuous* if its scalar-valued spectral measure is absolutely continuous with respect to Lebesgue measure on the line.

1.1 THEOREM (Putnam [1963]). *If S is a pure hyponormal operator and $S = A + iB$, where A and B are selfadjoint, then A and B are absolutely continuous.*

PROOF. Since iS is also hyponormal, it suffices to show that A is absolutely continuous. If $C = (1/2)(S^*S - SS^*)$, then

(1.2) $$AB - BA = -iC.$$

Let $A = \int t\,dE(t)$ be the spectral decomposition of A and let Δ be an open interval with midpoint α. From (1.2) we have that $(A - \alpha)B - B(A - \alpha) = -iC$. Multiplying both sides of this equation on the right as well as the left by $E(\Delta)$, we obtain

$$\left[\int_\Delta (t-\alpha)\,dE(t)\right] E(\Delta)BE(\Delta) - E(\Delta)BE(\Delta)\left[\int_\Delta (t-\alpha)\,dE(t)\right] = -iE(\Delta)CE(\Delta).$$

By hypothesis, C is a positive operator. Hence $\|E(\Delta)C^{1/2}\|^2 = \|E(\Delta)CE(\Delta)\| \leq 2\|E(\Delta)BE(\Delta)\|\,\|\int_\Delta (t-\alpha)dE(t)\| \leq \|B\|\,|\Delta|$, where $|\Delta|$ is used to denote the Lebesgue measure of the interval Δ. This can also be written as

(1.3) $$\|C^{1/2}E(\Delta)C^{1/2}\| \leq \|B\|\,|\Delta|.$$

If $\{\Delta_n\}$ is a sequence of pairwise disjoint open intervals and Δ is their union, then, using (1.3), we get

$$\|C^{1/2}E(\Delta)C^{1/2}\| = \left\|\sum_{n=1}^\infty C^{1/2}E(\Delta_n)C^{1/2}\right\|$$
$$\leq \sum_{n=1}^\infty \|C^{1/2}E(\Delta_n)C^{1/2}\|$$
$$\leq \|B\|\,|\Delta|$$

Thus (1.3) holds for any open set Δ.

Now let Δ be an arbitrary Borel set with $|\Delta| = 0$; so there is a decreasing sequence $\{U_n\}$ of open sets such that for all $n \geq 1$, $\Delta \subseteq U_n$ and $|U_n| < 2^{-n}$. Thus $\|C^{1/2}E(U_n)C^{1/2}\| \leq \|B\|2^{-n}$ and so $C^{1/2}E(U_n)C^{1/2} \to 0$ in norm. But $U_{n+1} \subseteq U_n$ and so $E(U_n) \to E(\bigcap_n U_n)$ (SOT). Therefore $0 \leq C^{1/2}E(\Delta)C^{1/2} \leq C^{1/2}E(\bigcap_n U_n)C^{1/2} = \lim C^{1/2}E(U_n)C^{1/2} = 0$. This implies that

(1.4) $$E(\Delta)C = 0 \quad \text{whenever } |\Delta| = 0.$$

Let $\mathscr{H}_a \equiv \{f \in \mathscr{H} : \|E(\cdot)f\|^2$ is absolutely continuous$\}$. By (1.4), $\operatorname{ran} C \subseteq \mathscr{H}_a$. Let $\mathscr{L}$ = the smallest reducing subspace of $\mathscr{H}$ that contains $\operatorname{ran} C$ and reduces S. Since $\operatorname{ran} C \subseteq \mathscr{L}$, $\mathscr{L}^\perp \subseteq (\operatorname{ran} C)^\perp = \ker C = \ker(S^*S - SS^*)$. Since $\mathscr{L}^\perp$ reduces S, it follows that $S|\mathscr{L}^\perp$ is a normal operator. But S is pure, so it must be that $\mathscr{L}^\perp = (0)$. That is, $\mathscr{L} = \mathscr{H}$.

Now to say that a space reduces S is equivalent to saying that it is invariant for both A and B, the real and imaginary parts of S. So the proof will be complete if it can be shown that $\mathscr{H}_a$ contains the smallest subspace that is invariant for both A and B and contains $\operatorname{ran} C$. Indeed, this smallest space is the space $\mathscr{L}$ of the last paragraph and so this will prove that $\mathscr{H} \subseteq \mathscr{H}_a$.

It is left to the reader to show that $\mathscr{H}_a$ is a closed linear subspace of $\mathscr{H}$. Thus to complete the proof, it suffices to show that $\mathscr{H}_a$ contains $\operatorname{ran}[A^{m_k} B^{n_k} \cdots A^{m_1} B^{n_1} C]$ for any choice of positive integers $m_1, n_1, \ldots, m_k, n_k$. To do this we will show the following.

1.5 CLAIM. $E(\Delta) A^{m_k} B^{n_k} \cdots A^{m_1} B^{n_1} C = 0$ if $|\Delta| = 0$.

The proof of this claim is by induction on k. Let s be any real number not in $\sigma(B)$. By (1.2) $A(B - s) - (B - s)A = -iC$. Multiplying both sides of this equation on the left as well as the right by $(B - s)^{-1}$, we obtain

$$(B - s)^{-1}A - A(B - s)^{-1} = -i(B - s)^{-1}C(B - s)^{-1}$$
$$= i[(B - s)^{-1}C^{1/2}][(B - s)^{-1}C^{1/2}]^*.$$

But this implies that $A - i(B - s)^{-1}$ is hyponormal. In the same way that (1.4) is derived from (1.2), we have that

$$E(\Delta)(B - s)^{-1}C = 0 \quad \text{whenever } |\Delta| = 0.$$

Differentiating this equation n times with respect to s gives $E(\Delta) \times (B - s)^{-n}C = 0$ if $|\Delta| = 0$. Hence for any polynomial p, if $|\Delta| = 0$ then $E(\Delta)p((B - s)^{-1})C = 0$. By taking an appropriate sequence of polynomials and passing to the limit, we conclude that $E(\Delta)B^n C = 0$ for all $n \geq 1$ provided $|\Delta| = 0$. Since A commutes with $E(\Delta)$,

(1.6) $$E(\Delta) A^m B^n C = 0$$

for all $m, n \geq 1$, whenever $|\Delta| = 0$. This is precisely Claim 1.5 for $k = 1$.

Suppose (1.5) is valid for some $k \geq 1$. Fix m_j and n_j for $1 \leq j \leq k$ and put $L = A^{m_k - 1} B^{n_k} \cdots A^{m_1} B^{n_1} C$. Fix Δ with $|\Delta| = 0$; so $E(\Delta)AL = 0$. From (1.2) it follows that $BA = AB + iC$. Hence $B^2 A = BAB + iBC = (AB + iC)B + iBC = AB^2 + i(CB + BC)$. An induction argument (left to the reader) shows that for every $n \geq 1$, there are integers $k_j \geq 0$ and operators T_j, $1 \leq j \leq n$, such that $B^n A = AB^n + i \sum_{j=1}^{n} B^{k_j} C T_j$. Hence $E(\Delta)B^n A = E(\Delta)AB^n + i \sum_{j=1}^{n} E(\Delta) B^{k_j} C T_j$. But (1.6) implies that the sum in this expression is 0 and so we get that $E(\Delta)B^n AL = E(\Delta)AB^n L = AE(\Delta)B^n L$. That is,

$$E(\Delta)B^n A^{m_k} B^{n_k} \cdots A^{m_1} B^{n_1} C = AE(\Delta)B^n A^{m_k-1} B^{n_k} \cdots A^{m_1} B^{n_1} C.$$

By iterating this procedure we obtain

$$E(\Delta)B^n A^{m_k} B^{n_k} \cdots A^{m_1} B^{n_1} C = A^{m_k} E(\Delta) B^{n+n_k} A^{m_{k-1}} \cdots A^{m_1} B^{n_1} C,$$

and this is 0 by the induction hypothesis. Since A^m commutes with $E(\Delta)$, it follows that $E(\Delta)A^m B^n A^{m_k} B^{n_k} \cdots A^{m_1} B^{n_1} C = 0$. This establishes Claim 1.5 and completes the proof. □

For further results along the same lines, see Putnam [**1967**]. In particular, he shows (page 42) that

(1.7) $$\|S^*S - SS^*\| \leq \frac{2}{\pi} \|B\| |\sigma(A)|$$

for any hyponormal operator S.

To assume that S is a subnormal operator does not seem to simplify the proof of Theorem 1.1. Actually, in spite of the convoluted induction argument used to establish Claim 1.5, the proof is elementary so that it is hard to see any significant simplification in the case of a subnormal operator. Nevertheless it would be interesting to obtain a special proof in this case since it might reveal a conceptual difference between subnormal and hyponormal operators. In particular, can (1.7) be more easily deduced for subnormal operators?

Exercise

Give a direct proof (without using Theorem 1.1) that the real part of a pure subnormal operator has no point spectrum.

§2 The Berger-Shaw Theorem. An operator T on a Hilbert space $\mathscr{H}$ is *finitely multicyclic* if there are a finite number of vectors $g_1, \ldots, g_m$ in $\mathscr{H}$ such that $\mathscr{H}$ is the closed linear span of $\{f(T)g_j : 1 \leq j \leq m$ and $f \in \text{Rat}(\sigma(T))\}$. The vectors $g_1, \ldots, g_m$ are called *generating vectors*. If T is finitely multicyclic and m is the smallest number of generating vectors, then T is said to be *m-multicyclic*.

Denote the *commutator* of two operators A and B by $[A, B] \equiv AB - BA$. The *self-commutator* of an operator T is the operator $[T^*, T] = T^*T - TT^*$. Note that an operator T is hyponormal exactly when its self-commutator is positive.

The purpose of this section is to prove the following theorem.

2.1 THE BERGER-SHAW THEOREM. *If T is an m-multicyclic hyponormal operator, then $[T^*, T]$ is a trace class operator and*

$$\text{tr}[T^*, T] \leq \frac{m}{\pi} \text{Area}(\sigma(T)).$$

By a consideration of the unilateral shift of multiplicity 1, we see that this inequality is sharp.

The proof given here is far simpler than the original and is from Hadwin and Nordgren [1988]; their proof is, in turn, based on an earlier simplification of the original in Voiculescu [1980]. The proof requires several lemmas. Recall that the Hilbert-Schmidt norm of an operator X is defined by $\|X\|_2 \equiv [\operatorname{tr}(X^*X)]^{1/2}$.

2.2 LEMMA. *If $T \in \mathscr{B}(\mathscr{H})$ and P is a finite rank projection, then*
$$\operatorname{tr}(P[T^*, T]P) \leq \|P^\perp TP\|_2^2.$$

PROOF. Write $\mathscr{H} = (\operatorname{ran} P) \oplus (\operatorname{ran} P^\perp)$ and $T = \begin{bmatrix} A & B \\ C & D \end{bmatrix}$ with respect to this decomposition. Since $P = \begin{bmatrix} 1 & 0 \\ 0 & 0 \end{bmatrix}$, a routine computation shows that $P[T^*, T]P$ is reduced by $\operatorname{ran} P$ and on $\operatorname{ran} P$, $P[T^*, T]P = [A^*, A] + C^*C - BB^*$. Thus $\operatorname{tr}(P[T^*, T]P) = \operatorname{tr}[A^*, A] + \|C\|_2^2 - \|B\|_2^2$. Since A operates on a finite dimensional space, $\operatorname{tr}[A^*, A] = 0$. Hence $\operatorname{tr}(P[T^*, T]P) \leq \|C\|_2^2 = \|P^\perp TP\|_2^2$. □

2.3 LEMMA. *If T is an m-multicyclic operator, then there is a sequence of finite rank projections $\{P_k\}$ such that $P_k \uparrow 1(\operatorname{SOT})$ and $\operatorname{rank}[P_k^\perp T P_k] \leq m$ for all $k \geq 1$.*

PROOF. Let $g_1, \ldots, g_m$ be the generating vectors for T and let $\{\lambda_j\}$ be a countable dense subset of $\mathbf{C} \setminus \sigma(T)$; for convenience, arrange the sequence $\{\lambda_j\}$ so that each point is repeated infinitely often. Let $P_k =$ the projection of $\mathscr{H}$ onto
$$\bigvee \{T^j(T - \lambda_1)^{-1} \cdots (T - \lambda_k)^{-1} g_i : 0 \leq j \leq 2k, 1 \leq i \leq m\}.$$
Thus P_k is finite rank and $P_k \leq P_{k+1}$. Also, the only vectors in the spanning set that defines $\operatorname{ran} P_k$ that can possibly be mapped by T outside of $\operatorname{ran} P_k$ are the vectors $T^{2k}(T - \lambda_1)^{-1} \cdots (T - \lambda_k)^{-1} g_i$, $1 \leq i \leq m$. Since there are only m of these, $\operatorname{rank}[P_k^\perp T P_k] \leq m$ for all $k \geq 1$. In order to complete this proof it must be shown that $P_k \to 1(\operatorname{SOT})$.

Since the sequence $\{P_k\}$ is increasing, $\mathscr{L} = \operatorname{cl}[\bigcup_k \operatorname{ran} P_k]$ is a closed linear space. To show that $P_k \to 1(\operatorname{SOT})$, it suffices to show that $\mathscr{L} = \mathscr{H}$. To do this it suffices to show that $f(T)\mathscr{L} \subseteq \mathscr{L}$ for every f in $\operatorname{Rat}(\sigma(T))$. Since the sequence $\{\lambda_j\}$ is dense in the complement of $\sigma(T)$, it is only necessary to show that $f(T)\mathscr{L} \subseteq \mathscr{L}$ when f is a rational function with poles in the set $\{\lambda_j\}$. Hence we must show that $T\mathscr{L} \subseteq \mathscr{L}$ and $(T - \lambda_j)^{-1}\mathscr{L} \subseteq \mathscr{L}$ for all $j \geq 1$. From the definition of $\mathscr{L}$ we see that these two conditions are equivalent, respectively, to showing that for all $k \geq 1$:

(2.4) $\quad T[T^j(T - \lambda_1)^{-1} \cdots (T - \lambda_k)^{-1} g_i] \in \mathscr{L} \quad \text{for } 0 \leq j \leq 2k;$

(2.5) $\quad (T - \lambda_m)^{-1}[T^j(T - \lambda_1)^{-1} \cdots (T - \lambda_k)^{-1} g_i] \in \mathscr{L}$

for $0 \leq j \leq 2k$ and all m. To prove (2.4), we need only consider the case where $j = 2k$. Now $T^{2k+1}(T - \lambda_1)^{-1} \cdots (T - \lambda_{2k})^{-1} g_i \in \operatorname{ran} P_{2k}$ and

$A = (T - \lambda_{k+1}) \cdots (T - \lambda_{2k})$ is a polynomial in T of degree $2k - k$. Hence $T^{2k+1}(T - \lambda_1)^{-1} \cdots (T - \lambda_k)^{-1} g_i = AT^{2k+1}(T - \lambda_1)^{-1} \cdots (T - \lambda_{2k})^{-1} g_i \in \operatorname{ran} P_{2k} \subseteq \mathscr{L}$, proving (2.4).

Since (2.4) implies that $\mathscr{L}$ is an invariant subspace for T, to show (2.5) it suffices to show that $(T - \lambda_m)^{-1}[(T - \lambda_1)^{-1} \cdots (T - \lambda_k)^{-1} g_i] \in \mathscr{L}$ for all m. Since λ_m is repeated infinitely often, we may assume that $m \geq k + 2$. If $B = (T - \lambda_{k+1}) \cdots (T - \lambda_{m-1})$, then B is a polynomial in T of degree $m + k - 1$. Hence $(T - \lambda_m)^{-1}[(T - \lambda_1)^{-1} \cdots (T - \lambda_k)^{-1} g_i] = B(T - \lambda_1)^{-1} \cdots (T - \lambda_m)^{-1} g_i \in \operatorname{ran} P_m \subseteq \mathscr{L}$, proving (2.5). $\square$

2.6 LEMMA. *If T is an m-multicyclic hyponormal operator, then*
$$\operatorname{tr}[T^*, T] \leq m\|T\|^2.$$

PROOF. By Lemma 2.3 there is an increasing sequence of finite rank projections $\{P_k\}$ such that $P_k \uparrow 1(\mathrm{SOT})$ and $\operatorname{rank}[P_k^\perp T P_k] \leq m$ for all $k \geq 1$. Note that $\|P_k^\perp T P_k\|_2^2 \leq m\|P_k^\perp T P_k\|^2$. Since $\{P_k\}$ is an increasing sequence, $\operatorname{tr}[T^*, T] = \lim_k \operatorname{tr}(P_k[T^*, T]P_k)$. Using Lemma 2.2 we get
$$\operatorname{tr}[T^*, T] \leq \limsup \|P_k^\perp T P_k\|_2^2$$
$$\leq \limsup[m\|P_k^\perp T P_k\|^2]$$
$$\leq m\|T\|^2.$$

This completes the proof. $\square$

PROOF OF THEOREM 1.1. Let $R = \|T\|$ and put $D = \overline{B}(0; R)$. If $\varepsilon > 0$, let $D_1, \ldots, D_n$ be pairwise disjoint closed disks contained in $D \setminus \sigma(T)$ such that $\operatorname{Area}(D) < \operatorname{Area}(\sigma(T)) + \sum_j \operatorname{Area}(D_j) + \varepsilon$. If $D_j = \overline{B}(a_j; r_j)$, this inequality says that
$$\pi R^2 - \pi \sum_j r_j^2 < \operatorname{Area}(\sigma(T)) + \varepsilon.$$

If S is the unilateral shift of multiplicity 1, let $S_j = (a_j + r_j S)^{(m)}$. Note that each S_j is m-multicyclic. Exercise 1 implies that $A = T \oplus \bigoplus_1^n S_j$ is an m-multicyclic hyponormal operator since the spectra of the operator summands are pairwise disjoint. Also $\|A\| = R$. By Lemma 2.6, $\operatorname{tr}[A^*, A] \leq mR^2$. But
$$\operatorname{tr}[A^*, A] = \operatorname{tr}[T^*, T] + \sum_{j=1}^n \operatorname{tr}[S_j^*, S_j]$$
$$= \operatorname{tr}[T^*, T] + m\sum_{j=1}^n r_j^2.$$

Combining these facts we get
$$\pi \operatorname{tr}[T^*, T] \leq m\left(\pi R^2 - \pi \sum_{j=1}^n r_j^2\right)$$
$$\leq m(\operatorname{Area}(\sigma(T)) + \varepsilon).$$

Since ε was arbitrary, this completes the proof. □

This theorem first appeared in Berger and Shaw [**1973a** and **1973b**] and, in spite of the elementary proof given here, it is one of the deepest results about hyponormal operators. It is one of those unfortunate incidents that the original manuscript Berger-Shaw [**preprint**] was never published because there is information there that is not contained in Voiculescu [**1980**] and Hadwin and Nordgren [**1988**]. The original proof is more complicated but it contains a number of interesting ideas and constructions. The serious reader would find it worthwhile to look at the original papers Berger and Shaw [**1973a** and **1973b**] as well as [**preprint**] if possible.

It is also true that the simplified proofs produced more than the original. Voiculescu [**1980**] obtained the following generalization whose proof will not be given here. (See Exercise 5.)

2.7 THEOREM. *If T is an operator such that $[T^*, T]$ is the sum of a positive operator and a trace class operator and if there exists a Hilbert-Schmidt operator X such that $T + X$ is m-multicyclic, then*

$$\mathrm{tr}[T^*, T] \leq \frac{m}{\pi} \mathrm{Area}(\sigma(T + X)).$$

Hadwin and Nordgren [**1988**] obtained an analogous result which we state after a definition.

2.8 DEFINITION. For any operator T, let

$$\tau(T) \equiv \liminf_{P \to 1}[\mathrm{rank}(P^\perp T P)],$$

where the lim inf is taken as the finite rank projection P converges strongly to the identity. This is interpreted as follows. Let $\mathscr{P}_0(\mathscr{H})$ denote the finite rank projections on $\mathscr{H}$. For each SOT neighborhood U of 1, let

$$\tau_U(T) = \inf\{[\mathrm{rank}(P^\perp T P)] : P \in \mathscr{P}_0(\mathscr{H}) \text{ and } P \in U\}.$$

Note that $\tau_U(T) \leq \tau_V(T)$ if $V \subseteq U$. Since the neighborhoods of 1 form a directed set, $\{\tau_U(T)\}$ is a net that must converge to its supremum. Thus

$$\tau(T) = \lim \tau_U(T) = \sup\{\tau_U(T) : U \text{ is an SOT neighborhood of } 1\}.$$

Thus $\tau(T)$ is an extended positive integer and was christened by them as the *modulus of triangularity*.

The theorem of Hadwin and Nordgren [**1988**] is as follows.

2.9 THEOREM. *If S is an operator such that $[S^*, S]$ is the sum of a positive operator and a trace class operator and if X is any Hilbert-Schmidt operator, then*

$$\mathrm{tr}[S^*, S] \leq \tau(S + X) \mathrm{Area}(\sigma(S + X)).$$

Exercises

1. For $j = 1, 2$ let T_j be an m_j-multicyclic operator. Show that if $\sigma(T_1) \cap \sigma(T_2) = \varnothing$, then $T_1 \oplus T_2$ is m-multicyclic, where $m = \max(m_1, m_2)$.

2. If S is the Bergman operator for the unit disk $\mathbb{D}$, $\phi \in H^\infty(\mathbb{D})$, and $\phi(z) = \sum a_n z^n$ is its power series expansion, then

$$\text{tr}[\phi(S)^*, \phi(S)] = \sum_{n=1}^{\infty} n|a_n|^2 = \frac{1}{\pi} \iint_{\mathbb{D}} |\phi'|^2.$$

3. If G is a bounded simply connected region with smooth boundary, ϕ is the Riemann map of $\mathbb{D}$ onto G, and S is the Bergman operator for $\mathbb{D}$, then $\text{tr}[\phi(S)^*, \phi(S)] = (1/\pi)\,\text{Area}(G)$.
4. State and prove a version of Exercise 2 for arbitrary hyponormal weighted unilateral shifts of norm 1.
5. Modify the proof of the Berger-Shaw Theorem to obtain a proof of Theorem 2.7.

§3 The area of the spectrum of a hyponormal operator. The reference for the next result is Putnam [**1970**].

3.1 PUTNAM'S INEQUALITY. *If S is a hyponormal operator, then*

$$\|[S^*, S]\| \leq \frac{1}{\pi} \text{Area}(\sigma(S)).$$

PROOF. Fix a vector f with $\|f\| \leq 1$ and let $\mathscr{K} \equiv \bigvee\{r(S)f : r \in \text{Rat}(\sigma(S))\}$. If $T = S|\mathscr{K}$, then T is a 1-multicyclic hyponormal operator. By the Berger-Shaw Theorem and the fact that $\|T^*f\| \leq \|S^*f\|$, we get that $\langle [S^*, S]f, f \rangle = \|Sf\|^2 - \|S^*f\|^2 \leq \|Tf\|^2 - \|T^*f\|^2 = \langle [T^*, T]f, f \rangle \leq \text{tr}[T^*, T] \leq (1/\pi)\text{Area}(\sigma(T)) \leq (1/\pi)\text{Area}(\sigma(S))$. Since f was chosen arbitrarily amongst the unit vectors, the result follows. □

3.2 COROLLARY. *If S is a hyponormal operator whose spectrum has zero area, then S is normal.*

For related results see Putnam [**1967, 1970, 1971a, 1971b,** and **1972**].

As was mentioned in the introduction to this chapter, just after Theorem V.3.8 below there is a proof of Putnam's Inequality which does not depend on the Berger-Shaw Theorem but only for the case of subnormal operators.

§4 The self-commutator of the Bergman operator. Let's begin this section with an open problem.

4.1 OPEN QUESTION. If G is a bounded open subset of $\mathbb{C}$ and S is the Bergman operator for G, is S m-multicyclic for some integer m.

It has been shown in Akeroyd, Khavinson, and Shapiro [**preprint**] that if G is the slit disk, then the Bergman operator is not cyclic, and hence not 1-multicyclic.

In spite of the fact that this question remains unsolved, it is possible to give an estimate of the trace of the self-commutator of a Bergman operator.

4.2 THEOREM (Berger and Shaw [**preprint**]). *If G is a bounded open subset of $\mathbb{C}$ and S is the Bergman operator for G, then*

$$\text{tr}[S^*, S] \leq \frac{1}{\pi} \text{Area}(G).$$

PROOF. Let $\{U_n\}$ be a sequence of open subsets of G such that $\operatorname{cl} U_n \subseteq U_{n+1}$, U_n has only a finite number of components, and ∂U_n consists of a finite number of pairwise disjoint smooth Jordan curves. (It is left as an exercise for the reader to show that this is the case.) Let $\mathscr{H}_n = \{f \in L^2(G) : f$ is analytic on $U_n\}$. So

$$\mathscr{H}_n = L_a^2(U_n) \oplus L^2(G \setminus U_n). \tag{4.3}$$

Let P_n be the projection of $L^2(G)$ onto $\mathscr{H}_n$ and let Q_n be the projection of $L^2(G)$ onto $L^2(G \setminus U_n)$. Clearly

$$L_a^2(G) \subseteq \mathscr{H}_{n+1} \subseteq \mathscr{H}_n \subseteq L^2(G) \tag{4.4}$$

and

$$P_n Q_n = Q_n P_n = Q_n. \tag{4.5}$$

Denote the minimal normal extension of S by $N : Nf = zf$ for all f in $L^2(G)$. Observe that each of the direct summands in (4.3) is invariant for N and $L^2(G \setminus U_n)$ reduces N, so that $NQ_n = Q_n N$. Now the operator NP_n is practically the same as the restriction of N to $\mathscr{H}_n$. Since the restriction of N to $L^2(G \setminus U_n)$ is normal, we can expect that

$$[P_n N^*, NP_n] = [(P_n - Q_n)N^*, N(P_n - Q_n)]. \tag{4.6}$$

In fact this can be verified by direct computation and use of (4.5) as well as the fact that N commutes with Q_n.

But $P_n - Q_n$ is the projection of $L^2(G)$ onto $L_a^2(U_n) \oplus (0)$. If S_n is the Bergman operator for the open set U_n, then $N(P_n - Q_n)$ is unitarily equivalent to $S_n \oplus 0$. But by Theorem II.8.16, S_n is 1-multicyclic. Thus the Berger-Shaw Theorem and (4.6) imply that

$$\operatorname{tr}[P_n N^*, NP_n] \leq \frac{1}{\pi} \operatorname{Area}(U_n). \tag{4.7}$$

Note that $\bigcap_n \mathscr{H}_n = L_a^2(G)$. Thus by (4.4) $P_n \to P(\operatorname{SOT})$, where P is the projection of $L^2(G)$ onto $L_a^2(G)$. Fix an orthonormal basis $\{e_k\}$ for $L_a^2(G)$. Since $L_a^2(G) \subseteq \mathscr{H}_n$, $P_n e_k = e_k$ for all n and k. Using (4.7), we have

$$\sum_{k=1}^{\infty} [\|Ne_k\|^2 - \|P_n N^* e_k\|^2] \leq \operatorname{tr}[P_n N^*, NP_n] \leq \frac{1}{\pi} \operatorname{Area}(U_n).$$

But since $P_n \to P(\operatorname{SOT})$, $\|Ne_k\|^2 - \|P_n N^* e_k\|^2 \to \|Se_k\|^2 - \|S^* e_k\|^2$ as

$n \to \infty$. By Fatou's Lemma,

$$\begin{aligned}
\operatorname{tr}[S^*, S] &= \sum_{k=1}^{\infty} \|Se_k\|^2 - \|S^*e_k\|^2 \\
&= \sum_{k=1}^{\infty} \lim_{n \to \infty} [\|Ne_k\|^2 - \|P_n N^* e_k\|^2] \\
&\leq \liminf_{n \to \infty} \sum_{k=1}^{\infty} [\|Ne_k\|^2 - \|P_n N^* e_k\|^2] \\
&\leq \liminf_{n \to \infty} \frac{1}{\pi} \operatorname{Area}(U_n) \\
&= \frac{1}{\pi} \operatorname{Area}(G).
\end{aligned}$$

This completes the proof. □

Berger and Shaw [**preprint**] actually prove more than Theorem 4.2. First, the inequality in (4.2) is an equality. Moreover, they show that if $\phi \in H^\infty(G)$, then

$$\operatorname{tr}[\phi(S)^*, \phi(S)] = \frac{1}{\pi} \int |\phi'|^2 \, d\mathscr{A}.$$

Using some results of Hadwin and Nordgren [**1988**], Elias [**1988**] extends Theorem 4.2 to the case of the Bergman operator on a weighted Bergman space. These results have also been extended to certain unbounded Bergman operators in Jin [**1989**].

§5 The decomposition of essentially normal operators.

In this section it will be shown that an essentially normal operator (defined below) is the direct sum of a normal operator and a countable family of irreducible essentially normal operators. By the Berger-Shaw Theorem, every finitely multicyclic hyponormal operator is essentially normal. In the decomposition just mentioned, each of the irreducible summands are finitely multicyclic hyponormal operators. This also applies to subnormal operators. In particular, it will be shown that a pure cyclic subnormal operator is the direct sum of irreducible cyclic subnormal operators.

5.1 DEFINITION. An operator S is *essentially normal* if its image in the Calkin algebra, $\mathscr{B}/\mathscr{B}_0$, is normal.

Equivalently, S is essentially normal if the self-commutator $[S^*, S] = S^*S - SS^*$ is compact.

Clearly every normal operator is essentially normal. The direct sum of a finite number of essentially normal operators is essentially normal and the norm limit of a sequence of essentially normal operators is essentially normal. It is also not difficult to see that if f is analytic in a neighborhood of $\sigma(S)$ and S is essentially normal, then $f(S)$ is essentially normal and essential normality is a unitary invariant. Also, if S is a finitely multicyclic hyponormal operator, then $S^*S - SS^*$ is trace class and so S is essentially normal.

5.2 DEFINITION. An operator is *irreducible* if the only reducing subspaces for it are (0) and the whole space. The proof of the next proposition is left to the reader.

5.3 PROPOSITION. *For an operator S the following statements are equivalent.*

(a) *S is irreducible.*
(b) *If P is a projection and $PS = SP$, then $P = 0$ or 1.*
(c) *$W^*(S) = \mathscr{B}(\mathscr{H})$.*
(d) *$\{S^*, S\}' = \mathbb{C}$.*

The next result was obtained by Behncke [1968]. Gilfeather [1971] extends this theorem to operators S for which there is a polynomial $p(x, y)$ in two noncommuting variables such that $p(S, S^*)$ is compact and gives more information. These results go back to a paper of Suzuki [1966] in which he examined operators S for which $S - S^*$ is compact.

5.4 THEOREM. *If $S \in \mathscr{B}(\mathscr{H})$ and S is essentially normal, then $\mathscr{H} = \mathscr{H}_0 \oplus \mathscr{H}_1 \oplus \cdots$, where*:

(a) *each of the spaces $\mathscr{H}_n$ reduces S;*
(b) *$S_0 \equiv S|\mathscr{H}_0$ is a normal operator;*
(c) *for $n \geq 1$, $\mathscr{H}_n$ is infinite dimensional and $S_n \equiv S|\mathscr{H}_n$ is an irreducible essentially normal operator.*

This decomposition is unique except for the order of the summands.

PROOF. Let $\mathscr{A} = C^*(S)$ be the C^*-algebra generated by S and the identity and let $\mathscr{I} = \mathscr{A} \cap \mathscr{B}_0(\mathscr{H})$. Because $\mathscr{I}$ is an ideal of $\mathscr{A}$, $\vee[\mathscr{I}\mathscr{H}]$ is an invariant subspace for every operator in $\mathscr{A}$. Because $A^* \in \mathscr{A}$ whenever $A \in \mathscr{A}$, $\mathscr{H}_0 \equiv [\mathscr{I}\mathscr{H}]^\perp$ is a reducing subspace for the algebra $\mathscr{A}$. Let $S_0 \equiv S|\mathscr{H}_0$ and note that $(S^*S - SS^*)|\mathscr{H}_0 = S_0^*S_0 - S_0S_0^*$. But $S^*S - SS^* \in \mathscr{I}$ and so $(S^*S - SS^*)\mathscr{H}_0 \subseteq \vee[\mathscr{I}\mathscr{H}] = \mathscr{H}_0^\perp$. Thus $(S^*S - SS^*)\mathscr{H}_0 \subseteq \mathscr{H}_0 \cap \mathscr{H}_0^\perp = (0)$, and so S_0 is normal.

Since $\mathscr{I}|\mathscr{H}_0^\perp$ is a C^*-algebra of compact operators, a standard result for C^*-algebra theory (see, for example, page 21 in Arveson [1976]) which asserts that there are reducing spaces $\mathscr{H}_1, \mathscr{H}_2, \ldots$ for $\mathscr{I}$ such that $\mathscr{H}_0^\perp = \mathscr{H}_1 \oplus \mathscr{H}_2 \oplus \cdots$ and $\mathscr{I}|\mathscr{H}_n = \mathscr{B}_0(\mathscr{H}_n)$ for all $n \geq 1$. Since $\mathscr{H}_n = \text{cl}[\mathscr{B}_0(\mathscr{H}_n)\mathscr{H}_n] = \vee[\mathscr{I}\mathscr{H}_n]$ and $\mathscr{I}$ is an ideal of $\mathscr{A}$, each $\mathscr{H}_n$ reduces $\mathscr{A}$. Put $S_n = S|\mathscr{H}_n$. Now $A \to A|\mathscr{H}_n$ is a *-homomorphism of $\mathscr{A}$ into $\mathscr{B}(\mathscr{H}_n)$ and hence it has closed range. That is, $\mathscr{A}|\mathscr{H}_n$ is a C^*-subalgebra of $\mathscr{B}(\mathscr{H}_n)$. Also, if $p(x, y)$ is a polynomial in two noncommuting symbols x and y, it is easy to check that $p(S^*, S)|\mathscr{H}_n = p(S_n^*, S_n)$; thus $\mathscr{A}|\mathscr{H}_n = C^*(S_n)$ and S_n is essentially normal. But this implies that $C^*(S_n) \supseteq \mathscr{I}|\mathscr{H}_n = \mathscr{B}_0(\mathscr{H}_n)$. Hence $W^*(S_n) \supseteq W^*(\mathscr{B}_0(\mathscr{H}_n)) = \mathscr{B}(\mathscr{H}_n)$ and S_n is an irreducible operator.

This is not quite the end of the proof of existence; it may be that some of the spaces $\mathscr{H}_n$, $n \geq 1$, are not infinite dimensional. If this is the case, however, it must be that $\dim \mathcal{H}_n = 1$. Indeed, if $\dim \mathcal{H}_n < \infty$, then the fact that S_n is essentially normal implies that S_n is normal. Since S_n is also irreducible, the Spectral Theorem implies that $\dim \mathcal{H}_n = 1$. Just combine such one dimensional summands with the operator S_0 and renumber the sequence.

The proof of uniqueness is left to the reader (see Exercise 5). □

5.5 COROLLARY. *If S is an m-multicyclic hyponormal operator, then $S = S_0 \oplus S_1 \oplus \cdots$, where S_0 is normal and, for $n \geq 1$, S_n is an irreducible hyponormal operator acting on an infinite dimensional space such that S_n is m_n-multicyclic for some integer $m_n \leq m$.*

Except for the ordering of the summands, this decomposition is unique.

PROOF. Use the Berger-Shaw Theorem and Theorem 5.4. □

5.6 COROLLARY. *If μ is a compactly supported positive measure on the plane, then there exist pairwise disjoint Borel sets $\Delta_0, \Delta_1, \ldots$ such that if $\mu_n = \mu | \Delta_n$ ($n \geq 0$), then:*

(a) $\mu = \sum_{n=0}^{\infty} \mu_n$;
(b) $P^2(\mu) = L^2(\mu_0) \oplus P^2(\mu_1) \oplus P^2(\mu_2) \oplus \cdots$;
(c) *for $n \geq 1$, $P^2(\mu_n)$ is infinite dimensional and $P^2(\mu_n) \cap L^\infty(\mu_n)$ contains no nontrivial characteristic functions.*

Except for the ordering of the summands, this decomposition is unique.

PROOF. Because $S = S_\mu$ is a cyclic subnormal operator, Corollary 5.5 applies. So $P^2(\mu) = \mathcal{H}_0 \oplus \mathcal{H}_1 \oplus \cdots$ and $S = S_0 \oplus S_1 \oplus \cdots$ as in (5.5). It follows that S_0 is normal and for $n \geq 1$, S_n is a cyclic subnormal operator that is irreducible and acts on an infinite dimensional space. Let P_n be the projection of $P^2(\mu)$ onto $\mathcal{H}_n$.

Since $\mathcal{H}_n$ reduces S, $P_n \in \{S\}'$. But Yoshino's Theorem implies that $\{S\}' = \{M_\phi : \phi \in P^2(\mu) \cap L^\infty(\mu)\}$. Because P_n is a projection, it corresponds to M_ϕ for ϕ a characteristic function in $P^2(\mu) \cap L^\infty(\mu)$. Thus there are Borel sets $\Delta_0, \Delta_1, \ldots$ such that $P_n = M_{\chi_{\Delta_n}}$; let $\chi_n = \chi_{\Delta_n}$. Since the projections are pairwise orthogonal, the sets $\{\Delta_n\}$ are pairwise disjoint; since $\sum_{n=0}^{\infty} P_n = 1$, (a) holds.

Clearly $\mathcal{H}_n \subseteq L^2(\mu_n)$. Since $\mathcal{H}_n \subseteq P^2(\mu)$, it follows that $\mathcal{H}_n \subseteq P^2(\mu_n)$. Thus $P^2(\mu) \subseteq P^2(\mu_0) \oplus P^2(\mu_1) \oplus \cdots$. On the other hand, if $f \in P^2(\mu_n)$, there is a sequence of polynomials $\{p_j\}$ such that $\int |f - p_j|^2 d\mu_n \to 0$ as $j \to \infty$. But $\chi_n \in P^2(\mu) \cap L^\infty(\mu)$ and so $\chi_n p_j \in P^2(\mu)$ for each $j \geq 1$. Defining f to be 0 off Δ_n, we obtain that $\int |\chi_n p_j - f|^2 d\mu = \int |p_j - f|^2 d\mu_n \to 0$. Hence $f \in P^2(\mu)$ and so $P^2(\mu) = P^2(\mu_0) \oplus P^2(\mu_1) \oplus \cdots$.

Now in addition, $S_0 = S | P^2(\mu_0) = S_{\mu_0}$ is normal. For any one of a variety of reasons, this implies that $P^2(\mu_0) = L^2(\mu_0)$ and so (b) holds.

Since the characteristic functions in $P^2(\mu_n) \cap L^\infty(\mu_n)$ are in one-to-one correspondence with the projections in $\{S_n\}'$, part (c) follows from the fact that S_n is irreducible and acts on an infinite dimensional space. □

The preceding corollary is a curious phenomenon. This corollary, as stated, only involves functions and measures. However the proof involves operator theory. For some time this was the only proof of this result. In his proof of the bounded point evaluation problem, Thomson [**preprint**] gets this corollary as a (minor) byproduct of his work. See Theorem VIII.5.1 below.

5.7 COROLLARY. *With the notation of the preceding corollary, if $\mathcal{M}$ is a reducing subspace for S_μ and $n \geq 1$, then either $P^2(\mu_n) \subseteq \mathcal{M}$ or $P^2(\mu_n) \perp \mathcal{M}$.*

PROOF. Let Q be the projection of $P^2(\mu)$ onto $\mathcal{M}$; so $QS = SQ$. By Yoshino's Theorem there is a function χ in $P^2(\mu) \cap L^\infty(\mu)$ such that $Q = M_\chi | P^2(\mu)$. As in (5.6b), $\chi = \chi_0 \oplus \chi_1 \oplus \cdots$. Clearly χ and each χ_n is a characteristic function. But part (c) of the preceding corollary implies that for each $n \geq 1$, either $\chi_n = 0$ or $\chi_n = 1$. This corresponds to the two alternatives in the conclusion of this corollary. □

In light of the Berger-Shaw Theorem and the results of this section, hyponormal operators and subnormal operators with finite rank self-commutator seem like natural objects for study. Here a sharp distinction between subnormal operators and hyponormal operators arises. There are many hyponormal operators with a rank one self-commutator (see page 11 in Clancey [**1979**]) however the unilateral shift is essentially the only subnormal operator with this property (Morrel [**1973**]; see Exercise 6 below).

More recently D. Xia [**1987a** and **1987b**] has succeeded in characterizing the subnormal operators with finite rank self-commutators and, independently, Olin, Thomson, and Trent [**preprint**] have obtained this characterization in the cyclic case by entirely different methods.

To say that S is irreducible is to say that $W^*(S) = \mathcal{B}(\mathcal{H})$. Thus $\mathcal{B}(\mathcal{H})$, as a von Neumann algebra, has a subnormal generator. What other von Neumann algebras or C^*-algebras have subnormal generators? See Behncke [**1970**] and Wogen [**1971**] for answers.

Some additional papers that discuss irreducible subnormal operators and sufficient conditions for the existence of nontrivial reducing subspaces for a subnormal operator are Conway and Putnam [**1985**], Lautzenheiser [**1973** and **1979**], McGuire [**1988b**], Mlak [**1972b**], Olin and Thomson [**1980c**], Putnam [**1974a, 1974b, 1976a, 1976b,** and **1977a**].

Exercises

1. Show that if S and T are irreducible operators and X is a bounded operator such that $SX = XT$ and $S^*X = XT^*$, then either $X = 0$ or S and T are unitarily equivalent.
2. If S is an irreducible operator, what is $W^*(S \oplus S)$?

3. If S and T are irreducible operators, show that either $W^*(S \oplus T)' = W^*(S)' \oplus W^*(T)'$ or S and T are unitarily equivalent. What is $W^*(S \oplus S)'$?

4. If $\{S_n\}$ is a sequence of irreducible operators, no two of which are unitarily equivalent, then $W^*(\bigoplus_{n=1}^{\infty} S_n) = \bigoplus_{n=1}^{\infty} W^*(S_n)$. What happens if some of these operators are unitarily equivalent to others?

5. Adopt the notation of Theorem 5.4.
   (a) Show that $W^*(S)' = W^*(S_0)' \oplus W^*(\bigoplus_{n=1}^{\infty} S_n)'$ and $W^*(S) = W^*(S_0) \oplus W^*(\bigoplus_{n=1}^{\infty} S_n)$.
   (b) If P_n is the projection of $\mathcal{H}$ onto $\mathcal{H}_n$, show that for $n \geq 1$, P_n is a minimal projection in $W^*(S)'$.
   (c) Prove the uniqueness statement in Theorem 5.4.

6. (Morrel [1973]) Let S be a pure subnormal operator with a rank one self-commutator.
   (a) Show that there are constants α and β such that if $T = \alpha S + \beta$, then there is a unit vector e_0 such that $[T^*, T] = e_0 \otimes e_0$ and $Te_0 \in \ker[T^*, T]$. (Actually it suffices to only assume that S is hyponormal for this part.)
   (b) Recall that $\ker[T^*, T]$ is invariant for T (Exercise II.2.7) and so $\operatorname{ran} T \subseteq \ker[T^*, T]$.
   (c) Prove that T is quasinormal.
   (d) Show that T is the unilateral shift of multiplicity 1.

7. If $\phi \in H^{\infty}$ and T_ϕ is the corresponding Toeplitz operator on H^2, show that $[T_\phi^*, T_\phi]$ has finite rank if and only if there is a finite Blaschke product b such that $\phi \in H^2 \cap (z\phi H^2)^{\perp}$. Also, show that such functions ϕ are rational functions and that b can be chosen so that $\operatorname{ran}[T_\phi^*, T_\phi] = H^2 \cap (z\phi H^2)^{\perp}$.

8. If S is essentially normal, what is the commutator ideal of $C^*(S)$? (See Lemma II.12.9.)

9. Compare the ideal $\mathcal{I}$ in the proof of Theorem 5.4 and the commutator ideal of $C^*(S)$.

Chapter V

Uniform Rational Approximation

Throughout this chapter K will denote a compact subset of $\mathbb{C}$. Recall that the uniform closure of the rational functions with poles off K, $\text{Rat}(K)$, is denoted by $R(K)$; so $R(K)$ is a closed subalgebra of $C(K)$, the continuous complex-valued functions defined on K. $P(K)$ is the closure in $C(K)$ of the polynomials and $A(K)$ is the algebra of functions in $C(K)$ that are analytic on $\text{int}\, K$. For any set X and any function $f: X \to \mathbb{C}$, $\|f\|_X \equiv \sup\{|f(z)| z \in X\}$.

We have already seen that for a von Neumann operator S, $f(S)$ can be defined for any f in $R(\sigma(S))$ (II.9.6). Moreover, if S is a von Neumann operator and $R(\sigma(S)) = C(\sigma(S))$, then S is normal (II.9.9). These are good reasons to study the algebra $R(K)$, but if subnormal operators are considered, there are even more reasons for such an endeavor.

If S is a subnormal operator and μ is a scalar-valued spectral measure for S, then $\phi(S)$ can be defined for any ϕ in $R^\infty(\sigma(S), \mu)$, the weak*-closure of $R(\sigma(S))$ in $L^\infty(\mu)$. In order to fully utilize this functional calculus for subnormal operators, it is necessary to understand the function algebra $R^\infty(K, \mu)$ for an arbitrary compact subset K of $\mathbb{C}$ and an arbitrary positive measure μ supported on K. This is the subject of the next chapter and it clearly depends on the present one.

The treatment of rational approximation here is far from complete, though more information will be given than the minimum amount needed to undertake a discussion of $R^\infty(K, \mu)$. A more thorough treatment can be found in Gamelin [**1969**] or Stout [**1971**], from which much of the present treatment is derived. Another source are the notes of Gamelin [**notes**] which have never been published and Dudziak [**1981**].

§1 Function algebras: examples and elementary properties

1.1 DEFINITION. A *function algebra* on a compact space X is a closed subalgebra A of $C(X)$ that contains the constant functions and separates the points of X.

The condition that A separates points means that for distinct points x and y of X there is a function f in A such that $f(x) \neq f(y)$. If $g = [f - f(x)]/[f(x) - f(y)]$, then $g \in A$ and $g(x) = 0$ and $g(y) = 1$. Thus the condition that A separates points is equivalent to the requirement that for distinct points x and y in X there is a g in A with $g(x) = 0$ and $g(y) = 1$.

1.2 EXAMPLES. (a) If K is a compact subset of $\mathbb{C}$, then $P(K)$, $R(K)$, and $A(K)$ are all function algebras on K.

(b) If A is a function algebra on X and B is any closed subalgebra of $C(X)$ that contains A, then B is a function algebra.

(c) If A, B are function algebras on X, Y, then the norm closed linear span in $C(X \times Y)$ of $\{(x, y) \to f(x)g(y) : f \in A \text{ and } g \in Y\}$ is a function algebra on $X \times Y$.

1.3 PROPOSITION. *If A is a function algebra and $\rho \colon A \to \mathbb{C}$ is a nonzero homomorphism, then $\|\rho\| = 1$.*

PROOF. Since $\rho(f) = \rho(1 \cdot f) = \rho(1)\rho(f)$ and $\rho(f) \neq 0$ for at least one f in A, we have that $\rho(1) = 1$. Thus $\|\rho\| \geq 1$. On the other hand if $f \in A$ and $\|f\| = 1$, then for $|\lambda| > 1$ the expansion of $(\lambda - f)^{-1}$ in a geometric series shows that $(\lambda - f)^{-1} \in A$. But then $1 = \rho((\lambda - f)(\lambda - f)^{-1}) = \rho(\lambda - f)\rho((\lambda - f)^{-1})$ and so $\rho(\lambda - f) \neq 0$. That is, $\rho(f) \neq \lambda$ for any λ with $|\lambda| > 1 = \|f\|$. Hence it must be that $|\rho(f)| \leq 1$ when $\|f\| = 1$. Therefore $\|\rho\| \leq 1$. □

The proof of the next proposition follows traditional lines and is left to the reader. The part of the statement that there is a natural homeomorphic embedding of X into $\mathcal{M}_A$, the maximal ideal space of A, refers to the map $x \to \delta_x$, where $\delta_x(f) = f(x)$ for all f in A. For this reason we will refer to points x in X as homomorphisms.

1.4 PROPOSITION. *If A is a function algebra on X and $\mathcal{M}_A$ is the maximal ideal space of A, then the Gelfand transform is an isometric isomorphism of A onto a function algebra on $\mathcal{M}_A$. Also, there is a natural homeomorphic embedding of X into $\mathcal{M}_A$.*

We now focus our attention on the specific function algebras $P(K)$, $R(K)$, and $A(K)$ for a compact subset K in the plane. Clearly $P(K) \subseteq R(K) \subseteq A(K) \subseteq C(K)$; and if K is a subset of $\mathbb{R}$, all these algebras are equal.

The next proposition is a straightforward application of Runge's Theorem. The details are left to the reader.

1.5 PROPOSITION. *If f is analytic in a neighborhood of K, then $f \in R(K)$.*

1.6 PROPOSITION. *For any compact set K, $P(K) = R(\widehat{K}) = P(\widehat{K}) = P(\partial \widehat{K})$. Precisely speaking, the identity map on polynomials extends to an isometric isomorphism between the algebras $P(K)$, $P(\widehat{K})$, $R(\widehat{K})$, and $P(\partial \widehat{K})$.*

PROOF. If $\rho(p) = p$ for every polynomial p, then the Maximum Modulus Theorem implies that ρ is isometrically defined on a dense subalgebra of $P(K)$. Hence ρ extends to an isometric isomorphism $\rho \colon P(K) \to P(\widehat{K})$. That is, $P(K) = P(\widehat{K})$.

Since $\widehat{K} = \widehat{\partial K}$, the preceding paragraph shows that $P(\widehat{K}) = P(\partial \widehat{K})$.

It remains to show that $P(K) = R(K)$ when $K = \hat{K}$. Clearly $P(K) \subseteq R(K)$, so it suffices to show that $\text{Rat}(K) \subseteq P(K)$ if $K = \hat{K}$. But this is also an easy exercise in the application of Runge's Theorem. □

1.7 PROPOSITION. *If K is a closed disk, then $P(K) = R(K) = A(K)$.*

PROOF. From the preceding proposition and elementary observations, we know that $P(K) = R(K) \subseteq A(K)$. The equality of $P(K)$ and $A(K)$ was shown in (III.9.3). □

1.8 PROPOSITION. *If K is an annulus, then $R(K) = A(K)$.*

PROOF. Without loss of generality we may assume that $K = \{z: r \leq |z| \leq R\}$. If $f \in A(K)$, then f has a Laurent series development $f(z) = \sum_n a_n z^n$ and the convergence is uniform on compact subsets of $\text{int } K$. Let $f_0(z) \equiv \sum_{n \geq 0} a_n z^n$ and $f_1(z) \equiv \sum_{n < 0} a_n z^n$. So $f_0(z)$ is analytic for $|z| < R$ and $f_1(z)$ is analytic for $|z| > r$, as well as in a neighborhood of ∞ where $f_1(\infty) = 0$. Since $f \in A(K)$, $f_0 = f - f_1 \in A(\{z: |z| \leq R\})$. By the preceding proposition, $f_0 \in P(\{z: |z| \leq R\})$ and so $f_0 \in R(K)$. If $g(z) = f_1(z^{-1})$, then $g \in A(\{z: |z| \leq R^{-1}\})$. Again the preceding proposition implies there is a sequence of polynomial $\{p_k\}$ such that $p_k(z) \to g(z)$ uniformly for $|z| \leq R$. Moreover, since $g(0) = 0$, p_k may be replaced by the polynomial $p_k - p_k(0)$ and so we may assume that $p_k(0) = 0$ for all k. If $u_k(z) = p_k(z^{-1})$, then $u_k \in \text{Rat}(K)$ and $\|u_k - f_1\|_K \to 0$. □

Actually this result can be generalized to any set K such that $\mathbb{C}\setminus K$ has a finite number of components. See Corollary 19.3 below.

A consideration of the annulus gives an example of a set K with $P(K) \neq R(K)$. To find an example of a K with $R(K) \neq A(K)$ is a little more difficult.

1.9 EXAMPLE. A *Swiss Cheese* is a compact set K obtained as follows. Let $\{\Delta_j\}$ be a sequence of open disks of radius r_j having the following properties:

(i) $\text{cl}\,\Delta_j \subseteq \mathbb{D}$ and $\text{cl}\,\Delta_j \cap \text{cl}\,\Delta_i = \emptyset$ for $i \neq j$;
(ii) $\sum_j r_j < \infty$;
(iii) $K = \text{cl}\,\mathbb{D}\setminus \bigcup_j \Delta_j$ has no interior.

The fact that such disks can be found is left as an exercise for the reader.

It is claimed that for this choice of K, $R(K) \neq A(K)$. In fact, $A(K) = C(K)$ since $\text{int } K = \emptyset$. So to see that $R(K) \neq C(K)$ we will exhibit a nonzero measure μ on K with $\mu \perp R(K)$. For $j \geq 1$, let γ_j be the boundary of Δ_j with positive orientation and let ν_j be the measure on K such that $\int f \, d\nu_j = -\int_{\gamma_j} f$ for every f in $C(K)$. Note that $\|\nu_j\| = r_j$. Let γ_0 be the positively oriented boundary of $\mathbb{D}$ and let ν_0 be the measure on K such that $\int f \, d\nu_0 = \int_{\gamma_0} f$ for every f in $C(K)$. Because of condition (ii) above, $\mu \equiv \nu_0 + \sum_j \nu_j$ converges in $M(K)$. It follows from Cauchy's Theorem that $\int f \, d\mu = 0$ for every f in $\text{Rat}(K)$. Hence $\mu \perp R(K)$. □

Is it always the case that $\operatorname{int} K = \varnothing$ (and so $A(K) = C(K)$) when $R(K) \neq A(K)$? No.

1.10 EXAMPLE. Let E be a compact subset of $\mathbb{D}$ with positive area and void interior. Now choose a sequence of open disks $\{\Delta_j\}$ having the following properties:

(i) $\operatorname{cl}\Delta_j \subseteq \mathbb{D}\setminus E$ and $\operatorname{cl}\Delta_j \cap \operatorname{cl}\Delta_i = \varnothing$ for $i \neq j$;
(ii) if r_j is the radius of Δ_j, then $\sum_j r_j < \infty$;
(iii) $E = \operatorname{cl}[(\bigcup_j \Delta_j)]\setminus \bigcup_j \operatorname{cl}\Delta_j$; that is, a point is an accumulation point of the disks Δ_j if and only if it belongs to E.

Let $K = \operatorname{cl}\mathbb{D}\setminus\bigcup_j \Delta_j$ and define the measure μ as in the preceding example. Once again $\mu \perp R(K)$. Since $\operatorname{int} K \neq \varnothing$, $A(K) \neq C(K)$. We will show that $R(K) \neq A(K)$ by exhibiting a function f in $A(K)$ such that $\int f \, d\mu \neq 0$.

Let

$$f(w) = \int_E (z-w)^{-1} \, d\mathscr{A}(z).$$

From Proposition 2.2 in the next section, f is a continuous function on $\mathbb{C}_\infty$, f is analytic off E, $f(\infty) = 0$, and $f'(\infty) = -\operatorname{Area} E \neq 0$. Thus $f \in A(K)$.

The power series expansion of f near ∞ is given by $f(z) = a_1 z^{-1} + a_2 z^{-2} + \cdots$, where $a_1 = -\operatorname{Area} E$. Since f is analytic in a neighborhood of $\operatorname{cl}\Delta_j$ for all j, $\int f \, d\nu_j = 0$ for $j \geq 1$ (using the notation of the preceding example). Hence $\int f \, d\mu = \int_{\gamma_0} f = -2\pi i \operatorname{Area} E \neq 0$.

Exercises

1. Verify the statements made in Example 1.2.
2. Prove Proposition 1.4.
3. Show that if $f \in R(K)$ and $a \in \operatorname{int} K$, then there is a g in $R(K)$ with $f - f(a) = (z-a)g$. If $R(K)$ is replaced by $P(K)$ or $A(K)$, is this still true?
4. Is the group of invertible elements in $R(K)$ connected?
5. Give the details in the construction of the Swiss Cheese.
6. Give the details in the construction of the set K in Example 1.10.

§2 Distributions and some results from analysis. This section contains a number of miscellaneous results that many consider basic facts of mathematical analysis, but that don't seem to have found a home in the usual elementary courses in the graduate curriculum. There will also appear some results that are found in such places but that are recalled here to be rephrased in the form needed in this book.

We begin with an elementary computation.

2.1 Lemma. *If K is a compact subset of $\mathbb{C}$, then for every z*

$$\int_K |z-\zeta|^{-1} \, d\mathscr{A}(\zeta) < \infty.$$

Proof. If z is fixed, then $\int_K |z-\zeta|^{-1} d\mathscr{A}(\zeta) = \int \chi_K(\zeta)|z-\zeta|^{-1} d\mathscr{A}(\zeta) = \int \chi_K(z-\zeta)|\zeta|^{-1} d\mathscr{A}(\zeta) = \int_{z-K} |\zeta|^{-1} d\mathscr{A}(\zeta)$. If R is sufficiently large that $z - K \subseteq B(0; R)$, then $\int_K |z-\zeta|^{-1} d\mathscr{A}(\zeta) \leq \int_{B(0;R)} |\zeta|^{-1} d\mathscr{A}(\zeta) = 2\pi R$ by using polar coordinates. $\square$

With this established, we can give the properties of a construction that is basic for much of what will be done.

2.2 Proposition. *If K is a compact set having positive area and*

$$f(z) \equiv \int_K (\zeta - z)^{-1} \, d\mathscr{A}(\zeta)$$

for all z in $\mathbb{C}$ and $f(\infty) = 0$, then $f: \mathbb{C}_\infty \to \mathbb{C}$ is a continuous function that is analytic on $\mathbb{C}_\infty \setminus K$ with $f'(\infty) = -\operatorname{Area} K$. In addition,

$$|f(z)| \leq [\pi \operatorname{Area}(K)]^{1/2}.$$

Proof. The fact that f is a continuous function on $\mathbb{C}_\infty$ is left as an exercise for the reader. It is also easy to see that f is analytic on $\mathbb{C} \setminus K$ and since f is continuous at ∞, ∞ is a removable singularity. Because $f(\infty) = 0$, $f'(\infty)$ is the limit of $zf(z)$ as $z \to \infty$. But $zf(z) = \int_K (\zeta/z - 1)^{-1} d\mathscr{A}(\zeta) \to -\operatorname{Area} K$ as $z \to \infty$ since $\zeta/z \to 0$ uniformly for ζ in K.

It remains to prove the inequality for $|f(z)|$. This inequality is due to Ahlfors and Beurling [**1950**], though the proof here is from Gamelin and Khavinson [**preprint**]. From the properties we have already established and the Maximum Modulus Theorem, f attains its maximum value at some point of K. By translating the set K, we may assume that $0 \in K$ and f attains its maximum at 0. In addition, if f is replaced by a suitable unimodular multiple of itself, we may assume that $f(0) > 0$. Thus

$$|f(z)| \leq f(0) = \int_K \operatorname{Re} \frac{1}{\zeta} \, d\mathscr{A}(\zeta).$$

Let $c = \frac{1}{2}[\frac{\pi}{\operatorname{Area} K}]^{1/2}$ and let $a = 1/2c$. It is elementary to see that the closed disk $D = \bar{B}(a; a)$ is $\{z: \operatorname{Re}(1/z) \geq c\}$ and that D and K have the same area. Thus $\mathscr{A}(D \cap K) + \mathscr{A}(D \setminus K) = \mathscr{A}(D) = \mathscr{A}(K) = \mathscr{A}(D \cap K) + \mathscr{A}(K \setminus D)$; hence $\mathscr{A}(D \setminus K) = \mathscr{A}(K \setminus D)$. On the other hand, $\operatorname{Re}(1/\zeta) \leq c$ for ζ in $K \setminus D$

and $\operatorname{Re}(1/\zeta) \geq c$ for ζ in $D\setminus K$. Therefore

$$\begin{aligned}
f(0) &\leq \int_{K\cap D} \operatorname{Re}\frac{1}{\zeta}\, d\mathscr{A} + c\mathscr{A}(K\setminus D) \\
&= \int_{K\cap D} \operatorname{Re}\frac{1}{\zeta}\, d\mathscr{A} + c\mathscr{A}(D\setminus K) \\
&\leq \int_{K\cap D} \operatorname{Re}\frac{1}{\zeta}\, d\mathscr{A} + \int_{D\setminus K} \operatorname{Re}\frac{1}{\zeta} \\
&= \int_D \operatorname{Re}\frac{1}{\zeta}\, d\mathscr{A}.
\end{aligned}$$

We leave it to the reader to show that for $0 < r < a$,

$$\int_0^{2\pi} \frac{1}{a + re^{i\theta}}\, d\theta = \frac{2\pi}{a},$$

by converting this to an integral around the circle $z = a + re^{i\theta}$. Hence converting to polar coordinates, we get that

$$\begin{aligned}
\int_D \operatorname{Re}\frac{1}{\zeta}\, d\mathscr{A} &= \int_0^a \int_0^{2\pi} \operatorname{Re}\frac{1}{a + re^{i\theta}}\, d\theta\, r\, dr \\
&= \int_0^a \operatorname{Re}\int_0^{2\pi} \frac{1}{a + re^{i\theta}}\, d\theta\, r\, dr \\
&= \int_0^a \frac{2\pi}{a} r\, dr \\
&= \pi a \\
&= \pi \left[\frac{\operatorname{Area} K}{\pi}\right]^{1/2} \\
&= [\pi \operatorname{Area} K]^{1/2}. \quad \square
\end{aligned}$$

For an open subset of $\mathbb{C}$ and $n \geq 1$, $C^n(G)$ is the space of those complex-valued functions on G whose partial derivatives up to and including the nth order exist and are continuous. $C_c^n(G)$ is the subalgebra of $C^n(G)$ consisting of those functions with compact support in G. (The *support* of a function ϕ is defined as the closure of $\{z: \phi(z) \neq 0\}$ and is abbreviated by $\operatorname{supp}\phi$.) Also $C^n \equiv C^n(\mathbb{C})$ and $C_c^n \equiv C_c^n(\mathbb{C})$.

Functions on $\mathbb{C}$ can, of course, be thought of as functions defined on $\mathbb{R}^2$. Because we are interested in making connections with analytic function theory, it is more convenient not to think of such functions as functions of two real variables. But we cannot think of a function u defined on an open subset of $\mathbb{C}$ as a function of a single complex variable since we want to discuss the differentiability of such functions even when they are not analytic. Hence it is necessary to introduce the two complex variables z and $\bar{z}$ that are related to the two real variables in the usual way: $z = x + iy$ and $\bar{z} = x - iy$. With this change of variables and the corresponding change in the basis of the space of differentials, the following formulas relating the corresponding derivatives hold:

$$\partial u \equiv \partial_z u \equiv \frac{\partial u}{\partial z} \equiv \frac{1}{2}\left(\frac{\partial u}{\partial x} - i\frac{\partial u}{\partial y}\right),$$

$$\overline{\partial} u \equiv \partial_{\bar{z}} u \equiv \frac{\partial u}{\partial \bar{z}} \equiv \frac{1}{2}\left(\frac{\partial u}{\partial x} + i\frac{\partial u}{\partial y}\right).$$

The usual rules for differentiating algebraic combinations of functions as well as the chain rule remain in force for these new derivatives. This can be verified by direct computation or appeal to the theory of differential forms.

Using this notation, the Cauchy-Riemann Equations become $\overline{\partial} u = 0$. That is, a function u in $C^1(G)$ is analytic if and only if $\overline{\partial} u = 0$. In words, this says that a continuously differentiable function is analytic precisely when it is a function of z alone and not $\bar{z}$. This notation also produces the following form of Green's Theorem.

2.3 GREEN'S THEOREM. *Let Γ be a smooth positively oriented Jordan system of curves and let $G = \text{ins}\,\Gamma$. If $u \in C(\text{cl}\,G)$ such that $u \in C^1(G)$ with $\overline{\partial} u$ integrable over G, then*

$$\int_\Gamma u = 2i \int_G \overline{\partial} u \, d\mathcal{A}.$$

You may derive this formulation of Green's Theorem from your favorite version of that result. This allows the following extension of Cauchy's Integral Formula.

2.4 THEOREM. *Let Γ be a smooth positively oriented Jordan system of curves and let $G = \text{ins}\,\Gamma$. If $u \in C(\text{cl}\,G)$ and $u \in C^1(G)$ with $\overline{\partial} u$ bounded, then for every w in G,*

$$u(w) = \frac{1}{2\pi i}\int_\Gamma u(z)(z-w)^{-1} dz - \frac{1}{\pi}\int_G (z-w)^{-1}\overline{\partial} u(z) \, d\mathcal{A}(z).$$

PROOF. Fix w in G and choose $\varepsilon > 0$ sufficiently small that $\overline{B}(w;\varepsilon) \subseteq G$. Put $\Delta_\varepsilon = B(w;\varepsilon)$ and let $G_\varepsilon = G\setminus\text{cl}\,\Delta_\varepsilon$. Now apply Green's Theorem to G_ε and the function $(z-w)^{-1}u$. On G_ε we have $\overline{\partial}[(z-w)^{-1}u] = (z-w)^{-1}\overline{\partial} u$ since $(z-w)^{-1}$ is analytic on G_ε. Hence Green's Theorem implies

(2.5) $$\int_\Gamma \frac{u(z)}{z-w} dz - \int_{\partial\Delta_\varepsilon} \frac{u(z)}{z-w} dz = 2i \int_{G_\varepsilon} \frac{1}{z-w} \overline{\partial} u \, d\mathcal{A}(z).$$

But

$$\lim_{\varepsilon\to 0}\int_{\partial\Delta_\varepsilon} \frac{u(z)}{z-w} dz = \lim_{\varepsilon\to 0} i\int_0^{2\pi} u(w + \varepsilon e^{i\theta})\,d\theta = 2\pi i u(w).$$

Because $(z-w)^{-1}$ is locally integrable (Lemma 2.1), the limit, as $\varepsilon \to 0$, of the right-hand side of (2.5) also exists. Thus letting $\varepsilon \to 0$ in (2.5) yields

$$\int_\Gamma \frac{u(z)}{z-w} dz - 2\pi i u(w) = 2i \int_G \frac{1}{z-w} \overline{\partial} u \, d\mathcal{A}(z).$$

This completes the proof. □

2.6 COROLLARY. *If $u \in C_c^1$, then for all w in $\mathbb{C}$*

$$u(w) = -\frac{1}{\pi} \int_G \frac{1}{z-w} \overline{\partial} u(z) \, d\mathscr{A}(z).$$

2.7 DEFINITION. If G is an open subset of $\mathbb{C}$, a *distribution* on G is a linear functional $L: C_c^\infty(G) \to \mathbb{C}$ with the property that if K is any compact subset of G and $\{\phi_k\}$ is a sequence in $C_c^\infty(G)$ with $\operatorname{supp} \phi_k \subseteq K$ for every $k \geq 1$ and if for all $m, n \geq 1$, $\partial^n \overline{\partial}^m \phi_k(z) \to 0$ uniformly for z in K as $k \to \infty$, then $L(\phi_k) \to 0$. The functions in $C_c^\infty(G)$ are referred to as *test functions*.

It is possible to define a topology on $C_c^\infty(G)$ such that $C_c^\infty(G)$ becomes a locally convex topological vector space and the distributions are precisely the continuous linear functionals on this space. See §IV.5 in [ACFA].

2.8 EXAMPLES. (a) If $L_{\text{loc}}^1(G)$ denotes the space of locally integrable functions on G (that is, the measurable functions u on G such that $\int_K |u| \, d\mathscr{A} < \infty$ for every compact subset K of (G) and $u \in L_{\text{loc}}^1(G)$, then u defines a distribution L_u via the formula $L_u(\phi) = \int u\phi \, d\mathscr{A}$.

(b) If μ is an extended complex-valued regular Borel measure on G, then μ defines a distribution L_μ via the formula $L_\mu(\phi) = \int \phi \, d\mu$.

The verification of these statements is left to the reader. Also, note that if L is a distribution on G then $\phi \to L(\partial \phi)$ and $\phi \to L(\overline{\partial}\phi)$ also define distributions on G. This justifies the following definition.

2.9 DEFINITION. If L is a distribution on G, let ∂L and $\overline{\partial} L$ be the distributions defined by $\partial L(\phi) = -L(\partial \phi)$ and $\overline{\partial} L(\phi) = -L(\overline{\partial}\phi)$.

The minus signs are placed here in the definition so that if u is a continuously differentiable function on G, then $\partial L_u = L_{\partial u}$ and $\overline{\partial} L_u = L_{\overline{\partial} u}$.

For the most part we will be concerned with distributions that are defined by locally integrable functions, measures, and the derivatives of such distributions. Be aware, however, that the derivative of a distribution defined by a function is not necessarily a distribution defined by a function. If $u \in L_{\text{loc}}^1(G)$, we will often consider ∂u as the distributional derivative of u. That is, ∂u is the distribution ∂L_u and the caution just expressed is the reminder that ∂u is not necessarily a function. Similar statements hold for $\overline{\partial} L$ and all higher derivatives.

Corollary 2.6 can now be given a distributional interpretation.

2.10 PROPOSITION. *If $w \in G$, then $\overline{\partial}[(z-w)^{-1}] = \pi \delta_w$, where δ_w is the unit point mass at w.*

PROOF. If $u(z) = (z-w)^{-1}$ and $L = L_u$, then Corollary 2.6 implies that for any test function ϕ, $\overline{\partial} u(\phi) = -\int (z-w)^{-1} \overline{\partial} \phi \, d\mathscr{A}(z) = \pi \phi(w)$. □

Let ϕ be an infinitely differentiable function on $\mathbb{C}$ such that $\phi \geq 0$, $\phi(z) = 0$ off $\mathbb{D}$, and $\int \phi \, d\mathscr{A} = 1$. (An example of such a function is obtained by letting $\phi(z) = c\exp[-1(1-|z|^2)^{-1}]$ for $|z| < 1$ and $\phi(z) = 0$ for $|z| \geq 1$, where the constant c is chosen so that $\int \phi \, d\mathscr{A} = 1$.) For $\varepsilon > 0$, let $\phi_\varepsilon(z) = \varepsilon^{-2}\phi(z/\varepsilon)$. Note that ϕ_ε is still infinitely differentiable, $\phi_\varepsilon(z) = 0$ for $|z| \geq \varepsilon$, and $\int \phi_\varepsilon \, d\mathscr{A} = 1$. This net $\{\phi_\varepsilon\}$ is called a *mollifier* or *regularizer* and for f in L^1_{loc},

$$\phi_\varepsilon * f(w) = \int \phi_\varepsilon(w-z) f(z) \, d\mathscr{A}(z)$$

is called the *mollification* or *regularization* of f.

2.11 PROPOSITION. *Let $f \in L^1_{\text{loc}}$ and let K be a compact subset of $\mathbb{C}$.*

(a) *$\phi_\varepsilon * f \in C^\infty$ for every $\varepsilon > 0$.*
(b) *If $f = 0$ off K and U is an open set containing K, then $\phi_\varepsilon * f \in C^\infty_c(U)$ for $0 < \varepsilon < \text{dist}(K, \partial U)$.*
(c) *If f is continuous on an open set that contains K, then $\phi_\varepsilon * f \to f$ uniformly on K.*
(d) *If $f \in L^p_{\text{loc}}$, $1 \leq p < \infty$, then*

$$\lim_{\varepsilon \to 0} \int_K |\phi_\varepsilon * f(w) - f(w)|^p \, d\mathscr{A}(w) = 0.$$

PROOF. (a) Since ϕ_ε is infinitely differentiable, the fact that $\phi_\varepsilon * f$ is infinitely differentiable follows by applying Leibniz's rule for differentiating under the integral sign.

(b) Let $0 < \varepsilon < \text{dist}(K, \mathbb{C}\setminus U)$. If $\text{dist}(w, K) \geq \varepsilon$, then $\phi_\varepsilon(z-w) = 0$ for all z in K. Hence if $\text{dist}(w, K) \geq \varepsilon$, then $\phi_\varepsilon * f(w) = \int_K \phi_\varepsilon(w-z) f(z) \, d\mathscr{A}(z) = 0$.

(c) Only consider $\varepsilon < \text{dist}(K, \mathbb{C}\setminus U)$. Because $\int \phi_\varepsilon \, d\mathscr{A} = 1$ and $\phi_\varepsilon = 0$ off $B(0;\varepsilon)$, for w in K we have that $|\phi_\varepsilon * f(w) - f(w)| = |\int \phi_\varepsilon(w-z)[f(z) - f(w)] \, d\mathscr{A}(z)| \leq \sup\{|f(z) - f(w)| : |w - z| < \varepsilon\}$. But f must be uniformly continuous in a neighborhood of K and so the right-hand side of this inequality can be made arbitrarily small uniformly for w in K.

(d) Let U be a bounded open set containing K and let $\alpha > 0$. Let g be a continuous function with support contained in U such that $\int_U |f-g|^p \, d\mathscr{A} < \alpha^p$. If $0 < \varepsilon < \text{dist}(K, \mathbb{C}\setminus U)$, then

$$\int_K |\phi_\varepsilon * f(w) - \phi_\varepsilon * g(w)|^p \, d\mathscr{A}(w)$$

$$= \int_K \left| \int_U \phi_\varepsilon(z-w)^{1/q} \phi_\varepsilon(z-w)^{1/p} [f(z) - g(z)] \, d\mathscr{A}(z) \right|^p d\mathscr{A}(w)$$

$$\leq \int_K \left(\int_U \phi_\varepsilon(z-w) \, d\mathscr{A}(z) \right)^{p/q} \left(\int_U \phi_\varepsilon(z-w) |f(z) - g(z)|^p d\mathscr{A}(z) \right) d\mathscr{A}(w)$$

$$= \int_U |f(z) - g(z)|^p \left[\int_K \phi_\varepsilon(z-w) \, d\mathscr{A}(w) \right] d\mathscr{A}(z)$$

$$< \alpha^p.$$

Therefore

$$\left[\int_K |\phi_\varepsilon * f(w) - f(w)|^p \, d\mathscr{A}(w)\right]^{1/p}$$
$$< 2\alpha + \left[\int_K |\phi_\varepsilon * g(w) - g(w)|^p \, d\mathscr{A}(w)\right]^{1/p}.$$

By part (c), the right-hand side of this inequality can be smaller than 3α if ε is chosen sufficiently small. □

2.12 WEYL'S LEMMA. *If $u \in L^1_{\text{loc}}(G)$ and $\bar{\partial}u = 0$ as a distribution, then there is an analytic function f on G such that $u = f$ a.e.* [Area].

PROOF. Let $\{\phi_\varepsilon\}$ be a mollifier. Fix δ with $0 < \varepsilon < \delta$ and put $G_\delta = \{z \in G : \text{dist}(z, \partial G) > \delta\}$. Note that $(\partial/\partial\bar{z})(\phi_\varepsilon(w-z)) = -(\partial/\partial\bar{w})(\phi_\varepsilon(w-z))$. Hence if $\psi \in C_c^\infty(G_\delta)$, then

$$\int \psi \bar{\partial}(\phi_\varepsilon * u) \, d\mathscr{A} = -\int \bar{\partial}\psi(w)(\phi_\varepsilon * u)(w) \, d\mathscr{A}(w)$$
$$= -\int u(z) \left[\int \bar{\partial}\psi(w)\phi_\varepsilon(w-z) \, d\mathscr{A}(w)\right] d\mathscr{A}(z)$$
$$= +\int u(z) \left[\int \psi(w) \frac{\partial}{\partial \bar{w}}[\phi_\varepsilon(w-z)] \, d\mathscr{A}(w)\right] d\mathscr{A}(z)$$
$$= -\int u(z) \left[\int \psi(w) \frac{\partial}{\partial \bar{z}}[\phi_\varepsilon(w-z)] \, d\mathscr{A}(w)\right] d\mathscr{A}(z)$$
$$= -\int u(z) \frac{\partial}{\partial \bar{z}} \left[\int \psi(w)\phi_\varepsilon(w-z) \, d\mathscr{A}(w)\right] d\mathscr{A}(z)$$
$$= -\int u(z)\bar{\partial}(\phi_\varepsilon * \psi)(z) \, d\mathscr{A}(z)$$
$$= \bar{\partial}L_u(\phi_\varepsilon * \psi)$$
$$= 0.$$

That is, $\bar{\partial}(\phi_\varepsilon * u) = 0$ on G_δ. Since $\phi_\varepsilon * u \in C^\infty$, $\phi_\varepsilon * u$ is analytic on G_δ when $0 < \varepsilon < \delta$. By part (d) of the preceding lemma, $\int_K |\phi_\varepsilon * u - u| \, d\mathscr{A} \to 0$ as $\varepsilon \to 0$ for any compact subset K of G_δ. Since Bergman spaces are complete, $u \in L^1_a(U)$ for any open set U with $\text{cl}\, U \subseteq G_\delta$. Since δ was arbitrary, the result follows. □

This section closes with the construction of a C^∞ partition of unity which uses the properties of mollifiers. The existence of a continuous partition of unity is presumed to be familiar to the reader. (See V.6.5 in [ACFA].)

2.13 THEOREM. *If K is a compact subset of $\mathbb{C}$ and $\{G_1, \ldots, G_n\}$ is an open cover of K, then there are functions $\phi_1, \ldots, \phi_n$ in C^∞ such that:*

(a) $0 \leq \phi_k \leq 1$ for $1 \leq k \leq n$;
(b) support $\phi_k \subseteq G_k$ for $1 \leq k \leq n$;
(c) $\sum_{k=1}^n \phi_k(z) = 1$ for all z in K.

PROOF. Let L be a compact set such that $K \subseteq \text{int } L \subseteq \bigcup_{k=1}^{n} G_n$. Let $f_1, \ldots, f_n$ be continuous functions such that support $f_k \subseteq G_k$, $0 \leq f_k \leq 1$, and $\sum_{k=1}^{n} f_k = 1$ on L. Choose a positive number ε with $\varepsilon < \text{dist}(K, \mathbb{C} \backslash L)$ and $\varepsilon < \text{dist}(\text{support } f_k, \mathbb{C} \backslash G_k)$ for $1 \leq k \leq n$. Let ϕ_ε be a mollifier and put $\phi_k = \phi_\varepsilon * f_k$. Since ϕ_ε and f_k are both positive functions, ϕ_k is also positive. Also, $\phi_k(w) = \int \phi_\varepsilon(w-z) f_k(z) \, d\mathscr{A}(z) \leq \int \phi_\varepsilon \, d\mathscr{A} = 1$. By Proposition 2.11 (b) and the choice of ε, $\phi_k \in C^\infty$ and has its support in G_k.

Finally, because $\phi_\varepsilon(z) = 0$ if $|z| \geq \varepsilon$, if $w \in K$, then

$$\sum_{k=1}^{n} \phi_k(w) = \int_{B(w;\varepsilon)} \phi_\varepsilon(w-z) \sum_{k=1}^{n} f_k(z) \, d\mathscr{A}(z).$$

But with w in K and $|z-w| < \varepsilon$, we have that $z \in L$. Hence $\sum_{k=1}^{n} f_k(z) = 1$. Therefore $\sum_{k=1}^{n} \phi_k(w) = 1$. □

Say that the partition of unity $\{\phi_k\}$ is *subordinate to the cover* $\{G_k\}$ when it satisfies the properties listed in the preceding theorem.

Excercises

1. Show that a continuously differentiable function $u: G \to \mathbb{C}$ is analytic if and only if $\overline{\partial} u = 0$.
2. Verify the statements in Example 2.8.
3. Use integration by parts to verify that for a continuously differentiable function u on G, $\partial L_u = L_{\partial u}$ and $\overline{\partial} L_u = L_{\overline{\partial} u}$.
4. State and prove a product rule for differentiating distributions.
5. When does equality occur in the inequality in Proposition 2.2?
6. Using the method used in proving Proposition 2.2, show that for any compact set K, $\int_K |z-w|^{-1} \, d\mathscr{A}(w) \leq 2[\pi \, \text{Area}(K)]^{1/2}$. When does equality occur?

§3 The Cauchy transform.

In this section we introduce and give the elementary properties of the Cauchy transform of a compactly supported measure on the plane. This in turn will be used to obtain a few results on rational approximation and these results will, in turn, be used to obtain some results about subnormal operators. In the next section we will use some of this information to show that every subnormal operator has a nontrivial invariant subspace.

If μ is any compactly supported measure on $\mathbb{C}$, let

$$\tilde{\mu}(w) = \int \frac{1}{|w-z|} \, d|\mu|(z)$$

when the integral converges, and let $\tilde{\mu}(w) = \infty$ when the integral diverges. It happens that $\tilde{\mu}$ is finite a.e. [Area]. Indeed if K is the support of μ and

$R > 0$, then

$$\int_{B(0;R)} \tilde{\mu}(w)\, d\mathscr{A}(w) = \int_{B(0;R)} \int_K \frac{1}{|w-z|}\, d|\mu|(z)\, d\mathscr{A}(w)$$
$$= \int_K \int_{B(0;R)} \frac{1}{|w-z|}\, d\mathscr{A}(w)\, d|\mu|(z)$$
$$\leq C_R \|\mu\|$$

where the constant C_R exists by Lemma 2.1. Thus $\tilde{\mu}$ is locally integrable and hence finite a.e. [Area]. This discussion legitimizes the following definition.

3.1 DEFINITION. If μ is a compactly supported measure on the plane, the *Cauchy transform* of μ is the function $\hat{\mu}$ defined a.e. [Area] by the equation

$$\hat{\mu}(w) = \int \frac{1}{z-w}\, d\mu(z).$$

3.2 PROPOSITION. *If μ is a compactly supported measure, the following statements hold.*

(a) $\hat{\mu}$ *is locally integrable.*
(b) $\hat{\mu}$ *is analytic on* $\mathbb{C}_\infty \setminus \text{support}(\mu)$.
(c) *For w in* $\mathbb{C} \setminus \text{support}(\mu)$ *and* $n \geq 0$,

$$\partial^n \hat{\mu}(w) = (-1)^n n! \int (z-w)^{n+1}\, d\mu(z).$$

(d) $\hat{\mu}(\infty) = 0$ *and the power series of $\hat{\mu}$ near ∞ is given by*

$$\hat{\mu}(w) = \sum_{n=0}^{\infty} \left(\int z^n\, d\mu(z) \right) \frac{1}{w^{n+1}}.$$

PROOF. The proof of (a) follows the lines of the discussion preceding the definition. Part (c), and hence the proof that $\hat{\mu}$ is analytic on $\mathbb{C} \setminus \text{support}(\mu)$, follows by differentiating under the integral sign. Note that as $w \to \infty$, $(z-w)^{-1} = w^{-1}(z/w - 1)^{-1} \to 0$ uniformly for z in any compact set. Hence $\hat{\mu}$ has a removable singularity at ∞ and $\hat{\mu}(\infty) = 0$.

It remains to establish (d). This is done by choosing R so that $\text{support}(\mu) \subseteq B(0;R)$, expanding $(z-w)^{-1} = -w^{-1}(1 - z/w)^{-1}$ in a geometric series for $|w| > R$, and integrating term-by-term. □

It is worth observing that the function f defined in Proposition 2.2 is the Cauchy transform of $\mu = \text{Area}|K$. This function f is continuous but in general the Cauchy transform of a measure need not be continuous or even finite-valued. For example, if $\mu = \delta_a$, then $\hat{\mu}(z) = (a-z)^{-1}$.

Since $\hat{\mu}$ is locally integrable it defines a distribution on $\mathbb{C}$ and so it can be differentiated.

3.3 THEOREM. *If μ is a compactly supported measure on $\mathbb{C}$, then*

$$\overline{\partial} \hat{\mu} = -\pi \mu.$$

Moreover, $\hat{\mu}$ is the unique solution to this differential equation in the sense that if $h \in L^1_{\text{loc}}$ such that $\bar{\partial} h = -\pi \mu$, h is analytic in a neighborhood of ∞, and $h(\infty) = 0$, then $h = \hat{\mu}$ a.e. [Area].

PROOF. If $\phi \in C_c^\infty$, then

$$\bar{\partial}\hat{\mu}(\phi) = -\int \hat{\mu}\bar{\partial}\phi \, d\mathscr{A} = -\int \bar{\partial}\phi(z)\left[\int (w-z)^{-1} d\mu(w)\right] d\mathscr{A}(z)$$
$$= -\int \left[\int \bar{\partial}\phi(z)(w-z)^{-1} d\mathscr{A}(z)\right] d\mu(z).$$

By Corollary 2.6 this becomes $\bar{\partial}\hat{\mu}(\phi) = -\pi \int \phi \, d\mu$, whence the first part of the theorem.

For the uniqueness statement, suppose h is such a function. It follows that $\bar{\partial}(\hat{\mu} - h) = 0$. By Weyl's Lemma, $\hat{\mu} - h$ is almost everywhere equal to an entire function f. But $f = \hat{\mu} - h$ has a removable singularity at ∞ and is 0 there. Hence $f \equiv 0$. □

3.4 COROLLARY. *If G is an open set and $\hat{\mu} = 0$ a.e. [Area] on G, then $|\mu|(G) = 0$.*

PROOF. If $\phi \in C_c^\infty(G)$, then $\int \phi \, d\mu = -\pi^{-1}\bar{\partial}\hat{\mu}(\phi) = 0$. It follows (by an application of Proposition 2.11) that $|\mu|(G) = 0$. □

The Cauchy transform is the premier tool in uniform rational approximation. The next result is the reason for this because, by means of the Hahn-Banach Theorem, it reduces questions of rational approximation in the supremum norm to questions of weak approximation.

3.5 THEOREM. *If K is a compact subset of $\mathbb{C}$ and μ is a measure on K, then $\mu \perp R(K)$ if and only if $\hat{\mu}(w) = 0$ a.e. [Area] on $\mathbb{C}\setminus K$.*

PROOF. If $\mu \perp R(K)$, then for $w \notin K$, $(z-w)^{-1} \in R(K)$. Hence $\hat{\mu}(w) = 0$ off K.

Conversely, assume that $\hat{\mu} = 0$ a.e. [Area] off K; since $\hat{\mu}$ is analytic off K, $\hat{\mu}$ is identically 0 off K. This implies that all the derivatives of $\hat{\mu}$ vanish on $\mathbb{C}_\infty\setminus K$. From (3.2c) and (3.2d) we get that μ annihilates all polynomials and all rational functions with poles off K. □

As an illustration of the power of this method and to convince the reader that we have indeed accomplished something, even though each step along the way has been almost effortless, we present the following classical result.

3.6 HARTOGS-ROSENTHAL THEOREM. *If $\text{Area}(K) = 0$, then $R(K) = C(K)$.*

PROOF. Let $\mu \in M(K)$ such that $\mu \perp R(K)$; so $\hat{\mu} = 0$ off K. Since $\text{Area}(K) = 0$, this implies that $\hat{\mu} = 0$ a.e. [Area] on $\mathbb{C}$. By Corollary 3.4, $\mu = 0$. By the Hahn-Banach Theorem $R(K) = C(K)$. □

If the Hartogs-Rosenthal Theorem is applied to Theorem II.9.9, the next result is immediate.

3.7 COROLLARY. *If S is a von Neumann operator and* $\text{Area}(\sigma(S)) = 0$, *then S is normal.*

This corollary was already proved for subnormal operators by virtue of Putnam's Inequality and, indeed, it is valid for hyponormal operators.

There is a quantitative version of the Hartogs-Rosenthal Theorem due to Alexander [1973] that is easy to prove at this point and has some interesting applications. Note that if the function $\bar{z}$ belongs to $R(K)$, then $R(K) = C(K)$. Indeed, since $R(K)$ is an algebra, the presence of $\bar{z}$ in $R(K)$ implies that $R(K)$ contains the polynomials in z and $\bar{z}$. Thus we arrive at the observation that $R(K) = C(K)$ if and only if the distance (in $C(K)$) from $\bar{z}$ to $R(K)$ is 0.

3.8 THEOREM (Alexander [1973]). *If K is a compact subset of $\mathbb{C}$, then*

$$\text{dist}(\bar{z}, R(K)) \leq \left[\frac{\text{Area } K}{\pi}\right]^{1/2}.$$

PROOF. Choose a function ψ in C_c^∞ such that $\psi(z) = \bar{z}$ in a neighborhood of K. From the preceding corollary, $\psi(z) = -\pi^{-1} \int (w-z)^{-1} \bar{\partial}\psi(w) \, d\mathscr{A}(w)$. Now $\bar{\partial}\psi \equiv 1$ near K so that for z in K,

$$\bar{z} = -\frac{1}{\pi}\int_K \frac{1}{w-z} d\mathscr{A}(w) - \frac{1}{\pi}\int_{\mathbb{C}\setminus K} \frac{1}{w-z} \bar{\partial}\psi(w) \, d\mathscr{A}(w).$$

Define the function $f: K \to \mathbb{C}$ by

$$f(z) = -\frac{1}{\pi}\int_{\mathbb{C}\setminus K} \frac{1}{w-z}\bar{\partial}\psi(w) \, d\mathscr{A}(w).$$

If $\mu \in M(K)$ and $\mu \perp R(K)$, then

$$\int f \, d\mu = -\frac{1}{\pi}\int_{\mathbb{C}\setminus K} \hat{\mu}(w)\bar{\partial}\psi(w) \, d\mathscr{A}(w) = 0.$$

Hence $f \in R(K)$. On the other hand, by Proposition 2.2

$$\begin{aligned}\text{dist}(\bar{z}, R(K)) &\leq \|\bar{z} - f\|_K \\ &= \left\|\frac{1}{\pi}\int_K \frac{1}{w-z} d\mathscr{A}(w)\right\|_K \\ &\leq \left[\frac{\text{Area } K}{\pi}\right]^{1/2}. \quad \square\end{aligned}$$

At this point we can give another proof, due to Axler and Shapiro [1985], of Putnam's Inequality (IV.3.1) for the case of a subnormal operator. The proof that was first given of Putnam's Inequality was based on the Berger-Shaw Theorem. Even though the simplified proof of that theorem somewhat lessens the desire for a special proof for subnormal operators, such a specialized proof helps to throw a little more light on the inequality. Actually, a

somewhat better result is obtained as it will be shown that for any subnormal operator S,
$$\|S^*S - SS^*\| \leq \operatorname{dist}_{C(\sigma(S))}[\bar{z}, R(\sigma(S))],$$
which, by virtue of the preceding theorem, captures Putnam's Inequality.

So assume that S is a subnormal operator and, initially, also assume that S is rationally cyclic. Thus we may assume that $\mathscr{H} = R^2(\sigma(S), \mu)$ for some compactly supported measure μ and $Sf = zf$ for all f in $\mathscr{H}$. If P is the projection of $L^2(\mu)$ onto $\mathscr{H}$ and $g \in \mathscr{H}$ with $\|g\| = 1$, then

$$\begin{aligned}
\langle (S^*S - SS^*)g, g \rangle &= \|Sg\|^2 - \|S^*g\|^2 \\
&= \|zg\|^2 - \|P(\bar{z}g)\|^2 \\
&= \|\bar{z}g\|^2 - \|P(\bar{z}g)\|^2 \\
&= \operatorname{dist}_{L^2(\mu)}[\bar{z}g, \mathscr{H}]^2 \\
&= \inf\{\|\bar{z}g - h\|^2 : h \in \mathscr{H}\} \\
&\leq \inf\{\|\bar{z}g - hg\|^2 : h \in R(\sigma(S))\} \\
&\leq \inf\{\|\bar{z} - h\|_{\sigma(S)} : h \in R(\sigma(S))\} \\
&= \operatorname{dist}_{C(\sigma(S))}[\bar{z}, R(\sigma(S))].
\end{aligned}$$

In the case of an arbitrary subnormal operator, fix a unit vector f in $\mathscr{H}$ and let $\mathscr{L}$ be the rationally invariant subspace for S generated by f. From the preceding paragraph, if $T = S|\mathscr{L}$

$$\|T^*T - TT^*\| \leq \operatorname{dist}_{C(\sigma(T))}[\bar{z}, R(\sigma(T))].$$

Now since $r(S)\mathscr{L} \subseteq \mathscr{L}$ for every r in $\operatorname{Rat}(\sigma(S))$, it follows that $\sigma(T) \subseteq \sigma(S)$ and so

$$\begin{aligned}
\operatorname{dist}_{C(\sigma(T))}[\bar{z}, R(\sigma(T))] &= \inf\{\|\bar{z} - r\|_{\sigma(T)} : r \in \operatorname{Rat}(\sigma(T))\} \\
&\leq \inf\{\|\bar{z} - r\|_{\sigma(S)} : r \in \operatorname{Rat}(\sigma(S))\} \\
&= \operatorname{dist}_{C(\sigma(S))}[\bar{z}, R(\sigma(S))].
\end{aligned}$$

Also, if Q is the projection of $\mathscr{H}$ onto $\mathscr{L}$, then $T^* = QS^*|\mathscr{L}$. Thus for the fixed vector f, $\langle (S^*S - SS^*)f, f \rangle = \|Sf\|^2 - \|S^*f\|^2 \leq \|Tf\|^2 - \|T^*f\|^2 = \langle (T^*T - TT^*)f, f \rangle \leq \|T^*T - TT^*\| \leq \operatorname{dist}_{C(\sigma(S))}[\bar{z}, R(\sigma(S))]$.

In addition to the paper of Axler and Shapiro [**1985**], the reader might also look at Gamelin [**1985**].

The next result about Cauchy transforms has great value, even though its proof is modest. This fact will also be of importance in §6 when we examine some additional function algebras.

3.9 Proposition. *If μ is a compactly supported measure and $g \in C_c^1$, then the measure*
$$\nu = g\mu - \pi^{-1}\bar{\partial}g\hat{\mu} \cdot (\text{Area})$$

satisfies
$$\hat{\nu} = g\hat{\mu}.$$

PROOF. Because g has compact support, $g\hat{\mu}$ is analytic at ∞ and vanishes there. Also, in the sense of distributions $\overline{\partial}(g\hat{\mu}) = g\overline{\partial}\hat{\mu} + \overline{\partial}g\hat{\mu} = -\pi g\mu + \overline{\partial}g\hat{\mu} = -\pi\nu = \overline{\partial}\nu$. By the uniqueness part of Theorem 3.3, $\hat{\nu} = g\hat{\mu}$. □

The next theorem says that a continuous function belongs to $R(K)$ if it belongs to $R(K)$ locally.

3.10 BISHOP'S LOCALIZATION THEOREM. *If $f \in C(K)$ and for each a in K there is a neighborhood U of a such that $f|(K \cap \mathrm{cl}\, U) \in R(K \cap \mathrm{cl}\, U)$, then $f \in R(K)$.*

PROOF. From the hypothesis we can find an open cover $\{U_1, \ldots, U_n\}$ of K such that $f|(K \cap \mathrm{cl}\, U_j) \in R(K \cap \mathrm{cl}\, U_j)$ for $1 \leq j \leq n$. Let $\{\phi_1, \ldots, \phi_n\}$ be a C^∞ partition of unity subordinate to this cover (2.13).

Let $\mu \in M(K)$ such that $\mu \perp R(K)$. By the preceding proposition, the measures
$$\mu_j = \phi_j \mu - \pi^{-1}\overline{\partial}\phi_j \hat{\mu} \cdot (\text{Area})$$
satisfy $\hat{\mu}_j = \phi_j \hat{\mu}$. Since ϕ_j vanishes outside U_j and $\hat{\mu}$ vanishes off K, $\hat{\mu}_j$ vanishes off $K \cap \mathrm{cl}\, U_j$. Thus $\mu_j \perp R(K \cap \mathrm{cl}\, U_j)$ and so $\int f\, d\mu_j = 0$ by hypothesis. But $\sum_j \phi_j = 1$ in a neighborhood of K and so $\sum_j \mu_j = \mu$. Thus $\int f\, d\mu = \sum_j \int f\, d\mu_j = 0$. Therefore $f \in R(K)$. □

3.11 LEMMA. *If μ and ν are compactly supported measures such that $\hat{\mu}$ and $\hat{\nu}$ are continuous functions, then $\hat{\mu}\hat{\nu}$ is the Cauchy transform of the measure $\hat{\nu}\mu + \hat{\mu}\nu$.*

PROOF. Note that $\overline{\partial}(\hat{\mu}\hat{\nu}) = \hat{\nu}\overline{\partial}\hat{\mu} + \hat{\mu}\overline{\partial}\hat{\nu} = -\pi(\hat{\nu}\mu + \hat{\mu}\nu)$. The result now follows by Weyl's Lemma. □

(Where was the continuity of $\hat{\mu}$ and $\hat{\nu}$ used in the preceding proof?)

Note that if $h \in L_c^\infty$ (the bounded Borel functions on $\mathbb{C}$ with compact support), then $\hat{h}(w) = \int (z-w)^{-1} h(z)\, d\mathscr{A}(z)$ is a continuous function on $\mathbb{C}$ (2.2). So the preceding lemma implies that if h and $k \in L_c^\infty$, $\hat{h}\hat{k}$ is the Cauchy transform of $\hat{h}k + h\hat{k}$ and, moreover, $\hat{h}k + h\hat{k} \in L_c^\infty$. These algebraic observations lead us to a new class of function algebras. But first we give an equivalent formulation of the algebra $R(K)$.

3.12 PROPOSITION. *If K is a compact subset of $\mathbb{C}$, then $\{\hat{h} : h \in L_c^\infty$ and $h = 0$ on $K\}$ is a norm dense subalgebra of $R(K)$.*

PROOF. Let $\mathscr{B} = \{\hat{h} : h \in L_c^\infty$ and $h = 0$ on $K\}$; so Lemma 3.11 and the comments preceding this proposition show that $\mathscr{B}$ is a subalgebra of $C(K)$. (To be completely precise, we should say that $\{\hat{h}|K : h \in L_c^\infty$ and $h = 0$ on $K\}$ is a subalgebra of $C(K)$. Nevertheless, we will proceed without making a distinction between $\hat{h}$ and $\hat{h}|K$.)

Let $\mu \in M(K)$ such that $\mu \perp R(K)$. So $\hat{\mu} = 0$ on $\mathbb{C} \setminus K$. Hence if $h \in L_c^\infty$ and $h = 0$ on K, then $h\hat{\mu} = 0$ everywhere on $\mathbb{C}$. Therefore

$$\int \hat{h} \, d\mu = \int \left[\int (z-w)^{-1} h(z) \, d\mathscr{A}(z) \right] d\mu(w)$$

$$= \int h(z) \left[\int (z-w)^{-1} \, d\mu(w) \right] d\mathscr{A}(z)$$

$$= -\int h(z) \hat{\mu}(z) \, d\mathscr{A}(z)$$

$$= 0.$$

That is, $\mu \perp \mathscr{B}$ and so $\mathscr{B} \subseteq R(K)$.

Now let $f_0 \in \text{Rat}(K)$ and let $\varepsilon > 0$. There is a function f in C_c^∞ that is analytic in a neighborhood of K and satisfies $\|f - f_0\|_\infty < \varepsilon$. If $h = -\pi^{-1} \bar{\partial} f$, then $h \in L_c^\infty$ and h vanishes on K. Thus $\hat{h} \in \mathscr{B}$. But $\bar{\partial} \hat{h} = -\pi h = \bar{\partial} f$. By Weyl's Lemma, $\hat{h} = f$. □

3.13 DEFINITION. If K is a compact subset of $\mathbb{C}$ and E is a measurable subset of K, let $R(K, E)$ be the uniform closure in $C(K)$ of $\{\hat{h} : h \in L_c^\infty$ and $h = 0$ on $E\}$. If E is any bounded measurable set, let $R(E) = R(\text{cl}\, E, E)$.

From the discussions that have already occurred, $R(K, E)$ is a subalgebra of $C(K)$. Moreover, if $E =$ the compact set K, the preceding proposition shows that this new definition of $R(K)$ agrees with the previous one as the closure of rational functions. The next result contains some elementary properties of $R(K, E)$. The proof is left to the reader.

3.14 PROPOSITION. *Let E be a Borel subset of K.*
(a) $R(K) \subseteq R(K, E)$.
(b) *A measure μ on K is orthogonal to $R(K, E)$ if and only if $\hat{\mu} = 0$ a.e. [Area] on $\mathbb{C} \setminus E$.*
(c) *If $\text{Area}(E) = 0$, then $R(K, E) = C(K)$.*

The algebras $R(E)$ and $R(K, E)$ will have a prominent role in the course of this book. They are being introduced here merely as an extra set of examples. We conclude this section with an application of Bishop's Localization Theorem to subnormal operators. But first a lemma is needed.

3.15 LEMMA. *Let S be a subnormal operator with $N = \text{mne}(S)$ and let $N = \int z \, dE(z)$ be the spectral decomposition of N. If X is any Borel set such that $\mathscr{H}_1 = \text{cl}[E(X)\mathscr{H}] \neq (0)$ and $\mathscr{K}_1 = E(X)\mathscr{K}$, then:*
(a) $S_1 \equiv N | \mathscr{H}_1$ *is a subnormal operator;*
(b) $N_1 \equiv N | \mathscr{K}_1$ *is the minimal normal extension of S_1;*
(c) $\sigma(S_1) \subseteq \sigma(S) \cap \hat{X}$.

PROOF. Part (a) is clear and the proof of part (b) is left as an exercise. Since $\sigma(N_1) \subseteq \text{cl}\, X$, $\sigma(S_1) \subseteq \hat{X}$. Fix λ with $\lambda \notin \sigma(S)$. If $\lambda \in \hat{X}$, then

$(N-\lambda)^{-1}\mathcal{H} = \mathcal{H}$. Therefore $(N-\lambda)^{-1}E(X)\mathcal{H} = E(X)(N-\lambda)^{-1}\mathcal{H} = E(X)\mathcal{H}$ and so $(N-\lambda)^{-1}\mathcal{H}_1 = \mathcal{H}_1$. This implies that $\lambda \notin \sigma(S_1)$. □

3.16 THEOREM (Clancey and Putnam [1972]). *If S is a subnormal operator and D is an open disk such that $\sigma(S) \cap D \neq \varnothing$ and $R(\sigma(S) \cap \operatorname{cl} D) = C(\sigma(S) \cap \operatorname{cl} D)$, then S is not pure.*

PROOF. Let $K = \sigma(S)$, put $X = D \cap K$, and let's apply the preceding lemma. If $N = \operatorname{mne}(S)$, then $D \cap \sigma(N) \neq \varnothing$. In fact, if $D \cap \sigma(N) = \varnothing$, then $D \cap \sigma(S) = D \cap [\sigma(S) \backslash \sigma(N)]$, which is an open subset of $\mathbb{C}$ since $\partial \sigma(S) \subseteq \sigma(N)$. Because $R(\operatorname{cl} X) = C(\operatorname{cl} X)$, $\operatorname{cl} X$ can have no interior.

If $N = \int z\, dE(z)$, then $E(X) \neq 0$ and, because N is the minimal normal extension of S, $\mathcal{H}_1 = \operatorname{cl}[E(X)\mathcal{H}] \neq (0)$. Let $S_1 = N|\mathcal{H}_1$; so $\sigma(S_1) \subseteq \operatorname{cl} X$ by the preceding lemma. But $R(\operatorname{cl} X) = C(\operatorname{cl} X)$, and so S_1 must be a normal operator (II.9.9). Hence $\mathcal{H}_1$ reduces N_1.

It remains to show that $\mathcal{H}_1 \subseteq \mathcal{H}$. To this end, suppose a is the center of D and r is its radius. Choose $R > r$ so that $K \subseteq B(a; R)$, let $A = \{z : r \leq |z - a| \leq R\}$, and $Y = A \cup \operatorname{cl} X = A \cup X$. Choose increasing sequences $\{t_n\}$ and $\{r_n\}$ such that $0 < t_n < r_n < r$ and both sequences converge to r. Define functions $f_n : Y \to [0, 1]$ by

$$f_n(z) = \begin{cases} 1 & \text{if } |z - a| < t_n \\ \frac{|z-a|-r_n}{t_n - r_n} & \text{if } t_n \leq |z - a| \leq r_n \\ 0 & \text{if } |z - a| \geq r_n \end{cases}.$$

Clearly $f_n \in C(Y)$. If $z \in X$, then there is a $t > 0$ such that $B(z; t) \subseteq D$. Hence by hypothesis $f_n|[Y \cap \overline{B}(z; t)] \in R(Y \cap \overline{B}(z; t))$. Since z was arbitrarily chosen from X and f_n vanishes in a neighborhood of A, Bishop's Localization Theorem implies that $f_n \in R(Y)$. Because $Y \supseteq \sigma(S)$, $f_n(N)\mathcal{H} = f_n(S)\mathcal{H} \subseteq \mathcal{H}$. But $\{f_n(z)\}$ increases monotonically to $\chi_X(z)$. Hence $f_n(N) \to E(X)$ (WOT) and so $\mathcal{H}_1 \subseteq \mathcal{H}$. □

Clancey and Putnam [1972] also showed that if K is a compact subset of $\mathbb{C}$ such that $R(K \cap D) \neq C(K \cap D)$ for every closed disk D whose interior meets K, then there is a pure subnormal operator S such that $\sigma(S) = K$.

Exercises

1. Improve Proposition 1.1 by showing that if G is an open set containing K and $f \in C^1(G)$ such that $\overline{\partial} f = 0$ on K, then $f \in R(K)$. (*Hint*. Use Theorem 3.5.)
2. (Wermer [1955]) If S is a subnormal operator and $\operatorname{Area}(\sigma_n(S)) = 0$, then S has a nontrivial invariant subspace.
3. If E and F are measurable subsets of K and $\operatorname{Area}[(E \backslash F) \cup (F \backslash E)] = 0$, what can be said about $R(E)$ and $R(F)$?
4. If μ and ν are measures with compact support and μ is absolutely continuous, show that $\int \hat{\mu}\, d\nu = -\int \hat{\nu}\, d\mu$.
5. What is $R(K, \varnothing)$?

§4 Invariant subspaces for subnormal operators.

In this section we present the remarkably simple proof of the existence of invariant subspaces for subnormal operators discovered by James Thomson. The only tools of any substance used are the properties of the Cauchy transform developed in the last section.

One key idea in this proof is to consider L^p spaces for $p > 2$, even though the theorem concerns operators on a Hilbert space. In particular, we want to briefly discuss bounded point evaluations for the spaces $R^p(K, \mu)$. The definition is clear: a complex number λ is a *bounded point evaluation* for $R^p(K, \mu)$ is there is a constant $C > 0$ such that $|f(\lambda)| \leq C\|f\|_p$ for every f in $\text{Rat}(K)$. If p is finite, this implies the existence of a function g in $L^q(\mu)$, where $(1/p) + (1/q) = 1$, such that $f(\lambda) = \int f g \, d\mu$ for all f in $\text{Rat}(K)$. Note that if $p \neq 2$, we cannot necessarily find the function g in $R^q(K, \mu)$ since this latter space is not the dual of $R^p(K, \mu)$.

We begin this discussion with a lemma. In this section, μ will always be a compactly supported positive Borel measure.

4.1 LEMMA. *If $1 < q < 2$ and $g \in L^q(\mu)$, then $(z - \lambda)^{-1} g \in L^q(\mu)$, except possibly for λ belonging to a set of area 0.*

PROOF. Let K be the support of μ and let $R > 0$ be sufficiently large that $K \subseteq B(0; R)$. If $z \in K$, then

$$\int_{|\lambda| \leq R} |z - \lambda|^{-q} \, d\mathscr{A}(\lambda) = \int_{|z-\lambda| \leq R} |\lambda|^{-q} \, d\mathscr{A}(\lambda)$$
$$\leq \int_{|\lambda| \leq 2R} |\lambda|^{-q} \, d\mathscr{A}(\lambda)$$
$$= \frac{2\pi}{2-q} (2R)^{2-q}.$$

Hence

$$\int_K \left[\int |(z-\lambda)^{-1} g(z)|^q \, d\mu(z) \right] d\mathscr{A}(\lambda)$$
$$= \int \left[\int_K |z - \lambda|^{-q} \, d\mathscr{A}(\lambda) \right] |g(z)|^q \, d\mu(z)$$
$$\leq \int \left[\int_{|\lambda| \leq R} |z - \lambda|^{-q} \, d\mathscr{A}(\lambda) \right] |g(z)|^q \, d\mu(z)$$
$$\leq \frac{2\pi}{2-q} (2R)^{2-q} \int_K |g|^q \, d\mu < \infty.$$

From here the lemma follows immediately. □

Unlike the Hilbert space case, bounded point evaluations always exist for $p > 2$ unless $R^p(K, \mu)$ is as big as it can possibly be. Namely, if $R^p(K, \mu) = L^p(\mu)$, then the only way that $L^p(\mu)$ can have a bounded point evaluation at λ is if μ has an atom at λ.

4.2 THEOREM (Brennan [**1971a**]). *If $p > 2$ and $R^p(K, \mu) \neq L^p(\mu)$, then $R^p(K, \mu)$ has a bounded point evaluation.*

PROOF. If $(1/p) + (1/q) = 1$, then by hypothesis there is a nonzero function g in $L^q(\mu)$ such that $\int fg \, d\mu = 0$ for every f in Rat(K). Thus there is a Borel set E of positive area such that the Cauchy transform of the measure $g\mu$ is well defined on E and does not vanish there. (Note that since $g\mu \perp R(K)$, $E \subseteq K$.) By the preceding lemma, there is a λ in E for which $(z - \lambda)^{-1} g \in L^q(\mu)$. But for every f in Rat(K),

$$0 = \int \frac{f(z) - f(\lambda)}{z - \lambda} g \, d\mu$$

and so

$$f(\lambda) = \frac{1}{\widehat{g\mu}(\lambda)} \int \frac{f(z)}{z - \lambda} g(z) \, d\mu$$

for all f in Rat(K). But this implies that for f in Rat(K), $|f(\lambda)| \leq C\|f\|_p$ with $C = |\widehat{g\mu}(\lambda)|^{-1} \|(z - \lambda)^{-1} g\|_q$. Hence λ is a bounded point evaluation. □

In view of our comments preceding the proof, it might be worthwhile observing that if the bounded point evaluation λ in the preceding proof is an atom of μ, then $g(\lambda) = 0$.

The technique in the preceding proof of manufacturing a bounded point evaluation by using the Cauchy transform is one we will see again when we discuss representing measures later in this chapter as well as Thomson's solution of the bounded point evaluation problem in Chapter VIII.

4.3 LEMMA. *Assume that k is a positive integer, $p = 2k+1$, and $q = p/2k$ (p and q are conjugate indices). If $h \in L^q(\mu)$ and $\|h\|_q = \sup\{|\int fh \, d\mu| : f \in P^p(\mu) \text{ and } \|f\|_p \leq 1\}$, then there is a function x in $P^2(\mu)$ such that $|h| = |x|^2$ a.e. $[\mu]$.*

PROOF. Without loss of generality we may assume that $\|h\|_q = 1$. Since the closed unit ball of $P^p(\mu)$ is weakly compact, there is a function u in $P^p(\mu)$ such that $\|u\|_p = 1$ and $1 = \int uh \, d\mu \leq \|u\|_p \|h\|_q = 1$. Thus equality holds in Hölder's Inequality. This implies that $|u|^p = |h|^q$ a.e. $[\mu]$. Substituting the values of p and q and performing the required arithmetic, we obtain that $|u|^{2k} = |h|$ a.e. $[\mu]$. Put $x = u^k$.

To show that $x \in P^2(\mu)$, let $\{p_j\}$ be a sequence of polynomials such that $0 = \lim \int |p_j - u|^p \, d\mu = \lim \int |p_j - u|^{2k+1} \, d\mu$. Let $M = \sup_j \|p_j\|_p$; by passing to a subsequence if necessary, we may assume that $p_j \to u$ a.e. $[\mu]$. Now Hölder's Inequality implies that $\|p_j^k\|_2 = [\int |p_j^{2k}| \, d\mu]^{1/2} \leq [(\int |p_j^{2k}|^q \, d\mu)^{1/q} (\int 1 \, d\mu)^{1/p}]^{1/2} = C[(\int |p_j|^{2k+1} \, d\mu)^{2k/2k+1}]^{1/2} = C\|p_j\|_p^k \leq CM^k$ for all $j \geq 1$. Therefore, since $p_j \to u$ a.e. $[\mu]$, $p_j^k \to u^k = x$ in L^2 norm. That is, $x = u^k \in P^2(\mu)$. □

As surprising as it may seem, we can now prove the main theorem of this section.

4.4 THEOREM (S. W. Brown [1978]). *A subnormal operator on a Hilbert space of dimension at least 2 has a nontrivial invariant subspace.*

PROOF. (Thomson [1986]). To avoid trivialities, we may assume that the Hilbert space on which the subnormal operator S acts is infinite dimensional. Since for any vector $f \neq 0$, $\bigvee\{S^n f; n \geq 0\}$ is a nonzero invariant subspace for S, we may assume that S is a cyclic subnormal operator. Thus there is a compactly supported positive measure μ on $\mathbb{C}$ such that $S \cong S_\mu$ acting on $P^2(\mu)$; let's assume that $S = S_\mu$. Also, since we might as well restrict our attention to pure subnormal operators, we can certainly assume that $P^2(\mu) \neq L^2(\mu)$. Let g be a nonzero function in $L^2(\mu)$ such that $\int fg\, d\mu = 0$ for all f in $P^2(\mu)$. It follows that $g \in L^q(\mu)$ for $1 \leq q \leq 2$ (Why?) and hence $P^p(\mu) \neq L^p(\mu)$ for $p \geq 2$. Let's take $p = 3$ and apply the preceding lemma with $k = 1$; so the corresponding value of q is $3/2$. By Theorem 4.2 there is a λ in $\mathbb{C}$ that is a bounded point evaluation for $P^3(\mu)$. Hence there is a function h in $L^{3/2}(\mu)$ such that

$$(4.5) \qquad \int ph\, d\mu = p(\lambda)$$

for every polynomial p and

$$\|h\|_{3/2} = \sup\{|p(\lambda)|: p \text{ is a polynomial and } \|p\|_3 \leq 1\}.$$

The preceding lemma implies there is a function x in $P^2(\mu)$ such that $|h| = |x|^2$ a.e. $[\mu]$. Define the function k by letting $k(z) = h(z)/x(z)$ when $x(z) \neq 0$ and $k(z) = 0$ otherwise. It follows that $|k|^2 = |h|$ and hence $k \in L^2(\mu)$. Let $\mathscr{M} = $ the closed linear span of $\{z^n(z-\lambda)x: n \geq 0\}$. Since $x \in P^2(\mu)$, $\mathscr{M} \subseteq P^2(\mu)$; clearly $\mathscr{M}$ is invariant for S. It remains to show that $\mathscr{M}$ is not trivial.

First $\mathscr{M} \neq (0)$ since $(z-\lambda)x \in \mathscr{M}$ and $(z-\lambda)x \neq 0$ (Why?). Also, for every polynomial p, $\int (z-\lambda)pxk\, d\mu = \int (z-\lambda)ph\, d\mu = 0$ by (4.5). Hence $\bar{k} \perp \mathscr{M}$ in $L^2(\mu)$. On the other hand, $x \in P^2(\mu)$ and $\langle x, \bar{g}\rangle = \int xk\, d\mu = \int h\, d\mu = 1$ by (4.5); so $\bar{k}$ is not orthogonal to $P^2(\mu)$. Thus $\mathscr{M} \neq P^2(\mu)$. □

Brown's original proof of Theorem 4.4 was quite long and complicated. Its strategy bore a slight resemblance to that of Thomson's proof in that it produced a complex number λ and functions f and g in $P^2(\mu)$ such that $p(\lambda) = \langle pf, g\rangle$ for all polynomials p. From here the existence of the invariant subspace for S_μ follows as in the previous proof. This strategy was far more difficult to execute than Thomson's.

It is one of those incidents that justify the maxim that a proof is worth a thousand theorems. Here this is almost literally true since what has become known as "the Scott Brown technique" has been used repeatedly to prove invariant subspace theorems in a wide variety of situations. A recent survey of some of these results is found in Bercovici, Foias, and Pearcy [**1985**]. Also Brown [**1987**] has used this technique together with some additional operator theory results to show that if T is a hyponormal operator and $R(\sigma(T)) \neq C(\sigma(T))$, then T has a nontrivial invariant subspace.

Actually Thomson [**1986**] proved a better theorem than (4.4).

4.6 THEOREM (Thomson [**1986**]). *If μ is a compactly supported positive measure on $\mathbb{C}$, $\mathcal{H}$ is a closed subspace of $L^2(\mu)$ containing the constant functions, and A is a subalgebra of $L^\infty(\mu)$ containing the function z that leaves $\mathcal{H}$ invariant, then there is a nontrivial subspace $\mathcal{M}$ of $\mathcal{H}$ such that $A\mathcal{M} \subseteq \mathcal{M}$.*

The proof of this theorem follows the lines of the previous proof and is left to the reader.

The last theorem has some significant advantages over Theorem 4.4. Here are some corollaries, none of which were known at the time of Brown's result.

4.7 COROLLARY. *Every rationally cyclic subnormal operator has a nontrivial hyperinvariant subspace.*

4.8 COROLLARY. *If S is a subnormal operator acting on $\mathcal{H}$, then there is a nontrivial subspace $\mathcal{M}$ of $\mathcal{H}$ such that $r(S)\mathcal{M} \subseteq \mathcal{M}$ for every r in $\mathrm{Rat}(\sigma(S))$.*

The following invariant subspace problem remains open.

4.9 OPEN PROBLEM. Does every subnormal operator have a nontrivial hyperinvariant subspace?

The difficulty here is that unlike the invariant subspace problem for subnormal operators, the hyperinvariant problem cannot be reduced to the cyclic case.

There have been uses of Thomson's Theorem 4.6 and his technique. Trent [**1987**] shows that it is not necessary to assume that the space $\mathcal{H}$ contains the constants or that the algebra A contains the polynomials. Yan [**preprint**] uses Thomson's techniques to show that any system of subnormal operators with commuting normal extensions has a nontrivial common invariant subspace. Trent (private communication) has made improvements in this result of Yan similar to those he made in Thomson's Theorem.

§5 Vitushkin localization operators. Let $g \in C_c^1$ and let f be a bounded Borel function on $\mathbb{C}$. Define $T_g f \colon \mathbb{C} \to \mathbb{C}$ by

$$(5.1) \qquad T_g f(w) = f(w)g(w) + \frac{1}{\pi} \int \frac{f(z)}{z-w} \bar{\partial} g(z) \, d\mathcal{A}(z).$$

As a consequence of Corollary 2.6, we can also write

$$(5.2) \qquad T_g f(w) = \frac{1}{\pi} \int \frac{f(z) - f(w)}{z-w} \bar{\partial} g(z) \, d\mathcal{A}(z).$$

This operator T_g is called the *Vitushkin localization operator*.

5.3 PROPOSITION. *If f is a bounded Borel function and $g \in C_c^1$, then the following statements hold.*

(a) $\bar{\partial}(T_g f) = g \bar{\partial} f$ *(in the sense of distributions).*
(b) *If E is a nonempty subset of $\mathbb{C}$, $a \in E$, and $f|E$ is continuous at a, then $T_g f|E$ is also continuous at a.*
(c) $T_g f$ *is analytic wherever f is analytic.*
(d) $T_g f$ *is analytic on $\mathbb{C}_\infty \setminus \mathrm{supp}(g)$ and vanishes at ∞.*
(e) $f - T_g f$ *is analytic on* $\mathrm{int}\{z : g(z) = 1\}$.

PROOF. (a) First note that the integral that appears in (5.1) is the Cauchy transform of $f \bar{\partial} g$. (Of course when we speak of the Cauchy transform of a function we really mean the Cauchy transform of that function times planar Lebesgue measure.) Thus Theorem 3.3 implies that $\bar{\partial}(T_g f) = \bar{\partial}(fg) + \pi^{-1} \bar{\partial}(\widehat{f \bar{\partial} g}) = f \bar{\partial} g + g \bar{\partial} f - f \bar{\partial} g = g \bar{\partial} f$.

(b) This is an easy consequence of the Lebesgue Dominated Convergence Theorem.

(c) If f is analytic on an open set G, then $\bar{\partial} f = 0$ on G and so (c) follows from part (a) and Weyl's Lemma.

(d) Again part (a) implies that $\bar{\partial} T_g f = 0$ off the support of g. Also since g has compact support, the fact that $T_g f$ vanishes at ∞ is a consequence of the same fact for Cauchy transforms.

(e) Note that $\bar{\partial}(f - T_g f) = \bar{\partial} f - g \bar{\partial} f \equiv 0$ on the interior of $\{z : g(z) = 1\}$. □

Let K be a compact subset of $\mathbb{C}$ and let $f \in C(K)$. Extend f to be defined on all of $\mathbb{C}$ by letting $f(z) = 0$ for z not in K. According to part (b) of the preceding proposition, $T_g f | K \in C(K)$. The next result improves this and gives some indication why these operators are of value in the study of approximation.

5.4 PROPOSITION. *If $g \in C_c^1$, then $T_g : C(K) \to C(K)$ is a bounded linear operator and its adjoint $T_g^* : M(K) \to M(K)$ is given by $T_g^* \mu =$ the unique measure ν supported on K such that $\hat{\nu} = g \hat{\mu}$. That is, $T_g^* \mu = g\mu - \pi^{-1} \bar{\partial} g \hat{\mu} \cdot$ (Area).*

PROOF. If $w \in K$ and $f \in C(K)$, then with $L = \mathrm{supp}(g)$

$$\left| \int (z-w)^{-1} f(z) \bar{\partial} g(z) \, d\mathscr{A}(z) \right| \leq \|f\|_\infty \|\bar{\partial} g\|_\infty \int_L |z-w|^{-1} \, d\mathscr{A}(z)$$
$$\leq 2[\pi \, \mathrm{Area}(L)]^{1/2} \|f\|_\infty \|\bar{\partial} g\|_\infty$$

by Proposition 2.14. From here it follows that T_g is a bounded operator. Thus $T_g^* : M(K) \to M(K)$ is well defined and bounded. Let $\mu \in M(K)$ and put $\nu = T_g^*\mu$. If $f \in C(K)$, then

$$\int f\,d\nu = \int f\,dT_g^*\mu$$
$$= \int (T_g f)\,d\mu$$
$$= \int fg\,d\mu + \pi^{-1} \iint f(z)(z-w)^{-1}\overline{\partial}g(z)\,d\mathscr{A}(z)\,d\mu(w)$$
$$= \int fg\,d\mu + \pi^{-1} \int f(z)\left[\int (z-w)^{-1}\,d\mu(w)\right]\overline{\partial}g(z)\,d\mathscr{A}(z)$$
$$= \int fg\,d\mu - \pi^{-1} \int f\hat{\mu}\overline{\partial}g\,d\mathscr{A}(z).$$

An examination of this equation shows that $T_g^*\mu = g\mu - \pi^{-1}\overline{\partial}g\hat{\mu}\cdot(\text{Area}) = \nu$. By Proposition 3.9, $\hat{\nu} = g\hat{\mu}$. □

Different estimates are needed for the norm of $T_g f$ than can be gotten from the proof of the preceding proposition. These will be obtained to conclude this technical section, but first an elementary lemma is required.

5.5 LEMMA. *If $a \in \mathbb{C}$ and $\delta > 0$, then there is a function g in C_c^1 such that $0 \leq g \leq 1$, $g \equiv 1$ on $B(a;\delta/2)$, $g \equiv 0$ off $B(a,\delta)$, and $\|\overline{\partial}g\|_\infty \leq 4/\delta$.*

PROOF. In fact let $h: \mathbb{R} \to \mathbb{R}$ be a continuously differentiable function such that $0 \leq h \leq 1$, $h(t) \equiv 1$ for $|t| \leq \delta/2$, $h(t) \equiv 0$ for $|t| \geq \delta$, and $|h'| \leq 4/\delta$. Put $g(z) = h(|z-a|)$. It is left to the reader to check that g has the desired properties. □

Recall that if $f: X \to \mathbb{C}$, the *oscillation* of f over X is defined by $\text{osc}(f;X) \equiv \sup\{|f(x) - f(y)|: x, y \in X\}$. That is, $\text{osc}(f;X)$ is the diameter of $f(X)$.

5.6 PROPOSITION. *If $g \in C_c^1$ and f is a bounded Borel function, then*

$$\|T_g f\|_\infty \leq \frac{2}{\sqrt{\pi}} \text{osc}(f;\text{supp}(g))\|\overline{\partial}g\|_\infty[\text{Area}(\text{supp}(g))]^{1/2}.$$

PROOF. Using (5.2), if $w \in \text{supp}(g)$, then

$$|T_g f(w)| \leq \pi^{-1}\int \frac{|f(z) - f(w)|}{|z-w|}|\overline{\partial}g(z)|\,d\mathscr{A}(z)$$
$$\leq \pi^{-1}\|\overline{\partial}g\|_\infty \text{osc}(f;\text{supp}(g))\int_{\text{supp}(g)} |z-w|^{-1}\,d\mathscr{A}(z)$$
$$\leq \frac{2}{\sqrt{\pi}}\text{osc}(f;\text{supp}(g))\|\overline{\partial}g\|_\infty[\text{Area}(\text{supp}(g))]^{1/2}$$

by Proposition 2.14. Thus the desired estimate holds on $\text{supp}(g)$. Now if $w_0 \in \partial(\text{supp } g)$ and $\{w_n\}$ is a sequence in the complement of $\text{supp}(g)$ that converges to w_0, then there is a constant M such that on $\text{supp}(g)$, $|[f - f(w_n)]\overline{\partial}g||z - w_n|^{-1}$ is dominated by $M||z - w_0| - |w_0 - w_n||^{-1}$, an integrable function. Thus $T_g f(w_n) \to T_g f(w_0)$. Since $T_g f$ is analytic on $\mathbb{C}_\infty \setminus \mathrm{supp}(g)$, the estimate holds everywhere by the Maximum Modulus Theorem. □

5.7 PROPOSITION. *Let $g \in C_c^1$ such that $0 \le g \le 1$, $g \equiv 1$ on $B(a; \delta/2)$, $g \equiv 0$ off $B(a, \delta)$, and $\|\overline{\partial} g\|_\infty \le 4/\delta$. If f is a bounded Borel function, then the following estimates are valid.*
  (a) $\|T_g f\|_\infty \le 8 \,\mathrm{osc}(f; B(a; \delta)) \le 16 \|f\|_\infty$.
  (b) *For $|z - a| > \delta$, $|T_g f(z)| \le 16\delta \|f\|_\infty / (|z - a|)$.*
  (c) *For $|z - a| < \delta/2$,*
$$|(f - T_g)(z) - (f - T_g)(a)| \le \frac{68 |z - a| \|f\|_\infty}{\delta}$$

PROOF. (a) Since $\mathrm{supp}(g) \subseteq \overline{B}(a; \delta)$, $\mathrm{Area}(\mathrm{supp}(g)) \le \pi \delta^2$ and $\mathrm{osc}(f; \mathrm{supp}(g)) \le 2\|f\|_\infty$. Substituting these estimates into (5.6) gives
$$\|T_g f\|_\infty \le \frac{2}{\sqrt{\pi}} \mathrm{osc}(f; B(a; \delta)) \|\overline{\partial} g\|_\infty \sqrt{\pi \delta^2}$$
$$\le 2\delta \, \mathrm{osc}(f; B(a; \delta)) \frac{4}{\delta}$$
$$\le 8 \, \mathrm{osc}(f; B(a; \delta))$$
$$\le 16 \|f\|_\infty.$$

(b) For $|z - a| = \delta$, part (a) implies that $|(z-a) T_g f(z)| \le 16 |z - a| \|f\|_\infty = 16 \delta \|f\|_\infty$. But $(z - a) T_g f(z)$ is analytic on $\mathbb{C}_\infty \setminus \overline{B}(a; \delta)$. As in the proof of (5.6), this inequality holds for $|z - a| \ge \delta$.

(c) Once again we will use the Maximum Modulus Theorem and the fact that $f - T_g f$ is analytic on $\mathrm{int}\{z : g(z) = 1\} \supseteq B(a; \delta/2)$. For $|z-a| = \delta/2$,
$$\frac{|(f - T_g)(z) - (f - T_g)(a)|}{|z - a|} \le \frac{1}{|z - a|}[\|f\| + 16\|f\| + \|f\| + 16\|f\|]$$
$$\le 68 \|f\|/\delta. \quad \square$$

Exercises

1. Show that if f is analytic in a neighborhood of $\mathrm{supp}(g)$, then $T_g f = 0$.
2. Suppose that K is a compact subset of $\mathbb{C}$ and f is a continuous function on $\mathbb{C}_\infty$ that is analytic off K with $f(\infty) = 0$. Let $\{U_k\}$ be a locally finite cover of $\mathbb{C}$ and let $\{g_k\}$ be a C^1 partition of unity subordinate to this cover.
   (a) Show that $T_{g_k} f = 0$ for all but a finite number of k.
   (b) Show that $f = \sum_k T_{g_k} f$.

§6 T-invariant algebras

6.1 DEFINITION. If K is a compact subset of $\mathbb{C}$ and A is a closed subalgebra of $C(K)$, then A is said to be *T-invariant* if $R(K) \subseteq A$ and for f

in A and g in C_c^1, $T_g f \in A$, where f is extended to $\mathbb{C}$ by letting it be identically 0 off K and T_g is the Vitushkin localization operator discussed in the preceding section.

Note that since $R(K) \subseteq A$, every T-invariant algebra is a function algebra. Clearly $R(K)$ is a T-invariant algebra since, in light of (5.3c), $T_g f$ is analytic in a neighborhood of K whenever f is. Of course if K is not polynomially convex, $P(K)$ will not be a T-invariant algebra since it does not contain $R(K)$. However if we consider $P(K)$ as a subalgebra of the space of continuous functions on its polynomially convex hull, then $P(K)$ is a T-invariant subalgebra of $C(\widehat{K})$.

Before giving additional examples of T-invariant algebras, let's look at some equivalent formulations. The part of the definition of a T-invariant algebra that requires that $T_g f \in A$ whenever $f \in A$ assumes that f is extended to $\mathbb{C}$ by letting $f(z) = 0$ for z in $\mathbb{C}\setminus K$. What happens if some other extension of f is used? For example, suppose $f \in \operatorname{Rat}(K)$ and we take its natural extension to $\mathbb{C}$? Condition (b) in the next proposition clears this up.

6.2 PROPOSITION. *If K is a compact subset of $\mathbb{C}$ and A is a function algebra on K, the following statements are equivalent.*

(a) *A is a T-invariant algebra.*
(b) *If $g \in C_c^1$ and f is a bounded Borel function on $\mathbb{C}$ such that $f|K \in A$, then $T_g f|K \in A$.*
(c) *If $\mu \in M(K)$ and $\mu \perp A$, then $\hat{\mu} = 0$ off K and if $g \in C_c^1$, then $\nu \in A^\perp$ for the measure ν on K that satisfies $\hat{\nu} = g\hat{\mu}$.*

PROOF. (a) *implies* (b). If g and f are as in (b), then for w in K

$$T_g f(w) = (T_g(f|K))(w) + \int_{\mathbb{C}\setminus K} \frac{f(z)}{z-w} \bar{\partial} g(z) \, d\mathscr{A}(z).$$

Since A is T-invariant and $f|K \in A$, $T_g(f|K) \in A$. The integral in the above formula is the Cauchy transform of $\chi_{\mathbb{C}\setminus K} f \bar{\partial} g$, a bounded Borel function with compact support that vanishes on K. By Proposition 3.11, the function defined by this integral belongs to $R(K)$ and hence to A.

(b) *implies* (a). It is only necessary to show that $R(K) \subseteq A$. Let $h \in L_c^\infty$ such that $h \equiv 0$ on K. So $h|K \in A$ and, by (b), $T_g h \in A$ for any function g in C_c^1. Construct such a function g with $g(z) = \pi \bar{z}$ in a neighborhood of the support of h. For w in K

$$T_g h(w) = \frac{1}{\pi} \int_{\mathbb{C}\setminus K} \frac{h(z)}{z-w} \bar{\partial} g(z) \, d\mathscr{A}(z)$$
$$= \hat{h}(w).$$

Thus $\hat{h} \in A$ and so $R(K) \subseteq A$ by Proposition 3.12.

(a) *is equivalent to* (c). Note that the first part of (c) is equivalent to the requirement that $R(K) \subseteq A$. If $g \in C_c^1$, then $T_g : C(K) \to C(K)$ is a bounded linear operator and if $\mu \in M(K)$, then $T_g^*(\mu) = \nu$ where ν is the measure on K such that $\hat{\nu} = g\hat{\mu}$ (Proposition 5.4). Thus the second half of (c) is equivalent to the requirement that $T_g^*(A^\perp) \subseteq A^\perp$. But abstract duality theory for operators on a Banach space states that $T_g^*(A^\perp) \subseteq A^\perp$ if and only if $T(A) \subseteq A$, so that (a) and (c) are equivalent. □

Here is a collection of function algebras that generalize the concept of the algebra $A(K)$.

6.3 DEFINITION. If K is a compact subset of $\mathbb{C}$ and U is an open set that is included in K, then $A(K, U) \equiv \{f \in C(K) : f \text{ is analytic on } U\}$. So $A(K, \text{int } K) = A(K)$.

6.4 PROPOSITION. (a) *For any compact subset K of $\mathbb{C}$ and open set U included in K, $A(K, U)$ is a T-invariant algebra.*

(b) *For any measurable subset E of a compact set K, $R(K, E)$ is a T-invariant subalgebra of $C(K)$.*

PROOF. (a) Since $R(K) \subseteq A(K, U)$, the fact that $A(K, U)$ is a T-invariant algebra is an immediate consequence of Proposition 5.3 (b) and (c).

(b) A measure μ on K annihilates $R(K, E)$ if and only if $\hat{\mu} = 0$ off E (3.14). Thus the fact that $R(K, E)$ is a T-invariant subalgebra of $C(K)$ follows readily from part (c) of the preceding proposition. □

We wish to determine the Banach algebra properties of T-invariant algebras, but to do so we need some more information about the type of functions that can belong to them. The next few results are of this nature.

6.5 PROPOSITION. *Suppose A is a T-invariant subalgebra of $C(K)$ and $a \in K$. If $f \in A$ and f has an analytic extension to a neighborhood of a, then*

$$\frac{f - f(a)}{z - a} \in A.$$

PROOF. Let $F : \mathbb{C} \to \mathbb{C}$ be a continuous function with compact support such that $F|K = f$ and F is analytic in a neighborhood of a. Choose $\delta > 0$ such that F is analytic on $B_\delta = B(a; \delta)$ and let $g \in C_c^1$ such that $g(z) = (z - a)^{-1}$ on $\text{supp}(F) \setminus B_\delta$. Since A is a T-invariant algebra, $T_g F \in A$. Also $\bar{\partial}(T_g F) = g\bar{\partial} F$. But $\bar{\partial} F = 0$ on B_δ and off $\text{supp}(F)$. Hence

$$\bar{\partial}(T_g F) = g\bar{\partial} F = (z - a)^{-1} \bar{\partial} F = \bar{\partial}\left[\frac{F - F(a)}{z - a}\right].$$

So $T_g F - (F - F(a))/(z - a)$ is an entire function by Weyl's Lemma. Since both $T_g F$ and $(F - F(a))/(z - a)$ vanish at ∞, $T_g F = (F - F(a))/(z - a) = (f - f(a))/(z - a) \in A$. □

6.6 PROPOSITION. *If A is a T-invariant subalgebra of $C(K)$ and $a \in K$, then the subalgebra of A consisting of those functions in A that have analytic extensions to a neighborhood of a is dense in A.*

PROOF. Let $\varepsilon > 0$ and fix f in A. Use the same symbol f to denote a continuous function on $\mathbb{C}$ with compact support that extends this element of A. Choose $\delta > 0$ such that $\operatorname{osc}(f; B(a; \delta)) < \varepsilon/8$. Let $g \in C_c^1$ such that $g \equiv 1$ on $B(a; \delta/2)$, $\operatorname{supp}(g) \subseteq \overline{B}(a; \delta)$, and $\|\bar{\partial} g\|_\infty \leq 4/\delta$ (5.5). It follows that $F \equiv f - T_g f$ is analytic on $B(a; \delta/2)$ and $F \in A$.

Also (5.7(a)) implies that $\|F - f\|_K = \|T_g f\|_K \leq 8 \operatorname{osc}(f; B(a; \delta)) < \varepsilon$. □

The preceding proposition is quite useful. Indeed, we will use it to show that $R(K) = A(K)$ for certain sets K. A preview of this application can be seen in the exercise at the end of this section.

6.7 ARENS'S THEOREM. *If A is a T-invariant subalgebra of $C(K)$, then the maximal ideal space of A is $\{\delta_z : z \in K\}$.*

PROOF. Let $\mathscr{M}$ denote the maximal ideal space of A and for ρ in $\mathscr{M}$, put $a = \rho(z)$. So $z - a \in \ker \rho$ and hence is not invertible in A. Since $R(K) \subseteq A$, it must be that $a \in K$. Suppose $f \in A$ and f is analytic in a neighborhood of a. By (6.5), $h = (f - f(a))/(z - a) \in A$. Hence $\rho(f) = f(a) + \rho(h)\rho(z - a) = f(a)$. By Proposition 6.6, $\rho(f) = f(a)$ for all f in A. □

6.8 COROLLARY. *If E is a bounded subset of K, the maximal ideal space of $R(K, E)$ is K.*

REMARKS. The reader might consult Gamelin [**1973**] for some additional information about T-invariant algebras. The idea goes back to the original work of Vitushkin, but the person who introduced the concept is unknown to me. Also see Cole and Gamelin [**1982** and **1985**].

Exercise

Let $\{\Delta_n\}$ be a sequence of pairwise disjoint closed disks in $\mathbb{D}$ such that $\operatorname{cl}(\bigcup_n \Delta_n) = \bigcup_n \Delta_n \cup \{0\}$. (For example, let $\Delta_n = \overline{B}(1/n; 10^{-n})$.) If $K = \operatorname{cl} \mathbb{D} \setminus \bigcup_n (\operatorname{int} \Delta_n)$, then $A(K) = R(K)$.

§7 The Shilov boundary. Let A be an abelian Banach algebra with identity and let $\mathscr{M}$ be its maximal ideal space, identified with the space of nonzero multiplicative linear functionals from A into $\mathbb{C}$, and topologized in the usual way. (If we want to more closely associate the maximal ideal space with its parent Banach algebra, it will be denoted by $\mathscr{M}_A$.) For f in A, $\hat{f}: \mathscr{M} \to \mathbb{C}$ is the usual Gelfand transform, $\hat{f}(\rho) = \rho(f)$ for all ρ in $\mathscr{M}$. So $\hat{f}$ is a continuous function on the compact space $\mathscr{M}$ and hence $|\hat{f}|$ attains its maximum value.

§7 THE SHILOV BOUNDARY

7.1 DEFINITION. A subset B of $\mathcal{M}$ is called a *boundary* for the Banach algebra A if for every f in A, $|\hat{f}|$ attains its maximum value somewhere on B.

Of course $\mathcal{M}$ is itself a boundary and so the collection of boundaries is a nonempty collection of subsets of $\mathcal{M}$. Also note that if B is a boundary, so is any larger set (like $\operatorname{cl} B$).

7.2 EXAMPLE. If $A = A(K)$, then $\mathcal{M} = K$ and ∂K is a boundary for A. Indeed, this is immediate from the Maximum Modulus Theorem. It is also possible to show that any closed boundary for A must contain ∂K. (This will be done later.) Thus ∂K is the smallest closed boundary for $A(K)$. Since $R(K) \subseteq A(K)$, it follows that ∂K is a boundary for $R(K)$ and it is also true that ∂K is the smallest closed boundary for $R(K)$. It is possible to give examples of compact sets for which there are proper (nonclosed) subsets of ∂K that are boundaries for $R(K)$. See Exercise 11.3 below.

It turns out that just like $R(K)$ every Banach algebra has a unique minimal closed boundary.

7.3 LEMMA. *If $f_1, \ldots, f_n \in A$ and $V = \{\rho \in \mathcal{M} : |\rho(f_j)| < 1 \text{ for } 1 \leq j \leq n\}$, then either $V \cap B \neq \varnothing$ for every boundary B or $F \setminus V$ is a boundary for every closed boundary F.*

PROOF. It suffices to show that if F is a closed boundary for A such that $F \setminus V$ is not a boundary, then $V \cap E \neq \varnothing$ for every boundary E.

Since $F \setminus V$ is not a boundary, there is an f in A such that $\|\hat{f}\|_\infty = 1 > \|\hat{f}\|_{F \setminus V}$. Let E be any boundary. Thus there is a ρ_0 in E for which $|\rho_0(f)| = 1$. Now $\hat{f}^k \to 0$ uniformly on $F \setminus V$ while $|\rho_0(f^k)| = 1$ for all $k \geq 1$. So if f is replaced by f^k for sufficiently large k, we may assume that $\|\hat{f}\hat{f}_j\|_{F \setminus V} < 1$ for $1 \leq j \leq n$. If $\rho \in V$, then $|\hat{f}(\rho)\hat{f}_j(\rho)| \leq |\hat{f}_j(\rho)| < 1$. Thus $|\hat{f}\hat{f}_j| < 1$ on F. Since F is a boundary for A, this implies that $\|\hat{f}\hat{f}_j\|_\infty < 1$. In particular, for $1 \leq j \leq n$, $1 > |\rho_0(ff_j)| = |\rho_0(f)\rho_0(f_j)| = |\rho(f_j)|$ and so $\rho_0 \in V$. That is, $\rho_0 \in V \cap E$. □

7.4 THEOREM. *The intersection of all closed boundaries for A is a closed boundary for A.*

PROOF. Let $\partial_A \equiv$ the intersection of all closed boundaries for A; notice that at this point we do not know that ∂_A is a nonempty set. This and more, however, will follow from the following claim.

CLAIM. If $f \in A$ such that $\|\hat{f}\|_\infty = 1$ and $Q = \{\rho \in \mathcal{M} : |\hat{f}(\rho)| = 1\}$, then $Q \cap \partial_A \neq \varnothing$.

In fact, suppose $Q \cap \partial_A = \varnothing$. So for every ρ in Q there is a closed boundary F_ρ such that $\rho \notin F_\rho$. There is an open set V_ρ of the form described in the preceding lemma (see Exercise 1) such that $\rho \in V_\rho$ and $V_\rho \cap F_\rho = \varnothing$. By the compactness of Q, there are finitely many such open sets $V_1, \ldots, V_m$ and a corresponding number of closed boundaries $F_1, \ldots, F_m$

having the following properties.
  (i) $F_j \cap V_j = \emptyset$.
  (ii) $Q \subseteq \bigcup_j V_j$.

Since V_1 is disjoint from the boundary F_1, Lemma 7.3 implies $\mathscr{M} \setminus V_1$ is a boundary. Similarly, $V_2 \cap F_2 = \emptyset$ implies $\mathscr{M} \setminus (V_1 \cup V_2) = (\mathscr{M} \setminus V_1) \setminus V_2$ is a boundary. Continuing we get that $\mathscr{M} \setminus (\bigcup_j V_j)$ is a closed boundary for A. But property (ii) above implies that $|\hat{f}| < 1$ on $\mathscr{M} \setminus (\bigcup_j V_j)$, a contradiction. This establishes the claim.

Indeed the claim proves the whole theorem since it says that if $f \in A$, there is a ρ in ∂_A for which $|\hat{f}(\rho)| = \|\hat{f}\|_\infty$. □

The unique smallest closed boundary for A is called the *Shilov boundary* of A and is denoted, as in the proof, by ∂_A.

7.5 LEMMA. *Let K be a compact subset of $\mathbb{C}$.*

(a) *If $a \in \partial K$ and there is a point b different from a such that the line segment $[a, b]$ meets K only at the point a, then there is a function f in $R(K)$ with $f(a) = 1$ and $|f(z)| < 1$ for all z in K, $z \neq a$.*

(b) *The set of such points a that satisfy the condition in (a) is dense in ∂K.*

PROOF. (a) First let $f_1(z) = (z-a)/(z-b)$; $f_1 \in R(K)$, $f_1(K) \subseteq \mathbb{C} \setminus (-\infty, 0)$, and $f_1(a) = 0$. Let $f_2(z) = f_1(z)^{1/2}$. We want to show that $f_2 \in R(K)$. In fact, let $\{a_n\}$ be a sequence in the segment $[a, b]$ such that $a_n \to a$ and put $g_n(z) = [(z - a_n)/(z - b)]^{1/2}$. Since each g_n is analytic in a neighborhood of K, $g_n \in R(K)$. If $z \neq a$, $g_n(z) \to f_2(z)$ and $g_n(a) \to f_2(a) = 0$. Also the sequence $\{g_n\}$ is uniformly bounded on K (Why?). Thus for any measure μ in $M(K)$ that annihilates $R(K)$, $\int g_n d\mu \to \int f_2 d\mu$. It follows that $f_2 \in R(K)$.

Now put $f(z) = \exp(-f_2(z))$. Since $R(K)$ is a Banach algebra, $f \in R(K)$. Since $\text{Re} f_2(z) > 0$ on $K \setminus \{a\}$, $|f(z)| < 1$ for z in K and $z \neq a$. It is also clear that $f(a) = 1$.

(b) Let $a \in \partial K$ and let $\varepsilon > 0$. Pick any point b in $\mathbb{C} \setminus K$ with $|a - b| < \varepsilon$. Let $t_0 = \max\{t \in [0, 1]: tb + (1-t)a \in K\}$; so $0 \leq t_0 < 1$. Put $a_0 = t_0 b + (1 - t_0)a$. It is easy to check that a_0 satisfies the condition in (a) and $|a - a_0| < \varepsilon$. □

7.6 COROLLARY. *The Shilov boundary of $P(K)$ is the outer boundary of K. (That is, $\partial \hat{K}$.)*

7.7 COROLLARY. *If A is a T-invariant subalgebra of $C(K)$, then $\partial_A \supseteq \partial K$.*

The inclusion in the last corollary can be proper. A demonstration that this can happen is outlined in Exercise 2 below, but it may be best to postpone this exercise until we finish gathering information about the Shilov boundary.

7.8 Proposition. *If A is an abelian Banach algebra with identity and $f \in A$, then $\hat{f}(\partial_A) \supseteq$ the topological boundary of $\hat{f}(\mathcal{M}_A)$.*

Proof. Since ∂_A is compact, so is $\hat{f}(\partial_A)$. Suppose α belongs to the topological boundary of $\hat{f}(\mathcal{M})$; that is, $\alpha \in \partial \hat{f}(\mathcal{M})$. Assume that $\alpha \notin \hat{f}(\partial_A)$; so there is a $\delta > 0$ such that $\text{dist}(\alpha, \hat{f}(\partial_A)) > \delta$. Let $\lambda \in \mathbb{C} \backslash \hat{f}(\mathcal{M})$ such that $|\lambda - \alpha| < \delta/2$. Now $\sigma_A(f) = \hat{f}(\mathcal{M})$ so that $f - \lambda$ is invertible in A; let $g = (f - \lambda)^{-1}$. If $\rho \in \partial_A$, then $\delta < |\hat{f}(\rho) - \alpha| \leq |\hat{f}(\rho) - \lambda| + |\lambda - \alpha| < |\hat{f}(\rho) - \lambda| + \delta/2$. Hence $|\hat{g}(\rho)| = |1/(\hat{f}(\rho) - \lambda)| < 2/\delta$. Since ∂_A is a boundary for A, $\|\hat{g}\|_\infty < 2/\delta$. But $\alpha \in \hat{f}(\mathcal{M})$ and so there is a ρ_0 in $\mathcal{M}$ with $\hat{f}(\rho_0) = \alpha$ and thus $|\hat{g}(\rho_0)| = |\hat{f}(\rho_0) - \lambda|^{-1} > 2/\delta$, a contradiction. □

It turns out that T-invariant algebras have a lot of analytic structure as long as the Shilov boundary is avoided.

7.9 Proposition. *If A is a T-invariant subalgebra of $C(K)$, then each f in A is analytic on $K \backslash \partial_A$.*

Proof. By Corollary 7.7, $\partial_A \supseteq \partial K$, so $K \backslash \partial_A = (\text{int } K) \backslash \partial_A$ is an open set. Let $a \in K \backslash \partial_A$ and choose $r > 0$ such that $\overline{B}(a; r) \subseteq K \backslash \partial_A$. Let $f \in A$ and let $g \in C_c^1(B(a; r))$ so that $g(z) = 1$ for $|z - a| < r/2$; hence $T_g f \in A$. But $\partial_A \subseteq \mathbb{C}_\infty \backslash \overline{B}(a; r)$ and ∂_A is a boundary for A. Hence $\|T_g f\|_{\mathbb{C}_\infty \backslash \overline{B}(a; r)}$ is attained at some point of ∂_A. By (5.3) $T_g f$ is analytic on $\mathbb{C}_\infty \backslash \overline{B}(a; r)$. It follows that $T_g f$ must be constant on $\mathbb{C}_\infty \backslash \overline{B}(a; r)$; since $T_g f$ vanishes at ∞, $T_g f \equiv 0$ off $\overline{B}(a; r)$. In particular, $T_g f \equiv 0$ on ∂_A. Again the fact that ∂_A is a boundary implies that $T_g f \equiv 0$ on K. Therefore $f = f - T_g f$ is analytic in $\text{int}\{z: g(z) = 1\} \supseteq B(a; r/2)$. Since a was arbitrary, f is analytic on $K \backslash \partial_A$. □

Exercises

1. If A is an abelian Banach algebra and $\mathcal{V}$ is the collection of all sets of the form $\{\rho \in \mathcal{M}: |\rho(f_j)| < 1 \text{ for } 1 \leq j \leq n\}$, where $\{f_1, \ldots, f_n\}$ is an arbitrary finite subset of A, then $\mathcal{V}$ is a base for the topology for $\mathcal{M}$.
2. Let E be a compact subset of $\mathbb{D}$ with positive area such that the support of the restriction of area measure to E is the set E itself. Put $K = \text{cl } \mathbb{D}$ and $U = \mathbb{D} \backslash E$ and consider the algebra $A = A(K, U)$. Let $f(z) = \int_E (w - z)^{-1} d\mathcal{A}(w)$ and prove that $f \in A$ and cannot be analytically continued to any open set that meets E. Prove that $\partial_A = \partial \mathbb{D} \cup E$.

§8 Representing measures

8.1 Definition. If A is a function algebra on X and $\rho \in \mathcal{M}_A$, then a *representing measure* for ρ is a probability measure μ on X such that $\rho(f) = \int f d\mu$ for all f in A. If $a \in X$ and $\rho(f) = f(a)$ for f in A, then a representing measure for ρ will be called a representing measure for a.

For the most part in this section, A will be a function algebra on a compact space X and $\mathscr{M} = \mathscr{M}_A$.

8.2 PROPOSITION. *If A is a function algebra on X and $\rho \in \mathscr{M}$, then there is a representing measure μ for ρ. In fact μ can be chosen with its support contained in ∂_A.*

PROOF. By the Hahn-Banach Theorem, ρ has a norm preserving extension to $C(X)$ and this extension is given by a regular Borel measure μ. Thus $\|\mu\| = \|\rho\| = 1$. But also $\mu(X) = \rho(1) = 1 = \|\mu\|$; therefore μ must be positive. That is, μ is a representing measure for ρ.

The restriction map $f \to f|\partial_A$ defines an isometric monomorphism of A onto a function algebra B on the Shilov boundary ∂_A. The homomorphism ρ can be identified with a homomorphism on the algebra B. If the same argument as in the preceding paragraph is used, we see that the representing measure μ can be chosen with support contained in ∂_A. □

If $\rho \in \mathscr{M}$, let M_ρ denote the set of representing measures for ρ. The set M_ρ is a weak* compact convex subset of $M(X)$ (Exercise 1).

8.3 EXAMPLES. (a) If $a \in X$, then δ_a, the unit point mass at a, is a representing measure for a. Note that if $a \notin \partial_A$, then this representing measure is not supported by the Shilov boundary. If $A = C(X)$, then $\partial_A = X$ and δ_a is the only representing measure for evaluation at a.

(b) If $A = P(\text{cl}\,\mathbb{D})$, $|a| < 1$, and P_a is the Poisson kernel at a, then $\mu = P_a\,dm$ is a representing measure for a that is supported on the Shilov boundary of A.

(c) If $A = A(K)$ and $a \in \text{int}\,K$, let $B(a;r) \subseteq K$. The measure $\mu = (\pi r^2)^{-1}\,\text{Area}|B(a;r)$ is a representing measure for evaluation at a.

This superabundance of representing measures is not a drawback. In fact we will eventually see that most measures are related to representing measures and this fact has a certain advantage.

The next easy lemma will be used often.

8.4 LEMMA. *Let A be a function algebra on X and let $\rho \in \mathscr{M}$. If $\ker \rho \equiv \{f \in A : \rho(f) = 0\}$, then $M_\rho = \{\mu : \mu$ is a probability measure and $\mu \perp \ker \rho\}$.*

PROOF. If $\mu \in M_\rho$, then clearly $\int f\,d\mu = 0$ for every function f in $\ker \rho$. For the other inclusion, just observe that for any f in A, $f - \rho(f) \in \ker \rho$. □

Often we can represent a homomorphism by means of a complex-valued measure rather than a probability measure. This extra degree of freedom has certain advantages.

8.5 DEFINITION. If A is a function algebra on X and $\rho \in \mathscr{M}$, a *complex representing measure for ρ* is a measure μ in $M(X)$ such that $\rho(f) = \int f\,d\mu$ for all f in A.

The proof of the next result is like that of Lemma 8.4 and is left to you.

8.6 LEMMA. *If A is a function algebra on X and $\rho \in \mathscr{M}$, then the set of complex representing measures for ρ is $\{\mu \in M(X): \mu(X) = 1$ and $\mu \perp \ker \rho\}$.*

Suppose K is a compact subset of $\mathbb{C}$, $A = A(K)$, and $\overline{B}(a;r) \subseteq \operatorname{int} K$. If $\gamma(t) = a + re^{it}$, $0 \le t \le 2\pi$, and σ is normalized arc length measure on γ, then $rz\sigma$ is a complex representing measure for a that is not a representing measure.

In general if $\rho \in \mathscr{M}$ and $\mu \in M_\rho$, then $\mu + \eta$ is a complex representing measure for ρ whenever $\eta \in A^\perp$. Another construction of complex representing measures is to let $\mu \in M_\rho$ and let $g \in A$ such that $\rho(g) = 0$; then $(g+1)\mu$ is a complex representing measure. The next theorem can be interpreted as saying that for every complex representing measure there is a representing measure lurking in the background.

8.7 THEOREM. *If A is a function algebra on X, $\rho \in \mathscr{M}$, and μ is a complex representing measure for ρ, then there is a representing measure ν that is absolutely continuous with respect to μ.*

PROOF. Let $\mathscr{H}$ = the closure of $\ker \rho = \{f \in A : \int f \, d\mu = 0\}$ in $L^2(|\mu|)$. If $f \in \ker \rho$, then $1 = |\int (1-f)^2 \, d\mu| \le \int |1-f|^2 \, d|\mu|$. Hence $1 \notin \mathscr{H}$ and so there is a function F in the closure of A in $L^2(|\mu|)$ such that $\int |F|^2 \, d|\mu| = 1$ and $F \perp \mathscr{H}$. An easy limiting argument shows that $(\ker \rho)F \subseteq \mathscr{H}$ and so $F \perp (\ker \rho)F$. That is, for every f in $\ker \rho$, $0 = \langle fF, F \rangle = \int f|F|^2 \, d|\mu|$. Let $\nu = |F|^2|\mu|$. So ν is a probability measure and if $f \in \ker \rho$, $\int f \, d\nu = 0$. By Lemma 8.4, ν is a representing measure. □

We now turn our attention to T-invariant algebras. The first result here is the analogue of Lemma 8.4 and Lemma 8.6

8.8 LEMMA. *Let A be a T-invariant algebra on K and let $a \in K$.*
  (a) *$M_a = \{\mu : \mu$ is a probability measure on K and $(z-a)\mu \perp A\}$.*
  (b) *The set of complex representing measures for a is $\{\mu \in M(K) : \mu(K) = 1$ and $(z-a)\mu \perp A\}$.*

PROOF. The proof is rather easy if Propositions 6.5 and 6.6 are properly utilized. □

The proof of the extremely useful result that follows is a straightforward application of the preceding lemma.

8.9 PROPOSITION. *If A is a T-invariant subalgebra of $C(K)$, $\mu \in A^\perp$, and $a \in K$ such that $\tilde{\mu}(a) < \infty$ and $\hat{\mu}(a) \ne 0$, then*
$$\frac{1}{\hat{\mu}(a)} \frac{1}{z-a} \mu$$
is a complex representing measure for a.

8.10 COROLLARY. *If A is a T-invariant subalgebra of $C(K)$, $\mu \in A^\perp$, and $a \in K$ such that $\tilde{\mu}(a) < \infty$ and $\hat{\mu}(a) \ne 0$, then there is a ν in M_a such that $\nu \ll \mu$.*

Exercises

1. If A is a function algebra and $\rho \in M_A$, show that M_ρ is a weak* compact convex subset of $M(X)$.

2. Let A be a T-invariant algebra on K, let $a \in K$, and let μ be a representing measure for a. If $\mathscr{H}$ is the closure of A in $L^2(\mu)$ and $S: \mathscr{H} \to \mathscr{H}$ is defined as multiplication by z, show that S is a subnormal operator, $a \in \sigma(S)$, and $a \notin \sigma_{ap}(S)$. What is $[\operatorname{ran}(S-a)]^\perp$?

§9 Harmonic measure. This section begins with some basic facts about harmonic functions that may not be familiar to many of the readers, though the reader is assumed to be acquainted with the most elementary properties of harmonic functions as found in [**FOCV**]. Later harmonic measure will be introduced and some of its properties deduced.

We will deal here with complex valued harmonic functions; that is, functions whose real and imaginary parts are harmonic.

9.1 DEFINITION. A *Dirichlet set* is a bounded open subset of $\mathbb{C}$ with the property that for each continuous function $u: \partial G \to \mathbb{C}$ there is a continuous function $h: \operatorname{cl} G \to \mathbb{C}$ such that h is harmonic on G and $h|\partial G = u$. The function h is called the (classical) solution of the Dirichlet Problem with boundary values u. A connected Dirichlet set is called a *Dirichlet region*.

In [**FOCV**] it is shown that if each component of ∂G consists of more than one point, then G is a Dirichlet set. It is also shown there that the punctured disk is not a Dirichlet set.

Even though there are open subsets of $\mathbb{C}$ that are not Dirichlet sets, each bounded function on ∂G gives rise to a "candidate" for the solution of the Dirichlet Problem.

Recall (III.5.1) that a function $\phi: G \to [-\infty, \infty)$ is subharmonic if ϕ is upper semicontinuous and for every closed disk $\overline{B}(a; r)$ contained in G, we have the inequality

$$\phi(a) \leq \frac{1}{2\pi} \int \phi(a + re^{i\theta}) \, d\theta.$$

A function $\psi: G \to (-\infty, \infty]$ is superharmonic if $-\psi$ is subharmonic.

Say that a function $\phi: G \to [-\infty, \infty)$ *satisfies the Maximum Principle* if for every compact set K contained in G, $\phi \leq h$ on K whenever h is a continuous function on K that is harmonic on $\operatorname{int} K$ and satisfies $\phi \leq h$ on ∂K. It is known (see [**FOCV**]) that an upper semicontinuous ϕ on G is subharmonic if and only if ϕ satisfies the Maximum Principle.

9.2 PROPOSITION. *If G is a region in $\mathbb{C}$ and $\phi: G \to \mathbb{R}$ is a C^2 function such that $\Delta \phi \geq 0$ on G, then ϕ is subharmonic.*

PROOF. Assume for the moment that G_1 is a bounded region with $\operatorname{cl} G_1 \subseteq G$ and that ϕ is not constant. It will be shown that ϕ cannot attain its maximum value on $\operatorname{cl} G_1$ at a point of G_1. From here it follows that ϕ satisfies the Maximum Principle (Exercise 2). Since it is assumed that ϕ is continuous on the compact set cl G_1, what we want to show is equivalent to proving the statement that $\sup\{\phi(z): z \in G_1\} = \sup\{\phi(z): z \in \partial G_1\}$.

Note that we can assume that $\Delta\phi > 0$ on G. Indeed, consider $\phi_\varepsilon = \phi + \varepsilon|z|^2$. So $\Delta\phi_\varepsilon > 0$. So if it is proved that $\sup\{\phi_\varepsilon(z): z \in G_1\} = \sup\{\phi_\varepsilon(z): z \in \partial G_1\}$ for each ϕ_ε, then

$$\sup\{\phi(z): z \in G_1\} \leq \sup\{\phi_\varepsilon(z): z \in G_1\}$$
$$= \sup\{\phi_\varepsilon(z): z \in \partial G_1\}$$
$$\leq \varepsilon M + \sup\{\phi(z): z \in \partial G_1\},$$

where $M = \sup\{|z|^2: z \in G_1\}$. Letting $\varepsilon \to 0$ gives the desired conclusion for ϕ.

Suppose there is a point ζ in G such that $\phi(\zeta) > \sup\{\phi(z): z \in \partial G\}$. Let $\zeta = x + iy$ and put

$$a = \inf\{\alpha: x + it \in G \text{ for } \alpha < t \leq y\},$$
$$b = \sup\{\beta: x + it \in G \text{ for } y \leq t < \beta\}.$$

Define $g(t) = \phi(x + it) = \phi(x, t)$ for $a \leq t \leq b$. By assumption, g attains its maximum value at the interior point y. Therefore $0 \geq g''(y) = ((\partial^2\phi)/(\partial y^2))(x, y)$. Similarly, $(\partial^2\phi/\partial x^2)(x, y) \leq 0$. This implies that $\Delta\phi(\zeta) \leq 0$, a contradiction. □

9.3 DEFINITION. If $u: \partial G \to \mathbb{R}$ is any bounded function, let

$$\hat{\mathscr{P}}(u, G) = \{\phi: \phi \text{ is subharmonic on } G \text{ and } \limsup_{z \to a} \phi(z) \leq u(a)$$
$$\text{for every } a \text{ in } \partial G\},$$

$$\check{\mathscr{P}}(u, G) = \{\psi: \psi \text{ is superharmonic on } G \text{ and } \liminf_{z \to a} \psi(z) \geq u(a)$$
$$\text{for every } a \text{ in } \partial G\}.$$

Also, define functions $\hat{u}$ and $\check{u}$ on G by

$$\hat{u}(z) = \sup\{\phi(z): \phi \in \hat{\mathscr{P}}(u, G)\},$$
$$\check{u}(z) = \inf\{\psi(z): \psi \in \check{\mathscr{P}}(u, G)\}.$$

9.4 PROPOSITION. (a) $\hat{u}$ and $\check{u}$ are harmonic functions on G.
(b) If c is a non-negative real number, then $\widehat{(cu)} = c\hat{u}$ and $\widecheck{(cu)} = c\check{u}$.
(c) If c is a nonpositive real number, then $\widehat{(cu)} = c\check{u}$ and $\widecheck{(cu)} = c\hat{u}$.
(d) $\hat{u} + \hat{v} \leq \widehat{(u+v)} \leq \widecheck{(u+v)} \leq \check{u} + \check{v}$.
(e) $-\|u\|_{\partial G} \leq \hat{u} \leq \check{u} \leq \|u\|_{\partial G}$.

The proof of part (a) of this proposition can be found in the proof of Theorem X.3.11 of [**FOCV**]. The proof of the remaining parts is an easy exercise in manipulating definitions.

If $u \in \mathbb{C}_\mathbb{R}(\partial G)$ and there is a solution h of the Dirichlet Problem with boundary values u, then $\hat{u} = \check{u} = h$ (Exercise 4). So in order to solve the Dirichlet Problem, it must be that $\hat{u} = \check{u}$. Suppose that this does indeed happen. That is, assume that $\hat{u} = \check{u}$. Define $h: \operatorname{cl} G \to \mathbb{R}$ by letting $h = \hat{u} = \check{u}$ on G and $h = u$ on ∂G. There are two difficulties here. First, how do we know that h is continuous on $\operatorname{cl} G$? Indeed, it will not always be so since we cannot always solve the Dirichlet Problem. Second, how can we decide whether $\hat{u} = \check{u}$? We have already alluded to the first question and referred to [**FOCV**]. A necessary and sufficient condition in terms of logarithmic capacity for the solution of the Dirichlet Problem can be found on page 104 of Tsuji [**1975**]. The second question always has an affirmative answer.

9.5 DEFINITION. Say that a bounded function $u: \partial G \to \mathbb{R}$ is *solvable* if $\hat{u} = \check{u}$.

9.6 PROPOSITION. *If $\mathscr{S}$ is the collection of all solvable functions on ∂G and $\mathscr{S}$ is endowed with the supremum norm, then $\mathscr{S}$ is a real Banach space.*

PROOF. The fact that $\mathscr{S}$ is a real linear space is a consequence of (b), (c), and (d) of the preceding proposition. To show that $\mathscr{S}$ is complete, let $\{u_n\}$ be a sequence of functions in $\mathscr{S}$ and assume $u: \partial G \to \mathbb{R}$ is a bounded function such that $\|u_n - u\|_{\partial G} \to 0$. Using parts (d) and (e) of Proposition 9.4, we have that $\check{u} = \widetilde{(u - u_n + u_n)} \leq \widetilde{(u - u_n)} + \widetilde{u}_n \leq \|u - u_n\| + \widetilde{u}_n$; also $\hat{u} = \widehat{(u - u_n + u_n)} \geq \widehat{(u - u_n)} + \widehat{u}_n \geq -\|u - u_n\| + \widehat{u}_n$. Since $\widehat{u}_n = \widetilde{u}_n$, $0 \leq \check{u} - \hat{u} \leq 2\|u - u_n\|$. Hence $u \in \mathscr{S}$. □

The next result is due to Wiener.

9.7 THEOREM. *Every continuous real-valued function on ∂G is solvable.*

PROOF. Let u be a real-valued continuous function on ∂G and let $\{p_n(z, \bar{z})\}$ be a sequence of polynomials in z and $\bar{z}$ that converges to u uniformly on ∂G. Let c_n be a positive constant that is sufficiently large that $\Delta(p_n + c_n|z|^2) > 0$ on $\operatorname{cl} G$. By Proposition 9.2, $p_n + c_n|z|^2$ is subharmonic on G. But it is not difficult to see that any continuous function that is subharmonic on G is solvable. Thus both $p_n + c_n|z|^2$ and $c_n|z|^2$ are solvable. Therefore p_n is solvable by Proposition 9.6. This same proposition implies that u, the limit of $\{p_n\}$, is solvable. □

9.8 COROLLARY. *If u and $v \in C_{\mathbb{R}}(\partial G)$ and $\alpha, \beta \in \mathbb{R}$, then $\widehat{\alpha u + \beta v} = \alpha \hat{u} + \beta \hat{v}$.*

9.9 COROLLARY. *If $a \in G$, the map $u \to \hat{u}(a)$ is a positive linear functional of norm 1.*

PROOF. If u is a positive function in $C_{\mathbb{R}}(\partial G)$, then $0 \in \hat{\mathscr{P}}(u, G)$; hence $\hat{u} \geq 0$. Corollary 9.8 implies that $u \to \hat{u}$ is linear. Therefore $u \to \hat{u}(a)$ is a positive linear functional. Since $\hat{1} = 1$, this functional has norm 1. □

According to the preceding corollary, for each a in G there is a unique probability measure ω_a supported on ∂G such that

$$(9.10) \qquad \hat{u}(a) = \int_{\partial G} u \, d\omega_a.$$

9.11 Definition. For any bounded open set G and any point a in G, the unique probability measure ω_a supported on ∂G and satisfying (9.10) for every u in $C_{\mathbb{R}}(\partial G)$ is called *harmonic measure* for G at a. If K is a compact subset of $\mathbb{C}$ and $a \in \text{int } K$, then *harmonic measure* for K at a is the same as harmonic measure for $\text{int } K$ at a.

If $G = \mathbb{D}$ and $a \in \mathbb{D}$, then $d\omega_a = P_a\, dm$, where P_a is the Poisson kernel at a and m is normalized arc length measure om $\partial \mathbb{D}$. Thus harmonic measure for $\mathbb{D}$ and arc length measure on $\partial \mathbb{D}$ are mutually absolutely continuous. This is not an isolated incident.

9.12 Theorem. *If G is a simply connected region such that ∂G is a rectifiable Jordan curve, then harmonic measure for G and arc length measure on ∂G are mutually absolutely continuous.*

Proof. Fix a in G and let $\tau: \mathbb{D} \to G$ be the Riemann map such that $\tau(0) = a$ and $\tau'(0) > 0$. A result from function theory (Koosis [**1980**]) says that τ extends to a homeomorphism $\tau: \text{cl } \mathbb{D} \to \text{cl } G$. Thus $\theta \to \tau(e^{i\theta})$ is a parameterization of ∂G. Since ∂G is a rectifiable curve, Theorem III.9.7 and its corollaries imply that $\theta \to \tau(e^{i\theta})$ is absolutely continuous and the derivative with respect to θ is $ie^{i\theta}\tau'(e^{i\theta})$ (here $\tau'(e^{i\theta})$ denotes the radial limit of τ'). So if $u: \partial G \to \mathbb{R}$ is a continuous function, the measure μ defined on ∂G by

$$\int u\, d\mu = \frac{1}{2\pi} \int_0^{2\pi} u(\tau(e^{i\theta}))|\tau'(e^{i\theta})|\, d\theta$$

is (non-normalized) arc length measure.

Claim. If E is a Borel subset of $\partial \mathbb{D}$, then $m(E) = 0$ if and only if $\mu(\tau(E)) = 0$.

From the above formula, we know that for any Borel set Δ contained in ∂G, $\mu(\Delta) = \int_{\tau^{-1}(\Delta)} |\tau'|\, dm$. Assume that $m(E) = 0$. Since τ is a homeomorphism, $E = \tau^{-1}(\tau(E))$ and so $\mu(\tau(E)) = \int_E |\tau'|\, dm = 0$. Conversely, if $\mu(\tau(E)) = 0$, then $\tau' = 0$ a.e. $[m]$ on E. Since $\tau' \in H^1$ (III.9.7), this implies that $m(E) = 0$.

On the other hand, if $u \in C_{\mathbb{R}}(\partial G)$, $u \circ \tau \in C_{\mathbb{R}}(\partial \mathbb{D})$; let h be the solution of the Dirichlet Problem on $\mathbb{D}$ with boundary values $u \circ \tau$. It is easy to see that $h \circ \tau^{-1}$ is the solution of the Dirichlet Problem on G with boundary values u. That is, $h \circ \tau^{-1} = \hat{u}$. Hence

$$\int u\, d\omega_a = \hat{u}(a) = h(0) = \int u \circ \tau\, dm.$$

Since u was arbitrary, this implies that $\omega_a = m \circ \tau^{-1}$. Thus for a Borel subset Δ of ∂G, $\omega_a(\Delta) = 0$ if and only if $m(\tau^{-1}(\Delta)) = 0$. In light of the claim, this proves the theorem. $\square$

The above result extends to any region whose boundary consists of a finite number of rectifiable Jordan curves. In fact, for smooth curves this can be obtained from Green's Theorem and an invocation of the normal derivative of the Green function for the region.

In general, we will not be so concerned with the exact form of harmonic measure but rather with its measure class; that is, with the sets of harmonic measure 0. Moreover, the focus here will be on harmonic measure for compact sets rather than open sets. Indeed, if $a \in \text{int}\, K$ and $f \in A(K)$, then f is the solution of the Dirichlet Problem with boundary values $f|\partial K$ and so $f(a) = \int f\, d\omega_a$. That is, $\omega_a \in M_a$, the set of representing measures for a. Moreover, ω_a is supported on ∂K.

The remainder of the section will be devoted to developing the basic properties of harmonic measure. The first result makes explicit something that is implicit in the definition of harmonic measure for a compact set. Indeed, harmonic measure for K at a is the same as harmonic measure for $\text{cl}[\text{int}\, K]$ at a.

9.13 PROPOSITION. *If $a \in \text{int}\, K$ and ω_a is harmonic measure for K at a, then $\omega_a(K \setminus \text{cl}[\text{int}\, K]) = 0$.*

Recall that two measures μ and ν are said to be mutually absolutely continuous if they have the same sets of measure 0. When this is the case, we can form the two Radon-Nikodym derivatives $d\mu/d\nu$ and $d\nu/d\mu$. Say that μ and ν are *boundedly mutually absolutely continuous* if they are mutually absolutely continuous and the two Radon-Nikodym derivatives are bounded functions.

The next result will be stated for open sets since this is the form needed in later sections.

9.14 PROPOSITION. *If G is an open subset of the plane and a and b belong to the same component of G, then the harmonic measures for G at a and b, ω_a and ω_b, are boundedly mutually absolutely continuous. Moreover, there is a constant $\rho > 0$ such that if H is any open set containing G and μ_a and μ_b are the harmonic measures for H at a and b, then $\rho\mu_a \leq \mu_b \leq \rho^{-1}\mu_a$.*

PROOF. This follows from Harnack's Inequality. If $\overline{B}(a;R) \subseteq G$, $|b-a| = r < R$, and u is a positive continuous function on ∂G, then Harnack's Inequality implies that $\rho \hat{u}(a) \leq \hat{u}(b) \leq \rho^{-1}\hat{u}(a)$, where $\rho = (R-r)/(R+r)$. Thus $\rho \int u\, d\omega_a \leq \int u\, d\omega_b \leq \rho^{-1} \int u\, d\omega_a$. But this implies that $\rho\omega_a(\Delta) \leq \omega_b(\Delta) \leq \rho^{-1}\omega_a(\Delta)$ for every Borel set Δ contained in ∂G. Thus ω_a and ω_b are boundedly mutually absolutely continuous and

$$\rho \leq \frac{d\omega_b}{d\omega_a} \leq \rho^{-1}.$$

Note that this constant ρ depends only on r and R. Thus the same inequalities hold for μ_a and μ_b:

$$\rho \leq \frac{d\mu_b}{d\mu_a} \leq \rho^{-1}.$$

If a and b belong to the same component of G, then there are points $a_0, \ldots, a_n$ and positive numbers $R_0, \ldots, R_n$ such that: (i) $a = a_0$, $b = a_n$; (ii) $|a_j - a_{j-1}| < R_j$, $1 \leq j \leq n$; (iii) $\overline{B}(a_j; R_j) \subseteq G$. By the preceding paragraph, for $1 \leq j \leq n$, ω_{a_j} and $\omega_{a_{j-1}}$ are boundedly mutually absolutely continuous with constant ρ_j. Thus ω_a and ω_b are mutually absolutely continuous and

$$\frac{d\omega_a}{d\omega_b} = \frac{d\omega_{a_0}}{d\omega_{a_1}} \cdots \frac{d\omega_{a_{n-1}}}{d\omega_{a_n}};$$

hence

$$\rho \leq \frac{d\omega_b}{d\omega_a} \leq \rho^{-1},$$

where $\rho = \rho_1 \rho_2 \cdots \rho_n$. Similarly the same inequality holds for the measures μ_a and μ_b. □

If $u \in C(\partial K)$, we have defined $\hat{u}$ on $\text{int } K$. Extend $\hat{u}$ to K by letting $\hat{u} = u$ on ∂K. If the Dirichlet Problem can be solved, then $\hat{u}$ is continuous on K. If there is no such solution, then $\hat{u}$ is still a bounded Borel function. Hopefully no confusion will arise by defining $\hat{u}$ in this way.

If $\mu \in M(K)$, then $u \to \int \hat{u} \, d\mu$ defines a bounded linear functional on $C(\partial K)$. Thus there is a measure $\hat{\mu}$ in $M(\partial K)$ such that

$$(9.15) \qquad \int_K \hat{u} \, d\mu = \int_{\partial K} u \, d\hat{\mu}$$

for every u in $C(\partial K)$.

9.16 DEFINITION. If μ is a measure on K, the measure $\hat{\mu}$ defined by (9.15) is called the *sweep* of μ.

(It is one of those unfortunate consequences of the fact that typewriters are designed by people with a total lack of interest in mathematics that there are so few symbols such as $\hat{}$ and $\sim$ available. This results in our having the same notation for the sweep of a measure and the Cauchy transform of a measure. This coincidence is historical and not limited to this book.)

Note that if $a \in \text{int } K$, then the sweep of δ_a is ω_a.

The next two proofs are left to the reader.

9.17 PROPOSITION. *The map $\mu \to \hat{\mu}$ is a contractive linear map of $M(K)$ into $M(\partial K)$. If $\mu \in M(K)$ and μ is supported on ∂K, then $\hat{\mu} = \mu$.*

9.18 PROPOSITION. *If $a \in K$ and μ is a representing measure for a and the algebra $A(K)$, then $\hat{\mu}$ is also a representing measure.*

In line with the previous comments that we will not be so interested in harmonic measure itself but rather with its measure class (that is, the collection of measures that have the same sets of measure 0 as harmonic measure), let us introduce the following idea.

9.19 DEFINITION. If K is a compact subset of $\mathbb{C}$, then *harmonic measure* for K is any measure that is mutually absolutely continuous with a measure

ω obtained in the following way. Let $G_1, G_2, \ldots$ be the components of int K and for each $n \geq 1$ pick a_n in G_n. Let ω_n be harmonic measure for K at a_n and put $\omega = \sum_n 2^{-n} \omega_n$.

Thus harmonic measure for a compact set is actually a measure class $[\omega]$, rather than a specific measure. This means it is impossible to define the L^p space of the class if $1 < p < \infty$. We can, however, define the L^1 space and the L^∞ space since the definitions of these spaces actually only depend on the collection of sets of measure 0. Even though we will not need these concepts until later, it is appropriate to define them here. Here are the details.

9.20 DEFINITION. If K is a compact subset of $\mathbb{C}$ and ω is harmonic measure for K, define
$$L^\infty(\partial K) = \{f : f \text{ is } \omega\text{-essentially bounded}\},$$
$$L^1(\partial K) = \{\mu \in M(\partial K) : \mu \ll \omega\}.$$

As usual, functions in $L^\infty(\partial K)$ are identified if they agree a.e. $[\omega]$. Thus $L^\infty(\partial K) = L^\infty(\omega)$ and the definition of the norm on $L^\infty(\partial K)$ is independent of which form of harmonic measure for K we choose. The same is not quite true for $L^1(\partial K)$. If ω and ω' are two harmonic measures for K that are not boundedly mutually absolutely continuous, then $L^1(\omega)$ and $L^1(\omega')$ can be significantly different *if* we define these spaces as spaces of integrable functions. In fact, these spaces may not be equal as sets let alone isometric as Banach spaces. But the definition of $L^1(\partial K)$ given above as a subspace of $M(\partial K)$ with the total variation norm removes this ambiguity.

9.21 PROPOSITION. *If $\mu \in M(K)$ and $|\mu|(\partial K) = 0$, then $\hat{\mu}$, the sweep of μ, belongs to $L^1(\partial K)$.*

PROOF. Adopt the notation of Definition 9.19 and let Δ be a compact subset of ∂K with $\omega(\Delta) = 0$. It suffices to assume that $\mu \geq 0$ and show that $\hat{\mu}(\Delta) = 0$. Let $\{u_n\}$ be a sequence of continuous functions on ∂K such that $\chi_\Delta \leq u_n \leq 1$ for all $n \geq 1$ and $\{u_n(z)\}$ decreases monotonically to $\chi_\Delta(z)$ for each z in ∂K. If $a \in \text{int } K$, then $\hat{u}_n(a) = \int u_n d\omega_a \to \omega_a(\Delta) = 0$. Since $0 \leq \hat{u}_n(a) \leq 1$ for all $n \geq 1$ and for all a in int K, $\hat{\mu}(\Delta) = \lim \int u_n d\hat{\mu} = \lim \int_{\text{int } K} \hat{u}_n d\mu = 0$. □

Additional information about the topics covered in this section can be found in Ohtsuka [**1970**] and Tsuji [**1975**].

Exercises

1. If $\{\phi_n\}$ is a sequence of subharmonic functions on G that is decreasing, show that $\phi(z) \equiv \lim_n \phi_n(z)$ is a subharmonic function.
2. Complete the proof of Proposition 9.2. See page 265 in [**FOCV**] for some of the details.
3. If $u: \partial G \to \mathbb{R}$ is any bounded function and $a \in G$ such that $\hat{u}(a) = \check{u}(a)$, then $\hat{u}(z) = \check{u}(z)$ for all z in the component of G that contains a.

4. If $u \in C_{\mathbb{R}}(\partial G)$ and there is a solution h of the Dirichlet Problem with boundary values u, prove that $\hat{u} = \check{u} = h$.
5. Suppose $\{u_n\}$ is a uniformly bounded sequence in $\mathscr{S}$, the space of solvable functions on ∂G, such that $u_n \leq u_{n+1}$ for all $n \geq 1$. If $u(z) = \lim_n u_n(z)$ on ∂G, show that $u \in \mathscr{S}$ and $\hat{u}_n(z) \to \hat{u}(z)$ for all z in G. What does this tell you about $\mathscr{S}$?
6. If $K = \operatorname{cl} \mathbb{D} \cup [1, 2]$ and $a \in \mathbb{D}$, what is harmonic measure for K at a?
7. Suppose that $\operatorname{int} K$ is a Dirichlet set and show that $u \to \hat{u}$ is a bounded linear map of $C(\partial K)$ into $C(K)$. What is the dual of this map?
8. If μ is a positive measure carried by $\mathbb{D}$, define $R_\mu : \partial \mathbb{D} \to \mathbb{R}$ by $R_\mu(\zeta) = \int P_\zeta \, d\mu$, where P_ζ is the Poisson kernel. Let m be normalized arc length measure on $\partial \mathbb{D}$.
   (a) Show that R_μ is a non-negative function in $L^1(m)$.
   (b) If $f \in H^\infty$, show that $\int_{\mathbb{D}} |f|^2 \, d\mu \leq \int |f|^2 R_\mu \, dm$.
   (c) Prove that $\hat{\mu} = R_\mu m$.
9. Let G be a bounded simply connected region and let $\tau : \mathbb{D} \to G$ be the Riemann map with $\tau(0) = a$ and $\tau'(0) > 0$. If m is normalized arc length measure on $\partial \mathbb{D}$ and τ is extended to $\partial \mathbb{D}$ by means of its radial limits, prove that $m \circ \tau^{-1} = \omega_a$, harmonic measure for G at a.
10. In Proposition 9.12, what is the Radon-Nikodym derivative of harmonic measure with respect to arc length measure?

§10 Hardy spaces for an arbitrary region. This section is not prerequisite for any of the remaining sections in this book and can be skipped by the reader who is in a hurry. Another collection of examples of subnormal operators is introduced here and the basic properties of these operators are explored. We begin by proving another characterization of functions that belong to the Hardy space H^p of the unit disk.

10.1 PROPOSITION. *If $1 \leq p < \infty$ and $f: \mathbb{D} \to \mathbb{C}$ is an analytic function, then $f \in H^p$ if and only if there is a harmonic function $u: \mathbb{D} \to [0, \infty)$ such that $|f|^p \leq u$ on $\mathbb{D}$.*

PROOF. Suppose f is analytic and there exists such a harmonic function u. By the Mean Value Property, for $0 < r < 1$

$$M_p(r, f) = \left[\frac{1}{2\pi} \int_0^{2\pi} |f(re^{i\theta})|^p \, d\theta \right]^{1/p}$$

$$\leq \left[\frac{1}{2\pi} \int_0^{2\pi} u(re^{i\theta}) \, d\theta \right]^{1/p}$$

$$= u(0)^{1/p}.$$

Hence $f \in H^p$. Conversely, if $f \in H^p$, let u be the Poisson integral of the boundary values of $|f|^p$. So u is harmonic and $u \geq 0$. Using the factorization of functions in H^p (III.8.10)

$$|f(z)|^p \leq \exp\left\{\int P_z(w)\log|f(w)|^p\,dm(w)\right\}$$
$$\leq u(z)$$

by Proposition III.5.2, since the exponential is a convex function. □

Even though our real interest only lies in Hilbert spaces, we will continue for awhile to consider all the values of p, $1 \leq p < \infty$, as there is no extra effort involved.

10.2 DEFINITION. If G is an arbitrary region in the complex plane and $1 \leq p < \infty$, $H^p(G)$ consists of all the analytic functions f defined on G such that there is a harmonic function $u: G \to [0,\infty)$ with $|f|^p \leq u$ on G. As before, $H^\infty(G)$ denotes the algebra of bounded analytic functions on G. These spaces are called *Hardy spaces* for G.

It is not necessary to assume that G is connected to define $H^p(G)$, but later we will see a good reason for this assumption. In light of Proposition 10.1, this definition of $H^p(\mathbb{D})$ is equivalent to the one given in Chapter III. The harmonic function u that dominates $|f|^p$ on G is said to be a *harmonic majorant* of $|f|^p$.

What are some examples of functions in $H^p(G)$? $H^p(G)$ always contains the constant functions; more generally, $H^\infty(G) \subseteq H^p(G)$. It is not hard to see that $H^p(\mathbb{C})$ only contains the constant functions (Exercise 3) and there are other examples where $H^p(G)$ is trivial. There are also examples of unbounded regions G such that $H^p(G)$ is finite dimensional but not trivial (see Hedjal [1973]).

Before stating the properties of these Hardy spaces, some function theoretic background is required. The first result is standard. The reader can use Proposition VII.4.4 from [ACFA] to fashion a proof.

10.3 LEMMA. *If G is a region in $\mathbb{C}$ and K is a compact subset of G, there is a sequence of regions $\{G_n\}$ having the following properties.*

(a) *For each n, $K \subseteq G_n$, $\operatorname{cl} G_n \subseteq G_{n+1}$, and $G = \bigcup_n G_n$.*
(b) *For each n, ∂G_n is a finite smooth Jordan system of curves.*

Call a collection of regions $\{G_n\}$ such as in the preceding lemma a *smooth exhaustion* of G.

10.4 THEOREM. *If G is a region in $\mathbb{C}$ and ϕ is a subharmonic function on G that is not identically $-\infty$ and if ϕ has a harmonic majorant, then there is a unique harmonic function u on G such that:*

(a) $u \geq \phi$;
(b) *if v is any harmonic function on G such that $v \geq \phi$, then $v \geq u$.*

Moreover, if $\{G_n\}$ is any smooth exhaustion of G and if for each $n \geq 1$, ω_n^z

denotes harmonic measure for G_n evaluated at z, then

(10.5) $$u(z) = \sup_n \int_{\partial G_n} \phi \, d\omega_n^z = \lim_n \int_{\partial G_n} \phi \, d\omega_n^z.$$

PROOF. Let Φ be a harmonic function on G such that $\Phi \geq \phi$ and for each n let $u_n : G \to \mathbb{C}$ be defined by $u_n(z) = \int_{\partial G_n} \phi \, d\omega_n^z$.

Fix n. Since ϕ is upper semicontinuous, there is a decreasing sequence of continuous functions $\{\phi_k\}$ on ∂G_n such that for every z in ∂G_n, $\phi_k(z) > \phi(z)$ and $\phi_k(z) \to \phi(z)$ (Tsuji [1975] page 36). Let w_k be the solution for the Dirichlet Problem for G_n with boundary values ϕ_k. Thus $w_k(z) = \int \phi_k \, d\omega_n^z$ for z in G_n and, if $w_k = \phi_k$ on ∂G_n, w_k is continuous on cl G_n. By the Maximum Principle, $w_k(z) > \phi(z)$ on G_n. By the Monotone Convergence Theorem, $w_k(z) \to u_n(z)$ for z in G_n. Thus $u_n \geq \phi$ on G_n.

If there is a point z in G such that $\phi(z) = \Phi(z)$, then the Maximum Principle implies that $\phi = \Phi$ and the proof of the theorem is complete (just take $u = \phi = \Phi$). So we can assume that $\phi(z) < \Phi(z)$ for all z in G. Thus for each z in ∂G_n there is an integer $k(z)$ such that $\Phi(z) > w_k(z)$ for $k \geq k(z)$. By continuity, this same inequality holds in a neighborhood of z. A compactness argument implies there is an integer k_0 such that whenever $k \geq k_0$, $\Phi(z) > w_k(z)$ for all z in ∂G_n. Thus $\Phi \geq w_k$ on G_n and so $\Phi \geq u_n$. A similar argument shows that $u_n \leq u_{n+1} \leq \Phi$ on G_n. Thus the supremum and the limit in (10.5) are equal and, by Harnack's Theorem, the function u defined there is harmonic on G. Clearly $u \geq \phi$ on G.

Now assume that v is a harmonic function on G such that $v \geq \phi$. Again the Maximum Principle implies that $v \geq u_n$ on G_n. Consequently $v \geq u$ on G. The uniqueness of u is an easy consequence of (b). □

The harmonic function u obtained in the preceding proposition is called the *least harmonic majorant* of ϕ.

10.6 COROLLARY. *If $f \in H^p(G)$, then $|f|^p$ has a least harmonic majorant on G.*

PROOF. Recall that $|f|^p$ is a subharmonic function on G whenever f is analytic (III.5.6). □

10.8 COROLLARY. *If f is an analytic function on G and $\{G_n\}$ is a smooth exhaustion of G, then the following statements are equivalent.*

(a) $f \in H^p(G)$.
(b) *For every z in G, $\sup_n \int_{\partial G_n} |f|^p \, d\omega_n^z < \infty$.*
(c) *There is a point a in G such that $\sup_n \int_{\partial G_n} |f|^p \, d\omega_n^a < \infty$.*

PROOF. The equivalence of (a) and (b) is clear from Theorem 10.4. For the equivalence of (b) and (c) suppose that $a, z \in G_n$. Proposition 9.14 implies there is one nonzero constant c_n such that $\omega_m^z \leq c_n \omega_m^a$ for all $m \geq n$. This leads quickly to the proof that (c) implies (b). □

10.9 THEOREM. *If G is any region in $\mathbb{C}$ and $1 \leq p < \infty$, then $H^p(G)$ is a linear space. If a is any fixed point in G and $\|f\|$ is defined for each f in $H^p(G)$ by $\|f\| \equiv u(a)^{1/p}$, where u is the least harmonic majorant of $|f|^p$, then $\|\cdot\|$ defines a norm on $H^p(G)$ and relative to this norm $H^p(G)$ is a Banach space. If $\{G_n\}$ is a smooth exhaustion of G such that each G_n contains a and ω_n is harmonic measure for G_n evaluated at a, then*

$$(10.10) \qquad \|f\|^p = \sup_n \int |f|^p \, d\omega_n = \lim_n \int |f|^p \, d\omega_n.$$

PROOF. The formula (10.10) for $\|f\|^p$ is just an interpretation of (10.5).

If $f \in H^p(G)$ and $\|f\| = 0$, then for each n, $\int |f|^p \, d\omega_n = 0$. Using Corollary 10.8, this implies that $\int |f|^p \, d\omega_n^z = 0$ for z in G_n and $n \geq 1$. Since G_n is a Dirichlet region, this implies that $f = 0$ on G_n for each $n \geq 1$, and hence $f \equiv 0$.

If f and $g \in H^p(G)$, then, with the notation of the statement of the theorem, $\sup_n [\int |f+g|^p \, d\omega_n]^{1/p} \leq \sup_n [\int |f|^p \, d\omega_n]^{1/p} + \sup_n [\int |g|^p \, d\omega_n]^{1/p} < \infty$, so that $f + g \in H^p(G)$ and $\|f + g\| \leq \|f\| + \|g\|$. The remainder of the proof that $H^p(G)$ is a linear space and $\|\cdot\|$ is a norm is clear.

Now assume that $\{f_k\}$ is a Cauchy sequence in $H^p(G)$. If K is a compact subset of G, then there is an integer n_1 such that $K \subseteq G_n$ for $n \geq n_1$. As in the proof of Corollary 10.8, there is a constant $c_K > 0$ such that $\omega_n^z \leq c_K \omega_n$ for all z in K and all $n \geq n_1$. Therefore if $z \in K$ and $n \geq n_1$, then for every j and k,

$$\sup\{|f_j(z) - f_k(z)| : z \in K\} = \sup\left\{\left|\int (f_j - f_k) \, d\omega_n^z\right| : z \in K\right\}$$
$$\leq c_K \int |f_j - f_k| \, d\omega_n$$
$$\leq c_K \|f_j - f_k\|.$$

Thus $\{f_k\}$ is a uniformly Cauchy sequence on each compact subset of G and so there is an analytic function f on G such that $f_k \to f$ uniformly on compact subsets of G.

If n is fixed, then $\operatorname{cl} G_n$ is a compact subset of G so that $f_j \to f$ uniformly on $\operatorname{cl} G_n$. Thus $\int |f - f_j|^p \, d\omega_n \to 0$. If $\varepsilon > 0$, choose k_0 such that if $j, k \geq k_0$, then $\|f_j - f_k\| < \varepsilon$. Therefore for $j, k \geq k_0$,

$$\left[\int |f - f_k|^p \, d\omega_n\right]^{1/p} \leq \left[\int |f - f_j|^p \, d\omega_n\right]^{1/p} + \left[\int |f_j - f_k|^p \, d\omega_n\right]^{1/p}$$
$$\leq \left[\int |f - f_j|^p \, d\omega_n\right]^{1/p} + \varepsilon.$$

Letting $j \to \infty$, we see that $[\int |f - f_k|^p \, d\omega_n]^{1/p} \leq \varepsilon$ for $k \geq k_0$. Since k_0 does not depend on n, $\|f - f_k\| \leq \varepsilon$ for $k \geq k_0$. It follows, therefore, that $f \in H^p(G)$ and $f_k \to f$ in $H^p(G)$. □

It should be emphasized here that the connectedness of G was used to show that $\|\cdot\|$ is a norm. Indeed, if G is not connected and Λ is a component of G that does not contain the norming point, then $\|f\| = 0$ for $f = \chi_\Lambda$. Precisely where in the proof of the preceding theorem was the connectedness of G used?

The astute reader has by now asked about the dependence of the definition of the norm of $H^p(G)$ on the point a. This point a will be referred to as the *norming point*. What happens if the norming point is changed? Because of Proposition 9.14, changing the norming point gives an equivalent norm, though it is not equal.

How does this compare with the norm for the Hardy spaces of the disk? Recall that for f in $H^p(\mathbb{D})$, $\|f\| = \sup M_p(r, f)$. Since the Poisson kernel gives harmonic measure for $\mathbb{D}$, a little reflection by the reader, keeping in mind the formula for the Poisson kernel of an arbitrary disk, will show that $M_p(r, \phi)$ is almost equal to $[\int |f|^p \, d\omega_r]^{1/p}$, where ω_r is harmonic measure for the disk $r\mathbb{D}$ at 0. (They aren't quite equal as there is a constant c_r involved, though $c_r \to 1$ as $r \to 1$.) Hence the formula (10.10) when applied to $G = \mathbb{D}$ with $a = 0$ gives the usual formula for the norm of a function in H^p.

Let's also remark that $H^2(G)$ is a Hilbert space, where for f and g in $H^2(G)$ the inner product is defined by

$$(10.11) \qquad \langle f, g \rangle = \lim_n \int f \overline{g} \, d\omega_n.$$

There is a temptation to think of $H^p(G)$ as a subspace of $L^p(\omega)$, where ω is harmonic measure for G evaluated at a, the norming point for $H^p(G)$. As a source of intuition, this is fine. Indeed, in many cases, for example when ∂G consists of a finite number of disjoint smooth Jordan curves, this is a mathematical truth. However, there are some reasonable regions where this is not the case. An example occurs when G is a slit disk. In this case if $f \in H^p(G)$, then $f(z)$ has two nontangential limits at the points of the slit; one from the top of the slit and one from the bottom, and these two limits may not agree. (Example?) This makes it impossible to embed $H^p(G)$ into $L^p(\omega)$. There is a way to partially compensate for this, however. If the abstract concept of the Martin boundary Δ for G is introduced, then functions in $H^p(G)$ can be defined on Δ and there is a corresponding harmonic measure ω on Δ. In this situation, $H^p(G)$ does embed in $L^p(\Delta, \omega)$. See Hasumi [1983] page 73, for details. In the case of a "nice" region, this Martin boundary coincides with the topological boundary and this assignment of boundary values is the process of taking nontangential limits. In the case of a slit disk, the Martin boundary associates two points with each point of the slit: one corresponding to each side of the slit.

Now we restrict our attention to the case of a Hilbert space, $p = 2$, and define the operator on $H^2(G)$ for a bounded region G. In all that follows the notation used in this section will prevail. In particular, a is the norming point and ω is harmonic measure for G at a.

10.12 DEFINITION. If G is a bounded region in the plane, the *Hardy operator* for G is the operator S defined on $H^2(G)$ by $Sf = zf$ for all f in $H^2(G)$.

Note that if we could embed $H^2(G)$ into $L^2(\omega)$, then it is immediate that the Hardy operator for G is subnormal. Since $H^2(G)$ contains the analytic polynomials, it would also be the case that the minimal normal extension of S is multiplication by z on $L^2(\omega)$. Since such an embedding is not always possible, an alternative will have to be found. Before giving the details concerning the Hardy operator, here is a function theory result.

10.13 LEMMA. *If $\lambda \in G$ and $d = \text{dist}(\lambda, \partial G)$, then for every f in $H^2(G)$ such that $f(\lambda) = 0$ there is a function g in $H^2(G)$ such that $f = (z - \lambda)g$ and $\|g\| \leq d^{-1}\|f\|$.*

PROOF. Let u be the least harmonic majorant of $|f|^2$ on G and let g be the analytic function on G such that $f = (z-\lambda)g$. For $0 < \varepsilon < d$, put $G_\varepsilon = \{z \in G: \text{dist}(z, \partial G) > \varepsilon\}$. On ∂G_ε, $|g|^2 = |z - \lambda|^{-2}|f|^2 \leq (d - \varepsilon)^{-2}u$. But $|g|^2$ is subharmonic on G so that $|g|^2 \leq |z - \lambda|^{-2}u$ on G_ε. Letting $\varepsilon \to 0$ shows that $|g|^2 \leq d^{-2}u$ on G. Hence $g \in H^2(G)$ and $\|g\|^2 \leq d^{-2}\|f\|$. □

The statement of the next theorem as well as most of its proof is analogous to Theorem II.8.5, which concerns the Bergman operator.

10.14 THEOREM. *If S is the Hardy operator for the bounded region G, then S is a bounded subnormal operator and the following statements hold.*

(a) *$\sigma(S) = \text{cl } G$ and $\sigma_p(S) = \varnothing$.*

(b) *For λ in G there is a k_λ in $H^2(G)$ such that $\langle f, k_\lambda \rangle = f(\lambda)$ for all f in $H^2(G)$.*

(c) *For λ in G, $\text{ran}(S - \lambda) = \{f \in H^2(G): f(\lambda) = 0\}$ is closed, $S - \lambda$ is Fredholm, and $\text{ind}(S - \lambda) = -1$.*

PROOF. It is clear that S is bounded. To see that S is subnormal, let $f_0, \ldots, f_m \in H^2(G)$. Using the formula (10.11) for the inner product on $H^2(G)$, we see that

$$\sum_{j,k=0}^{m} \langle S^j f_k, S^k f_j \rangle = \lim_n \sum_{j,k=0}^{m} \int z^j \bar{z}^k f_k \bar{f}_j \, d\omega_n$$

$$= \lim_n \int \left| \sum_{j=0}^{m} z^j \bar{f}_j \right|^2 d\omega_n.$$

By Theorem II.1.9, S is subnormal.

The proof of parts (a), (b), and (c) and similar to the corresponding parts in Theorem II.8.3 and are left to the reader. □

The determination of the essential spectrum of the Hardy operator is found in Conway [**1987**]. In general $\partial[\operatorname{cl} G] \subseteq \sigma_e(S) \subseteq \partial G$.

The proof of the next theorem is also left to the reader because of its similarity to the proof of Theorem II.8.17.

10.15 THEOREM. *If S is the Hardy operator for G, then $\{S\}' = \{\phi(S): \phi \in H^\infty(G)\}$.*

10.16 COROLLARY. *The Hardy operator is irreducible and hence pure.*

PROOF. Since G is connected, $H^\infty(G)$ has no nontrivial idempotents. □

What is the minimal normal extension of the Hardy operator? The answer is not the normal operator N_ω. In fact, if G is the slit disk then the minimal normal extension is $N_\omega \oplus N_\lambda$, where λ is Lebesgue measure on the slit. If we think of normal operators as being determined by their scalar-valued spectral measure and the multiplicity function, then the next result is the first step in this determination.

10.17 PROPOSITION. *The scalar-valued spectral measure for the Hardy operator on G is harmonic measure for G.*

The only proof of this proposition that I know involves the use of the universal analytic covering map $\tau: \mathbb{D} \to G$. The details won't be given here because this would require us to go too far into function theory for the sake of this one fact. But here is an outline. If this map in chosen with $\tau(0) = a$, the norming point for $H^2(G)$, then $\omega = m \circ \tau^{-1}$. That is, harmonic measure ω for G at a is given by $\omega(\Delta) = m(\tau^{-1}(\Delta))$ for all Borel subsets Δ of ∂G, where τ is given its usual boundary values by means of radial limits (G is bounded). Let $\mathscr{G}_\tau$ denote the group of Möbius transformations T of $\mathbb{D}$ onto itself such that $\tau \circ T = T$. So $\mathscr{G}_\tau$ is the group of deck transformations for the covering map τ. If $H_\tau^2 \equiv \{g \in H^2(\mathbb{D}): g \circ T = g$ for all T in $\mathscr{G}_\tau\}$, then $Uf = f \circ \tau$ defines an isomorphism $U: H^2(G) \to H_\tau^2$. If S is the Hardy space operator for G, then $USU^{-1} = S_\tau$, multiplication by τ on H_τ^2. If $L_\tau^2 = \{g \in L^2: g \circ T = g$ for all T in $\mathscr{G}_\tau\}$, then it can be shown that the minimal normal extension of S_τ is N_τ, multiplication by τ on L_τ^2. (This gives another proof that S is subnormal.) Thus multiplication by τ on L^2 is a normal extension of S_τ and standard normal operator theory implies the scalar-valued spectral measure of this last normal operator is $m \circ \tau^{-1} = \omega$. If μ is the scalar-valued spectral measure for S, this shows that $\mu \ll \omega$. It is now necessary to use a separate argument to show that $[\mu] = [\omega]$. See Abrahamse and Douglas [**1976**] for proofs in the case G is bounded by a smooth Jordan system. It would be nice to have a direct proof of Proposition 10.17 that does not depend on the analytic covering map.

To complete the determination of the minimal normal extension of S, we need to find the multiplicity function. This study was carried out in Spraker [**preprint a** and **preprint b**]. He shows that if m_N is the multiplicity function for S, the Hardy operator for G, then $m_N \leq 2$ and he gives a necessary and sufficient condition on a subset Δ of ∂G having positive harmonic measure that $m_N = 2$ on Δ. In particular, for the case of the slit disk, m_N is 2 on the slit and 1 on the circle.

The result for Hardy spaces analogous to (II.8.15) is also valid (Exercise 5). That is, for a Carathéodory region G the polynomials are dense in $H^2(G)$. To obtain the analogue of (II.8.16), that $R^2(\operatorname{cl} G, \omega) = H^2(G)$ where ∂G is a finite Jordan system and ω is harmonic measure for G, is an undertaking that is also similar to the proof of that result. There is, however, some function theory that must precede the proof. See Exercise 6 for a sketch of this. Actually, in this case if σ denotes arc length measure on ∂G, $H^2(G)$ and $R^2(\operatorname{cl} G, \sigma)$ are equal as sets and their norms are equivalent since ω and σ are boundedly absolutely continuous. Thus the Hardy operator is similar to multiplication by z on $R^2(\operatorname{cl} G, \sigma)$.

REMARKS. Hardy spaces for arbitrary regions in $\mathbb{C}$ were first introduced in Parreau [**1951**] and Rudin [**1955**]. Two additional early works on the subject are Heins [**1969**] and Yamashita [**1968**]. This is not the place to sort out priorities in the results presented above, but the interested reader can consult Hasumi [**1983**] where this is done. Moreover, Hasumi [**1983**] is the only book devoted to Hardy spaces though the elementary theory can also be found in Fisher [**1983**]. For a sampling of papers on hardy spaces where some of their peculiarities can be found, see Bañuelos and Wolff [**1985**], Conway, Dudziak, and Straube [**1987**], Hedjal [**1973**], and Kobayashi [**1978**] as well as the previously cited works.

In addition to the references to the Hardy operator for a region G already cited, we must also include Akeroyd [**1987, 1989**, and **preprint**], where the question of the density of the polynomials in $H^2(G)$ is essentially settled for crescents G (simply connected regions G for which $\operatorname{cl} G$ is homeomorphic to the closure of the region bounded by two circles that are internally tangent at a point). Also see Remark 5 following Theorem II.8.16 for connections between the question of the density of polynomials and the existence of cyclic vectors. The paper Akeroyd, Khavinson, and Shapiro [**preprint**] also contains results about cyclic vectors for the Hardy operators.

Abrahamse [**1974**] contains a theory of Toeplitz operators on Hardy spaces of finitely connected regions. In this vein, the reader should also consult Ahern and Sarason [**1967a** and **1967b**] and Sarason [**1965a**].

There is also the class of subnormal operators known as bundle shifts. These generalize the Hardy operators. See Abrahamse and Douglas [**1976**], Abrahamse and Bastian [**1978**], Rudol [**1982** and **1988**].

Exercises

1. Give the details of the proof that if a and b are two norming points for G, these norms are equivalent on $H^p(G)$.

2. For $1 \leq p < \infty$, show that a sequence in $H^p(G)$ converges weakly if and only if it is uniformly bounded and converges uniformly on compact subsets.
3. (Conway [**1987**]) Show that if λ is an isolated point of ∂G, then every function f in $H^p(G)$ has a removable singularity at λ and the extended function belongs to $H^p(G \cup \{\lambda\})$ and has the same norm. Use this to show that if K is a compact subset of $\mathbb{C}$, then every function in $H^p(\mathbb{C}\setminus K)$ has a removable singularity at ∞. Show that the only functions in $H^p(\mathbb{C})$ are the constants.
4. If S is the Hardy operator for the bounded region G, show that $\partial[\operatorname{cl} G] \subseteq \sigma_e(S) \subseteq \partial G$.
5. If G is a bounded Carathéodory region and ω is harmonic measure for G, show that $P^2(\omega) = H^2(G)$, so that the Hardy operator for G is S_ω.
6. (Conway [**1987**]) This is the only reference I know where these results appear, though it is the case that most of the exercise was at least "folklore" before the appearance of this paper.)
    (a) If K is a compact subset of the open set W and $u: W\setminus K \to \mathbb{R}$ is a harmonic function, then there is a harmonic function $u_0: \mathbb{C}\setminus K \to \mathbb{R}$ such that for some constant a, $u_0(z) - a\log|z| \to 0$ as $z \to \infty$ and there is a harmonic function $u_1: W \to \mathbb{R}$ such that $u = u_0 + u_1$ on $W\setminus K$. These functions are unique.
    (b) Let G be a region such that ∂G is a finite Jordan system. Let $K_1, \ldots, K_m$ be the bounded components of the complement of G, let $G_j = \mathbb{C}_\infty \setminus K_j$, and let G_0 be the Carathéodory region $G \cup K_1 \cup \cdots \cup K_m$. If $f \in H^2(G)$, then there are analytic functions $f_j: G_j \to \mathbb{C}$ with $f_j(\infty) = 0$ and an analytic function f_0 on G_0 such that the following hold:
    (i) $f = f_0 + f_1 + \cdots + f_m$;
    (ii) for $0 \leq j \leq m$, $f_j \in H^2(\mathbb{C}\setminus K_j)$;
    (iii) for $0 \leq j \leq m$ there is a constant M_j such that $\|f_j\| \leq M_j\|f\|$;
    (iv) the functions f_j are unique;
    (v) if $P_j: H^2(G) \to H^2(G)$ is defined by $P_j f = f_j|G$, then P_j is a bounded idempotent. (Also see Lemma 12.19 below.)
    (c) With the notation as in (b), if $a_j \in K_j$ for $1 \leq j \leq m$, then $\{f : f$ is a rational function whose only poles are at $\infty, a_1, \ldots,$ and $a_m\}$, is dense in $H^2(G)$. (Hint: Use Exercise 5 above.)

§**11 Peak points.** Remember that unless otherwise stipulated A is a function algebra on a compact space X and $\mathscr{M} = \mathscr{M}_A$.

11.1 DEFINITION. A point x in X is a *peak point* if there is an f in A such that $f(x) = 1$ and $|f(y)| < 1$ for $y \neq x$. Such a function f is said to peak at f.

A compact subset E of X is a *peak interpolating set* for A if for every g in $C(E)$ there is a function f in A such that $f|E = g$ and $|f(x)| < \|g\|_E$ for x in $X\backslash E$.

Some observations and remarks are in order here. First, if x is a peak point, then $\{x\}$ is a peak interpolating set. Next if E is a peak interpolating set, then E is a G_δ set. In fact, take $g = 1$ in the definition and let $f \in A$ such that $f|E = 1$ and $|f(x)| < 1$ for x in $X\backslash E$. Clearly $\bigcap_n \{y \in X : |f(y)| > 1 - n^{-1}\} = E$. In particular, peak points have a countable neighborhood base. For this reason we will limit our attention to the case where X is metrizable. There is a concept of a "generalized peak point" that avoids this difficulty, but the only algebras we will consider live on the plane so this generalization is not germane to our considerations. The interested reader can consult Gamelin [**1969**].

Also note that if f peaks at x, then $|f|$ attains its maximum value on X only at the point x. Thus the set of peak points is contained in ∂_A. Also, if X is a metric space and E is a peak interpolation set for A, then each point of E is a peak point so that E is also contained in the Shilov boundary.

11.2 Proposition. *For a compact subset K of $\mathbb{C}$, the peak points for $R(K)$ are dense in ∂K.*

Proof. In fact this was proved in Lemma 7.5. □

11.3 Theorem. *If X is a metric space, A is a function algebra on X, and $E \subseteq X$, then the following statements are equivalent.*

(a) *E is a peak interpolating set for A.*
(b) *If $g \in C(E)$, then there is a uniformly bounded sequence $\{f_n\}$ in A with $f_n|E = g$ for all $n \geq 1$ and such that $f_n(x) \to 0$ uniformly on compact subsets of $X\backslash E$.*
(c) *If μ is a measure on X such that $\mu \perp A$, then $|\mu|(E) = 0$.*
(d) *For every neighborhood U of E and every function g in $C(E)$, there is a function f in A such that $f|E = g$, $\|f\|_X \leq 2\|g\|_E$, and $|f(x)| \leq (1/4)\|g\|_E$ for x in $X\backslash U$.*
(e) *There are constants c and M with $0 < c < 1$ and $M \geq 1$, such that for every neighborhood U of E and every function g in $C(E)$, there is a function f in A such that $f|E = g$, $\|f\|_X \leq M\|g\|_E$, and $|f(x)| \leq c\|g\|_E$ on $X\backslash U$.*

The proof of this theorem needs a lemma which we prove for a linear space contained in $C(X)$ rather than an algebra. The assumption that X is a metric space remains in effect.

11.4 Lemma. *Suppose that B is a closed linear subspace of $C(X)$, E is a closed subset of X, and $g \in C(E)$ such that $\|g\|_E \leq 1$. If there are constants c and M with $0 < c < 1$ and $M \geq 1$ such that for every neighborhood U of E there is a function f in B with $f|E = g$, $\|f\|_X \leq M$, and $|f(x)| \leq c$*

on $X \backslash U$, then there is a function f in B such that $f|E = g$ and $|f(x)| < 1$ for x in $X \backslash E$.

PROOF. Let $0 < s < 1$ and let $\{\varepsilon_n\}_{n \geq 0}$ be a sequence of positive numbers that decrease to 0. (Both s and the sequence $\{\varepsilon_n\}$ will be further specified later in the proof.) Let $\{F_n\}_{n \geq 0}$ be a sequence of closed subsets of X such that $X \backslash E = \bigcup_n F_n$ and $F_0 = \varnothing$. Let $f_0 \in B$ such that $f_0|E = g$ and $\|f_0\|_X \leq M$. Inductively define a sequence $\{f_n\}_{n \geq 0}$ in B and a sequence $\{K_n\}_{n \geq 0}$ of closed sets in X by

$$K_n = \{x \in X : \max_{1 \leq k \leq n} |f_k(x)| \geq 1 + \varepsilon_n\},$$

and choose the functions f_n so that $f_n|E = g$, $\|f_n\|_X \leq M$, and

$$|f_n(x)| \leq c \quad \text{for } x \text{ in } K_{n-1} \cup F_{n-1} \quad \text{for } n \geq 1.$$

It is left to the reader to show that the hypothesis can be used to perform the required induction argument.

Let $f = (1-s) \sum_{n=0}^{\infty} s^n f_n$. Thus $f \in B$ and $f|E = g$. Fix x in $X \backslash E$. If $x \notin \bigcup_m K_m$, then, because $\varepsilon_m \to 0$, $|f_n(x)| \leq 1$ for all n. On the other hand, there is at least one integer n such that $x \in F_{n-1}$ and so $|f_n(x)| \leq c < 1$. Hence $|f(x)| < 1$. Now if x does belong to some set K_m, pick m to be the smallest such integer. For the moment assume $m \geq 1$; so $x \notin K_{m-1}$. Therefore $|f_n(x)| < 1 + \varepsilon_{m-1}$ for $0 \leq n \leq m-1$; $|f_m(x)| \leq M$; and $|f_n(x)| \leq c$ for $n \geq m+1$. Therefore

$$|f(x)| \leq (1-s)\left[(1 + \varepsilon_{m-1}) \sum_{n=0}^{m-1} s^n + Ms^m + c \sum_{n=m+1}^{\infty} s^n\right]$$
$$= (1-s)\left[(1 + \varepsilon_{m-1})(1 - s^m)(1-s)^{-1} + Ms^m + cs^m[(1-s)^{-1} - 1]\right]$$
$$= 1 + \varepsilon_{m-1}(1 - s^m) + s^m[(M-1) + s(c - M)].$$

But s can be chosen sufficiently close to 1 that $(M-1) + s(c - M) < 0$. Once this is done, then numbers ε_m can be chosen so that

$$\varepsilon_{m-1}(1 - s^m) + s^m[(M-1) + s(c - M)] < 0.$$

The choice of these numbers only depends on c and M. Thus $|f(x)| < 1$.

Now return to the possibility that $x \in K_0 (m = 0)$. Here $|f_0(x)| \leq M$ and $|f_n(x)| \leq c$ for all $n \geq 1$. Performing the same manipulations as in the preceding paragraph we get that $|f(x)| \leq M + s(c - M)$. By the choice of s already made, $|f(x)| < 1$.

Thus $|f(x)| < 1$ for all x in $X \backslash E$. □

PROOF OF THEOREM 11.3. (a) *implies* (b). Let $f \in A$ such that $f|E = g$ and $|f(x)| < \|g\|_E$ for x in $X \backslash E$. Also, there is a function h in A such that $h(x) = 1$ for all x in E and $|h(x)| < 1$ for x in $X \backslash E$. The sequence $f_n = fh^n$ satisfies condition (b).

(b) *implies* (c). Fix a measure μ on X such that $\mu \perp A$. Let g be any continuous function on E and let $\{f_n\}$ be as in part (b). So $\int_E g \, d\mu = \lim_n \int_E f_n \, d\mu = \lim_n \int f_n \, d\mu = 0$. Thus $|\mu|(E) = 0$.

(c) *implies* (d). Let U any open neighborhood of E and let $p: X \to [1/8, 1]$ be a continuous function such that $p(x) = 1$ for x in E and $p(x) = 1/8$ for x in $X \setminus U$. Let B be the closed linear subspace of $C(E)$ defined by $B = \{f/p : f \in A\}$. Define $R: B \to C(E)$ to be the restriction map $Rh = h|E$. Clearly R is a contraction. Also if $\nu \in M(E)$ and $h \in B$, then $\langle h, R^*(\nu) \rangle = \langle Rh, \nu \rangle = \int h \, d\nu$. That is, $R^*\nu = \nu$ considered as a measure on X rather than E. Thus R^* is an isometry. This implies two things.

First ran R^* is closed and so ran R is closed (See VI.1.10 in [ACFA]). But (c) implies that ran R is dense. Indeed, ran $R = \{f|E : f \in A\}$ since $p|E \equiv 1$. So if $\nu \in M(E)$ and $\nu \perp$ ran R, then $\nu \perp A$ and so $\nu = 0$ by (c). Thus R is surjective.

Next the induced map $\tilde{R}: B/\ker R \to C(E)$ is an isometry so that if $g \in C(E)$, then there is a function h in B such that $Rh = g$ and $\|h\|_X \leq 2\|g\|_E$. But $h = f/p$ for some f in A. It follows that $f|E = g$ and for any x in X, $|f(x)| \leq p(x)|h(x)| \leq 2p(x)\|g\|_E$. In particular, $\|f\|_X \leq 2\|g\|_E$ and for x in $X \setminus U$, $|f(x)| \leq 2(1/8)\|g\|_E = (1/4)\|g\|_E$.

(d) *implies* (e). Let $M = 2$ and $c = 1/4$.

(e) *implies* (a). This follows from Lemma 11.4. □

There is an analogous characterization of peak points.

11.5 THEOREM. *If X is a metric space, A is a function algebra on X, and $a \in X$, then the following statements are equivalent.*

(a) *a is a peak point.*
(b) *There is a uniformly bounded sequence $\{f_n\}$ in A with $f_n(a) = 1$ for all $n \geq 1$ and such that $f_n(x) \to 0$ uniformly on compact subsets of $X \setminus \{a\}$.*
(c) *$M_a = \{\delta_a\}$.*
(d) *For every neighborhood U of a, there is a function f in A such that $\|f\| \leq 1$, $f(a) > 3/4$, and $|f(x)| < 1/4$ for x in $X \setminus U$.*
(e) *There are constants c and M with $0 < c < 1$ and $M \geq 1$, such that for every neighborhood U of a there is an f in A with $f(a) = 1$, $\|f\| \leq M$, and $|f| \leq c$ on $X \setminus U$.*

The proof of this theorem is not a straightforward specialization of Theorem 11.3 and in fact requires an additional lemma.

11.6 LEMMA. *If $\rho \in \mathscr{M}_A$ and $u: \mathscr{M} \to \mathbb{R}$ is an upper semicontinuous function, then*

$$\sup\left\{\int u \, d\mu : \mu \in M_\rho\right\} = \inf\{\operatorname{Re} \rho(f) : f \in A \text{ and } \operatorname{Re} f > u \text{ on } X\}.$$

PROOF. Let β be the above supremum and α the infimum. If $u < \operatorname{Re} f$ on X and $\mu \in M_p$, then $\int u \, d\mu \leq \int \operatorname{Re} f \, d\mu = \operatorname{Re} \int f \, d\mu = \operatorname{Re} p(f)$; thus $\beta \leq \alpha$. For the reverse inequality, a measure μ must be found in M_p with $\int u \, d\mu \geq \alpha$.

Let $V = \{v \in C_{\mathbb{R}}(X): v(x) > u(x) - \alpha \text{ for all } x \text{ in } X\}$. Note that V is open in $C_{\mathbb{R}}(X)$ since u is upper semicontinuous.

CLAIM. $V \cap [\operatorname{Re}(\ker p)] = \varnothing$.

In fact, if there is a function f in $\ker p$ with $\operatorname{Re} f$ in V, then, because V is open, there is a positive number ε such that $\operatorname{Re} f - \varepsilon \in V$. That is, $\operatorname{Re} f - \varepsilon > u - \alpha$ everywhere on X. But $\operatorname{Re}(p(f) + \alpha - \varepsilon) = \alpha - \varepsilon < \alpha$, contradicting the definition of α. This proves the claim.

Because V and $\ker p$ are disjoint convex sets and V is open, the Hahn-Banach Theorem implies there is a real-valued measure μ on X and a constant c in $\mathbb{R}$ such that $\sup\{\int \operatorname{Re} f \, d\mu : f \in \ker p\} = c \leq \inf\{\int v \, d\mu : v \in V\}$. Because $\ker p$ is a linear space, $c = 0$ and so $\int f \, d\mu = 0$ for every f in $\ker p$. Now if $v \in C_{\mathbb{R}}(X)$ and $v(x) > 0$ for all x, then $tv \in V$ for sufficiently large $t > 0$. Hence $\int v \, d\mu \geq 0$; that is, μ is a positive measure. By normalizing, it may be assumed that μ is a probability measure. This implies that $\mu \in M_p$. Since u is upper semicontinuous, there is a decreasing sequence $\{v_n\}$ in V which converges to $u - \alpha$ (Tsuji [1975] page 36). This shows that $0 \leq \int(u - \alpha) \, d\mu$ and so $\int u \, d\mu \geq \alpha$. □

PROOF OF THEOREM 11.5. (a) *implies* (b). If $f \in A$ such that f peaks at a, then the sequence $f_n = f^n$ satisfies condition (b).

(b) *implies* (c). Given the sequence $\{f_n\}$ in (b) and μ in M_a, $1 = \int f_n \, d\mu \to \mu(\{a\})$. Hence $\mu = \delta_a$.

(c) *implies* (d). Let U be any open neighborhood of a and let $u: X \to [0, \infty)$ be a continuous function such that $u(a) < -\log 3/4$ and $u(x) > -\log 1/4$ for x in $X \setminus U$. By the preceding lemma and the fact that we are assuming (c) to be true, $u(a) = \inf\{\operatorname{Re}(a): g \in A \text{ and } \operatorname{Re} g(x) > u(x) \text{ on } X\}$. Hence we can find a function g in A with $\operatorname{Re} g(x) > u(x)$ for all x in X and $u(a) < \operatorname{Re} g(a) < -\log 3/4$. Now for any θ, $f = e^{-g + i\theta} \in A$ and for a suitable choice of θ, $f(a) = |f(a)| = \exp(-\operatorname{Re} g(a)) > 3/4$. If $x \in X$, then $|f(x)| = \exp(-\operatorname{Re} g(x)) < \exp(-u(x)) \leq 1$. Finally, if $x \in X \setminus U$, then $|f(x)| < \exp(-u(x)) \leq 1/4$.

(d) *implies* (e). Let $M = 4/3$ and $c = 1/3$. According to (d), if U is a neighborhood of a, then there is a function h in A with $\|h\| \leq 1$, $h(a) > 3/4$, and $|h| < 1/4$ on $X \setminus U$. If $f = |h(a)|^{-1} h$, then it is left to the reader to check that f has the desired properties.

(e) *implies* (a). This follows from Lemma 11.4 by taking $E = \{a\}$ and $g = 1$. □

11.7 COROLLARY. *The set of peak points of a function algebra on a metric space X is a G_δ-set.*

PROOF. Denote the metric on X by d. For each integer $n \geq 1$, let $U_n = \{x \in X :$ there is a function f in A with $\|f\|_X \leq 1$, $f(x) > 3/4$, and

$|f(y)| < 1/4$ when $d(x, y) \geq 1/n\}$. It is not hard to see that U_n is open. According to Theorem 11.3, $\bigcap_n U_n$ is precisely the set of peak points. □

We now restrict ourselves to T-invariant algebras.

11.8 Proposition. *If A is a T-invariant subalgebra of $C(K)$ and $\nu \in A^\perp$, then $\{z \in K : \tilde{\nu}(z) < \infty \text{ and } \hat{\nu}(z) \neq 0\}$ contains no peak points for A.*

PROOF. Suppose $z \in K$ with $\tilde{\nu}(z) < \infty$ and $\hat{\nu}(z) \neq 0$. According to Corollary 8.10 there is a representing measure μ for z such that $\mu \ll \nu$. But since $\tilde{\nu}(z) < \infty$, ν puts no mass at z. Hence μ does not have an atom at z. By Theorem 11.3, z is not a peak point for A. □

Recall the definition of $R(K, E)$ from (3.13).

11.9 Theorem. *If A is a T-invariant algebra on K and Q is the set of nonpeak points of A, then $A \supseteq R(K, Q)$.*

PROOF. Let $\nu \in A^\perp$; it will be shown that $\nu \in R(K, Q)^\perp$. But since $R(K) \subseteq A$, $\hat{\nu} = 0$ on $\mathbb{C}\backslash K$. In addition, the preceding proposition and the fact that $\tilde{\nu}$ is finite a.e. [Area] imply that $\tilde{\nu}$ vanishes a.e. [Area] on $\mathbb{C}\backslash Q$. By Proposition 3.14, $\nu \perp R(K, Q)$. □

11.10 Corollary. *If A is a T-invariant algebra on K and almost every point of K is a peak point, then $A = C(K)$.*

We will return to peak points in §13.

Exercises

1. Define a subset E of X to be a *peak set* if there is an f in A such that $f(z) = 1$ for all z in E and $|f(z)| < 1$ for z in $X\backslash E$. Show that the countable intersection of peak sets is a peak set.
2. Show that if E is a peak set for A and $\mu \in M(X)$ such that $\mu \perp A$, then $\mu|E \perp A$. (The converse of this is also true. That is, if E is a closed subset of X and $\mu|E \perp A$ whenever $\mu \perp A$, then E is a peak set for A. Glicksberg [**1962**])
3. Give an example of a compact subset K of $\mathbb{C}$ such that there is a point a in ∂K for which every line segment $[a, b] \cap K$ properly contains $\{a\}$. Is this point a a peak point? Give an example of a compact set K such that ∂K has at least one nonpeak point for $R(K)$.

§12 Capacity. Capacity is a measure of smallness. There are different concepts of capacity littering the analytical landscape, each invented in an attempt to quantify an idea of negligibility. Some capacities are better than others in that they possess more properties. There is, in addition, an abstract theory of capacity (Choquet [**1955**] and Carleson [**1967**]). Unfortunately the two capacities we will encounter here do not fit into this general framework and thus cannot enjoy the fringe benefits of belonging to a larger theory.

12.1 Definition. If F is a subset of $\mathbb{C}$, let $H(F)$ be the collection of all analytic functions defined on $\mathbb{C}_\infty \backslash K$ for some compact subset K of F

such that $f(\infty) = 0$ and $\|f\|_{\mathbb{C}\setminus K} \leq 1$. $HC(F)$ consists of all the continuous functions defined on $\mathbb{C}_\infty$ that belong to $H(F)$ and satisfy $\|f\|_\mathbb{C} \leq 1$.

Recall that if f is analytic in a neighborhood of ∞, then

$$f(z) = a_0 + a_1 z^{-1} + a_2 z^{-2} + \cdots,$$

where the convergence is uniform for $|z| \geq R$ for some large value of R. Thus

$$f(\infty) = a_0,$$

and

$$f'(\infty) = a_1 = \lim_{z \to \infty} z[f(z) - f(\infty)].$$

12.2 DEFINITION. If F is a subset of $\mathbb{C}$,

$$\gamma(F) = \sup\{|f'(\infty)|: f \in H(F)\}$$

is called the *analytic capacity* of F. The number

$$\alpha(F) = \sup\{|f'(\infty)|: f \in HC(F)\}$$

is called the *continuous analytic capacity* of F.

These concepts of capacity originated in the work of Ahlfors and Beurling [1950].

Clearly $\alpha(F) \leq \gamma(F)$ for every set F and by elementary complex analysis $\gamma(F) = 0$ for any discrete set F. Also, an easy exercise using Morera's Theorem shows that if F is a closed line segment, then $\alpha(F) = 0$. We will see later that for a closed line segment F, $\gamma(F) = l/4$, where l is the length of F (Corollary 12.10 below).

12.3 PROPOSITION. (a) *If* $F_1 \subseteq F_2$, *then* $\gamma(F_1) \leq \gamma(F_2)$ *and* $\alpha(F_1) \leq \alpha(F_2)$.

(b) *If* a *and* $b \in \mathbb{C}$, *then* $\gamma(a + bF) = |b|\gamma(F)$ *and* $\alpha(a + bF) = |b|\alpha(F)$ *for all sets* F.

(c) *If* K *is a compact subset of* $\mathbb{C}$, *then* $\gamma(K) = \gamma(\partial K) = \gamma(\widehat{K}) = \gamma(\partial \widehat{K})$.

PROOF. The proof of part (a) is clear. For part (b) it suffices to consider the case where $b \neq 0$ (Why?). Note that $H(a + bF) = \{f(b^{-1}(z - a)); f \in H(F)\}$ and $HC(a + bF) = \{f(b^{-1}(z - a)); f \in HC(F)\}$.

To prove (c) first observe that (a) implies that $\gamma(\partial \widehat{K}) \leq \gamma(\partial K) \leq \gamma(K) \leq \gamma(\widehat{K})$. Thus it suffices to assume that K is polynomially convex and show that $\gamma(K) \leq \gamma(\partial K)$. Let $\{f_n\} \subseteq H(K)$ such that $f'_n(\infty) \to \gamma(K)$. Since the sequence $\{f_n\}$ is uniformly bounded by 1 on $\mathbb{C}\setminus K$, by passing to a subsequence if necessary we may assume that $\{f_n\}$ converges uniformly on compact subsets of $\mathbb{C}_\infty \setminus K$ to an analytic function f. It follows that $f \in H(K)$ and $f'(\infty) = \gamma(K)$. Now define $g: \mathbb{C}_\infty \setminus \partial K \to \mathbb{C}$ by letting $g(z) = f(z)$ for z in $\mathbb{C}_\infty \setminus K$ and $g(z) = 0$ for z in $\text{int } K$. Then $g \in H(\partial K)$ and so $\gamma(\partial K) \geq g'(\infty) = f'(\infty) = \gamma(K)$. □

The statement for continuous capacity that is analogous to part (c) above is not true. For example if D is a closed disk, then an application of Morera's Theorem shows that $\alpha(\partial D) = 0$ while $\alpha(D)$ will be shown below (Corollary 12.9) to be the radius of D. Where does the proof of (c) break down for α?

Part of the proof of the preceding proposition is worth noting.

12.4 PROPOSITION. *If K is a compact set, there is a function f in $H(K)$ such that $f'(\infty) = \gamma(K)$. If g is another function in $H(K)$ such that $g'(\infty) = \gamma(K)$, then f and g agree on $\mathbb{C}\setminus\widehat{K}$.*

PROOF. The existence of f was shown in the preceding proof. It remains to establish uniqueness. If f and g are functions in $H(K)$ with $f'(\infty) = g'(\infty) = \gamma(K)$, then $h = (1/2)(f+g) \in H(K)$ and $h'(\infty) = \gamma(K)$; let $k = (1/2)(f-g)$. It is easy to check that $k \in H(K)$ and $k'(\infty) = 0$. It will be shown that $k \equiv 0$ in $\mathbb{C}\setminus\widehat{K}$.

Now $h + k = f$ and $h - k = g$. So $1 \geq |h \pm k|^2 = |h|^2 + |k|^2 \pm 2\operatorname{Re}(h\overline{k})$. Thus $|h|^2 + |k|^2 \leq 1$ and so $(1/2)|k|^2 \leq (1/2)[1 - |h|^2] = (1/2)(1 - |h|)(1 + |h|) \leq 1 - |h|$. Hence $(1/2)|k|^2 + |h| \leq 1$ on $\mathbb{C}\setminus K$. If k is not identically 0 in a neighborhood of ∞, we can write $(1/2)k^2 = a_n z^{-n} + a_{n+1} z^{-n-1} + \cdots$, where $a_n \neq 0$ and $n > 1$ since $k(\infty) = k'(\infty) = 0$.

Let U be an open neighborhood of $\widehat{K}$ and choose $\varepsilon > 0$ such that $\varepsilon|a_n z^{n-1}| < 1$ on U. Put $f_1(z) = h(z) + (1/2)\varepsilon \overline{a}_n z^{n-1} k(z)^2$. Clearly f_1 is analytic off $\widehat{K}$ and $f_1(\infty) = 0$. Also, for z in $U\setminus K$, $|f_1(z)| \leq |h(z)| + (1/2)\varepsilon|a_n z^{n-1}||k(z)|^2 \leq |h(z)| + (1/2)|k(z)|^2 \leq 1$. By the Maximum Modulus Theorem $|f_1| \leq 1$ on $\mathbb{C}\setminus\widehat{K}$. Therefore $f_1 \in H(\widehat{K})$. But $f_1'(\infty) = h'(\infty) + \varepsilon|a_n|^2 > h'(\infty) = \gamma(K) = \gamma(\widehat{K})$, a contradiction. Therefore it must be that $k \equiv 0$ on $\mathbb{C}\setminus\widehat{K}$. □

12.5 COROLLARY. *If K is a compact set, there is a unique function f in $H(\widehat{K})$ such that $f'(\infty) = \gamma(K)$.*

12.6 DEFINITION. For a compact set K, the unique function f in $H(\widehat{K})$ such that $f'(\infty) = \gamma(K)$ is called the *Ahlfors function* for K.

There may be parts of the preceding proofs that are reminiscent of the reader's past experiences in complex analysis. Indeed, there are similarities with one of the standard proofs of the Riemann Mapping Theorem (for example, the one given in [**FOCV**]). An examination of that proof will yield the following.

12.7 PROPOSITION. *If K is a compact connected subset of $\mathbb{C}$, then the Ahlfors function for K is the Riemann mapping of $\mathbb{C}_\infty \setminus \widehat{K}$ onto $\mathbb{D}$ that takes ∞ to 0.*

Note that K must be assumed connected in this proposition for otherwise $\mathbb{C}_\infty \setminus \widehat{K}$ may not be simply connected.

Proposition 12.7 allows us to compute the capacity of certain sets provided we know the corresponding Riemann map. For example, what is the Riemann map of $\mathbb{C}_\infty \setminus \overline{B}(a;r)$ onto $\mathbb{D}$? The answer gives the next corollary.

§12 CAPACITY

12.8 COROLLARY. *If D is a closed disk of radius r, then $\gamma(D) = r$.*

12.9 COROLLARY. *If D is a closed disk of radius r, then $\alpha(D) = r$.*

PROOF. We have $\alpha(D) \leq \gamma(D) = r$. But if f is the Ahlfors function for D, then this Riemann map has a continuous extension to $\mathbb{C}_\infty \setminus (\operatorname{int} D)$ and hence to $\mathbb{C}_\infty$. That is, $f \in HC(D)$. □

12.10 COROLLARY. *If K is a straight line segment of length l, $\gamma(K) = l/4$.*

PROOF. By part (a) of Proposition 12.3 we may assume that $K = [-l/2, l/2]$. But $w \to (l/4)(w + w^{-1})$ is the Riemann map of $\mathbb{D}$ onto $\mathbb{C}_\infty \setminus K$ that takes 0 to ∞. By uniqueness, if f is the Ahlfors function for K, then $z = (l/4)[f(z) + f(z)^{-1}] = (l/4)[(f(z)^2 + 1)/(f(z))]$. So $\gamma(K) = \lim_{z \to \infty} z f(z) = \lim_{z \to \infty} (l/4)[1 + f(z)^2] = (l/4)$. □

The emphasis on computing the capacity of compact sets is not accidental because the capacity of every set can be estimated in terms of the capacity of its compact subsets. The proof of this is left to the reader.

12.11 PROPOSITION. *If F is any set, then*

$$\gamma(F) = \sup\{\gamma(K): K \text{ is a compact subset of } F\},$$
$$\alpha(F) = \sup\{\alpha(K): K \text{ is a compact subset of } F\}.$$

12.12 COROLLARY. *If D is an open disk of radius r, $\gamma(D) = r$. If F is an open or half-open line segment of length l, $\gamma(F) = (l/4)$.*

12.13 PROPOSITION. *If K is compact and connected, then*

$$\gamma(K) \leq \operatorname{diam} K \leq 4\gamma(K).$$

PROOF. If d is the diameter of K and $a \in K$, then $K \subseteq D = \overline{B}(a; d)$ and so $\gamma(K) \leq \gamma(D) = d$. To prove the other inequality, we may assume that K is not a singleton.

Without loss of generality we may assume that K is polynomially convex. If g is the Ahlfors function for K and h is the inverse of g, then $h: \mathbb{D} \to \mathbb{C}_\infty \setminus K$ is the Riemann map with $h(0) = \infty$. Fix a in K and put $f = \gamma(K)(h - a)^{-1}$ on $\mathbb{D}$. So f is one-to-one on $\mathbb{D}$ with $f(0) = 0$.

Now $g(z) = \gamma(K)z^{-1} + a_2 z^{-2} + \cdots$ in a neighborhood of ∞. Hence $w = g(h(w)) = \gamma(K)h(w)^{-1} + a_2 h(w)^{-2} + \cdots$ for w in a neighborhood of 0 and so $wh(w) = \gamma(K) + h(w)^{-1}[a_2 h(w)^{-1} + \cdots]$. Since $h(0) = \infty$, we get that $\gamma(K) = \lim_{w \to 0} wh(w)$. Therefore $f'(0) = \lim_{w \to 0} f(w)/w = \lim_{w \to 0} \gamma(K)/(wh(w) - wa) = 1$. But now we can apply the Köebe 1/4 Theorem to conclude that $f(\mathbb{D}) \supseteq \{z: |z| < 1/4\}$. If $b \notin K$, then $b \notin h(\mathbb{D})$ and so $\gamma(K)(b - a)^{-1} \notin f(\mathbb{D})$. Hence $1/4 \leq \gamma(K)|b - a|^{-1}$, or $\gamma(K) \geq (1/4)|b - a|$. Since a and b were arbitrary points of K, we have the other inequality. □

12.14 Proposition. *If F is any bounded measurable set, then*

$$\gamma(F) \geq \alpha(F) \geq \sqrt{\frac{\text{Area}(F)}{\pi}}.$$

PROOF. If K is any compact subset of F and $f(z) = \int_K (w-z)^{-1} d\mathscr{A}(w)$, then f is continuous on $\mathbb{C}_\infty$, analytic on $\mathbb{C}_\infty \setminus K$, $f(\infty) = 0$, $f'(\infty) = -\text{Area}(K)$, and $|f(z)| \leq [\pi \text{Area}(K)]^{1/2}$ (2.2). So $g = [\pi \text{Area}(K)]^{-1/2} f \in HC(K)$. Therefore, $\alpha(K) \geq g'(\infty) = (\text{Area}(K)/\pi)^{1/2}$. The result follows from Proposition 12.11. □

12.15 Proposition. *If $\{K_n\}$ is a decreasing sequence of compact sets and $K = \bigcap_n K_n$, then $\gamma(K) = \lim_n \gamma(K_n)$. Moreover, if f_n is the Ahlfors function for K_n, then $\{f_n\}$ converges uniformly on compact subsets of $\mathbb{C} \setminus \hat{K}$ to the Ahlfors function for K.*

PROOF. Exercise 3. □

12.16 Corollary. *If K is a compact set, then $\gamma(K) = \inf\{\gamma(U) : U \text{ is an open set containing } K\}$.*

PROOF. Clearly $\gamma(K) \leq \inf\{\gamma(U) : U \text{ is an open set containing } K\}$. On the other hand, let $\{U_n\}$ be a sequence of bounded open sets such that $\text{cl}\, U_{n+1} \subseteq U_n$ for every $n \geq 1$ and $K = \bigcap_n U_n$. By the preceding proposition, $\gamma(\text{cl}\, U_n) \to \gamma(K)$. But $\gamma(U_{n+1}) \leq \gamma(\text{cl}\, U_{n+1}) \leq \gamma(U_n)$. □

12.17 Proposition. *If U is an open set, then $\gamma(U) = \alpha(U)$.*

PROOF. Let K be an arbitrary compact subset of U and choose another compact set L such that $K \subseteq \text{int}\, L$ and $L \subseteq U$. If $f \in H(K)$, then there is a g in $HC(L)$ such that $g = f$ on $\mathbb{C}_\infty \setminus L$. Hence $\gamma(K) \leq \alpha(L) \leq \alpha(U)$. By (12.11), $\gamma(U) \leq \alpha(U)$. □

This last result shows that there cannot be propositions analogous to (12.15) and (12.16) that are valid for continuous capacity.

We now conclude this section with a discussion of the significance of a set with analytic capacity 0. Call such sets *Painlevé null sets*. First an elementary result.

12.18 Proposition. (a) *A compact set K is a Painlevé null set if there is no nonconstant bounded analytic function on $\mathbb{C} \setminus K$.*

(b) *If K is a compact Painlevé null set, then K is totally disconnected.*

PROOF. The proof of (a) is a triviality. Part (b) is a direct consequence of (12.13). □

The preceding proposition definitely justifies the claim that analytic capacity measures the smallness of a set. Before proving the next proposition, we present a lemma that is used frequently over the course of this book. (Also see Exercise 10.6.)

12.19 Lemma. *If G is an open subset of $\mathbb{C}$, K is a compact subset of G, and $f: G\backslash K \to \mathbb{C}$ is an analytic function, then there are unique analytic functions $f_0: G \to \mathbb{C}$ and $f_\infty: \mathbb{C}_\infty \backslash K \to \mathbb{C}$ such that $f_\infty(\infty) = 0$ and $f(z) = f_0(z) + f_\infty(z)$ for z in $G\backslash K$. If f is bounded, so are f_0 and f_∞.*

Proof. If $z \in G$, there is a smooth Jordan system Γ_0 in $G\backslash K$ such that $K \cup \{z\}$ is included in the inside of Γ_0 (see (10.3)). Let

$$f_0(z) = \frac{1}{2\pi i} \int_{\Gamma_0} \frac{f(w)}{w-z} \, dw.$$

Cauchy's Theorem implies that the definition of $f_0(z)$ is independent of the Jordan system Γ_0. Also, it is easy to see that f_0 is an analytic function on G (differentiate under the integral sign).

If $z \in \mathbb{C}\backslash K$, let Γ_∞ be a smooth Jordan system in $G\backslash K$ such that it contains K in its inside and has the point z in its outside. Let

$$f_\infty(z) = -\frac{1}{2\pi i} \int_{\Gamma_\infty} \frac{f(w)}{w-z} \, dw.$$

Once again the definition of $f_\infty(z)$ is independent of the choice of Jordan system Γ_∞ and $f_\infty: \mathbb{C}\backslash K \to \mathbb{C}$ is an analytic function. It is also easy to see that $f_\infty(z) \to 0$ as $z \to \infty$, so that ∞ is a removable singularity. Therefore if we define $f_\infty(\infty) = 0$, f_∞ becomes an analytic function on $\mathbb{C}_\infty \backslash K$. If $z \in G\backslash K$, then Γ_0 and Γ_∞ can be chosen above so that $\Gamma \equiv \Gamma_0 - \Gamma_\infty$ is also a Jordan system with winding number about the point z equal to 0. Thus Cauchy's Integral Formula implies that $f(z) = f_0(z) + f_\infty(z)$.

It is left to the reader to show that f_0 and f_∞ are bounded when f is. □

12.20 Proposition. *A compact set K is a Painlevé null set if and only if for every open set G containing K, each function in $H^\infty(G\backslash K)$ has an analytic extension to G.*

Proof. If K has this property, take $G = \mathbb{C}$ and let f be the Ahlfors function for K. So f has an extension to an entire function vanishing at ∞, and so $f \equiv 0$. Therefore $\gamma(K) = 0$.

Conversely, assume $\gamma(K) = 0$, let G be an open set containing K, and let $f \in H^\infty(G\backslash K)$. By the preceding lemma, $f = f_0 + f_\infty$, where $f_0 \in H^\infty(G)$ and $f_\infty \in H^\infty(\mathbb{C}_\infty \backslash K)$ and $f_\infty(\infty) = 0$. Since $\gamma(K) = 0$, Proposition 12.18 implies $f_\infty \equiv 0$. Thus $f = f_0$ and we have an analytic extension to G. □

We finish with a result about continuous analytic capacity, whose proof is left as an exercise for the interested reader.

12.21 Proposition. *A compact set K has $\alpha(K) = 0$ if and only if for every open set G containing K, every continuous function $f: \operatorname{cl}[G\backslash K] \to \mathbb{C}$ that is analytic on $G\backslash K$ has an extension to a continuous function on $\operatorname{cl} G$ that is analytic on G.*

We will make use of analytic capacity, and more thorough treatments of rational approximation make even further use of it. For additional information, consult Ahlfors and Beurling [1950], Gamelin [1969], and Garnett [1972].

Exercises

1. Give a direct proof of Proposition 12.7 by using Schwarz's Lemma.
2. Show that if Γ is a Jordan curve, $\gamma(\widehat{\Gamma}) = \alpha(\widehat{\Gamma})$.
3. Prove Proposition 12.15.
4. Prove Proposition 12.21.

§13 Some applications of analytic capacity. In this section we will see some applications of capacity as well as applications of the applications. Capacity can be used to give a necessary and sufficient condition for a point in ∂K to be a peak point of $R(K)$, though this will not be proved here (see Gamelin [1969]). We begin by proving a sufficient condition for a peak point of $R(K)$.

13.1 CURTIS'S PEAK POINT CRITERION. *Let K be a compact subset of $\mathbb{C}$ and let $a \in K$.*

(a) *If*
$$\limsup_{r \to 0} \frac{\gamma(B(a;r) \setminus K)}{r} > 0,$$
then a is a peak point of $R(K)$.

(b) *If*
$$\limsup_{r \to 0} \frac{\alpha(B(a;r) \setminus \operatorname{int} K)}{r} > 0,$$
then a is a peak point of $A(K)$.

PROOF. Only a proof of (a) is presented. Using this proof the reader can fashion a proof of (b).

If the condition in (a) is satisfied, then note that a must be a boundary point of K and there is a positive number β and a sequence of positive numbers $\{r_n\}$ such that $r_n \to 0$ and $\gamma(B(a;r_n) \setminus K) > \beta r_n$ for all $n \geq 1$. Thus for each $n \geq 1$ there is a compact subset L_n contained in $B(a;r_n) \setminus K$ and a function f_n in $H(L_n)$ with $f_n'(\infty) = \gamma(L_n) > \beta r_n$. Define g_n on $\mathbb{C}_\infty \setminus L_n$ by putting $g_n(z) = ((z-a)f_n(z))/f_n'(\infty)$ for z in $\mathbb{C}_\infty \setminus \widehat{L}_n$ and letting g_n be identically 0 on $\widehat{L}_n \setminus L_n$. Note that g_n is analytic and $g_n(\infty) = \lim_{z \to \infty} (z-a)f_n(z)/f_n'(\infty) = 1$.

Now for z in $\mathbb{C} \setminus L_n$ and $|z - a| < r_n$, $|g_n(z)| \leq r_n |f_n(z)|/f_n'(\infty) \leq r_n/\gamma(L_n) \leq 1/\beta$. Since $\beta \leq 1$ and $g_n(\infty) = 1$, the Maximum Modulus Theorem implies that $|g_n| \leq 1/\beta$ on $\mathbb{C}_\infty \setminus L_n$. By Montel's Theorem there is an analytic function g defined of $\mathbb{C}_\infty \setminus \{a\}$ such that, replacing $\{g_n\}$ by a subsequence if necessary, $g_n \to g$ uniformly on compact subsets of $\mathbb{C}_\infty \setminus \{a\}$. But $|g| \leq 1/\beta$ and so a must be a removable singularity. By Liouville's Theorem, g is a constant function. Since $g(\infty) = 1$, $g \equiv 1$.

Let $h_n = 1 - g_n$. Since h_n is analytic in a neighborhood of K, $h_n \in R(K)$. Also, $h_n \to 0$ uniformly on compact subsets of $\mathbb{C}_\infty \setminus \{a\}$ and $\|h_n\|_K \leq 1 + 1/\beta = M$. So if U is any open set containing a, there is an $n \geq 1$ such that $\|h_n\|_{K \setminus U} < 1/2 = c$. Since $h_n(a) = 1$ for all n, Theorem 11.5 implies that a is a peak point. □

The preceding theorem and its consequent corollaries below are a good illustration of the fact that even though it is somewhere between intolerably difficult and impossible to compute the analytic capacity of a given set, this does not negate the usefulness of the concept. That is, computability is not the only criterion for the usefulness of an idea. Indeed, operator theory knows this because of its experience with the spectrum. From Curtis's Criterion several easily applicable conditions for points to be peak points can be obtained, as we do presently.

13.2 GONCHAR'S CRITERION. *Let $a \in \partial K$ and for $r > 0$ let $d(r)$ be the supremum of the diameters of the components of $B(a; r) \setminus K$. If*

$$\limsup_{r \to 0} \frac{d(r)}{r} > 0,$$

then a is a peak point of $R(K)$.

PROOF. If C is a component of $B(a; r) \setminus K$, then $\gamma(C) \geq \sup\{\gamma(L): L$ is a compact, connected subset of $C\} \geq (1/4) \sup\{\operatorname{diam} L: L$ is a compact, connected subset of $C\} = (1/4) \operatorname{diam} C$. Hence $\gamma(B(a; r) \setminus K) \geq \sup\{\gamma(C): C$ is a component of $B(a; r) \setminus K\} \geq (1/4) d(r)$. By Curtis's Criterion, a is a peak point. □

13.3 COROLLARY. *If G is a component of $\mathbb{C} \setminus K$ and $a \in \partial G$, then a is a peak point of $R(K)$.*

A topological lemma is needed for the proof of this corollary.

13.4 LEMMA. *Let G be a component of $\mathbb{C} \setminus K$ and let $a \in \partial G$. If $0 < \varepsilon < r < \operatorname{diam} G$, then there is a component V of $B(a; r) \setminus K$ with $\operatorname{diam} V \geq r - \varepsilon$.*

PROOF. Let $b \in G$ with $|a - b| < \varepsilon$ and let a_1 be on the line segment $[a, b]$ such that $[a_1, b] \cap K = \{a_1\}$. Hence $|a - a_1| < \varepsilon$ and $|b - a_1| < \varepsilon$. Since $(a_1, b] \subseteq \mathbb{C} \setminus K$, $b \in G$, and G is connected, we have that $(a_1, b] \subseteq G$. Also $\varepsilon < r$ and so $(a_1, b] \subseteq B(a; r) \setminus K$. Let V be the component of $B(a; r) \setminus K$ that contains $(a_1, b]$.

It is claimed that $(\operatorname{cl} V) \cap \partial B(a; r) \neq \varnothing$. Once this is established, the proof of the lemma comes to a speedy conclusion. In fact, suppose there is a c in $(\operatorname{cl} V) \cap \partial B(a; r)$. Since $a_1 \in \operatorname{cl} V$, $\operatorname{diam} V = \operatorname{diam}[\operatorname{cl} V] \geq |a_1 - c| \geq |a - c| - |a - a_1| \geq r - \varepsilon$.

So to prove that $(\operatorname{cl} V) \cap \partial B(a; r) \neq \varnothing$, suppose the contrary. Let $b_1 \in G \setminus \overline{B}(a; r)$ (possible since $r < \operatorname{diam} G$) and let $\gamma: [0, 1] \to G$ be a continuous path with $\gamma(0) = b$ and $\gamma(1) = b_1$. Let $t^* = \sup\{t: \gamma(t) \in V\}$. It follows that $0 < t^* < 1$ and $\gamma(t^*) \in \partial V$. But $(\operatorname{cl} V) \cap \partial B(a;r) = \varnothing$, so it must be that $\operatorname{cl} V \subseteq B(a;r)$. Also V is a component of $B(a;r)\setminus K$ and so it must be that $\partial V \subseteq K$. But then this implies that $\gamma(t^*) \in K$, contradicting the fact that $\gamma(t^*) \in G$. This proves that $(\operatorname{cl} V) \cap \partial B(a;r) \neq \varnothing$ and completes the proof of the lemma. □

PROOF OF COROLLARY 13.3. Let $d(r)$ be defined as in Gonchar's Criterion. The preceding lemma shows that $d(r) \geq r$ and so $\limsup_{r \to 0} d(r)/r \geq 1$. □

13.5 COROLLARY. *If there is a positive number d such that each component of $\mathbb{C}\setminus K$ has a diameter at least d, then each point of ∂K is a peak point of $R(K)$.*

PROOF. Let $a \in \partial K$ and let $0 < r/2 < d$. Let G be a component of $\mathbb{C}\setminus K$ that meets $B(a;r/2)$. Pick a_1 in ∂G with $|a - a_1| < r/2$. Since $\operatorname{diam} G \geq d > r/2$, Lemma 13.4 implies that for $0 < \varepsilon < r/2$ there is a component V_1 of $B(a;r/2)\setminus K$ with $\operatorname{diam} V_1 \geq r/2 - \varepsilon$. Let V be the component of $B(a;r)\setminus K$ that contains V_1. If $d(r)$ is defined as in Gonchar's Criterion, then $d(r) \geq \operatorname{diam} V \geq \operatorname{diam} V_1 \geq r/2 - \varepsilon$ for all small ε. Hence $d(r) \geq r/2$ and a must be a peak point. □

Say that the compact set K is *finitely connected* if $\mathbb{C}\setminus K$ has a finite number of components.

13.6 COROLLARY. *If K is finitely connected, then every point of ∂K is a peak point.*

Another sufficient condition for a point to be a peak point can be obtained from the idea of area density.

13.7 DEFINITION. If F is a measurable subset of $\mathbb{C}$ and $a \in \mathbb{C}$, then F is said to have *positive upper area density* at a if

$$\limsup_{r \to 0} \frac{\operatorname{Area}[B(a;r) \cap F]}{\pi r^2} > 0.$$

If F does have positive upper area density at a, then there is a sequence of positive numbers $\{r_n\}$ and a positive constant β such that $r_n \to 0$ and

$$\frac{\operatorname{Area}[B(a;r_n) \cap F]}{\pi r_n^2} > \beta.$$

We now turn our attention to the algebras $R(K, E)$ for a bounded measurable set E.

13.8 PROPOSITION. *If E is a measurable subset of K, $a \in K$, and $\mathbb{C}\setminus E$ has positive upper area density at a, then a is a peak point of $R(K, E)$.*

PROOF. Let $B(r) = B(a;r)$ for $r > 0$. By hypothesis there is a sequence of positive numbers $\{r_n\}$ and a positive constant α such that $r_n \to 0$ and

Area$(B(r_n)\backslash E) \geq \alpha \pi r_n^2$ for all $n \geq 1$. Let $c_n = \int_{B(r_n)\backslash E} |w - a|^{-1} d\mathscr{A}(w) \geq r_n^{-1}$ Area$(B(r_n)\backslash E) \geq \alpha \pi r_n$.

Let $\sigma(z) = z/|z|$ for $z \neq 0$ and $\sigma(0) = 1$. Define

$$f_n(z) = \frac{1}{c_n} \int_{B(r_n)\backslash E} \frac{\sigma(w-a)}{w-z} d\mathscr{A}(w).$$

So $f_n \in R(K, E)$ by definition and $f_n(a) = 1$. Also Proposition 2.14 implies

$$|f_n(z)| \leq \frac{1}{\alpha \pi r_n} \int_{B(r_n)\backslash E} |w - z|^{-1} d\mathscr{A}(w)$$

$$\leq \frac{1}{\alpha \pi r_n} 2[\pi \text{Area}(B(r_n)\backslash E)]^{1/2}$$

$$\leq \frac{1}{\alpha \pi r_n} 2\pi r_n$$

$$= \frac{2}{\alpha} \equiv M.$$

For $|z - a| = r_n$, $|(z-a)f_n(z)| \leq M r_n$. But f_n is analytic on $\mathbb{C}_\infty \backslash [\text{cl } B(r_n)]$ and continuous on $\mathbb{C}_\infty$. By the Maximum Modulus Theorem, $|f_n(z)| \leq M r_n/|z - a|$ for $|z - a| \geq r_n$.

So if $r > 0$, fix n such that $r_n < r/2M$. If $z \notin B(r)$, then $|f_n(z)| \leq M r_n/|z-a| \leq M r_n/r < 1/2 = c$. By Theorem 11.5, a is a peak point of $R(K, E)$. □

13.9 COROLLARY. *If K is a compact subset of $\mathbb{C}$, $a \in K$, and $\mathbb{C}\backslash K$ has positive upper area density at a, then a is a peak point of $R(K)$.*

Note that a point a in K that does not belong to cl E satisfies the hypothesis of the preceding proposition and is thus a peak point of $R(K, E)$. So if Q is the set of nonpeak points of $R(K, E)$, then $Q \subset \text{cl } E$ but Q may contain points that do not belong to E. Since [cl $E]\backslash E$ may have positive area, this has a potential for causing trouble. But it doesn't.

13.10 PROPOSITION. *If E is a measurable set and Q is the set of nonpeak points for $R(K, E)$, then Area$(Q\backslash E) = 0$.*

PROOF. We know that the set of peak points of $R(K, E)$ is a G_δ set, so that Q and $Q\backslash E$ are measurable sets. For each a in $\mathbb{C}$, let

$$d(a) = \limsup_{r \to 0} \frac{\text{Area}(B(a; r)\backslash E)}{\pi r^2}.$$

The Lebesgue Density Theorem implies that Area$(\{a: d(a) = 0\}) = 0$. But Proposition 13.8 implies that $Q\backslash E \subseteq \{a: d(a) = 0\}$. □

It is not necessarily true, however, that Area$(E\backslash Q) = 0$. For example, there is a compact set E of positive area such that $R(E) = C(E)$ (Exercise 2). Hence $Q = \varnothing$.

13.11 PROPOSITION. *If E is a bounded measurable set and Q is the set of nonpeak points, then $R(K, E) = R(K, Q)$.*

PROOF. Let $h \in L_c^\infty$ such that $h = 0$ on E. By the preceding proposition, $h = 0$ a.e. on $(Q \cap E) \cup (Q \setminus E) = Q$. Hence $\hat{h}$, the Cauchy transform of h, belongs to $R(K, Q)$. That is, $R(K, E) \subseteq R(K, Q)$. On the other hand, Theorem 11.9 implies that $R(K, Q) \subseteq R(K, E)$. □

Exercises

1. Give examples of a polynomially convex set K with a boundary point a such that $\mathbb{C} \setminus K$ does not have positive upper area density at a. What does this say about Proposition 13.8?
2. Construct a compact set E of positive area such that $R(E) = C(E)$. (If you cannot do this directly, see Lavrentiev's Theorem (14.20 below.))

§14 Dirichlet algebras. In this section we will study a class of function algebras that are of particular importance for subnormal operators.

14.1 DEFINITION. If A is a function algebra on X, then A is a *Dirichlet algebra* on X if $\operatorname{Re} A \equiv \{\operatorname{Re} f : f \in A\}$ is dense in $C_\mathbb{R}(X)$.

This imposes a stringent requirement on the relation between A and X. To see exactly what is entailed here, suppose A is a Dirichlet algebra on X, let $a \in X$, and let U be an open neighborhood of a. If $\varepsilon > 0$ and $\delta > 0$ is chosen so that $e^\delta < 1 + \varepsilon$, then there is a continuous function u on X such that $u \leq 0$, $u(a) = 0$, and $u(x) \equiv \log \varepsilon$ off U. By hypothesis there is a function f in A with $f(a) = 0$ and $\|\operatorname{Re} f - u\| < \delta$. Put $g = e^f$. So $g \in A$, $g(a) = 1$, and $\|g\| \leq e^\delta \leq 1 + \varepsilon$. Also off U, $|g| \leq 2\varepsilon$. By Theorem 11.5, a is a peak point for A. We have, therefore, proved the following.

14.2 PROPOSITION. *If A is a Dirichlet algebra on X, then every point of X is a peak point for A.*

For any function algebra A on X, the restriction of A to ∂_A, the Shilov boundary of A, is also a function algebra on ∂_A. Bowing to the reality of the preceding proposition, we adopt the following definition.

14.3 DEFINITION. If A is a function algebra on X, then A is a *Dirichlet algebra* if $A|\partial_A$ is a Dirichlet algebra on ∂_A.

The difference between (14.1) and this definition is that the last one is independent of the host space X. The next result is immediate from Proposition 14.2.

14.4 PROPOSITION. *If A is a Dirichlet algebra, then every point of ∂_A is a peak point.*

14.5 EXAMPLE. If $X = \operatorname{cl} \mathbb{D}$, then $P(X)$ is a Dirichlet algebra. In fact, $\{\operatorname{Re} p|\partial \mathbb{D} : p \text{ is a polynomial}\}$ is precisely the real-valued trigonometric polynomials which are dense in $C_\mathbb{R}(\partial \mathbb{D})$.

14.6 Proposition. *If A is a function algebra, then A is a Dirichlet algebra if and only if the only real-valued measure μ on ∂_A that satisfies $\int f \, d\mu = 0$ for all f in A is the zero measure.*

PROOF. This is an immediate consequence of the Hahn-Banach Theorem. □

Here is another proof that $A = P(\operatorname{cl} \mathbb{D})$ is a Dirichlet algebra; it is this proof that will be generalized to show that $P(K)$ is a Dirichlet algebra. Let $\mu \in M_\mathbb{R}(\partial \mathbb{D})$ such that $\mu \perp A$. By the F. and M. Riesz Theorem, $\mu = h \, dm$ for some function h in H_0^1. But μ is real-valued, so h is real-valued and analytic. Hence $h = 0$.

14.7 Proposition. *Suppose A is a Dirichlet algebra.*

(a) *If $p \in \mathscr{M}_A$, then there is a unique representing measure for p that is supported on ∂_A.*

(b) *If E is a closed subset of ∂_A that is also relatively open, then there is an f in A such that $f|\partial_A = \chi_E$.*

PROOF. (a) If μ_1 and μ_2 are two representing measures for p, then $\mu = \mu_1 - \mu_2$ is a real-valued measure that satisfies $\int f \, d\mu = 0$ for every f in A. So if both μ_1 and μ_2 are carried by ∂_A, then $\mu = 0$ by (14.6).

(b) Since E is closed and relatively open in ∂_A, χ_E is a real-valued continuous function on ∂_A. By hypothesis there is a function g in A such that $\|\operatorname{Re} g - \chi_E\|_{\partial_A} < 1/4$. Thus $3/4 < \operatorname{Re} g < 5/4$ on E and $|\operatorname{Re} g| < 1/4$ on $\partial_A \setminus E$. Let G be the union of these two open strips; we have that $g(\partial_A) \subseteq G$. By Proposition 7.8 the topological boundary of $\hat{g}(\mathscr{M}_A) \subseteq g(\partial_A) \subseteq G$. It follows that $\hat{g}(\mathscr{M}_A) \subseteq G$. Let h be the characteristic function of the open strip $\{w : 3/4 < \operatorname{Re} w < 5/4\}$. Because h is analytic in a neighborhood of $\sigma(g) = \hat{g}(\mathscr{M}_A)$, $f = h \circ g \in A$. Clearly $\operatorname{Re} f = \chi_E$. □

Recall that the Shilov boundary of $R(K)$ is ∂K. If $R(K)$ is a Dirichlet algebra and $a \in \operatorname{int} K$, then harmonic measure at a is the unique representing measure for a supported on ∂K that is guaranteed by the preceding proposition. If $a \in \partial K$, then δ_a is that representing measure. Let's record this here as well as an additional consequence.

14.8 Proposition. *Let K be a compact subset of $\mathbb{C}$ and assume that $R(K)$ is a Dirichlet algebra.*

(a) *If $a \in \operatorname{int} K$, then the unique representing measure for a that is supported on ∂K is harmonic measure.*

(b) $R(\partial K) = C(\partial K)$.

PROOF. The proof of (a) has already been given. To prove (b) observe that since every point of ∂K is a peak point of $R(K)$, every point of ∂K is also a peak point of $R(\partial K)$. By Corollary 11.10, $R(\partial K) = C(\partial K)$. □

Proposition 14.7(b) shows that if K is an annulus, then $R(K)$ is not a Dirichlet algebra since the outer boundary of K is a closed and relatively open subset of ∂K and there is no f in $R(K)$ that f is identically 1 on the outer boundary and identically 0 on the inner boundary. Nevertheless, $R(\partial K) = C(\partial K)$ for an annulus (13.6).

We now embark on a series of results aimed at proving that $P(K)$ is a Dirichlet algebra. Along the way we will encounter several results that have importance and interest on their own.

14.9 LEMMA. *Suppose K is a compact subset of $\mathbb{C}$ and $K = \bigcap_n K_n$, where each K_n is compact, $K_{n+1} \subseteq \operatorname{int} K_n$, and $\operatorname{int} K_n$ is a Dirichlet set for each n. Let a be a peak point of $R(K)$. If ω_n is harmonic measure for K_n at a, then $\omega_n \to \delta_a$ weak* in $M(K_1)$.*

PROOF. In fact, $\{\omega_n\}$ is a sequence of probability measures on K_1 and so there is a weak* cluster point μ in $M(K_1)$; of necessity, μ is also a probability measure. Let f be a function that is analytic on an open set U that contains K.

Thus there is an integer n_0 such that $K_n \subseteq U$ for $n \geq n_0$. From the definition of harmonic measure, $f(a) = \int f \, d\omega_n$ for $n \geq n_0$; thus $f(a) = \int f \, d\mu$.

It also follows that μ is supported by ∂K. In fact, if $\phi \in C(K_1)$ and $(\operatorname{supp} \phi) \cap \partial K = \varnothing$, then $(\operatorname{supp} \phi) \cap \partial K_n = \varnothing$ for all sufficiently large n. Thus $\int \phi \, d\omega_n = 0$ for all large n and so $\int \phi \, d\mu = 0$.

Therefore μ is a representing measure for $R(K)$ at a. Since a is a peak point, $\mu = \delta_a$. But this means that δ_a is the only possible weak* cluster point of the sequence $\{\omega_n\}$ and so $\omega_n \to \delta_a$ weak* in $M(K_1)$. □

14.10 THEOREM. *If K is a compact subset of $\mathbb{C}$ and every point of ∂K is a peak point for $R(K)$, then $\operatorname{int} K$ is a Dirichlet set. Moreover, if $u \in C(\partial K)$, then there is a sequence $\{u_n\}$ of functions harmonic on a neighborhood of K such that $u_n(z) \to u(z)$ uniformly on ∂K.*

PROOF. Construct a sequence of compact sets $\{K_n\}$ such that for each $n \geq 1$, $K_{n+1} \subseteq K_n$, $\operatorname{int} K_n$ is a Dirichlet set, and $K = \bigcap_n K_n$. Assume that u is real-valued. By Tietze's Extension Theorem, there is a continuous function $f: K_1 \to \mathbb{R}$ that extends u and also satisfies $f(K_1) \subseteq [\min u, \max u]$. For each n let $h_n \in C(K_n)$ such that h_n is harmonic on $\operatorname{int} K_n$ and $h_n|\partial K_n = f|\partial K_n$. Now fix a point a in ∂K and let ω_n be harmonic measure for K_n at a. By the preceding lemma, $\omega_n \to \delta_a$ weak* in $M(K_1)$. Thus $h_n(a) = \int h_n \, d\omega_n = \int f \, d\omega_n \to \int f \, d\delta_a = u(a)$. Also $\|h_n\|_{\partial K} \leq \|h_n\|_{\partial K_n} \leq \|f\|_{\partial K_1} = \|u\|_{\partial K}$.

Therefore, $\{h_n\}$ is a sequence of functions that are harmonic in a neighborhood of K, are uniformly bounded on K, and converge pointwise on ∂K to u. Thus for any measure μ in $M(\partial K)$, $\int h_n \, d\mu \to \int u \, d\mu$; equivalently, $h_n \to u$ weakly in the Banach space $C(\partial K)$. Therefore, there is a sequence of functions $\{u_k\}$ such that each u_k is a convex combination of elements of the sequence $\{h_n\}$ and $u_k \to u$ uniformly on ∂K (see V.1.4 in [ACFA]). □

§14 DIRICHLET ALGEBRAS

It is well known (see VIII.2.2 in [**FOCV**]) that a region is simply connected if and only if each harmonic function on G has a harmonic conjugate and this is also equivalent to the assertion that every analytic function on G has a primitive. The existence of harmonic conjugates and the existence of primitives for analytic functions turns out to be related on an individual basis, independent of the supporting region G. To be specific, note that an easy consequence of the Cauchy-Riemann equations is the following result.

14.11 PROPOSITION. *If $u: G \to \mathbb{C}$ is a C^2 function, then u is a harmonic function on G if and only if $f = u_x - iu_y = 2\partial u$ is an analytic function on G.*

It turns out that there is a close relation between the harmonic function u and the analytic function $f = u_x - iu_y$. Indeed, one function can often be studied with the help of the other. A key to this is the following computation. If γ is any closed rectifiable curve in G, then

$$(14.12) \qquad \int_\gamma f = i \int_\gamma (u_x\, dy - u_y\, dx).$$

In fact, $\int_\gamma f = \int_\gamma (u_x - iu_y)(dx + i\, dy) = \int_\gamma (u_x\, dx + u_y\, dy) + i\int_\gamma (u_x\, dy - u_y\, dx)$ and $\int_\gamma (u_x\, dx + u_y\, dy) = 0$ since this is the integral of an exact differential.

We are now ready to present a direct relation between the existence of a harmonic conjugate and the existence of a primitive.

14.13 THEOREM. *If G is a region in $\mathbb{C}$ and $u: G \to \mathbb{R}$ is a harmonic function, then the following statements are equivalent.*

(a) *u has a harmonic conjugate.*
(b) *The analytic function $f = \partial u$ has a primitive in G.*
(c) *For every closed rectifiable curve γ in G, $\int_\gamma (u_x\, dy - u_y\, dx) = 0$.*
(d) *For every closed rectifiable curve γ in G, $\int_\gamma f = 0$.*

PROOF. Parts (c) and (d) are equivalent by (14.12). Standard analytic function theory shows that (b) and (d) are equivalent. It remains to show that (a) and (b) are equivalent.

(a) *implies* (b). If g is an analytic function on G such that $g = u + iv$, then the fact that the Cauchy-Riemann equations hold implies that $g' = u_x + iv_x = u_x - iu_y = f$.

(b) *implies* (a). Suppose g is an analytic function on G such that $g' = f$ and let U and V be the real and imaginary parts of g. Thus $g' = U_x + iV_x = f = u_x - iu_y$. So $U_x = u_x$ and $U_y = -V_x = u_y$ in G. That is, the gradient of $U - u$ vanishes in G. By Exercise X.1.9 of [**FOCV**], U and u differ by a constant. Hence V is a harmonic conjugate of u. □

The reader might question whether Theorem 14.13 actually characterizes the harmonic functions that have a conjugate since it merely states that this problem is equivalent to another problem of equal difficulty: whether a given analytic function has a primitive. There is some validity in this criticism, though this does not diminish the value of (14.13); it is a criticism of the result as it relates to the originally stated objective rather than any internal defect. In a sense, the theorem says that to check whether a function has a conjugate you must still check an infinite number of conditions (part (c)). For certain regions, having a harmonic conjugate is equivalent to a finite collection of conditions.

Say that a region G in $\mathbb{C}$ is *n-connected* if $\mathbb{C}_\infty \backslash G$ has n components. Thus a 1-connected region is simply connected. A region G is finitely connected if it is n-connected for some positive integer n. Note that if G is any region in $\mathbb{C}$ and K is any component of $\mathbb{C}_\infty \backslash G$ that does not contain ∞, then K must be a compact subset of $\mathbb{C}$; in fact, such components of $\mathbb{C}_\infty \backslash G$ are precisely the bounded components of $\mathbb{C} \backslash G$.

If G is an $n+1$-connected region in $\mathbb{C}$, E is a compact subset of G, and $K_1, \ldots, K_n$ are the bounded components of $\mathbb{C} \backslash G$, then there is a positive Jordan system $\Gamma = \{\gamma_0, \gamma_1, \ldots, \gamma_n\}$ in G having the following properties:

(14.14) $\begin{cases} \text{(i)} & \text{inside } \Gamma \subseteq G; \\ \text{(ii)} & \text{for } 1 \leq j \leq n, K_j \subseteq \text{inside } \gamma_j; \\ \text{(iii)} & \text{cl(inside } \gamma_j) \cap \text{cl(inside } \gamma_k) = \emptyset \text{ for } j \neq k. \end{cases}$

The idea here is that for $1 \leq j \leq n$ and a in K_j, $n(\gamma_j; a) = -1$ while $n(\gamma_0; a) = 1$. (Also see (10.3).)

A positive Jordan system Γ satisfying (14.14) will be called a *curve generating system* for G. In fact, the curves $\gamma_1, \ldots, \gamma_n$ are a set of generators for the first homology group of G as well as the fundamental group of G.

14.15 THEOREM. *Let G be an $(n+1)$-connected region and let $K_1, \ldots, K_n$ be the bounded components of $\mathbb{C} \backslash G$. If Γ is a positive Jordan system satisfying (14.14) and if $u: G \to \mathbb{R}$ is a harmonic function, then u has a harmonic conjugate if and only if $\int_{\gamma_j} (u_x \, dy - u_y \, dx) = 0$ for $1 \leq j \leq n$.*

PROOF. From Theorem 14.13 it is easy to see that it suffices to assume that $\int_{\gamma_j} f = 0$ for $1 \leq j \leq n$, where $f = \partial u$, and prove that f has a primitive. Fix a point a_j in K_j for $1 \leq j \leq n$. If γ is any closed rectifiable curve in G, put $m_j = n(\gamma; a_j)$. It follows that the system of curves $\Gamma \equiv \{\gamma, -m_1\gamma_1, \ldots, -m_n\gamma_n\}$ is homologous to 0 in G. By Cauchy's Theorem,

$$0 = \int_\gamma f - \sum m_j \int_{\gamma_j} f.$$

But this implies that $\int_\gamma f = 0$ by assumption. Hence f has a primitive. □

What is going on in the preceding lemma is that the first homology group of G is a free abelian group on n generators and the curves $\{\gamma_j\}$ form a system of generators for this group. Moreover, if Γ is an element of the first homology group, then Γ corresponds to a system of closed curves in G and the map $\Gamma \to \int_\Gamma f$ is a homomorphism of this group into the additive group

ℂ. Thus the condition of the preceding theorem is that this homomorphism vanishes on the generators, and hence vanishes identically.

Theorem 14.15 can be used to describe exactly how a harmonic function differs from one that has a harmonic conjugate. The next result is classical, though the proof here and the development preceding it are taken from Axler [1986].

14.16 Theorem. *Let G be an $(n+1)$-connected region with $K_1, \ldots, K_n$ the bounded components of its complement; for $1 \leq j \leq n$, let $a_j \in K_j$. If u is a harmonic function on G, then there are real scalars $c_1, \ldots, c_n$ and an analytic function h on G such that*

$$u = \operatorname{Re} h + \sum_j c_j \log|z - a_j|.$$

PROOF. Let $\Gamma = \{\gamma_0, \gamma_1, \ldots, \gamma_n\}$ be a positive Jordan system satisfying (14.14) and put

$$c_j = (1/2\pi) \int_{\gamma_j} (u_x\,dy - u_y\,dx).$$

Note that $c_j \in \mathbb{R}$. Consider the harmonic function

$$U = u - \sum_j c_j \log|z - a_j|.$$

Now $l(z) = \log|z|$ is harmonic on $\mathbb{C}\setminus\{0\}$ and an elementary computation shows that $l_x - il_y = \bar{z}/|z|^2 = z^{-1}$. Thus (14.12) implies that for any closed rectifiable curve γ not passing through 0,

$$\int_\gamma (l_x\,dy - l_y\,dx) = -i\int_\gamma z^{-1}\,dz = 2\pi n(\gamma; 0).$$

So if $L_k(z) = \log|z - a_k|$, $\int_{\gamma_j}(L_{k_x}\,dy - L_{k_y}\,dx) = 2\pi$ if $k = j$ and 0 otherwise. It is straightforward to see that the choice of c_j gives that $\int_{\gamma_j}(U_x\,dy - U_y\,dx) = 0$ for $1 \leq j \leq n$. By Theorem 14.15, there is an analytic function h on G such that $U = \operatorname{Re} h$. □

14.17 Corollary. *Assume K is finitely connected and let $G_1, \ldots, G_n$ be the bounded components of $\mathbb{C}\setminus K$; choose a_j from G_j, $1 \leq j \leq n$. If u is harmonic in a neighborhood of K, then for every $\varepsilon > 0$ there is a function h analytic in a neighborhood of K and real constants $c_1, \ldots, c_n$ such that*

$$\|u - \operatorname{Re} h - \sum_{j=1}^n c_j \log|z - a_j|\|_K < \varepsilon.$$

PROOF. Since K is finitely connected, there is a $\delta > 0$ such that if $G = \{z : \operatorname{dist}(z, K) < \delta\}$, then u is harmonic on G, G is finitely connected, and each bounded component of $\mathbb{C}\setminus G$ contains one of the points a_j. Now apply the theorem. □

The next result is actually another corollary of Theorem 14.16 but it is of sufficient importance to merit a more significant appellation.

14.18 THEOREM. *Assume K is finitely connected and let $G_1, \ldots, G_n$ be the bounded components of $\mathbb{C} \backslash K$; choose a_j from G_j, $1 \le j \le n$. If $f \in C_{\mathbb{R}}(\partial K)$ and $\varepsilon > 0$, then there is a function g in $R(K)$ and real constants $c_1, \ldots, c_n$ such that*

$$\left\| f - \operatorname{Re} g - \sum_{j=1}^{n} c_j \log|z - a_j| \right\|_{\partial K} < \varepsilon.$$

PROOF. Corollary 13.6 implies each point of ∂K is a peak point. Theorem 14.10 implies that f can be approximated uniformly on ∂K by functions that are harmonic in a neighborhood of K. This theorem now follows by an application of the preceding corollary. □

Since polynomially convex sets have connected complements, our objective is achieved as a consequence of this theorem.

14.19 COROLLARY. *If K is polynomially convex, $P(K)$ is a Dirichlet algebra.*

If K is the closure of the cornucopia together with the unit disk, then K is polynomially convex and so $R(K)$ is a Dirichlet algebra. The same holds if K is the union of two tangent disks. Note that in both these examples the interior of K is not connected.

The preceding corollary also gives a classical result as an easy consequence.

14.20 LAVRENTIEV'S THEOREM. *If K is a compact subset of $\mathbb{C}$, then $P(K) = C(K)$ if and only if K is polynomially convex and $\operatorname{int} K = \varnothing$.*

PROOF. If $P(K) = C(K)$ then it is clear that $\operatorname{int} K = \varnothing$ and K is polynomially convex. For the converse, the assumption that K is polynomially convex implies that $P(K)$ is a Dirichlet algebra. Thus every point of $K = \partial K$ is a peak point. By Corollary 11.10, $P(K) = C(K)$. □

We will return to the study of Dirichlet algebras periodically in the course of this book. Indeed, these algebras have great significance for the study of operator theory. We conclude this section with some applications to operator theory of the results of this section.

14.21 THEOREM (Bram [1955]). *If N is a $*$-cyclic normal operator, then N has a cyclic vector. Moreover, if N is represented as multiplication by z on $L^2(\mu)$ for some compactly supported measure μ on $\mathbb{C}$, then N has a cyclic vector in $L^\infty(\mu)$.*

PROOF. We may assume that $\mathscr{H} = L^2(\mu)$ and $N = N_\mu$, where μ is a probability measure with compact support X. Let D be a closed disk that contains X in its interior. The proof consists in constructing a sequence $\{F_k\}$ of compact sets having the following properties:

(i) F_k is polynomially convex and $\operatorname{int} F_k = \varnothing$;
(ii) $F_k \subseteq F_{k+1}$;
(iii) $\mu(F_k) \to 1 = \|\mu\|$.

Indeed once this is done, we can complete the proof by an appeal to Lemma II.10.7 with $E_k = F_k$, $K = D$, and $f(z) = \bar{z}$. To see this, note that Lavrentiev's Theorem and (i) imply that $\bar{z} \in P(F_k) = R(F_k, D)$, so Lemma II.10.7 implies there is a measure ν with $[\nu] = [\mu]$, $\bar{z} \in P^2(\nu)$, and $\phi = (d\nu/d\mu)^{1/2}$ a bounded function. It follows that $\bar{z}q \in P^2(\nu)$ for every polynomial q and hence, by a limiting argument, $\bar{z}f \in P^2(\nu)$ for every f in $P^2(\nu)$; thus $P^2(\nu) = L^2(\nu)$. So for any f in $L^2(\mu)$ there is a sequence of polynomials $\{p_n\}$ such that $p_n \to f/\phi$ in $L^2(\nu)$. But

$$\int |p_n \phi - f|^2 d\mu = \int |p_n \phi - f|^2 \phi^{-2} d\nu = \int |p_n - f/\phi|^2 d\nu.$$

Therefore, $p_n \phi \to f$ in $L^2(\mu)$ and so ϕ is a cyclic vector.

Now to construct the sets F_k. Let $R = \{r_n\}$ be a countable dense subset of D that contains no atoms of μ. Since r_n is not an atom of μ, for each integer $k \geq 1$, there is an open disk Δ_n^k with center r_n such that $\mu(\Delta_n^k) < (k2^n)^{-1}$. Arrange these disks so that $\Delta_n^{k+1} \subseteq \Delta_n^k$. Put $\Delta^k = \bigcup_{n=1}^\infty \Delta_n^k$; so $\mu(\Delta^k) < k^{-1}$, Δ^k is open, and $R \subseteq \Delta^k$.

Now consider all the open rectangles having $[r_n, r_{n+1}]$ as one on the oriented sides of their closure. Since the intersection of all such rectangles is empty, for each $k \geq 1$ we can find one of them, T_n^k, such that $\mu(T_n^k) < (k2^n)^{-1}$. Arrange these rectangles so that $T_n^{k+1} \subseteq T_n^k$. Let $T^k = \bigcup_{n=1}^\infty T_n^k$. So T^k is open and $\mu(T^k) < k^{-1}$. Now fix some point z_0 in $\mathbb{C}\setminus D$ and let T_0^k be an open rectangle with z_0 and r_1 as two of its vertices such that $\mu(T_0^k) < k^{-1}$. Let s_1^k and s_2^k be the other vertices of T_0^k such that $\partial T_0^k = [z_0, r_1, s_1^k, s_2^k]$ is the positive orientation. Also, arrange the rectangles T_0^k so that $T_0^{k+1} \subseteq T_0^k$.

Let $U^k = \Delta^k \cup T^k \cup T_0^k$ and put $F_k = D\setminus U^k$. By arrangement, (ii) is satisfied and F_k is compact. Also, $\mu(U^k) < 3/k$, so that (iii) is also satisfied. Since U^k is open, includes the set R, and R is dense in D, we have int $F_k = \varnothing$. It remains to show that F_k is polynomially convex.

Now each of the disks Δ_n^k and Δ_{n+1}^k have nonempty intersection with the rectangle T_n^k, and so $\Delta_n^k \cup T_n^k \cup \Delta_{n+1}^k$ is connected. Thus U^k can be written as the union of a sequence of connected sets, consecutive pairs of which have nonempty intersection. This implies that U^k is connected. But $\mathbb{C}\setminus F_k = (\mathbb{C}\setminus D) \cup U^k$, both U^k and $\mathbb{C}\setminus D$ are connected and $z_0 \in (\mathbb{C}\setminus D) \cap U^k$; therefore $\mathbb{C}\setminus F_k$ is connected and so F_k is polynomially convex. □

14.22 COROLLARY. *If μ is a compactly supported measure on the plane and ψ is a Borel function, then there is a sequence of polynomials $\{p_n\}$ such that $p_n(z) \to \psi(z)$ a.e. $[\mu]$.*

14.23 COROLLARY. *If S is a subnormal operator whose minimal normal extension is $*$-cyclic, then S^* has a cyclic vector.*

PROOF. If $N = \text{mne}(S)$, Theorem 14.21 implies that N^* has a cyclic vector ϕ. Let P be the projection of $\mathcal{K}$ onto $\mathcal{H}$ and put $f = P\phi$. It will be shown that f is a cyclic vector for S^*. In fact, if $g \in \mathcal{H}$ and $g \perp S^{*n}f$ for all $n \geq 0$, then $0 = \langle f, S^n g \rangle = \langle \phi, N^n g \rangle = \langle N^{*n}\phi, g \rangle$. Since ϕ is cyclic for N^*, $g = 0$. □

Wogen [1978] has shown that if $\phi \in H^\infty$ and ϕ is not constant, then T_ϕ^* is cyclic. In fact, he has shown that there is one function f in H^2 that is a cyclic vector for every operator T_ϕ^* with ϕ nonconstant. He also has shown that the adjoint of a pure quasinormal operator is cyclic and raised the following question.

14.24 OPEN QUESTION. Is there a pure subnormal operator S such that S^* is not cyclic?

Exercises

1. If K is a compact subset of $\mathbb{C}$, $\text{int}\, K$ is connected, and ∂K is not connected, then $R(K)$ is not a Dirichlet algebra.
2. If $R(K)$ is a Dirichlet algebra, prove that $\text{int}\, K$ is a Dirichlet set.
3. Give an example of a compact set K such that $R(K)$ is a Dirichlet algebra and $\text{int}\, K$ has an infinite number of components.
4. Characterize the cyclic vectors for the bilateral shift of multiplicity 1.
5. If $K = \{z\colon |z| \leq 1 \text{ and } |z - 1/2| \geq 1/2\}$, show that $R(K)$ is a Dirichlet algebra. (This will be easier later in the book.)

§15 Gleason parts. We have already seen several applications of the theory of harmonic functions to the study of function algebras and especially to $R(K)$. Indeed the fully informed reader has noticed several similarities between concepts from harmonic functions and several of the topics that have been covered so far. In this section we will explore yet another analogy with harmonic functions, an abstraction of Harnack's Inequality.

Fix the following notation for this section. Let A be a function algebra on X with maximal ideal space $\mathcal{M}$ and define $\rho\colon \mathcal{M} \times \mathcal{M} \to [0, \infty]$ by

$$\rho(\alpha, \beta) = \inf\left\{c > 0\colon \frac{1}{c} \leq \frac{\operatorname{Re}\alpha(f)}{\operatorname{Re}\beta(f)} \leq c \text{ for every } f \text{ in } A \text{ with } \operatorname{Re}\hat{f} > 0\right\}$$

when such constants c exist, and $\rho(\alpha, \beta) = \infty$ otherwise. The proof of the next result is left as an exercise.

15.1 PROPOSITION. *Let A be a function algebra and let α, β, and $\gamma \in \mathcal{M}$.*

(a) $\rho(\alpha, \beta) \geq 1$ *and* $\rho(\alpha, \beta) = 1$ *if and only if* $\alpha = \beta$.
(b) $\rho(\alpha, \beta) = \rho(\beta, \alpha)$.
(c) $\rho(\alpha, \beta) \leq \rho(\alpha, \gamma)\rho(\gamma, \beta)$.

Using this proposition it is easy to see that the relation $\alpha \sim \beta$ defined by $\rho(\alpha, \beta) < \infty$ is an equivalence relation on $\mathscr{M}$.

15.2 DEFINITION. The *Gleason parts* for A are the equivalence classes of the equivalence relation just defined. A Gleason part is *trivial* if it consists of a single homomorphism; otherwise the Gleason part is said to be *nontrivial*.

The next proposition is a consequence of the definition and some rudimentary analysis. The next inequality is the generalization of Harnack's Inequality alluded to previously.

15.3 PROPOSITION. *If α and β belong to the same Gleason part for A, then for every f in A with $\operatorname{Re} \hat{f} > 0$,*

$$\frac{1}{\rho(\alpha, \beta)} \leq \frac{\operatorname{Re} \alpha(f)}{\operatorname{Re} \beta(f)} \leq \rho(\alpha, \beta).$$

To minimize the complexity of the notation, we will often consider functions in A as functions on $\mathscr{M}_A$ as well as functions on X. So in particular, if μ is a representing measure on $\mathscr{M}_A$ for some homomorphism α, we will write $\int f d\mu$ rather than $\int \hat{f} d\mu$. In the applications that are of interest to us it is always the case that $\mathscr{M}_A = X$, so that ultimately there will be no confusion.

15.4 THEOREM. *If α and β belong to the same Gleason part for A, then there exist representing measures μ and ν for α and β, respectively, such that μ and ν are supported on ∂_A and μ and ν are boundedly mutually absolutely continuous. Moreover,*

$$\frac{1}{\rho(\alpha, \beta)} \leq \frac{d\mu}{d\nu} \leq \rho(\alpha, \beta).$$

PROOF. Let $\rho = \rho(\alpha, \beta)$. If $u \in \operatorname{Re} \hat{A}$ and $u \geq 0$, then $\rho u(\beta) - u(\alpha) \geq 0$. Thus $u \to \rho u(\beta) - u(\alpha)$ is a positive linear functional on $\operatorname{Re} \hat{A}|\partial_A$ and the constant function 1 belongs to this space. Hence there is a positive measure σ supported on ∂_A such that $\rho u(\beta) - u(\alpha) = \int u d\sigma$ for all u in $\operatorname{Re} \hat{A}$ and $\|\sigma\| = \int 1 d\sigma = \rho - 1$. Similarly there is a positive measure τ on ∂_A such that $\rho u(\alpha) - u(\beta) = \int u d\tau$ for all u in $\operatorname{Re} \hat{A}$ and $\|\tau\| = \rho - 1$. Define μ and ν by the equations $\mu = (\rho^2 - 1)^{-1}(\sigma + \rho\tau)$ and $\nu = (\rho^2 - 1)^{-1}(\tau + \rho\sigma)$. The reader can verify that μ and ν have the desired properties. (The idea behind this proof is to think of the equations $\rho u(\beta) - u(\alpha) = \int u d\sigma$ and $\rho u(\alpha) - u(\beta) = \int u d\tau$ as the equations $\rho \nu - \mu = \sigma$ and $\rho \mu - \nu = \tau$ and solve for μ and ν.) □

15.5 COROLLARY. *If α and β belong to the same Gleason part for A and $\eta \in M_\alpha$, then there is a λ in M_β such that $\eta \ll \lambda$ and $d\eta/d\lambda$ is bounded.*

PROOF. Let μ and ν be as in the preceding proposition. If $0 < \varepsilon < \rho^{-1}$, then $\nu - \varepsilon\mu \geq 0$. Hence $\lambda = \nu + \varepsilon(\eta - \mu) = \varepsilon\eta + (\nu - \varepsilon\mu) \geq 0$. Also since $\eta - \mu \perp \hat{A}$, $\lambda \in M_\beta$. Finally, $\lambda \geq \varepsilon\eta$, so $\eta \ll \lambda$ and $d\eta/d\lambda \leq \varepsilon^{-1}$. □

15.6 COROLLARY. *If $a \in X$ and a is a peak point, then $\{a\}$ is a trivial Gleason part.*

PROOF. Let G be the Gleason part for A that contains a. If $\beta \in G$, then we know that there are representing measures μ and ν for a and β, respectively, that are supported on the Shilov boundary and that are boundedly mutually absolutely continuous. But a is a peak point and so it must be that $\mu = \delta_a$. Hence $\nu = \delta_a$ and so $\beta = a$. □

15.7 COROLLARY. *Let K be a compact subset of $\mathbb{C}$ and for a in $\operatorname{int} K$, let ω_a be harmonic measure for K at a. If $R(K)$ is a Dirichlet algebra and a and b belong to the same Gleason part for $R(K)$, then ω_a and ω_b are boundedly mutually absolutely continuous.*

PROOF. Let G be the Gleason part for $R(K)$ that contains a and b. Since $R(K)$ is a Dirichlet algebra, each point of ∂K is a peak point for $R(K)$. Since G is a nontrivial part, the preceding corollary implies that $G \subseteq \operatorname{int} K$. The corollary now follows by Proposition 15.5 and the fact that because $R(K)$ is a Dirichlet algebra, points in K have unique representing measures supported on ∂K. □

If α and $\beta \in \mathcal{M}$, let $\|\alpha - \beta\|$ be the norm of $\alpha - \beta$ in A^*. We know that $\|\alpha - \beta\| \leq 2$ for all α and β.

15.8 THEOREM. *If α and $\beta \in \mathcal{M}$, the following statements are equivalent.*
  (a) *α and β belong to distinct Gleason parts.*
  (b) *$\|\alpha - \beta\| = 2$.*
  (c) *If $A_\beta \equiv \{f \in A: \beta(f) = 0\}$, then $\|\alpha|A_\beta\| = 1$.*
  (d) *There is a sequence $\{f_n\}$ in A with $\|f_n\| \leq 1$ for all $n \geq 1$ such that $|\alpha(f_n)| \to 1$ and $\limsup |\beta(f_n)| < 1$.*

PROOF. (a) *implies* (b). Since $\rho(\alpha, \beta) = \infty$, the definition of $\rho(\alpha, \beta)$ implies there is a sequence $\{g_n\}$ in A such that $\operatorname{Re} g_n > 0$, $\operatorname{Re} \alpha(g_n) \to 0$, and $\operatorname{Re} \beta(g_n) \geq s > 0$ for all n. Replacing g_n by $[\operatorname{Re} \alpha(g_n)]^{-1/2} g_n$, we may assume that $\operatorname{Re} \beta(g_n) \to \infty$. Moreover, replacing this new g_n by $g_n - i \operatorname{Im} \alpha(g_n)$, we may also assume that $\alpha(g_n) \to 0$. Since $\operatorname{Re} g_n > 0$ on $\mathcal{M}_A$, $f = (g_n - 1)/(g_n + 1) \in R(K)$ and $|f_n| < 1$ on $\mathcal{M}_A$. Finally, $\alpha(f_n) \to -1$ and $\beta(f_n) \to +1$. Hence $|(\alpha - \beta)(f_n)| \to 2$.

(b) *implies* (c). According to (b) there is a sequence $\{f_n\}$ in A such that $\|f_n\| \leq 1$ and $|\alpha(f_n) - \beta(f_n)| \to 2$. Note that it must be that $|\alpha(f_n)| \to 1$. So if each f_n is replaced by a unimodular multiple of itself, we obtain that $\alpha(f_n) \to 1$. It follows that $\beta(f_n) \to -1$. Let $g_n = [f_n - \beta(f_n)][1 - \overline{\beta(f_n)} f_n]^{-1}$. Then $g_n \in A$, $\beta(g_n) = 0$, $\|g_n\| \leq 1$, and $|\alpha(g_n)| \to 1$.

(c) *implies* (d). This is clear.

(d) *implies* (a). According to (d) there is a sequence $\{f_n\}$ in A such that $\|f_n\| < 1$, $\alpha(f_n) \to 1$, and $|\beta(f_n)| \leq s < 1$ for all $n \geq 1$. Thus

$\operatorname{Re}(1 - f_n) > 0$ on $\mathscr{M}_A$, $\operatorname{Re}\alpha(1 - f_n) \to 0$, and $\operatorname{Re}\beta(1 - f_n) \geq 1 - s$. Thus $\rho(\alpha, \beta) = \infty$. □

If α and β belong to distinct Gleason parts, there is a dramatic counterpoint to Theorem 15.4.

15.9 THEOREM. *If α and β belong to distinct Gleason parts, then there is a Borel subset E of X such that every representing measure for α is carried by E and every representing measure for β is carried by $X \setminus E$.*

To facilitate the proof of this theorem, we first prove a lemma that will also be used later in this book.

15.10 LEMMA. *If $\alpha \in \mathscr{M}$ and $\{f_n\}$ is a sequence in A such that $\|f_n\| \leq 1$ for all $n \geq 1$ and $\alpha(f_n) \to 1$, then for every μ in M_α, $f_n \to 1$ weak* in $L^\infty(\mu)$. Moreover, if $\sum_n |1 - \alpha(f_n)| < \infty$ and $E = \{x \in X : \lim_n f_n(x) = 1\}$, then $\mu(E) = 1$ for all μ in M_α.*

PROOF. Fix a measure μ in M_α. Since $\|f_n\| \leq 1$ for all $n \geq 1$, $\{f_n\}$ has a weak* cluster point, f, in ball $L^\infty(\mu)$. Since $\int f_n d\mu = \alpha(f_n) \to 1$, it must be that $\int f d\mu = 1$. But $\|f\|_\infty \leq 1$ so this can only happen if $f = 1$ a.e. $[\mu]$. That is, 1 is the unique weak* cluster point of $\{f_n\}$. Hence $f_n \to 1$ weak* in $L^\infty(\mu)$.

For the second half, assume that $\sum |1 - \alpha(f_n)| < \infty$ and define E as in the lemma. Thus $\int |1 - f_n|^2 d\mu = 1 + \int |f_n|^2 d\mu - 2\operatorname{Re}\int f_n d\mu \leq 2 - 2\operatorname{Re}\alpha(f_n) \leq 2|1 - \alpha(f_n)|$. Hence $\int \sum_n |1 - f_n|^2 d\mu = \sum_n \int |1 - f_n|^2 d\mu \leq 2 \sum_n |1 - \alpha(f_n)| < \infty$. Hence $|1 - f_n| \to 0$ a.e. $[\mu]$. □

PROOF OF THEOREM 15.9. Because $\|\alpha - \beta\| = 2$, there is a sequence $\{f_n\}$ in A such that $\|f_n\| \leq 1$, $\alpha(f_n) \to 1$, and $\beta(f_n) \to -1$. By passing to a subsequence if necessary, we may assume that $\sum_n |1 - \alpha(f_n)| < \infty$ and $\sum_n |1 + \beta(f_n)| < \infty$. Let $E = \{\rho \in \mathscr{M}_A : \lim \rho(f_n) = 1\}$ and $F = \{\rho \in \mathscr{M}_A : \lim \rho(f_n) = -1\}$. The preceding lemma implies that E carries every representing measure for α and F carries every representing measure for β. □

We now focus on T-invariant algebras. For the remainder of this section, A will be a T-invariant algebra on K, a compact subset of the plane. Recall that here $\mathscr{M}_A = K$.

15.11 LEMMA. *If A is a T-invariant algebra on K, Q is a Gleason part for A, $a \in Q$, μ is a representing measure for a, and $\nu = (z - a)\mu$, then $\{z : \tilde{\nu}(z) < \infty \text{ and } \hat{\nu}(z) \neq 0\} \subseteq Q$.*

PROOF. Note that for f in A, $\int f d\nu = \int (z - a) f d\mu = 0$. Thus if $\tilde{\nu}(z) < \infty$ and $\hat{\nu}(z) \neq 0$, then Corollary 8.10 implies there is a representing measure η for A at z such that $\eta \ll \nu$. Since μ cannot have an atom at a, $\eta \ll \mu$. By Theorem 15.9, z must belong to the same Gleason part as a. □

15.12 Proposition. *Suppose A is a T-invariant algebra on K and Q is a Gleason part for A. If $a \in Q$ and μ is a representing measure for a, then $\operatorname{supp}(\mu) \subseteq \operatorname{cl} Q$. Also if $f \in R(K, Q)$, then $f(a) = \int f \, d\mu$.*

Proof. If $\mu = \delta_a$, there is nothing to prove. If $\mu \neq \delta_a$, then $\mu = \eta + \alpha \delta_a$, where $0 \leq \alpha < 1$ and $\eta(\{a\}) = 0$. It is easy to see that $(1-\alpha)^{-1}\eta \in M_a$. Thus it suffices to assume that $\mu(\{a\}) = 0$.

Let $\nu = (z-a)\mu$; by assumption $\nu \neq 0$. Also the preceding lemma implies that $B \equiv \{z \colon \tilde{\nu}(z) < \infty$ and $\hat{\nu}(z) \neq 0\} \subseteq Q$. Hence $\hat{\nu} = 0$ a.e. [Area] off Q. Thus $|\nu|(\mathbb{C} \setminus \operatorname{cl} Q) = 0$ (3.4), from which it follows that μ is supported by $\operatorname{cl} Q$. Also by Proposition 3.14, $\nu \perp R(K, Q)$. Since $\tilde{\nu}(a) = \hat{\nu}(a) = 1$, it is easy to check that $\mu = (z-a)^{-1}\nu$ is a representing measure for $R(K, Q)$ at a. □

15.13 Corollary. *The trivial Gleason parts of a T-invariant algebra are precisely the singleton sets consisting of a peak point.*

Proof. We already know that for a peak point a, $\{a\}$ is a Gleason part (15.6). Conversely, if $\{a\}$ is a Gleason part, then the preceding proposition says that every representing measure for a is supported on $\{a\}$ so that δ_a is the only possibility. □

15.14 Corollary. *If A is a T-invariant algebra and Q is a nontrivial Gleason part, then Q has positive area.*

Proof. Let $a \in Q$ and let $\mu \in M_a$. As in the proof of (15.12), we may assume that $\mu(\{a\}) = 0$. Therefore $\nu = (z-a)\mu \neq 0$ and $\nu \perp A$. By Lemma 15.11, $B = \{z \colon \tilde{\nu}(z) < \infty$ and $\hat{\nu}(z) \neq 0\} \subseteq Q$. But because $\nu \neq 0$, $\operatorname{Area}(B) > 0$ (3.4) □

15.15 Corollary. *A T-invariant algebra has at most a countable number of nontrivial Gleason parts.*

We can now combine this information about T-invariant algebras with Theorem 15.9 to get an important structure theorem.

15.16 Theorem. *If A is a T-invariant algebra and $Q_1, Q_2, \ldots$ are the nontrivial Gleason parts, then there are pairwise disjoint Borel sets $E_1, E_2, \ldots$ such that $Q_n \subseteq E_n \subseteq \operatorname{cl} Q_n$ for every $n \geq 1$ and if μ is a representing measure for a point a in Q_n, then μ is carried by E_n.*

Proof. Temporarily fix $n \geq 1$. For each integer $k \neq n$, there is a Borel set F_n^k such that if $a \in Q_n$, $b \in Q_k$, $\mu \in M_a$, $\nu \in M_b$, then $\mu(F_n^k) = 0$ and $\nu(F_n^k) = 1$. Put $F_n = K \setminus \bigcup_{k \neq n} F_n^k$. So for μ and ν as described, $\mu(F_n) = 1$ and $\nu(F_n) = 0$. Let $E_n = F_n \setminus \bigcup_{k \neq n} F_k$. Clearly the sets $\{E_n\}$ are pairwise disjoint Borel sets. If $a \in Q_n$ and $\mu \in M_a$, then $\mu(E_n) = 1$ since $\mu(F_n) = 1$ and $\mu(F_k) = 0$ for all $k \neq n$. Since δ_a is one possible choice for μ, $Q_n \subseteq E_n$. By (15.12) we can replace this E_n by $E_n \cap (\operatorname{cl} Q_n)$ and all the properties desired for the sequence of sets are obtained. □

With the notation of the preceding theorem, the sets E_n are called the *carriers* of the Gleason parts.

15.17 Proposition. *If K is a compact subset of $\mathbb{C}$, Q is a nontrivial Gleason part for $R(K)$, and Q contains an interior point a of K, then Q contains the component of $\operatorname{int} K$ that contains a.*

Proof. Let G be the component of $\operatorname{int} K$ that contains a. If $b \in G$, then ω_a and ω_b, the harmonic measures for the two points, are mutually absolutely continuous. By Theorem 15.9, $b \in Q$. □

15.18 Corollary. *If K is finitely connected and $\operatorname{int} K$ is connected, then $\operatorname{int} K$ is the only nontrivial Gleason part for $R(K)$.*

Proof. By Corollary 13.6 the peak points for $R(K)$ are the points on ∂K. Thus the nontrivial Gleason parts must be contained in $\operatorname{int} K$. But the preceding proposition and the hypothesis imply that $\operatorname{int} K$ is contained in a single Gleason part. □

Later we will see that for a finitely connected set K the nontrivial Gleason parts are precisely the components of $\operatorname{int} K$.

15.19 Example. A *string of beads* is a compact set

$$K = [\operatorname{cl} \mathbb{D}] \setminus \bigcup_{n=1}^{\infty} \Delta_n,$$

where $\{\Delta_n\}$ is a sequence of open disks in $\mathbb{D}$ having the following properties:

(i) $\operatorname{cl} \Delta_n \cap \operatorname{cl} \Delta_m = \varnothing$ for $n \neq m$;
(ii) the center of each Δ_n lies on the interval $[-1, 1]$;
(iii) $[-1, 1] \setminus \bigcup_n \Delta_n$ contains no interval;
(iv) $[-1, 1] \cap K$ has positive one-dimensional Lebesgue measure.

The details of the construction of this set are left to the reader.

Notice that $\operatorname{int} K$ has two components, each of which is simply connected; denote these components by Q_+ and Q_-. Let $a_\pm \in Q_\pm$ and let $\omega_\pm =$ harmonic measure for K at the point $a_\pm$. Since $\partial Q_\pm$ is a rectifiable Jordan curve (Why?), $\omega_\pm$ and arc length measure on $\partial Q_\pm$ are mutually absolutely continuous (9.12). Since $\omega_\pm \in M_{a_\pm}$ and ω_+ and ω_- are not mutually singular, a_+ and a_- cannot belong to distinct Gleason parts (15.9). Thus Q_+ and Q_- are contained in the same Gleason part. That is, it is possible for distinct components of $\operatorname{int} K$ to be included in the same Gleason part.

§16 The Wermer Embedding Theorem. It is a surprising phenomenon that under relatively mild assumptions a function algebra exhibits not only properties analogous to algebras of analytic functions, but in fact possesses actual analytic structure. The precise meaning of this statement is connected with the following concept and will become clearer as we progress.

16.1 Definition. If A is a function algebra, a subset Δ of its maximal ideal space is called an *analytic disk* if there is a continuous one-to-one map

$\tau: \mathbb{D} \to \Delta$ such that $\tau(\mathbb{D}) = \Delta$ such that for every f in A, $f \circ \tau$ is an analytic function on $\mathbb{D}$. The map τ and its inverse will sometimes be called the Riemann map.

It is clear that for the algebra $R(K)$ analytic disks abound. The following theorem is, however, far from obvious.

16.2 WERMER EMBEDDING THEOREM (Wermer [1960]). *If A is a Dirichlet algebra, then every nontrivial Gleason part is an analytic disk.*

Two lemmas are needed for the proof which are of independent interest. But first let's set some notation that will remain in force throughout this section. As usual, A will denote a function algebra on X and we will always assume that A is a Dirichlet algebra. Since A is a Dirichlet algebra, for each α in $\mathcal{M}_A$ there is a unique representing measure for α that is supported on ∂_A, the Shilov boundary of A. This unique measure will be denoted by ω_α.

Note that if α and β belong to the same Gleason part for A and $\rho = \rho(\alpha, \beta)$ (as in §15), then Theorem 15.4 implies that ω_α and ω_β are boundedly mutually absolutely continuous and $\rho^{-1} \leq d\omega_\alpha/d\omega_\beta \leq \rho$. This implies that the identity mapping $L^2(\omega_\alpha) \to L^2(\omega_\beta)$ is a bounded bijection. Thus the identity map between $L^2(\omega_\alpha)$ and $L^2(\omega_\beta)$ is a Banach space isomorphism, though it is not an isometry for $\alpha \neq \beta$. On the other hand, the identity map between $L^\infty(\omega_\alpha)$ and $L^\infty(\omega_\beta)$ is an isometric isomorphism and weak* homeomorphism. With these observations set, we will readily and frequently talk of functions as simultaneously belonging to the Lebesgue spaces of these two measures.

16.3 LEMMA. *Suppose A is a Dirichlet algebra, $\alpha \in \mathcal{M}_A$, and α is not a peak point. If $\mathcal{H}$ is the closure of A in $L^2(\omega_\alpha)$ and $\mathcal{H}_0$ is the closure of $\ker \alpha$, then $\mathcal{H} \cap \mathcal{H}_0^\perp = $ the constant functions and there is a Z in $\mathcal{H}_0$ such that $|Z| = 1$ a.e. $[\omega_\alpha]$ and $\mathcal{H}_0 = Z\mathcal{H}$.*

PROOF. If $f \in \ker \alpha$, then $\langle 1, f \rangle = \int f \, d\omega_\alpha = 0$. Hence $1 \in \mathcal{H} \cap \mathcal{H}_0^\perp$. Since the closed linear span of $\mathcal{H}_0$ and the constant functions is easily seen to be all of $\mathcal{H}$, we have that $\mathcal{H} \cap \mathcal{H}_0^\perp = \mathbb{C}$.

Let β be a point in the same Gleason part as α, $\beta \neq \alpha$. From the remarks preceding this lemma, $\mathcal{H}$ and $\mathcal{H}_0$ are both closed subspaces of $L^2(\omega_\beta)$.

Put $c = \|\beta|\ker \alpha\|$; so $0 < c < 1$ since α and β belong to the same Gleason part. Let μ be a measure on ∂_A such that $\|\mu\| = c$ and $\beta(f) = \int f \, d\mu$ for all f in $\ker \alpha$ and let $\{f_n\} \subseteq \ker \alpha$ such that $\|f_n\| < 1$ and $\beta(f_n) \to c$. By passing to a subsequence, we may assume that there is a function Z in $L^\infty(|\mu|)$ such that $f_n \to Z$ weak*; thus $|Z| \leq 1$ a.e. $[|\mu|]$. On the other hand, $\int Z \, d\mu = \lim \int f_n \, d\mu = \lim \beta(f_n) = c = \|\mu\|$. Hence $Z\mu = |\mu|$ and $|Z| = 1$ a.e. $[|\mu|]$.

CLAIM. $\omega_\beta = c^{-1} Z \mu$.

§16 THE WERMER EMBEDDING THEOREM

In fact, if $g \in \ker \beta$, then $\int g \, d|\mu| = \int gZ \, d\mu = \lim \int g f_n \, d\mu = \lim \beta(g f_n) = 0$. Hence $c^{-1} Z \mu = c^{-1} |\mu|$ is a probability measure that is orthogonal to $\ker \beta$ and therefore must be a representing measure for β (8.4). But μ is supported by ∂_A and so the fact that A is a Dirichlet algebra implies such representing measures are unique, whence the claim.

In light of the claim, we have that $|Z| = 1$ a.e. $[\omega_\beta]$. But $[\omega_\beta] = [\omega_\alpha]$ since α and β belong to the same Gleason part, so $Z\mathscr{H}$ is a closed subspace of $L^2(\omega_\alpha)$. But the fact that $f_n \to Z$ weak* in $L^\infty(|\mu|)$ implies that $f_n \to Z$ weak* in $L^\infty(\omega_\alpha)$. It follows that $f_n \to Z$ weakly in $L^2(\omega_\alpha)$. Thus $Z \in \mathscr{H}_0$ and it is easy to show that $Z\mathscr{H} \subseteq \mathscr{H}_0$.

To show that $Z\mathscr{H} = \mathscr{H}_0$, let $g \in \mathscr{H}_0$ such that $g \perp Z\mathscr{H}$ in $L^2(\omega_\beta)$. Thus

$$(16.4) \qquad \int f Z \bar{g} \, d\omega_\beta = 0 \quad \text{for all } f \text{ in } A.$$

Now fix a function f in $\ker \beta$. We have, in light of the claim, $\int g f \bar{Z} \, d\omega_\beta = c^{-1} \int g f \, d\mu$. Let $\{g_n\}$ be a sequence in $\ker \alpha$ such that $\int |g - g_n|^2 \, d\omega_\alpha \to 0$; this also gives that $\int g f \, d\mu = \lim \int g_n f \, d\mu = \lim \beta(g_n f) = 0$, since $f \in \ker \beta$. Hence $\int g f \bar{Z} \, d\omega_\beta = 0$ for all f in $\ker \beta$. Combining this with (16.4) we get that for all functions f in the algebra A,

$$\int \bar{g} Z (\operatorname{Re} f) \, d\omega_\beta = \frac{1}{2} \int \bar{g} Z \bar{f} \, d\omega_\beta$$
$$= \frac{1}{2} \int \bar{g} Z (\bar{f} - \overline{\beta(f)}) \, d\omega_\beta + \frac{\overline{\beta(f)}}{2} \int \bar{g} Z \, d\omega_\beta$$
$$= 0.$$

But since A is a Dirichlet algebra, $\{\operatorname{Re} f : f \in A\}$ is dense in $C_\mathbb{R}(\partial_A)$. Thus $\bar{g} Z = 0$ a.e. $[\omega_\beta]$. But since $|Z| = 1$ a.e. $[\omega_\beta]$, it must be that $g = 0$. □

16.5 LEMMA. *If α, $\mathscr{H}$, and Z are as in the preceding lemma, then for $|\lambda| < 1$, $(Z - \lambda)\mathscr{H}$ is a closed subspace of $\mathscr{H}$ and*

$$[(1 - \bar{\lambda} Z)^{-1}]^\perp = (Z - \lambda)\mathscr{H}.$$

PROOF. Since $|Z| = 1$ a.e. $[\omega_\alpha]$, multiplication by Z on $\mathscr{H}$ is an isometry; hence $(Z - \lambda)\mathscr{H}$ is a closed subspace of $\mathscr{H}$ whenever $|\lambda| < 1$. If $f \in \mathscr{H}$, then $fZ \in \mathscr{H}_0$ and so

$$\int f \left(\frac{Z - \lambda}{1 - \lambda \bar{Z}} \right) d\omega_\alpha = \int f \frac{Z}{Z} \left(\frac{Z - \lambda}{1 - \lambda \bar{Z}} \right) d\omega_\alpha$$
$$= \int f Z \, d\omega_\alpha$$
$$= 0.$$

Thus $(1 - \bar{\lambda} Z)^{-1} \perp (Z - \lambda)\mathscr{H}$.

If $g \in \mathscr{H} \cap [(Z-\lambda)\mathscr{H}]^\perp$, then for all f in $\mathscr{H}$, $0 = \int \overline{g}(Z-\lambda)f\,d\omega_\alpha = \int \overline{g}(1-\lambda\overline{Z})Zf\,d\omega_\alpha$. Thus $g(1-\overline{\lambda}Z) \perp Z\mathscr{H} = \mathscr{H}_0$. From the preceding lemma we get that $g(1-\overline{\lambda}Z)$ is a constant. □

PROOF OF THEOREM 16.2. Fix a homomorphism α on A and assume that α is not a peak point. Let Q be the Gleason part that contains α. Let ω_α, $\mathscr{H}$, $\mathscr{H}_0$, and Z be as in Lemma 16.3. We first make the observation that

$$\text{(16.6)} \qquad \int Z^m\,d\omega_\alpha = 0 \quad \text{for } m \neq 0.$$

In fact, this is easy for $m = 1$ since $1 \perp \mathscr{H}_0 = Z\mathscr{H}$. Using Exercise 3, (16.6) follows for all positive integers m. The statement also holds for negative m by taking complex conjugates and remembering that $|Z| = 1$ a.e. $[\omega_\alpha]$.

For $|\lambda| < 1$, define $\phi_\lambda: A \to \mathbb{C}$ by

$$\phi_\lambda(f) = \int f(1-\lambda\overline{Z})^{-1}\,d\omega_\alpha.$$

Since $(1-\lambda\overline{Z})^{-1}$ is a bounded function, ϕ_λ is a bounded linear functional on A. We want to show that ϕ_λ is a homomorphism on A. Now Lemma 16.5 implies that $\ker\phi_\lambda = A \cap (Z-\lambda)\mathscr{H}$. From this and the fact $\mathscr{H}$ is invariant for multiplication by functions in A, it is easy to see that $\ker\phi_\lambda$ is an ideal of A. Moreover, since $|\lambda| < 1$ and $|Z| = 1$, $(1-\lambda\overline{Z})^{-1} = \sum_0^\infty \lambda^n \overline{Z}^n$, where the convergence is in the $L^\infty(\omega_\alpha)$ norm. Therefore

$$\phi_\lambda(1) = \sum_{n=0}^\infty \lambda^n \int \overline{Z}^n\,d\omega_\alpha = 1,$$

by (16.6).

Thus $\lambda \to \phi_\lambda$ is a map from $\mathbb{D}$ into $\mathscr{M}_A$. If $f \in A$, then an examination of the definition of ϕ_λ and an application of differentiation under the integral sign shows the function $\lambda \to \phi_\lambda(f)$ to be analytic on $\mathbb{D}$. That is $\{\phi_\lambda : |\lambda| < 1\}$ is an analytic disk in $\mathscr{M}_A$. We must now show that this is the Gleason part Q.

Let $\beta = \phi_\lambda$ for some λ in $\mathbb{D}$. Now $\eta = (1-\lambda Z)^{-1}\omega_\alpha$ is a complex representing measure for β and so there is a representing measure μ for β with $\mu \ll \omega_\alpha$. By Theorem 15.9, $\beta \in Q$.

Now fix a β in Q and put $\lambda = \int Z\,d\omega_\beta$. Since $|Z| = 1$ a.e. $[\omega_\beta]$ and Z is not constant (Why?), $|\lambda| = |\int Z\,d\omega_\beta| < \int |Z|\,d\omega_\beta = 1$. We want to show that $\beta = \phi_\lambda$. In fact, if $f \in A$, then (16.5) implies there is a constant c and a function g in $\mathscr{H}$ such that $f = c(1-\overline{\lambda}Z)^{-1} + (Z-\lambda)g$. Hence, using Exercise 3,

$$\beta(f) = \int f \, d\omega_\beta$$
$$= c \int (1 - \bar{\lambda}Z)^{-1} d\omega_\beta + \int (Z - \lambda) g \, d\omega_\beta$$
$$= c \sum_{n=0}^{\infty} \bar{\lambda}^n \int Z^n d\omega_\beta + \left(\int (Z - \lambda) d\omega_\beta \right) \left(\int g \, d\omega_\beta \right)$$
$$= c \sum_{n=0}^{\infty} |\lambda|^{2n}$$
$$= c(1 - |\lambda|^2)^{-1}.$$

But we also have that

$$\phi_\lambda(f) = \int f(1 - \lambda \bar{Z})^{-1} d\omega_\alpha$$
$$= c \int \frac{1}{|1 - \lambda \bar{Z}|^2} d\omega_\alpha + \int \frac{Z - \lambda}{1 - \lambda \bar{Z}} g \, d\omega_\alpha$$
$$= c \sum_{m,n=0}^{\infty} \bar{\lambda}^n \lambda^m \int Z^n \bar{Z}^m d\omega_\alpha + \int Z g \, d\omega_\alpha$$
$$= c(1 - |\lambda|^2)^{-1}$$
$$= \beta(f).$$

This completes the proof. □

16.7 COROLLARY. *If Q, a, and Z are as in the preceding proof, then*

$$b \to \int Z \, d\omega_b$$

is the Riemann map of Q onto $\mathbb{D}$ that takes a to 0.

16.8 COROLLARY. *If $R(K)$ is a Dirichlet algebra, then the nontrivial Gleason parts of K are the components of $\operatorname{int} K$ and these components are simply connected.*

PROOF. Let Q be a nontrivial Gleason part for $R(K)$. Since $R(K)$ is a Dirichlet algebra, every point of ∂K is a peak point; thus $Q \subseteq \operatorname{int} K$. Fix a point a in Q and let G be the component of $\operatorname{int} K$ containing a. For any point b in G, ω_a and ω_b are boundedly mutually absolutely continuous; thus $G \subseteq Q$. But the Wermer Embedding Theorem implies that there is a function $\phi: \mathbb{D} \to Q$ that is bijective and has the property that for every f in $R(K)$, $f \circ \phi$ is analytic on $\mathbb{D}$. Taking $f = z$, we see that ϕ is a Riemann map. In particular, Q is connected and so $Q = G$ and the simple connectedness of Q is immediate. □

Exercises

1. Suppose $R(K)$ is a Dirichlet algebra and $a \in \text{int } K$. Define S on $R^2(K, \omega_a)$ by $Sf = zf$. Find $\sigma(S)$ and $\sigma_e(S)$.
2. Repeat Exercise 1, but this time assume that K is finitely connected.
3. Let ω_α, ω_β, $\mathscr{H}$, and Z be as in Lemma 16.3 and show that if f and g are functions in $\mathscr{H} \cap L^\infty(\omega_\alpha)$, then $fg \in \mathscr{H}$ and $\int fg\, d\omega_\beta = (\int f\, d\omega_\beta)(\int g\, d\omega_\beta)$.

§17 Bands of measures. In this section we will develop the elementary properties of bands of measures. This concept will be the basis for an abstract F. and M. Riesz Theorem which will be proved in the next section.

17.1 DEFINITION. If X is a compact space, a *band of measures* is a norm closed linear subspace $\mathscr{B}$ of $M(X)$ such that if $\mu \in \mathscr{B}$ and ν is a measure on X that is absolutely continuous with respect to μ, then $\nu \in \mathscr{B}$.

Perhaps a word to the cautious is worth uttering here. Let's agree (as all sane mathematicians do) that to say that two complex-valued measure ν and μ satisfy $\nu \ll \mu$ means that $|\nu| \ll |\mu|$. That is, $|\nu|(\Delta) = 0$ for every Borel set E with $|\mu|(\Delta) = 0$.

17.2 EXAMPLES. (a) $M(X)$ and (0) are bands. Call these the *trivial bands*.

(b) If μ is a positive measure on X, then $L^1(\mu)$ can be identified with a closed subspace of $M(X)$ by means of the Radon-Nikodym Theorem. That is, $L^1(\mu) = \{\nu \in M(X): \nu \ll \mu\}$. With this identification, $L^1(\mu)$ is a band.

(c) The collection of purely atomic measures on X is a band.

(d) The collection of completely nonatomic measures is a band.

The proof of the first result is an easy exercise.

17.3 PROPOSITION. *If $\mathscr{B}$ is a band of measures on X and $\mu \in \mathscr{B}$, then the following statements hold.*

(a) *If Δ is any Borel set, $\mu|\Delta \in \mathscr{B}$.*
(b) *If $\mu = (\mu_1 - \mu_2) + i(\mu_3 - \mu_4)$ is the Jordan decomposition of μ, then $\mu_j \in \mathscr{B}$ for $1 \leq j \leq 4$.*
(c) $|\mu| \in \mathscr{B}$.

The next theorem is a generalization of the Lebesgue Decomposition Theorem. Indeed, the proof is the same.

17.4 THEOREM. *If $\mathscr{B}$ is a band of measures on X and $\nu \in M(X)$, then $\nu = \nu_a + \nu_s$, where $\nu_a \in \mathscr{B}$ and $\nu_s \perp \mu$ for every μ in $\mathscr{B}$. The measures ν_a and ν_s are unique.*

PROOF. If $\nu \perp \mu$ for every μ in $\mathscr{B}$, then we are done. So assume the contrary. Thus there is a Borel set F such that $|\nu|(F) > 0$ and $\nu|F \ll \mu$ for some μ in $\mathscr{B}$. Note that this implies that $\nu|F \in \mathscr{B}$. Let $c = \sup\{|\nu|(F): F \text{ is a Borel subset of } X \text{ and } \nu|F \in \mathscr{B}\}$; by our assumption,

$c > 0$. It follows that there is an increasing sequence $\{F_n\}$ of Borel sets such that $\nu|F_n \in \mathscr{B}$ and $|\nu|(F_n) \to c$. If $F = F_1 \cup F_2 \cdots$, then the fact that $\mathscr{B}$ is norm closed implies $\nu|F \in \mathscr{B}$ and $|\nu|(F) = c$.

Let $\nu_a = \nu|F$ and $\nu_s = \nu - \nu_A = \nu|(X\backslash F)$. Clearly $\nu_a \in \mathscr{B}$. If E is any Borel set disjoint from F and $\nu|E \ll \mu$ for some μ in $\mathscr{B}$, then $\nu|E \in \mathscr{B}$ and $|\nu|(F \cup E) = c + |\nu|(E)$. By the definition of c, it must be that $|\nu|(E) = 0$. Therefore $\nu_s \perp \mathscr{B}$.

The proof of uniqueness is left to the reader. □

Call the above decomposition of ν the *Lebesgue decomposition* of ν with respect to $\mathscr{B}$.

For convenience, let's agree that for a nonempty subset $\mathscr{S}$ of $M(X)$, the notation $\mu \perp \mathscr{S}$ means that $\mu \perp \sigma$ for every σ in $\mathscr{S}$.

17.5 PROPOSITION. *If $\mathscr{S}$ is a nonempty subset of measures on X and $\mathscr{S}' \equiv \{\nu \in M(X): \nu \perp \mathscr{S}\}$, then $\mathscr{S}'$ is a band of measures.*

In fact this proof is an immediate consequence of the definitions. If $\mathscr{B}$ is a band of measures, then the band $\mathscr{B}'$ is called the *complementary band* to $\mathscr{B}$. Note that $\mathscr{B} \cap \mathscr{B}' = (0)$. This terminology is justified by the next proposition.

17.6 PROPOSITION. *Let $\mathscr{B}$ be a band of measures on X.*

(a) $(\mathscr{B}')' = \mathscr{B}$.
(b) $M(X) = \mathscr{B} \oplus_1 \mathscr{B}'$, *where $\oplus_1$ denotes the Banach space l^1-direct sum. (That is, $\|\nu \oplus \mu\| = \|\nu\| + \|\mu\|$.)*

PROOF. (a) From the definition we have that $\mathscr{B} \subseteq (\mathscr{B}')'$. If $\nu \in (\mathscr{B}')'$, then the preceding theorem implies that $\nu = \mu + \eta$, where $\mu \in \mathscr{B}$ and $\eta \in \mathscr{B}'$. But $(\mathscr{B}')'$ is a band, so μ and $\eta \in (\mathscr{B}')'$. But then $\eta \in \mathscr{B}' \cap (\mathscr{B}')'$ and hence $\eta = 0$. Thus $\nu = \mu \in \mathscr{B}$.

(b) This is a straightforward reformulation of Theorem 17.4. □

Note that the intersection of any nonempty collection of bands in $M(X)$ is again a band. Thus for any nonempty subset $\mathscr{S}$ of $M(X)$, define the *band generated by S* to be the intersection of all bands that contain $\mathscr{S}$. So the band generated by $\mathscr{S}$ is the smallest band containing $\mathscr{S}$.

17.7 PROPOSITION. *If $\mathscr{S}$ is a nonempty subset of $M(X)$ and $\mathscr{B} = \{\mu \in M(X): \mu \perp \mathscr{S}'\}$, then $\mathscr{B}$ is the band generated by $\mathscr{S}$.*

PROOF. Let $\mathscr{A}$ be any band containing $\mathscr{S}$. It is easy to see that $\mathscr{A}' \subseteq \mathscr{S}'$ and so $\mathscr{B} = (\mathscr{S}')' \subseteq (\mathscr{A}')' = \mathscr{A}$. □

17.8 LEMMA. *Let $\{\nu_n\}$ be a sequence in $M(X)$ such that $\sum_n \nu_n$ converges in norm to a measure ν. If $\eta \in M(X)$ and $\eta \ll \nu$, then $\eta = \sum_n \eta_n$ where the measures $\{\eta_n\}$ are pairwise singular and $\eta_n \ll \nu_n$ for every $n \geq 1$.*

PROOF. Let $\eta = \eta_1 + \sigma_1$ be the Lebesgue decomposition of η with respect to ν_1. So there is a Borel partition $\{E_1, F_1\}$ of X such that $\eta_1 = \eta|E_1$, $\sigma_1 =$

$\eta|F_1$, and $|\nu_1|(F_1) = 0$. Let $\sigma_1 = \eta_2 + \sigma_2$ be the Lebesgue decomposition of σ_1 with respect to ν_2. This produces a Borel partition $\{E_2, F_2\}$ of the set F_1 such that $\eta_1 = \eta|E_2$, $\sigma_2 = \eta|F_2$, and $|\nu_2|(F_2) = 0$. Continue and we obtain a sequence $\{E_n\}$ of pairwise disjoint Borel sets and a decreasing sequence $\{F_n\}$ of Borel sets having the following properties:

  (i) $\eta_n = \eta|E_n \ll \nu_n$;
  (ii) $E_1 \cup \cdots \cup E_n \cup F_n = X$;
  (iii) $(E_1 \cup \cdots \cup E_n) \cap F_n = \varnothing$;
  (iv) $|\nu_n|(F_n) = 0$

If $\sigma_n = \eta|F_n$, then $\{\sigma_n\}$ converges in norm to a measure σ in $M(X)$. In fact, $\sigma = \eta|F$, where $F = \bigcap_n F_n$. But $|\nu_n|(F) = 0$ for every $n \geq 1$. Thus $|\nu|(F) = 0$ and so $|\eta|(F) = 0$; that is, $\sigma = 0$ and so $\eta = \sum_n \eta_n$. □

17.9 Proposition. *If $\mathscr{S}$ is a nonempty subset of $M(X)$ and $\mathscr{B}$ is the band generated by $\mathscr{S}$, the following statements are equivalent for a measure ν in $M(X)$.*

  (a) *$\nu \in \mathscr{B}$.*
  (b) *$\nu = \sum_n \nu_n$, where this series is norm convergent, $\nu_n \perp \nu_m$ for $n \neq m$, and for each n there is a μ_n in $\mathscr{S}$ with $\nu_n \ll \mu_n$.*
  (c) *$\nu = \sum_n \nu_n$, where this series is norm convergent and for each n there is a μ_n in $\mathscr{S}$ with $\nu_n \ll \mu_n$.*

Proof. Clearly (b) implies (c) and the preceding lemma gives that (c) implies (b). Let $\mathscr{A}$ be the set of measures described in (b) (or (c)). It will be shown that $\mathscr{A}$ is a band. Once this is established, the equivalence of (a) and (b) will follow. Indeed, $\mathscr{S}$ is clearly a subset of $\mathscr{A}$ and so $\mathscr{A}' \subseteq \mathscr{S}'$. On the other hand, if $\eta \in \mathscr{S}'$ and $\nu = \sum_n \nu_n$ with $\nu_n \ll \mu_n$ for some μ_n in $\mathscr{S}$, then $\eta \perp \mu_n$ for every n and hence $\eta \perp \nu_n$ for every n. Thus $\eta \perp \nu$ and so $\eta \in \mathscr{A}'$. That is, $\mathscr{A}' = \mathscr{S}'$ and so $\mathscr{B} = (\mathscr{S}')' = (\mathscr{A}')' = \mathscr{A}$.

To show that $\mathscr{A}$ is a band we first establish that $\mathscr{A}$ is a closed subspace of $M(X)$. The fact that $\mathscr{A}$ is a linear space follows easily by using (c). Now suppose that $\{\nu^k\} \subseteq \mathscr{A}$ and $\nu^k \to \nu$. Let $\nu^k = \sum_n \nu_n^k$ where $\nu_n^k \ll \mu_n^k$ and $\mu_n^k \in \mathscr{S}$. Then $\nu = \nu^1 + \sum_k (\nu^{k+1} - \nu^k) = \sum_n \nu_n^1 + \sum_k \sum_n (\nu_n^{k+1} - \nu_n^k)$. From here the proof that ν belongs to $\mathscr{A}$ becomes a test of expository skills; this test is left to the reader.

If $\nu \in \mathscr{A}$ and $\eta \ll \nu$, then Lemma 17.8 shows that η satisfies the necessary conditions to belong to $\mathscr{A}$. Therefore $\mathscr{A}$ is a band and the proof is complete. □

Before proceeding, we must have another measure theoretic interlude. This lemma will also be used later in this book in a different context.

17.10 Lemma (Chaumat [**1974**]). *Let (X, Ω, μ) be a finite measure space and let C be a closed bounded convex subset of $L^p(\mu)$, $1 \leq p \leq \infty$. If $h \in C$ and $\varepsilon > 0$, then there is a function f in C such that $|g|\mu \ll |f|\mu$ for every g in C and $\|f - h\|_p < \varepsilon$.*

PROOF. It suffices to assume that $C \subseteq \text{ball } L^p(\mu)$. At first we will ignore the requirement that f be close to h and wait until the end to take care of that.

CLAIM 1. If f_1 and $f_2 \in C$ and $\varepsilon > 0$, then there is an α with $0 < \alpha < \varepsilon$ such that $|f_1|\mu$ and $|f_2|\mu$ are absolutely continuous with respect to $|f_1 + \alpha f_2|\mu$. To see this let $f_1\mu = kf_2\mu + h\mu$ be the Lebesgue decomposition of $f_1\mu$ with respect to $f_2\mu$. So $f_2 h = 0$ a.e. $[\mu]$. If $E_\alpha = \{x: k(x) = -\alpha\}$, then $\mu(E_\alpha) > 0$ for at most a countable number of α. Pick an α with $0 < \alpha < \varepsilon$ and $\mu(E_\alpha) = 0$. Thus $f_1 + \alpha f_2 = (\alpha + k)f_2 + h$ a.e. $[\mu]$. It follows that $|f_1|\mu$ and $|f_2|\mu$ are absolutely continuous with respect to $|f_1 + \alpha f_2|\mu$.

CLAIM 2. If $f_0, f_1, \ldots, \in C$, then there is an f in C with $|f_n|\mu \ll |f|\mu$ for every $n \geq 0$. For any function g put $S(g) = \{x: g(x) \neq 0\}$. The reason for the introduction of this set is the observation that $|g_1|\mu \ll |g_2|\mu$ if and only if $\mu(S(g_2) \setminus S(g_1)) = 0$. Let $g_0 = f_0$. There are positive numbers ε_0 and A_1 such that if $U_1 = \{x: |g_0(x)| > \varepsilon_0$ and $|f_1(x)| < A_1\}$, then $\mu(S(g_0) \setminus U_1) < 1/2$. By Claim 1 there is an α_1 with $0 < \alpha_1 < \min\{1/2, \varepsilon_0/4A_1\}$ such that if $g_1 = g_0 + \alpha_1 f_1 = f_0 + \alpha_1 f_1$, then $|f_0|\mu$ and $|f_1|\mu \ll |g_1|\mu$. Note that this last condition is equivalent to $\mu([S(f_0) \cup S(f_1)] \setminus S(g_1)) = 0$.

Continue; by induction we can choose positive constants ε_n, A_n, and α_n and functions g_n in $L^1(\mu)$ such that:

(i) $\varepsilon_{n+1} < \varepsilon_n/2$, $\alpha_n < \min\{2^{-n}, \varepsilon_{n-1}/4A_n\}$;
(ii) if $U_{n+1} = \{x: |g_n(x)| > \varepsilon_n$ and $|f_{n+1}(x)| < A_{n+1}\}$ and $g_{n+1} = g_n + \alpha_{n+1} f_{n+1}$, then
$$\mu(S(g_n) \setminus U_{n+1}) < 2^{-(n+1)} \text{ and}$$
$$\mu([S(g_n) \cup S(f_{n+1})] \setminus S(g_{n+1})) = 0.$$

Define $g = f_0 + \sum_{n \geq 1} \alpha_n f_n = \lim g_n$. (Because $\alpha_n < 2^{-n}$, this limit exists in $L^p(\mu)$.) It is left as an exercise to show that $S(g) \supseteq \bigcup_{n \geq 1} \bigcap_{k \geq n} U_k$ and for $m < n$, $\mu(S(g_m) \setminus \bigcap_{k \geq n+1} U_k) < 2^{-n}$. From these relations it follows that $\mu(S(g_m) \setminus S(g)) = 0$ and hence $|f_m|\mu \ll |g|\mu$. Put $\alpha_0 = 1$; if $f = (\sum \alpha_n)^{-1} \sum \alpha_n f_n = (\sum \alpha_n)^{-1} g$, then $f \in C$ and Claim 2 is established.

Now let $\gamma = \sup\{\mu(S(f)): f \in C\}$ and choose $\{f_n\}$ contained in C such that $\mu(S(f_n)) \to \gamma$. Let $f \in C$ that f satisfies Claim 2 for this sequence. If $g \in C$, then Claim 1 implies there is a g_1 in C such that $\mu([S(f) \cup S(g)] \setminus S(g_1)) = 0$ and g_1 is a convex combination of f and g. Hence $S(g_1) = S(f) \cup S(g)$. Also, $\mu(S(f)) = \gamma \geq \mu(S(g_1)) = \mu(S(f)) + \mu(S(g) \setminus S(f)) \geq \mu(S(f))$. So $\mu(S(g) \setminus S(f)) = 0$ and hence $|g|\mu \ll |f|\mu$.

Now to adjust f so that it is close to h. In fact, Claim 1 implies that there is an α, $0 < \alpha < \varepsilon/2$, such that $|f|\mu$ and $|h|\mu \ll |h + \alpha f|\mu$. Put $f_1 = (1+\alpha)^{-1}(h + \alpha f)$. So $f_1 \in C$ and hence $[|f|\mu] = [|f_1|\mu]$ so that $|g|\mu \ll |f_1|\mu$ for all g in C. Also, $\|f_1 - h\|_p = \alpha(1+\alpha)^{-1}\|f - h\|_p < 2\alpha < \varepsilon$. □

Also see Theorem 5 in Helson **[1983]** for a result related to the preceding lemma.

17.11 PROPOSITION. *If $\mathscr{S}$ is a closed convex set of measures and $\mathscr{B}$ is the band generated by $\mathscr{S}$, then $\nu \in \mathscr{B}$ if and only if there is a η in $\mathscr{S}$ such that $\nu \ll \eta$.*

PROOF. It suffices to show that if $\nu \in \mathscr{B}$, then $\nu \ll \eta$ for some η in $\mathscr{S}$. By Proposition 17.9, $\nu = \sum_n \nu_n$ where $\nu_n \ll \mu_n$ for some $\mu_n \in \mathscr{S}$ for each n. Let $\mu = \sum_n |\mu_n|$ and put $C = \{f \in L^1(\mu) : f\mu \in \mathscr{S} \text{ and } \|f\mu\| \leq 1\}$. It is easy to deduce from the hypothesis that C is a closed bounded convex subset of $L^1(\mu)$. Moreover, $\mu_n = f_n \mu \in C$ for every $n \geq 1$. Thus Lemma 17.10 implies that there is an f in C such that $\mu_n \ll f\mu \equiv \eta$ in $\mathscr{S}$. Clearly $\nu \ll \eta$. □

17.12 PROPOSITION. *If M is a weak* closed convex set of probability measures on X and if $\nu \perp M$, then there is a Borel set E such that E carries ν and $|\mu|(E) = 0$ for every μ in M.*

The proof of this proposition requires a result from general functional analysis. For a proof see Gamelin **[1969]** page 40.

17.13 THE MINIMAX THEOREM. *Let $\mathscr{V}$ be a vector space over $\mathbb{R}$ and let $\mathscr{X}$ be a real topological vector space. If C is a convex subset of $\mathscr{V}$, M is a compact convex subset of $\mathscr{X}$, and $F: C \times M \to \mathbb{R}$ is a function such that:*

(a) *$\inf\{F(c, m) : c \in C \text{ and } m \in M\} > -\infty$;*
(b) *for every m in M, $c \to F(c, m)$ is a convex function;*
(c) *for every c in C, $m \to F(c, m)$ is a continuous concave function;*

then
$$\sup_m \inf_c F(c, m) = \inf_c \sup_m F(c, m).$$

PROOF OF PROPOSITION 17.12. Let $C = \{u \in C_\mathbb{R}(X) : 0 < u < 1\}$ and let M be as in the statement of the proposition. Define $F: C \times M \to \mathbb{R}$ by $F(u, m) = \int u\, dm + \int (1-u)\, d|\nu|$. It is routine to check that F is convex linear in each variable and $F \geq 0$. Thus the Minimax Theorem applies. Since $\nu \perp m$ for every m in M, $\inf_u F(u, m) = 0$. Hence $\inf_u \sup_m F(u, m) = 0$. This says that there is a sequence $\{u_n\} \subseteq C_\mathbb{R}(X)$ with $0 < u_n < 1$ such that

$$\sup_{m \in M} \left\{ \int u_n\, dm + \int (1 - u_n)\, d|\nu| \right\} < n^{-2}.$$

Let $E = \{x : u_n(x) \to 1\}$. If $m \in M$, then $n^{-2} \geq \int_E u_n\, dm + \int_E (1 - u_n)\, d|\nu| \to m(E)$. Hence $m(E) = 0$ for every m in M. But $\sum_n \int (1 - u_n)\, d|\nu| \leq \sum_n n^{-2}$ and so $u_n \to 1$ a.e. $[|\nu|]$. Thus $\nu = \nu|E$. □

This concludes this introduction to the theory of bands. Now let's prove a result about subnormal operators that is a direct consequence of Lemma 17.10 and is of use in the theory of subnormal operators. Recall [ACFA] page 288, that if N is a normal operator on $\mathscr{K}$, a vector f in $\mathscr{K}$ is a *separating vector* for N if the only operator A in $W^*(N)$ that satisfies $Af = 0$ is $A = 0$. If $N = \int z\, dE$, then f is a separating vector if and only if $\langle E(\cdot)f, f \rangle$ is a scalar-valued spectral measure for N.

17.14 PROPOSITION. *If S is a subnormal operator on $\mathscr{H}$ with minimal normal extension N acting on $\mathscr{K}$, $h \in \mathscr{H}$, and $\varepsilon > 0$, then there is a vector f in $\mathscr{H}$ that is separating for N and satisfies $\|f - h\| < \varepsilon$.*

PROOF. As in the proof of (17.10), we initially ignore the requirement that f be found close to h. Let $N = \int z\, dE$ be the spectral decomposition of N and let e be a separating vector for N in $\mathscr{K}$ with $\|e\| = 1$. Put $\mu(\Delta) = \langle E(\Delta)e, e \rangle$ and for each f in $\mathscr{H}$ let $\mu_f(\Delta) = \langle E(\Delta)f, e \rangle$. So $C = \{\mu_f : f \in \text{ball}\,\mathscr{H}\}$ is a bounded convex subset of ball $L^1(\mu)$. Moreover, $f \to \mu_f$ is a contractive linear map of $\mathscr{H}$ into $L^1(\mu)$ and is thus weakly continuous. Therefore C is weakly compact and thus norm closed. By Lemma 17.10, there is an f in ball $\mathscr{H}$ such that $\mu_g \ll |\mu_f|$ for every g in $\mathscr{H}$. We claim that f is a separating vector for N. In fact, suppose Δ is a Borel set such that $E(\Delta)f = 0$; we want to show that $E(\Delta) = 0$. If $\Delta_1 \subseteq \Delta$, then $|\mu_f|(\Delta_1) = 0$ and so $\mu_g(\Delta_1) = 0$ for every vector g in $\mathscr{H}$. Hence $|\mu_g|(\Delta) = 0$ for every g in $\mathscr{H}$. If n and k are non-negative integers, then $0 = \int_\Delta z^n \bar{z}^k\, d\mu_g = \langle N^n N^{*k} g, e \rangle = \langle N^n N^{*k} g, E(\Delta)e \rangle$. But $\bigvee \{N^n N^{*k} g : g \in \mathscr{H}$ and $n, k \geq 0\} = \mathscr{K}$. Thus it must be that $E(\Delta)e = 0$. Since e is a separating vector for N, $E(\Delta) = 0$.

To get a separating vector for N that is close to the given vector h, we proceed as in the proof of Claim 1 of the proof of (17.10). Since $\mu_h \ll \mu_f$, there is a Borel function ϕ such that $\mu_h = \phi \mu_f$. Now select α with $0 < \alpha < \varepsilon$ such that the function $\phi + \alpha \neq 0$ a.e. $[\mu_f]$; put $f_1 = h + \alpha f$. Now $\mu_{f_1} = (\phi + \alpha)\mu_f$ and $|\mu_{f_1}|$ and $|\mu_f|$ are mutually absolutely continuous. Thus f_1 is a separating vector for N and $\|f_1 - h\| \leq \alpha < \varepsilon$. □

Actually, a fact is contained in the last paragraph of the preceding proof that is worth recording because it will be used later in this book.

17.15 COROLLARY. *If f is a separating vector for N and h is any vector, then for all but a countable number of scalars α, $h + \alpha f$ is a separating vector for N.*

Exercises

1. Show that there is a band $\mathscr{B}$ and a measure ν in $\mathscr{B}'$ such that no Borel set E exists with the property that $\nu = \nu|E$ and $|\mu|(E) = 0$ for every μ in $\mathscr{B}$.
2. Let $\mathscr{B}$ be a band and define $P: M(X) \to M(X)$ by $P\nu = \nu_a$ as in Theorem 17.4. Show that P is a linear idempotent with $\|P\| = 1$, ran $P = \mathscr{B}$, and ker $P = \mathscr{B}'$.
3. Show that the map Θ in Theorem II.12.6 is injective.

§18 Annihilating measures. In this section we will study measures that annihilate a function algebra. In particular, we will prove an abstract version of the F. and M. Riesz Theorem, for which the central concept is the following.

18.1 DEFINITION. If A is a function algebra on X and $\mathscr{B}$ is a band of measures on X, then $\mathscr{B}$ is a *reducing band* (for A) if for every μ in $A^\perp$ with Lebesgue decomposition $\mu = \mu_a + \mu_s$, μ_a in $\mathscr{B}$ and μ_s in $\mathscr{B}'$, it follows that μ_a and μ_s both belong to $A^\perp$.

18.2 THE ABSTRACT F. AND M. RIESZ THEOREM. *If A is a function algebra and $\rho \in \mathscr{M}_A$, then the band generated by the representing measures for ρ is a reducing band.*

We won't try to trace the history of the abstract F. and M. Riesz Theorem, but it is a result that passed through several evolutionary stages before reaching the above form. The above formulation is due to Koenig and Seever [**1969**], which was strongly influenced by Glicksberg [**1967**]. For more on reducing bands, the reader can see Cole and Gamelin [**1982 and 1985**] and Gamelin [**1973**].

The classical F. and M. Riesz Theorem can be deduced from this abstract version, once an additional result about T-invariant algebras is proved. This will be done after (18.5) below.

PROOF OF THEOREM 18.2. Let $\rho \in \mathscr{M}_A$ and let $\mathscr{B}$ be the band generated by M_ρ. Let $\mu \in A^\perp$ and let $\mu = \mu_a + \mu_s$ be the Lebesgue decomposition of μ with respect to $\mathscr{B}$. By (17.11), $\nu \in \mathscr{B}$ if and only if there is an η in M_ρ such that $\nu \ll \eta$. Thus Proposition 17.12 implies there is a Borel subset E of X such that $|\nu|(E) = 0$ for all ν in $\mathscr{B}$ and E carries μ_s. Without loss of generality it may be assumed that $E = \bigcup_n E_n$, where each E_n is compact and $E_n \subseteq E_{n+1}$.

By Lemma 11.6 we get that for every $n \geq 1$ there is a function f_n in A such that $\operatorname{Re} f_n > n\chi_{E_n}$ and $0 < \rho(f_n) < \delta_n$, where the numbers δ_n will be specified later. (Actually, the function f_n obtained in Lemma 11.6 must be replaced by $f_n - i \operatorname{Im} \rho(f_n)$.) Let $g_n = e^{-f_n}$. So $g_n \in A$, $\|g_n\| \leq 1$, $|g_n| < e^{-n}$ on E_n, and $|1 - \rho(g_n)| < 1/n^2$ if the δ_n are chosen appropriately. Since $\sum_n |1 - \rho(g_n)| < \infty$, Lemma 15.10 implies $g_n \to 1$ a.e. $[\mu_a]$. However, $|g_n| < e^{-n}$ on E_n and so $g_n \to 0$ a.e. $[\mu_s]$. Therefore, $g_n \mu \to \mu_a$ weak* in $M(X)$. So if $f \in A$, $\int f d\mu_a = \lim_n \int f g_n d\mu = 0$, since $\mu \perp A$ and $fg_n \in A$. Thus $\mu_a \in A^\perp$. □

Before giving some corollaries of this theorem and further information about annihilating measures, let's take a little time to rephrase some of our previous results in terms of bands.

Let Q be a nontrivial Gleason part for the algebra A and let $\rho \in Q$. From Corollary 3.5 we have that the band generated by M_ρ is the same as the band generated by the representing measures for any other element of Q.

Define the *band generated by* Q to be the band of measures on X generated by M_ρ for any ρ in Q. Denote this band by $\mathscr{B}_Q$.

Let $\{Q_i\}$ be the nontrivial Gleason parts for A. (In general, there may be an uncountable number of these parts, but not, of course, for T-invariant algebras.) According to Theorem 15.9, $\mathscr{B}_{Q_i} \subseteq \mathscr{B}'_{Q_j}$ for $Q_i \neq Q_j$. We therefore arrive at the following result whose proof is straightforward.

18.3 PROPOSITION. *If $\{Q_i\}$ are the nontrivial Gleason parts of the function algebra A on X and $\mathscr{B}_{Q_i}$ is the band of measures generated by Q_i, then*

$$M(X) = \mathscr{S} \oplus \bigoplus_i \mathscr{B}_{Q_i},$$

where $\mathscr{S}$ is the band $\bigcap_i \mathscr{B}'_{Q_j}$ consisting of the measures that are singular to every representing measure for every homomorphism belonging to a nontrivial Gleason part, and the direct sum is an l^1 direct sum. Thus every measure ν in $M(X)$ can be written as $\nu = \nu_0 + \sum_i \nu_i$, where $\nu_0 \in \mathscr{S}$, $\nu_i \in \mathscr{B}_{Q_i}$ for each i, and $\|\nu\| = \|\nu_0\| + \sum_i \|\nu_i\|$.

Call the band $\mathscr{S}$ that appears in the preceding proposition the *singular band* for A. Note that $\mathscr{S}$ at least contains the point masses δ_a for each peak point a for A. The presence of this singular band is something of a nuisance, but, as we shall see, for T-invariant algebras and measures in the annihilator of A, it can be ignored. But first an explicit combination of Proposition 18.3 and the abstract F. and M. Riesz Theorem.

18.4 COROLLARY. *If $\mu \in A^\perp$, then $\mu = \mu_0 + \sum_i \mu_i$, where for all i, $\mu_0 \perp \mathscr{B}_{Q_i}$ and $\mu_i \in A^\perp \cap \mathscr{B}_{Q_i}$, $\|\mu\| = \|\mu_0\| + \sum_i \|\mu_i\|$, and $\mu_i \perp \mu_j$ for $i \neq j$.*

Now focus your attention on T-invariant algebras.

18.5 WILKIN'S THEOREM. *If A is a T-invariant algebra on K, then there is no nonzero annihilating measure that belongs to the singular band for A.*

PROOF. Suppose $\nu \in A^\perp$ and $\nu \perp M_a$ for every nonpeak point a in K; it must be shown that $\nu = 0$. But since $R(K) \subseteq A$, $\hat{\nu}$, the Cauchy transform of ν, vanishes off K. If $a \in K$ such that $\tilde{\nu}(a) < \infty$ and $\hat{\nu}(a) \neq 0$, then a is not a peak point and there is a μ in M_a such that $\mu \ll \nu$ (8.10). So it must be that $\hat{\nu} = 0$ a.e. [Area] and hence $\nu = 0$. $\square$

We can now derive the classical F. and M. Riesz Theorem as a consequence of the results of this section. Let A be the disk algebra, the uniform closure of the polynomials in $C(\partial \mathbb{D})$. So $\mathscr{M}_A = \operatorname{cl} \mathbb{D}$. If ρ is evaluation at 0 and M_ρ = the representing measures for ρ that are supported on $\partial \mathbb{D}$, then $M_\rho = \{m\}$. Thus the band generated by M_ρ is precisely $L^1(m) = L^1$. If μ is a measure on $\partial \mathbb{D}$ that annihilates A, then Theorem 18.2 implies that if $\mu = \mu_a + \mu_s$ is the Lebesgue decomposition of μ with respect to m, then μ_a and $\mu_s \in A^\perp$. On the other hand, Wilkin's Theorem implies that $\mu_s = 0$. Thus $\mu = \mu_a$ and is thus absolutely continuous.

18.6 COROLLARY. *If A is a T-invariant algebra on K and Q is the set of nonpeak points for A, then $\mathscr{B}_Q$, the band generated by the representing measures for points in Q, is the same as the band generated by the set of annihilating measures of A.*

PROOF. Let $\mathscr{C}$ be the band generated by the annihilating measures of A. By Wilkins's Theorem, $\mathscr{C} \subseteq \mathscr{B}_Q$. On the other hand, if $a \in Q$, then there is a representing measure μ for a such that $\mu \neq \delta_a$. Thus $\mu - \delta_a \perp A$ and both μ and δ_a are absolutely continuous with respect to $\mu - \delta_a$. Hence $\mathscr{C}$ contains all representing measures for nonpeak points and so $\mathscr{B}_Q \subseteq \mathscr{C}$. □

In light of Wilkin's Theorem, Proposition 18.3 can be combined with Theorem 15.16 to produce a good structure theorem for the annihilator of a T-invariant algebra.

18.7 THEOREM. *Let A be a T-invariant algebra on K with nontrivial Gleason parts $Q_1, Q_2, \ldots$ and carriers $E_1, E_2, \ldots$. If ν is an annihilating measure of A, then $\nu = \sum_n \nu|E_n$, $\nu|E_n \in A^\perp$, and for each $n \geq 1$ and for every choice of a_n from Q_n, there is a μ_n in M_{a_n} such that $\nu|E_n \ll \mu_n$ (and thus $\nu|E_n \in \mathscr{B}_{Q_n}$). Also the Cauchy transform of $\nu|E_n$ is the function $\chi_{Q_n} \hat{\nu}$ and $\nu_n \perp R(K, Q_n)$.*

PROOF. The carriers E_n exist from Theorem 15.16 and for each choice of a_n in Q_n, every representing measure for a_n is carried by E_n. Therefore, by Proposition 17.11, every measure in $\mathscr{B}_{Q_n}$ is carried by E_n. Let ν_n be the projection of ν into $\mathscr{B}_{Q_n}$. Proposition 18.3 and Wilkin's Theorem imply that $\nu = \sum_n \nu_n$ and $\nu_n \in A^\perp$. Since the sets $\{E_n\}$ are pairwise disjoint, it must be that $\nu_n = \nu|E_n$. This establishes the first part of the theorem.

To see that $\hat{\nu}_n = \chi_{Q_n} \hat{\nu}$, observe that if w is a point with $\infty > \tilde{\nu}(w) = \sum_n \int |z-w|^{-1} d|\nu_n|(z)$, then $\hat{\nu}(w) = \sum_n \hat{\nu}_n(w)$. But $\nu_n \in \mathscr{B}_{Q_n}$ and so Corollary 8.10 implies that $\{w : \tilde{\nu}_n(w) < \infty \text{ and } \hat{\nu}_n(w) \neq 0\} \subseteq Q_n$. Hence $\hat{\nu}_n(w) = 0$ a.e. [Area] off Q_n. Thus $\hat{\nu}_n = \chi_{Q_n} \hat{\nu}$. Finally, since $\hat{\nu}_n = 0$ a.e. off Q_n, $\nu_n \perp R(K, Q_n)$ by Proposition 3.14. □

The following corollary is a shorthand formulation of the preceding theorem though, like all shorthand versions, it contains less information.

18.8 COROLLARY. *If A is a T-invariant algebra on K with nontrivial Gleason parts $Q_1, Q_2, \ldots$, then*

$$A^\perp = \bigoplus_i [A^\perp \cap \mathscr{B}_{Q_i}].$$

The next result generalizes the classical F. and M. Riesz Theorem.

18.9 COROLLARY. *If $R(K)$ is a Dirichlet algebra and ν is an annihilating measure of $R(K)$, then $\nu|\partial K$ is absolutely continuous with respect to harmonic measure for K.*

PROOF. Adopt the notation of Theorem 18.7 for the T-invariant algebra $R(K)$. So $\nu_n \ll \mu_n$ for some representing measure μ_n of a_n in Q_n. It follows that $\hat{\mu}_n$, the sweep of μ_n, is also a representing measure for a_n that is concentrated on ∂K. Since $R(K)$ is a Dirichlet algebra, $\hat{\mu}_n = \omega_n$, harmonic measure for K at a_n. But $\hat{\mu}_n = \mu_n|\partial K + (\mu|\operatorname{int} K)\hat{}$ so that $\mu_n|\partial K \ll \omega_n$. Hence $\nu_n|\partial K \ll \omega_n$ for every $n \geq 1$. □

18.10 COROLLARY. *If $R(K)$ is a Dirichlet algebra and E is a compact subset of ∂K, then E is a peak interpolating set for $R(K)$ if and only if E has zero harmonic measure.*

PROOF. Combine the preceding corollary with Theorem 11.3. □

§**19 Mergelyan's Theorem.** In this section some of the techniques that have been developed will be applied to determine sufficient conditions for $R(K)$ and $A(K)$ to coincide. Necessary and sufficient conditions for this can be found but this would take us too far from our goal of studying subnormal operators. The interested reader can find this material in Gamelin [**1969**] page 217.

We begin with one of the oldest theorems of this type.

19.1 MERGELYAN'S THEOREM. *If K is polynomially convex, then $P(K) = R(K) = A(K)$*

PROOF. Of course the first equality, $P(K) = R(K)$, is a direct consequence of Runge's Theorem and we were already aware of this. Since $P(K) \subseteq A(K)$, it remains to show that $A(K) \subseteq P(K)$. Actually we will prove that $A(K)|\partial K \subseteq P(K)|\partial K$ which will also complete the proof of the theorem. Let $\nu \in M(\partial K)$ such that $\nu \perp P(K)$; it must be shown that $\nu \perp A(K)$.

Let $Q_1, Q_2, \ldots$ be the nontrivial Gleason parts and let $E_1, E_2, \ldots$ be the corresponding carriers. Put $\nu_n = \nu|E_n$. By Theorem 18.7, $\nu_n \perp P(K)$. By Corollary 18.8, $\nu_n \ll \omega_n$, harmonic measure for a point a_n in Q_n. It must be shown that $\nu_n \perp A(K)$ for each $n \geq 1$.

Fix $n \geq 1$; also fix f in $A(K)$ and let $C = 2\|f\|$. Hence $\operatorname{Re}(f + C) > 0$ on K and so $g = \log(f + C) \in A(K)$. Since $P(K)$ is a Dirichlet algebra, there is a sequence of polynomials $\{p_k\}$ such that for every $k \geq 1$, $\|\operatorname{Re} p_k - \operatorname{Re} g\|_{\partial K} < 2^{-k}$. By the Maximum Principle we have that $\|\operatorname{Re} p_k - \operatorname{Re} g\|_K < 2^{-k}$ for all k. Also the polynomials p_k can be chosen so that $p_k(a_n) = g(a_n)$. Now $(p_k - g)^2$ is harmonic on $\operatorname{int} K$ and vanishes at a_n, so $0 = \int (p_k - g)^2 d\omega_n$. Since the real part of this integral must vanish, we have that

$$\int (\operatorname{Im} p_k - \operatorname{Im} g)^2 d\omega_n = \int (\operatorname{Re} p_k - \operatorname{Re} g)^2 d\omega_n$$
$$< (2^{-k})^2.$$

Hence
$$\int \sum_{k=1}^{\infty} |p_k - g|^2 d\omega_n = \sum_{k=1}^{\infty} \int |p_k - g|^2 d\omega_n < \infty.$$

Therefore $p_k(z) \to g(z)$ a.e. $[\omega_n]$ and hence a.e. $[\nu_n]$. Thus $e^{p_k} \to f + C$ a.e. $[\nu_n]$. But $|e^{p_k}| = e^{\operatorname{Re} p_k} \leq e^{(\operatorname{Re} g + 1)}$. By the Lebesgue Dominated Convergence Theorem, $\int f\, d\nu_n = \int (f + C)\, d\nu_n = \lim_k \int e^{p_k} d\nu_n = 0$ since $e^{p_k} \in P(K)$ for each $k \geq 1$. □

Mergelyan's Theorem can be used, in conjunction with Bishop's Localization Theorem, to show that $R(K) = A(K)$ whenever K is finitely connected. The same proof technique can be used to obtain an even better result.

19.2 THEOREM. *If K is a compact subset of $\mathbb{C}$ with the property that there is a $\delta > 0$ such that each component of $\mathbb{C} \setminus K$ has diameter at least δ, then $R(K) = A(K)$.*

PROOF. Fix f in $A(K)$. Let $a \in K$ and put $U = B(a; \delta/3)$; by Bishop's Theorem (3.10), if it can be shown that $f|(K \cap \operatorname{cl} U) \in R(K \cap \operatorname{cl} U)$, then $f \in R(K)$. But clearly $f|(K \cap \operatorname{cl} U) \in A(K \cap \operatorname{cl} U)$, so if we can show that $K \cap \operatorname{cl} U$ is polynomially convex, then the result will follow by Mergelyan's Theorem. This is precisely what will be shown.

In fact, suppose to the contrary that $\mathbb{C} \setminus (K \cap \operatorname{cl} U)$ has a bounded component W. Now $\mathbb{C} \setminus (K \cap \operatorname{cl} U) = (\mathbb{C} \setminus K) \cup (\mathbb{C} \setminus \operatorname{cl} U)$ and the fact that $\mathbb{C} \setminus \operatorname{cl} U$ is connected and unbounded implies that $W \cap (\mathbb{C} \setminus \operatorname{cl} U) = \varnothing$. Hence $W \subseteq \operatorname{cl} U$ and so $\operatorname{diam} W \leq 2\delta/3$. We also have that $W \subseteq \mathbb{C} \setminus K$ and so there is a component W_1 of $\mathbb{C} \setminus K$ that contains W. But $\mathbb{C} \setminus K \subseteq \mathbb{C} \setminus (K \cap \operatorname{cl} U)$ and so $W = W_1$. Since W has diameter $< \delta$, this is a contradiction. □

19.3 COROLLARY. *If K is finitely connected, $R(K) = A(K)$.*

§20 The double dual of a T-invariant algebra. In this section we will characterize the second dual space of a T-invariant algebra. If A is a function algebra on X, then general Banach space theory tells us that $A^* = M(X)/A^\perp$ and $A^{**} = A^{\perp\perp}$ = the weak* closure of A in $M(X)^{**}$. For T-invariant algebras we can improve this and relate A^{**} to the structure of the algebra. Clearly this will involve the annihilator of the algebra and Theorem 18.7 will be of value. We begin by determining the dual of a band of measures.

20.1 DEFINITION. If X is a compact metric space and $\mathscr{B}$ is a band of measures on X, define $L^\infty(\mathscr{B})$ to be the collection of all $F = \{F_\mu\}$ in the Cartesian product $\prod\{L^\infty(\mu): \mu \in \mathscr{B}\}$ such that if μ and $\nu \in \mathscr{B}$ and $\mu \ll \nu$, then $F_\mu = F_\nu$ a.e. $[\mu]$.

It is easy to see that $L^\infty(\mathscr{B})$ is a linear subspace of $\prod\{L^\infty(\mu): \mu \in \mathscr{B}\}$. Also if f is a bounded Borel function on X and $F_\mu = f$ for all μ in $\mathscr{B}$, then this defines an element of $L^\infty(\mathscr{B})$. It will be shown that $L^\infty(\mathscr{B})$ is the Banach space dual of $\mathscr{B}$, but first we must attend to a few amenities like defining the norm on $L^\infty(\mathscr{B})$.

20.2 LEMMA. *If $\mathscr{B}$ is a band of measures on X and $F \in L^\infty(\mathscr{B})$, then* $\sup\{\|F_\mu\|_\infty : \mu \in \mathscr{B}\} < \infty$.

PROOF. If this supremum is infinite, then there is a sequence $\{\mu_n\}$ in $\mathscr{B}$ such that $\|F_{\mu_n}\|_\infty \to \infty$. We may assume that $\|\mu_n\| \le 1$ for all $n \ge 1$ and so $\mu = \sum_n 2^{-n}|\mu_n| \in \mathscr{B}$. Since $\mu_n \ll \mu$, $F_{\mu_n} = F_\mu$ a.e. $[\mu_n]$ for all $n \ge 1$. Thus $\|F_\mu\|_\infty \ge \|F_{\mu_n}\|_\infty$ for all $n \ge 1$, contradicting the fact that $F_\mu \in L^\infty(\mu)$. □

The proof of the next proposition is left to the reader.

20.3 PROPOSITION. *If $\mathscr{B}$ is a band of measures on X and $\|F\| \equiv \sup\{\|F_\mu\|_\infty : \mu \in \mathscr{B}\}$ for F in $L^\infty(\mathscr{B})$, then $L^\infty(\mathscr{B})$ is an abelian C^*-algebra.*

20.4 EXAMPLES. (a) If μ is a positive measure on X and $\mathscr{B} = L^1(\mu)$, then for each f in $L^\infty(\mu)$ and ν in $L^1(\mu)$ define f_ν to be the element of $L^\infty(\nu)$ naturally associated with f. That is, the inclusion map $L^1(\nu) \to L^1(\mu)$ is an isometry (but possibly not surjective) and f_ν is the image of f under the dual of this map. In a certain sense f_ν is a restriction of f. It follows that $f \to \{f_\nu\}$ defines an isometric isomorphism of $L^\infty(\mu)$ onto $L^\infty(\mathscr{B})$.

(b) If $\mathscr{B}$ is the band of all purely atomic measures on X, then $L^\infty(\mathscr{B})$ "$=$" $l^\infty(X)$.

20.5 THEOREM. *If $\mathscr{B}$ is a band of measures on X and for F in $L^\infty(\mathscr{B})$, $\Phi_F : \mathscr{B} \to \mathbb{C}$ is defined by*

$$\Phi_F(\mu) = \int F_\mu \, d\mu,$$

then the map $F \to \Phi_F$ defines an isometric isomorphism of $L^\infty(\mathscr{B})$ onto $\mathscr{B}^*$.

PROOF. The fact that each Φ_F is linear is left to the reader. Also $|\Phi_F(\mu)| \le \int |F_\mu| \, d|\mu| \le \|F\| \|\mu\|$. Hence $\Phi_F \in \mathscr{B}^*$ and $\rho : L^\infty(\mathscr{B}) \to \mathscr{B}^*$ defined by $\rho(F) = \Phi_F$ is a linear contraction. It remains to show that ρ is isometric and surjective. As often happens in these situations, both these properties will be demonstrated simultaneously.

Fix Φ in $\mathscr{B}^*$. If $\mu \in \mathscr{B}$ and $g \in L^1(\mu)$, then $g\mu \in \mathscr{B}$. Thus $g \to \Phi(g\mu)$ is a well-defined linear functional on $L^1(\mu)$. Moreover, $|\Phi(g\mu)| \le \|\Phi\| \|g\mu\| = \|\Phi\| \|g\|_1$; hence there is an F_μ in $L^\infty(\mu)$ with $\|F_\mu\|_\infty \le \|\Phi\|$ such that $\Phi(g\mu) = \int g F_\mu \, d\mu$ for all g in $L^1(\mu)$.

Now suppose $\mu, \nu \in \mathscr{B}$ and $\mu \ll \nu$. Hence $\mu = g\nu$ for some g in $L^1(\nu)$. Thus for every h in $L^1(\mu)$, $hg \in L^1(\nu)$ and $h\mu = hg\nu$. This gives that $\int h F_\mu \, d\mu = \Phi(h\mu) = \Phi(hg\nu) = \int hg F_\nu \, d\nu$ for all h in $L^1(\mu)$. It follows that $F_\mu = F_\nu$ a.e. $[\mu]$. Therefore $F = \{F_\mu\} \in L^\infty(\mathscr{B})$. Also $\|F_\mu\|_\infty \le \|\Phi\|$ and so $\|F\| \le \|\Phi\|$. This completes the proof. □

20.6 COROLLARY. *The Banach space dual of $M(X)$ is isometrically isomorphic to $L^\infty(M(X))$.*

Now that we know that $L^\infty(\mathscr{B})$ is the dual of the Banach space $\mathscr{B}$, $L^\infty(\mathscr{B})$ has a weak* topology.

20.7 PROPOSITION. *A net $\{F_i\}$ in $L^\infty(\mathscr{B})$ converges to F in the weak* topology if and only if $(F_i)_\mu \to F_\mu$ weak* in $L^\infty(\mu)$ for every μ in $\mathscr{B}$.*

The proof of this proposition is left as an exercise for the reader. Notice that this says that the weak* topology on $L^\infty(\mathscr{B})$ is the relative product topology it has as a subset of $\prod\{L^\infty(\mu) : \mu \in \mathscr{B}\}$, where each coordinate has its natural weak* topology.

If $\mathscr{S} \subseteq L^\infty(\mathscr{B})$, define $\mathscr{S}_\mu \equiv \{F_\mu : F \in \mathscr{S}\}$.

20.8 PROPOSITION. *If $\mathscr{S} \subseteq L^\infty(\mathscr{B})$ and $F \in L^\infty(\mathscr{B})$, then F belongs to the weak* closure of $\mathscr{S}$ if and only if for every μ in $\mathscr{B}$, F_μ belongs to the weak* closure of $\mathscr{S}_\mu$ in $L^\infty(\mu)$.*

PROOF. It is easy to check that if $F \in \text{wk}^*\text{-cl}\,\mathscr{S}$, then for every μ in $\mathscr{B}$, $F_\mu \in wk^* - \text{cl}\,\mathscr{S}_\mu$. So let's concentrate on the converse; so assume that $F \in L^\infty(\mathscr{B})$ and $F_\mu \in wk^* - \text{cl}\,\mathscr{S}_\mu$ for every μ in $\mathscr{B}$. Let I = the collection of all pairs (ε, M), where $\varepsilon > 0$ and M is a finite subset of $\mathscr{B}$. Define an order on I as follows: $(\varepsilon, M) \leq (\delta, N)$ if $\delta \leq \varepsilon$, and $M \subseteq N$. Clearly with this definition of order, I becomes a directed set.

If $\alpha = (\varepsilon, \{\mu_1, \ldots, \mu_m\}) \in I$, then $\mu = |\mu_1| + \cdots + |\mu_m| \in \mathscr{B}$ and $\mu_1, \ldots, \mu_m \in L^1(\mu)$. By hypothesis there is a G is $\mathscr{S}$ such that $|\int(G_\mu - F_\mu)\,d\mu_j| < \varepsilon$ for $1 \leq j \leq m$. (Let's remark that $\int(G_\mu - F_\mu)\,d\mu_j = \int(G_{\mu_j} - F_{\mu_j})\,d\mu_j$.) Denote any such element G of $\mathscr{S}$ by F_α. Thus $\{F_\alpha : \alpha \in I\}$ is a net in $\mathscr{S}$. It is claimed that $F_\alpha \to F$ weak* in $L^\infty(\mathscr{B})$.

By Proposition 20.7 it must be shown that $(F_\alpha)_\mu \to F_\mu$ weak* in $L^\infty(\mu)$ for every μ in $\mathscr{B}$. That is, it must be shown that for μ in $\mathscr{B}$ and g in $L^1(\mu)$, $\int (F_\alpha)_\mu g\,d\mu \to \int F_\mu g\,d\mu$. So fix μ in $\mathscr{B}$, g in $L^1(\mu)$, and $\varepsilon > 0$. Let $\alpha_0 = (\varepsilon, \{\mu, g\mu\})$ in I. If $\alpha \geq \alpha_0$, then $\alpha = (\delta, M)$, where $\delta \leq \varepsilon$ and $\{\mu, g\mu\} \subseteq M$. By definition of F_α, we have $|\int((F_\alpha)_\mu - F_\mu)g\,d\mu| < \delta \leq \varepsilon$, precisely what we had to show. □

Note that $\mathscr{B}$ is a module over $L^\infty(\mathscr{B})$. That is, if $F \in L^\infty(\mathscr{B})$ and $\mu \in \mathscr{B}$, then define $\mu F \equiv (F_\mu)\mu \in \mathscr{B}$. It is easy to check that the desired distributive laws are satisfied and also $\|\mu F\| \leq \|\mu\|\|F\|$. (Write μF rather than $F\mu$ to avoid the possible confusion of this product with the usual notation of F_μ for the μth coordinate of the element F of $L^\infty(\mathscr{B})$.)

The algebra $L^\infty(\mathscr{B})$ has several idempotents. For example, if E is a Borel set and $F_\mu = \chi_E$ for every μ in $\mathscr{B}$, then $F \in L^\infty(\mathscr{B})$ and $F^2 = F$. If F is any idempotent of $L^\infty(\mathscr{B})$, not just one of the preceding type, then it can be verified by the reader that $\mathscr{B}F$ is a norm closed subband of $\mathscr{B}$.

The converse is also true. (Recall that the set of idempotents in any abelian ring forms a lattice.)

20.9 PROPOSITION. *If $\mathscr{B}$ is a band of measures on X, then the map $F \to \mathscr{B}F$ defines a lattice isomorphism between the lattice of idempotents of $L^\infty(\mathscr{B})$ and the lattice of subbands of $\mathscr{B}$.*

PROOF. Only a sketch of the proof is given here. The reader can fill in the details. Suppose $\mathscr{A}$ is a subband of $\mathscr{B}$ and for μ in $\mathscr{B}$ let $\mu = \mu_a + \mu_s$ be the Lebesgue decomposition of μ with respect to the band $\mathscr{A}$. Define F_μ by letting it be 1 a.e. $[\mu_a]$ and 0 a.e. $[\mu_s]$. Then $F \in L^\infty(\mu)$, $F^2 = F$, and $\mathscr{B}F = \mathscr{A}$. □

20.10 DEFINITION. If A is a function algebra on X and $\mu \in M(X)$, let $A^\infty(\mu)$ be the weak* closure of A in $L^\infty(\mu) \equiv L^\infty(|\mu|)$. If $\mathscr{B}$ is a band of measures on X, then there is a natural inclusion of A inside $L^\infty(\mathscr{B})$. Let $A^\infty(\mathscr{B})$ be the weak* closure of A in $L^\infty(\mathscr{B})$.

20.11 PROPOSITION. *The double dual of the function algebra A on X is naturally isometrically isomorphic to $A^\infty(M(X))$.*

REMARK. The word "naturally" in the preceding proposition means that the diagram

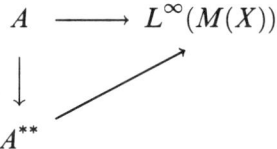

is commutative, where the horizontal and vertical arrows are the natural embeddings of A.

The proof of this proposition is just a specific instance of a Banach space phenomenon. Namely, if $\mathscr{X}$ is a Banach space and $\mathscr{Y}$ is a closed subspace of $\mathscr{X}$, then $\mathscr{Y} \subseteq \mathscr{X} \subseteq \mathscr{X}^{**}$ and $\mathscr{Y}^{**}$ "is" the closure of $\mathscr{Y}$ in the weak* topology of $\mathscr{X}^{**}$.

For the remainder of this book we will identify A^{**} with $A^\infty(M(X))$.

20.12 PROPOSITION. *If A is a function algebra on X and $F \in L^\infty(M(X))$, the following statements are equivalent.*

(a) $F \in A^\infty(M(X))$.
(b) *If* $\nu \in A^\perp$, *then* $\int F_\nu \, d\nu = 0$.
(c) $A^\perp F \subseteq A^\perp$.

PROOF. The equivalence of (a) and (b) is the consequence of general Banach space theory and is left to the reader.

(a) *implies* (c). Let $\nu \in A^\perp$; we want to show that $\nu F \in A^\perp$. Since $F \in A^\infty(M(X))$, there is a net $\{f_i\}$ in A such that $f_i \to F$ weak* in $L^\infty(M(X))$.

That is, for every μ in $M(X)$, $\int f_i d\mu \to \int F_\mu d\mu$. In particular, if $g \in A$, then $\int f_i g \, d\nu \to \int F_\nu g \, d\nu$. But $g\nu \in A^\perp$ for g in A and so $\int f_i g \, d\nu = 0$ for all i. Thus $\int F_\nu g \, d\nu = 0$ for all g in A and so $\nu F \in A^\perp$.

(c) implies (b). Fix ν in $A^\perp$. By hypothesis $\nu F \in A^\perp$; that is, $F_\nu \nu \in A^\perp$. Thus $\int F_\nu \, d\nu = \int 1 F_\nu \, d\nu = 0$ and (b) is proved. □

Return, momentarily, to the consideration of an arbitrary band $\mathscr{B}$ of measures on X with no reference to a function algebra living on X. If F is the idempotent in $L^\infty(M(X))$ such that $\mathscr{B} = M(X)F$ (20.9), then $1 - F$ is also an idempotent and it is easy to check that $M(X)(1 - F) = \mathscr{B}'$, the complementary band for $\mathscr{B}$ (Exercise 6). These observations allow us to show that under a natural map

(20.13) $$L^\infty(M(X)) = L^\infty(\mathscr{B}) \oplus_\infty L^\infty(\mathscr{B}').$$

In fact, the natural map in question is given by $G \to FG \oplus (1 - F)G$. To see that this map is surjective, note that if $G \in L^\infty(\mathscr{B})$, then G can be extended to an element of $L^\infty(M(X))$ by letting $G_\mu = 0$ for all μ in $\mathscr{B}'$. If G denotes this natural extension, then $FG = G$. The reader can supply the details.

If A is a given function algebra, then there is a distinguished class of bands associated with A (the reducing ones) and there is a distinguished class of idempotents (the ones that belong to $A^\infty(M(X))$). The next proposition says that these respective subclasses correspond under the identification between subbands and idempotents. It will also settle the following question.

Let $\mathscr{B}$ be a band and let F be the idempotent in $L^\infty(M(X))$ such that $\mathscr{B} = M(X)F$. So $F_\nu = 1$ for all ν in $\mathscr{B}$ and $F_\nu = 0$ for ν in $\mathscr{B}'$. By Proposition 20.8, $F \in A^\infty(\mathscr{B})$ and $1 - F \in A^\infty(\mathscr{B}')$. Is $F \in A^\infty(M(X))$? In other words, let $\nu \in M(X)$ and write $\nu = \nu_a + \nu_s$ with $\nu_a \in \mathscr{B}$ and $\nu_s \in \mathscr{B}'$. Even though we know that $F_{\nu_a} \in A^\infty(\nu_a)$ and $F_{\nu_s} \in A^\infty(\nu_s)$, can we conclude that $F_\nu \in A^\infty(\nu)$? The next theorem says that the reducing bands are the ones for which the answer to this question is yes.

20.14 THEOREM. *If A is a function algebra on X, F is an idempotent in $L^\infty(M(X))$, and $\mathscr{B}$ is the band $M(X)F$, then the following statements are equivalent.*

(a) *$\mathscr{B}$ is a reducing band for A.*
(b) *$A^\infty(\mathscr{B}) \subseteq A^\infty(M(X))$.*
(c) *$F \in A^\infty(M(X))$.*

Moreover, when $\mathscr{B}$ is a reducing band for A,

$$FA^\infty(M(X)) = A^\infty(\mathscr{B})$$

and

$$A^\infty(M(X)) = A^\infty(\mathscr{B}) \oplus_\infty A^\infty(\mathscr{B}').$$

PROOF. That (a) and (c) are equivalent is just a restatement of the fact that conditions (a) and (c) in (20.12) are equivalent.

(b) *implies* (c). By definition, for each ν in $\mathcal{B}$, $F_\nu = 1$ a.e. $[\nu]$ (since in this case $L^1(\nu) \subseteq \mathcal{B}$ and so $L^1(\nu)F = L^1(\nu)$). Thus $F_\nu \in (A)_\nu$ for every ν in $\mathcal{B}$. By (20.8), $F \in A^\infty(\mathcal{B})$ which, by hypothesis, is contained in $A^\infty(M(X))$.

(c) *implies* (b). Fix G in $A^\infty(\mathcal{B})$ and let $\{g_i\}$ be a net in A such that $g_i \to G$ weak* in $L^\infty(\mathcal{B})$. Thus for every ν in $\mathcal{B}$, $g_i \to G_\nu$ weak* in $L^\infty(\nu)$. In light of (20.13), G has a natural extension to an element of $L^\infty(M(X))$, still denoted by G, and $G = FG$; so $G_\eta = 0$ for all η in $\mathcal{B}'$. But for ν in $\mathcal{B}$, $(Fg_i)_\nu = g_i \to G_\nu$ weak* in $L^\infty(\nu)$ and for η in $\mathcal{B}'$, $(Fg_i)_\eta = 0 = G_\eta$. Hence it follows that $(Fg_i)_\mu \to G_\mu$ weak* in $L^\infty(\mu)$ for every μ in $M(X)$. By Proposition 20.7, $Fg_i \to G$ weak* in $L^\infty(M(X))$. But since $F \in A^\infty(M(X))$ by (c) and each $g_i \in A$, it easily follows that $Fg_i \in A^\infty(M(X))$ (Verify!). Therefore $G \in A^\infty(M(X))$, proving (b).

It remains to establish the two equations in the theorem. But observe that the preceding paragraph proved that for each G in $A^\infty(\mathcal{B})$, $G = FG \in A^\infty(M(X))$. Thus $A^\infty(\mathcal{B}) \subseteq FA^\infty(M(X))$. But in general, $FL^\infty(M(X)) = L^\infty(\mathcal{B})$ and so $FA \subseteq FA^\infty(\mathcal{B}) = A^\infty(\mathcal{B})$. It is now immediate that $FA^\infty(M(X)) \subseteq A^\infty(\mathcal{B})$, establishing the first of the two equations.

For the second equation, note that the first also establishes that $(1-F)A^\infty(M(X)) = A^\infty(\mathcal{B}')$ from which it follows that $A^\infty(M(X)) = FA^\infty(M(X)) \oplus_\infty (1-F)A^\infty(M(X)) = A^\infty(\mathcal{B}) \oplus_\infty A^\infty(\mathcal{B}')$. □

Now we conclude this section by interpreting these results for T-invariant algebras. Recall that Wilkin's Theorem (18.5) says that there can be no nonzero annihilating measures in the singular band for a T-invariant algebra, though this does not imply that this singular band does not exist.

20.15 THEOREM. *Let A be a T-invariant algebra on K, let $Q_1, Q_2, \ldots$ be the nontrivial Gleason parts, and let $Q = \bigcup_j Q_j =$ the set of nonpeak points for A. If $\mathcal{S}$ is the singular band for A, then*

$$A^\infty(M(X)) = L^\infty(\mathcal{S}) \oplus_\infty \bigoplus_j A^\infty(\mathcal{B}_{Q_j})$$

$$= L^\infty(\mathcal{S}) \oplus_\infty A^\infty(\mathcal{B}_Q),$$

where $\mathcal{B}_Q = \bigoplus_j \mathcal{B}_{Q_j}$.

PROOF. The proof of the fact that $A^\infty(\mathcal{B}_Q) = \bigoplus_j A^\infty(\mathcal{B}_{Q_j})$ is left to the reader. From the abstract F. and M. Riesz Theorem we know that $\mathcal{B}_Q$ is a reducing band and Proposition 18.3 implies that $M(X) = \mathcal{S} \oplus_1 \mathcal{B}_Q$. Hence the preceding theorem implies that $A^\infty(M(X)) = A^\infty(\mathcal{B}'_Q) \oplus_\infty A^\infty(\mathcal{B}_Q)$. Since $\mathcal{B}'_Q = \mathcal{S}$ (Why?), it remains to show that $A^\infty(\mathcal{S}) = L^\infty(\mathcal{S})$.

Now $L^\infty(\mathcal{S})$ is the Banach space dual of $\mathcal{S}$ and $A^\infty(\mathcal{S})$ is, by definition, the weak* closure of A in $L^\infty(\mathcal{S})$. So to show that $A^\infty(\mathcal{S}) = L^\infty(\mathcal{S})$, we must show that if $\mu \in \mathcal{S}$ and $\mu \perp A$, then $\mu = 0$. But this is the precise content of Wilkin's Theorem. □

Exercises

1. Verify the statements made in Example 20.4.
2. Prove Proposition 20.7.
3. Give the details of the proof of Proposition 20.10.
4. If $\mathscr{B}$ is a band, E is a Borel set, and F is the idempotent in $L^\infty(\mathscr{B})$ defined by $F_\mu = \chi_E$ for every μ in $\mathscr{B}$, then $\mathscr{B}F$ is the subband of $\mathscr{B}$ consisting of those μ in $\mathscr{B}$ carried by E.
5. If A is a function algebra on X show that the collection of bands that are reducing for A form a complete sublattice of the lattice of all bands of measures on X.
6. If F is the idempotent in $L^\infty(M(X))$ and $\mathscr{B} = M(X)F$, show that $1 - F$ is also an idempotent and $M(X)(1-F) = \mathscr{B}'$.
7. Prove (20.13).

§21 The Lautzenheiser-Mlak Theorem. In this section some of the ideas of this chapter will be applied to obtain a direct sum decomposition of an operator T with a spectral set K.

21.1 THE LAUTZENHEISER-MLAK DECOMPOSITION THEOREM. *If T is a bounded operator on a Hilbert space, K is a spectral set for T, and $Q_1, Q_2, \ldots$ are the nontrivial Gleason parts for $R(K)$, then $T = T_0 \oplus T_1 \oplus T_2 \oplus \cdots$, where:*

(a) *T_0 is a normal operator with spectrum contained in ∂K and if $\mathscr{M}$ is a subspace that is invariant for $f(T_0)$ for every function f in $R(K)$, then $\mathscr{M}$ reduces T_0;*

(b) *the spectrum of T_n is contained in $\operatorname{cl} Q_n$ and $\operatorname{cl} Q_n$ is a spectral set for T_n.*

Some of these summands may be trivial.

This decomposition theorem was first proved in Sarason [**1965b**] for the case where the spectral set K is polynomially convex. His proof will also work if $R(K)$ is only assumed to be a Dirichlet algebra. It was extended in Mlak [**1972b**] and independently in Lautzenheiser [**1973**]. The approach here is closer to the treatment of Seever [**preprint**] which was never published.

Most of the hard work for the proof of this theorem has been done, but we still need a few preliminary facts to relate the function theory to the operator theory. In addition, there is a uniqueness statement associated with Theorem 21.1, but this must wait until this preliminary material is developed before it is stated. See Theorem 21.12 below.

Recall (II.9.5) that if K is a spectral set for T, then we have a contractive functional calculus $\rho: R(K) \to \mathscr{B}(\mathscr{H})$, $\rho(f) = f(T)$, which extends the Riesz-Dunford functional calculus. We want to extend this functional calculus still further.

Consider the dual maps $\rho^*: \mathscr{B}(\mathscr{H})^* \to R(K)^*$ and $\rho^{**}: R(K)^{**} \to \mathscr{B}(\mathscr{H})^{**}$. In light of Theorem 20.15 we have rather good information about

$R(K)^{**}$. In fact, $R(K)^{**} = R^\infty(M(K)) = L^\infty(\mathscr{S}) \oplus R^\infty(\mathscr{B}_Q)$, where Q is the set of nonpeak points for $R(K)$ and $R^\infty(\mathscr{B}_Q)$ is the weak* closure of $R(K)$ in $L^\infty(\mathscr{B}_Q)$. In the next section we will obtain even more information about $R^\infty(\mathscr{B}_Q)$ when we realize this algebra as an algebra of functions. Thus ρ^{**} will extend the functional calculus ρ to a larger class of functions. A difficulty, or impending sacrifice, seems to present itself in that the range of ρ^{**} is contained in $\mathscr{B}(\mathscr{H})^{**}$ rather than $\mathscr{B}(\mathscr{H})$, and so this extended functional calculus will not yield an operator for each function in $R^\infty(M(K))$. The next proposition overcomes this.

21.2 PROPOSITION. *There is a bounded linear map* $\tau: \mathscr{B}(\mathscr{H})^{**} \to \mathscr{B}(\mathscr{H})$ *such that* $\|\tau\| = 1$ *and* $\tau(A) = A$ *for every* A *in* $\mathscr{B}(\mathscr{H})$.

PROOF. If C is a trace class operator on $\mathscr{H}$, define $\omega_C: \mathscr{B}(\mathscr{H}) \to \mathbb{C}$ by $\omega_C(A) = \operatorname{tr}(AC)$. That is, $C \to \omega_C$ is the natural embedding of $\mathscr{B}_1(\mathscr{H}) \to \mathscr{B}(\mathscr{H})^*$. Define $\tau: \mathscr{B}(\mathscr{H})^{**} \to \mathscr{B}(\mathscr{H})$ to be the dual of the map $C \to \omega_C$. Since the mapping $C \to \omega_C$ is an isometry, τ is contractive. But it is an easy exercise to check that $\tau(A) = A$ for A in $\mathscr{B}(\mathscr{H})$ and so $\|\tau\| = 1$. □

It is worthwhile to make the mapping τ in the preceding proposition a little more explicit. For x and y in $\mathscr{H}$, define $\omega_{x,y}: \mathscr{B}(\mathscr{H}) \to \mathbb{C}$ by $\omega_{x,y}(A) = \langle Ax, y \rangle$ for A in $\mathscr{B}(\mathscr{H})$. Note that if $F \in \mathscr{B}(\mathscr{H})^{**}$, then

(21.3) $$\langle \tau(F)x, y \rangle = F(\omega_{x,y})$$

for all x and y in $\mathscr{H}$.

Return to the operator T on $\mathscr{H}$ with spectral set K and functional calculus $\rho: R(K) \to \mathscr{B}(\mathscr{H})$. Define

(21.4) $$\rho: R^\infty(M(K)) \to \mathscr{B}(\mathscr{H})$$

by $\rho = \tau \circ \rho^{**}$, where τ is the contraction from Proposition 21.2 (The reader must be forgiving here for the author's abuse of notation in using ρ to denote its own extension.) Thus for every F in $R^\infty(M(K))$ and all x and y in $\mathscr{H}$,

$$\langle \rho(F)x, y \rangle = \rho^{**}(F)(\omega_{x,y}).$$

Also recall that $R^\infty(M(K))$ is a Banach algebra. The next proposition contains the germane information about this extension of ρ.

21.5 PROPOSITION. *If T is a bounded operator on $\mathscr{H}$ and K is a spectral set for T, then the mapping $\rho: R^\infty(M(K)) \to \mathscr{B}(\mathscr{H})$ defined in (21.4) is a Banach algebra homomorphism that extends the $R(K)$ functional calculus for T, $\|\rho\| = 1$, and ρ is weak* continuous.*

PROOF. The fact that ρ is a contraction is immediate from the fact that it is the composition of two contractions; since $\rho(z) = T$, $\|\rho\| = 1$. The fact that the two maps whose composition is ρ are both weak* continuous implies that ρ is weak* continuous. It remains to show that ρ is multiplicative.

We already know that ρ is multiplicative on $R(K)$. Let $F \in R^\infty(M(K))$ and let $g \in R(K)$. Let $\{f_i\}$ be a net in $R(K)$ such that $f_i \to F$ weak* in $R^\infty(M(K))$. It's easy to see that $f_i g \to Fg$ weak* in $R^\infty(M(K))$. Hence $\rho(f_i g) \to \rho(Fg)$ weak* in $\mathscr{B}(\mathscr{H})$. But $\rho(f_i g) = f_i(T)g(T) \to \rho(F)g(T)$ weak* in $\mathscr{B}(\mathscr{H})$. Therefore $\rho(Fg) = \rho(F)g(T) = \rho(F)\rho(g)$ for F in $R^\infty(M(K))$ and g in $R(K)$.

Maintaining the notation of the preceding paragraph, let $G \in R^\infty(M(K)) \subseteq L^\infty(M(K))$. If $\mu \in M(K)$ and F_μ and G_μ are the μ-coordinates of F and G, then F_μ and $G_\mu \in L^\infty(\mu)$. Since $f_i \to F$ weak* in $R^\infty(M(K))$, $f_i = (f_i)_\mu \to F_\mu$ weak* in $L^\infty(\mu)$ (20.7). Hence $(f_i G)_\mu = f_i G_\mu \to F_\mu G_\mu$ weak* in $L^\infty(\mu)$. Since μ was arbitrary, $f_i G \to FG$ weak* in $R^\infty(M(K))$. Therefore $\rho(f_i G) \to \rho(FG)$ weak* in $\mathscr{B}(\mathscr{H})$. But, by the preceding paragraph, $\rho(f_i G) = \rho(f_i)\rho(G) \to \rho(F)\rho(G)$ weak*. Hence $\rho(FG) = \rho(F)\rho(G)$ and ρ is a homomorphism. □

21.6 DEFINITION. If K is a spectral set for T and $F \in R^\infty(M(K))$, then $F(T)$ denotes the operator $\rho(F)$, where ρ is the homomorphism in (21.4).

This defines an extension of the functional calculus for the operator T. In Conway and Dudziak [**1990**] this functional calculus is refined (using the results of the next section) and an application is given.

The next result is a variation on previous themes.

21.7 THEOREM. *Let A be a T-invariant algebra on K with nontrivial Gleason parts $Q_1, Q_2, \ldots$ and let $\mathscr{B}_n = \mathscr{B}_{Q_n}$ for $n \geq 1$. If A_n is the uniform closure in $C(K)$ of $A + R(K, Q_n)$, then the following statements hold.*

(a) *Each A_n is a T-invariant algebra on K.*
(b) $A_n^\perp = A^\perp \cap \mathscr{B}_n$.
(c) *Q_n is the only nontrivial Gleason part for A_n.*
(d) *If $a \in Q_n$, then the set of representing measures for a and the algebra A is the same as the set of representing measures for a and the algebra A_n.*
(e) $A = \bigcap_{n=1}^\infty A_n$.

PROOF. We begin by proving (b). From Banach space generalities, $A_n^\perp = A^\perp \cap R(K, Q_n)^\perp$. Let $\nu \in A_n^\perp$; since $\nu \in R(K, Q_n)^\perp$, $\hat{\nu} = 0$ a.e. [Area] off Q_n. If E_m is the carrier for Q_m, then Theorem 18.7 implies $\nu_m = \nu|E_m \in A^\perp \cap \mathscr{B}_m$ and $\hat{\nu}_m = \hat{\nu}\chi_{E_m}$. Thus for $m \neq n$, $\hat{\nu}_m = 0$ a.e. [Area] on $\mathbb{C}$ and it must be that $\nu_m = 0$. Therefore $\nu = \nu_n \in A^\perp \cap \mathscr{B}_n$.

Conversely, if $\nu \in A^\perp \cap \mathscr{B}_n$, then Corollary 8.10 implies that $\hat{\nu}$ must vanish a.e. [Area] off Q_n. Hence $\nu \in R(K, Q_n)^\perp$ and so $\nu \in A_n^\perp$.

(a) Let $f \in A$ and $g \in R(K, Q_n)$. By (b) if $\nu \in A_n^\perp$, then $\nu \in A^\perp \cap \mathscr{B}_n$. Hence $f\nu \in A^\perp \cap \mathscr{B}_n = A_n^\perp$ and it follows that $\nu \perp fg$ since $R(K, Q_n) \subseteq A_n$. Since ν was an arbitrary element of $A_n^\perp$, $fg \in A_n$. It is now immediate that

A_n is a subalgebra of $C(K)$. Since A and $R(K, Q_n)$ are both T-invariant algebras and the Vitushkin operators are bounded and linear, it follows that A_n is a T-invariant algebra.

(d) Let $a \in Q_n$ and let μ be a representing measure for a with respect to A; so $\mu \in \mathscr{B}_n$. Thus $(z - a)\mu \in A^\perp \cap \mathscr{B}_n = A_n^\perp$. By Lemma 8.8, μ is a representing measure for a with respect to A_n. Since $A \subseteq A_n$, this finishes the proof of (d).

(c) Fix a in Q_n and let R_n be the Gleason part of A_n that contains a. If $b \in Q_n$, then a and b have mutually absolutely continuous representing measures with respect to A. By (d), a and b have mutually absolutely continuous representing measures with respect to A_n. Hence $Q_n \subseteq R_n$. But since $A \subseteq A_n$, each Gleason part for A_n is contained in a Gleason part for A. Thus $Q_n = R_n$. But almost every point of $K \setminus Q_n$ is a peak point of $R(K, Q_n)$ (13.10) and hence a peak point of A_n. Since nontrivial Gleason parts of a T-invariant algebra must have positive area, Q_n is the only nontrivial Gleason part for A_n.

(e) Since $A \subseteq \bigcap_{n=1}^\infty A_n$, it suffices to prove that $A^\perp \subseteq [\bigcap_{n=1}^\infty A_n]^\perp$. If $\nu \in A^\perp$, then $\nu = \sum_n \nu_n$ with $\nu_n \in A^\perp \cap \mathscr{B}_n = A_n^\perp$ (18.7). So if $f \in A_n$ for every $n \geq 1$, then $\int f \, d\nu_n = 0$ for all $n \geq 1$. $\square$

Let's establish some notation that will remain in force for the remainder of the section. Let $Q_1, Q_2, \ldots$ be the nontrivial Gleason parts for $R(K)$, let $\mathscr{B}_n = \mathscr{B}_{Q_n}$, and let $\mathscr{S}$ be the singular band for $R(K)$. According to Theorem 20.15

$$(21.8) \quad R^\infty(M(K)) = L^\infty(\mathscr{S}) \oplus R^\infty(K, \mathscr{B}_1) \oplus R^\infty(K, \mathscr{B}_2) \oplus \cdots.$$

Also for every $n \geq 1$ there is an idempotent E_n in $R^\infty(M(K))$ such that

$$(21.9) \quad \mathscr{B}_n = M(K)E_n \text{ and } R^\infty(K, \mathscr{B}_n) = E_n R^\infty(M(K)).$$

(The fact that there may arise some confusion between this idempotent E_n and the carrier of the Gleason part Q_n is intentional. The idempotent E_n can be thought of as the characteristic function of the carrier.) In fact E_n is defined by $(E_n)_\mu = 1$ if $\mu \in \mathscr{B}_n$ and $(E_n)_\mu = 0$ if $\mu \in \mathscr{B}_n'$. Also let E_0 be the idempotent in $R^\infty(M(K))$ such that $M(K)E_0 = \mathscr{S}$ and $E_0 R^\infty(M(K)) = L^\infty(\mathscr{S})$. So $E_0 + E_1 + E_2 + \cdots = 1$ (where the sum converges in the weak* topology of $L^\infty(M(K))$.

21.10 LEMMA. *If $x \in \mathscr{H}$, then there is a measure μ on K such that $\langle f(T)x, x \rangle = \int f \, d\mu$ for every f in $R(K)$ and $\|\mu\| = \|x\|^2$. If μ is any such measure, then μ must be positive.*

PROOF. If $L: R(K) \to \mathbb{C}$ is defined by $L(f) = \langle f(T)x, x \rangle$, then L is a linear functional on $R(K)$ and, since K is a spectral set for T, $\|L\| \leq \|x\|^2$. But $L(1) = \|x\|^2$, so $\|L\| = \|x\|^2$. This guarantees the existence of the measure μ. The fact that such a measure must be positive is a consequence of the fact that $\mu(K) = L(1) = \|L\| = \|\mu\|$. $\square$

21.11 LEMMA. *If $x \in \mathcal{H}$ and μ is as in the preceding lemma, then for every F in $R^\infty(M(K))$.*

$$\langle F(T)x, x \rangle = \int F_\mu \, d\mu$$

and

$$\langle F(T)^* x, x \rangle = \int \overline{F}_\mu \, d\mu.$$

PROOF. Let $\{f_i\}$ be a net in $R(K)$ such that $f_i \to F$ (weak*). Thus $\int f_i d\mu \to \int F_\mu d\mu$. On the other hand, $f_i(T) \to F(T)$ weak* in $\mathcal{B}(\mathcal{H})$. This establishes the first equation. The second equation follows from the fact that μ is a positive measure. □

PROOF OF THEOREM 21.1. For $n \geq 0$ let $P_n = E_n(T)$. Since ρ is a contractive homomorphism, P_n is an idempotent with $\|P_n\| = 1$; hence each P_n is a selfadjoint projection. Put $\mathcal{H}_n = P_n\mathcal{H}$; since $E_n E_m = 0$ for $n \neq m$, the spaces $\{\mathcal{H}_n\}$ are pairwise orthogonal. Since the sum of all the idempotents E_n is 1, we get $\mathcal{H} = \mathcal{H}_0 \oplus \mathcal{H}_1 \oplus \mathcal{H}_2 \oplus \cdots$.

Since T commutes with each P_n, T is reduced by each of the spaces $\mathcal{H}_n$. Let $T_n \equiv T|\mathcal{H}_n$ so that $T = T_0 \oplus T_1 \oplus T_2 \oplus \cdots$.

Now fix $n \geq 1$ and suppose $\lambda \notin \operatorname{cl} Q_n$. Hence $F = E_n(z-\lambda)^{-1} \in R^\infty(K, \mathcal{B}_n)$ by (2.19). Also $E_n = E_n F(z-\lambda)$. Since ρ is a homomorphism, $P_n = P_n F(T)(T-\lambda) = P_n F(T)(T_n - \lambda) = (T_n - \lambda)P_n F(T)$. That is, $F(T)|\mathcal{H}_n = (T_n - \lambda)^{-1}$ and $\lambda \notin \sigma(T_n)$. Thus $\sigma(T_n) \subseteq \operatorname{cl} Q_n$. To finish the proof that Q_n is a spectral set for T_n, let $f \in R(\operatorname{cl} Q_n)$. Once again $E_n f \in R^\infty(K, \mathcal{B}_n)$ and so $\|f(T_n)\| = \|E_n(T)f(T)\| \leq \|E_n f\| \leq \|E_n\| \|f\| \leq \|f\|_{\operatorname{cl} Q_n}$. Therefore $\operatorname{cl} Q_n$ is a spectral set for T_n.

It remains to prove part (a). Let $F = \overline{z}E_0 (\in L^\infty(\mathcal{S}))$. We want to show that $F(T)|\mathcal{H}_0 = T_0^*$, from which it is immediate that T_0 is normal. Let $x \in \mathcal{H}_0$ and let μ be the positive measure on K corresponding to x as in Lemma 21.10. Let $\mu_n = \mu E_n$ for $n \geq 0$, so that $\mu_0 \in \mathcal{S}$ and $\mu_n \in \mathcal{B}_n$ for $n \geq 1$. If $f \in R(K)$, then $\int f d\mu = \langle f(T)x, x \rangle = \langle f(T)P_0 x, x \rangle = \langle (E_0 f)(T)x, x \rangle = (21.11) \int (E_0 f)_\mu d\mu = \int f d\mu_0$. Thus we may assume that $\mu = \mu_0 \in \mathcal{S}$. But then the other half of Lemma 21.11 shows that $\langle F(T)^* x, x \rangle = \int \overline{F}_\mu d\mu = \int z \, d\mu = \langle Tx, x \rangle$. That is, $F(T)^*|\mathcal{H}_0 = T_0$ and so $F(T)|\mathcal{H}_0 = T_0^*$ and T_0 is normal.

Since each component of the interior of K is contained in some nontrivial Gleason part of $R(K)$, it follows that each measure in $\mathcal{S}$ is supported on ∂K. So if $\lambda \in \operatorname{int} K$, $F = E_0(z-\lambda)^{-1} \in L^\infty(\mathcal{S})$ and a routine calculation shows that $F(T)|\mathcal{H}_0 = (T_0 - \lambda)^{-1}$. Hence $\sigma(T_0) \subseteq \partial K$.

Finally, if $\mathcal{M}$ is an invariant subspace for $f(T_0)$ for every f in $R(K)$, then it is easy to see that $F(T)\mathcal{M} \subseteq \mathcal{M}$ for every F in $R^\infty(M(K))$. In particular, if $F_0 = \overline{z}E_0$, $F_0(T)\mathcal{M} = T_0^*\mathcal{M} = F_0(T)\mathcal{M} \subseteq \mathcal{M}$. □

21.12 THEOREM (Uniqueness for the Lautzenheiser-Mlak decomposition).

Let K be a spectral set for the operator T with nontrivial Gleason parts $Q_1, Q_2, \ldots$ and corresponding reducing bands $\mathscr{B}_1, \mathscr{B}_2, \ldots$. Let $\mathscr{B}_0$ be the singular band. Let $T = T_0 \oplus T_1 \oplus T_2 \oplus \cdots$ be the Lautzenheiser-Mlak decomposition on $\mathscr{H} = \mathscr{H}_0 \oplus \mathscr{H}_1 \oplus \mathscr{H}_2 \oplus \cdots$. If $x \in \mathscr{H}_n$ and μ is a positive measure on K with $\int f\, d\mu = \langle f(T)x, x\rangle$ for all f in $R(K)$, then $\mu \in \mathscr{B}_n$.

Conversely, if $\mathscr{L}_0, \mathscr{L}_1, \mathscr{L}_2 \ldots$ are reducing subspaces for T such that $\mathscr{H} = \mathscr{L}_0 \oplus \mathscr{L}_1 \oplus \mathscr{L}_2 \oplus \cdots$ and $\mu \in \mathscr{B}_n$ whenever μ is a positive measure on K with $\int f\, d\mu = \langle f(T)x, x\rangle$ for all f in $R(K)$ and some $x \in \mathscr{L}_n$, then $\mathscr{L}_n = \mathscr{H}_n$ for all $n \geq 0$.

PROOF. First fix x in $\mathscr{H}_n$ and let μ be as in the statement of the theorem. For $k \geq 0$, let E_k be the idempotent in $H^\infty(M(K))$ such that $M(K)E_k = \mathscr{B}_k$. If $\mu_k = \mu E_k$, then $\mu = \sum_k \mu_k$. Also for every f in $R(K)$, $\int f\, d\mu_k = \int (fE_k)_\mu\, d\mu = \langle f(T)E_k(T)x, x\rangle = 0$ for $k \neq n$, since $E_k(T)x \perp x$ in this case. But μ, and hence μ_k, is positive, so taking $f = 1$ shows that $\mu_k = 0$ for $k \neq n$. Thus $\mu = \mu_n \in \mathscr{B}_n$.

For the converse, fix $n \geq 0$ and let $x \in \mathscr{L}_n$; put $x_k = E_k(T)x$. Let μ be any positive measure such that $\langle f(T)x, x\rangle = \int f\, d\mu$ for all f in $R(K)$. By hypothesis $\mu \in \mathscr{B}_n$. Thus for $k \neq n$, $\mu E_k = 0$. Therefore if $k \neq n$ and $f \in R(K)$, $0 = \int (fE_k)_\mu\, d\mu = \langle f(T)E_k(T)x, x\rangle = \langle f(T)x_k, x_k\rangle$. Taking $f = 1$, we get that $x_k = 0$ for $k \neq n$. This proves that $\mathscr{L}_n \subseteq \mathscr{H}_n$ for all $n \geq 0$. But $\mathscr{H} = \mathscr{L}_0 \oplus \mathscr{L}_1 \oplus \mathscr{L}_2 \oplus \cdots = \mathscr{H}_0 \oplus \mathscr{H}_1 \oplus \mathscr{H}_2 \oplus \cdots$ and so it follows that $\mathscr{L}_n = \mathscr{H}_n$. □

A criticism of the preceding theorem might be that this uniqueness result is in terms of the measures μ rather than strictly in terms of the properties of the operator T. But this seems to be one of the difficulties of life rather than a difficulty with the result. After all, the decomposition is a function of the set K as well as the operator T. For the case of a pure operator, however, an improvement can be made and the uniqueness can be stated solely in terms of operator theoretic properties.

21.13 THEOREM. *Adopt the notation of Theorem* 21.1. *If T is a pure operator and $T = S_1 \oplus S_2 \oplus \cdots$ where $\sigma(S_n) \subseteq \operatorname{cl} Q_n$ and $\operatorname{cl} Q_n$ is a spectral set for S_n, then $S_n = T_n$ for all $n \geq 1$.*

PROOF. Let $\mathscr{L}_n$ be the space on which S_n acts. Applying Theorem 21.1 to the operator S_n and the spectral set $\operatorname{cl} Q_n$ and then the first half of the preceding theorem, we obtain that for every x in $\mathscr{L}_n$ and for any positive measure μ on $\operatorname{cl} Q_n$ such that $\int f\, d\mu = \langle f(T)x, x\rangle$ for all f in $R(\operatorname{cl} Q_n)$, it must be that μ belongs to the band $\mathscr{A}_n$ of measures on $\operatorname{cl} Q_n$ generated by the annihilating measures of $R(\operatorname{cl} Q_n)$. By using Theorem 21.7, this implies that each such measure μ belongs to $\mathscr{B}_n$. According to the preceding theorem, $S_n = T_n$ for all $n \geq 1$. □

Some of the summands T_n in the decomposition of Theorem 21.1 may not appear. This will become clear in the following collection of examples. It will also become clear that the theorem is no panacea, as the very first example demonstrates.

21.14 EXAMPLE. Let T be a strict contraction; that is, $\|T\| < 1$. So $K = \operatorname{cl}\mathbb{D}$ is a spectral set for T. But here $R(K) = A(K) = A(\operatorname{cl}\mathbb{D})$ and there is only one summand, $T_1 = T$.

21.15 EXAMPLE. Again let T be a strict contraction and let $K = \operatorname{cl}\mathbb{D} \cup \overline{B}(2;1)$. Here $R(K)$ has 2 nontrivial Gleason parts $\mathbb{D}$ and $B(2;1)$ although the decomposition still only has one summand, the one corresponding to the part $\mathbb{D}$.

21.16 EXAMPLE. Let K be as in the preceding example, let $G = \operatorname{int} K$, and let T be the Bergman operator for G. In this case there are two summands and they are the Bergman operators for $\mathbb{D}$ and $B(2;1)$.

21.17 EXAMPLE. Let K and G be as in the preceding example, let $\mu =$ area measure on K, let S be the Bergman operator for G, and put $T = N_\mu \oplus S$. In this case $T = T_1 \oplus T_2$, with $T_1 = S_1 \oplus N_{\mu_1}$, $T_2 = S_2 \oplus N_{\mu_2}$, where S_1 and S_2 are the Bergman operators for $\mathbb{D}$ and $B(2;1)$, respectively, and μ_1 and μ_2 are the restrictions of area measure to $\mathbb{D}$ and $B(2;1)$, respectively.

21.18 EXAMPLE. Let K be the closed unit disk and let $\{a_n\}$ be a dense sequence in $\partial \mathbb{D}$. Put $\mu = \operatorname{Area}|\mathbb{D} + \sum_n 2^{-n}\delta_{a_n}$ and $T = N_\mu$. Then the decomposition of T becomes $T = T_0 \oplus T_1$, where T_0 is multiplication by z on $L^2(\sum_n 2^{-n}\delta_{a_n})$ and T_1 is multiplication by z on $L^2(\operatorname{Area}|\mathbb{D})$.

This last example shows that the T_0 summand is not the only one that can be normal.

Exercises

1. Prove (21.3).
2. Let $\{a_n\}$ be a countable dense subset of $\operatorname{cl}\mathbb{D}$, put $\mu = \sum_n 2^{-n}\delta_{a_n}$, and let $T = N_\mu$. If $K = \operatorname{cl}\mathbb{D}$, find the Lautzenheiser-Mlak decomposition of T.
3. Let μ be as in the preceding exercise and let $\nu = \mu + m$, where m is normalized arc length measure on $\partial \mathbb{D}$. Interpret Theorem 21.1 when $K = \operatorname{cl}\mathbb{D}$ and $T = N_\nu$.
4. Let K be the string of beads and let $\mu = \omega_+ + \omega_-$ as in (15.19). Interpret Theorem 21.1 when $T = N_\mu$.

§22 Davie's Theorem. Let $\mathscr{B}$ be a band of measures on the compact space X and fix a measure μ in $\mathscr{B}$. Thus $L^1(\mu)$ is a subband of $\mathscr{B}$ and so, by Proposition 20.9, there is an idempotent F in $\mathscr{B}$ such that $L^1(\mu) = \mathscr{B}F$. Thus $\nu \to \nu F$ defines a projection of $\mathscr{B}$ onto $\mathscr{B}F = L^1(\mu)$. Also $G \to FG$ defines a projection of $L^\infty(\mathscr{B})$ onto $L^\infty(\mu)$. In this situation, moreover, $FG = G_\mu$. In this section we will focus on this latter projection which is called the *natural projection* of $L^\infty(\mathscr{B})$ onto $L^\infty(\mu)$. Clearly the natural projection has norm 1.

Here is another approach to this projection. If $i: L^1(\mu) \to \mathscr{B}$ is the inclusion map, then the dual map $i^*: \mathscr{B}^* \to L^\infty(\mu)$ is, once the appropriate identifications are made, the projection $i^*(G) = FG = G_\mu$ of $L^\infty(\mathscr{B})$ onto $L^\infty(\mu)$ (Exercise 1). Note that looking at it this way shows that every natural projection is weak* continuous.

If A is a function algebra on X, then we can restrict the natural projection $L^\infty(\mathscr{B}) \to L^\infty(\mu)$ to the subspace $A^\infty(\mathscr{B})$, the weak* closure of A in $L^\infty(\mathscr{B})$. It is easy to see that the image of this restriction is contained in $A^\infty(\mu)$.

Recall that if A is a T-invariant algebra on K and Q is the set of nonpeak points for A, then $\mathscr{B}_Q$ is the band generated by $\bigcup \{M_a : a \in Q\}$. Also, the set Q is an F_σ set and so for any Borel measure μ, μ_Q is a well-defined Borel measure. The purpose of this section is to prove the following theorem.

22.1 DAVIE'S THEOREM. *If A is a T-invariant algebra on K, Q is the set of nonpeak points, and λ_Q is the restriction of planar Lebesgue measure to Q, then $\lambda_Q \in \mathscr{B}_Q$ and the natural projection*

$$A^\infty(\mathscr{B}_Q) \to A^\infty(\lambda_Q)$$

is a dual algebra isomorphism.

Before beginning the proof of this theorem, one of its corollaries is worth stating. Indeed, this was Davie's original statement of the result in Davie **[1972a]** and is an equivalent formulation of the theorem (see Exercise 4).

22.2 COROLLARY. *If A is a T-invariant algebra and $f \in A^\infty(\lambda_Q)$, then there is a sequence of functions $\{f_n\}$ in A such that $\|f_n\| \leq \|f\|_\infty$ and $f_n \to f$ a.e. $[\lambda_Q]$.*

In order to prove this corollary, we first state a result that will be used here as well as in the next chapter.

22.3 PROPOSITION. *Let (X, Ω, μ) be a finite measure space. If there is a sequence $\{\phi_n\}$ in $L^\infty(\mu)$ such that $\phi_n \to \phi$ weak* in $L^\infty(\mu)$, then there is a uniformly bounded sequence $\{\psi_n\}$ such that $\psi_n(z) \to \phi(z)$ a.e. $[\mu]$ and each ψ_n is a convex combination of the functions $\phi_1, \phi_2, \ldots$. Thus $\sup_n \|\psi_n\|_\infty \leq \sup_n \|\phi_n\|_\infty$.*

PROOF. Since $\phi_n \to \phi$ weak* in $L^\infty(\mu)$, there is a constant M such that $\|\phi_n\|_\infty \leq M$ for all $n \geq 1$. Since μ is a finite measure, $L^2(\mu) \subseteq L^1(\mu)$ and so $\phi_n \to \phi$ weakly in $L^2(\mu)$. From general Hilbert space theory, ϕ belongs to the norm closed convex hull of $\{\phi_1, \phi_2, \ldots\}$ in $L^2(\mu)$. Thus there is a sequence $\{\psi_n\}$ of convex combinations of the functions $\phi_1, \phi_2, \ldots$ such that $\int |\psi_n - \phi|^2 d\mu \to 0$ (See V.1.4 in [ACFA]). But now measure theory assures us that a subsequence of $\{\psi_n\}$ can be found that converges to ϕ a.e. $[\mu]$. □

The next result comes from combining the preceding proposition with the Krein-Millman Theorem.

22.4 COROLLARY. *Let (X, Ω, μ) be a finite measure space. If $L^1(\mu)$ is separable and S is a convex subset of $L^\infty(\mu)$, then S is weak* closed if and only if whenever $\{\phi_n\} \subseteq S$, $\{\phi_n\}$ is uniformly bounded, and $\phi_n(z) \to \phi(z)$ a.e. $[\mu]$, then $\phi \in S$.*

PROOF OF COROLLARY 22.2. Assume $\|f\|_\infty = 1$. Since $A^\infty(\mathscr{B}_Q) \cong A^{**}$, the unit ball of A is weak* dense in ball $A^\infty(\mathscr{B}_Q)$. Therefore by Davie's Theorem ball A is weak* dense in ball $A^\infty(\lambda_Q)$. Therefore there is a sequence in ball A that converges to f weak* in $A^\infty(\lambda_Q)$. By Proposition 22.3 we can find a sequence $\{f_n\}$ in ball A such that $f_n \to f$ a.e. $[\lambda_Q]$. □

It can be taken as a maxim that behind every beautiful theorem there is a grubby lemma. This beautiful theorem has several grubby lemmas behind it. The first guarantees that λ_Q belongs to $\mathscr{B}_Q$, but it also says more.

The notation of the theorem will remain in effect for the rest of the section. In addition let $Q_1, Q_2, \ldots$ be the nontrivial Gleason parts of A; so $Q = Q_1 \cup Q_2 \cup \cdots$.

22.5 PROPOSITION. *If $\mu \in M(K)$ and $|\mu|(K \backslash Q) = 0$, then $\mu \in \mathscr{B}_Q$.*

PROOF. Let E be the idempotent in $L^\infty(M(K))$ such that $M(K)E = \mathscr{B}_Q$. Since $\mathscr{B}_Q$ is a reducing band for A, $E \in A^\infty(M(K))$.

From general Banach space theory, ball A is weak* dense in ball $A^\infty(M(K))$. Hence ball A is weak* dense in (ball $A^\infty(M(K)))_\nu$ for every ν in $M(K)$ (20.8). In particular, E_ν belongs to the weak* closure of ball A in $L^\infty(\nu)$ for every ν in $M(K)$. For $n \geq 1$, fix a_n in Q_n and let $\eta = |\mu| + \sum_n 2^{-n} \delta_{a_n}$. By (22.3) there is a sequence $\{f_k\}$ in ball A such that $\int |f_k - E_\eta|^2 \, d\eta \to 0$.

Now $\delta_{a_n} \in M_{a_n} \subseteq \mathscr{B}_Q$ and so $E_\eta(a_n) = 1$ for all $n \geq 1$. Thus for $n \geq 1$, $f_k(a_n) \to 1$ as $k \to \infty$. Fix n for the moment and let $b \in Q_n$. Since $f_k \in \text{ball } A$, $\operatorname{Re}(1 - f_k) \geq 0$ and so the definition of a Gleason part shows that there is a constant $c > 0$ such that for all $k \geq 1$, $c^{-1} \operatorname{Re}(1 - f_k(b)) \leq \operatorname{Re}(1 - f_k(a_n)) \leq c \operatorname{Re}(1 - f_k(b))$. Combined with similar estimates for the imaginary parts, this shows that $f_k(b) \to 1$ for all b in Q. Since η is carried by Q, $f_k \to 1$ a.e. $[\eta]$. Thus $E_\eta = 1$ a.e. $[\eta]$ and hence $E_\mu = 1$ a.e. $[\mu]$. Therefore $\mu = \mu E \in M(K)E = \mathscr{B}_Q$. □

Now we know that $\lambda_Q \in \mathscr{B}_Q$ and so the natural projection $\pi \colon A^\infty(\mathscr{B}_Q) \to A^\infty(\lambda_Q)$ is a well-defined linear contraction that is weak* continuous.

22.6 LEMMA. *The natural projection $\pi \colon A^\infty(\mathscr{B}_Q) \to A^\infty(\lambda_Q)$ is an isometry.*

This lemma is more than just a lemma; it is the heart of the theorem. In fact, once this lemma is established, the proof of the theorem follows rather quickly, as we now see.

PROOF OF DAVIE'S THEOREM (ASSUMING LEMMA 22.6). Let R be the range of π. Since π is an isometry, ball $R = \pi[\text{ball}\,A^\infty(\mathscr{B}_Q)]$. By Alaoglu's Theorem and the fact that a natural projection is weak* continuous, ball R is weak* compact in $A^\infty(\lambda_Q)$ and hence weak* closed. By the Krein-Smulian Theorem, R is weak* closed in $A^\infty(\lambda_Q)$. But $R \supseteq A$ and hence π is surjective. The fact that π is a weak* homeomorphism is an exercise in the application of the Krein-Smulian Theorem. □

To begin the road to the proof of Lemma 22.6, we introduce a class of operators related to the Vitushkin operators.

22.7 DEFINITION. If $g \in C_c^1$, define $R_g : L^\infty(\lambda_Q) \to C(K)$ by

$$R_g f = T_g f - fg,$$

where T_g is the Vitushkin operator. Call R_g a *remainder operator*.

Note that for f in $L^\infty(\lambda_Q)$,

$$R_g f(z) = \frac{1}{\pi} \int \frac{f(w)}{w-z} \overline{\partial} g(w)\, d\lambda_Q(w),$$

whence the continuity of $R_g f$ not only on K but on all of $\mathbb{C}$.

22.8 LEMMA. *If $\{f_i\}$ is a net in $L^\infty(\lambda_Q)$ and $f_i \to f$ weak*, then $R_g f_i \to R_g f$ weakly in $C(K)$.*

PROOF. Let $\mu \in M(K)$; it must be shown that $\int R_g f_i\, d\mu \to \int R_g f\, d\mu$. In fact

$$\int R_g f_i\, d\mu = \int \left[\frac{1}{\pi} \int \frac{f(w)}{w-z} \overline{\partial} g(w)\, d\lambda_Q(w)\right] d\mu(z)$$

$$= \frac{1}{\pi} \int f_i(w) \overline{\partial} g(w) \left[\int \frac{1}{w-z}\, d\mu(z)\right] d\lambda_Q(w)$$

$$= -\frac{1}{\pi} \int f_i(w) \overline{\partial} g(w) \hat{\mu}(w)\, d\lambda_Q(w).$$

But $\overline{\partial} g \hat{\mu} \in L^1(\lambda_Q)$ and so

$$\int R_g f_i\, d\mu \to -\frac{1}{\pi} \int f(w) \overline{\partial} g(w) \hat{\mu}(w)\, d\lambda_Q(w) = \int R_g f\, d\mu. \quad \square$$

The next lemma is a consequence of (22.6) but must be proved separately since it will be used in the proof of that lemma.

22.9 LEMMA. *The natural projection $\pi : A^\infty(\mathscr{B}_Q) \to A^\infty(\lambda_Q)$ is injective.*

PROOF. Suppose $F \in A^\infty(\mathscr{B}_Q)$ and $F_{\lambda_Q} = 0$; it must be shown that $F = 0$. Since $\{\mu \in \mathscr{B}_Q : F_\mu = 0\}$ is a subband of $\mathscr{B}_Q$, Corollary 18.6 implies that it suffices to show that $F_\mu = 0$ for every μ in $A^\perp$.

Let $\{f_i\}$ be a net in A such that $f_i \to F$ weak* in $L^\infty(\mathscr{B}_Q)$ and temporarily fix $\mu \perp A$. If $g \in C_c^1$, then $T_g f_i \in A$ since A is a T-invariant algebra. Thus

$$0 = \int T_g f_i \, d\mu = \int f_i g \, d\mu + \int R_g f_i \, d\mu.$$

Now $g\mu \ll \mu$ and so $g\mu \in \mathscr{B}_Q$. Since $f_i \to F$ weak* in $L^\infty(\mathscr{B}_Q)$, $\int f_i g \, d\mu \to \int F_\mu g \, d\mu$. It also follows that $f_i \to F_{\lambda_Q} = 0$ weak* in $L^\infty(\lambda_Q)$ and so, by the preceding lemma, $\int R_g f_i \, d\mu \to 0$. Combining these facts with the above equation, we get that $\int F_\mu g \, d\mu = 0$ for every g in C_c^1. Thus $F_\mu = 0$ a.e. $[\mu]$. Since μ was an arbitrary measure in $A^\perp$ we have that $F = 0$ in $A^\infty(\mathscr{B}_Q)$. □

The next several lemmas are useful for other parts of rational approximation. They deal with covers of the plane and partitions of unity subordinate to these covers. Similar results will appear in Chapter VIII when Thomson's Theorem on bounded point evaluations is proved.

22.10 LEMMA. *There are positive constants c_1 and $c_2 > 1$ such that for every $\delta > 0$ and every integer $k \geq 1$, there exists a disk $\Delta_{k\delta}$ and a positive function $g_{k\delta}$ in $C_c^1(\Delta_{k\delta})$ with the following properties.*

(a) *For δ fixed, $\sum_k g_{k\delta} = 1$ on $\mathbb{C}$.*
(b) *The radius of $\Delta_{k\delta}$ is δ.*
(c) *$\|\bar{\partial} g_{k\delta}\|_\infty \leq c_1/\delta$.*
(d) *For δ fixed, no point of $\mathbb{C}$ lies in more than c_2 of the disks $\Delta_{k\delta}$.*

PROOF. Fix a function g in $C_c^1(B(0; 1/2))$ such that $g \geq 0$ and $\iint g = 1$. Let $\{E_k\}_{k \geq 1}$ be a partition of $\mathbb{C}$ into disjoint open squares with sides of length $1/2$; let a_k be the center of E_k and put $g_k = \chi_{E_k} * g$. For each $\delta > 0$, let $\Delta_{k\delta} = B(\delta a_k; \delta)$, $g_{k\delta}(z) = g_k(z/\delta)$. It is left to the reader to verify that the conditions of the lemma are satisfied with $c_1 = \|\bar{\partial} g\|_\infty$ and $c_2 = 25$. □

22.11 DEFINITION. If $\{\Delta_{k\delta}\}$ and $\{g_{k\delta}\}$ are as in the preceding lemma, call $\{\Delta_{k\delta}\}$ a *Vitushkin cover* of $\mathbb{C}$ and call $\{\Delta_{k\delta}, g_{k\delta}\}$ a *Vitushkin partition of unity*.

For the rest of the section the constants c_1 and c_2 obtained in the preceding lemma will remain fixed.

22.12 LEMMA. *There is a positive constant c_3 such that if $\{\Delta_{k\delta}\}$ is a Vitushkin cover of $\mathbb{C}$, then for every a in $\mathbb{C}$, $\delta > 0$, and integer $n \geq 0$, the number $N(n)$ of disks $\Delta_{k\delta}$ whose centers lie in the annulus $\{z: n\delta \leq |z-a| \leq (n+1)\delta\}$ satisfies $N(n) \leq c_3(n+1)$.*

PROOF. Let I be the set of positive integers k such that the center of the disk $\Delta_{k\delta}$ lies in the annulus in question; so $N(n)$ is the cardinality of I. Note that if $k \in I$, then $\Delta_{k\delta} \subseteq \{z: (n-1)\delta \leq |z-a| \leq (n+2)\delta\} \equiv R$. Since no point in $\mathbb{C}$ can belong to more than c_2 of the disks $\Delta_{k\delta}$, we have that $\sum\{\chi_{\Delta_{k\delta}} : k \in I\} \leq c_2 \chi_R$. Thus

$$N(n)\pi\delta^2 \leq \int \sum\{\chi_{\Delta_{k\delta}} : k \in I\}\, d\mathscr{A} \leq c_2 \int \chi_R\, d\mathscr{A}$$
$$= c_2\pi\delta^2[(n+2)^2 - (n-1)^2]$$
$$= c_2\pi\delta^2[6n+3],$$

from which it follows that $N(n) \leq c_2(6n+3) \leq 6c_2(n+1)$. So $c_3 = 6c_2$ works. □

22.13 LEMMA. *There is a positive constant c_4 such that if $\{\Delta_{k\delta}\}$ is a Vitushkin cover, $\delta > 0$, and $M \geq 0$, then the following hold.*

(a) *If for each $k \geq 1$, $h_{k\delta}$ is a Borel function with $|h_{k\delta}| \leq M$ and such that $h_{k\delta}$ is analytic on $\mathbb{C}_\infty \setminus \operatorname{cl} \Delta_{k\delta}$ with a triple 0 at ∞, then for every a in $\mathbb{C}$*

$$\sum_{k=1}^{\infty} |h_{k\delta}(a)| \leq c_4 M.$$

(b) *If L is a compact subset of $\mathbb{C}$ such that $h_{k\delta} = 0$ whenever $L \cap \Delta_{k\delta} = \varnothing$, then for every a in $\mathbb{C} \setminus L$*

$$\sum_{k=1}^{\infty} |h_{k\delta}(a)| \leq c_4 M \delta[\operatorname{dist}(a, L)]^{-1}.$$

PROOF. Let $a_{k\delta}$ be the center of $\Delta_{k\delta}$. By hypothesis, $(z - a_{k\delta})^3 h_{k\delta}$ is analytic on $\mathbb{C}_\infty \setminus [\operatorname{cl} \Delta_{k\delta}]$. Also for z in $\operatorname{cl} \Delta_{k\delta}$, $|(z - a_{k\delta})^3 h_{k\delta}(z)| \leq \delta^3 M$. The Maximum Modulus Theorem implies that

$$|h_{k\delta}(z)| \leq \delta^3 M |z - a_{k\delta}|^{-3}$$

for all z in $\mathbb{C}$.

Fix $\delta > 0$ and a in $\mathbb{C}$ and let $N(n)$ be as in preceding lemma. That is, $N(n)$ is the number of disks $\Delta_{k\delta}$ for which $n\delta \leq |a_{k\delta} - a| \leq (n+1)\delta$. If $n \geq 1$ and $a_{k\delta}$ satisfies this relation, then

(22.14) $$|h_{k\delta}(a)| \leq \delta^3 M |a - a_{k\delta}|^{-3} \leq M/n^3.$$

Hence

$$\sum_{k=1}^{\infty} |h_{k\delta}(a)| \leq M \left[N(0) + \sum_{n=1}^{\infty} \frac{N(n)}{n^3} \right]$$
$$\leq Mc_3 \left[1 + \sum_{n=1}^{\infty} \frac{n+1}{n^3} \right]$$
$$\leq 5c_3 M.$$

Thus any constant $c_4 \geq 5c_3$ will work for part (a).

For part (b) let $a \in \mathbb{C} \setminus L$ and $d = \text{dist}(a, L)$; we consider two possibilities for δ. If $\delta \geq d/6$, then

$$\sum_{k=1}^{\infty} |h_{k\delta}(a)| \leq 5c_3 M \leq \frac{30 c_3 M \delta}{d}.$$

Thus a choice of $c_4 \geq 30 c_3$ will satisfy (b) for these δ as well as part (a).

Suppose $\delta < d/6$. If $n\delta \leq |a_{k\delta} - a| \leq (n+1)\delta$ with $n \leq (d/\delta) - 2$, then for every z in $\Delta_{k\delta}$, $|z-a| \leq |z-a_{k\delta}| + |a_{k\delta}-a| < \delta + (n+1)\delta \leq d = \text{dist}(a, L)$ and so $L \cap \Delta_{k\delta} = \emptyset$; in this case the assumption in (b) implies $h_{k\delta}(a) = 0$. So in estimating $\sum_k |h_{k\delta}(a)|$, the only values of k that matter are those in the set $J = \{k \geq 1 : n\delta \leq |a_{k\delta} - a| \leq (n+1)\delta \text{ with } n > (d/\delta) - 2\}$. From (22.14) we get

$$\sum_{k=1}^{\infty} |h_{k\delta}(a)| = \sum \{|h_{k\delta}(a)| : k \in J\}$$

$$\leq M \sum_k \left\{ \frac{N(n)}{n^3} : k \in J \right\}$$

$$\leq c_3 M \sum \left\{ \frac{n+1}{n^3} : k \in J \right\}$$

$$\leq 2 c_3 M \sum_k \left\{ \frac{1}{n^2} : k \in J \right\}$$

$$\leq 2 c_3 M \int_{\frac{d}{\delta} - 3}^{\infty} t^{-2} \, dt$$

$$= \frac{2 c_3 M \delta}{d - 3\delta}.$$

Now $2(d - 3\delta) = d + (d - 6\delta) \geq d$ and so we have that

$$\sum_{k=1}^{\infty} |h_{k\delta}(a)| \leq \frac{4 c_3 M \delta}{d}.$$

So here we need that $c_4 \geq 4 c_3$. Hence the choice of $c_4 = 30 c_3$ satisfies all our needs.

22.15 LEMMA. *If Q is a Borel set of finite area and λ_Q is the restriction of planar Lebesgue measure to Q, then there is a constant c_5 such that for f in $L^{\infty}(\lambda_Q)$ and g in C_c^1.*

$$\|R_g f\|_{\infty} \leq c_5 \|\overline{\partial} g\|_{\infty} \|f\|_{L^3(\lambda_Q)}.$$

PROOF. By modifying some of the estimates used in our discussion of Cauchy transforms, it can be shown that

(22.16) $$\int_Q |z - a|^{-3/2} \, d\lambda(z) \leq 4 \pi^{3/4} \lambda(Q)^{1/4}$$

for all a in $\mathbb{C}$. Thus $(z-a)^{-1} \in L^{3/2}(\lambda_Q)$ and $\|(z-a)^{-1}\|_{L^{3/2}(\lambda_Q)} \le 4^{2/3} \pi^{1/2} \lambda(Q)^{1/6}$. Hölder's Inequality implies that for every a in $\mathbb{C}$

$$\begin{aligned}|R_g f(a)| &\le \frac{1}{\pi} \int \frac{|f(z)|}{|z-a|} |\bar{\partial} g(z)| \, d\lambda_Q(z) \\ &\le \frac{1}{\pi} \|(z-a)^{-1}\|_{3/2} \|f \bar{\partial} g\|_3 \\ &\le 4^{2/3} \pi^{-1/2} \lambda(Q)^{1/6} \|\bar{\partial} g\|_\infty \|f\|_3.\end{aligned}$$

So we can take $c_5 = 4^{2/3} \pi^{-1/2} \lambda(Q)^{1/6}$. □

It is now time to assemble the pieces.

PROOF OF LEMMA 22.6. Suppose the natural projection $\pi: A^\infty(\mathscr{B}_Q) \to A^\infty(\lambda_Q)$ is not an isometry. That is, assume there is an F in $A^\infty(\mathscr{B}_Q)$ such that $\|F\| = 1$ and $\|F_{\lambda_Q}\| < 1$. Put $F_1 = F_{\lambda_Q}$ and let $0 < \varepsilon < 1/2$. (The reason for choosing $\varepsilon < 1/2$ will surface shortly. Later ε will be further restricted.) If F is replaced by F^n for large n, we may assume that $\|F_1\| \le \varepsilon$.

For each integer n, there is a measure μ_n in $\mathscr{B}_Q$ such that $\|F_{\mu_n}\| > 1 - n^{-1}$; there is no loss in generality in assuming that $\|\mu_n\| = 1$ and $\mu_n \ge 0$. Put $\mu = \lambda_Q + \sum_n 2^{-n} \mu_n$; so $\mu \in \mathscr{B}_Q$ and $\|F_\mu\| = 1$. Replacing F by $e^{i\theta} F$ for suitable θ, we may assume that $1 \in$ the μ-essential range of F_μ. Thus there is a compact set L such that $\mu(L) > 0$ and $|F_\mu(z) - 1| < \varepsilon$ for z in L. Because $\varepsilon < 1/2$, it must be that $\lambda_Q(L) = 0$.

Let $\mathscr{S} = \mathscr{B}_Q'$ be the band of measures complementary to $\mathscr{B}_Q$. Hence $A^\infty(M(K)) = L^\infty(\mathscr{S}) \oplus A^\infty(\mathscr{B}_Q)$ and $0 \oplus F \in$ ball $A^\infty(M(K)) =$ ball $A^{**} =$ the weak* closure of ball A. By (20.8), $F_\mu \in$ the weak* closure of ball A in $L^\infty(\mu)$. But $L^{3/2}(\mu) \subseteq L^1(\mu)$ and so F_μ belongs to the weak closure of ball A in $L^3(\mu)$. By an argument like that used to prove (22.3), there is a sequence $\{f_n\}$ in A with $\|f_n\| \le 1$ such that $\int |f_n - F_\mu|^3 d\mu \to 0$. By passing to a subsequence if necessary, we may assume that $f_n \to F_\mu$ a.e. $[\mu]$. By Egoroff's Theorem, L can be replaced by a smaller compact set and a finite number of the functions f_n can be discarded to achieve the fact that

$$|f_n(z) - 1| \le 2\varepsilon$$

for all z in L and all $n \ge 1$.

Let $\{\Delta_{k\delta}, g_{k\delta}\}$ be a Vitushkin partition and put $R_{k\delta} =$ the operator $R_{g_{k\delta}}$. By Lemma 22.12, for every $\delta > 0$ the number of k for which the disk $\Delta_{k\delta}$ meets L is finite. So for z in L, $\sum_k g_{k\delta}(z)$ is a finite sum. We now want to estimate $\|R_{k\delta}F_1\|$. For a in $\mathbb{C}$,

$$|R_{k\delta}F_1(a)| = \left| \frac{1}{\pi} \int_{\Delta_{k\delta}} \frac{F_1(z)}{z-a} \bar{\partial} g_{k\delta}(z)\, d\lambda_Q(z) \right|$$

$$\leq \frac{1}{\pi} \int_{\Delta_{k\delta}} \frac{|F_1(z)|}{|z-a|} |\bar{\partial} g_{k\delta}(z)|\, d\lambda_Q(z)$$

$$\leq \frac{1}{\pi} \|F_1\| \|\bar{\partial} g_{k\delta}\|_\infty \int_{\Delta_{k\delta}} \frac{1}{|z-a|}\, d\lambda_Q(z)$$

(by 2.2) $\leq \dfrac{1}{\pi} \|F_1\| \|\bar{\partial} g_{k\delta}\|_\infty [\pi \operatorname{Area}(\Delta_{k\delta})]^{1/2}$

(by 22.10) $\leq \dfrac{1}{\pi} \pi \delta \varepsilon \dfrac{c_1}{\delta}$

$= c_1 \varepsilon.$

Thus $\|R_{k\delta}F_1\|_\infty \leq c_1 \varepsilon$.

Since $f_n \to F_\mu$ in $L^3(\mu)$, $\|f_n - F_1\|_3 \to 0$. ($\|\cdot\|_3$ will be used for the norm in $L^3(\lambda_Q)$.) Also by Lemma 22.15,

$$\|R_{k\delta}f_n - R_{k\delta}F_1\|_\infty \leq c_5 \|\bar{\partial} g_{k\delta}\| \|f_n - F_1\|_3$$
$$\leq \frac{c_5 c_1}{\delta} \|f_n - F_1\|_3.$$

Since the right-hand side of this inequality is independent of k,

$$\lim_{n \to \infty} \sup_k \|R_{k\delta}f_n - R_{k\delta}F_1\|_\infty = 0.$$

Hence there is an integer $n_1 = n_1(\delta)$ such that

(22.17) $$\|R_{k\delta}f_n\|_\infty \leq 3c_1 \varepsilon$$

for all $n \geq n_1$ and for all $k \geq 1$. Henceforth we will assume that $n \geq n_1(\delta)$.

Let

$$F_{n\delta} = \sum_k (T_{g_{k\delta}} f_n)^3 = \sum_k (f_n g_{k\delta} + R_{k\delta}f_n)^3.$$

By Lemma 22.12, at any given point in $\mathbb{C}$ this is actually only a finite sum and is hence well defined. Because of this and the fact that A is a T-invariant algebra on the compact set K, $F_{n\delta} \in A$. Now $T_{g_{k\delta}} f_n$ is analytic off $\Delta_{k\delta}$ and 0 at ∞; hence $(T_{g_{k\delta}} f_n)^3$ has a triple zero at ∞. Also (22.17) implies that $\|T_{g_{k\delta}} f_n\|_\infty = \|f_n g_{k\delta} + R_{k\delta}f_n\| \leq 1 + 3c_1 \varepsilon$. By Lemma (22.13),

(22.18) $$\begin{cases} \|F_{n\delta}\|_\infty \leq c_4(1 + 3c_1 \varepsilon)^3, \\ |F_{n\delta}(a)| \leq \dfrac{c_4(1 + 3c_1 \varepsilon)^3}{\operatorname{dist}(a, L)} & \text{for } a \notin L. \end{cases}$$

Now let's estimate $|F_{n\delta} - 1|$ on L. In fact

(22.19)
$$\begin{aligned}
F_{n\delta} - 1 &= \sum_k (f_n g_{k\delta} + R_{k\delta} f_n)^3 - 1 \\
&= \sum_k (R_{k\delta} f_n)^3 + 3 f_n \sum_k g_{k\delta} (R_{k\delta} f_n)^2 \\
&\quad + 3 f_n^2 \sum_k g_{k\delta}^2 (R_{k\delta} f_n) \\
&\quad + f_n^3 \left[\left(\sum_k g_{k\delta}^3 \right) - 1 \right] + (f_n^3 - 1).
\end{aligned}$$

We will now estimate each summand in (22.19), starting with the first.
By (22.13) and (22.17),
$$\left| \sum_k (R_{k\delta} f_n)^3 \right| \le c_4 (3 c_1 \varepsilon)^3$$
on $\mathbb{C}$. For the second summand, the fact that $\|f_n\| \le 1$, $g_{k\delta} \ge 0$, and $\sum_k g_{k\delta} = 1$ on $\mathbb{C}$ implies that
$$\left| 3 f_n^2 \sum_k g_{k\delta} [R_{k\delta} f_n]^2 \right| \le 3(3 c_1 \varepsilon)^2 \sum_k g_{k\delta} = 3(3 c_1 \varepsilon)^2.$$

Also
$$\left| 3 f_n^2 \sum_k g_{k\delta}^2 [R_{k\delta} f_n] \right| \le 3(3 c_1 \varepsilon) \sum_k g_{k\delta}^2 \le 3(3 c_1 \varepsilon),$$
estimating the third summand.

By Lemma 22.10, for any given value of a in $\mathbb{C}$, $g_{k\delta}(a) \ne 0$ for at most c_2 values of k. Since $\sum_k g_{k\delta}(a) = 1$, $\max_k g_{k\delta}(a) \ge 1/c_2$. Hence in the fourth summand of (22.19),
$$\left| f_n^3 \left[\left(\sum_k g_{k\delta}^3 \right) - 1 \right] \right| \le 1 - \sum_k g_{k\delta}^3$$
$$\le 1 - \frac{1}{c_2^3}.$$

Finally, $|f_n^3 - 1| = |f_n^2 + f_n + 1| |f_n - 1| \le 3(2\varepsilon)$ on L. Now combine all these estimates to derive from (22.19) that on L

(22.20)
$$|F_{n\delta} - 1| \le c_4 (3 c_1 \varepsilon)^3 + 3(3 c_1 \varepsilon)^2 + 3(3 c_1 \varepsilon) \\ + (1 - c_2^{-3}) + 3(2\varepsilon) \equiv b(\varepsilon).$$

Note that as $\varepsilon \to 0$, $b(\varepsilon) \to 1 - c_2^{-3} < 1$. Also $b(\varepsilon)$ is independent of δ. So by choosing ε sufficiently small and putting $b_1 = c_4 (1 + 3 c_1 \varepsilon)^3$ and $b_2 = b(\varepsilon)$, we have that for every $\delta > 0$ there is a function F_δ in A (let $F_\delta = F_{n\delta}$ for any $n \ge n_1(\delta)$) with
    (i) $\|F_\delta\| \le b_1$;

(ii) $|F_\delta(a)| \le b_1 \delta [\text{dist}(a, L)]^{-1}$ if $a \notin L$;
(iii) $|F_\delta(z) - 1| \le b_2 < 1$ for z in L.

From (ii) we see that as $\delta \to 0$, $F_\delta \to 0$ pointwise off L. But $\lambda_Q(L) = 0$. Therefore because of (i), we have that $F_\delta \to 0$ weak* in $A^\infty(\lambda_Q)$. But $\{F_\delta : \delta > 0\}$ is a bounded net in $A^\infty(\mathscr{B}_Q)$ and thus has at least one weak* cluster point. But any weak* cluster point of $\{F_\delta\}$ in $A^\infty(\mathscr{B}_Q)$ is mapped by the natural projection onto a weak* cluster point of $\{F_\delta\}$ in $A^\infty(\lambda_Q)$. Since $F_\delta \to 0$ weak* in $A^\infty(\lambda_Q)$ and the natural projection is one-to-one (22.9), we have that 0 is the unique weak* cluster point of $\{F_\delta\}$ in $A^\infty(\mathscr{B}_Q)$. Hence $F_\delta \to 0$ weak* in $A^\infty(\mathscr{B}_Q)$. In particular, $\int F_\delta d\mu \to 0$ for the measure μ in $\mathscr{B}_Q$ defined at the beginning of this proof.

But from (iii) above, $\text{Re}\, F_\delta \ge 1 - b_2 > 0$ on L. Hence $|\int_L F_\delta d\mu| \ge \text{Re} \int_L F_\delta d\mu = \int_L \text{Re}\, F_\delta d\mu \ge (1 - b_2)\mu(L) > 0$ for all $\delta > 0$, contradicting the fact that $\int F_\delta d\mu \to 0$. This contradiction completes the proof. □

REMARKS. The development of Davie's Theorem in this section was taken from Dudziak [**1981**] which was taken from Gamelin [**notes**]. Also see Cole and Gamelin [**1982** and **1985**] and Gamelin [**1973**].

Exercises

1. Let $\mathscr{B}$ be a band of measures, let $\mathscr{A}$ be a subband, and let F be the idempotent in $L^\infty(\mathscr{B})$ such that $\mathscr{A} = \mathscr{B}F$. If $i: \mathscr{A} \to \mathscr{B}$ is the inclusion map, show that the dual map $i^*: L^\infty(\mathscr{B}) \to L^\infty(\mathscr{A})$ is given by $i^*(G) = FG$.
2. If $\mathscr{B}$ is a band of measures and $\mu \in \mathscr{B}$, restrict the Cartesian projection $\prod\{L^\infty(\nu): \nu \in \mathscr{B}\} \to L^\infty(\mu)$ to the band $L^\infty(\mathscr{B})$ to get a map $\tau: L^\infty(\mathscr{B}) \to L^\infty(\mu)$. Compare this with the natural projection of $L^\infty(\mathscr{B})$ onto $L^\infty(\mu)$.
3. Prove the estimate (22.16).
4. Derive Davie's Theorem as a consequence of Corollary 22.2.
5. Show that for any measure μ in $\mathscr{B}_Q$, the natural projection $A^\infty(\mathscr{B}_Q) \to A^\infty(\mu + \lambda_Q)$ is a dual algebra isomorphism.
6. Assume that K is a compact subset of $\mathbb{C}$ such that Q, the set of nonpeak points for $R(K)$, is the interior of K. If ω is harmonic measure for K, show that $\omega \in \mathscr{B}_Q$ and the natural projection $H^\infty(\mathscr{B}_Q) \to H^\infty(\omega)$ is an isometric isomorphism and a weak* homeomorphism.

CHAPTER VI

Weak-Star Rational Approximation

In this chapter we will investigate $R^\infty(K, \mu)$, the weak-star closure of $R(K)$ in $L^\infty(\mu)$ for a measure μ supported on the compact subset K in the plane. Specifically, we will prove Chaumat's Theorem characterizing $R^\infty(K, \mu)$ as $R^\infty(K, \lambda)$, where λ is the restriction of Lebesgue measure to an appropriate set. This will shed light on the functional calculus for subnormal operators. We will also derive Sarason's Theorem characterizing $P^\infty(\mu)$, the weak-star closure of the polynomials in $L^\infty(\mu)$. In the next chapter this information will be exploited to derive some structure theorems for subnormal operators.

§1 Weak-star closed subalgebras of $L^\infty(\mu)$. In this section (X, Ω, μ) will always denote a finite measure space. It is not necessary to restrict our attention to Borel measures here. Also it is pointless to allow μ to be σ-finite since if (X, Ω) carries a σ-finite measure ν, there is an equivalent measure μ that is finite. Since $L^\infty(\mu) = L^\infty(\nu)$ for mutually absolutely continuous measures and we are only interested in the L^∞ spaces, we might as well assume that the measure μ is finite. We will consider subalgebras $\mathscr{A}$ of $L^\infty(\mu)$ and it will always be assumed that $\mathscr{A}$ contains the constant functions.

1.1 DEFINITION. A subalgebra $\mathscr{A}$ of $L^\infty(\mu)$ is *antisymmetric* if $1 \in \mathscr{A}$ and the constant functions are the only real-valued functions in $\mathscr{A}$. The algebra $\mathscr{A}$ is *pseudosymmetric* if $1 \in \mathscr{A}$ and for any measurable set Δ with $\mu(\Delta) > 0$, there is a function φ in $\mathscr{A}$ with φ real-valued on Δ and φ not constant on Δ.

Algebras that are antisymmetric are easy to find. For example, let G be a bounded region (connected) in $\mathbb{C}$, let $\mu =$ the restriction of area measure to G, and let $\mathscr{A} = H^\infty(G) \subseteq L^\infty(\mu)$. It is easy to see that $\mathscr{A}$ is antisymmetric. Notice that $L^\infty(\mu)$ is pseudosymmetric provided μ has no atoms. Nontrivial pseudosymmetric subalgebras are less common.

1.2 EXAMPLE. If μ is the restriction of area measure to $\mathbb{D}$ and $\mathscr{A}$ is the weak* closed subalgebra of $L^\infty(\mu)$ generated by z and $|z|$, then $\mathscr{A} \neq L^\infty(\mu)$ and $\mathscr{A}$ is pseudosymmetric. In fact, $\int z^n |z|^m \bar{z}\, d\mu = (\int_0^1 r^{n+m+2} dr) \times (\int_0^{2\pi} e^{i(n+1)\theta} d\theta) = 0$. That is, the $L^1(\mu)$ function $\bar{z}$ annihilates $\mathscr{A}$ and so $\mathscr{A} \neq L^\infty(\mu)$. But note that for any set Δ with $\mu(\Delta) > 0$, $|z|$ is real-valued and nonconstant on Δ.

1.3 PROPOSITION. *If $\mathscr{A}$ is a weak* * closed subalgebra of $L^\infty(\mu)$, $\varphi \in \mathscr{A}$, φ is real-valued, and $\Delta = \varphi^{-1}(F)$ for some measurable subset F of $\mathbb{R}$, then $\chi_\Delta \in \mathscr{A}$.*

PROOF. We first show this for a closed subset F of $\mathbb{R}$. If F is a closed subset of $\mathbb{R}$, there is a sequence $\{u_n\}$ of continuous functions on $\mathbb{R}$ such that $0 \le u_n \le 1$ and $\{u_n(t)\}$ decreases to $\chi_F(t)$ for every t in $\mathbb{R}$. Since each function u_n can be uniformly approximated by polynomials on $[-\|\varphi\|_\infty, \|\varphi\|_\infty]$, $u_n \circ \varphi \in A$. It can be checked that $u_n \circ \varphi \to \chi_\Delta$ weak* in $L^\infty(\mu)$. If $\mathscr{F} = \{F: F \text{ is a Borel subset of } \mathbb{R} \text{ and } \chi_{\varphi^{-1}(F)} \in \mathscr{A}\}$, then the fact that $\mathscr{A}$ is an algebra with identity shows that $\mathscr{F}$ is closed under the operations of union, intersection, and complementation. The fact that $\mathscr{A}$ is weak* closed shows that $\mathscr{A}$ is closed under countable unions and intersections. That is, $\mathscr{F}$ is a σ-algebra. Since $\mathscr{F}$ contains the closed sets, $\mathscr{F}$ must be the Borel sets. □

It turns out that both pseudosymmetric and antisymmetric algebras can be characterized in terms of the projections, or characteristic functions, they contain.

1.4 PROPOSITION. *If $\mathscr{A}$ is a weak* closed subalgebra of $L^\infty(\mu)$, then $\mathscr{A}$ is pseudosymmetric if and only if $1 \in \mathscr{A}$ and $\mathscr{A}$ contains no nonzero minimal projections.*

PROOF. Suppose $\mathscr{A}$ is pseudosymmetric. If $\chi_\Delta \in \mathscr{A}$ and $\chi_\Delta \ne 0$, let $\varphi \in A$ such that $\varphi|\Delta$ is real-valued but not constant. We can assume that $\varphi = \varphi \chi_\Delta$ and so $\varphi(x) = 0$ for $x \notin \Delta$. Since φ is not constant on Δ, there is a real number α such that measure theoretically, $\Sigma = \varphi^{-1}([\alpha, \infty)) \cap \Delta$ is neither empty nor all of Δ. By the preceding proposition, $\chi_\Sigma \in \mathscr{A}$. Hence χ_Δ is not minimal.

Now assume that $\mathscr{A}$ has no nonzero minimal projections. Suppose Δ_0 is a measurable set with $\mu(\Delta_0) > 0$ and put $\mathscr{S} = \{\Sigma: \chi_\Sigma \in \mathscr{A} \text{ and } \Sigma \supseteq \Delta_0\}$. Since $1 \in \mathscr{A}$, $\mathscr{S} \ne \varnothing$. Let $\{\Sigma_n\} \subseteq \mathscr{S}$ such that $\mu(\Sigma_n) \to \alpha \equiv \inf\{\mu(\Sigma): \Sigma \in \mathscr{S}\}$. Since $\mathscr{A}$ is an algebra, we may assume that the sets $\{\Sigma_n\}$ are decreasing. Put $\Sigma = \bigcap_n \Sigma_n$. Because $\mathscr{A}$ is weak* closed, $\chi_\Sigma \in \mathscr{A}$; clearly $\mu(\Sigma) = \alpha$ and $\chi_\Sigma \ne 0$ in $\mathscr{A}$. Since χ_Σ cannot be a minimal projection in $\mathscr{A}$, there is a measurable set $\Delta \subseteq \Sigma$ such that $\chi_\Delta \in \mathscr{A}$ and $0 < \mu(\Delta) < \alpha$. Thus $\Delta \notin \mathscr{S}$ and so $\mu(\Delta_0 \setminus \Delta) > 0$. Also, $\alpha > \mu(\Sigma \setminus \Delta)$ and so $\Sigma \setminus \Delta \notin \mathscr{S}$; thus $\mu(\Delta \cap \Delta_0) > 0$. This implies that χ_Δ is not constant on Δ_0. □

1.5 PROPOSITION. *If $\mathscr{A}$ is a weak* closed subalgebra of $L^\infty(\mu)$, then $\mathscr{A}$ is antisymmetric if and only if the only projections in $\mathscr{A}$ are 0 and 1.*

PROOF. Since projections are real-valued functions, it is clear that the only projections in an antisymmetric algebra are 0 and 1. So assume that $\mathscr{A}$ is a weak* closed subalgebra of $L^\infty(\mu)$ such that the only projections in $\mathscr{A}$ are

0 and 1. If $\varphi \in \mathscr{A}$ and φ is real-valued, then Proposition 1.3 implies that for every real number α, $\chi_{\Delta_\alpha} \in \mathscr{A}$ where $\Delta_\alpha = \varphi^{-1}([\alpha, \infty))$. But for each α, $\chi_{\Delta_\alpha} = 0$ or 1. It is left to the reader to show that φ must be constant. □

1.6 THEOREM (Conway and Olin [1977]). *If (X, Ω, μ) is a finite measure space and $\mathscr{A}$ is a weak* closed subalgebra of $L^\infty(\mu)$ containing the constant functions, then there is a measurable partition $\{\Delta_0, \Delta_1, \Delta_2, \ldots\}$ of X having the following properties.*

(a) $\chi_{\Delta_n} \in \mathscr{A}$ *for all* $n \geq 0$.
(b) *For* $n \geq 1$, χ_{Δ_n} *is a minimal projection in* $\mathscr{A}$.
(c) $\mathscr{A}|\Delta_0$ *is a pseudosymmetric subalgebra of* $L^\infty(\mu|\Delta_0)$.
(d) *For* $n \geq 1$, $\mathscr{A}|\Delta_n$ *is an antisymmetric subalgebra of* $L^\infty(\mu|\Delta_n)$.
(e) $\mathscr{A} = \mathscr{A}|\Delta_0 \oplus \mathscr{A}|\Delta_1 \oplus \mathscr{A}|\Delta_2 \oplus \cdots$.

Moreover, this decomposition $\{\Delta_n : n \geq 0\}$ is unique except for the ordering of the terms.

PROOF. Let $\mathscr{P}$ = the collection of minimal projections in $\mathscr{A}$. Because each function in $\mathscr{P}$ corresponds to a set in Ω, distinct elements of $\mathscr{P}$ correspond to disjoint sets in Ω. Since μ is a finite measure, $\mathscr{P}$ can be at most countable. Let $\mathscr{P} = \{\chi_1, \chi_2, \ldots\}$, where $\chi_n = \chi_{\Delta_n}$ for some Δ_n in Ω and $\Delta_n \cap \Delta_m = \varnothing$ for $n \neq m$. Let Δ_0 be the complement of $\bigcup_{n \geq 1} \Delta_n$. Since $\chi_1 + \cdots + \chi_n \in \mathscr{A}$ for every $n \geq 1$ and $\mathscr{A}$ is weak* closed, $\sum_{n \geq 1} \chi_n \in \mathscr{A}$; thus $\chi_{\Delta_0} \in \mathscr{A}$.

Condition (b) is satisfied by definition. Condition (c) holds by Proposition 1.4 and the construction of Δ_0 and part (d) holds by Proposition 1.5. Condition (e) follows directly from the fact that $\chi_n \in \mathscr{A}$, $\{\Delta_n\}$ is a partition, and $\mathscr{A}$ is weak* closed.

It remains to prove the uniqueness statement. But this is essentially immediate from the fact that if $\Delta \in \Omega$ and χ_Δ is a minimal projection, then $\Delta = \Delta_n$ for some $n \geq 1$. □

Note that the set Δ_0 in the preceding theorem contains no atoms of μ since every function in $L^\infty(\mu)$ is constant on an atom. Hence each atom of μ is contained in one of the antisymmetric parts, Δ_n with $n \geq 1$. The following might be worthwhile recording.

1.7 EXAMPLE. If $\mathscr{A} = L^\infty(\mu)$, then the decomposition in Theorem 1.6 becomes the decomposition of the measure space (X, Ω, μ) into the atoms of μ, $\Delta_1, \Delta_2, \ldots$, and the continuous part of the measure μ, Δ_0.

1.8 EXAMPLE. Let μ be a compactly supported measure on $\mathbb{C}$ and put $\mathscr{A} = P^2(\mu) \cap L^\infty(\mu)$. The decomposition of $\mathscr{A}$ obtained in Theorem 1.6 is the same as the decomposition of $\mathscr{A}$ obtained in Corollary IV.5.6 except that some of the antisymmetric summands in Theorem 1.6 can be finite dimensional (and hence 1-dimensional). In (IV.5.6) these 1-dimensional summands were thrown together with the algebra $L^\infty(\mu_0)$. It was also shown in (IV.5.6) that $P^2(\mu|\Delta_0) \cap L^\infty(\mu|\Delta_0) = L^\infty(\mu|\Delta_0)$.

If, in Theorem 1.6, one of the sets Δ_n is an atom, then $1 = \dim(\mathscr{A}|\Delta_n)$. It can be shown (Exercise 2) that if $\mathscr{A}$ is an antisymmetric algebra that is finite dimensional, then $\dim \mathscr{A} = 1$ and so $\mathscr{A} = \mathbb{C}$. It does not follow, however, that the set Δ_n is an atom when $\dim(\mathscr{A}|\Delta_n) = 1$ as can be seen in Example 1.10 below.

1.9 EXAMPLE. Let G be a bounded open subset of $\mathbb{C}$, let $\mu = \text{Area}|G$, and let $G_1, G_2, \ldots$ be the components of G. If $\mathscr{A} = H^\infty(G) \subseteq L^\infty(\mu)$, then the sets $\Delta_n = G_n$ for $n \geq 1$ and $\Delta_0 = \varnothing$.

1.10 EXAMPLE. Let G, μ, and $G_1, G_2, \ldots$ be as in the preceding example and let $\mathscr{A} = \{f \in H^\infty(G) : f \text{ is constant on } G_1\}$. Once again the sets Δ_n, $n \geq 1$ obtained by Theorem 1.6 are $\Delta_n = G_n$, but this time $\mathscr{A}|\Delta_1$ is one-dimensional.

The question arises as to when the pseudosymmetric part of the decomposition in Theorem 1.6, $\mathscr{A}|\Delta_0$, equals $L^\infty(\mu|\Delta_0)$. Indeed, this question will be of interest to us when we study $R^\infty(K, \mu)$ and the resulting functional calculus for subnormal operators. In fact, in this situation it is the case that the pseudosymmetric part is the entire L^∞ space (Theorem 1.14, below). Now another decomposition of a weak* closed subalgebra of $L^\infty(\mu)$ is given that can be combined with Theorem 1.6 to give a finer decomposition of the pseudosymmetric summand in (1.6). The main tool used to obtain this decomposition is Lemma V.17.10.

1.11 PROPOSITION. *If (X, Ω, μ) is a finite measure space and $\mathscr{A}$ is a weak* closed subalgebra of $L^\infty(\mu)$ containing the constant functions, then there is a measurable partition $\{\Delta_1, \Delta_2\}$ of X such that χ_{Δ_1} and $\chi_{\Delta_2} \in \mathscr{A}$,*

$$\mathscr{A} = L^\infty(\mu|\Delta_1) \oplus \mathscr{A}|\Delta_2,$$

and $\mathscr{A}|\Delta_2$ contains no nontrivial weak closed ideal of $L^\infty(\mu)$. Moreover, if $1 \leq p < \infty$, then there is a measure ν on (X, Ω) that is carried by Δ_2 such that $\int f \, d\nu = 0$ for all f in $\mathscr{A}$, $[\nu] = [\mu|\Delta_2]$, and the Radon-Nikodym derivative of ν with respect to μ belongs to $L^p(\mu)$.*

PROOF. According to Proposition II.11.6, there is no loss in generality in assuming that $\mathscr{A}^q(\mu) \cap L^\infty(\mu) = \mathscr{A}$, where q is dual to p. Let $C \equiv \{f \in L^p(\mu) : \int f\varphi \, d\mu = 0 \text{ for all } \varphi \text{ in } \mathscr{A}\}$; according to Lemma V.17.10, there is a function f in C such that $g\mu \ll f\mu$ for all g in C. Put $\Delta_1 = \{x : f(x) = 0\}$ and $\Delta_2 = \{x : f(x) \neq 0\}$. Clearly $\mathscr{A} \subseteq L^\infty(\mu|\Delta_1) \oplus \mathscr{A}|\Delta_2$. On the other hand, if $\varphi \in L^\infty(\mu|\Delta_1)$, then $\int g\varphi \, d\mu = 0$ for every g in C. By the Hahn-Banach Theorem, $L^\infty(\mu|\Delta_1) \subseteq \mathscr{A}$. Some algebraic manipulation shows that $\mathscr{A}|\Delta_2 \subseteq \mathscr{A}$ so that we get $\mathscr{A} = L^\infty(\mu|\Delta_1) \oplus \mathscr{A}|\Delta_2$.

Now every weak* closed ideal of $L^\infty(\mu)$ is of the form $L^\infty(\mu|\Delta)$ for some Δ in Ω (Exercise 3). So if $\mathscr{A}|\Delta_2$ contains a weak* closed ideal this would contradict the definition of Δ_2. Finally, put $\nu = f\mu$. By the definition of Δ_2, $[\nu] = [\mu|\Delta_2]$. □

1.12 COROLLARY. *If $\mathscr{A}$ is an antisymmetric subalgebra of $L^\infty(\mu)$, there is a measure ν such that $[\nu] = [\mu]$ and ν annihilates $\mathscr{A}$. Moreover, ν can be chosen so that if $1 \leq p < \infty$, then the Radon-Nikodym derivative of ν with respect to μ belongs to $L^p(\mu)$.*

1.13 DEFINITION. A weak* closed subalgebra of $L^\infty(\mu)$ *contains no L^∞ summand* if the set Δ_1 in Proposition 1.11 is void.

Equivalently, a weak* closed subalgebra of $L^\infty(\mu)$ contains no L^∞ summand provided there is a measure equivalent to μ that annihilates $\mathscr{A}$.

Pseudosymmetric algebras appear to be quite mysterious. However, if the algebra is the weak* closure of a T-invariant algebra, the mystery disappears.

1.14 THEOREM. *If A is a T-invariant algebra on K and μ is a positive Borel measure on K such that $A^\infty(\mu)$, the weak* closure of A in $L^\infty(\mu)$, is pseudosymmetric, then $A^\infty(\mu) = L^\infty(\mu)$.*

To prove this theorem we need a preliminary result, which sheds a little more light on pseudosymmetric algebras and thus has some interest of its own.

1.15 PROPOSITION. *If $\mathscr{A}$ is a pseudosymmetric subalgebra of $L^\infty(\mu)$, then $\mathscr{A}$ has no weak* continuous multiplicative linear functionals.*

PROOF. Suppose $\rho: \mathscr{A} \to \mathbb{C}$ is a weak* continuous homomorphism and $\rho \neq 0$. Since ρ is a homomorphism, ρ assumes the value 1 at every idempotent of $\mathscr{A}$. Put $\mathscr{E} = \{\Delta \in \Omega: \chi_\Delta \in \mathscr{A} \text{ and } \rho(\chi_\Delta) = 1\}$. Since $\rho \neq 0$, $X \in \mathscr{E}$ and so $\mathscr{E}$ is a nonempty collection of measurable sets. If Δ_1 and $\Delta_2 \in \mathscr{E}$, the $\rho(\chi_{\Delta_1 \cap \Delta_2}) = \rho(\chi_{\Delta_1})\rho(\chi_{\Delta_2}) = 1$ and so $\Delta_1 \cap \Delta_2 \in \mathscr{E}$. Thus there is a sequence $\{\Delta_n\} \subseteq \mathscr{E}$ such that $\Delta_n \supseteq \Delta_{n+1}$ and $\mu(\Delta_n) \to \alpha \equiv \inf\{\mu(\Delta): \Delta \in \mathscr{E}\}$. If $\Delta = \bigcap_n \Delta_n$, then $\chi_{\Delta_n} \to \chi_\Delta$ weak* in $L^\infty(\mu)$ and so $\chi_\Delta \in \mathscr{A}$. Since ρ is weak* continuous, $\rho(\chi_\Delta) = 1$. That is, $\Delta \in \mathscr{E}$ and clearly $\mu(\Delta) = \alpha$.

If $\chi_\Sigma \in \mathscr{A}$ and $\Sigma \subseteq \Delta$, then $1 = \rho(\chi_\Delta) = \rho(\chi_{\Delta \setminus \Sigma}) + \rho(\chi_\Sigma)$. Now either $\rho(\chi_\Sigma) = 1$ or $\rho(\chi_\Sigma) = 0$. Hence either Σ or $\Delta \setminus \Sigma \in \mathscr{E}$. But by the construction of Δ, this means that either $\chi_\Sigma = \chi_\Delta$ or $\chi_\Sigma = 0$. That is, χ_Δ is a minimal projection in $\mathscr{A}$. According to Proposition 1.4, this is impossible since $\mathscr{A}$ is pseudosymmetric. □

PROOF OF THEOREM 1.14. It will be shown that if $\nu \in M(K)$, $\nu \ll \mu$, and $\nu \perp A$, then $\nu = 0$. Suppose $\nu \neq 0$. Then there is a point a in K such that $\hat{\nu}(a) < \infty$ and $\hat{\nu}(a) \neq 0$. According to Corollary V.7.9 there is a representing measure λ for the point a with $\lambda \ll \nu \ll \mu$. It follows that $\rho(f) = \int f d\lambda$ defines a weak* continuous homomorphism on $A^\infty(\mu)$, contradicting the preceding proposition. □

Theorem 1.14 can be combined with Theorem 1.6 to give the following decomposition theorem which will be of particular interest to us for the development of the functional calculus for subnormal operators.

1.16 PROPOSITION. *If K is a compact subset of $\mathbb{C}$ and μ is a positive* measure in $M(K)$, then there is a Borel partition $\{\Delta_0, \Delta_1, \Delta_2, \cdots\}$ of K such that

$$R^\infty(K, \mu) = L^\infty(\mu|\Delta_0) \oplus R^\infty(K, \mu|\Delta_1) \oplus R^\infty(K, \mu|\Delta_2) \oplus \cdots$$

where $R^\infty(K, \mu|\Delta_n)$ is antisymmetric for $n \geq 1$.

Exercises

1. Let $\mathscr{A}$ be a weak* closed subalgebra of $L^\infty(\mu)$ and let K be a compact subset of $\mathbb{C}$ such that $\mathbb{C}\setminus K$ is connected and int $K = \emptyset$.
   (a) Show that for every f in $\mathscr{A}$, $\chi_{f^{-1}(K)} \in \mathscr{A}$.
   (b) If E is a Borel set contained in K and $f \in \mathscr{A}$, show that $\chi_{f^{-1}(E)} \in \mathscr{A}$.
2. Show that a finite dimensional antisymmetric subalgebra of $L^\infty(\mu)$ is one-dimensional and consists of the constant functions.
3. Prove that every weak* closed ideal of $L^\infty(\mu)$ is of the form $\{f \in L^\infty(\mu) : f = f\chi_\Delta\}$ for some measurable set Δ.
4. If $\mathscr{B}$ is a subalgebra of $L^\infty(\mu)$ and $\mathscr{A}$ is the weak* closure of $\mathscr{B}$, show that $\mathscr{A}$ is an algebra. If $\Delta \in \Omega$ and $\chi_\Delta \in \mathscr{A}$, show that $\mathscr{A}|\Delta$ is the weak* closure of $\mathscr{B}|\Delta$.
5. If $\mathscr{A}$ is a pseudosymmetric subalgebra of $L^\infty(\mu)$ and φ is a real-valued function in $\mathscr{A}$, show that $\mathscr{A}$ contains the von Neumann algebra generated by φ.
6. Give the details of the proof of Proposition 1.16.

§2 The envelope. Starting with this section we will focus our attention on the algebra $R^\infty(K, \mu)$, the weak* closure of $R(K)$ in $L^\infty(\mu)$. Throughout this section, K will, as usual, denote a compact subset of $\mathbb{C}$ and μ is a positive Borel measure carried by K. Here and throughout the book, $\|\varphi\|_\mu$ will denote the μ-essential supremum norm of a function φ.

2.1 DEFINITION. The *envelope* of μ with respect to K is the set $E(K, \mu)$ of all complex numbers a for which there is a measure μ_a such that μ_a is absolutely continuous with respect to μ, $\mu_a(\{a\}) = 0$, and $f(a) = \int f d\mu_a$ for all f in $R(K)$. At times we may also use E_μ or just E to denote the envelope, if there is no possibility of a misconception.

2.2 PROPOSITION. *If E is the envelope of μ with respect to K, then the following hold.*
   (a) *E contains no peak points of $R(K)$.*
   (b) *If $\nu \in L^1(\mu)$ and $\nu \perp R(K)$, then E contains $\{z: \{z\}$ is not an atom of μ and $\check{\nu}(z) < \infty$ and $\hat{\nu}(z) \neq 0\}$. Hence if $\nu \in L^1(\mu)$ and $\nu \perp R(K)$, then $\hat{\nu}(z) = 0$ a.e. [Area] off E.*

PROOF. Exercise. (There is a small difficulty in the second statement of part (b) since we do not as yet know that E is measurable. This will be remedied in Proposition 2.8 below.) □

Another set of complex numbers associated with K and μ, which we will see to be the same as the envelope in all the cases that matter, is the set of weak* continuous homomorphisms. If $\rho: R^\infty(K, \mu) \to \mathbb{C}$ is a homomorphism, then $\rho | R(K)$ is a homomorphism on $R(K)$. Hence there is a point a in K such that $\rho(f) = f(a)$ for all f in $R(K)$. If, in addition, ρ is weak* continuous on $R^\infty(K, \mu)$, then ρ is completely determined by its action on $R(K)$. That is, the set of weak* continuous homomorphisms on $R^\infty(K, \mu)$ is in one-to-one correspondence with a subset of K.

2.3 DEFINITION. The set $\Sigma(K, \mu) = \{a \in K:$ there is a weak* continuous homomorphism $\rho: R^\infty(K, \mu) \to \mathbb{C}$ such that $\rho(f) = f(a)$ for all f in $R(K)\}$.

The reader can probably already see a relation between $E(K, \mu)$ and $\Sigma(K, \mu)$. Before making this explicit and perhaps proving a little more than the reader can see at this moment, we need an elementary lemma.

2.4 LEMMA. *If (F_1, F_2) is a measurable partition of K such that $\chi_{F_j} \in R^\infty(K, \mu)$, then $R^\infty(K, \mu) = R^\infty(K, \mu|F_1) \oplus R^\infty(K, \mu|F_2)$.*

PROOF. Since $\chi_{F_1} \in R^\infty(K, \mu)$, it is easy to see that $R^\infty(K, \mu) = \chi_{F_1} R^\infty(K, \mu) \oplus \chi_{F_2} R^\infty(K, \mu)$. So it is only necessary to show that $\chi_{F_1} R^\infty(K, \mu) = R^\infty(K, \mu|F_1)$. The fact that $\chi_{F_1} R^\infty(K, \mu) \subseteq R^\infty(K, \mu|F_1)$ is an easy consequence of the fact that $\chi_{F_1} \in R^\infty(K, \mu)$ and is left to the reader. For the other inclusion, let $\varphi \in R^\infty(K, \mu|F_1)$ and let $\{f_i\}$ be a net in $R(K)$ such that $f_i \to \varphi$ weak* in $L^\infty(\mu|F_1)$. Since $\chi_{F_1} \in R^\infty(K, \mu)$, $f_i \chi_{F_1} \in \chi_{F_1} R^\infty(K, \mu)$ and it is immediate that $f_i \chi_{F_1} \to \varphi \chi_{F_1}$ weak* in $L^\infty(\mu)$. Thus $\varphi \chi_{F_1} \in R^\infty(K, \mu)$; but $\varphi \chi_{F_1} = \varphi$ and so the proof is complete. □

2.5 PROPOSITION. *If $E = E(K, \mu)$ and $\Sigma = \Sigma(K, \mu)$, then $E \subseteq \Sigma$ and $\Sigma \backslash E$ is at most a countable set consisting of atoms of μ. Moreover for each a in $\Sigma \backslash E$, $\chi_{\{a\}} \in R^\infty(K, \mu)$. In particular, if $R^\infty(K, \mu)$ contains no L^∞ summand, then $E = \Sigma$.*

PROOF. The fact that $E \subseteq \Sigma$ is immediate from the definition of the envelope since for each a in E the homomorphism $f \to f(a)$ on $R(K)$ is represented by a measure in $L^1(\mu)$. If $a \in \Sigma \backslash E$, then there is a measure ν in $L^1(\mu)$ such that $f(a) = \int f \, d\nu$ for all f in $R(K)$. Since $a \notin E$, it must be that $\nu(\{a\}) > 0$. Thus a is an atom of μ.

Let $a \in \Sigma \backslash E$, put $\mu_0 = \mu - \mu(\{a\})\delta_a$, and let $\tau: R^\infty(K, \mu) \to R^\infty(K, \mu_0)$ be the natural homomorphism defined by restriction. Clearly τ is a weak* continuous contraction and τ has weak* dense range since $\operatorname{ran} \tau \supseteq R(K)$. If τ is an isometry, then τ is a dual algebra isomorphism (I.2.10). Therefore $\Sigma(K, \mu_0) = \Sigma$. Since $a \in \Sigma$, this implies there is a measure $\nu \ll \mu_0$ such that $f(a) = \int f \, d\nu$ for all f in $R(K)$. But then $\nu \ll \mu$ and $\nu(\{a\}) = 0$

and so $a \in E(K, \mu)$, a contradiction. Therefore it must be that τ is not an isometry.

Since τ is not an isometry, there is a function φ in $R^\infty(K, \mu)$ with $\|\varphi\|_\mu = 1$ and $\|\tau(\varphi)\|_{\mu_0} < 1$. But this says that $\varphi(a) = 1$ and $|\varphi(z)| < 1$ for μ almost every $z \neq a$. Thus $\varphi^n \to \chi_{\{a\}}$ weak* in $L^\infty(\mu)$. Hence $\chi_{\{a\}} \in R^\infty(K, \mu)$.

The last statement is an immediate consequence of the preceding lemma. □

2.6 PROPOSITION. *Let K be a compact subset of $\mathbb{C}$ and let μ be a positive Borel measure on K. If for each z in $\Sigma(K, \mu)$, ρ_z is the unique homomorphism on $R^\infty(K, \mu)$ such that $\rho_z(f) = f(z)$ for all f in $R(K)$, then the following statements hold.*

(a) *For each φ in $R^\infty(K, \mu)$, $\varphi(z) = \rho_z(\varphi)$ a.e. $[\mu]$ on $\Sigma(K, \mu)$.*
(b) *For each φ in $R^\infty(K, \mu)$, the function $z \to \rho_z(\varphi)$ is a bounded analytic function on $\text{int}\{\Sigma(K, \mu)\}$.*
(c) $\text{int}\{\text{supp}\,\mu\} \subseteq E(K, \mu)$.

PROOF. Let $\mathscr{M} = \{\varphi \in R^\infty(K, \mu): z \to \rho_z(\varphi)$ is a bounded analytic function on $\text{int}\,\Sigma(K, \mu)$ and $\varphi(z) = \rho_z(\varphi)$ a.e. $[\mu]$ on $\Sigma(K, \mu)\}$. Clearly $\mathscr{M} \supseteq R(K)$. So to prove both (a) and (b), it suffices to prove that $\mathscr{M}$ is weak* closed. Let $\{\varphi_n\}$ be a sequence in $\mathscr{M}$ such that $\|\varphi_n\|_\infty \le 1$ for all $n \ge 1$, and assume that $\varphi_n(z) \to \varphi(z)$ a.e. $[\mu]$ in $L^\infty(\mu)$. Since $\|\rho_z\| = 1$ for all z, $|\rho_z(\varphi_n)| \le 1$ for all $n \ge 1$. Also $\varphi_n \to \varphi$ weak* and so $\rho_z(\varphi_n) \to \rho_z(\varphi)$ for each z in $\Sigma(K, \mu)$. By Montel's Theorem, $z \to \rho_z(\varphi)$ is a bounded analytic function on $\text{int}\{\Sigma(K, \mu)\}$. It also follows that $\varphi(z) = \rho_z(\varphi)$ a.e. $[\mu]$ on $\Sigma(K, \mu)$. Thus $\varphi \in \mathscr{M}$ and so, by Corollary V.22.4, $\mathscr{M}$ is weak* closed. This proves (a) and (b).

(c) Let $G = \text{int}\{\text{supp}\,\mu\}$. As in the preceding paragraph, the linear manifold $\{\varphi \in R^\infty(K, \mu):$ there is a function g in $H^\infty(G)$ such that $\varphi = g$ a.e. $[\mu]$ on $G\}$ is weak* closed and contains $R(K)$; hence this is all of $R^\infty(K, \mu)$. Notice that for each φ in $R^\infty(K, \mu)$ the function g in $H^\infty(G)$ such that $\varphi = g$ a.e. $[\mu]$ on G is unique. (Why?)

Now let $a \in G$ and define $\rho: R^\infty(K, \mu) \to \mathbb{C}$ by letting $\rho(\varphi) = g(a)$ where g is the unique function in $H^\infty(G)$ such that $\varphi = g$ a.e. $[\mu]$ on G. It is left to the reader to show that ρ is a homomorphism. It is also left as an exercise in the application of Corollary V.22.4 to show that $\{\varphi \in R^\infty(K, \mu): \rho(\varphi) = 0\}$ is weak* closed. Hence ρ is weak* continuous. Thus $G \subseteq \Sigma(K, \mu)$.

Suppose $a \notin E$. By Proposition 2.5, $\{a\}$ must be an atom of μ and $\chi = \chi_{\{a\}} \in R^\infty(K, \mu)$. Let U be the component of G that contains a. Now $z \to \rho_z(\chi)$ is a bounded analytic function on U and, since χ is idempotent, must equal either 0 or 1 at all points of U. Thus either $\rho_z(\chi) \equiv 0$ or $\rho_z(\chi) \equiv 1$ on U. But $\rho_z(\chi) = \chi$ a.e. $[\mu]$. This implies that $\rho_z(\chi) = 0$ for

$z \neq a$ and $p_z(\chi) = 1$ since $\{a\}$ is an atom of μ. This contradiction shows that $a \in E$. □

This section concludes with a proof that the envelope is a measurable set. This proof begins with a lemma.

2.7 LEMMA. *If $a \in E(K, \mu)$, there is a measurable subset F of E such that F has full area density at a.*

PROOF. Fix a in $E(K, \mu)$ and let μ_a be as in the definition of the envelope. Put $F = \{z : |z - a|\tilde{\mu}_a(z) < 1\}$ and let $B_\delta = B(a; \delta)$. First we show that F has full area density at a. That is, we want to show that $\text{Area}(F \cap B_\delta)/\pi\delta^2 \to 1$ as $\delta \to 0$. Equivalently, we want to show that $\text{Area}(F \backslash B_\delta)/\pi\delta^2 \to 0$.

$$\frac{\text{Area}(B_\delta \backslash F)}{\pi\delta^2} = \frac{1}{\pi\delta^2} \int_{B_\delta} \chi_{\mathbb{C}\backslash F}(z) \, d\mathscr{A}(z)$$
$$\leq \frac{1}{\pi\delta^2} \int_{B_\delta} |z - a|\tilde{\mu}_a(z) \, d\mathscr{A}(z)$$
$$= \frac{1}{\pi\delta^2} \int_{B_\delta} \left[\int \left| \frac{z-a}{w-z} \right| d|\mu_a|(w) \right] d\mathscr{A}(z)$$
$$= \int \left[\frac{1}{\pi\delta^2} \int_{B_\delta} \left| \frac{z-a}{w-z} \right| d\mathscr{A}(z) \right] d|\mu_a|(w).$$

But if $w \neq a$, then for $0 < \delta < |w - a|$ and z in B_δ, $|w - z| \geq |w - a| - \delta$ and so $|(z-a)/(w-z)| \leq \delta/(|w - a| - \delta)$. Hence $g_\delta(w) = (1/\pi\delta^2) \int_{B_\delta} |(z-a)/(w-z)| \, d\mathscr{A}(z) \leq \delta/(|w-a| - \delta)$ for all $w \neq a$. This says two things: for $w \neq a$, $g_\delta(w) \to 0$ as $\delta \to 0$; and there is a constant C such that $|g_\delta(w)| \leq C|w - a|^{-1}$ for all δ and $w \neq a$. For $w = a$, $g_\delta(a) = 1/\pi\delta^2 \int_{B_\delta} |(z-a)/(a-z)| \, d\mathscr{A}(z) = 1$. Hence the Lebesgue Dominated Convergence Theorem implies that

$$\frac{1}{\pi\delta^2} \int_{B_\delta} \left| \frac{z-a}{w-z} \right| d\mathscr{A}(z) \to \chi_{\{a\}}(w)$$

for all w. Thus

$$\limsup_{\delta \to 0} \frac{\text{Area}(B_\delta \backslash F)}{\pi\delta^2} \leq |\mu_a|(\{a\}) = 0,$$

from which it follows that F has full area density at a.

It remains to prove that $F \subseteq E$. Let $\nu = (z - a)\mu_a$; so $\nu \perp R(K)$. Note that if C is the distance from a to $\text{supp}\,\mu_a$, then $\tilde{\nu}(z) = \int |w - z|^{-1} d|\nu|(w) = \int |(w - a)/(w - z)| \, d|\mu_a|(w) \leq C\tilde{\mu}_a(z)$. Fix z in

F. So $\hat{v}(z) < \infty$ and

$$|1 - \hat{v}(z)| \leq \left|1 - \int \left[\frac{w-a}{w-z}\right] d\mu_a(w)\right|$$

$$= \left|\int \left[\frac{a-z}{w-z}\right] d\mu_a(w)\right|$$

$$\leq |z-a|\tilde{\mu}_a(z)$$

$$< 1.$$

Hence $\hat{v}(z) \neq 0$ and so $z \in E(K, \mu)$ by (2.2b). □

2.8 PROPOSITION. *$E(K, \mu)$ is a measurable set with density 1 at each of its points.*

PROOF. Let $E = E(K, \mu)$ and fix a bounded open set U that contains E. If $a \in E$, then the preceding lemma implies there is a measurable set F with full area density at a and $F \subseteq E$. Thus if $n \geq 1$, there is an $r_0 > 0$ such that $\text{Area}(B(a; r) \setminus F) \leq (1/n) \text{Area}(B(a; r))$ whenever $r < r_0$. This implies that the outer measure of $B(a; r) \setminus E$ satisfies $\text{Area}^*(B(a; r) \setminus E) \leq (1/n) \text{Area}(B(a; r))$ for every $r < r_0$. Thus if $\mathscr{V}_n$ is the collection of all closed disks D contained in U with center in E and such that $\text{Area}^*(D \setminus E) \leq (1/n) \text{Area}(D)$, then $\mathscr{V}_n$ forms a Vitali cover of E. By the Vitali Covering Theorem there is a sequence of such disks $\{D_m^n : m \geq 1\}$ that are pairwise disjoint and such that $E \setminus \bigcup_m D_m^n$ is a set of outer measure 0. Put $C = \bigcap_{n=1}^{\infty} \bigcup_{m=1}^{\infty} D_m^n$.

Now $E = (E \setminus C) \cup (E \cap C)$. But $E \setminus C = \bigcup_{n=1}^{\infty} (E \setminus \bigcup_{m=1}^{\infty} D_m^n)$ and so $E \setminus C$ is a set of measure 0. So it suffices to show that $E \cap C$ is measurable. But $E \cap C = C \setminus [C \setminus E]$ so it suffices to show that $C \setminus E = \bigcap_{n=1}^{\infty} (\bigcup_{m=1}^{\infty} (D_m^n \setminus E))$ is measurable. But

$$\text{Area}^* \left(\bigcup_{m=1}^{\infty} (D_m^n \setminus E)\right) \leq \sum_{m=1}^{\infty} \text{Area}^*(D_m^n \setminus E)$$

$$\leq \sum_{m=1}^{\infty} \frac{1}{n} \text{Area}(D_m^n)$$

$$= \frac{1}{n} \text{Area} \left(\bigcup_{m=1}^{\infty} D_m^n\right)$$

$$\leq \frac{1}{n} M,$$

where $M = \text{Area}(U)$ for the open set U fixed at the start of the proof. Hence $\text{Area}^*(C \setminus E) \leq M/n$ for all $n \geq 1$ and so $C \setminus E$ is a set of measure 0 and hence must be measurable.

The fact that E has density 1 at each of its points is an immediate consequence of Lemma 2.7. □

§3 Chaumat's Theorem.

Once again, throughout this section, K will denote a fixed compact subset of $\mathbb{C}$ and μ will be a positive measure supported on K. Also fix further notation for this section by letting $E = E(K, \mu)$, the envelope introduced in the preceding section.

As we saw there, E is a Lebesgue measurable set and so we can consider the T-invariant algebra $R(K, E) \subseteq C(K)$. In this section we will prove Chaumat's Theorem characterizing $R^\infty(K, \mu)$. In later sections we will further explore this result and apply it to special classes of algebras $R(K)$ to obtain more specific information. To state Chaumat's Theorem we need more notation and a few preliminary results.

3.1 PROPOSITION. (a) *If $\nu \ll \mu$ and $\nu \perp R(K)$, then $\operatorname{supp} \nu \subseteq \operatorname{cl} E$ and $\nu \perp R(K, E)$.*

(b) *If $\eta \ll \mu$ and η represents evaluation at a on $R(K)$, then $\operatorname{supp} \eta \subseteq \operatorname{cl} E$ and η represents evaluation at a on $R(K, E)$.*

PROOF. (a) From Proposition V.8.9, $\hat{\nu} = 0$ a.e. [Area] off E. By Corollary V.3.4, ν is supported by $\operatorname{cl} E$. Also Proposition V.3.14 implies that $\nu \perp R(K, E)$.

(b) Put $\nu = (z - a)\eta$; so $\nu \perp R(K)$. By part (a) $\operatorname{supp} \nu \subseteq \operatorname{cl} E$ and $\nu \perp R(K, E)$. A computation shows that $\hat{\nu}(a) = \|\eta\|$ and $\hat{\nu}(a) = 1$. By (V.8.9), $\eta = [\hat{\nu}(a)]^{-1}(z-a)^{-1}\nu$ represents evaluation at a on $R(K, E)$. □

3.2 COROLLARY. $R(E) \subseteq R^\infty(K, \mu)$.

Let $Q = Q(K, \mu)$ be the collection of nonpeak points of $R(K, E)$. Actually this notation will be short-lived as we will see that $Q(K, \mu) = E(K, \mu)$. Let $\mathscr{B}_Q$ be the band of measures on K generated by the representing measures for the nonpeak points of $R(K, E)$ and denote by μ_Q the projection of μ into $\mathscr{B}_Q$. The measure μ_Q is not merely the restriction of μ to Q. (What is an example?) Let $\mathscr{S}$ be the singular band for $R(K, E)$ (V.18.3) and denote by $\mu_\mathscr{S}$ the projection of μ into $\mathscr{S}$; so $\mu_\mathscr{S} = \mu - \mu_Q$.

3.3 LEMMA. (a) $E \subseteq Q$ and $\operatorname{Area}(Q \setminus E) = 0$.

(b) *E and Q have the same closure and $R(K, E) = R(K, Q)$.*

(c) $E = E(K, \mu_Q) = \Sigma(K, \mu_Q)$.

PROOF. (a) If $a \in E$, then there is a representing measure ν for evaluation at a on $R(K)$ that is absolutely continuous with respect to μ and has no atom at a. By Proposition 3.1(b), ν is a representing measure for evaluation at a on $R(K, E)$ and so $a \in Q$. The fact that $\operatorname{Area}(Q \setminus E) = 0$ is just a restatement of Proposition V.13.10.

The proofs of (b) and (c) are left as an exercise. □

If ν is a measure supported on K, then, in analogy with previous notation, let $R^\infty(E, \nu)$ be the weak* closure of $R(K, E)$ in $L^\infty(\nu)$. We can now state Chaumat's Theorem originally proved in Chaumat [1974].

3.4 CHAUMAT'S THEOREM. *If K is a compact subset of $\mathbb{C}$ and μ is a positive measure supported on K, then $\mu = \mu_0 + \mu_a$, $\mu_a \perp \mu_0$, and the following conditions are satisfied.*

   (a) *$\mu_a \in \mathscr{B}_Q$ and $R^\infty(K, \mu_a)$ has no L^∞-summand.*
   (b) *$R^\infty(K, \mu) = L^\infty(\mu_0) \oplus R^\infty(K, \mu_a)$.*
   (c) *The identity map $R(K) \to R(K, E)$ extends to a dual algebra isomorphism of $R^\infty(K, \mu_a)$ onto $R^\infty(E, \mathscr{A}|E)$.*

Note that the measure μ_a in Chaumat's Theorem may not be the projection of μ into the band $\mathscr{B}_Q$. An easy example of this is to let K be the closed unit disk, $m = $ Lebesgue measure on ∂K, and $\mu = m|A$, where $A = \{z \in \partial K : \operatorname{Re} z \geq 0\}$. In this case, $\mu = \mu_Q$ but $\mu_a = 0$.

The proof cannot be given without a few preliminary lemmas. In particular, it is necessary to define the mapping between $R^\infty(K, \mu_a)$ and $R^\infty(E, \mathscr{A}|Q)$. Before beginning this process, let's look at one important corollary.

3.5 COROLLARY. *If $\varphi \in R^\infty(K, \mu_a)$, then there exists a sequence $\{f_n\}$ in $R(K, E)$ such that $\|f_n\| \leq \|\varphi\|_\infty$ and $f_n(z) \to \varphi(z)$ a.e. $[\mu_a]$.*

PROOF. Identify $R^\infty(K, \mu_a)$ and $R^\infty(E, \mathscr{A}|E)$ as in Chaumat's Theorem. Using the Corollary to Davie's Theorem (V.22.2), there is a sequence $\{f_n\}$ in $R(K, E)$ such that $\|f_n\| \leq \|\varphi\|_\infty$ and $f_n(z) \to \varphi(z)$ a.e. [Area] on E. Hence $f_n \to \varphi$ weak* in $R^\infty(E, \mathscr{A}|E)$. This implies that $f_n \to \varphi$ weak* in $R^\infty(K, \mu_a)$. Thus if ρ_z is the weak* continuous homomorphism on $R^\infty(K, \mu_a)$ corresponding to the point z in E, $f_n(z) = \rho_z(f_n) \to \rho_z(\varphi)$ for each z in E. But $\rho_z(\varphi) = \varphi(z)$ a.e. $[\mu_a]$ by Proposition 2.6. □

3.6 LEMMA. (a) *There exist measures μ_a in $\mathscr{B}_Q$ such that if $\mu_0 = \mu - \mu_a$, then $\mu_a \perp \mu_0$, $R^\infty(K, \mu) = R^\infty(K, \mu_a) \oplus L^\infty(\mu_0)$, and $R^\infty(K, \mu_a)$ has no L^∞-summand.*

(b) *The measure μ_a has the property that if $\nu \ll \mu$ and $\nu \perp R(K)$, then $\nu \ll \mu_a$.*

(c) *$R^\infty(K, \mu_a) = R^\infty(E, \mu_a)$.*

PROOF. (a) The existence of measures μ_a and μ_0 such that $R^\infty(K, \mu) = R^\infty(K, \mu_a) \oplus L^\infty(\mu_0)$ and $R^\infty(K, \mu_a)$ has no L^∞-summand follows from Proposition 1.11. This same proposition implies there is a measure $\eta \perp R(K)$ with $[|\eta|] = [\mu_a]$. Since η is an annihilating measure, $\eta \in \mathscr{B}_Q$. Hence $\mu_a \in \mathscr{B}_Q$.

(b) This is clear from (a).

(c) Since $E \subseteq K$, $R(K) \subseteq R(K, E)$ and hence $R^\infty(K, \mu_a) \subseteq R^\infty(E, \mu_a)$. The other inclusion follows from part (b) and Proposition 3.1. □

Recall that for any measure ν and any Borel function φ, $\|\varphi\|_\nu$ denotes the ν-essential supremum norm of φ. Observe that if $\varphi \in R^\infty(E, \mu_a)$, then $\|\varphi\|_\mu = \|\varphi\|_{\mu_a}$.

If $a \in E$, then there is a weak* continuous homomorphism $\rho_a: R^\infty(K, \mu) \to \mathbb{C}$ such that $\rho_a(f) = f(a)$ for all f in $R(K)$. In fact, $\rho_a(\varphi) = \int \varphi \, d\mu_a$ for all φ in $R^\infty(K, \mu)$, where μ_a is as in the definition of $E(K, \mu) = E$. Since $R^\infty(K, \mu) = R^\infty(E, \mu)$, we can formulate the following definition.

3.7 DEFINITION. If $\varphi \in R^\infty(E, \mu_a)$, define $\check{\varphi}: E(K, \mu) \to \mathbb{C}$ by $\check{\varphi}(a) = \rho_a(\varphi)$ for all a in $E = E(K, \mu)$, where ρ_a is the unique weak* continuous homomorphism from $R^\infty(E, \mu_a)$ into $\mathbb{C}$ that satisfies $\rho_a(f) = f(a)$ for all f in $R(K)$. $\check{\varphi}$ is called the *Chaumat transform* of φ.

Note that $\check{\varphi}$ is a well-defined function and $|\check{\varphi}(a)| \leq \|\varphi\|_\infty$. Also if $f \in R(K, E)$, write $f = f(a) + (z - a)g$ for some g in $R(K, E)$. Thus we see that $\check{f}(a) = f(a)$ for a in E and f in $R(K, E)$.

PROOF OF CHAUMAT'S THEOREM. In light of Lemma 3.6, μ_0 and μ_a exist and $R^\infty(K, \mu_a) = R^\infty(E, \mu_a)$. It only remains to demonstrate that the Chaumat transform defines an isomorphism of $R^\infty(E, \mu_a)$ onto $R^\infty(E, \mathscr{A}|Q)$. Let $\mathscr{M} = \{\varphi \in R^\infty(E, \mu_a): \check{\varphi} \in R^\infty(E, \mathscr{A}|E)\}$. Clearly $\mathscr{M} \supseteq R(K)$. We now show that $\mathscr{M}$ is weak* closed in $R^\infty(E, \mu_a)$. In fact, if $\{\varphi_n\} \subseteq \mathscr{M}$ and $\varphi_n \to \varphi$ weak*, then for every z in E, $\check{\varphi}_n(z) \to \check{\varphi}(z)$ and $\sup_n \|\check{\varphi}_n\|_E < \infty$. Thus $\varphi \in \mathscr{M}$ and $\mathscr{M}$ is indeed weak* closed. Therefore, we can define $\gamma: R^\infty(E, \mu_a) \to R^\infty(E, \mathscr{A}|E)$ by $\gamma(\varphi) = \check{\varphi}$. The preceding argument shows that γ is weak* sequentially continuous and hence must be weak* continuous by the Krein-Smulian Theorem. It is also easy to see that γ is contractive, $\gamma(f) = f$ for each f in $R(K)$, and γ is a multiplicative linear map. It remains to show that γ is surjective and an isometry.

Let $D: R^\infty(E, \mathscr{B}_Q) \to R^\infty(E, \mathscr{A}|E)$ be the map from Davie's Theorem and let $P: R^\infty(E, \mathscr{B}_Q) \to R^\infty(E, \mu_a)$ be the natural projection. Since, by the discussion following (3.7), $D(f) = \gamma \circ P(f)$ for every f in $R(K, E)$ and all the maps are weak* continuous, it follows that $D = \gamma \circ P$. But Davie's Theorem implies that for Φ in $R^\infty(E, \mathscr{B}_Q)$, $\|\Phi\| = \|D(\Phi)\|_\infty = \|\gamma(P(\Phi))\|_\infty \leq \|P(\Phi)\| \leq \|\Phi\|$. Hence P is an isometry and γ is an isometry on ran P. But both ran P and ran γ are weak* dense in their host spaces and so the usual application of the Krein-Smulian Theorem (see Proposition I.2.10) shows that both P and γ are dual algebra isomorphisms. □

We will conclude this section with a few facts about the envelope $E(K, \mu)$. We maintain the notation that $\gamma: R^\infty(K, \mu_a) \to R^\infty(E, \mathscr{A}|Q)$ is the isometric isomorphism found in Chaumat's Theorem. In fact, this map will henceforward be called the *Chaumat map*.

3.8 PROPOSITION. $E(K, \mu) = Q(K, \mu)$.

PROOF. We already know that $E \subseteq Q$ (3.3). Let $w \in Q$ and put $\nu = \mu_a - \mu_a(\{w\})$. Since $\nu \ll \mu_a$, $\nu \in \mathscr{B}_Q$. It follows from the definitions and Proposition 2.5 that $E = E(K, \mu_a) \subseteq E(K, \nu) \subseteq \Sigma(K, \nu) \subseteq \Sigma(K, \mu_a) = E$. Hence $E(K, \nu) = E$ and so $Q(K, \nu) = Q$. Thus the Chaumat map $\gamma_\nu: R^\infty(K, \nu) \to R^\infty(E, \mathscr{A}|E)$ is an isometry. But if $\rho: R^\infty(E, \mu_a) \to R^\infty(E, \nu)$ is the restriction map, then $\gamma_\nu \circ \rho = \gamma$ and hence ρ is an isometry.

It follows by (I.2.10) that $\rho: R^\infty(E, \mu_a) \to R^\infty(E, \nu)$ is a dual algebra isomorphism. Thus $\varphi \to \rho_w(\rho^{-1}(\varphi))$ (see Definition 3.6 for the notation) is a weak* continuous homomorphism. Therefore, there is a measure $\nu_w \ll \nu$ such that $\rho_w(\rho^{-1}(\varphi)) = \int \varphi\, d\nu_w$ for every φ in $R^\infty(E, \nu)$. But then $\nu_w \ll \mu$ and ν_w is a complex representing measure for evaluation at w on $R(K)$. Since $\nu_w(\{w\}) = 0$, $w \in E(K, \mu)$ by definition. □

3.9 PROPOSITION. *$E(K, \mu)$ and its closure have the same interior.*

PROOF. Let $\eta = \mu_a + \mathscr{A}|Q$; since $\mu_a \ll \eta$, the restriction map $R^\infty(K, \eta) \to R^\infty(K, \mu_a)$ is a well-defined contractive homomorphism that is the identity on $R(K)$. Now Chaumat's Theorem implies that $\|\varphi\|_{\mu_a} = \|\varphi\|_{\mathscr{A}|Q}$ for each φ in $R^\infty(K, \mu_a)$. But $\|\varphi\|_\eta = \max\{\|\varphi\|_{\mu_a}, \|\varphi\|_{\mathscr{A}|Q}\}$ for each φ in $R^\infty(K, \eta)$ and so the restriction map $R^\infty(K, \eta) \to R^\infty(K, \mu_a)$ is a dual algebra isomorphism. Thus $\Sigma(K, \eta) = \Sigma(K, \mu_a) = E$ by Proposition 2.5. Since these algebras have no L^∞-summand, $E(K, \eta) = \Sigma(K, \eta) = E$. But Proposition 2.6 implies that $E \supseteq \text{int}\{\text{supp}\, \eta\} = \text{int}\{\text{cl}\, E\}$. Hence $\text{int}\{\text{cl}\, E\} \subseteq \text{int}\, E$. The other inclusion is elementary. □

3.10 PROPOSITION. (a) *If $\varphi \in R^\infty(K, \mu_a)$, then φ is equal a.e. $[\mu]$ to a bounded analytic function on $\text{int}\{E(K, \mu)\}$.*

(b) *If $\varphi \in R^\infty(K, \mu_a)$ and $w \in \text{int}\, E$, then there is a function ψ in $R^\infty(K, \mu_a)$ such that $\varphi - \check{\varphi}(w) = (z - w)\psi$.*

PROOF. (a) If $\varphi \in R^\infty(K, \mu_a)$, then Corollary 3.5 implies there is a sequence $\{f_n\}$ in $R(K, E)$ such that $\|f_n\| \leq \|\varphi\|_\infty$ and $f_n \to \varphi$ a.e. $[\mu_a]$. Since each f_n is analytic on $\text{int}\, E$, an easy application of Montel's Theorem shows that (a) is true.

(b) Fix an open neighborhood $B = B(w; r)$ such that $\text{cl}\, B \subseteq \text{int}\, E$ and let $\{f_n\}$ be as in the proof of part (a). Since each f_n is analytic on $\text{int}\, E$, there is a function g_n in $R(K, E)$ such that $f_n - f_n(w) = (z - w)g_n$. Using the Maximum Modulus Theorem we get that $\|g_n\| \leq r^{-1}\|f_n\| \leq \|\varphi\|_\infty$. Thus $g_n \to (z - w)^{-1}(\varphi - \check{\varphi}(w)) \equiv \psi$ boundedly a.e. $[\mu_a]$; hence $\psi \in R^\infty(K, \mu_a)$. □

3.11 PROPOSITION. (a) $\text{Supp}\, \mu_a \subseteq \text{cl}\, E(K, \mu) \subseteq (\text{supp}\, \mu_a)\hat{\,}$.

(b) *If U is a bounded component of the complement of $\text{supp}\, \mu_a$, then either $U \subseteq \text{cl}\, E$ or $U \cap (\text{cl}\, E) = \varnothing$.*

Consequently, $\text{cl}\, E$ is the union of $\text{supp}\, \mu_a$ together with some collection of bounded components of $\mathbb{C} \setminus \text{supp}\, \mu_a$.

PROOF. (a) Every measure in the band $\mathscr{B}_Q$ is supported on $\text{cl}\, Q$ and so the first inclusion is a consequence of the fact that $Q = E$. On the other hand, if p is a polynomial, then Chaumat's Theorem says that $\|p\|_{\mu_a} = \|p\|_{\mathscr{A}|Q} = \|p\|_E$. Thus we have the second inclusion.

(b) Let U be as in the proposition; it suffices to show that if $U \cap (\operatorname{cl} E) \neq \varnothing$, then $U \subseteq \operatorname{cl} E$. Since U meets $\operatorname{cl} E$, there is a point a in $U \cap E$. Let $\mu_a \ll \mu$ such that μ_a is a representing measure for evaluation at a on $R(K)$ and put $\nu = (z - a)\mu_a$. Thus $\nu \ll \mu$ and $\nu \perp R(K)$. By Proposition 3.1, $\nu \perp R(K, E)$ and hence $\nu \in \mathscr{B}_Q$. But this says that $\nu \ll \mu_a$ (3.6) and so $\operatorname{supp}\nu \subseteq \operatorname{supp}\mu_a$. Hence the Cauchy transform of ν, $\hat{\nu}$, is analytic on U. Since $\hat{\nu}(a) = 1$, $\hat{\nu}$ is not identically 0 on U and so its zeros are isolated. Thus $U \subseteq \operatorname{cl}\{z : \hat{\nu}(z) \neq 0\}$. But Proposition 2.2 implies that $\{z : \hat{\nu}(z) \neq 0\} \subseteq E$ so that $U \subseteq \operatorname{cl} E$. □

By combining Chaumat's Theorem with some results from Chapter V, we can achieve a finer decomposition of $R^\infty(K, \mu)$. The details of the proof are left to the reader.

3.12 THEOREM. *Let $R^\infty(K, \mu) = L^\infty(\mu_0) \oplus R^\infty(K, \mu_a)$ as in Chaumat's Theorem and let $E = E(K, \mu)$ be the envelope. If $G_1, G_2, \ldots$ are the nontrivial Gleason parts of $R(K, E)$, then there are pairwise singular measures $\mu_1, \mu_2, \ldots$, such that the following hold.*
   (a) $\mu_a = \sum_n \mu_n$.
   (b) $R^\infty(K, \mu_a) = \oplus_n R^\infty(G_n, \mu_n)$.
   (c) *Each $R^\infty(G_n, \mu_n)$ is antisymmetric.*

Exercises

1. With the notation as in Chaumat's Theorem, show that the spectrum of z as an element of the Banach algebra $R^\infty(K, \mu)$ is $(\operatorname{supp}\mu_0) \cup \operatorname{cl} E$.
2. Show that if $R^\infty(K, \mu)$ has no L^∞-summand, then $w \in E(K, \mu)$ if and only if $w \in K$ and $\{f \in R(K) : f(w) = 0\}$ is not weak* dense in $R^\infty(K, \mu)$.
3. Use Corollary 3.5 to show that if $\varphi \in R^\infty(K, \mu_a)$, then $\varphi = \check{\varphi}$ a.e. $[\mu_a]$ on E.
4. Use Corollary 3.5 to show that if L is a weak* continuous linear functional on $R^\infty(K, \mu_a)$, then $\|L\| = \|L|R(K, E)\|$.
5. Show that $R^\infty(K, \nu)$ as defined just prior to Chaumat's Theorem is the same as the weak* closure of $R(\operatorname{cl} E \cup \operatorname{supp}\nu, E)$ in $L^\infty(\nu)$.
6. Give the details of the proof of Theorem 3.12.

§4 $H^\infty(\partial K)$ for a Dirichlet algebra. As before (V.9.19), for a compact subset K of $\mathbb{C}$, define harmonic measure ω for K as $\omega = \sum_n 2^{-n}\omega_n$, where $G_1, G_2, \ldots$ is an enumeration of the components of $\operatorname{int} K$, $a_n \in G_n$, and ω_n is harmonic measure for K evaluated at a_n. Since the definition of $L^\infty(\omega)$ depends only on the sets of ω-measure 0, it is unambiguous to define $L^\infty(\partial K) = L^\infty(\omega)$ as in (V.9.20). That is, this definition of $L^\infty(\partial K)$ is independent of the choice of points a_n. Similarly define $L^1(\partial K)$ as the set of all Borel measures on ∂K that are absolutely continuous with respect to ω and give $L^1(\partial K)$ the total variation norm (V.9.20). Hence $L^1(\partial K)^* = L^\infty(\partial K)$. (Note that $L^p(\partial K)$ cannot be defined independent of ω.)

4.1 DEFINITION. If K is a compact subset of $\mathbb{C}$ with nonempty interior, define $H^\infty(\partial K)$ as the weak* closure of $R(K)$ in $L^\infty(\partial K)$. Equivalently, $H^\infty(\partial K) = R^\infty(K, \omega)$.

In this section we will concretely characterize $H^\infty(\partial K)$, when $R(K)$ is a Dirichlet algebra, as the bounded analytic functions on int K.

Before giving the next proposition, let's introduce the following notation. For a measure μ and a function f, $\|f\|_\mu$ denotes the μ-essential supremum norm of f. In this and succeeding sections we will often have need to discuss several measures at once and this notation will be convenient. Also, if G is an open subset of $\mathbb{C}$, define the weak* topology on $H^\infty(G)$ to be the relative weak* topology it inherits as a subspace of $L^\infty(G, \mathscr{A}|G)$. By an application of the Krein-Smulian Theorem and Montel's Theorem it is easy to see that $H^\infty(G)$ is weak* closed in $L^\infty(G, \mathscr{A}|G)$. Thus a sequence in $H^\infty(G)$ converges weak* if and only if it is uniformly bounded and converges pointwise on G (or equivalently, converges uniformly on compact subsets of G).

4.2 DEFINITION. If $f \in H^\infty(\partial K)$, define $\hat{f}$: int $K \to \mathbb{C}$ by $\hat{f}(z) = \int f \, d\omega_z$, where ω_z is harmonic measure for K evaluated at z.

Because $\omega_z \in L^1(\partial K)$, this definition makes sense. We start with the following result that is true for arbitrary sets K. (We make the tacit assumption in this section that only compact sets K are considered that have nonempty interiors. Otherwise harmonic measure is meaningless.)

4.3 PROPOSITION. *If $f \in H^\infty(\partial K)$, then $\hat{f} \in H^\infty(\text{int } K)$ and the map $\rho: H^\infty(\partial K) \to H^\infty(\text{int } K)$ defined by $\rho(f) = \hat{f}$ is a contractive homomorphism that is weak* continuous and equals the identity on $R(K)$.*

PROOF. It is easy to see that $\|\hat{f}\|_{\text{int } K} \leq \|f\|_\omega$ and, by the definition of harmonic measure, $\hat{f} = f$ when $f \in R(K)$. To see that $\hat{f}$ is analytic for all f in $H^\infty(\partial K)$, let $\mathscr{M} = \{f \in H^\infty(\partial K): \hat{f} \in H^\infty(\text{int } K)\}$. We have that $R(K) \subseteq \mathscr{M}$ so we need only show that $\mathscr{M}$ is weak* closed in $H^\infty(\text{int } K)$. Let $\{f_n\}$ be a sequence in $\mathscr{M}$ such that $f_n \to f$ weak* in $H^\infty(\partial K)$. Since $\{f_n\}$ is uniformly bounded in $H^\infty(\partial K)$, it is easy to see that $\{\hat{f}_n\}$ is uniformly bounded in $H^\infty(\text{int } K)$. Since $\omega_z \in L^1(\partial K)$ for each z in int K, it follows that $\hat{f}_n(z) \to \hat{f}(z)$ for all z in int K. By the remarks preceding the proposition, $\hat{f}_n \to \hat{f}$ weak* in $H^\infty(\text{int } K)$. Thus $f \in \mathscr{M}$ and, by the Krein-Smulian Theorem, $\mathscr{M}$ is weak* closed and hence equal to $H^\infty(\partial K)$. The remainder of the proof is left to the reader. □

The first theorem of this section focuses on a specific harmonic measure ω_a rather than $H^\infty(\partial K)$. It will be used to study $H^\infty(\partial K)$.

4.4 THEOREM. *If $R(K)$ is a Dirichlet algebra, $a \in \text{int } K$, and ω_a is harmonic measure for K evaluated at a, then for every φ in $R^2(K, \omega_a) \cap$*

$L^\infty(\omega_a)$, there is a sequence $\{f_n\}$ in $R(K)$ such that for all $n \geq 1$

$$\|f_n\|_K \leq \|\varphi\|_{\omega_a} \text{ and } f_n \to \varphi \text{ a.e. } [\omega_a].$$

In particular, $R^2(K, \omega_a) \cap L^\infty(\omega_a) = R^\infty(K, \omega_a)$.

PROOF. First note that the last assertion of the theorem follows immediately from the first. To prove the first assertion, let $\varphi \in R^2(K, \omega_a) \cap L^\infty(\omega_a)$ and assume that $\|\varphi\|_{\omega_a} = 1$. Let $\{\varphi_n\} \subseteq R(K)$ such that $\int |\varphi_n - \varphi|^2 \, d\omega_a \to 0$. By passing to a subsequence if necessary, we may assume that $\varphi_n \to \varphi$ a.e. $[\omega_a]$.

Put $E_n = \{z \in \partial K : |\varphi_n(z)| \geq 1\}$. Note that E_n is closed. Define $h_n : \partial K \to \mathbb{R}$ by letting $h_n = -\log|\varphi_n|$ on E_n and $h_n = 0$ off E_n. Since φ_n is continuous on ∂K, $h_n \in C_{\mathbb{R}}(\partial K)$. Because $R(K)$ is a Dirichlet algebra there are functions $u_n + iv_n \in R(K)$ such that $v_n(a) = 0$ and $h_n - 1/n < u_n < h_n$ on ∂K. (Justify!) Put $g_n = \exp(u_n + iv_n)$. Hence $g_n \in R(K)$ and

(4.5i) $$g_n(a) = e^{u_n(a)},$$

(4.5ii) $$|g_n| = |\varphi_n|^{-1} \text{ on } E_n,$$

(4.5iii) $$|g_n| \leq 1 \text{ off } E_n.$$

Thus $g_n \varphi_n \in R(K)$, $\|g_n \varphi_n\|_K \leq 1$ and $0 \geq u_n(a) = \int u_n \, d\omega_a \geq \int h_n \, d\omega_a - 1/n = -1/n - \int_{E_n} \log|\varphi_n| \, d\omega_a$.

CLAIM. $\int_{E_n} \log|\varphi_n| \, d\omega_a \to 0$.

Once this claim is established, the proof can be concluded. Namely, the claim and the preceding inequalities show that $u_n(a) \to 0$. Hence (4.5i) implies that $g_n(a) \to 1$. Also (4.5) and the Maximum Modulus Theorem show that $\|g_n\|_K \leq 1$ and so $\int |g_n - 1|^2 \, d\omega_a = \int |g_n|^2 \, d\omega_a - 2\text{Re} \int g_n \, d\omega_a + 1 \leq 2[1 - \text{Re } g_n(a)] \to 0$ as $n \to \infty$. Therefore, there is a subsequence $\{g_{n_k}\}$ such that $g_{n_k} \to 1$ a.e. $[\omega_a]$. If $f_k = g_{n_k} \varphi_{n_k}$, then $f_k \in R(K)$, $\|f_k\|_K \leq 1$, and $f_k \to \varphi$ a.e. $[\omega_a]$, proving the theorem.

To prove the claim, let $\varepsilon > 0$ and put $F_n = \{z \in \partial K : |\varphi_n(z)| \geq 1 + \varepsilon\}$. Since $\varphi_n \to \varphi$ a.e. $[\omega_a]$ and $\|\varphi_n\|_{\omega_a} \leq 1$, it follows that $\limsup \omega_a(F_n) \leq \omega_a(\limsup F_n) \equiv \omega_a(\bigcap_n \bigcup_{m \geq n} F_m) = 0$. Hence $\omega_a(F_n) \to 0$ as $n \to \infty$. Also

$$0 \leq \int_{F_n} \log|\varphi_n| \, d\omega_a \leq \int_{F_n} |\varphi_n| \, d\omega_a$$

$$\leq \int |\varphi_n - \varphi| \, d\omega_a + \int_{F_n} |\varphi| \, d\omega_a$$

$$\leq \|\varphi_n - \varphi\|_{L^2(\omega_a)} + \omega_a(F_n).$$

Thus $\int_{F_n} \log|\varphi_n| d\omega_a \to 0$. But

$$0 \le \int_{E_n} \log|\varphi_n| d\omega_a = \int_{E_n \setminus F_n} \log|\varphi_n| d\omega_a + \int_{F_n} \log|\varphi_n| d\omega_a$$

$$\le \log(1+\varepsilon) + \int_{F_n} \log|\varphi_n| d\omega_a,$$

whence the claim. □

The next two lemmas are the means by which the consideration of $H^\infty(\partial K)$ when $R(K)$ is a Dirichlet algebra can be reduced to a consideration of the case where int K is connected, thus allowing the application of the preceding theorem. The first of these could be derived as a consequence of Theorem 3.12, but a direct proof will be given.

4.6 LEMMA. *If $R(K)$ is a Dirichlet algebra, $\{G_n\}$ are the components of int K, $a_n \in G_n$, and ω_n is the harmonic measure for K evaluated at a_n, then*

$$H^\infty(\partial K) = \bigoplus_{n=1}^\infty R^\infty(\operatorname{cl} G_n, \omega_n).$$

PROOF. Since the measures $\{\omega_n\}$ are pairwise singular, $L^\infty(\omega) = \bigoplus_n L^\infty(\omega_n)$; it is thus easy to see that $H^\infty(\partial K) \subseteq \bigoplus_n R^\infty(\operatorname{cl} G_n, \omega_n)$. To show the other inclusion, let ν be a measure that is absolutely continuous with respect to ω such that $\nu \perp H^\infty(\partial K)$. It must be shown that $\nu \perp \bigoplus_n R^\infty(\operatorname{cl} G_n, \omega_n)$. Since $\nu \perp R(K)$, Theorem V.18.7 implies that $\nu = \sum_n \nu_n$, where $\nu_n \perp \nu_m$ for $m \ne n$, ν_n belongs to the band $\mathscr{B}_n$ generated by the representing measures for points in G_n, and $\nu_n \perp R(\operatorname{cl} G_n)$. Since $\nu \ll \omega$ and ω_n is the projection of ω into $\mathscr{B}_n$, it follows that $\nu_n \ll \omega_n$ for each n. Therefore, $\nu_n \perp R^\infty(\operatorname{cl} G_n, \omega_n)$ and so $\nu \perp \bigoplus_n R^\infty(\operatorname{cl} G_n, \omega_n)$. □

In the next section we will see a result more general than the next lemma, but the present result will suffice for our immediate needs and its proof is completely elementary (and thus left to the reader).

4.7 LEMMA. *If $R(K)$ is a Dirichlet algebra, $\{G_n\}$ are the components of int K, and $\{n_k\}$ is any sequence of integers, then $R(K \setminus \bigcup_k G_{n_k})$ is a Dirichlet algebra.*

Now for the main result of this section.

4.8 THEOREM. *If $R(K)$ is a Dirichlet algebra and ω is harmonic measure for K, then the following properties hold.*

(a) *The envelope $E(K, \omega)$ equals the interior of K and $R(K, E(K,\omega)) = R(K)$.*

(b) *The map $f \to \hat{f}$, where $\hat{f}(z) = \int f d\omega_z$ for z in int K, is a dual algebra isomorphism of $H^\infty(\partial K)$ onto $H^\infty(\operatorname{int} K)$.*

(c) *If $\varphi \in H^\infty(\operatorname{int} K)$, then there is a sequence $\{f_n\}$ in $R(K)$ such that $\|f_n\|_K \le \|\varphi\|_{\operatorname{int} K}$ for all n and $f_n(z) \to \varphi(z)$ for all z in int K.*

PROOF. (a) Since points in the envelope $E = E(K, \omega)$ cannot be peak points and $R(K)$ is a Dirichlet algebra, $E \subseteq \text{int } K$. On the other hand, for each a in $\text{int } K$, harmonic measure for K at a, ω_a, is absolutely continuous with respect to ω, represents $R(K)$ at a, and puts no mass at a. Hence $\text{int } K \subseteq E$. Since E is the set of nonpeak points of $R(K)$, it follows that $R(K, E) = R(K)$ (V.13.11).

(b) Let γ be Chaumat's map. In this situation, $\omega \in \mathscr{B}_Q$ and so

$$\gamma : H^\infty(\partial K) \to R^\infty(E, \mathscr{A}|E).$$

Moreover for each f in $H^\infty(\partial K)$, $\gamma(f) = \hat{f}$, where $\hat{f}$ is as in Definition 4.2. Thus $R^\infty(E, \mathscr{A}|E) \subseteq H^\infty(\text{int } K)$ and so $\gamma : H^\infty(\partial K) \to H^\infty(\text{int } K)$ is an isometry. We must show that γ maps $H^\infty(\partial K)$ onto $H^\infty(\text{int } K)$.

In light of the preceding two lemmas, it suffices to consider the case where $\text{int } K$ is connected and $\omega = \omega_a$ for some point a in $E = \text{int } K$. By Lemma V.16.3, there is a function Z in $R^2(K, \omega)$ such that $|Z| = 1$ a.e. $[\omega]$ and $ZR^2(K, \omega) = $ the closure of $\{f \in R(K) : f(a) = 0\}$ in $R^2(K, \omega)$. By Theorem 4.4, $Z \in H^\infty(\partial K)$. Also Corollary V.16.7 implies that $\hat{Z} : E \to \mathbb{D}$ is the Riemann map such that $\hat{Z}(a) = 0$. If $h \in H^\infty(E)$, then $h \circ \hat{Z}^{-1} \in H^\infty(\mathbb{D})$. Hence there is a sequence of polynomials $\{p_n\}$ such that $\|p_n\|_\mathbb{D} \leq \|h \circ \hat{Z}^{-1}\|_\mathbb{D} = \|h\|_E$ and $p_n(\zeta) \to h(\hat{Z}^{-1}(\zeta))$ for all ζ in $\mathbb{D}$. But since $Z \in H^\infty(\partial K)$, $p_n \circ Z = p_n(Z) \in H^\infty(\partial K)$. Also $\|p_n \circ Z\|_\omega \leq \|h\|_E$. Thus there is a subsequence $\{p_{n_k} \circ Z\}$ such that $p_{n_k} \circ Z \to \varphi$ weak* in $H^\infty(\partial K)$. For any z in E, $\hat{\varphi}(z) = \lim(p_{n_k} \circ Z)\hat{\,}(z) = h(z)$. Thus $\gamma(\varphi) = h$ and γ is surjective.

(c) This follows by applying Corollary 3.5. □

The way to think of the preceding theorem is as a way of assigning boundary values to bounded analytic functions on $\text{int } K$ when $R(K)$ is a Dirichlet algebra. That is, to each φ in $H^\infty(\text{int } K)$ there is a unique function f in $H^\infty(\partial K)$ such that $\hat{f} = \varphi$ and $\|\varphi\|_\omega = \|f\|_{\text{int } K}$. If $K = \text{cl } \mathbb{D}$, this assignment of boundary values is precisely the process of finding the radial limits of a bounded analytic function on $\mathbb{D}$.

Also note that if $\{G_n\}$ are the components of $\text{int } K$, then $H^\infty(\partial G_n) \cong H^\infty(G_n)$ (see Exercise 1) and so $H^\infty(\partial G_n)$ is antisymmetric. The decomposition in Lemma 4.6 is precisely the decomposition of $H^\infty(\partial K)$ into its antisymmetric summands (1.6).

The final result of this section will be used later and, in its part (c), presents a somewhat bizarre twist.

4.9 PROPOSITION. *Let $R(K)$ be a Dirichlet algebra, let ω be harmonic measure for K, and let ν be a positive measure on ∂K that is singular with ω. If $\mu = \omega + \nu$, then the following statements hold.*

(a) $E(K, \mu) = E(K, \omega) = \text{int } K$.
(b) $R^\infty(K, \mu) \cong H^\infty(\partial K) \oplus L^\infty(\nu)$.

(c) If $\varphi \in H^\infty(\partial K)$ and $\psi \in L^\infty(\nu)$ such that $\|\varphi\|_\omega \le 1$ and $\|\psi\|_\nu \le 1$, then there is a sequence $\{f_n\} \subseteq R(K)$ with $\|f_n\|_K \le 1$, $f_n \to \varphi$ weak* in $H^\infty(\partial K)$, and $f_n \to \psi$ weak* in $L^\infty(\nu)$.

PROOF. Part (a) follows because $R(K)$ is a Dirichlet algebra and so the only representing measure for points in int K that is supported by ∂K is the relevant harmonic measure.

Since Q, the set of nonpeak points of $R(K)$, is int K in this case, it follows that the measures in $\mathscr{B}_Q$ that are supported by ∂K are precisely those in $L^1(\omega)$. Hence $\mu_Q = \omega$ and $\mu_\mathscr{S} = \nu$. Part (b) now follows by Chaumat's Theorem.

Part (c) is a straightforward interpretation of Corollary 3.5. □

This is not an appropriate place to unravel the origins of the results in this section; they are standard results from function algebras. The interested reader can consult Gamelin [1969] and Stout [1971]. The proofs here are slight modifications, made to incorporate Chaumat's Theorem, of those that appear in Wermer [1964].

Exercises

1. If $R(K)$ is a Dirichlet algebra, G is a component of int K, $a \in G$, and ω is harmonic measure for K evaluated at a, show that $R^\infty(\operatorname{cl} G, \omega) = H^\infty(\partial G)$.
2. Prove Lemma 4.6 by using Theorem 3.12.
3. Prove Lemma 4.7.
4. (Merril [1968]). If $R(K)$ is a Dirichlet algebra and int K is connected, show that $H^\infty(\partial K)$ is a maximal weak* closed subalgebra of $L^\infty(\partial K)$.

§5 Dirichlet algebras revisited. In this section several facts concerning Dirichlet algebras are collected. The first result, stated without proof, characterizes Dirichlet algebras in terms of analytic capacity. In the remainder of the section this theorem is used to show how to get new Dirichlet algebras from old ones and derive other characterizations of Dirichlet algebras. These results will then be used in the next section to study $R^\infty(K, \mu)$ when $R(K)$ is a Dirichlet algebra and μ is an arbitrary measure.

We begin by agreeing to call an open set *simply connected* if each of its components is simply connected. That is, simply connected sets do not have to be connected. (This is in fact the situation in [**FOCV**].)

5.1 PROPOSITION. *If K is a compact subset of $\mathbb{C}$, the following statements are equivalent.*

(a) *The interior of K is simply connected.*
(b) *The complement of* int K *is connected.*
(c) *The boundary of each component of* int K *is connected.*

The proof that (a) and (b) in the preceding proposition are equivalent can be found in [**FOCV**]; the proof of the equivalence of these statements with (c) is left to the reader's mathematical or bibliographic skills. This last condition won't be used much here.

The proof of the next theorem can be found as Theorem 9.3 in Gamelin and Garnett [**1971**].

5.2 THEOREM. *If K is a compact subset of $\mathbb{C}$, the following statements are equivalent.*

(a) *$R(K)$ is a Dirichlet algebra.*

(b) *The interior of K is simply connected and for every a in ∂K and for every $r > 0$,*

$$\gamma(B(a;r)\setminus K) \geq r/4.$$

(c) *The interior of K is simply connected and for all but a countable number of a in ∂K,*

$$\liminf_{\delta \to 0} \frac{\gamma(B(a;\delta)\setminus K)}{\delta} > 0.$$

5.3 THEOREM (Sarason [**1972**]). *Assume that $R(K)$ is a Dirichlet algebra and L is a compact subset of K. If each component V of $\mathbb{C}\setminus L$ satisfies $[\operatorname{cl} V] \cap [\mathbb{C}\setminus \operatorname{int} K] \neq \varnothing$, then $R(L)$ is a Dirichlet algebra.*

PROOF. We know that $\mathbb{C}\setminus \operatorname{int} K$ is connected since $\operatorname{int} K$ is simply connected. Now if V is a component of $\mathbb{C}\setminus L$, then by hypothesis $\operatorname{cl} V$ meets $\mathbb{C}\setminus \operatorname{int} K$. Hence $[\operatorname{cl} V] \cup [\mathbb{C}\setminus \operatorname{int} K]$ is connected. Therefore, $[\mathbb{C}\setminus \operatorname{int} K] \cup \bigcup \{\operatorname{cl} V : V$ is a component of $\mathbb{C}\setminus L\}$ is connected. But the closure of this set is precisely $\mathbb{C}\setminus \operatorname{int} L$. Thus $\operatorname{int} L$ is simply connected.

Now let $a \in \partial L$. If $a \in \partial K$ and $r > 0$, then $\gamma(B(a;r)\setminus L) \geq \gamma(B(a;r)\setminus K) \geq r/4$. If $a \in \partial L$ but $a \notin \partial K$, then $a \in \partial L \cap \operatorname{int} K$. Let $r > 0$ such that $B(a;r) \subseteq \operatorname{int} K$. If V is any component of $\mathbb{C}\setminus L$ such that V meets $B(a;r/2)$, then the assumption that $\operatorname{cl} V$ meets $\mathbb{C}\setminus \operatorname{int} K$ implies that $V\setminus B(a;r) \neq \varnothing$. Hence there is a curve in V from a point in $B(a;r/2)$ to a point not in $B(a;r)$. So this curve is a connected set of diameter at least $r/2$. By (V.12.13), $\gamma(B(a;r)\setminus L) \geq r/8$. By the preceding theorem, $R(L)$ is a Dirichlet algebra. □

5.4 COROLLARY. *If $R(K)$ is a Dirichlet algebra and W is an open subset of $\operatorname{int} K$ such that $\operatorname{int}(K\setminus W)$ is simply connected, then $R(K\setminus W)$ is a Dirichlet algebra.*

PROOF. Let $L = K\setminus W$; so $\mathbb{C}\setminus L = (\mathbb{C}\setminus K) \cup W$. If V is a component of $\mathbb{C}\setminus L$, then either V is a component of $\mathbb{C}\setminus K$ or V is a component of W. In either case $\operatorname{cl} V$ meets ∂K and the preceding theorem applies. □

Call a compact set K a *crescent* if $K = \operatorname{cl}\mathbb{D}\setminus W$, where W is a simply connected region contained in $\mathbb{D}$ and $\operatorname{cl} W$ meets $\partial\mathbb{D}$ at precisely one point. It follows from the preceding result that if K is a crescent, then $R(K)$ is a Dirichlet algebra.

5.5 THEOREM. *Let $\{K_\alpha\}$ be a decreasing chain of compact sets and let $K = \bigcap_\alpha K_\alpha$. If $R(K_\alpha)$ is a Dirichlet algebra for each α, then $R(K)$ is a Dirichlet algebra.*

PROOF. In fact, $\mathbb{C}\setminus \operatorname{int} K = \operatorname{cl}[\bigcup_\alpha (\mathbb{C}\setminus K_\alpha)]$ and so $\mathbb{C}\setminus \operatorname{int} K$ is connected. Now let $a \in \partial K$ and let $r > 0$. Since $B(a;r/2)\setminus K \neq \varnothing$, there is an α such that $B(a;r/2)\setminus K_\alpha \neq \varnothing$. Since $a \in K_\alpha$, there is a point b on ∂K_α with $|a-b| < r/2$. But Theorem 5.2 implies that $\gamma(B(a;r)\setminus K) \geq \gamma(B(b;r/2)) \geq r/8$. Again by Theorem 5.2, $R(K)$ is a Dirichlet algebra. □

5.6 EXAMPLE. Let $D_n = B(3/2^{n+1}; 1/2^{n+1})$ for $n \geq 1$. So successive disks in this sequence $\{D_n\}$ are tangent. Let $K = (\operatorname{cl}\mathbb{D})\setminus \bigcup_n D_n$. Then $R(K)$ is a Dirichlet algebra.

The next result follows by using Theorem 5.2 and the method in the proof of Gonchar's criterion (V.13.2). The details are left to the reader.

5.7 THEOREM. *For a in ∂K and $r > 0$, let $d(a;r)$ be the supremum of the diameters of the components of $B(a;r)\setminus K$. If $\limsup_{r \to 0} d(a;r)/r > 0$ for all but a countable number of points a on ∂K and $\operatorname{int} K$ is simply connected, then $R(K)$ is a Dirichlet algebra.*

Note that the preceding theorem can also be used to show that $R(K)$ is a Dirichlet algebra for the set K in Example 5.6.

5.8 COROLLARY. *If all but a countable number of points of ∂K are boundary points of some component of $\mathbb{C}\setminus K$, then $R(K)$ is a Dirichlet algebra if and only if $\operatorname{int} K$ is simply connected.*

5.9 COROLLARY. *If there is a number $d > 0$ such that each component of $\mathbb{C}\setminus K$ has diameter at least d, then $R(K)$ is a Dirichlet algebra if and only if $\operatorname{int} K$ is simply connected.*

5.10 COROLLARY. *If K is finitely connected, then $R(K)$ is a Dirichlet algebra if and only if $\operatorname{int} K$ is simply connected.*

Here is another result in the same spirit, though we won't have much need for it here. It is presented more for the reader's cultural edification.

5.11 THEOREM (Gamelin and Garnett [1971]). *For a compact subset K of $\mathbb{C}$, $R(K)$ is a Dirichlet algebra if and only if $\operatorname{int} K$ is simply connected, $R(\partial K) = C(\partial K)$, and $R(K)$ is boundedly pointwise dense in $H^\infty(\operatorname{int} K)$.*

We have already shown that these conditions necessarily hold if $R(K)$ is a Dirichlet algebra. For the proof of sufficiency, see the original source.

§6 The weak-star closure of a Dirichlet algebra.
The purpose of this section is to prove the following characterization of $R^\infty(K,\mu)$ when $R(K)$ is a Dirichlet algebra. For simplicity we will use the algebras $R(E)$ rather than $R(K,E)$. (Why is this permissible?)

§6 WEAK-STAR CLOSURE OF A DIRICHLET ALGEBRA

6.1. Theorem. *If $R(K)$ is a Dirichlet algebra and μ is a positive measure supported by K, then $\mu = \mu_a + \mu_s$, where $\mu_a \perp \mu_s$, $R^\infty(K, \mu) = R^\infty(K, \mu_a) \oplus L^\infty(\mu_s)$, $R^\infty(K, \mu_a)$ has no L^∞-summand, and the following properties are enjoyed.*

(a) *The envelope $E = E(K, \mu)$ is open and $R(\operatorname{cl} E)$ is a Dirichlet algebra that is equal to $R(E)$.*
(b) *The support of μ_a is contained in $\operatorname{cl} E$ and $R(E)$ is contained in $R^\infty(K, \mu_a)$.*
(c) *There is a dual algebra isomorphism of $R^\infty(K, \mu_a)$ onto $H^\infty(E)$ that is the identity on $R(E)$.*
(d) *If $\varphi \in R^\infty(K, \mu_a)$, there is a sequence $\{f_n\}$ in $R(E)$ such that $\|f_n\|_E \leq \|\varphi\|_{\mu_a}$ and $f_n \to \varphi$ a.e. $[\mu_a]$.*

Note that the first part of the proof is immediate from Chaumat's Theorem. The remainder of the proof will require a few lemmas that examine a transfinite process for obtaining the envelope $E(K, \mu)$ and deriving its properties.

Since the existence of the measures μ_a and μ_s come from Chaumat's Theorem, in the succeeding lemmas it will usually be assumed that $R^\infty(K, \mu)$ has no L^∞-summand. That is, it will be assumed that there is a measure ν with $[\nu] = [\mu]$ and $\nu \perp R(K)$. So fix such a measure μ and let $U_0, U_1, U_2, \ldots$ be the components of $\mathbb{C} \setminus \operatorname{supp} \mu$, where U_0 is the unbounded component. Recall (3.11) that for each $n \geq 1$, either $U_n \subseteq \operatorname{cl} E$ or $U_n \cap \operatorname{cl} E = \varnothing$ and $\operatorname{cl} E$ consists of $\operatorname{supp} \mu$ together with some collection of its holes. Thus we are led to define the collection $\mathscr{F}$ of all compact sets having the following properties.

(i) $F = K \cap [\operatorname{supp} \mu \cup \bigcup \{U_n : n \in I\}]$, for some nonempty subset I of the positive integers.
(ii) $E(K, \mu) \subseteq F$.
(iii) $R(F)$ is a Dirichlet algebra.

We seek a minimal element of $\mathscr{F}$ which will be shown to be $\operatorname{cl}[E(K, \mu)]$. But first things first.

6.2 Lemma. $\mathscr{F} \neq \varnothing$.

Proof. In fact, let $F_0 = K \cap [\operatorname{supp} \mu \cup \bigcup \{U_n : n \geq 1\}] = K \cap (\operatorname{supp} \mu)\hat{\ }$. By Proposition 3.11, $E \subseteq (\operatorname{supp} \mu)\hat{\ }$ and so $E \subseteq F_0$. Thus F_0 satisfies (i) and (ii). If V is a component of $\mathbb{C} \setminus F_0 = U_0 \cup (\mathbb{C} \setminus K)$, then either $V \supseteq U_0$ or $V \cap U_0 = \varnothing$. In the latter case V is a bounded component of $\mathbb{C} \setminus K$ and so $\partial V \cap \partial K \neq \varnothing$. If $V \supseteq U_0$, it is also clear that $\operatorname{cl} V \cap \partial K \neq \varnothing$. By Theorem 5.3, $R(F_0)$ is a Dirichlet algebra. Hence $F_0 \in \mathscr{F}$. □

6.3 Lemma. *If $R(K)$ is a Dirichlet algebra and $R^\infty(K, \mu)$ has no L^∞-summand, then the collection of sets $\mathscr{F}$ has a minimal element.*

Proof. Apply Zorn's Lemma. If $\{F_\alpha\}$ is a decreasing chain in $\mathscr{F}$ and $F = \bigcap_\alpha F_\alpha$, then Theorem 5.5 implies that $R(F)$ is a Dirichlet algebra and clearly $F \supseteq E$. It is easy to see that F also satisfies condition (i) and hence that $F \in \mathscr{F}$. □

Note that if $F \in \mathscr{F}$, then it must be that E is contained in the interior of F, not just F. In fact, since $R(F)$ is a Dirichlet algebra and a point of E cannot be a peak point for $R(E)$, and thus not a peak point for $R(F)$, it follows that $E \subseteq \text{int } F$.

The next lemma is the linchpin in this series.

6.4 LEMMA. *Assume that $R(K)$ is a Dirichlet algebra and $R^\infty(K, \mu)$ has no L^∞-summand. If $F \in \mathscr{F}$ and there is a function φ in $H^\infty(\text{int } F)$ with $\|\varphi\|_E < \|\varphi\|_{\text{int } F}$, then there is a component U of $\mathbb{C} \setminus \text{supp } \mu$ such that $F \setminus U \neq F$ and $F \setminus U \in \mathscr{F}$.*

PROOF. Let W be a component of $\{z \in \text{int } F : \text{there is a } \varphi \text{ in } H^\infty(\text{int } F) \text{ with } |\varphi(z)| > \|\varphi\|_E\}$; so the hypothesis implies that $W \neq \emptyset$. Clearly W is open and disjoint from E. Thus $W \cap \text{cl } E = \emptyset$ and so $W \cap (\text{supp } \mu) = \emptyset$ (we are using Proposition 3.11 and the fact that $R^\infty(K, \mu)$ has no L^∞-summand). Let U be the component of $\mathbb{C} \setminus \text{supp } \mu$ that contains W and put $F_1 = F \setminus U$.

If V is a component of $\mathbb{C} \setminus F_1 = U \cup (\mathbb{C} \setminus F)$, then either $V \supseteq U$ or V is a component of $\mathbb{C} \setminus F$. In the latter case, $\partial V \cap \partial F \neq \emptyset$. If $V \supseteq U$, then $V \supseteq W$ and the Maximum Modulus Theorem implies that $\partial W \cap \partial F \neq \emptyset$. Hence $\text{cl } V$ always meets $\mathbb{C} \setminus \text{int } F$. By Theorem 5.3, $R(F_1)$ is a Dirichlet algebra. It is clear that F_1 has the other two properties that forces it to belong to $\mathscr{F}$ and $F_1 \neq F$. □

PROOF OF THEOREM 6.1. By an application of Chaumat's Theorem we may assume that $R^\infty(K, \mu)$ has no L^∞-summand. Let F be a minimal element of $\mathscr{F}$ (6.3). If $\varphi \in H^\infty(\text{int } F)$, then Theorem 4.8 implies there is a sequence $\{f_n\}$ in $R(F)$ with $\|f_n\|_F \leq \|\varphi\|_{\text{int } F}$ and $f_n(z) \to \varphi(z)$ for all z in $\text{int } F$. Since $F \in \mathscr{F}$, $E \subseteq \text{int } F$ and so $\|f_n\|_E \leq \|\varphi\|_{\text{int } F}$. Hence $f_n \to \varphi$ weak* in $L^\infty(\mathscr{A}|E)$. But $E \subseteq F$ and so $R(F) \subseteq R(E)$. Therefore, $\varphi|E \in R^\infty(E, \mathscr{A}|E) \cong R^\infty(K, \mu)$ and $f_n \to \varphi|E$ weak* in $R^\infty(K, \mu)$. Define $\rho: H^\infty(\text{int } F) \to R^\infty(K, \mu)$ by $\rho(\varphi) = \varphi|E$. (The algebras $R^\infty(K, \mu)$ and $R^\infty(E, \mathscr{A}|E)$ are being identified here. The reader is welcomed, if so inclined, to keep them distinct and incorporate the Chaumat map in the following line of reasoning.) Note that ρ is the identity on $R(F)$.

Note that we have actually shown that ρ is weak* sequentially continuous and hence weak* continuous. By Lemma 6.4, ρ is an isometry. Thus, by the Krein-Smulian Theorem, the range of ρ is a weak* closed subalgebra of $R^\infty(K, \mu)$. Also $F \subseteq K$ so that $R(K) \subseteq R(F)$ and the range of ρ contains $R(K)$ and is, consequently, weak* dense. Therefore ρ is surjective.

Now since $E = \Sigma(K, \mu)$, the weak* continuous homomorphisms on $R_\infty(K, \mu)$, and it is easy to see that the weak* continuous homomorphisms on $H^\infty(\text{int } F)$ are the evaluations at points of $\text{int } F$, it follows that $E =$

int F. This establishes parts (a), (b), and (c) of the theorem. Part (d) follows by Corollary 3.5. □

6.5 EXAMPLE. See Chaumat [**1974**] page 114, for an example of a K and a μ such that $R(\operatorname{cl} E)$ is not pointwise boundedly dense in $R^\infty(K, \mu)$. In this example, $R(K)$ is, of course, not a Dirichlet algebra.

§7 The weak-star closure of the polynomials.

In this section the results of the previous section will be applied to characterize $P^\infty(\mu)$, the weak* closure of the polynomials in $L^\infty(\mu)$ for a compactly supported positive measure μ on $\mathbb{C}$. To begin we will merely interpret the main result from §6 since $P^\infty(\mu) = R^\infty(K, \mu)$ for $K = $ the polynomially convex hull of the support of μ. Afterwards we will examine the Sarason process for determining $P^\infty(\mu)$.

Sarason's characterization of $P^\infty(\mu)$ predates Chaumat's work and the Sarason process referred to is actually a transfinite recipe for finding $P^\infty(\mu)$ that was the original proof of this result. Thus we will give a proof of Sarason's Theorem that is (essentially) independent of Chaumat's Theorem. This is done for two reasons. First the algebra $P^\infty(\mu)$ is more important for the study of subnormal operators than the general $R^\infty(K, \mu)$, since the weak* closed algebra generated by a subnormal operator is isomorphic to $P^\infty(\mu)$, where μ is a scalar-valued spectral measure for the subnormal operator. Indeed, it is the knowledge of $P^\infty(\mu)$ that will allow us to prove that subnormal operators are reflexive in the next chapter.

The second reason is that this second proof using the Sarason process is actually a constructive technique to determine $P^\infty(\mu)$. (The word "constructive" is not being used here in the technical sense, since there is a transfinite process involved.) Even though the process is a transfinite one it can be readily applied to specific examples to find the precise answer. Chaumat's Theorem is, on the other hand, an existence theorem and does not afford a method of determining the envelope and the measure μ_a. (Once the measure is determined, however, the proof of Chaumat's Theorem, especially in the case that $R(K)$ is a Dirichlet algebra, is somewhat constructive.)

The first version of Sarason's Theorem (Sarason [**1972**]) is a direct application of Theorem 6.1 to the Dirichlet algebra $R(K)$ for $K = $ the polynomially convex hull of the support of μ.

7.1 SARASON'S THEOREM (First Version). *For any positive measure μ on $\mathbb{C}$ with compact support, there is a compact set K and measures μ_a and μ_s having the following properties.*

(a) $\mu = \mu_a + \mu_s$, $\mu_a \perp \mu_s$, *and* $P^\infty(\mu) = P^\infty(\mu_a) \oplus L^\infty(\mu_s)$.
(b) K *contains the support of* μ_a, $\mu_a | \partial K$ *is absolutely continuous with respect to harmonic measure, and* $R(K) \subseteq P^\infty(\mu_a)$.
(c) $R(K)$ *is a Dirichlet algebra.*
(d) *There is a dual algebra isomorphism* $\rho \colon H^\infty(\operatorname{int} K) \to P^\infty(\mu_a)$ *such that* $\rho(f) = f$ *for every f in $R(K)$.*

(e) If $\varphi \in P^\infty(\mu_a)$, there is a sequence $\{f_n\}$ in $R(K)$ such that $\|f_n\|_K \leq \|\varphi\|_{\mu_a}$ and $f_n \to \varphi$ a.e. $[\mu_a]$.

PROOF. Let K be the closure of the envelope $E((\operatorname{supp}\mu)\hat{}, \mu)$ and apply Theorem 6.1. The fact that $\mu_a|\partial K$ is absolutely continuous with respect to harmonic measure for K follows from Corollary V.18.9. The details are left to the reader. □

The second version of the theorem consists in obtaining a chain of compact sets K_α indexed by the ordinal numbers as well as corresponding measures μ_α. It will be shown that for a countable ordinal α, $K_\beta = K_\alpha$ and $\mu_\beta = \mu_\alpha$ for all $\beta \geq \alpha$. It will then be the case that μ_α is the measure μ_a from the preceding theorem and K_α is the desired set K.

Before defining the sets K_α, it should be remarked that the definition must be made simultaneously with the proving of a lemma. That is, there is part of the definition that must be justified for each ordinal α. This can be done with transfinite induction but will be postponed until after the definition is completed.

7.2 DEFINITION. For a positive compactly supported measure μ on $\mathbb{C}$, the *Sarason process* is the following transfinite sequence of pairs (K_α, μ_α), consisting of a compact set K_α and a positive measure μ_α.

(a) K_1 is the polynomially convex hull of the support of μ.

(b) If K_α is defined, then

$$\mu_\alpha = \mu|\operatorname{int} K_\alpha + (\mu|\partial K)_a,$$

where for any compact set L and any measure ν, $(\nu|\partial L)_a$ is the part of $\nu|\partial L$ that is absolutely continuous with respect to harmonic measure for L. The set $K_{\alpha+1}$ is defined by

$$K_{\alpha+1} = (\operatorname{supp}\mu_\alpha) \cup \{z \in \operatorname{int} K_\alpha : |\varphi(z)| \leq \|\tilde{\varphi}\|_{\mu_\alpha} \text{ for every } \varphi$$
$$\text{in } H^\infty(\operatorname{int} K_\alpha)\}.$$

(The term $\|\tilde{\varphi}\|_{\mu_\alpha}$ must be explained.)

(c) If α is a limit ordinal,

$$K_\alpha = \bigcap_{\beta < \alpha}\{K_\beta : \beta < \alpha\}.$$

The meaning to the term $\|\tilde{\varphi}\|_{\mu_\alpha}$ appearing in part (b) of the definition is as follows. It will be shown (see Lemma 7.3 below) that for every α, K_α is compact and $R(K_\alpha)$ is a Dirichlet algebra. Thus $H^\infty(\partial K_\alpha) \cong H^\infty(\operatorname{int} K_\alpha)$ (Theorem 4.8). As remarked there, this last result is to be regarded as a natural way of assigning boundary values to functions φ in $H^\infty(\operatorname{int} K_\alpha)$. So for each φ in $H^\infty(\operatorname{int} K_\alpha)$, $\tilde{\varphi}: \operatorname{cl} K_\alpha \to \mathbb{C}$ is defined by letting $\tilde{\varphi} = \varphi$ on $\operatorname{int} K_\alpha$ and $\tilde{\varphi}$ = the corresponding boundary values of φ in $H^\infty(\partial K_\alpha)$.

Thus $\tilde{\varphi}$ is defined everywhere on $\operatorname{int} K_\alpha$ and a.e. $[\omega]$ on ∂K_α. Since $\mu_\alpha | \partial K_\alpha$ is absolutely continuous with respect to harmonic measure on K_α, $\tilde{\varphi}$ is defined a.e. $[\mu_\alpha]$. Therefore, $\|\tilde{\varphi}\|_{\mu_\alpha}$ is well-defined.

Of course, to make complete sense of this we must show that K_α is compact and $R(K_\alpha)$ is a Dirichlet algebra. This is done forthwith. Since the proof is by transfinite induction, no logical contretemps occurs.

7.3 LEMMA. *For each ordinal number α, the set K_α in the Sarason process is compact and $R(K_\alpha)$ is a Dirichlet algebra.*

PROOF. For $\alpha = 1$ the conclusion is true since K_1 is polynomially convex. If α is a limit ordinal and for each $\beta < \alpha$, K_β is compact and $R(K_\beta)$ is a Dirichlet algebra, then, because the sets are decreasing, $K_\alpha = \bigcap_\beta K_\beta$ is compact and $R(K_\alpha)$ is a Dirichlet algebra by Theorem 5.5.

Now assume that K_α is compact and $R(K_\alpha)$ is a Dirichlet algebra. Let $\{z_n\} \subseteq K_{\alpha+1}$ and suppose $z_n \to z_0$. So $z_n \in K_\alpha$. We can assume that $z_0 \notin \operatorname{supp} \mu_\alpha$ and thus that $z_n \notin \operatorname{supp} \mu_\alpha$ for all $n \geq 1$. If $z_0 \in \partial K_\alpha$, then the fact that $R(K_\alpha)$ is a Dirichlet algebra implies that z_0 is a peak point for $R(K_\alpha)$. Thus there is a function f in $R(K_\alpha)$ such that $f(z_0) = 1$ and $|f(z)| < 1$ for all z in K_α, $z \neq z_0$. But $z_0 \notin \operatorname{supp} \mu_\alpha$ so that there is an open disk D about z_0 with $\operatorname{cl} D \cap \operatorname{supp} \mu_\alpha = \varnothing$. This shows that $\|f\|_{\mu_\alpha} \leq \|f\|_{K_\alpha \setminus D} < 1$. But $z_n \in D$ for all sufficiently large n and so $|f(z_n)| > \|f\|_{\mu_\alpha}$ for all large n. This contradicts the assumption that $z_n \in K_{\alpha+1}$. Hence it must be that $z_0 \in \operatorname{int} K_\alpha$. Since $|\varphi(z_n)| \leq \|\tilde{\varphi}\|_{\mu_\alpha}$ for all large n, it readily follows that $|\varphi(z_0)| \leq \|\tilde{\varphi}\|_{\mu_\alpha}$ and so $z_0 \in K_{\alpha+1}$. That is, $K_{\alpha+1}$ is closed.

Since $R(K_\alpha)$ is a Dirichlet algebra, $\|\tilde{\varphi}\|_{\mu_\alpha}$ makes sense for every φ in $H^\infty(\operatorname{int} K_\alpha)$. Let $W = \{z \in \operatorname{int} K_\alpha : \text{there is a } \varphi \text{ in } H^\infty(\operatorname{int} K_\alpha) \text{ with } |\varphi(z)| > \|\tilde{\varphi}\|_{\mu_\alpha}\}$. Note that W is an open set and $W \cap \operatorname{supp} \mu_\alpha = \varnothing$. Also the Maximum Modulus Theorem implies that the closure of any component of W must meet ∂K_α. Since $K_{\alpha+1} \subseteq K_\alpha \setminus W$, if V is any component of $\mathbb{C} \setminus K_{\alpha+1}$, then either V contains a component of W or V contains a component of $\mathbb{C} \setminus K_\alpha$. In either case $\operatorname{cl} V \cap \partial K_\alpha \neq \varnothing$. By Theorem 5.3, $R(K_{\alpha+1})$ is a Dirichlet algebra and the induction argument is complete. □

7.4 SARASON'S THEOREM (Second Version). *If μ is a positive compactly supported measure on $\mathbb{C}$ and $\{(K_\alpha, \mu_\alpha)\}$ are the pairs of compact sets and measures obtained by the Sarason process, then there is a countable ordinal α_0 such that the following statements hold.*

(a) $K_\alpha = K_{\alpha_0}$ and $\mu_\alpha = \mu_{\alpha_0}$ for all $\alpha \geq \alpha_0$.
(b) *If $K = K_{\alpha_0}$ and $\mu_a = \mu_{\alpha_0}$ and $\mu_s = \mu - \mu_a$, then $K = E((\operatorname{supp} \mu)\hat{\,}, \mu)$ and $P^\infty(\mu) = P^\infty(\mu_a) \oplus L^\infty(\mu_s)$.*
(c) $R(K) \subseteq P^\infty(\mu_a)$ *and the inclusion map $R(K) \to P^\infty(\mu_a)$ extends to a dual algebra isomorphism of $H^\infty(\operatorname{int} K)$ onto $P^\infty(\mu_a)$.*

(d) If $\varphi \in P^\infty(\mu_a)$, then there is a sequence $\{f_n\} \subseteq R(K)$ such that $\|f_n\|_K \le \|\varphi\|_{\mu_a}$ for all $n \ge 1$ and $f(z) \to \varphi(z)$ a.e. $[\mu_a]$.

7.5 COROLLARY. *For any compactly supported positive measure μ on $\mathbb{C}$, $P^\infty(\mu) = L^\infty(\mu)$ if and only if there is an ordinal number α such that the set K_α in the Sarason process is empty.*

Before proving this theorem, let's look at a few examples in order to familiarize ourselves with the Sarason process and to become convinced that this process can indeed be used to calculate $P^\infty(\mu)$.

7.6 EXAMPLE. Let $A = \{z \in \partial\mathbb{D}: \operatorname{Re} z \ge 0\}$ and $B = \{z \in \partial\mathbb{D}: \operatorname{Re} z \le 0\}$. Let $\{a_n\}$ be a dense sequence in A and put $\mu = m|B + \sum_n 2^{-n}\delta_{a_n}$, where m is arc length measure on $\partial\mathbb{D}$. In the Sarason process, it is easy to see that $K_1 = \operatorname{cl} \mathbb{D}$ and so $\mu_1 = m|B$. Thus $K_2 = B$ and so $\mu_2 = 0$. It follows that $K_\alpha = \varnothing$ and $\mu_\alpha = 0$ for all $\alpha \ge 3$. Thus $P^\infty(\mu) = L^\infty(\mu)$.

7.7 EXAMPLE. Let A, B, and $\{a_n\}$ be as in the preceding example. Let D_2 be the disk centered at 0 of radius $1/2$. Put $\mu = \mathscr{A}|D_1 + m|B + \sum_n 2^{-n}\delta_{a_n}$. Here $K_1 = \operatorname{cl} \mathbb{D}$ and $\mu_1 = \mathscr{A}|D_2 + m|B$. $K_2 = \operatorname{cl} D_2$ and $\mu_2 = \mathscr{A}|D_2$. It follows that $K_\alpha = \operatorname{cl} D_2$ and $\mu_\alpha = \mathscr{A}|D_2$ for all $\alpha \ge 2$. Thus $P^\infty(\mu) \cong H^\infty(D_2) \oplus L^\infty(m|B + \sum_n 2^{-n}\delta_{a_n})$.

By continuing this series of examples, it is possible to manufacture successively more complicated examples. See Exercise 1.

7.8 EXAMPLE. Let C be a Cantor set contained in $\partial\mathbb{D}$ of positive arc length. So $\partial\mathbb{D}\setminus C$ consists of a countable number of pairwise disjoint open arcs in $\partial\mathbb{D}$. Let G be the interior of the convex hull (not the polynomially convex hull but the convex hull) of $\partial\mathbb{D}\setminus C$. So each component of G is the open set whose boundary consists of one of the arcs that make up $\partial\mathbb{D}\setminus C$ and the corresponding chord. Let $\mu = \mathscr{A}|G$. Here $K_1 = \operatorname{cl} \mathbb{D}$ and $\mu_1 = \mu$; $K_\alpha = \operatorname{cl} G$ and $\mu_\alpha = \mu$ for $\alpha \ge 2$. Thus $P^\infty(\mu) \cong H^\infty(G)$.

7.9 EXAMPLE. Let G be as in the preceding example and put $G_1 = G$ and $G_2 = (1/2)G$. Let $\mu = \mathscr{A}|(G_1 \cup G_2)$. Then $K_1 = \operatorname{cl} \mathbb{D}$ and $\mu_1 = \mu$; $K_2 = \{z | z| \le 1/2\} \cup \operatorname{cl} G_1$ and $\mu_2 = \mu$; $K_\alpha = \operatorname{cl}[G_1 \cup G_2]$ and $\mu_\alpha = \mu$ for all $\alpha \ge 3$. Thus $P^\infty(\mu) \cong H^\infty(G_1 \cup G_2)$.

PROOF OF SARASON'S THEOREM. We begin by showing the existence of the ordinal number α_0. In fact, observe that the open sets $\{\operatorname{int}[K_\alpha \setminus K_{\alpha+1}]\}_\alpha$ are pairwise disjoint. Since $\mathbb{C}$ is separable, at most a countable number of them can be nonempty. Thus there is a countable ordinal α such that $K_\alpha \setminus K_{\alpha+1}$ has no interior. But from the definition of the sets K_α, if $\operatorname{int}[K_\alpha \setminus K_{\alpha+1}] = \varnothing$, then $\operatorname{int} K_\alpha = \operatorname{int} K_{\alpha+1}$ and $\mu_\alpha = \mu_{\alpha+1}$. It follows that if $\alpha_0 = \alpha + 1$, then $K_\beta = K_{\alpha_0}$ for all $\beta \ge \alpha_0$.

Let $K = K_{\alpha_0}$ and $\mu_a = \mu_{\alpha_0}$. For every ordinal α let $H^\infty(\operatorname{int} K_\alpha, \mu_\alpha) =$ the image of $H^\infty(\operatorname{int} K_\alpha)$ in $L^\infty(\mu_\alpha)$.

Note that for any α and φ in $H^\infty(\operatorname{int} K_\alpha)$, $\|\tilde\varphi\|_{\mu_\alpha} \le \|\varphi\|_{\operatorname{int} K_\alpha}$ and the map $\varphi \to \tilde\varphi$ is a weak* continuous homomorphism. For the ordinal $\alpha = \alpha_0$, this map, by virtue of the fact that $K_{\alpha_0} = K_{\alpha_0+1}$, is an isometry. Thus, with the aid of (I.2.10) we arrive at the following.

(7.10)   The map $\varphi \to \tilde{\varphi}$ is a dual algebra isomorphism of $H^\infty(\operatorname{int} K)$ onto its image $H^\infty(\operatorname{int} K, \mu_a)$ in $L^\infty(\mu_a)$.

CLAIM. For each ordinal α, $P^\infty(\mu) = R^\infty(K_\alpha, \mu_\alpha) \oplus L^\infty(\mu - \mu_\alpha)$ and $R^\infty(K_\alpha, \mu_\alpha) = P^\infty(\mu_\alpha)$.

Once this claim is established, the proof of the second version of Sarason's Theorem quickly follows. The details are left to the reader.

To prove the claim, first note that by definition, $\mu_\alpha \perp \mu - \mu_\alpha$ and so $P^\infty(\mu) \subseteq R^\infty(K_\alpha, \mu_\alpha) \oplus L^\infty(\mu - \mu_\alpha)$. Also, the second equality in the claim follows from the first. Thus to prove the claim it suffices to show

(7.11)   If η is a measure that is absolutely continuous with respect to μ and η annihilates the polynomials, then $\eta \ll \mu_\alpha$ and $\eta \perp R(K_\alpha)$.

This will be proved by induction. Let ω_α be harmonic measure for K_α. For $\alpha = 1$, the definition of μ_1 implies that $\mu - \mu_1 \perp \omega_1$. Since K_1 is polynomially convex, $\eta \perp R(K_1)$. Also, Proposition 4.9 shows that if $\psi \in L^\infty(\mu - \mu_1)$, there is a sequence $\{f_n\}$ in $R(K_1)$ such that $\|f_n\|_{K_1} \leq \|\psi\|_{\mu-\mu_1}$, $f_n \to \psi$ weak* in $L^\infty(\mu - \mu_1)$, and $f_n \to 0$ weak* in $H^\infty(\partial K_1)$. Hence $f_n \to 0$ weak* in $L^\infty(\mu_1)$ from which it follows that $f_n \to 0 \oplus \psi$ weak* in $L^\infty(\mu_1) \oplus L^\infty(\mu - \mu_1)$. Since $\eta \ll \mu$, $\int \psi\, d\eta = \lim \int f_n\, d\eta = 0$. Therefore $\eta \ll \mu_1$.

Now assume that (7.11) holds for some ordinal α; we show it holds for $\alpha + 1$. So assume that η is a measure such that $\eta \ll \mu_\alpha$ and $\eta \perp R(K_\alpha)$. We must show that $\eta \ll \mu_{\alpha+1}$ and $\eta \perp R(K_{\alpha+1})$. Fix a z_0 in $\mathbb{C} \setminus K_{\alpha+1}$ such that the Cauchy transform of η is defined at z_0. We know that the set of all such z_0 is a subset of $\mathbb{C} \setminus K_{\alpha+1}$ having full measure. Suppose $\hat{\eta}(z_0) \neq 0$. Since $\eta \perp R(K_\alpha)$ it must be that $z_0 \in K_\alpha$. If $z_0 \in \partial K_\alpha$, then there is a sequence $\{z_n\} \subseteq \mathbb{C} \setminus K_\alpha$ such that $z_n \to z_0$. Since $z_0 \notin K_{\alpha+1}$, $z_0 \notin \operatorname{supp} \mu_\alpha$ and so $(z - z_n)^{-1} \to (z - z_0)^{-1}$ uniformly on $\operatorname{supp} \mu_\alpha$. Since $\eta \ll \mu_\alpha$ it must be that $\hat{\eta}(z_0) = \lim \hat{\eta}(z_n) = 0$, contradicting the assumption. Hence it must be that $z_0 \in \operatorname{int} K_\alpha$. But $\eta \perp R(K_\alpha)$ and so $\hat{\eta}(z_0)^{-1}(z - z_0)^{-1} d\eta(z)$ is a complex representing measure for $R(K_\alpha)$. Thus there is a representing measure ν for $R(K_\alpha)$ at z_0 with $\nu \ll \eta$. Hence $\nu \ll \mu_\alpha$ and so $\varphi(z_0) = \int \tilde{\varphi}\, d\nu$ for all φ in $H^\infty(\operatorname{int} K_\alpha)$ by Theorem 4.8. But then by the definition of $K_{\alpha+1}$, this implies that $z_0 \in K_{\alpha+1}$, a contradiction.

Therefore, it must be that $\hat{\eta} = 0$ a.e. [Area] off $K_{\alpha+1}$. So η is supported by $K_{\alpha+1}$ and $\eta \perp R(K_{\alpha+1})$. Let $\eta = \eta_0 + \eta_1$ with $\eta_1 \ll \mu_{\alpha+1}$ and $\eta_0 \perp \mu_{\alpha+1}$. Since $\eta \ll \mu$, η_0 is carried by $\partial K_{\alpha+1}$ and $\eta_0 \perp \omega_{\alpha+1}$. Corollary V.18.9 implies that $\eta_0 = 0$ since $\eta \perp R(K_{\alpha+1})$. This proves the induction step for the case of a nonlimit ordinal.

Now assume that β is a limit ordinal and (7.11) holds for each $\alpha < \beta$. If $f \in \operatorname{Rat}(K_\beta)$, then there is an $\alpha < \beta$ such that $f \in \operatorname{Rat}(K_\alpha)$ and so $\int f\, d\eta = 0$. Thus $\eta \perp R(K_\beta)$. Let $\eta = \eta_0 + \eta_1$ with $\eta_1 \ll \mu_\beta$ and $\eta_0 \perp \mu_\beta$. It follows, as in the preceding paragraph, that $\eta_0 = 0$ and so $\eta \ll \mu_\beta$. This completes the proof. □

7.12 DEFINITION. For a compactly supported measure μ on $\mathbb{C}$, let $\Sigma(\mu)$ be the closure of the envelope $E((\operatorname{supp}\mu)\hat{\,}, \mu)$. The set $\Sigma(\mu)$ is called the *Sarason hull* of μ. If μ_a is as in Theorem 7.1, then $(\Sigma(\mu), \mu_a)$ is called the *Sarason pair* for μ.

7.13 EXAMPLE. Let $\{a_n\}$ be a sequence in $\mathbb{D}$ such that every point of $\partial \mathbb{D}$ is a limit point of $\{a_n\}$ and put $\mu = \sum_n 2^{-n} \delta_{a_n}$. Then $P^\infty(\mu) = L^\infty(\mu)$ if and only if there is a function φ in $H^\infty(\mathbb{D})$ such that $\|\varphi\|_\mathbb{D} > \sup_n |\varphi(a_n)|$. If $P^\infty(\mu) \neq L^\infty(\mu)$, then the Sarason pair for μ is $(\operatorname{cl}\mathbb{D}, \mu)$. Thus $P^\infty(\mu) = H^\infty(\mathbb{D}, \mu)$.

Sequences $\{a_n\}$ as in Example 7.13 with $\|\varphi\|_\mathbb{D} = \sup_n |\varphi(a_n)|$ for all φ in $H^\infty(\mathbb{D})$ were studied in Brown, Shields, and Zellar [1960]. In fact, this paper had an influence on Sarason [1972]. Such sequences are called *dominating sequences* and a necessary and sufficient condition for $\{a_n\}$ to be dominating is that a.e. point of $\partial \mathbb{D}$ be the nontangential limit of some subsequence of $\{a_n\}$. Also, see Bercovici [1984] for more on dominating sequences.

We conclude the section with a few miscellaneous results about $P^\infty(\mu)$. The proof of the next proposition is left as an exercise.

7.14 PROPOSITION. *For a compactly supported positive measure μ on $\mathbb{C}$ and a complex number a, the following statements are equivalent.*
   (a) $a \in \operatorname{int} \Sigma(\mu)$.
   (b) *The functional $p \to p(a)$ defined on the polynomials extends to a weak* continuous homomorphism of $P^\infty(\mu) \to \mathbb{C}$.*
   (c) $\{p: p \text{ is a polynomial and } p(a) = 0\}$ *is not weak* dense in $P^\infty(\mu)$.*

If $\mathscr{A}$ is a weak* closed subalgebra of $L^\infty(\mu)$ with 1 in $\mathscr{A}$, then a function φ in $\mathscr{A}$ is called a *weak* generator* of $\mathscr{A}$ if $\mathscr{A}$ is the smallest weak* closed algebra containing φ and 1; that is, $\mathscr{A}$ is the weak* closure of $\{p(\varphi): p \text{ is a polynomial}\}$.

7.15 PROPOSITION. *Let μ be a compactly supported positive measure on $\mathbb{C}$ and let G be a component of $\operatorname{int} \Sigma(\mu)$. If ψ is a Riemann map of $\mathbb{D}$ onto G, then ψ is a weak* generator of $H^\infty(\mathbb{D})$.*

PROOF. It suffices to assume that $\operatorname{int} \Sigma(\mu)$ is connected (so $G = \operatorname{int} \Sigma(\mu)$) and $P^\infty(\mu) \cong H^\infty(G)$ (Why?). Identifying $P^\infty(\mu)$ with $H^\infty(G)$, define $\rho: P^\infty(\mu) \to H^\infty(\mathbb{D})$ by $\rho(\varphi) = \varphi \circ \psi$. It is easy to see that ρ is a dual algebra isomorphism and $\rho(p) = p(\psi)$ for every polynomial p. The conclusion is now immediate. □

If ψ is a weak* generator of $H^\infty(\mathbb{D})$, it is easy to show that ψ is one-to-one. This condition is not sufficient, however, for ψ to be a generator.

Sarason [1966b] characterizes the weak* generators of $H^\infty(\mathbb{D})$ in terms of the geometry of the simply connected region $\psi(\mathbb{D})$. Also see Exercise 5.

7.16 PROPOSITION. *If μ is a compactly supported positive measure on $\mathbb{C}$ such that $P^\infty(\mu)$ is antisymmetric, then $\text{int}\,\Sigma(\mu)$ is connected; if τ is a Riemann map of $\text{int}\,\Sigma(\mu)$ onto $\mathbb{D}$, then τ is a weak* sequential generator of $P^\infty(\mu)$.*

PROOF. Let $G = \text{int}\,\Sigma(\mu)$. The fact that G is connected is immediate from the fact that $P^\infty(\mu)$ is antisymmetric. Also $P^\infty(\mu)$ can have no L^∞-summand and so we can identify $P^\infty(\mu)$ and $H^\infty(G)$. If $\varphi \in H^\infty(G)$, then $\varphi \circ \tau^{-1} \in H^\infty(\mathbb{D})$ and so there is a sequence of polynomials $\{p_n\}$ such that $\|p_n\|_\mathbb{D} \leq \|\varphi \circ \tau^{-1}\|_\mathbb{D} = \|\varphi\|_G$ and $p_n \to \varphi \circ \tau^{-1}$ weak* in $H^\infty(\mathbb{D})$. It follows that $\|p_n(\tau)\|_G \leq \|\varphi\|_G$ and $p_n(\tau) \to \varphi$ pointwise on G. Hence $p_n(\tau) \to \varphi$ weak* in $P^\infty(\mu)$. □

There are several applications of Sarason's Theorem to normal and subnormal operators. Indeed, this is the subject matter of the next chapter. One nice application that will not be presented is from Olin and Thomson [1977a]. Let $N = WP$ be the polar decomposition of the normal operator N, where W is a partial isometry. Olin and Thomson [1977a] give necessary and sufficient conditions that W and P belong to $P^\infty(N)$. For example, if $0 \in \sigma_p(N)$, then $P \in P^\infty(N)$ if and only if N is reductive (and so $P^\infty(N) = W^*(N)$). If $0 \notin \sigma_p(N)$, then $P \in P^\infty(N)$ if and only if $N = N_0 \oplus N_1$, where N_0 is reductive, $\sigma(N_1)$ is a circle centered at 0, and $P^\infty(N) = P^\infty(N_0) \oplus P^\infty(N_1)$. They also give necessary and sufficient conditions that W belongs to $P^\infty(N)$ in terms of the spectral properties of N.

Exercises

1. (a) (Sarason [1968]) Show that for any countable ordinal α_0 there is a positive compactly supported measure μ on $\mathbb{C}$ such that $K_\alpha = K_{\alpha_0}$ and $\mu_\alpha = \mu_{\alpha_0}$ for all $\alpha \geq \alpha_0$ but $K_\alpha \neq K_{\alpha_0}$ and $\mu_\alpha \neq \mu_{\alpha_0}$ for $\alpha < \alpha_0$.
   (b) Show that μ can be chosen so that in addition to the properties in (a) it is also true that $P^\infty(\mu) = L^\infty(\mu)$.
   (c) Show that μ can be chosen so that in addition to the properties in (a) it is also true that $P^\infty(\mu) \neq L^\infty(\mu)$.
2. Let E be a measurable subset of $\mathbb{C}$ having full area density at each of its points and let $\mu = \mathscr{A}|E$. Show that the support of μ is $\text{cl}\,E$. Can $P^\infty(\mu)$ have an L^∞-summand? Show that if α_0 is any countable ordinal, E can be chosen so that $K_\alpha = K_{\alpha_0}$ and $\mu_\alpha = \mu_{\alpha_0} = \mu$ for all $\alpha \geq \alpha_0$ but $K_\alpha \neq K_{\alpha_0}$ and $\mu_\alpha \neq \mu_{\alpha_0}$ for $\alpha < \alpha_0$.
3. Verify the statements in Example 7.13. Construct two sequences in $\mathbb{D}$ that give each of the two possibilities in Example 7.13.
4. Prove Proposition 7.14.

5. (a) Show that if ψ is a weak* generator of $H^\infty(\mathbb{D})$, then ψ is one-to-one on $\mathbb{D}$.
   (b) If G is the slit disk or an open crescent, show that the Riemann map of $\mathbb{D}$ onto G is not a weak* generator of $H^\infty(\mathbb{D})$.
   (c) Let G be a simply connected region and let ψ be a Riemann map of $\mathbb{D}$ onto G. If μ is the restriction of area measure to G, show that ψ is a weak* generator of $H^\infty(\mathbb{D})$ if and only if $G = \operatorname{int} \Sigma(\mu)$.
   (d) Maintaining the same notation, show that ψ is a weak* sequential generator of $H^\infty(\mathbb{D})$ if and only if $G = \operatorname{int} \Sigma(\mu)$ and the polynomials are weak* sequentially dense in $P^\infty(\mu)$.
6. Assume that $P^\infty(\mu)$ has no L^∞-summand and find a weak* sequential generator of $P^\infty(\mu)$.
7. (Conway [1977]) Let N and M be normal operators with scalar-valued spectral measures ν and μ, respectively, let $\lambda = \nu + \mu$, and let $(\Sigma(\nu), \nu_a)$, $(\Sigma(\mu), \mu_a)$, and $(\Sigma(\lambda), \lambda_a)$ be the corresponding Sarason pairs. Show that the following statements are equivalent.
   (a) $P^\infty(N \oplus M) = P^\infty(N) \oplus P^\infty(M)$.
   (b) $\operatorname{Lat}(N \oplus M) = \operatorname{Lat} N \oplus \operatorname{Lat} M$.
   (c) $\nu \perp \mu$, $[\operatorname{int} \Sigma(\nu)] \cap [\operatorname{int} \Sigma(\mu)] = \varnothing$, $\Sigma(\nu) \cup \Sigma(\mu) = \Sigma(\lambda)$, and $\lambda_a = \nu_a + \mu_a$.
8. (Conway [1977]) Let U and V be unitary operators with scalar-valued spectral measures μ and ν, respectively, let ω = harmonic measure on $\partial \mathbb{D}$, and let $\mu = \mu_a + \mu_s$ and $\nu = \nu_a + \nu_s$ be the Lebesgue decompositions of μ and ν with respect to ω. Show that the following statements are equivalent.
   (a) $P^\infty(U \oplus V) = P^\infty(U) \oplus P^\infty(V)$.
   (b) $\operatorname{Lat}(U \oplus V) = \operatorname{Lat} U \oplus \operatorname{Lat} V$.
   (c) $\mu \perp \nu$ and one of the following holds:
      (i) $[\mu_a] = [\omega]$ and $\nu_a = 0$;
      (ii) $[\nu_a] = [\omega]$ and $\mu_a = 0$;
      (iii) $[\mu_a + \nu_a] \neq [\omega]$.
   (d) $\mu \perp \nu$ and one of the following holds:
      (i) U has a bilateral shift as a direct summand and V is singular;
      (ii) V has a bilateral shift as a direct summand and U is singular;
      (iii) $U \oplus V$ does not have a bilateral shift as a direct summand.
9. Let R be the open ribbon that is part of the cornucopia, let μ be area measure restricted to R, and let $N = N_\mu$ on $L^2(\mu)$. Show that $\operatorname{Lat} N \subseteq \operatorname{Lat} N^{-1}$ but $\operatorname{Lat} N \neq \operatorname{Lat} N^{-1}$.
10. Let μ be a measure on $\partial \mathbb{D}$ and let m be normalized Lebesgue measure on $\partial \mathbb{D}$. Show that $P^\infty(\mu) \neq L^\infty(\mu)$ if and only if $m \ll \mu$.
11. If μ is a measure supported on the closed unit disk and $\Sigma(\mu) = \operatorname{cl} \mathbb{D}$ and $P^\infty(\mu) = H^\infty(\mathbb{D})$, show that for $0 < r < 1$ and $\mu_r = \mu | \{z : r \leq |z| \leq 1\}$, $P^\infty(\mu_r) \cong H^\infty(\mathbb{D})$.

CHAPTER VII

Some Structure Theory for Subnormal Operators

In this chapter we will see a number of applications of the preceding two chapters to the theory of subnormal operators. In the first section a decomposition theorem for subnormal operators is proved. The next four sections address questions concerning the functional calculus for subnormal operators: What is the minimal normal extension of $\phi(S)$ and what is its spectrum? Sections 5 and 6 prove some theorems on factoring weak* continuous linear functionals on the algebra $P^\infty(S)$. As one application it is shown that like normal operators, for every subnormal operator S, $P^\infty(S)$ is the same as the WOT closed algebra generated by S. In Section 7 these factorization theorems are applied to show that each subnormal operator has an abundance of cyclic subspaces on which there are as many analytic bounded point evaluations as is possible. In light of Thomson's Theorem in the next chapter, cyclic subspaces always have analytic bounded point evaluations, but this result says that there are cyclic subspaces such that the interior of the Sarason hull is the set of analytic bounded point evaluations. This is then used to show every pure subnormal operator S is reflexive; that is, S has sufficiently many invariant subspaces to completely determine $P^\infty(S)$. In Section 9, some results are presented on the collection of all subnormal operators having a fixed normal extension. The chapter concludes with an application of the functional calculus to the study of quasisimilarity for subnormal operators.

Though most of these sections depend on Section 1, there is some independence among those remaining. Section 4 is heavily dependent on Section 3 and Sections 5, 6, 7, and 8 form an interlocked quartet. But these two groups are essentially independent of the rest of the chapter.

§1 **A decomposition of subnormal operators.** The main theorem (1.4) of this section often permits a question concerning subnormal operators to be reduced to a question about a more elementary subnormal operator. This is especially true if the question is related to the algebra $P^\infty(S)$. Here questions can often be reduced to those involving subnormal operators where $P^\infty(S)$ is naturally isomorphic to $H^\infty(\mathbb{D})$.

One approach here would be to apply the Lautzenheiser-Mlak Decomposition Theorem (V.21.1) and obtain some refinements of this result for the special case of a subnormal operator. In light of the functional calculus available for subnormal operators, it is just as easy to obtain this decomposition directly.

We begin with a few preliminary results.

1.1 DEFINITION. For an operator T acting on $\mathcal{H}$ and a compact set K containing $\sigma(T)$, say that T is $R(K)$ *reductive* if every subspace of $\mathcal{H}$ that is invariant for $f(T)$ for every f in $\text{Rat}(K)$ is a reducing subspace of T. Say that T is *reductive* if every invariant subspace for T is reducing.

So an operator is reductive if and only if it is $R(\widehat{\sigma(T)})$ reductive.

1.2 PROPOSITION. *If N is a normal operator with scalar-valued spectral measure μ and K is a compact set containing $\sigma(N)$, then N is $R(K)$ reductive if and only if $R^\infty(K, \mu) = L^\infty(\mu)$.*

PROOF. If $\mathcal{M}$ is a subspace invariant for $f(N)$ for every f in $\text{Rat}(K)$, then it is easy to see that $\mathcal{M}$ is invariant for $\phi(N)$ for every ϕ in $R^\infty(K, \mu)$. So it is immediate that N is $R(K)$ reductive if $R^\infty(K, \mu) = L^\infty(\mu)$.

Now assume that N is $R(K)$ reductive. There is a reducing subspace $\mathcal{H}_1$ for N such that $N_1 \equiv N|\mathcal{H}_1$ is *-cyclic with scalar-valued spectral measure μ. It is straightforward that N_1 is $R(K)$ reductive. Thus the proposition will be proved if it can be proved under the additional hypothesis that N is *-cyclic. Make this assumption. That is, assume that $N = N_\mu$ acting on $L^2(\mu)$.

To show that $R^\infty(K, \mu) = L^\infty(\mu)$ it suffices to show that $\bar{z} \in R^\infty(K, \mu)$. So let $h \in L^1(\mu)$ such that $h \perp R^\infty(K, \mu)$ and let's show that $\int \bar{z} h \, d\mu = 0$. Write $h = f_1 \bar{f}_2$, where f_1 and f_2 are functions in $L^2(\mu)$. Let $\mathcal{M} = \bigvee \{gf_1 : g \in R(K)\}$. Clearly $\mathcal{M}$ is invariant for $g(N) = M_g$ for every g in $R(K)$. Thus $\bar{z} f_1 \in \mathcal{M}$ by hypothesis. But $f_2 \perp \mathcal{M}$ and so $0 = \langle \bar{z} f_1, f_2 \rangle = \int \bar{z} h \, d\mu$. □

1.3 COROLLARY. *If N is a normal operator with scalar-valued spectral measure μ, then N is reductive if and only if $P^\infty(\mu) = L^\infty(\mu)$.*

Here is the main theorem of the section.

1.4 THEOREM. *Let S be a subnormal operator on a Hilbert space $\mathcal{H}$ with minimal normal extension N acting on $\mathcal{K}$ and scalar-valued spectral measure μ and let K be a compact set containing $\sigma(S)$. Let $R^\infty(K, \mu) = L^\infty(\mu_0) \oplus R^\infty(K, \mu_a)$ as in Chaumat's Theorem, where $R^\infty(K, \mu_a)$ has no L^∞ summand. If $E = E(K, \mu) = E(K, \mu_a)$ is the envelope and $G_1, G_2, \ldots$ are the nontrivial Gleason parts of $R(K, E)$, then there are pairwise singular measures $\mu_1, \mu_2, \ldots$, pairwise orthogonal subspaces of $\mathcal{H}$, $\mathcal{H}_0, \mathcal{H}_1, \mathcal{H}_2, \ldots$, and pairwise orthogonal subspaces of $\mathcal{K}$, $\mathcal{K}_1, \mathcal{K}_2, \ldots$ such that the following conditions are satisfied.*

(a) $\mathcal{H}_0$ *reduces N and $N|\mathcal{H}_0$ is $R(K)$ reductive with scalar-valued spectral measure μ_0.*

(b) *For $n \geq 1$, $\mathcal{K}_n$ reduces S, $\mathcal{K}_n$ reduces N, $\mathcal{H}_n \leq \mathcal{K}_n$, and if $S_n \equiv S|\mathcal{H}_n$ and $N_n \equiv N|\mathcal{K}_n$, then $\sigma(S_n)$ and $\sigma(N_n)$ are contained in $\operatorname{cl} G_n$.*

(c) *For $n \geq 1$, $N_n = \operatorname{mne} S_n$ and μ_n is a scalar-valued spectral measure for N_n.*

(d) $\mathcal{K} = \mathcal{K}_0 \oplus \bigoplus_{n=1}^{\infty} \mathcal{K}_n$, $\mathcal{H} = \mathcal{H}_0 \oplus \bigoplus_{n=1}^{\infty} \mathcal{H}_n$, $N = N_0 \oplus \bigoplus_{n=1}^{\infty} N_n$, and $S = N_0 \oplus \bigoplus_{n=1}^{\infty} S_n$.

(e) $R^{\infty}(K; N) = W^*(N_0) \oplus \bigoplus_{n=1}^{\infty} R^{\infty}(\operatorname{cl} G_n; N_n)$ and $R^{\infty}(K; S) = W^*(N_0) \oplus \bigoplus_{n=1}^{\infty} R^{\infty}(\operatorname{cl} G_n; S_n)$.

(f) *For $n \geq 1$, both $R^{\infty}(\operatorname{cl} G_n; S_n)$ and $R^{\infty}(\operatorname{cl} G_n; N_n)$ are naturally isomorphic to $R^{\infty}(G_n, \mu_n) = R^{\infty}(\operatorname{cl} G_n, \mu_n)$.*

(g) *If $\mathcal{M}$ is an invariant subspace for $f(S)$ for every f in $R(K)$, then $\mathcal{M} = \mathcal{M}_0 \oplus \mathcal{M}_1 \oplus \mathcal{M}_2 \oplus \cdots$, where $\mathcal{M}_n \leq \mathcal{H}_n$, $\mathcal{M}_0$ reduces S, and, for $n \geq 1$, $\mathcal{M}_n$ is an invariant subspace for $f(S_n)$ for all f in $R(\operatorname{cl} G_n)$. A similar statement holds for the subspaces of $\mathcal{K}$ that are invariant for $f(N)$ for every f in $R(K)$.*

PROOF. First note that since $\sigma(S) \subseteq K$, $\phi(N)\mathcal{H} \subseteq \mathcal{H}$ for all ϕ in $R^{\infty}(K, \mu)$. As in Theorem VI.3.12, there are pairwise singular measures $\mu_0, \mu_1, \mu_2, \ldots$ such that $\mu = \sum_n \mu_n$ and

$$(1.5) \qquad R^{\infty}(K, \mu) = L^{\infty}(\mu_0) \oplus \bigoplus_{n=1}^{\infty} R^{\infty}(G_n, \mu_n),$$

where each $R^{\infty}(G_n, \mu_n)$ is antisymmetric. Since the measures μ_n are pairwise singular, there are characteristic functions $\{\chi_n\}$ such that $\chi_n \chi_m = 0$ for $n \neq m$ and $\mu_n = \chi_n \mu$ for all $n \geq 0$. Thus χ_n is the identity for $R^{\infty}(G_n, \mu_n)$. Let $P_n = \chi_n(N)$, $Q_n = \chi_n(S)$, $\mathcal{K}_n = P_n \mathcal{K}$, and $\mathcal{H}_n = Q_n \mathcal{H}$. Because $R^{\infty}(K, \mu_0) = L^{\infty}(\mu_0)$, $N|\mathcal{K}_0$ is an $R(K)$ reductive operator. Since $N = \operatorname{mne} S$, it must be that so that $\mathcal{H}_0 = \mathcal{K}_0 \leq \mathcal{H}$ and this proves (a).

Since $P_0 = Q_0$ and $1 = \sum_n \chi_n$, part (d) readily follows from (1.5).

It is left to the reader to verify that μ_n is the scalar-valued spectral measure for N_n. Note that $Q_n = P_n|\mathcal{H}$ and so $\mathcal{H}_n \leq \mathcal{K}_n$ and $S_n = N_n|\mathcal{H}_n$. Now $\sigma(N_n) = \operatorname{supp} \mu_n \subseteq \operatorname{cl} G_n$. Also if $\phi \in R^{\infty}(G_n, \mu_n)$, then $\phi = \phi \chi_n$ and so $\phi(N_n)\mathcal{H}_n = \chi_n(N)\phi(N)\mathcal{H} \subseteq \mathcal{H}$. In particular, if $\alpha \notin \operatorname{cl} G_n$, $(z-\alpha)^{-1} \in R^{\infty}(G_n, \mu_n)$ and so $(N_n - \alpha)^{-1} \mathcal{H}_n \subseteq \mathcal{H}_n$. Therefore $\sigma(S_n) \subseteq \operatorname{cl} G_n$, proving part (b).

If N_n is not the minimal normal extension of S_n, then the fact that $N = N_0 \oplus N_1 \oplus \cdots$ implies that N cannot be the minimal normal extension of S. This proves (c).

The verification of (e) and (f) are left to the reader.

Finally, to prove part (g), let $\mathcal{M}$ be as in the statement of (g). Thus $\mathcal{M}$ is an invariant subspace for $\phi(S)$ for every ϕ in $R^{\infty}(K, \mu)$ and so $\mathcal{M}_n = Q_n \mathcal{M} \subseteq \mathcal{M}$. It is easy to show that $\mathcal{M} = \oplus_n \mathcal{M}_n$ and $\mathcal{M}_n$ is invariant for $\phi(S_n)$ for every ϕ in $R^{\infty}(G_n, \mu_n)$. □

The next corollary follows from the preceding theorem by taking K to be the polynomially convex hull of $\sigma(S)$.

1.6 COROLLARY. *Let S be a subnormal operator on a Hilbert space $\mathcal{H}$ with minimal normal extension N acting on $\mathcal{K}$ and scalar-valued spectral measure μ. Let $\mu = \mu_s + \mu_a$ as in the statement of Sarason's Theorem so that $P^\infty(\mu_a)$ has no L^∞ summand. If $G_1, G_2, \ldots$ are the components of the interior of the Sarason hull of μ, then there are pairwise singular measures $\mu_1, \mu_2, \ldots$, pairwise orthogonal subspaces of $\mathcal{H}$, $\mathcal{H}_0, \mathcal{H}_1, \mathcal{H}_2, \ldots$, and pairwise orthogonal subspaces of $\mathcal{K}$, $\mathcal{K}_1, \mathcal{K}_2, \ldots$ such that $P^\infty(\mu) = L^\infty(\mu_s) \oplus \bigoplus_n P^\infty(\mu_n)$ and the following conditions are satisfied.*

(a) *$\mathcal{H}_0$ reduces N and $N|\mathcal{H}_0$ is reductive with scalar-valued spectral measures μ_0.*

(b) *For $n \geq 1$, $\mathcal{H}_n$ reduces S, $\mathcal{K}_n$ reduces N, $\mathcal{H}_n \leq \mathcal{K}_n$, and if $S_n \equiv S|\mathcal{H}_n$ and $N_n \equiv N|\mathcal{K}_n$, then $\sigma(S_n)$ and $\sigma(N_n)$ are contained in $\operatorname{cl} G_n$.*

(c) *For $n \geq 1$, $N_n = \operatorname{mne} S_n$ and μ_n is a scalar-valued spectral measure for N_n.*

(d) *$\mathcal{H} = \mathcal{H}_0 \oplus \bigoplus_{n=1}^\infty \mathcal{H}_n$, $\mathcal{K} = \mathcal{H}_0 \oplus \bigoplus_{n=1}^\infty \mathcal{K}_n$, $N = N_0 \oplus \bigoplus_{n=1}^\infty N_n$, and $S = N_0 \oplus \bigoplus_{n=1}^\infty S_n$.*

(e) *$P^\infty(N) = W^*(N_0) \oplus \bigoplus_{n=1}^\infty P^\infty(N_n)$ and $P^\infty(S) = W^*(N_0) \oplus \bigoplus_{n=1}^\infty P^\infty(S_n)$.*

(f) *For $n \geq 1$, both $P^\infty(S_n)$ and $P^\infty(N_n)$ are naturally isomorphic to $P^\infty(\mu_n)$ which is naturally isomorphic to $H^\infty(G_n)$. Hence $P^\infty(S_n)$ and $P^\infty(N_n)$ are antisymmetric.*

(g) *If $\mathcal{M}$ is an invariant subspace for S, then $\mathcal{M} = \mathcal{M}_0 \oplus \bigoplus_{n=1}^\infty \mathcal{M}_n$, where $\mathcal{M}_n \leq \mathcal{H}_n$, $\mathcal{M}_0$ reduces S, and for $n \geq 1$, $\mathcal{M}_n$ is an invariant subspace for S_n. A similar statement holds for invariant subspaces of N.*

Notice that nothing in the preceding theorem and its corollary precludes the possibility that $\mathcal{H}_n = \mathcal{K}_n$ for some integer $n \geq 1$. Of course this means that S is not pure. For example, the bilateral shift U is such that $P^\infty(U)$ is antisymmetric.

The preceding corollary when combined with the next proposition permits many questions concerning subnormal operators to be examined and solved by means of our knowledge of function theory on the open unit disk.

1.7 PROPOSITION. *Let S be a subnormal operator with scalar-valued spectral measure μ. If $P^\infty(\mu)$ is antisymmetric, $G = $ the interior of the Sarason hull of μ, and τ is the Riemann map of G onto $\mathbb{D}$, then the following statements hold.*

(a) *If N is the minimal normal extension of S, then $\tau(N)$ is the minimal normal extension of $\tau(S)$ and so $\tau(S)$ has $\nu = \mu \circ \tau^{-1}$ as a scalar-valued spectral measure.*

(b) *$P^\infty(\tau(S)) = P^\infty(S)$.*

(c) *$\operatorname{Lat} \tau(S) = \operatorname{Lat} S$.*

(d) *$P^\infty(\nu) \cong H^\infty(\mathbb{D})$.*

(e) *If $S = M_z$ on $P^2(\mu)$, then $\tau(S) \cong M_z$ on $P^2(\nu)$.*

PROOF. (a) Suppose S acts on $\mathscr{H}$ and N acts on $\mathscr{K}$. Let $\mathscr{L}$ be the smallest subspace of $\mathscr{K}$ that contains $\mathscr{H}$ and reduces $\tau(N)$; we want to show that $\mathscr{L} = \mathscr{K}$. According to Proposition VI.7.16, there is a sequence of polynomials $\{p_n\}$ such that $p_n(\tau) \to z$ weak* in $P^\infty(\mu)$. Hence $p_n(\tau(N)) \to N$ weak* in $\mathscr{B}(\mathscr{K})$ and so $p_n(\tau(N))^* \to N^*$. Since $\mathscr{L}$ reduces $\tau(N)$, this shows that $\mathscr{L}$ reduces N. But $N = \operatorname{mne} S$ and so it must be that $\mathscr{L} = \mathscr{K}$. It is a standard fact about normal operators that ν must be a scalar-valued spectral measure for $\tau(N)$.

(b) This is immediate from Proposition VI.7.16.

(c) Again, an easy application of (VI.7.16).

(d) By (a), $P^\infty(\nu) \cong P^\infty(\tau(S))$ via the map $\phi \to \phi(\tau(S))$. Also $f \to f \circ \tau$ defines an isomorphism of $H^\infty(\mathbb{D})$ onto $H^\infty(G)$ and $H^\infty(G) \cong P^\infty(S)$ via the map $\psi \to \psi(S)$. In light of part (b), this implies that $H^\infty(\mathbb{D}) \cong P^\infty(\nu)$ and, if the various maps whose composition forms this isometric isomorphism are examined, the isomorphism is the identity on polynomials.

(e) If $W: L^2(\nu) \to L^2(\mu)$ is defined by $Wf = f \circ \tau$, then W is a surjective isometry. For any polynomial p, $Wp = p \circ \tau \in P^\infty(\mu) \subseteq P^2(\mu)$. Thus $WP^2(\nu) \subseteq P^2(\mu)$. On the other hand, Proposition VI.7.16 and the fact that μ is a finite measure imply that $WP^2(\nu)$ must contain $P^\infty(\mu)$. Since $WP^2(\nu)$ is closed we get that W maps $P^2(\nu)$ onto $P^2(\mu)$.

Now for any polynomial p, $W(zp) = \tau W(p)$ and so we can conclude that $WM_zW^* = M_\tau$ on $P^2(\mu) = \tau(S)$. □

The following proposition is from Radjabalipour [1975].

1.8 PROPOSITION. *If S is a subnormal operator, $\bar{\alpha} \in \sigma_p(S^*)$, and α is a peak point of $\sigma(S)$, then $\alpha \in \sigma_p(S)$.*

PROOF. Let $N = \operatorname{mne} S$ and let x be a nonzero vector in $\ker(S^* - \bar{\alpha})$. So for every vector h in $\mathscr{H}$, $0 = \langle (S - \alpha)h, x \rangle$. Elementary algebra shows that $0 = \langle (r(S) - r(\alpha))h, x \rangle$ for all rational functions r in $\operatorname{Rat}(\sigma(S))$ and hence, by taking limits, $\langle (g(S) - g(\alpha))h, x \rangle = 0$ for all functions g in $R(\sigma(S))$. Let $f \in R(\sigma(S))$ such that $f(\alpha) = 1$ and $|f(z)| < 1$ for all $z \neq \alpha$. Thus $\langle (f(S)^n - 1)h, x \rangle = 0$ for all vectors h in $\mathscr{H}$ and all $n \geq 1$. Let χ denote the characteristic function of the singleton $\{\alpha\}$. If μ is a scalar-valued spectral measure for S, $f^n \to \chi$ weak* in $R^\infty(\sigma(S), \mu)$ and so $f(S)^n \to \chi(S)$ weak* in $\mathscr{B}(\mathscr{H})$. Thus $0 = \langle (\chi(S) - 1)h, x \rangle$ for all h in $\mathscr{H}$. But $\chi(S)$ is a projection and this implies that $x \in \ker(\chi(S) - 1)$. That is, $\chi(S) \neq 0$. But this implies that $\chi(N) \neq 0$ and it is easy to see that $\chi(N)$ must be the projection of $\mathscr{K}$ onto $\ker(N - \alpha)$. Thus $Sx = S\chi(S)x = N\chi(N)x = \alpha x$. □

1.9 PROPOSITION. *If S is a pure subnormal operator with $N = \operatorname{mne} S$ and U is a component of $\mathbb{C}\backslash\sigma(S)$, then $\sigma_p(N) \cap \partial U = \varnothing$.*

PROOF. Let α be any point in ∂U. By Corollary V.13.3, α is a peak point of $\sigma(S)$. Let f be a function in $R(\sigma(S))$ that peaks at α; so $f^n \to \chi_{\{\alpha\}}$ pointwise on $\sigma(S)$ and hence weak* in $L^\infty(\mu)$, where μ is the scalar-valued spectral measure for N. It follows that $f^n(N) \to E \equiv \chi_{\{\alpha\}}(N)$ (WOT). But E is the projection of $\mathscr{K}$ onto $\ker(N - \alpha)$. Since $f^n(N)\mathscr{H} \subseteq \mathscr{H}$ for all n, it must be that $E\mathscr{H} \subseteq \ker(S - \alpha)$. But S is pure and so $E\mathscr{H} = (0)$. That is, $\mathscr{H} \subseteq \ker E = [\ker(N - \alpha)]^\perp$. Since $N = \mathrm{mne}\, S$, this implies that $\ker(N - \alpha) = (0)$ so that $\alpha \notin \sigma_p(N)$. □

Putnam [1977b] shows that if $\alpha \in \sigma_p(N)$ and α is a peak point of $\sigma(S)$, then $\alpha \in \sigma_p(S)$. Olin [1977] shows that if K is a compact set and $\alpha \in K$ such that α is not a peak point of K, then there is a pure subnormal operator S such that $\sigma(S) \subseteq K$ and $\alpha \in \sigma_p(N)$.

This seems an appropriate place to present the following result.

1.10 PROPOSITION (Wermer [1952]). *If U is a unitary operator with scalar-valued spectral measure μ and m is normalized Lebesgue measure on $\partial \mathbb{D}$, then the following statements are equivalent.*

(a) *U is not reductive.*
(b) *$m \ll \mu$.*
(c) *There is a reducing subspace $\mathscr{M}$ for U such that $U|\mathscr{M}$ is unitarily equivalent to the unilateral shift of multiplicity 1.*

PROOF. The equivalence of (a) and (b) follows from Corollary 1.3 and Exercise VI.7.10. Since it is immediate that (c) implies (a), it remains to show that (b) implies (c). It is a standard result about normal operators that there is a vector e such that $m(\Delta) = \langle E(\Delta)e, e \rangle$ for all Borel sets Δ. (For example, see IX.8.6 in [ACFA]. If $\mathscr{M} = \bigvee\{U^n e : n \geq 0\}$, then it follows that $U|\mathscr{M}$ is a unilateral shift of multiplicity 1. □

What is an example of a reductive operator that is not normal? This natural and seemingly innocent question has much more to it than meets the eye. Dyer, Pedersen, and Porcelli [1972] have shown that there are reductive operators that are not normal if and only if there is an operator whose only invariant subspaces are the two trivial ones. In this vein, see Exercises 3 and 4 below.

An additional reference is Scroggs [1959].

Exercises

1. In Corollary 1.6 show that if $\sigma(S_n) \subseteq \partial G_n$, then $S_n = N_n$.
2. (Putnam [1977b]) Let $N = \mathrm{mne}\, S$ and let $N = \int z\, dE(z)$. If Δ is a peak set of $\sigma(S)$ and $E(\Delta) \neq 0$, then $E(\Delta)\mathscr{H}$ is a nonzero reducing subspace of S and the minimal normal extension of $S|E(\Delta)\mathscr{H}$ is $N|E(\Delta)\mathscr{H}$.
3. Prove that a reductive subnormal operator is normal.
4. Prove that every hyponormal operator has an invariant subspace if and only if every reductive hyponormal operator is normal.

5. Show that if N is a normal operator and $\operatorname{int} \sigma(N) \neq \varnothing$, then N is not reductive.
6. If N is a normal operator with $\operatorname{int} \sigma(N) = \varnothing$ and $\mathbb{C} \setminus \sigma(N)$ connected, show that N is reductive.
7. (Ando [1963c]) Every compact normal operator is reductive.
8. Determine all the invariant subspaces for a compact normal operator.
9. (Rosenthal [1968]) Show that a compact reductive operator is normal.
10. (Wermer [1952]) If N is a diagonalizable normal operator, then the following statements are equivalent.
    (a) N is reductive.
    (b) Every invariant subspace for N is the closed linear span of the eigenvectors it contains.
    (c) If $\mathscr{M} \in \operatorname{Lat} N$, then $\sigma_p(N|\mathscr{M}) \neq \varnothing$.
    (d) If $\{\lambda_n\}$ is a sequence of distinct eigenvalues of N and $\{\alpha_n\}$ is a sequence of scalars such that $\sum_n |\alpha_n| < \infty$ and $\sum_n \alpha_n \lambda_n^k = 0$ for all $k \geq 0$, then $\alpha_1 = \alpha_2 = \cdots = 0$.

§2 The minimal normal extension problem for subnormal operators.

Let S be a subnormal operator on a Hilbert space $\mathscr{H}$ and let N be its minimal normal extension acting on the Hilbert space $\mathscr{K}$. Let μ be a scalar-valued spectral measure for S. If ϕ is a function in the restriction algebra $\mathscr{R}(N, \mathscr{H})$, then $\phi(S) = \phi(N)|\mathscr{H}$ is a subnormal operator with $\phi(N)$ as a normal extension. The basic question addressed here is whether $\phi(N)$ is the minimal normal extension of $\phi(S)$.

It is easy to see the answer is no since if ϕ is constant, $\phi(S)$ is already normal. As we will see below, this is essentially the answer to the entire question. It will be shown that for suitable ϕ which are not constant on "big" sets, $\phi(N)$ is the minimal normal extension of $\phi(S)$.

2.1 DEFINITION. If K is a compact subset of $\mathbb{C}$ and μ is a measure supported on K, let $\mathscr{S}(K, \mu)$ denote the collection of all subnormal operators S with $\sigma(S) \subseteq K$ and scalar-valued spectral measure μ.

Note that the set K in the preceding definition need not equal the support of μ.

The point here is that for a function ϕ in $R^\infty(K, \mu)$, $\phi(S)$ is defined for each S in $\mathscr{S}(K, \mu)$. Thus if conditions are imposed solely on ϕ that aid in the determination of the minimal normal extension of $\phi(S)$, they apply to all the operators in $\mathscr{S}(K, \mu)$. Why not impose extra conditions on S as well as on ϕ so that "individual" results on the minimal normal extension of $\phi(S)$ are possible? This would be welcomed, but as yet no pertinent examples exist that demonstrate the utility of this approach. At this point the known examples can be covered by the existing results.

Once we agree to consider $\phi(S)$ for all S in $\mathscr{S}(K, \mu)$, the problem becomes a function theoretic one as we will see in the next result. The equivalence of (a), (c), and (d) in the next theorem is from Dudziak [1986], though a version for functions in $P^\infty(\mu)$ which includes an analogue of condition (b) was obtained in Conway and Olin [**1977**].

2.2 THEOREM. *If K is a compact subset of $\mathbb{C}$, μ is a measure supported on K, and $\phi \in R^\infty(K, \mu)$, then the following statements are equivalent.*

(a) *The algebra generated by $R^\infty(K, \mu)$ and $\overline{\phi}$ is weak* dense in $L^\infty(\mu)$.*
(b) *Whenever N is a normal operator with μ as its scalar-valued spectral measure, $(\operatorname{Lat} R^\infty(K; N)) \cap \operatorname{Lat} \phi(N)^*$ is the lattice of reducing subspaces for N.*
(c) *$\phi(\operatorname{mne} S) = \operatorname{mne} \phi(S)$ for every S in $\mathscr{S}(K, \mu)$.*
(d) *$\phi(\operatorname{mne} S) = \operatorname{mne} \phi(S)$ for each $R(K)$-cyclic operator S in $\mathscr{S}(K, \mu)$.*

PROOF. Let $\mathscr{A}$ = the subalgebra of $L^\infty(\mu)$ generated by $R^\infty(K, \mu)$ and $\overline{\phi}$.

(a) *implies* (b). Denote the reducing subspaces for N by $\operatorname{Red} N$ and note that we always have that $\operatorname{Red} N = \operatorname{Lat}\{\psi(N) : \psi \in L^\infty(\mu)\} \subseteq \operatorname{Lat}\{\psi(N) : \psi \in \mathscr{A}\}$. It is easily seen that $\operatorname{Lat}\{\psi(N) : \psi \in \mathscr{A}\} = (\operatorname{Lat} R^\infty(K; N)) \cap \operatorname{Lat} \phi(N)^*$, so if $\mathscr{A}$ is weak* dense in $L^\infty(\mu)$, then (b) follows.

(b) *implies* (c). If $S \in \mathscr{S}(K, \mu)$ and $N = \operatorname{mne} S$ acting on $\mathscr{K}$, then the minimal normal extension of $\phi(S)$ is $\phi(N)|\mathscr{L}$, where $\mathscr{L} = \bigvee\{\phi(N)^{*n}h : h \in \mathscr{H}$ and $n \geq 0\}$. Clearly $\mathscr{L}$ is invariant for $\psi(N)$ for every ψ in $\mathscr{A}$ and hence, according to (b), $\mathscr{L}$ reduces N and the restriction of N to $\mathscr{L}$ is normal. By minimality, $\mathscr{L} = \mathscr{K}$ and so $\phi(N) = \operatorname{mne} \phi(S)$.

(c) *implies* (d). Trivial.

(d) *implies* (a). By Proposition II.11.6, there is a measure ν such that $[\nu] = [\mu]$ and $\mathscr{A}^2(\nu) \cap L^\infty(\nu) = \mathscr{A}^\infty(\nu)$, the weak* closure of $\mathscr{A}$ in $L^\infty(\nu)$. Note that $\mathscr{A}^\infty(\nu) = \mathscr{A}^\infty(\mu)$. Let S = multiplication by z on $R^2(K, \nu)$; so $N = M_z$ on $L^2(\nu) = \operatorname{mne} S$. By (d), $M_\phi = \phi(N) = \operatorname{mne}\phi(S)$. By (II.2.4), this implies that $\mathscr{A}$ is dense in $L^2(\nu)$; that is, $\mathscr{A}^2(\nu) = L^2(\nu)$. Hence $\mathscr{A}^\infty(\mu) = \mathscr{A}^2(\nu) \cap L^\infty(\nu) = L^\infty(\mu)$. □

We can now state the main result of this section. Recall the definition of the envelope $E(K, \mu)$ (VI.2.1). Also recall that if $\phi \in R^\infty(K, \mu)$, there is a natural function $\check{\phi}: E(K, \mu) \to \mathbb{C}$ which agrees a.e. $[\mu]$ on $E(K, \mu)$ with ϕ (Exercise VI.3.3).

2.3 THEOREM. *Let $a_1, a_2, \ldots$ be an enumeration of the at most countable number of points ζ in $\mathbb{C}$ for which both $\mu(\phi^{-1}(\zeta)) > 0$ and $\operatorname{Area}(\check{\phi}^{-1}(\zeta)) > 0$. For $n \geq 1$, let $\mu_n = \mu|\phi^{-1}(a_n)$ and put $\mu_0 = \mu - \sum_n \mu_n$. If $\mathscr{A}^\infty(\phi, \mu)$ is the weak* closed subalgebra of $L^\infty(\mu)$ generated by $R(K)$ and $\overline{\phi}$, then*

$$\mathscr{A}^\infty(\phi, \mu) = L^\infty(\mu_0) \oplus \bigoplus_n R^\infty(\check{\phi}^{-1}(a_n), \mu_n).$$

This theorem, as stated here, is from Dudziak [**1989**]. It is almost the same as the main result of Gamelin, Russo, and Thomson [**1989**]. The first result of this type was proved in Conway and Olin [**1977**], where an analogous theorem was established for functions ϕ in $P^\infty(\mu)$ (see Corollary 2.6 below). In Dudziak [**1986**] some special cases of the preceding theorem were proved (which include the $P^\infty(\mu)$ case).

The proof here is from Dudziak [**1989**], though the application of disintegration of measures originated in Conway [**1989**], where another proof of the $P^\infty(\mu)$ case is given. For completeness, the existence and uniqueness of the disintegration of a measure with respect to a function will be established. But first let's see a few corollaries of Theorem 2.3. Additional corollaries will be given after the proof.

2.4 COROLLARY. *If μ is any measure supported on the compact set K and $\phi \in R^\infty(K, \mu)$ such that $\mu(\phi^{-1}(\zeta)) = 0$ for all ζ in $\mathbb{C}$, then the weak* closed subalgebra of $L^\infty(\mu)$ generated by $\overline{\phi}$ and $R(K)$ is all of $L^\infty(\mu)$.*

2.5 COROLLARY. *If μ is any measure supported on the compact set K and $\phi \in R^\infty(K, \mu)$ such that $\mu(\phi^{-1}(\zeta)) = 0$ for all ζ in $\mathbb{C}$, then $\phi(\mathrm{mne}\, S) = \mathrm{mne}\, \phi(S)$ for all S in $\mathscr{S}(K, \mu)$.*

2.6 COROLLARY. *If $\phi \in P^\infty(\mu)$, then $\phi(\mathrm{mne}\, S) = \mathrm{mne}\, \phi(S)$ for every subnormal operator with μ as a scalar-valued spectral measure if and only if ϕ is not constant on any component of the interior of the Sarason hull of μ.*

PROOF. We only consider the case where $P^\infty(\mu)$ has no L^∞ summand. Thus $P^\infty(\mu) \cong H^\infty(G)$, where G is the interior of the Sarason hull. Here G is the envelope $E((\mathrm{supp}\,\mu)^\wedge, \mu)$ and for ϕ in $P^\infty(\mu)$, $\check{\phi} = \phi$ considered as a bounded analytic function on G. Assume that ϕ is not constant on any component of G. Thus for each ζ, $\check{\phi}^{-1}(\zeta)$ consists of a countable number of points and thus $\mathrm{Area}(\check{\phi}^{-1}(\zeta)) = 0$. According to Theorem 2.3, $\mathscr{A}^\infty(\phi, \mu) = L^\infty(\mu)$ and this part of the corollary follows.

For the converse, suppose ϕ is constant on some component H of G; say $\phi|H \equiv a$. Let $\mathscr{A}$ be the subalgebra of $L^\infty(\mu)$ generated by $\overline{\phi}$ and $P^\infty(\mu)$. If D is a closed disk contained in H, then Sarason's Theorem implies that $\|\psi\|_D \leq \|\psi\|_\mu$ for all functions ψ in $P^\infty(\mu)$. So if $\psi \in \mathscr{A}$, then there are functions $\psi_1, \ldots, \psi_n$ in $P^\infty(\mu)$ such that for all z in H, $\psi(z) = \sum_k \bar{a}^k \psi_k(z)$. In particular, for z in D, $|\psi(z)| \leq \|\sum_k \bar{a}^k \psi_k\|_D \leq \|\sum_k \bar{a}^k \psi_k\|_\mu \leq \|\psi\|_\mu$. Therefore each function in $\mathscr{A}$ is analytic on H and so $\mathscr{A}^\infty(\phi, \mu) \neq L^\infty(\mu)$. By (2.2) there is a subnormal operator S having μ as a scalar-valued spectral measure such that $\phi(\mathrm{mne}\, S) \neq \mathrm{mne}\, \phi(S)$. □

The treatment of the disintegration of a measure given here is taken from Abrahamse and Kriete [**1973**]. Let X be a compact metric space and μ a positive regular Borel measure on X. It is important here to distinguish between an integrable Borel function and its equivalence class in $L^1(\mu)$. To facilitate this, let $\mathscr{L}^p(\mu)$ denote the collection of Borel functions f on X such that $\int |f|^p d\mu < \infty$; as usual, $L^p(\mu)$ is the corresponding Lebesgue space of the equivalence classes of these functions. Let Z be a compact metric space and let $\phi: X \to Z$ be a Borel mapping. Let ν be the Borel measure $\mu \circ \phi^{-1}$ on Z. That is, $\nu(\Delta) = \mu(\phi^{-1}(\Delta))$ for every Borel set Δ contained in Z.

If $f \in \mathscr{L}^1(\mu)$, then $\psi \to \int (\psi \circ \phi) f \, d\mu$ defines a bounded linear functional on $L^\infty(\nu)$. If attention is restricted to characteristic functions χ_Δ in $L^\infty(\nu)$, $\Delta \to \int (\chi_\Delta \circ \phi) f \, d\mu = \int_{\phi^{-1}(\Delta)} f \, d\mu$ is a countably additive measure defined on the Borel sets in Z that is absolutely continuous with respect to ν. Hence there is a unique element $E(f)$ in $L^1(\nu)$ such that $\int (\chi_\Delta \circ \phi) f \, d\mu = \int \chi_\Delta E(f) \, d\nu$ for all Borel subsets Δ of Z. An approximation argument shows that

(2.7) $$\int (\psi \circ \phi) f \, d\mu = \int \psi E(f) \, d\nu$$

for all ψ in $L^\infty(\nu)$. This defines a map $E: \mathscr{L}^1(\mu) \to L^1(\nu)$ called the *expectation operator*. The proof of the next proposition is left to the reader.

2.8 PROPOSITION. *The expectation operator $E: \mathscr{L}^1(\mu) \to L^1(\nu)$ has the following properties.*

(a) *E is linear and $\|E(f)\|_1 \leq \|f\|_1$ for all f in $\mathscr{L}^1(\mu)$.*
(b) *If $f, g \in \mathscr{L}^1(\mu)$ and $f = g$ a.e. $[\mu]$, then $E(f) = E(g)$. Thus E induces a contraction $E: L^1(\mu) \to L^1(\nu)$.*
(c) *If $f \in \mathscr{L}^\infty(\mu)$ and $\psi \in L^\infty(\nu)$, then $|\int \psi E(f) \, d\nu| \leq \|f\|_\mu \|\psi\|_1$. Thus E also defines a contraction $E: L^\infty(\mu) \to L^\infty(\nu)$.*

2.9 DEFINITION. A *disintegration of the measure μ with respect to ϕ* is a function $\zeta \to \lambda_\zeta$ from $Z \to M(X)$ such that:

(a) for each ζ in Z, λ_ζ is a probability measure;
(b) if $f \in \mathscr{L}^1(\mu)$, $E(f)(\zeta) = \int f \, d\lambda_\zeta$ a.e. $[\nu]$.

There is no claim made at present about the existence of a disintegration, though this will come shortly. Some easy examples of disintegration, however, can be found. For example, suppose we take X and Z to be the same space, let μ be any positive measure on Z, and let $\phi(\zeta) = \zeta$ for all ζ. Here $\nu = \mu$ and the disintegration is obtained by letting $\lambda_\zeta = \delta_\zeta$, the unit point mass at ζ. For another example, suppose Z is given, ν is any measure on Z, Y is any compact metric space, η is a probability measure on Y, and put $X = Y \times Z$, $\phi(y, \zeta) = \zeta$, and $\mu = \eta \times \nu$. So $\nu = \mu \circ \phi^{-1}$. Here $E(f)(\zeta) = \int f(y, \zeta) \, d\eta(y)$ and the disintegration arises by letting $\lambda_\zeta(\Delta) = \eta(\pi_1(\Delta \cap (Y \times \{\zeta\})))$, where π_1 is the projection of $Y \times Z$ onto Y.

Note that if in (2.7) we take $f = \chi_A$ for any Borel subset A of X and $\psi \equiv 1$, then the definition of disintegration shows that

$$\mu(A) = \int \lambda_\zeta(A) \, d\nu(\zeta).$$

So disintegration does indeed "disintegrate" the measure μ into the pieces λ_ζ. The measures λ_ζ are also carried by the level sets $\phi^{-1}(\zeta)$, as we see in the next result. Recall that ϕ is a function (in $\mathscr{L}^\infty(\mu)$) and not an equivalence class (in $L^\infty(\mu)$) so that it makes sense to talk about $\phi^{-1}(\zeta)$.

2.10 PROPOSITION. *If $\zeta \to \lambda_\zeta$ is a disintegration of μ with respect to ϕ, then λ_ζ is carried by $\phi^{-1}(\zeta)$ a.e. $[\nu]$.*

PROOF. Let $\mathscr{U}$ be a countable basis for the topology of Z. Note that $X \setminus \phi^{-1}(\zeta) = \bigcup \{\phi^{-1}(Z \setminus U) : U \in \mathscr{U}$ and $\zeta \in U\}$. So it suffices to show that $\lambda_\zeta(\phi^{-1}(Z \setminus U)) = 0$ except for ζ belonging to a set of ν measure 0.

For each U in $\mathscr{U}$, let $\Delta_U = \{\zeta \in U : \lambda_\zeta(\phi^{-1}(Z \setminus U)) > 0\}$. Using the definition of the disintegration (which insures the integrability of the functions below) and the definition of E, we get

$$\int_U \lambda_\zeta(\phi^{-1}(Z \setminus U)) \, d\nu(\zeta) = \int_U \int \chi_{\phi^{-1}(Z \setminus U)} \, d\lambda_\zeta \, d\nu(\zeta)$$
$$= \int_U E(\chi_{\phi^{-1}(Z \setminus U)}) \, d\nu$$
$$= \int \chi_U E(\chi_{\phi^{-1}(Z \setminus U)}) \, d\nu$$
$$= \int (\chi_U \circ \phi) \chi_{\phi^{-1}(Z \setminus U)} \, d\mu$$
$$= \int \chi_{\phi^{-1}(U)} \chi_{\phi^{-1}(Z \setminus U)} \, d\mu$$
$$= 0.$$

Since $\Delta_U \subseteq U$, this implies that $\nu(\Delta_U) = 0$. Thus the fact that $\mathscr{U}$ is countable implies that $\Delta \equiv \bigcup \{\Delta_U : U \in \mathscr{U}\}$ has $\nu(\Delta) = 0$. Fix ζ not in Δ. If $U \in \mathscr{U}$ and $\zeta \in U$, then the fact that $\zeta \notin \Delta_U$ implies that $\lambda_\zeta(\phi^{-1}(Z \setminus U)) = 0$. □

2.11 THEOREM. *Given a regular Borel measure μ on a compact metric space X and a Borel function ϕ from X into a compact metric space Z, there is a disintegration $\zeta \to \lambda_\zeta$ of μ with respect to ϕ. If $\zeta \to \lambda'_\zeta$ is another disintegration of μ with respect to ϕ, then $\lambda_\zeta = \lambda'_\zeta$ a.e. $[\nu]$.*

PROOF. It is easy to see from the definition of disintegration that it suffices to assume that X is the support of μ. Let $\mathscr{E}$ be a countable dense subset of $C(X)$ that is a complex-rational linear manifold containing the constant function 1. Since $\|E(f)\|_\nu \leq \|f\|_\mu = \|f\|_X$, for each f in $\mathscr{E}$ we can choose a representative $\tilde{f}$ of $E(f)$ in $\mathscr{L}^\infty(\nu)$ with $\|\tilde{f}\|_Z \leq \|f\|_X$; be sure to choose $\tilde{1} = 1$. If f and $g \in \mathscr{E}$ and α and β are complex-rational numbers, $\alpha \tilde{f} + \beta \tilde{g} = (\alpha f + \beta g)^\sim$ a.e. $[\nu]$. Since there are only a countable number of such α, β, f, and g, there is a set Δ contained in Z with $\nu(\Delta) = 0$ and such that $\alpha \tilde{f}(\zeta) + \beta \tilde{g}(\zeta) = (\alpha f + \beta g)^\sim(\zeta)$ whenever $\zeta \notin \Delta$.

Therefore for ζ not in Δ, $f \to \tilde{f}(\zeta)$ is a linear functional on $\mathscr{E}$ that is bounded by 1. Hence there is a unique Borel measure λ_ζ on X with $\|\lambda_\zeta\| \leq 1$ and $\tilde{f}(\zeta) = \int f\, d\lambda_\zeta$ for all f in $\mathscr{E}$. Since $\tilde{1} = 1$, we have that $1 = \int d\lambda_\zeta = \|\lambda_\zeta\|$; thus λ_ζ is a probability measure.

Note that by definition we have that $E(f)(\zeta) = \int f\, d\lambda_\zeta$ a.e. $[\nu]$ whenever $f \in \mathscr{E}$. We want this to hold whenever $f \in \mathscr{L}^1(\mu)$. To this end, let $\mathscr{L} = \{f \in \mathscr{L}^1(\mu) : E(f)(\zeta) = \int f\, d\lambda_\zeta \text{ a.e. } [\nu]\}$. Since E is linear, $\mathscr{L}$ is a linear manifold in $\mathscr{L}^1(\mu)$. We will establish the following facts about $\mathscr{L}$.

(2.12) If $\{f_n\} \subseteq \mathscr{L}$, $\sup_n \|f_n\|_X < \infty$, and $f_n(x) \to f(x)$ for all x in X, then $f \in \mathscr{L}$.

(2.13) If $\{f_n\} \subseteq \mathscr{L}$ such that $0 \leq f_n \leq f_{n+1}$ and $f \in \mathscr{L}^1(\mu)$ such that $f_n(x) \to f(x)$ for all x in X, then $f \in \mathscr{L}$.

Once these two facts are proved, it follows that $\mathscr{L} = \mathscr{L}^1(\mu)$, completing the proof of existence. Indeed, (2.12) and the density of $\mathscr{E}$ implies that $C(X) \subseteq \mathscr{L}$ and, since X is a compact metric space, that every bounded Borel function belongs to $\mathscr{L}$. But if $f \in \mathscr{L}^1(\mu)$ and f is non-negative, then $f_n(x) = \min\{f(x), n\}$ is a bounded Borel function, $0 \leq f_n \leq f_{n+1}$, and $f_n(x) \to f(x)$ for all x in X. By (2.13), $f \in \mathscr{L}$ from which we get that $\mathscr{L}^1(\mu) \subseteq \mathscr{L}$.

To establish (2.12), let $\{f_n\}$ and f be as described in (2.12). By the Lebesgue Dominated Convergence Theorem, $E(f_n)(\zeta) = \int f_n\, d\lambda_\zeta \to \int f\, d\lambda_\zeta$ a.e. $[\nu]$. Since the sequence $\{E(f_n)\}$ is also bounded a.e. $[\nu]$, the Lebesgue Dominated Convergence Theorem again implies that for every ψ in $\mathscr{L}^\infty(\nu)$, $\int \psi E(f_n)\, d\nu \to \int \psi(\zeta) \int f\, d\lambda_\zeta\, d\nu(\zeta)$. But using the definition of the expectation E, we also have that $\int \psi E(f_n)\, d\nu = \int (\psi \circ \phi) f_n\, d\mu \to \int (\psi \circ \phi) f\, d\mu = \int \psi E(f)\, d\nu$. Thus $\int \psi(\zeta) \int f\, d\lambda_\zeta\, d\nu(\zeta) = \int \psi E(f)\, d\nu$ for every ψ in $\mathscr{L}^\infty(\mu)$ and so $E(f)(\zeta) = \int f\, d\lambda_\zeta$ a.e. $[\nu]$. That is, $f \in \mathscr{L}$.

The proof of (2.13) is similar to that of (2.12), but with the monotone convergence theorem taking the place of the Lebesgue Dominated Convergence Theorem. The details are left to the reader.

To prove uniqueness, let $\zeta \to \lambda'_\zeta$ be a second disintegration of μ with respect to ϕ. Since $\int f\, d\lambda_\zeta = E(f)(\zeta) = \int f\, d\lambda'_\zeta$ a.e. $[\nu]$ and $\mathscr{E}$ is countable, the set $\Delta \equiv \{\zeta \in Z : \int f\, d\lambda_\zeta \neq \int f\, d\lambda'_\zeta \text{ for some } f \text{ in } \mathscr{E}\}$ has $\nu(\Delta) = 0$. So for ζ in $Z \setminus \Delta$, $\int f\, d\lambda_\zeta = \int f\, d\lambda'_\zeta$ for all f in $\mathscr{E}$. Since $\mathscr{E}$ is dense in $C(X)$, $\lambda_\zeta = \lambda'_\zeta$ for ζ in $Z \setminus \Delta$. □

The existence and uniqueness of the disintegration of a measure with respect to a measurable function can be proved in greater generality than in the preceding theorem. For one such generalization, see Bourbaki [1959].

Before applying the disintegration of a measure, observe that if $\phi' : X \to Z$ is another Borel mapping such that $\phi'(x) = \phi(x)$ a.e. $[\mu]$, then the disintegration of μ with respect to ϕ is the same as the disintegration of μ with respect to ϕ' (Exercise 2). Thus if μ is a measure on $\mathbb{C}$ with support contained in the compact set $K(=X)$, $\phi \in L^\infty(\mu)$, and Z is the essential range of ϕ, then we can talk of the disintegration of μ with respect to ϕ.

Let $E = E(K, \mu)$ be the envelope and recall that for any measure η supported on K, $R^\infty(E, \eta)$ is the weak* closure of $R(K, E)$ in $L^\infty(\eta)$. In this last piece of notation, the role of K is suppressed for convenience.

2.14 LEMMA. *If $\phi \in R^\infty(K, \mu)$, $\nu = \mu \circ \phi^{-1}$, and $\zeta \to \lambda_\zeta$ is a disintegration of μ with respect to ϕ, then*

$$\mathscr{A}^\infty(\phi, \mu) = \{f \in L^\infty(\mu) : f \in R^\infty(E, \lambda_\zeta) \text{ a.e. } [\nu]\}.$$

PROOF. Let $\mathscr{B} = \{f \in L^\infty(\mu) : f \in R^\infty(E, \lambda_\zeta)$ a.e. $[\nu]\}$ and note that $\mathscr{B}$ is a subalgebra of $L^\infty(\mu)$. Also observe that $R(K) \subseteq R(K, E) \subseteq \mathscr{B}$. In addition, for almost every ζ, $\bar{\phi}$ is constant a.e. $[\lambda_\zeta]$ (2.10) and hence $\bar{\phi} \in \mathscr{B}$. Thus $\mathscr{B}$ contains the algebra generated by $R(K)$ and $\bar{\phi}$. We want to show that $\mathscr{B}$ is weak* closed, from which it will follow that $\mathscr{A}^\infty(\phi, \mu) \subseteq \mathscr{B}$.

To show that $\mathscr{B}$ is weak* closed, we appeal to Corollary V.22.4. So let $\{f_n\} \subseteq \text{ball } \mathscr{B}$ and let $f \in \text{ball } L^\infty(\mu)$ such that $f_n \to f$ a.e. $[\mu]$. Fix representatives of these functions and let Δ be a Borel set such that μ is carried by Δ, $\|f_n\|_\Delta \leq 1$, and $f_n(z) \to f(z)$ for all z in Δ. Since $0 = \mu(K\setminus\Delta) = \int \lambda_\zeta(K\setminus\Delta) d\nu(\zeta)$, λ_ζ is carried by Δ for $[\nu]$ almost every ζ. Thus $f_n \to f$ weak* in $L^\infty(\lambda_\zeta)$ a.e. $[\nu]$. Since $f_n \in \mathscr{B}$, this implies that $f \in R^\infty(E, \lambda_\zeta)$ a.e. $[\nu]$ and hence $f \in \mathscr{B}$. This proves that $\mathscr{B}$ is weak* closed and establishes half of the lemma.

Let $h \in L^1(\mu)$ such that $h \perp \mathscr{A}^\infty(\phi, \mu)$. To prove that $\mathscr{B} \subseteq \mathscr{A}^\infty(\phi, \mu)$, it suffices to prove that $h \perp \mathscr{B}$. Fix a uniformly dense sequence $\{f_n\}$ in $R(K, E)$ and observe that if $p(\zeta, \bar{\zeta})$ is any polynomial in ζ and $\bar{\zeta}$, then $p(\phi, \bar{\phi})f_n \in A^\infty(\phi, \mu)$. Thus (2.7) implies

$$0 = \int p(\phi, \bar{\phi}) f_n h \, d\mu$$
$$= \int \left\{ \int p(\phi(z), \overline{\phi(z)}) f_n(z) h(z) \, d\lambda_\zeta(z) \right\} d\nu(\zeta)$$
$$= \int p(\zeta, \bar{\zeta}) \left\{ \int f_n h \, d\lambda_\zeta \right\} d\nu(\zeta),$$

since $p(\phi(z), \overline{\phi(z)}) \equiv p(\zeta, \bar{\zeta})$ a.e. $[\lambda_\zeta]$ by Proposition 2.10. Hence we have that the measure $(\int f_n h \, d\lambda_\zeta) d\nu(\zeta)$ annihilates all polynomials in ζ and $\bar{\zeta}$. Therefore for $[\nu]$ almost every ζ, $\int f_n h \, d\lambda_\zeta = 0$ by the Stone-Weierstrass Theorem. This implies there is a single Borel set Δ with $\nu(\Delta) = 0$ such that $\int f_n h \, d\lambda_\zeta = 0$ for all $n \geq 1$ and ζ not in Δ. But $h \in L^1(\lambda_\zeta)$ a.e. $[\nu]$ and so, since $\{f_n\}$ is uniformly dense in $R(K, E)$, $h \perp R^\infty(E, \lambda_\zeta)$ for ζ not in Δ. Now fix f in $\mathscr{B}$. By definition, $f \in R^\infty(E, \lambda_\zeta)$ a.e. $[\nu]$. So for almost every ζ not in Δ, $\int fh\, d\lambda_\zeta = 0$. Hence $\int fh\, d\mu = \int [\int fh\, d\lambda_\zeta]\, d\nu(\zeta) = 0$ and so $h \perp \mathscr{B}$. □

2.15 LEMMA. *If $\phi \in R^\infty(K, \mu)$, $\nu = \mu \circ \phi^{-1}$, and $\zeta \to \lambda_\zeta$ is a disintegration of μ with respect to ϕ, then $R^\infty(E, \lambda_\zeta) = R^\infty(\check\phi^{-1}(\zeta), \lambda_\zeta)$ a.e. $[\nu]$.*

PROOF. According to Corollary VI.3.5, there is a sequence $\{f_n\}$ in $R(K, E)$ such that $\|f_n\|_E \leq \|\phi\|_\mu$ and $f_n \to \phi$ a.e. $[\mu]$. Let Δ be a Borel set contained in K having full μ-measure such that $f_n(z) \to \phi(z)$ for all z in Δ. Now λ_ζ is carried by Δ for $[\nu]$ almost every ζ, so for $[\nu]$ almost every ζ,

(2.16) $$f_n \to \zeta \text{ a.e. } [\lambda_\zeta].$$

So it suffices to show that $R^\infty(E, \lambda_\zeta) = R^\infty(\check\phi^{-1}(\zeta), \lambda_\zeta)$ for those ζ for which (2.16) is valid.

Now $\check\phi^{-1}(\zeta) \subseteq E$ so $R^\infty(E, \lambda_\zeta) \subseteq R^\infty(\check\phi^{-1}(\zeta), \lambda_\zeta)$. Let $h \in L^1(\lambda_\zeta)$ such that $h \perp R^\infty(E, \lambda_\zeta)$. To prove that $h \perp R^\infty(\check\phi^{-1}(\zeta), \lambda_\zeta)$, it suffices to show that the measure $d\eta = h\, d\lambda_\zeta$ annihilates $R(K, \check\phi^{-1}(\zeta))$. To do this, it suffices to show that $\hat\eta$, the Cauchy transform of η, is 0 a.e. [Area] off $\check\phi^{-1}(\zeta)$. But $\eta \perp R(K, E)$ and so $\hat\eta = 0$ a.e. [Area] off E. So the lemma will be proved once we establish the following.

CLAIM. *If $a \in E \setminus \check\phi^{-1}(\zeta)$ and $\int |z-a|^{-1}\, d|\eta|(z) < \infty$, then $\hat\eta(a) = 0$.*

Using (V.6.6), we find a sequence $\{g_n\}$ in $R(K, E)$ such that each g_n has an extension that is analytic in a neighborhood of a and $\|g_n - f_n\|_E \to 0$. Since $f_n \to \phi$ boundedly a.e. $[\mu]$, $g_n \to \phi$ boundedly a.e. $[\mu]$; hence $\check g_n \to \check\phi$ pointwise on E. In particular $\check\phi(a) = \lim_n \check g_n(a) = \lim_n g_n(a)$. On the other hand, (2.16) implies that $g_n \to \zeta$ a.e. $[\eta]$, since $\eta \ll \lambda_\zeta$. But because each g_n is analytic in a neighborhood of a, the sequence $\{[g_n - g_n(a)]/(z-a)\}$ is uniformly bounded on cl E. Therefore the Lebesgue Dominated Convergence Theorem implies that

$$\lim_n \int \frac{g_n - g_n(a)}{z - a}\, d\eta = \int \frac{\zeta - \check\phi(a)}{z - a}\, d\eta.$$

But $[g_n - g_n(a)]/(z-a) \in R(K, E)$ (V.6.5) and so every term in this sequence of integrals is 0; thus $0 = \int (\zeta - \check\phi(a))/(z-a)\, d\eta = (\zeta - \check\phi(a))\hat\eta(a)$. But $\zeta - \check\phi(a)$ is constant a.e. $[\eta]$ and differs from 0 since $a \in E \setminus \check\phi^{-1}(\zeta)$. This implies that $\hat\eta(a) = 0$. □

PROOF OF THEOREM 2.3. If $f \in R(K)$, it is easy to see that $f \in L^\infty(\mu_0) \oplus \bigoplus_n R^\infty(\check\phi^{-1}(a_n), \mu_n)$. Since $\overline{\phi} \in L^\infty(\mu_0) \oplus \bigoplus_n R^\infty(\check\phi^{-1}(a_n), \mu_n)$ and this is clearly an algebra, $\mathscr{A}^\infty(\phi, \mu) \subseteq L^\infty(\mu_0) \oplus \bigoplus_n R^\infty(\check\phi^{-1}(a_n), \mu_n)$.

Now fix a function f in $L^\infty(\mu_0) \oplus \bigoplus_n R^\infty(\check\phi^{-1}(a_n), \mu_n)$. We'll show that $f \in \mathscr{A}^\infty(\phi, \mu)$ by appealing to Lemma 2.14. If ζ is a point such that $\mu(\phi^{-1}(\zeta)) > 0$, then it is easy to see that $\lambda_\zeta = [\mu(\phi^{-1}(\zeta))]^{-1}\mu|\phi^{-1}(\zeta)$ (Exercise 1). In particular, $\lambda_{a_n} = \|\mu_n\|^{-1}\mu_n$, so that $R^\infty(\check{\phi}^{-1}(a_n),\mu_n) = R^\infty(\check{\phi}^{-1}(a_n),\lambda_{a_n})$. But Lemma 2.15 implies that $R^\infty(E,\lambda_{a_n}) = R^\infty(\check{\phi}^{-1}(a_n),\lambda_{a_n})$; hence $f \in R^\infty(E,\lambda_{a_n})$ for each $n \geq 1$. Let $\Omega = \{\zeta :$ Area$(\check{\phi}^{-1}(\zeta)) > 0$ but $\mu(\phi^{-1}(\zeta)) = 0\}$; this set Ω is at most countable. Since $\nu(\{\zeta\}) = \mu(\phi^{-1}(\zeta))$, $\nu(\Omega) = 0$ and so we can ignore it when trying to verify the condition in Lemma 2.14.

This leaves us to consider those ζ for which Area$(\check{\phi}^{-1}(\zeta)) = 0$. Fix such a ζ that also satisfies $R^\infty(E,\lambda_\zeta) = R^\infty(\check{\phi}^{-1}(\zeta),\lambda_\zeta)$; by Lemma 2.15 this happens a.e. [ν]. Now if Δ is any Borel subset of K with zero area, $R(K,\Delta) = C(K)$ so that, in particular, $R^\infty(\Delta,\eta) = L^\infty(\eta)$ for any measure supported by K. Thus for this choice of ζ, $R^\infty(E,\lambda_\zeta) = L^\infty(\lambda_\zeta)$ and so $f \in R^\infty(E,\lambda_\zeta)$. □

The decomposition theorem for subnormal operators obtained in the preceding section can be combined with these results to give a complete answer, in the case of a function ϕ in $P^\infty(\mu)$, to the question "What is the minimal normal extension of $\phi(S)$?" The details of the proof of the next theorem are left to the reader.

2.17 THEOREM. *Using the notation of Corollary 1.6, if $\phi \in P^\infty(\mu)$ and $\mathbb{N}_1 = \{n \in \mathbb{N} : \phi$ is constant on $G_n\}$, then the minimal normal extension of $\phi(S)$ is*

$$\phi(N_0) \oplus \bigoplus \{\phi(N_n) : n \notin \mathbb{N}_1\} \oplus \bigoplus \{\alpha_n I_{\mathscr{H}_n} : n \in \mathbb{N}_1\},$$

where for n in $\mathbb{N}_1$, $\phi|G_n \equiv \alpha_n$ and $I_{\mathscr{H}_n}$ is the identity operator on $\mathscr{H}_n$.

The difficulty in extending this result to functions in $R^\infty(K,\mu)$ is that there is no nice criteria for $\phi(N_n)$ to be the minimal normal extension of $\phi(S_n)$ like there is for functions in $P^\infty(\mu)$.

We will conclude this section by giving an additional application to subnormal operators. This result forms a nice addition to Theorem 2.2.

2.18 PROPOSITION. *If S is a pure subnormal operator in $\mathscr{S}(K,\mu)$ and ϕ is a function in $R^\infty(K,\mu)$ such that the weak* closed algebra generated by $\bar{\phi}$ and $R(K)$ is $L^\infty(\mu)$, then $\phi(S)$ is a pure subnormal operator.*

PROOF. Let $N = $ mne S and put $T = N^*|\mathscr{H}^\perp$. Recall (II.2.10) that S is pure if and only if N^* is the minimal normal extension of T. For any set E, let $E^* = \{\bar{z} : z \in E\}$ and define $\mu^*(E) = \mu(E^*)$ for every Borel set E. Let $\phi^*(z) = \overline{\phi(\bar{z})}$. It is left to the reader to check that $\phi^* \in R^\infty(K^*,\mu^*)$ and $\mathscr{A}^\infty(\phi^*,\mu^*) = L^\infty(\mu^*)$. Since S is pure, Theorem 2.2 implies that $\phi^*(N^*)$ is the minimal normal extension of $\phi^*(T)$. But since $\mathscr{H}$ is invariant for $\phi(N)$, $\phi^*(T) = \phi(N)^*|\mathscr{H}^\perp$. By Proposition II.2.10, $\phi(S)$ is pure. □

Exercises

1. If $\zeta \to \lambda_\zeta$ is a disintegration of μ with respect to the function ϕ, show that if $\mu(\phi^{-1}(\zeta)) > 0$, then $\lambda_\zeta = [\mu(\phi^{-1}(\zeta))]^{-1}\mu|\phi^{-1}(\zeta)$.

2. Suppose X and Z are compact metric spaces and ϕ and ϕ' are two Borel mappings from X into Z; let $\zeta \to \lambda_\zeta$ and $\zeta \to \lambda'_\zeta$ be the corresponding disintegrations of a measure μ on X. Show that if $\phi(x) = \phi'(x)$ a.e. $[\mu]$, then $\nu = \mu \circ \phi^{-1} = \mu \circ \phi'^{-1}$ and $\lambda_\zeta = \lambda'_\zeta$ a.e. $[\nu]$.

3. Give the details of the proof of Proposition 2.18.

4. (Olin [1976]) Let K be a compact set such that either $R(K)$ is a Dirichlet algebra or Area$(\partial K) = 0$. If $f \in R(K)$ and f is not constant on any component of int K, then $f(\operatorname{mne} S) = \operatorname{mne} f(S)$ for every subnormal operator S with $\sigma(S) \subseteq K$.

5. (Dudziak [1986]) Adopt the notation of Theorem 2.3 and assume $R^\infty(K, \mu)$ has no L^∞ summand. Put $U = \{z : \check{\phi}$ has a nonconstant analytic extension to a neighborhood of $z\}$. Show that $\mathscr{A}^\infty(\phi, \mu) = \mathscr{A}^\infty(\phi, \mu|\mathbb{C}\backslash U) \oplus L^\infty(\mu|U)$.

§3 Spectral mapping theorems.

For any operator S on a Hilbert space and any function f analytic in a neighborhood of the spectrum of S, we know from the study of the Riesz-Dunford functional calculus that $\sigma(f(S)) = f(\sigma(S))$. If N is a normal operator, μ is a scalar-valued spectral measure for N, and $\phi \in L^\infty(\mu)$, then $\sigma(\phi(N))$ is the μ-essential range of ϕ, which has a legitimate interpretation as $\phi(\sigma(N))$. Each of these is the spectral mapping theorem for the respective functional calculus. The question addressed here is "Is there a spectral mapping theorem for the functional calculus for subnormal operators?" Specifically, if μ is a scalar-valued spectral measure for the subnormal operator S and $\phi \in R^\infty(\sigma(S), \mu)$, what is $\sigma(\phi(S))$? Is there any sense in which $\sigma(\phi(S)) = \phi(\sigma(S))$? These questions remain unanswered in full generality. (In the next section similar questions will be asked and partially answered about the essential spectrum of $\phi(S)$.)

Let's fix some notation. As usual, S will denote a subnormal operator and throughout this section μ will be a scalar-valued spectral measure for S. Let $E_S = E(\sigma(S), \mu)$, the envelope associated with $R^\infty(\sigma(S), \mu)$. For a compact set K containing $\sigma(S)$, let $E_K = E(K, \mu)$, the envelope associated with $R^\infty(K, \mu)$. This abuse of notation so introduced will only remain in effect in this section and the next and, since μ is always understood to have the same meaning, no confusion should arise. The idea here is that a spectral mapping theorem will be proved for functions in $R^\infty(K, \mu)$ when K satisfies a sufficiently strong condition to permit us to discover enough about the structure of such functions. The condition actually guarantees that the functions in $R^\infty(K, \mu)$ are bounded analytic functions on some open set.

We also make the following assumption throughout this section.

3.1 ASSUMPTION. $R^\infty(\sigma(S), \mu)$ has no L^∞ summand.

The reason for this assumption is that the proof of a spectral mapping theorem reduces to a consideration of pure operators. Indeed, by Theorem 1.4, $S = N_0 \oplus S_1$, where N_0 is an $R(\sigma(S))$ reductive normal operator and $R^\infty(\sigma(S_1), \mu_1)$ has no L^∞ summand. If $\phi \in R^\infty(\sigma(S), \mu)$, then $\phi = \phi_0 \oplus \phi_1$, where $\phi_0 \in L^\infty(\mu_0)$ and $\phi_1 \in R^\infty(\sigma(S_1), \mu_1)$. Thus $\phi(S) = \phi_0(N_0) \oplus \phi_1(S_1)$ and $\sigma(\phi(S)) = \sigma(\phi_0(N_0)) \cup \sigma(\phi_1(S_1))$. Now we already know the spectrum of $\phi_0(N_0)$ to be the usual μ_0-essential range of ϕ_0, so it remains to determine the spectrum of $\phi_1(S_1)$.

Note that since $\sigma(S) \subseteq K$, $R(K) \subseteq R(K, \sigma(S))$. Thus $R^\infty(K, \mu) \subseteq R^\infty(\sigma(S), \mu)$ and $E_S \subseteq E_K$. If $\phi \in R^\infty(K, \mu)$, then $\check\phi$, the Chaumat transform of ϕ, is defined on E_K. Since ϕ is also an element of $R^\infty(\sigma(S), \mu)$, it has a Chaumat transform defined on E_S. It is left as an exercise for the reader to show that this latter transform is the restriction of $\check\phi$ to E_S. Therefore there will be no notational distinction between the two.

Recall that by Chaumat's Theorem (VI.3.4), the map $\phi \to \check\phi$ defines an isometry of $R^\infty(K, \mu)$ onto $R^\infty(E_K, \mathscr{A}|E_K)$ as well as an isometry of $R^\infty(\sigma(S), \mu)$ onto $R^\infty(E_S, \mathscr{A}|E_S)$. The above remarks about the Chaumat transform can be interpreted as saying that there is a commutative diagram.

$$\begin{array}{ccc} R^\infty(K, \mu) & \longrightarrow & R^\infty(\sigma(S), \mu) \\ \downarrow & & \downarrow \\ R^\infty(E_K, \mathscr{A}|E_K) & \longrightarrow & R^\infty(E_S, \mathscr{A}|E_S) \end{array}$$

In this diagram the horizontal arrows are restriction maps and the vertical arrows are the isometric isomorphisms.

3.2 PROPOSITION. $\sigma(S) = \operatorname{cl} E_S$ and so μ is supported on $\operatorname{cl} E_S$.

PROOF. Clearly $\operatorname{cl} E_S \subseteq \sigma(S)$, so let a be a point not in $\operatorname{cl} E_S$. Thus $(z - a)^{-1} \in R(E_S) \subseteq R^\infty(E_S, \mathscr{A}|E_S)$. Thus $(S - a)^{-1}$ is defined and so $a \notin \sigma(S)$. □

Here is a spectral mapping theorem that holds for all subnormal operators except for a restricted class of functions. Note that the statement of the theorem makes sense in light of the preceding proposition.

3.3 THEOREM. If $f \in R(E_S)$, then $\sigma(f(S)) = f(\sigma(S))$.

PROOF. Let $\lambda \in f(\sigma(S))$; so $\lambda = f(a)$ for some a in $\sigma(S)$. By Proposition V.6.6 there is a sequence $\{f_n\}$ in $R(E_S)$ such that each f_n has an extension that is analytic in a neighborhood of a and $\|f_n - f\| \to 0$. By Proposition V.6.5, for each $n \geq 1$ there is a function g_n in $R(E_S)$ such that $f_n - f_n(a) = (z - a)g_n$. Thus $f_n(S) - f_n(a) = (S - a)g_n(S) = g_n(S)(S - a)$. Since $S - a$ is not invertible, $f_n(S) - f_n(a)$ is not invertible for each n. But $f_n(S) - f_n(a) \to f(S) - \lambda$ in norm and so it must be that $f(S) - \lambda$ is not invertible. Thus $f(\sigma(S)) \subseteq \sigma(f(S))$.

Conversely, suppose that $\lambda \notin f(\sigma(S))$; so $f - \lambda$ never vanishes on $\sigma(S) = \operatorname{cl} E_S$. Since $\sigma(S)$ is the maximal ideal space of $R(E_S)$, general Banach algebra theory says that $(f - \lambda)^{-1} \in R(E_S)$. It follows that $f(S) - \lambda$ is invertible. □

When faced with the problem of determining whether a spectral mapping theorem is valid for this functional calculus, it is necessary to decide what the appropriate definition of $\phi(\sigma(S))$ is for elements ϕ of $R^\infty(\sigma(S), \mu)$ since ϕ is not defined as a function on $\sigma(S)$. The preceding results suggests a possible definition. Since E_S is dense in $\sigma(S)$ and $\check{\phi}$ agrees with ϕ a.e. [μ] on E_S (Exercise VI.3.3 or (2.6)), the following seems appropriate.

3.4 DEFINITION. If $\phi \in R^\infty(\sigma(S), \mu)$, let
$$\phi(\sigma(S)) \equiv \operatorname{cl}[\check{\phi}(E_S)].$$

By the preceding theorem, this definition and the equation $\sigma(\phi(S)) = \phi(\sigma(S))$ are consistent for functions in $R(E_S)$.

In order to state the spectral mapping theorems it is necessary to introduce some additional terminology.

3.5 DEFINITION. Say that a measurable set Δ is *almost open* if $\Delta \setminus (\operatorname{int} \Delta)$ has 0 area. Say that a compact set is *essentially Dirichlet* if for all but a countable number of points a in ∂K,

(3.6) $$\liminf_{\delta \to 0} \frac{\gamma(B(a; \delta) \setminus K)}{\delta} > 0.$$

If K is essentially Dirichlet and the components of the interior of K are simply connected, then $R(K)$ is a Dirichlet algebra (VI.5.2). From the methods used in §V.13 it follows that if the components of the complement of K are bounded away from 0, then K is essentially Dirichlet. In particular, every finitely connected compact set is essentially Dirichlet.

Here is the spectral mapping theorem.

3.7 THEOREM (Dudziak [1984]). *Let S be a subnormal operator satisfying (3.1) and let K be a compact set containing $\sigma(S)$.*
  (a) *If the set of nonpeak points of $R(K)$ is almost open, then $\phi(\sigma(S)) \subseteq \sigma(\phi(S))$ for all ϕ in $R^\infty(K, \mu)$.*
  (b) *If K is essentially Dirichlet, then $\phi(\sigma(S)) = \sigma(\phi(S))$ for all ϕ in $R^\infty(K, \mu)$.*

The proof of the theorem requires a function theoretic detour. But first a corollary.

3.8 COROLLARY (Conway and Olin [1977]). *If S is any subnormal operator with scalar-valued spectral measure μ and $\phi \in P^\infty(\mu)$, then*
$$\sigma(\phi(S)) = \operatorname{cl}[\phi(\sigma(S) \cap \operatorname{int} \Sigma(\mu))] \cup \sigma(\phi(N_\mu)).$$

(*Remark*: $\Sigma(\mu)$ *is the Sarason hull of μ and ϕ is being considered as a bounded analytic function on* $\operatorname{int} \Sigma(\mu)$. *Also $\sigma(\phi(N_\mu))$ is known to be the μ-essential range of ϕ.*)

PROOF. It suffices to restrict our attention to the case where S satisfies (3.1). We apply Theorem 3.7 with $K =$ the polynomially convex hull of $\sigma(S)$. So $R^\infty(K, \mu) = P^\infty(\mu)$, $E_K = G$, the interior of the Sarason hull of μ, and for ϕ in $P^\infty(\mu)$, $\check{\phi} = \phi$ when $P^\infty(\mu)$ is identified with $H^\infty(G)$. Now $E_S \subseteq E_K = G$ and so $E_S \subseteq \sigma(S) \cap G$. So the preceding theorem implies that $\sigma(\phi(S)) = \phi(\sigma(S)) \subseteq \text{cl}[\phi(\sigma(S) \cap G)]$. It is left as an exercise for the reader to use the fact that $P^\infty(\mu) \cong H^\infty(G)$ to show that $\phi(\sigma(S) \cap G)$, and hence its closure, is contained in $\sigma(\phi(S))$, thus establishing equality. □

Now for the function theory. These results have independent interest in addition to having special use in the proof of the spectral mapping theorem.

Observe that Proposition V.13.9 implies that any point satisfying (3.6) is a peak point for $R(K)$. Since peak points are contained in ∂K, K and Q_K, the set of nonpeak points for $R(K)$, have the same interior. This gives the following implication.

3.9 PROPOSITION. *If K is essentially Dirichlet, then Q_K, the set of nonpeak points for $R(K)$, is almost open.*

The following lemma is from Browder [1969], page 157.

3.10 LEMMA. *If ν is any measure on $\mathbb{C}$ with compact support and a is any point, then*

$$\frac{1}{\pi r^2} \int_{B(a;r)} |a - z| \tilde{\nu}(z) \, d\mathscr{A}(z) \to |\nu(\{a\})|$$

as $r \to 0$.

PROOF. For $r > 0$ let

$$G_r(w) = \frac{1}{\pi r^2} \int_{B(a;r)} \left| \frac{z - a}{z - w} \right| d\mathscr{A}(z).$$

Note that $G_r(a) = 1$. If $w \neq a$, let $r < |w - a|$. So whenever $|z - a| < r$, $|w - a| \leq |w - z| + |z - a| < |w - z| + r$; hence, $|w - a| - r < |w - z|$. Thus $\sup\{|w - z|^{-1} : |z - a| < r\} \leq [|w - a| - r]^{-1}$ and so

$$G_r(w) \leq \frac{1}{\pi r^2} \int_{B(a;r)} \frac{r}{|w - a| - r} \, d\mathscr{A}(z)$$
$$= \frac{r}{|w - a| - r}.$$

Hence for $w \neq a$, $G_r(w) \to 0$ as $r \to 0$. In addition, Proposition V.2.14 implies that

$$G_r(w) \leq \frac{1}{\pi r^2} r \int_{B(a;r)} \frac{1}{|z - w|} \, d\mathscr{A}(z)$$
$$\leq \frac{1}{\pi r} 2[\pi \pi r^2]^{\frac{1}{2}}$$
$$= 2.$$

So the functions G_r are uniformly bounded and converge pointwise to $\chi_{\{a\}}$. Thus

$$|\nu(\{a\})| = \lim_{r \to 0} \int G_r(w) \, d|\nu|(w)$$

$$= \lim_{r \to 0} \int \left[\frac{1}{\pi r^2} \int_{B(a;r)} \left| \frac{z-a}{z-w} \right| d\mathscr{A}(z) \right] d|\nu|(w)$$

$$= \lim_{r \to 0} \frac{1}{\pi r^2} \int_{B(a;r)} |z-a| \widetilde{\nu}(z) \, d\mathscr{A}(z). \quad \square$$

3.11 LEMMA (Chaumat [1975]). *Let $a \in E_K$ and suppose ν is a complex representing measure for $R(K)$ at the point a with $\nu \ll \mu$ and $\nu(\{a\}) = 0$. If $0 < \rho < 1$, then $\{b \in \mathbb{C} : |b - a|\widetilde{\nu}(b) < \rho\}$ is a subset of E_K that has area density 1 at each of its points.*

PROOF. Let $\Delta \equiv \{b \in \mathbb{C} : |b - a|\widetilde{\nu}(b) < \rho\}$ and put $\eta = (z-a)\nu$. Note that $\eta \perp R(K)$ and if $b \in \Delta$, then $\widetilde{\eta}(b) = \int |z-b|^{-1} d|\eta|(z) = \int |(z-a)/(z-b)| \, d|\nu|(z) = \int |1 - (a-b)/(z-b)| \, d|\nu|(z) \leq \|\nu\| + \rho < \infty$. Also $\widehat{\eta}(b) = \int (z-b)^{-1} d\eta(z) = \int [1 - (a-b)/(z-b)] \, d\nu(z) = 1 - (a-b)\widehat{\nu}(b)$. Therefore, $|\widehat{\eta}(b)| \geq 1 - |a-b|\widetilde{\nu}(b) > 1 - \rho > 0$. By Proposition V.8.9,

$$(3.12) \qquad \nu_b \equiv \frac{1}{\widehat{\eta}(b)} \frac{1}{z-b} \eta = \frac{1}{\widehat{\eta}(b)} \frac{z-a}{z-b} \nu$$

is a complex representing measure for $R(K)$ at b. Also $\nu_b \ll \nu \ll \mu$ and $\nu_b(\{b\}) = 0$ since $\widehat{\nu}(b) < \infty$; thus $b \in E_K$. Since b was arbitrary, $\Delta \subseteq E_K$.

Now $\mathscr{A}(B(a;r) \backslash \Delta) = \int_{B(a;r) \backslash \Delta} 1 \, d\mathscr{A} \leq \int_{B(a;r)} |w-a|\widetilde{\nu}(w)\rho^{-1} \, d\mathscr{A}(w)$. Therefore

$$1 \geq \frac{\mathscr{A}(\Delta \cap B(a;r))}{\pi r^2} \geq 1 - \frac{1}{\rho} \frac{1}{\pi r^2} \int_{B(a;r)} |w-a|\widetilde{\nu}(w) \, d\mathscr{A}(w).$$

By the preceding lemma, the right-hand side of this inequality converges to 1 and so Δ has area density 1 at b. $\square$

3.13 LEMMA. *If $L: R^\infty(K, \mu) \to \mathbb{C}$ is a weak* continuous linear functional, then*

$$\|L\| = \sup\{|L(f)| : f \in R(E_K) \text{ and } \|f\| \leq 1\}.$$

The proof of this is immediate from Corollary VI.3.5.

If $a \in \operatorname{cl} E_K$, then we will identify a with the continuous linear functional $f \to f(a)$ defined on $R(E_K)$. Thus for two points a and b in $\operatorname{cl} E_K$, $\|a-b\|$ denotes the norm of the difference of these two evaluation functionals. If the point a belongs to E_K, then the weak* continuous linear functional on $R^\infty(K, \mu)$ defined by $\phi \to \phi(a)$ extends the evaluation functional on $R(E_K)$. Hence for a and b in E_K, the preceding lemma implies that $\|a-b\|$ designates the norm of the difference of these two linear functionals in $R^\infty(K, \mu)^*$ as well as in $R(E_K)^*$. We can now take comfort in the total lack of ambiguity in the notation.

3.14 PROPOSITION (Chaumat [**1975**]). (a) *For a in E_K and $\varepsilon > 0$, $\{b \in E_K : \|a - b\| < \varepsilon\}$ has area density 1 at a.*

(b) *If $a \in E_K \setminus (\operatorname{int} E_K)$, $0 < \varepsilon < 2$, and $\delta > 0$, then there is a point b in $E_K \cap B(a; \delta)$ such that $\|a - b\| = \varepsilon$.*

PROOF. (a) Fix ρ, $0 < \rho < 1$, let $\Delta = \{b \in \mathbb{C} : |b - a|\tilde{\nu}(b) < \rho\}$, and put $\eta = (z - a)\nu$ as in the proof of Lemma 3.11. Later in this proof we will make a judicious choice of ρ. Let $b \in \Delta$ and define ν_b as in (3.12). Thus

$$\|a - b\| \leq \|\nu - \nu_b\|$$
$$= \int \left| 1 - \frac{1}{\hat{\eta}(b)} \frac{z-a}{z-b} \right| d|\nu|(z)$$
$$= \frac{1}{|\hat{\eta}(b)|} \int \left| \hat{\eta}(b) - \frac{z-a}{z-b} \right| d|\nu|(z)$$
$$\leq \frac{1}{|\hat{\eta}(b)|} \left\{ |\hat{\eta}(b) - 1| \|\nu\| + \int \left| 1 - \frac{z-a}{z-b} \right| d|\nu|(z) \right\}$$
$$\leq \frac{1}{|\hat{\eta}(b)|} \{|\hat{\eta}(b) - 1| \|\nu\| + |b - a|\tilde{\nu}(b)\}$$
$$\leq \frac{1}{|\hat{\eta}(b)|} \{|\hat{\eta}(b) - 1| \|\nu\| + \rho\}.$$

As in the proof of Lemma 3.11, $|\hat{\eta}(b) - 1| \leq |b - a|\tilde{\nu}(b) < \rho$. Hence $\|a - b\| \leq |\hat{\eta}(b)|^{-1} \rho(\|\nu\| + 1) \leq \rho(\|\nu\| + 1)/(1 - \rho)$. Now choose ρ small enough that this last quantity is less than ε. Thus $a \in \Delta \subseteq \{b : \|b - a\| < \varepsilon\}$ and, by the Lemma 3.11, Δ has area density 1 at a.

(b) Suppose part (b) is false. That is, assume there is an ε and a δ such that $\{b \in \overline{B}(a; \delta) \cap E_K : \|a - b\| = \varepsilon\} = \varnothing$. Put $L = \{b \in \overline{B}(a; \delta) \cap \operatorname{cl} E_K : \|a - b\| \leq \varepsilon\}$.

Note that $a \in L$, $L \subseteq \operatorname{cl} E_K$, and L is compact. Now the fact that $\varepsilon < 2$ implies that L is contained in the same Gleason part of $R(E_K)$ as a (V.15.8). In particular, points in L are nonpeak points. By Proposition VI.3.8, $L \subseteq E_K$. Also the assumption implies that $L = \{b \in \overline{B}(a; \delta) \cap \operatorname{cl} E_K : \|a - b\| < \varepsilon\}$. According to part (a), for each b in L, $\{z \in E_K : \|b - z\| < \frac{1}{2}[\varepsilon - \|b - a\|]\}$ has area density 1 at b. But if z is in this set, then $\|z - a\| < \frac{1}{2}[\varepsilon - \|b - a\|] + \|b - a\| \leq \frac{1}{2}\|b - a\| + \frac{1}{2}\varepsilon < \varepsilon$. If, in addition, $|z - a| < \delta$, then $z \in L$. That is, $L \supseteq B(a; \delta) \cap \{z \in E_K : \|b - z\| < \frac{1}{2}[\varepsilon - \|b - a\|]\}$ and so L has area density 1 at each of its points b in the open disk $B(a; \delta)$.

Now let $z_0 \in B(a; \delta/2) \setminus \operatorname{cl} E_K$ (such a point exists since $\operatorname{int} E_K = \operatorname{int}(\operatorname{cl} E_K)$ and, by hypothesis, a belongs to the topological boundary of E_K). Since L is compact, there is a point z_1 in L with $|z_0 - z_1| = \operatorname{dist}(z_0, L)$. Thus $|z_0 - z_1| \leq |z_0 - a| < \delta/2$ and so $|z_1 - a| \leq |z_1 - z_0| + |z_0 - a| < \delta$. From the preceding paragraph this implies that L has area density 1 at z_1.

But $L \cap B(z_0; |z_0 - z_1|) = \varnothing$ and so

$$\limsup_{r \to 0} \frac{\mathscr{A}(L \cap B(z_1; r))}{\pi r^2} \leq \limsup_{r \to 0} \frac{\mathscr{A}(B(z_1; r) \setminus B(z_0; |z_0 - z_1|))}{\pi r^2}$$
$$= \frac{1}{2},$$

contradicting the fact that the area density is 1. □

We now follow a procedure like that used to prove Theorem VI.6.1. Let $U_0, U_1, U_2, \ldots$ be an enumeration of the components on $\mathbb{C} \setminus \operatorname{supp} \mu$, where U_0 is the unbounded one. Define $\mathscr{F}$ to be the collection of all compact nonempty sets F satisfying the following conditions.

(i) $F = K \cap [\operatorname{supp} \mu \cup \bigcup \{U_n : n \in I\}]$ for some nonempty subset I of the positive integers.
(ii) $E_K \subseteq F$.
(iii) $R(F \cap L)$ is a Dirichlet algebra whenever L is a compact set such that $R(K \cap L)$ is a Dirichlet algebra.

Note that this definition of $\mathscr{F}$ is the same as that of the collection $\mathscr{F}$ in §VI.6 except for condition (iii). The methods used to prove Lemmas 2, 3, and 4 in §VI.6 can be used to prove the following. The details are left to the reader (or see Dudziak [**1984**]).

3.15 LEMMA. (a) *The collection $\mathscr{F}$ is nonempty.*
(b) *$\mathscr{F}$ has a minimal element.*
(c) *If F is a minimal element of $\mathscr{F}$ and $\phi \in H^\infty(\operatorname{int} F)$, then $\|\phi\|_{\operatorname{int} F} = \|\phi\|_{E_K \cap \operatorname{int} F}$.*

The proof of the next proposition is similar to the proof of Theorem VI.6.1, save for one significant difference and so the details will be given.

3.16 PROPOSITION. *If F is a minimal element of the collection $\mathscr{F}$ above, then $F = \operatorname{cl} E_K$.*

PROOF. Let a be a point in $\operatorname{int} F$ and set $\mathscr{M} = \{\phi \in R^\infty(K, \mu):$ there exists a function g in $H^\infty(\operatorname{int} F)$ with $g(a) = 0$ and such that $\check{\phi} = g$ on $E_K \cap \operatorname{int} F\}$.

CLAIM. $\mathscr{M}$ is weak* closed in $R^\infty(K, \mu)$.

Let $\{\phi_n\}$ be a uniformly bounded sequence in $\mathscr{M}$ and let $\phi \in R^\infty(K, \mu)$ such that $\phi_n \to \phi$ a.e. $[\mu]$. So for each $n \geq 1$, there is a function g_n in $H^\infty(\operatorname{int} F)$ such that $g_n(a) = 0$ and $g_n = \check{\phi}_n$ on $E_K \cap \operatorname{int} F$. Part (c) of the preceding lemma implies that $\sup_n \|g_n\|_{\operatorname{int} F} \leq \sup_n \|\phi_n\|_\mu < \infty$. Using Montel's Theorem and replacing $\{g_n\}$ and $\{\phi_n\}$ by suitable subsequences, we may assume that there is a function g in $H^\infty(\operatorname{int} F)$ such that $g_n \to g$ uniformly on compact subsets of $\operatorname{int} F$. In particular, $g(a) = 0$ and $g = \check{\phi}$ on $E_K \cap \operatorname{int} F$. Thus $\phi \in \mathscr{M}$. By Corollary V.22.4, $\mathscr{M}$ is weak* closed.

Now $1 \notin \mathscr{M}$ and so the Hahn-Banach Theorem implies that there is a measure $\nu \ll \mu$ such that $\int d\nu = 1$ and $\int \phi \, d\nu = 0$ for all ϕ in $\mathscr{M}$. Thus for f in $R(K)$, $0 = \int (f - f(a))\,d\nu = \int f\,d\nu - f(a)$, so that ν is a complex representing measure for $R(K)$ at a. Now if $\mu(\{a\}) = 0$, then $\nu(\{a\}) = 0$ and so, by definition, $a \in E_K$. Since this excludes at most the countable number of atoms of μ, we conclude that $\text{int}\, F \subseteq \text{cl}\, E_K$.

On the other hand, the definition of $\mathscr{F}$ shows that $\partial F \subseteq \text{supp}\,\mu \subseteq \text{cl}\, E_K$ so that $F \subseteq \text{cl}\, E_K$. But $E_K \subseteq F$ by part (ii) of the definition of $\mathscr{F}$. □

3.17 COROLLARY. *If L is a compact set such that $R(K \cap L)$ is a Dirichlet algebra, then $R((\text{cl}\, E_K) \cap L)$ is a Dirichlet algebra.*

We now use these results to establish some "permanence" properties of the envelope.

3.18 PROPOSITION. (a) *If K is essentially Dirichlet, then $\text{cl}\, E_K$ is essentially Dirichlet.*

(b) *If Q_K is almost open, then E_K is almost open.*

PROOF. (a) Let $a \in \partial[\text{cl}\, E_K]$ and at first also assume that $a \in \text{int}\, K$; let $\delta_0 > 0$ such that $B(a; \delta_0) \subseteq K$. Thus $R(K \cap \overline{B}(a; \delta_0)) = R(\overline{B}(a; \delta_0))$ is a Dirichlet algebra and so, by the preceding corollary, $R(\text{cl}\, E_K \cap \overline{B}(a; \delta_0))$ is a Dirichlet algebra. Thus (VI.5.2) $\gamma(B(a; \delta) \setminus \text{cl}\, E_K) \geq \delta/4$ for $0 < \delta < \delta_0$ and so

$$(3.19) \qquad \liminf_{\delta \to 0} \frac{\gamma(B(a; \delta) \setminus \text{cl}\, E_K)}{\delta} > 0$$

whenever $a \in \partial[\text{cl}\, E_K] \cap \text{int}\, K$. But for a in $\partial K \cap \partial[\text{cl}\, E_K]$,

$$\liminf_{\delta \to 0} \frac{\gamma(B(a; \delta) \setminus \text{cl}\, E_K)}{\delta} \geq \liminf_{\delta \to 0} \frac{\gamma(B(a; \delta) \setminus K)}{\delta} > 0$$

except for at most a countable number of points. Thus $\text{cl}\, E_K$ is essentially Dirichlet.

(b) Now if Q is the set of nonpeak points for $R(\text{cl}\, E_K)$, then $E_K \subseteq Q$. But since E_K and its closure have the same interior (VI.3.9), $E_K \setminus (\text{int}\, E_K) \subseteq \partial[\text{cl}\, E_K]$. Thus $(\text{int}\, K) \cap (E_K \setminus (\text{int}\, E_K)) \subseteq (\text{int}\, K) \cap Q \cap \partial[\text{cl}\, E_K]$. But (3.19) was established for points in $(\text{int}\, K) \cap \partial[\text{cl}\, E_K]$ independent of the hypothesis that K is essentially Dirichlet and so such points are peak points of $R(\text{cl}\, E_K)$. Thus $(\text{int}\, K) \cap Q \cap \partial[\text{cl}\, E_K] = \varnothing$ and hence $E_K \setminus (\text{int}\, E_K) \subseteq E_K \setminus (\text{int}\, K)$. But $E_K \subseteq Q_K$ and $\text{int}\, K = \text{int}\, Q_K$. Therefore $E_K \setminus (\text{int}\, E_K) \subseteq Q_K \setminus (\text{int}\, Q_K)$, which has measure 0 by hypothesis. □

This result has an immediate benefit.

3.20 COROLLARY. *If Q_K is almost open, then the map $\phi \to \check{\phi}|\text{int}\, E_K$ is a dual algebra isomorphism of $R^\infty(K, \mu)$ onto a weak* closed subalgebra of $H^\infty(\text{int}\, E_K)$.*

PROOF. From Chaumat's Theorem we know that $R^\infty(K, \mu) \cong R^\infty(E_K, \mathscr{A}|E_K)$ via the map $\phi \to \check{\phi}$. But the preceding theorem shows that $\mathscr{A}|E_K = \mathscr{A}|(\text{int}\, E_K)$ so that $R^\infty(K, \mu) \cong R^\infty(E_K, \mathscr{A}|(\text{int}\, E_K))$. On the other hand, for ϕ in $R^\infty(K, \mu)$, $\check\phi$ is analytic on $\operatorname{int} E_K$ (VI.3.10) and thus the result. □

We would like to have the isomorphism in the preceding corollary surjective. But as in life, wanting and getting are not synonymous. We need an extra hypothesis, but it is at hand.

3.21 THEOREM (Gamelin and Garnett [1971]). *If U is an open subset of $\mathbb{C}$ such that for all but a countable number of points a in $(\operatorname{cl} Q_K) \setminus U$,*

$$\liminf_{\delta \to 0} \frac{\gamma(B(a; \delta) \setminus K)}{\delta} > 0,$$

then $H^\infty(U) \subseteq R^\infty(K, \mathcal{A}|Q_K)$. That is, for each h in $H^\infty(U)$ there is a function ψ in $R^\infty(K, \mathcal{A}|Q_K)$ such that $\psi = h|Q_K$ a.e..

Remark. Note that Theorem V.13.1 implies that U contains all but at most a countable number of points of Q_K so that $\mathcal{A}|Q_K \leq \mathcal{A}|U$.

The proof of this theorem will not be given. The interested reader can consult the reference. It does lead to the result we are seeking.

3.22 COROLLARY. *If K is essentially Dirichlet, then the map $\phi \to \check\phi|\operatorname{int} E_K$ is a dual algebra isomorphism of $R^\infty(K, \mu)$ onto $H^\infty(\operatorname{int} E_K)$.*

PROOF. In light of Corollary 3.20, it remains to show that the map is surjective. So fix a function g in $H^\infty(\operatorname{int} E_K)$. Since $\operatorname{cl} E_K$ is essentially Dirichlet, the preceding theorem implies that $g \in R^\infty(\operatorname{cl} E_K, \mathcal{A}|Q)$, where Q is the set of nonpeak points for $R(\operatorname{cl} E_K)$. But $E_K \subseteq Q$ and so $\mathcal{A}|E_K \leq \mathcal{A}|Q$. Since g is the weak* limit in $L^\infty(\mathcal{A}|Q)$ of a net of functions from $R(\operatorname{cl} E_K)$, this implies that g is the weak* limit in $L^\infty(\mathcal{A}|E_K)$ of this same net. Thus $g \in R^\infty(\operatorname{cl} E_K, \mathcal{A}|E_K)$. But by Chaumat's Theorem, $R^\infty(K, \mu) \cong R^\infty(\operatorname{cl} E_K, \mathcal{A}|E_K)$ and hence there is a function ϕ in $R^\infty(K, \mu)$ such that $g = \check\phi|\operatorname{int} E_K$. □

PROOF OF THEOREM 3.7. (a) We must show that $\check\phi(E_S) \subseteq \sigma(\phi(S))$. So fix a in E_S. By Proposition 3.14, $\{b \in E_S : \|a - b\| < \varepsilon\}$ has area density 1 at a. But E_K is almost open and so, since $E_S \subseteq E_K$, $\mathcal{A}(E_S \setminus \operatorname{int} E_K) = 0$. Thus $\{b \in E_S \cap (\operatorname{int} E_K) : \|a - b\| < \varepsilon\}$ has area density 1 at a. If $b \in E_S \cap (\operatorname{int} E_K)$ such that $\|a - b\| < \varepsilon$, then $|\check\phi(b) - \check\phi(a)| \leq \varepsilon \|\phi\|_\mu$ for all ϕ in $R^\infty(\sigma(S), \mu)$.

Now fix ϕ in $R^\infty(K, \mu)$; so $\phi \in R^\infty(\sigma(S), \mu)$ and $|\check\phi(b) - \check\phi(a)| \leq \varepsilon \|\phi\|_\mu$.

CLAIM. There is a function ψ in $R^\infty(K, \mu)$ such that $\phi - \check\phi(b) = (z - b)\psi$.

In fact, (VI.3.5) implies there is a sequence $\{f_n\}$ in $R(E_K)$ such that for all $n \geq 1$, $\|f_n\|_{E_K} \leq \|\phi\|_\mu$ and $f_n \to \phi$ a.e. $[\mu]$. By (V.6.3) for each n there is a function g_n in $R(E_K)$ such that $f_n - f_n(b) = (z - b)g_n$. If $\overline{B}(b; r) \subseteq \operatorname{int} E_K$, an application of the Maximum Modulus Theorem implies that $\|g_n\|_{E_K} \leq 2r^{-1}\|f_n\|_{E_K} \leq 2r^{-1}\|\phi\|_\mu$. By passing to a subsequence we may assume there is a function ψ in $R^\infty(K, \mu)$ such that $g_n \to \psi$ weak*. For this function ψ the claim holds.

Thus $\phi(S) - \check\phi(b) = (S - b)\psi(S) = \psi(S)(S - b)$. Since it also true that $b \in E_S \subseteq \sigma(S)$, $S - b$ is not invertible and hence $\phi(S) - \check\phi(b)$ cannot be invertible. That is, $\check\phi(b) \in \sigma(\phi(S))$. But $|\check\phi(a) - \check\phi(b)| \leq \varepsilon\|\phi\|_\mu$ and ε is arbitrary. Therefore $\check\phi(a) \in \sigma(\phi(S))$.

(b) Suppose $\lambda \notin \phi(\sigma(S))$. So there is an $\varepsilon > 0$ such that $|\check\phi - \lambda| \geq \varepsilon$ on E_S. Now $\check\phi$ is analytic on int E_K and so there is an open set U such that $(\operatorname{int} E_K) \cap \sigma(S) \subseteq U \subseteq \operatorname{int} E_K$ and $|\check\phi - \lambda| > \varepsilon/2$ on U. Hence $(\check\phi - \lambda)^{-1} \in H^\infty(U)$. (Recall that $\sigma(S) = \operatorname{cl} E_S$.)

Since K is essentially Dirichlet, $\operatorname{cl} E_K$ is also. Hence for all but a countable number of points a in $\partial[\operatorname{cl} E_K]$

$$\liminf_{\delta \to 0} \frac{\gamma(B(a;\delta)\setminus\sigma(S))}{\delta} \geq \liminf_{\delta \to 0} \frac{\gamma(B(a;\delta)\setminus \operatorname{cl} E_K)}{\delta} > 0.$$

Let Q_S be the set of nonpeak points for $R(\sigma(S))$. It is not hard to see that $(\operatorname{cl} Q_S)\setminus U \subseteq \sigma(S)\setminus(\operatorname{int} E_K) \subseteq (\operatorname{cl} E_K)\setminus(\operatorname{int} E_K)$. But $\operatorname{int} E_K = \operatorname{int}\{\operatorname{cl} E_K\}$ (VI.3.9) so that we get that $(\operatorname{cl} Q_S)\setminus U \subseteq \partial[\operatorname{cl} E_K]$. Hence

$$\liminf_{\delta \to 0} \frac{\gamma(B(a;\delta)\setminus\sigma(S))}{\delta} > 0$$

for all but at most a countable number of points in $(\operatorname{cl} Q_S)\setminus U$. By Theorem 3.21, $H^\infty(U) \subseteq R^\infty(\sigma(S), \mathscr{A}|Q_S)$.

In particular, $(\check\phi - \lambda)^{-1} \in R^\infty(\sigma(S), \mathscr{A}|Q_S)$. Thus there is a net $\{f_i\}$ in $R(\sigma(S))$ such that $f_i \to (\check\phi-\lambda)^{-1}$ weak* in $L^\infty(\mathscr{A}|Q_S)$. But $E_S \subseteq Q_S$ and so $f_i \to (\check\phi-\lambda)^{-1}$ weak* in $L^\infty(\mathscr{A}|E_S)$. That is, $(\check\phi-\lambda)^{-1} \in R^\infty(\sigma(S), \mathscr{A}|E_S)$. But by Chaumat's Theorem, $R^\infty(\sigma(S), \mathscr{A}|E_S) \cong R^\infty(\sigma(S), \mu)$ and so $(\phi - \lambda)^{-1} \in R^\infty(\sigma(S), \mu)$. Therefore $(\phi(S) - \lambda)^{-1}$ exists and so $\lambda \notin \sigma(\phi(S))$. The reverse inclusion holds by part (a). □

§4 Spectral mapping theorems for the essential spectrum. In this section Assumption 3.1 from the previous section will remain in force, as will the notation introduced there.

We begin with a spectral mapping theorem valid without restriction.

4.1 THEOREM (Dudziak [1984]). *Assume that condition* (3.1) *holds.*

(a) *If $f \in R(E_S)$, then $\sigma_e(f(S)) = f(\sigma_e(S))$.*

(b) *If $\lambda \notin \sigma_e(f(S))$, then $f^{-1}(\lambda) = \{a \in \sigma(S) : f(a) = \lambda\}$ is a finite subset of $\sigma(S)\setminus\sigma_e(S)$ and f is analytic in a neighborhood of each point of this set.*

(c) *If $\lambda \notin \sigma_e(f(S))$ and for each a in $f^{-1}(\lambda)$, $n_\lambda(f;a)$ is the multiplicity of the zero of $f - \lambda$ at a, then $n_\lambda(f;a) < \infty$ and*

$$\operatorname{ind}(f(S) - \lambda) = \sum_{a \in f^{-1}(\lambda)} n_\lambda(f;a)\operatorname{ind}(S - a).$$

PROOF. Suppose $\lambda \in f(\sigma_e(S))$ and put $\lambda = f(a)$ for some a in $\sigma_e(S)$. According to Proposition V.6.6, there is a sequence of functions $\{f_n\}$ in $R(E_S)$ such that each f_n has an analytic extension to a neighborhood of a and $\|f - f_n\| \to 0$. Also, Proposition V.6.5 implies that for every $n \geq 1$ there is a function g_n in $R(E_S)$ such that $f_n - f_n(a) = (z - a)g_n$. Thus $f_n(S) - f_n(a) = (S - a)g_n(S) = g_n(S)(S - a)$. Since $S - a$ is not a Fredholm operator, $f_n(S) - f_n(a)$ is not Fredholm for all $n \geq 1$. Since $f_n(S) - f_n(a) \to f(S) - \lambda$ in norm, $f(S) - \lambda$ is not a Fredholm operator and so $\lambda \in \sigma_e(f(S))$.

Now assume that $\lambda \notin f(\sigma_e(S))$. By Assumption 3.1, the boundary of the spectrum of S is contained in $\sigma_e(S)$. Thus the set $f^{-1}(\lambda)$ must be contained in int $\sigma(S)$. But f is analytic on int $\sigma(S)$ and so $f^{-1}(\lambda)$ is finite. Also if $a \in f^{-1}(\lambda)$ and $f - \lambda$ has a zero of infinite order at λ, then f is constantly equal to λ on the component of int $\sigma(S)$ containing the point a. But this would imply that f attains the value λ at a boundary point of $\sigma(S)$, contradicting the assumption that $\lambda \notin f(\sigma_e(S))$. So $n_\lambda(f;a) < \infty$ for all a in $f^{-1}(\lambda)$. Let $a_1, \ldots, a_m$ be the distinct points in $f^{-1}(\lambda)$ and let $n_j = n_\lambda(f; a_j)$. Repeated application of (V.6.5) shows that there is a function g in $R(E_S)$ that does not vanish on $\sigma(S)$ and such that $f - \lambda = \prod_j (z - a_j)^{n_j} g$. From here the formula in part (c) of the theorem follows from properties of the Fredholm index and the fact (3.3) that $g(S)$ is invertible. □

As in the preceding section we are faced with the problem of giving an appropriate meaning to the term $\phi(\sigma_e(S))$ when $\phi \in R^\infty(\sigma(S), \mu)$. Taking a lead from the definition of $\phi(\sigma(S))$ we are tempted to let $\text{cl}[\check{\phi}(E_S \cap \sigma_e(S))] = \phi(\sigma_e(S))$. This fails. Indeed, if S is the unilateral shift of multiplicity 1, then $E_S = \mathbb{D}$ and $\sigma_e(S) = \partial \mathbb{D}$. So in this instance, $\check{\phi}(E_S \cap \sigma_e(S)) = \varnothing$.

Some experience leads us to consider the set of cluster values of a function as the appropriate definition of $\phi(\sigma_e(S))$.

4.2 DEFINITION. If $\phi \in R^\infty(K, \mu)$, the *cluster set* of ϕ on $\sigma_e(S)$ is the set $\text{clu}(\phi; \sigma_e(S)) \equiv \{\lambda \in \mathbb{C}: \text{there is a sequence } \{z_n\} \text{ in } E_S \text{ converging to a point in } \sigma_e(S) \text{ such that } \check{\phi}(z_n) \to \lambda\}$.

Notice that $\text{clu}(\phi; \sigma_e(S))$ includes $\check{\phi}(E_S \cap \sigma_e(S))$.

4.3 THEOREM (Dudziak [1984]). *Let S be a subnormal operator satisfying* (3.1) *and let K be a compact set containing $\sigma(S)$.*

(a) *If Q_K is almost open and $\phi \in R^\infty(K, \mu)$, then $\text{clu}(\phi; \sigma_e(S)) \subseteq \sigma_e(\phi(S))$.*

(b) *If K is essentially Dirichlet and $\phi \in R^\infty(K, \mu)$, then $\text{clu}(\phi; \sigma_e(S)) = \sigma_e(\phi(S))$. Moreover, if $\lambda \notin \sigma_e(\phi(S))$, then the following hold*:

(i) *$Z_\lambda(\phi) \equiv \{a \in E_s : \check{\phi}(a) = \lambda\}$ is a finite subset of $\sigma(S) \setminus \sigma_e(S)$ and $\check{\phi}$ is analytic in a neighborhood of each point in $Z_\lambda(\phi)$;*
(ii) *if $a \in Z_\lambda(\phi)$, $\check{\phi} - \lambda$ has a zero of finite order $n_\lambda(\phi; a)$ at a;*
(iii) *$\text{ind}(\phi(S) - \lambda) = \sum_{a \in Z_\lambda(\phi)} n_\lambda(\phi; a) \, \text{ind}(S - a)$.*

We must do some preliminary work, including a small excursion into function theory, before proving this theorem. But one lemma can be given immediately. In fact, its proof is an application of (V.6.5) and (V.6.6) and uses the usual factorization techniques; the details are left to the reader.

4.4 LEMMA. *If $\phi \in R^\infty(K, \mu)$, then $\check\phi(\sigma_e(S)) \cap \operatorname{int} E_K \subseteq \sigma_e(\phi(S))$.*

The method used to prove Theorem 4.3 involves another concept. The next several results have no reference to the subnormal operator S. Indeed, we will apply them to the compact set $\sigma(S)$ as well as the generic set K.

4.5 DEFINITION. If K is any compact subset of $\mathbb{C}$, a sequence $\{z_n\}$ in E_K is an *interpolating sequence* for $R^\infty(K, \mu)$ if the map $\phi \to \{\check\phi(z_n)\}$ takes $R^\infty(K, \mu)$ onto l^∞.

The use of interpolating sequences in determining the essential spectrum of a function of an operator first appears in Axler [**1982**] where the essential spectrum of multiplication by a bounded analytic function on the Bergman space of an open subset of $\mathbb{C}$ is determined.

The following results about interpolating sequences have their analogues in Gamelin and Garnett [**1970**], which investigates interpolating sequences for $H^\infty(G)$; the techniques of proof used here are taken from this reference.

4.6 PROPOSITION. *If $\{z_n\}$ is a sequence in E_K, then $\{z_n\}$ is an interpolating sequence for $R^\infty(K, \mu)$ if and only if there are constants M and α with $M \geq 1$ and $0 < \alpha < 1$ such that for every s in l^∞ there is a ϕ in $R^\infty(K, \mu)$ with $\|\phi\| \leq M\|s\|$ and $\|\{\check\phi(z_n) - s_n\}\| < \alpha\|s\|$.*

PROOF. Of course if the sequence is an interpolating sequence, the condition holds by the Open Mapping Theorem. So assume that $\{z_n\}$ satisfies the stated conditions for some constants M and α and fix a sequence s in l^∞. By induction there is a sequence of functions $\{\phi_k\}$ in $R^\infty(K, \mu)$ such that for all $k \geq 1$, $\|\phi_k\| \leq M\alpha^{k-1}\|s\|$ and $\|\{\sum_{j=1}^k \check\phi_j(z_n) - s_n\}_n\| \leq \alpha^k\|s\|$. Thus $\sum_k \|\phi_k\| < \infty$ and so $\sum_k \phi_k$ converges in $R^\infty(K, \mu)$ to a function ϕ. It is routine to check that $\check\phi(z_n) = s_n$ for all n. □

Now we find a way to manufacture interpolating sequences.

4.7 PROPOSITION. *If $\{z_n\}$ is a sequence in E_K that converges to a point a that does not belong to E_K, then there is a subsequence of $\{z_n\}$ that is an interpolating sequence.*

PROOF. Observe that since $a \notin E_K$, Proposition VI.3.8 implies that a is a peak point for $R(E_K)$. So let $g \in R(E_K)$ such that g peaks at a. That is, $g(a) = 1$ and $|g(z)| < 1$ for z in $\operatorname{cl} E_K$ and $z \neq a$. A rather straightforward induction argument produces a subsequence $\{w_j\}$ of $\{z_n\}$, a decreasing sequence of open disks $\{\Delta_j\}$ whose intersection is $\{a\}$, and a sequence $\{g_j\}$ of increasing powers of the function g such that the following hold.

(i) $|g_j| \leq (\frac{1}{2})^{j+2}$ on $(\operatorname{cl} E_K) \setminus \Delta_j$;

(ii) $|g_j - 1| \leq (\frac{1}{2})^{j+2}$ on $(\operatorname{cl} E_K) \cap \Delta_{j+1}$;

(iii) $|g_j(w_k)| \leq (\frac{1}{2})^{j+2}$ for $k \leq j-1$;

(iv) $|g_j(w_k) - 1| \leq (\frac{1}{2})^{j+2}$ for $k \geq j$.

Now let $f_j = g_j - g_{j+1}$. So $f_j \in R(E_K)$ and this sequence of functions satisfies the following inequalities.

(i) $\|f_j\| \leq 2$;

(ii) $|f_j| \leq (\frac{1}{2})^{j+1}$ on $(\operatorname{cl} E_K) \setminus (\Delta_j \setminus \Delta_{j+1})$;

(iii) $|f_j(w_j) - 1| \leq (\frac{1}{2})^{j+1}$;

(iv) $|f_j(w_k)| \leq (\frac{1}{2})^{j+1}$ for $k \neq j$.

It is claimed that for any s in l^∞, $\phi = \sum_j s_j f_j \in R^\infty(K, \mu)$, where this series converges in the weak* topology. To see this first assume that $z \in \operatorname{cl} E_K$ and that z belongs to one of the annuli $\Delta_k \setminus \Delta_{k+1}$. Thus

$$\sum_{j=1}^\infty |s_j f_j(z)| = \sum_{j \neq k} |s_j f_j(z)| + |s_k f_k(z)|$$

$$\leq \|s\|_\infty \sum_{j \neq k} \left(\frac{1}{2}\right)^{j+1} + \|s\|_\infty |f_k(z)|$$

$$\leq \frac{1}{2}\|s\|_\infty + 2\|s\|_\infty$$

$$\leq 3\|s\|_\infty.$$

If the point does not belong to any of the disks Δ_j or if $z = a$, then the same inequality is even easier to prove. Thus the sequence of finite sums $\{\sum_1^n s_j f_j\}_n$ is uniformly bounded in $R^\infty(K, \mu)$ and is weak* Cauchy. Therefore the infinite series $\sum_j s_j f_j$ converges weak* to a function ϕ in $R^\infty(K, \mu)$.

It follows that for any w in E_K, $\check{\phi}(w) = \sum_j s_j \check{f}_j(w) = \sum_j s_j f_j(w)$. In particular, $\check{\phi}(w_k) = \sum_j s_j f_j(w_k)$. So for any $k \geq 1$,

$$|\check{\phi}(w_k) - s_k| = \left|\sum_{j \neq k} s_j f_j(w_k)\right| + |s_k(f_k(w_k) - 1)|$$

$$\leq \|s\| \sum_{j \neq k} \left(\frac{1}{2}\right)^{j+1} + \|s\| \left(\frac{1}{2}\right)^{k+1}$$

$$\leq \frac{1}{2}\|s\|.$$

Since $\|\phi\| \leq 3\|s\|$, $\{w_j\}$ must be an interpolating sequence by Proposition 4.6. □

We now want to investigate what happens when a sequence $\{z_n\}$ in E_K converges to a point that belongs to E_K. To do this we return to the Vitushkin localization operators (§V.5).

4.8 LEMMA. *Fix a point a in E_K and for each $\delta > 0$ let $B_\delta = B(a;\delta)$ and let g_δ be a function in C_c^1 such that $0 \leq g_\delta \leq 1$, $g_\delta \equiv 1$ on $B(a;\delta/2)$, support $g_\delta \subseteq B_\delta$, and $\|\bar\partial g_\delta\|_\infty \leq 4/\delta$. If T_δ is the Vitushkin localization operator T_{g_δ}, then*

$$0 = \lim_{\delta \to 0} \sup\{|T_\delta f(a)| : f \text{ is a bounded Borel function on } \mathbb{C},$$

$$\|f\|_{\mathbb{C}} \leq 1, \text{ and } f|\operatorname{cl} E_K \in R(E_K)\}.$$

PROOF. Let $\varepsilon > 0$ and put $E_\varepsilon = \{z \in E_K : \|z - a\| < \varepsilon\}$. Let f be a bounded Borel function on $\mathbb{C}$ with $\|f\|_{\mathbb{C}} \leq 1$ and $f|\operatorname{cl} E_K \in R(E_K)$. From (V.5.2) we get that

$$|T_\delta f(a)| \leq \frac{1}{\pi} \int_{B_\delta \cap E_\varepsilon} \left|\frac{f(z) - f(a)}{z - a}\right| |\bar\partial g_\delta(z)| \, d\mathscr{A}(z)$$

$$+ \frac{1}{\pi} \int_{B_\delta \setminus E_\varepsilon} \left|\frac{f(z) - f(a)}{z - a}\right| |\bar\partial g_\delta(z)| \, d\mathscr{A}(z)$$

$$\leq \frac{1}{\pi} \varepsilon \frac{4}{\delta} \int_{B_\delta \cap E_\varepsilon} \frac{1}{|z-a|} \, d\mathscr{A}(z) + \frac{1}{\pi} \frac{8}{\delta} \int_{B_\delta \setminus E} \frac{1}{|z-a|} \, d\mathscr{A}(z).$$

Using Proposition V.2.14 this implies that

$$|T_\delta f(a)| \leq \frac{1}{\pi} \varepsilon \frac{8}{\delta} \sqrt{\pi \operatorname{Area}(B_\delta \cap E_\varepsilon)} + \frac{1}{\pi} \frac{16}{\delta} \sqrt{\pi \operatorname{Area}(B_\delta \setminus E_\varepsilon)}$$

$$\leq \frac{1}{\pi} \frac{8\varepsilon}{\delta} \pi \delta + 16 \sqrt{\frac{\operatorname{Area}(B_\delta \setminus E_\varepsilon)}{\pi \delta^2}}$$

$$= 8\varepsilon + 16 \sqrt{\frac{\operatorname{Area}(B_\delta \setminus E_\delta)}{\pi \delta^2}}.$$

But E_ε has area density at 1 at a (3.14) and so the right-hand side of this inequality can be made less than 24ε for all sufficiently small δ. □

4.9 PROPOSITION. *If $\{z_n\}$ is a sequence in E_K and $z_n \to a$ with a in E_K, then either $\|z_n - a\| \to 0$ or $\{z_n\}$ has a subsequence that is an interpolating sequence for $R^\infty(K, \mu)$.*

PROOF. Assume that $\{z_n\}$ does not converge to a in $R^\infty(K, \mu)^*$. By passing to a subsequence if necessary, we may assume that there is a constant $c > 0$ such that $\|z_n - a\| > c^{-1}$ for all $n \geq 1$. Thus for each integer $n \geq 1$ we can find a function h_n in $R(E_K)$ such that $\|h_n\| \leq 1$ and $|h_n(a) - h_n(z_n)| \geq c^{-1}$. Extend h_n continuously to $\mathbb{C}$ without increasing its norm. Let

$$g_n = \frac{h_n}{h_n(z_n) - h_n(a_n)}.$$

So g_n is continuous on $\mathbb{C}$, $\|g_n\| \leq c$, $g_n|E_K \in R(E_K)$, and $g_n(z_n) - g_n(a) = 1$.

Let B_δ and T_δ be as in the preceding lemma and let $\varepsilon > 0$. According to Lemma 4.8, for all sufficiently small $\delta > 0$, $|T_\delta g_n(a)| < \varepsilon/4$ uniformly in n. Fix a value of $r > 0$ and let $B_r = B(a; r)$. By Proposition V.5.7(b), if $|z - a| > r$, then for all $\delta > 0$ and all $n \geq 1$, $|T_\delta g_n(z)| \leq 16\delta c/r$. So by choosing δ small enough we get that $|T_\delta g_n(z)| \leq \varepsilon/4$ for $|z - a| > r$ and $n \geq 1$.

On the other hand, (V.5.7(c)) implies that if $|z_n - a| < \delta/2$, then

$$|1 - [T_\delta g_n(z_n) - T_\delta g_n(a)]| = |(g_n - T_\delta g_n)(z_n) - (g_n - T_\delta g_n)(a)|$$
$$\leq \frac{68c|z_n - a|}{\delta}.$$

Therefore there is an integer N such that for $n \geq N$, $|T_\delta g_n(z_n) - T_\delta g_n(a)| \geq \frac{1}{2}$. Put

$$f_n = \frac{T_\delta g_n - T_\delta g_n(a)}{T_\delta g_n(z_n) - T_\delta g_n(a)}.$$

Note that $f_n \in R(E_K)$ since $R(E_K)$ is a T-invariant algebra. Also (V.5.7(a)) implies that $\|T_\delta g_n\| \leq 16c$ and so $\|f_n\| \leq 2 \cdot 2 \cdot 16c = 64c$.

To synopsize, we have that for any disk B_r there is a function f_n (for an appropriate choice of n and δ) such that

f_n is continuous on $\mathbb{C}$;
$f_n|(\operatorname{cl} E_K) \in R(E_K)$;
$\|f_n\| \leq 64c$;
$f_n(z_n) = 1$ and $f_n(a) = 0$;
$|f_n| \leq \varepsilon$ off the disk B_r.

We can now proceed inductively to choose a subsequence $\{w_j\}$ of $\{z_n\}$, a sequence of disks $\{\Delta_j\}$ decreasing to a, and a sequence of functions $\{f_j\}$ that are continuous on $\mathbb{C}$ and satisfy:

$f_j|\operatorname{cl} E_K \in R(E_K)$;
$\|f_j\| \leq 64c$;
$|f_j| < (\frac{1}{2})^{j+1}$ off $\Delta_j \backslash \Delta_{j+1}$;
$f_j(w_j) = 1$ and $f_j(a) = 0$;
$|f_j(w_k)| < (\frac{1}{2})^{j+1}$ for $k \neq j$.

As in the proof of Proposition 4.7, for any s in l^∞, $\phi = \sum_j s_j f_j$ converges weak* in $R^\infty(K, \mu)$ and $\check{\phi}(w_j) = s_j$ for all $j \geq 1$. Therefore $\{w_j\}$ is an interpolating sequence for $R^\infty(K, \mu)$. □

4.10 THEOREM. *If $\{z_n\}$ is a sequence in E_K that converges to some point a in $\operatorname{cl} E_K$, then one of the two following mutually exclusive statements hold.*

(a) *$a \in E_K$ and $\|z_n - a\| \to 0$.*
(b) *$\{z_n\}$ has a subsequence that is an interpolating sequence for $R^\infty(K, \mu)$.*

PROOF. Propositions 4.7 and 4.9 show that one of these two statements is valid. It must be shown that both cannot hold simultaneously. But if (a) holds, then $\check{\phi}(z_n) \to \check{\phi}(a)$ for every ϕ so that no subsequence of $\{z_n\}$ could possibly be an interpolating sequence. □

Now we return to the operator theory.

4.11 LEMMA. *If $\phi \in R^\infty(K, \mu)$ and $\{z_n\}$ is a sequence in $(\operatorname{int} E_K) \cap E_S$ that is an interpolating sequence for $R^\infty(\sigma(S), \mu)$ and $\check{\phi}(z_n) \to \lambda$, then $\lambda \in \sigma_e(\phi(S))$.*

PROOF. If $z_n \in \sigma_e(S)$ for an infinite number of n, then passing to a subsequence we may assume that $\{z_n\} \subseteq \sigma_e(S)$. But then $\check{\phi}(z_n) \in \sigma_e(\phi(S))$ by Lemma 4.4 and so $\lambda \in \sigma_e(\phi(S))$.

Now assume that only a finite number of the z_n belong to $\sigma_e(S)$; consequently we may assume that $z_n \notin \sigma_e(S)$ for all $n \geq 1$. But $z_n \in E_S \subseteq \sigma(S)$ and so it must be that for all $n \geq 1$, either $\ker(S - z_n) \neq (0)$ or $\ker(S - z_n)^* \neq (0)$. Hence for each $N \geq 1$, either $\bigvee\{\ker(S - z_n) : n \geq N\}$ or $\bigvee\{\ker(S - z_n)^* : n \geq N\}$ is infinite dimensional. Put this in your memory bank.

Since $\{z_n\}$ is an interpolating sequence, the Open Mapping Theorem implies that there is a constant $M > 0$ such that for each s in l^∞, there is a function ψ in $R^\infty(\sigma(S), \mu)$ such that $\{\check{\psi}(z_n)\} = s$ and $\|\psi\| \leq M\|s\|$. For each $m \geq 1$, let $\psi_m \in R^\infty(\sigma(S), \mu)$ such that

(i) $\|\psi_m\| \leq M \sup\{|\check{\phi}(z_n) - \lambda| : n \geq m\}$;
(ii) $\check{\psi}_m(z_n) = 0$ for $n < m$;
(iii) $\check{\psi}_m(z_n) = \check{\phi}(z_n) - \lambda$ for $n \geq m$.

Since $\check{\phi}(z_n) \to \lambda$ as $n \to \infty$, condition (i) implies that $\|\psi_m\| \to 0$ and so $\|\psi_m(S)\| \to 0$. Therefore $\phi(S) - \lambda - \psi_m(S) \to \phi(S) - \lambda$. Hence to show that $\lambda \in \sigma_e(\phi(S))$, it suffices to show that $\phi(S) - \lambda - \psi_m(S)$ is not a Fredholm operator for all large m.

Fix m and let $n \geq m$. So $z_n \in E_S \setminus \sigma_e(S) \subseteq \sigma(S) \setminus \partial \sigma_e(S) = \operatorname{int} \sigma(S) = \operatorname{int}\{\operatorname{cl} E_S\}(3.2) = \operatorname{int} E_S$ (VI.3.9). Thus by Proposition VI.3.10(b), there is a function ξ in $R^\infty(\sigma(S), \mu)$ such that $\phi - \lambda - \psi_m = (z - z_n)\xi$. Hence $\phi(S) - \lambda - \psi_m(S) = (S - z_n)\xi(S) = \xi(S)(S - z_n)$. But this implies that $\ker(S - z_n) \subseteq \ker(\phi(S) - \lambda - \psi_m(S))$ and $\ker(S - z_n)^* \subseteq \ker(\phi(S) - \lambda - \psi_m(S))^*$. But then information in your memory bank implies that either the kernel or the cokernel of $\phi(S) - \lambda - \psi_m(S)$ is infinite dimensional and so $\phi(S) - \lambda - \psi_m(S)$ cannot be a Fredholm operator. □

PROOF OF THEOREM 4.3. (a) Let $\lambda \in \operatorname{clu}(\phi; \sigma_e(S))$ and let $\{z_n\}$ be a sequence in E_S that converges to a point a in $\sigma_e(S)$ such that $\check{\phi}(z_n) \to \lambda$; it must be shown that $\lambda \in \sigma_e(\phi(S))$. We consider three cases.

CASE 1. $a \in \operatorname{int} E_S$. Here $\check{\phi}$ is an analytic function on $\operatorname{int} E_S$ and so $\check{\phi}(a) = \lambda$ and Lemma 4.4 implies that $\lambda \in \sigma_e(\phi(S))$.

CASE 2. $\{z_n\}$ has a subsequence that is interpolating for $R^\infty(\sigma(S), \mu)$. Here $\{z \in E_S : \|z - z_n\| < 1/n\}$ has area density 1 at z_n. But since Q_K is almost open, E_K is almost open. Because $E_S \subseteq E_K$, $\operatorname{Area}(E_S \setminus \operatorname{int} E_K) = 0$

and so $\{z \in E_S \cap \operatorname{int} E_K : \|z - z_n\| < 1/n\}$ has area density 1 at z_n. In particular, there is a point w_n in $B(z_n; 1/n) \cap E_S \cap \operatorname{int} E_K$ with $\|w_n - z_n\| < 1/n$. Now by Theorem 4.10, $\{z_n\}$ cannot converge to a in the norm of $R^\infty(\sigma(S), \mu)^*$ and so the same holds for the sequence $\{w_n\}$. Thus $\{w_n\}$ must have a subsequence that is interpolating for $R^\infty(\sigma(S), \mu)$. Since $\check{\phi}(w_n) \to \lambda$, Lemma 4.11 implies that $\lambda \in \sigma_e(\phi(S))$.

CASE 3. Neither Case 1 nor Case 2 holds. It follows by Theorem 4.10 that $a \in E_S \setminus \operatorname{int} E_S$ and $\|z_n - a\| \to 0$. Therefore, $\check{\phi}(z_n) \to \check{\phi}(a)$ and so $\check{\phi}(a) = \lambda$. Fix ε with $0 < \varepsilon < 2$. By Proposition 3.14(b) there is a sequence $\{w_n\}$ in E_S such that $w_n \to a$ and $\|w_n - a\| = \varepsilon$ for all $n \geq 1$. Without loss of generality we may assume that $\check{\phi}(w_n) \to \lambda'$ for some complex number λ'. By Theorem 4.10 $\{w_n\}$ has a subsequence that is interpolating for $R^\infty(\sigma(S), \mu)$ and so by Case 2, $\lambda' \in \sigma_e(\phi(S))$. But $|\check{\phi}(w_n) - \lambda| = |\check{\phi}(w_n) - \check{\phi}(a)| \leq \varepsilon \|\phi\|$ for all n and so $|\lambda' - \lambda| \leq \varepsilon \|\phi\|$. Since ε was arbitrary, $\lambda = \lambda' \in \sigma_e(\phi(S))$.

(b) To prove the containment left untouched by part (a), let $\lambda \notin \operatorname{clu}(\phi; \sigma_e(S))$. So there is a neighborhood U of $\sigma_e(S)$ such that $\lambda \notin \check{\phi}(U \cap E_S)$. Because $\partial \sigma(S) \subseteq \sigma_e(S)$, we have that $Z_\lambda(\phi) \subseteq E_S \setminus U \subseteq \sigma(S) \setminus \sigma_e(S) \subseteq \operatorname{int} \sigma(S) = \operatorname{int}\{\operatorname{cl} E_S\} = \operatorname{int} E_S$ by (VI.3.9). But ϕ is analytic on $\operatorname{int} E_S$. The assertions (i) and (ii) now follow as did the corresponding assertions in Theorem 4.1.

From Proposition VI.3.10(b) we obtain a function ψ in $R^\infty(K, \mu)$ and a polynomial p such that $\check{\psi}$ does not vanish at any point of $Z_\lambda(\phi)$, $Z_\lambda(\phi)$ are precisely the zeros of p and the order of the zero of p at a in $Z_\lambda(\phi)$ is $n_\lambda(\phi; a)$, and $\phi - \lambda = p\psi$. Hence $\phi(S) - \lambda = p(S)\psi(S) = \psi(S)p(S)$. Note that $\sigma_e(p(S)) = p(\sigma_e(S))$ and so $0 \notin \sigma_e(p(S))$; that is, $p(S)$ is a Fredholm operator. We will show that $\phi(S) - \lambda$ is Fredholm by showing that $\psi(S)$ is invertible.

Suppose $\psi(S)$ is not invertible so that $0 \in \sigma(\psi(S))$. By Theorem 3.7(b), $0 \in \check{\psi}(\sigma(S))$. Thus there is a sequence $\{z_n\}$ in E_S such that $\check{\psi}(z_n) \to 0$. Passing to a subsequence if necessary, we may assume that $\{z_n\}$ converges to some point a in $\operatorname{cl} E_S$. Observe that $\check{\psi}$ has no zeros on $\operatorname{int} E_S$ by the way that ψ was constructed. Thus it must be that $a \notin \operatorname{int} E_S$, for otherwise, since $\check{\psi}$ is analytic on $\operatorname{int} E_S$, $\check{\psi}(a) = 0$. Since $\operatorname{int} E_S = \operatorname{int}\{\operatorname{cl} E_S\}$, this implies that $a \in \partial[\operatorname{cl} E_S] = \partial \sigma(S) \subseteq \sigma_e(S)$. But then $\check{\phi}(z_n) - \lambda = p(z_n)\check{\psi}(z_n) \to p(a) \cdot 0 = 0$ and this says that $\lambda \in \operatorname{clu}(\phi; \sigma_e(S))$, a contradiction to our assumption. Therefore, it must be that $0 \notin \sigma(\psi(S))$ and so $\psi(S)$ is invertible.

From properties of the index, $\operatorname{ind}(\phi(S) - \lambda) = \operatorname{ind} p(S)$ and part (iii) is immediate. □

Exercise

1. Let $\{z_n\}$ be a sequence in E_K that converges to a point a in E_K. Show that if $\{z_n\}$ (considered as a sequence in $R^\infty(K, \mu)^*$) converges weak* to a, then $\|z_n - a\| \to 0$.

§5 A factorization theorem. This section is the first in a series that provides the beginning of a structure theory of subnormal operators and their invariant subspace lattices. The purpose of this section is to prove the following.

5.1 THEOREM (Olin and Thomson [**1980b**]). *There is a constant C such that if S is a subnormal operator acting on the Hilbert space $\mathscr{H}$ and $L: P^\infty(S) \to \mathbb{C}$ is a weak-star continuous linear functional, then there are vectors x and y in $\mathscr{H}$ with $\|x\|^2$ and $\|y\|^2$ bounded by $C\|L\|$ and*

$$L(T) = \langle Tx, y \rangle$$

for every T in $P^\infty(S)$.

Here is an application of this theorem.

5.2 THEOREM. *If S is a subnormal operator, then $P^\infty(S)$ equals the WOT closed algebra generated by S and the identity and the WOT and the weak* topology agree on $P^\infty(S)$.*

PROOF. It is immediate from Theorem 5.1 that the WOT and the weak* topology agree on $P^\infty(S)$. If $W(S)$ is the WOT closed algebra generated by S and 1, then $P^\infty(S) \subseteq W(S)$. Let $T \in W(S)$ and let $P^\infty(S)_*$ denote the collection of all weak* continuous linear functionals on $P^\infty(S)$ with the usual norm. Define $\Lambda: P^\infty(S)_* \to \mathbb{C}$ as follows. If $L \in P^\infty(S)_*$, let x and y be the vectors obtained in Theorem 5.1 and put $\Lambda(L) = \langle Tx, y \rangle$. It is routine to check that Λ is well-defined and linear. Also, if C is the constant in (5.1), then $|\Lambda(L)| \leq C^2\|L\|$. Thus Λ is a bounded linear functional on $P^\infty(S)_*$ and so there is an operator A in $P^\infty(S)$ such that $\Lambda(L) = L(A)$ for all L in $P^\infty(S)_*$. But this implies that $\langle Tx, y \rangle = \langle Ax, y \rangle$ for all vectors x and y so that $T = A \in P^\infty(S)$. □

Before beginning the rather involved process that leads to the proof of Theorem 5.1, a few bibliographical comments seem appropriate. The development presented here is from Thomson [**1988a** and **1988b**] with a few modifications that permit use of this material in the following section. If S is a contraction of class $\mathbb{A}$ (not defined here), then a similar conclusion has been obtained by Bercovici [**1988**]. In fact, using the results of Bercovici [**1988**] and the decomposition theorem from §1, it has been shown (Bercovici and Conway [**1988**]) that the constant C in the above theorem can be chosen to be $1 + \varepsilon$ for any positive ε. More recently, Conway and Dudziak [**1990**] have shown Theorem 5.1 to be valid for any operator whose spectrum is a spectral set by using a functional calculus for these operators in order to reduce the problem to that solved in Bercovici [**1988**]. By concentrating on subnormal operators and not insisting on the best value of the constant C, the proofs are simpler than the more abstract situation.

Now for some notation. If S is a subnormal operator on $\mathscr{H}$ with minimal normal extension N on $\mathscr{K}$ and scalar-valued spectral measure μ, then the algebras $P^\infty(S)$, $P^\infty(N)$, and $P^\infty(\mu)$ are all isomorphic in a natural way

(II.11.4) and their weak* topologies are all homeomorphic. Thus a linear functional on any one of them is naturally identified with one on any other of these algebras. We will often flit between these algebras and their functionals as though they were all equal.

If x and y are vectors in $\mathcal{H}$, define $x \otimes y \colon P^\infty(N) \to \mathbb{C}$ by $x \otimes y(T) = \langle Tx, y \rangle$ for all T in $P^\infty(N)$. Note that $x \otimes y$ is a weak* continuous linear functional on $P^\infty(N)$. By the identifications discussed in the preceding paragraph, we also have that $x \otimes y$ is a weak* continuous linear functional on $P^\infty(S)$. If $P^\infty(S)_*$ is the collection of all weak* continuous linear functionals on $P^\infty(S)$, then $P^\infty(S)_*$ is a Banach space whose dual is $P^\infty(S)$. Let $\|\cdot\|_*$ denote the norm on this Banach space. Observe that $\|x \otimes y\|_* \leq \|x\| \|y\|$.

Theorem 5.1 can therefore be rephrased by saying that every functional L in $P^\infty(S)_*$ can be factored as $L = x \otimes y$ with x and y in $\mathcal{H}$ and with control of the norms of x and y. If the vectors x and y were only required to belong to $\mathcal{K}$, then the conclusion would be easy (see Exercise 2). Once we have x in $\mathcal{H}$, however, the fact that y belongs to $\mathcal{H}$ rather than $\mathcal{K}$ is not significant since a factorization $L = x \otimes y$ with y in $\mathcal{K}$ produces such a factorization with y in $\mathcal{H}$ by projecting y onto $\mathcal{H}$.

So the essential part of Theorem 5.1 (aside from the condition on the norms) is that x belongs to $\mathcal{H}$. Indeed, this was the route to the original proof of existence of invariant subspaces for subnormal operators in Brown [**1978**] where Theorem 5.1 was proved for a special class of subnormal operators—a class that contains all the subnormal operators that do not clearly have nontrivial invariant subspaces. The idea is as follows. Let λ be a point in the interior of the Sarason hull of μ and let $e_\lambda \colon P^\infty(\mu) \to \mathbb{C}$ be evaluation at λ. Let x and y be vectors in $\mathcal{H}$ such that $e_\lambda = x \otimes y$. Put $\mathcal{M} = \bigvee \{(S - \lambda)^n x : n \geq 1\}$. Clearly $\mathcal{M}$ is invariant for $S - \lambda$ and hence for S. But $y \perp \mathcal{M}$ and so $\mathcal{M} \neq \mathcal{H}$. If $\mathcal{M} = (0)$, then $\ker(S - \lambda)$ is a nontrivial invariant subspace; otherwise $\mathcal{M}$ is the sought for invariant subspace.

Of course the proof of invariant subspaces for subnormal operators given in §V.4 is easier than the proof of the present theorem, but we will see other applications of Theorem 5.1 and the technique of its proof in subsequent sections.

Note that for vectors x and y in $L^2(\mu)$, the linear functional $x \otimes y \colon P^\infty(\mu) \to \mathbb{C}$ is given by

$$x \otimes y(\phi) = \langle \phi x, y \rangle = \int \phi x \bar{y} \, d\mu.$$

Here we have that $\|x \otimes y\|_* \leq \|xy\|_1 \leq \|x\|_2 \|y\|_2$.

For μ a compactly supported measure in the plane, let $P^\infty(\mu)_\perp \equiv \{f \in L^1(\mu) : \int f \phi \, d\mu = 0$ for all ϕ in $P^\infty(\mu)\}$ and let $P^\infty(\mu)_* \equiv L^1(\mu)/P^\infty(\mu)_\perp$, the predual of $P^\infty(\mu)$. Thus $(P^\infty(\mu)_*)^* = P^\infty(\mu)$.

The strategy here is that we will prove Theorem 5.1 under a special set of circumstances and then show how to reduce the general case to this special one using the decomposition theorem from §1. The reason for first attacking the special case is the great simplification of the function theory involved. The special assumptions are:

(5.3i) $\|\mu\| = 1$ and S is multiplication by z on $P^2(\mu)$;

(5.3ii) $P^\infty(\mu) = H^\infty(\mathbb{D})$.

Note that condition (5.3ii) is to be interpreted by means of Sarason's Theorem (VI.7.1). Thus μ is supported on $\operatorname{cl}\mathbb{D}$, $\mu|\partial\mathbb{D}$ is absolutely continuous with respect to m, Lebesgue measure, and the identity map on polynomials extends to a dual algebra isomorphism of $P^\infty(\mu)$ onto $H^\infty(\mathbb{D})$.

Thus for every ϕ in $P^\infty(\mu)$, it make sense to evaluate ϕ at points of $\mathbb{D}$. A sequence in $P^\infty(\mu)$ converges weak* to 0 if and only if it is uniformly bounded on $\mathbb{D}$ and converges pointwise on $\mathbb{D}$. Let $P_0^\infty(\mu) = \{\phi \in P^\infty(\mu) : \phi(0) = 0\}$. Hence $P_0^\infty(\mu)$ can be identified with $H_0^\infty = \{\phi \in H^\infty : \phi(0) = 0\}$. Recall that A denotes the disk algebra consisting of the functions that are continuous on $\operatorname{cl}\mathbb{D}$ and analytic on $\mathbb{D}$. Let $A_0 = \{f \in A : f(0) = 0\}$.

5.4 LEMMA. *Assume* (5.3). *If* $L \in P_0^\infty(\mu)_*$ *with* $\|L\|_* = 1$ *and if* $\nu \in M(\operatorname{cl}\mathbb{D})$ *such that* $L(f) = \int f\,d\nu$ *for all* f *in* A_0 *and* $\|\nu\| \leq 1$, *then* $\nu \ll m$ *and* $\|\nu\| = 1$.

PROOF. First let K be a compact subset of $\partial\mathbb{D}$ such that $m(K) = 0$. Thus, there is a function g in the disk algebra such that $g(z) = \bar{z}$ on K and $\|g\| = 1$ (V.18.10). Let $h \in A$ such that $h|K \equiv 1$ and $|h| < 1$ off K. Put $f = zhg$. So $f \in A_0$, $f|K \equiv 1$, and $|f| < 1$ off K. Since $\mu|\partial\mathbb{D} \ll m$, $f^n \to 0$ weak* in $P_0^\infty(\mu)$. Therefore, $0 = \lim_n L(f^n) = \lim_n \int f^n d\nu = \nu(K)$. Since K was an arbitrary compact subset of $\partial\mathbb{D}$ having Lebesgue measure 0, it follows that $\nu|\partial\mathbb{D} \ll m$.

Now the closed unit ball of $P_0^\infty(\mu)$ is weak* compact and so there is a function ϕ in $P_0^\infty(\mu)$ with $\|\phi\|_\mu = 1$ and $L(\phi) = 1$. Let $\{f_n\}$ be the sequence of Cesàro sums of ϕ. So $f_n \in A_0$, $\|f_n\| \leq \|\phi\|_\mathbb{D}$, $f_n \to \phi$ pointwise on $\mathbb{D}$, and $f_n \to \phi$ weak* in $L^\infty(m)$. Since $\nu|\partial\mathbb{D} \ll m$, $\int f_n d\nu \to \int \phi d\nu$. Also it follows from (5.3) that $f_n \to \phi$ weak* in $P_0^\infty(\mu)$ and so $L(f_n) \to L(\phi)$. Since $L(f_n) = \int f_n d\nu$ for each n, $L(\phi) = \int \phi d\nu$. Thus $1 = \int \phi d\nu \leq \|\phi\|_\mu \|\nu\| \leq 1$, yielding equality. But this implies that $\|\nu\| = 1$ and, since $\int \phi d\nu = 1$, $|\phi| = 1$ a.e. $[\nu]$. But ϕ is analytic on $\mathbb{D}$ and $\phi(0) = 0$. Since $\|\phi\|_\mu = 1$, $|\phi| < 1$ on $\mathbb{D}$ and so it must be that $\operatorname{supp}\nu \subseteq \partial\mathbb{D}$. □

Let $L \in P^\infty(\mu)_* = L^1(\mu)/P^\infty(\mu)_\perp$. Before proceeding further, it is worthwhile to apprise the reader of a certain limitation in the Hahn-Banach Theorem that may have escaped notice. We know that L has a norm preserving extension to $L^\infty(\mu)$; this extension, however, need not be weak* continuous and hence represented by a function in $L^1(\mu)$. We can, if we prefer, obtain a weak* continuous extension of L, but this extension need not have the same norm as L. The Hahn-Banach Theorem does not guarantee the existence of a weak* continuous extension of L with the same norm. Using the definition of the quotient norm, however, we can conclude that for L in $P^\infty(\mu)_*$, there is a sequence $\{u_n\}$ in $L^1(\mu)$ such that $\|u_n - L\|_* = 0$ and $\|u_n\|_1 \downarrow \|L\|_*$. Similar comments apply to functionals in $P_0^\infty(\mu)_*$.

5.5 LEMMA. *Assume (5.3). Let $L \in P_0^\infty(\mu)_*$ with $\|L\|_* = 1$. If $\{u_n\}$ is any sequence in $L^1(\mu)$ such that $\|u_n - L\|_* \to 0$ and $\limsup_n \|u_n\|_1 \leq 1$, then for any r, $0 < r < 1$, and $\Delta_r \equiv \{z: |z| < r\}$, we have that $\int_{\Delta_r} |u_n| d\mu \to 0$ as $n \to \infty$. Also, if ν is any weak* cluster point of $\{u_n \mu\}$ in $M(\mathrm{cl}\,\mathbb{D})$, then $\|\nu\| = 1$ and $\nu \ll m$.*

PROOF. First note that if $f \in A_0$, then $\int f u_n d\mu \to_{\mathrm{cl}} \int f d\nu$ while we also have that $\int f u_n d\mu \to L(f)$. Since the hypothesis clearly implies that $\|\nu\| \leq 1$, we have from Lemma 5.4 that $\|\nu\| = 1$ and $\nu \ll m$.

Now fix r, $0 < r < 1$, and assume that $\int_{\Delta_r} |u_n| d\mu \geq \varepsilon$ for infinitely many n; by passing to a subsequence it may be assumed that this inequality holds for all n. Let ν be a weak* cluster point of $\{u_n \mu\}$ in $M(\mathrm{cl}\,\mathbb{D})$; since $M(\mathrm{cl}\,\mathbb{D})$ is the dual of a separable Banach space, it may be supposed that $u_n \mu \to \nu$ weak* in $M(\mathrm{cl}\,\mathbb{D})$. From the first part of the proof we know that $\|\nu\| = 1$ and $\nu \ll m$; in particular, ν is supported by $\partial \mathbb{D}$. Let $A_r = (\mathrm{cl}\,\mathbb{D}) \setminus \Delta_r$. There is a subsequence $\{u_{n_k}\}$ and a measure η on $\mathrm{cl}\,\mathbb{D}$ such that $u_{n_k} \chi_{\Delta_r} \mu \to \eta$ weak* in $M(\mathrm{cl}\,\mathbb{D})$. Note that $\mathrm{supp}(\eta) \subseteq \mathrm{cl}\,\Delta_r$. Now $\chi_{A_r} u_{n_k} \mu = u_{n_k} \mu - \chi_{\Delta_r} u_{n_k} \mu \to \nu - \eta$ and $\|\chi_{A_r} u_{n_k} \mu\| = \int_{A_r} |u_{n_k}| d\mu = \int |u_{n_k}| d\mu - \int_{\Delta_r} |u_{n_k}| d\mu \leq \|u_{n_k}\|_1 - \varepsilon$. Therefore, $\|\nu - \eta\| \leq \limsup \|\chi_{A_r} u_{n_k} \mu\| \leq 1 - \varepsilon$. But η and ν have disjoint supports so that $\|\nu - \eta\| = \|\nu\| + \|\eta\| \geq \|\nu\| = 1$. This contradiction shows that it is impossible for there to be an $\varepsilon > 0$ and an infinity of n such that $\int_{\Delta_r} |u_n| d\mu \geq \varepsilon$. That is, $\int_{\Delta_r} |u_n| d\mu \to 0$ as $n \to \infty$. □

We now begin an approximation scheme. The general strategy of approaching the proof of a theorem such as Theorem 5.1 in this manner goes back to S. W. Brown [1978]. Indeed, this has become known as the Scott Brown technique. The tactics needed to execute this approach, however, are more than an easy modification of those used in Brown [1978] and, in this instance, originated in Olin and Thomson [1980b] with considerable simplification in Thomson [1988b]. We now return to functionals in $P^\infty(\mu)_*$ rather than $P_0^\infty(\mu)_*$.

5.6 LEMMA. *Assume (5.3). Let $r, \delta, \gamma_1, \ldots, \gamma_k, \eta_1, \ldots, \eta_m$ be positive constants with $0 < r < 1$ and put $A_r = \{z : r \leq |z| \leq 1\}$. If $L \in P^\infty(\mu)_*$ such that $\|L\|_* < \delta^2$, then for any $\varepsilon > 0$ and any functions $g_1, \ldots, g_k$, $h_1, \ldots, h_m$ in $L^2(\mu)$, there is a function x in $P^2(\mu)$ and a function y in $L^2(\mu | A_r)$ such that:*

(a) $\|x\| < \delta$ and $\|y\| < \delta$;
(b) $\|L - x \otimes y\|_* < \varepsilon$;

(c) *for* $1 \leq i \leq k$, $\|y g_i \chi_{\mathbb{D}}\|_1 < \gamma_i$;
(d) *for* $1 \leq j \leq m$, $\|x h_j \chi_{\mathbb{D}}\|_1 < \eta_j$.

(*Note that conditions* (c) *and* (d) *imply that* $\|g_i \otimes y \chi_{\mathbb{D}}\|_* < \gamma_i$ *and* $\|x \otimes h_j \chi_{\mathbb{D}}\|_* < \eta_j$.)

PROOF. First we prove the lemma under the assumption that $L \in P_0^\infty(\mu)_*$. Later we'll see how to derive the general lemma from this one. As observed prior to the statement of Lemma 5.5, there is a sequence $\{v_n\}$ in $L^1(\mu)$ such that $\|v_n\|_1 \downarrow \|L\|_* < \delta^2$ and $\|v_n - L\|_* \to 0$; it is assumed that $\|v_n\|_1 < \delta^2$ for all n. A minor approximation argument shows that the functions v_n can be chosen in $L^\infty(\mu)$. Fix $n \geq 1$ and observe that since $\|v_n\|_q \to \|v_n\|_1$ as $q \to 1$, we can pick a value of p so that p has the form considered in Lemma V.4.3 and is sufficiently large that for the conjugate index q, $\|v_n\|_q < \|L\|_* + 1/n$. (Realize that p, and hence q, depends on n but this dependence does not affect the argument. This dependence is suppressed in the notation.) By the Hahn-Banach Theorem, there is a function u_n in $L^q(\mu)$ such that $\int f u_n d\mu = \int f v_n d\mu$ for every f in $P^p(\mu)$ and $\|u_n\|_q = \sup\{|\int f v_n d\mu| : f \in P^p(\mu) \text{ and } \|f\|_p \leq 1\}$. By Lemma V.4.3 there is a function x_n in $P^2(\mu)$ such that $|x_n|^2 = |u_n|$ a.e. $[\mu]$. Now u_n and v_n define the same linear functional on $P_0^\infty(\mu)_*$, so that $\|u_n - L\|_* \to 0$. Also, the fact that $\|\mu\| = 1$ implies that $\|u_n\|_1 \leq \|u_n\|_q \leq \|v_n\|_q$; hence $\limsup_n \|u_n\|_1 \leq \|L\|_*$. Thus by Lemma 5.5, for any s, $0 < s < 1$, $\int_{\Delta_s} |u_n| d\mu \to 0$ as $n \to \infty$.

Put $y_n = (\bar{u}_n / \bar{x}_n) \chi_{A_r}$ when x_n is not 0 and $y_n = 0$ otherwise; so $y_n \in L^2(\mu|A_r)$. Observe that $|y_n|^2 = |x_n|^2 = |u_n|$ a.e. $[\mu]$ on A_r. Thus $\limsup_n \|x_n\|_2^2 = \limsup_n \|u_n\|_1 \leq \|L\|_* < \delta^2$; so $\|x_n\|_2 < \delta$ for sufficiently large n. Similarly, $\limsup_n \|y_n\|_2 < \delta$. Let n_1 be such that $\|x_n\|_2 < \delta$ and $\|y_n\|_2 < \delta$ for all $n \geq n_1$. We will take $x = x_n$ and $y = y_n$ for some $n \geq n_1$; we have just shown that no matter how n is chosen, condition (a) is always satisfied.

Now $\|L - x_n \otimes y_n\|_* \leq \|L - u_n\|_* + \|u_n - x_n \otimes y_n\|_*$ and $\|L - u_n\|_* \to 0$ as $n \to \infty$. In addition, if $\phi \in P_0^\infty(\mu)$, then

$$\langle u_n - x_n \otimes y_n, \phi \rangle = \int \phi u_n d\mu - \int \phi x_n \bar{y}_n d\mu$$
$$= \int \phi u_n d\mu - \int_{A_r} \phi x_n \frac{u_n}{x_n} d\mu$$
$$= \int_{\Delta_r} \phi u_n d\mu.$$

So

$$\|u_n - x_n \otimes y_n\|_* \leq \int_{\Delta_r} |u_n| d\mu.$$

Since $\int_{\Delta_r} |u_n| d\mu \to 0$ as $n \to \infty$, there is an $n_2 \geq n_1$ such that $\|x_n \otimes y_n - L\|_* < \varepsilon$ for $n \geq n_2$.

Now fix h in $L^2(\mu)$ and let $\eta > 0$. Choose s such that $\int_{\mathbb{D}\setminus\Delta_s} |h|^2 d\mu < \eta^2$. Thus

$$\|x_n h \chi_{\mathbb{D}}\|_1 = \int_{\mathbb{D}} |x_n| |h| d\mu$$

$$= \int_{\Delta_s} |x_n| |h| d\mu + \int_{\mathbb{D}\setminus\Delta_s} |x_n| |h| d\mu$$

$$\leq \|h\|_2 \left[\int_{\Delta_s} |x_n|^2 d\mu\right]^{\frac{1}{2}} + \|x_n\|_2 \left[\int_{\mathbb{D}\setminus\Delta_s} |h|^2 d\mu\right]^{\frac{1}{2}}$$

$$\leq \|h\|_2 \left[\int_{\Delta_s} |u_n| d\mu\right]^{\frac{1}{2}} + \delta\eta.$$

Thus $\|x_n h \chi_{\mathbb{D}}\|_1 \to 0$. Similarly, $\|y_n g \chi_{\mathbb{D}}\|_1 \to 0$ for any g in $L^2(\mu)$.

The proof is completed for the case where $L \in P_0^\infty(\mu)_*$ by letting $x = x_n$ and $y = y_n$ for an appropriate value of $n \geq n_2$.

Now suppose that $L \in P^\infty(\mu)_*$. If $\phi \in P_0^\infty(\mu)$ and $\|\phi\|_\mu \leq 1$, then ϕ is an analytic function on $\mathbb{D}$ vanishing at 0 and bounded by 1. By Schwarz's Lemma, ϕ/z is also bounded by 1. Thus if we define $L_1 : P_0^\infty(\mu) \to \mathbb{C}$ by $L_1(\phi) = L(\phi/z)$, then $L_1 \in P_0^\infty(\mu)_*$ and $\|L_1\| \leq \|L\| < \delta^2$. From the preceding part of the proof there is a function x_1 in $P^2(\mu)$ and a function y_1 in $L^2(\mu|A_r)$ such that $\|x_1\|_2 < \delta$, $\|y_1\|_2 < \delta$, $|L_1(\phi) - \int \phi x_1 \bar{y}_1 d\mu| < \varepsilon \|\phi\|$ for all ϕ in $P_0^\infty(\mu)$, $\|g_i y_1 \chi_{\mathbb{D}}\|_1 < \gamma_i$, and $\|h_j x_1 \chi_{\mathbb{D}}\|_1 < \eta_j$ for all i and j. Put $x = x_1$ and $y = \bar{z} y_1$. If $\phi \in P^\infty(\mu)$, then

$$\left|L(\phi) - \int \phi x \bar{y} d\mu\right| = \left|L_1(z\phi) - \int \phi z x_1 \bar{y}_1 d\mu\right|$$

$$< \varepsilon \|z\phi\|$$

$$\leq \varepsilon \|\phi\|.$$

It is not difficult to show that with this definition of x and y, the remaining inequalities in the lemma are also satisfied. □

The next lemma is Lemma 6 from Olin and Thomson [1980b]. The full strength of this lemma will not be employed until the next section.

5.7 Lemma. *Assume* (5.3). *Let* $\varepsilon, \gamma_1, \ldots, \gamma_m > 0$, $0 < r < 1$, $0 < \delta < \frac{1}{3}$, *and* $h_1, \ldots, h_m \in L^2(\mu)$. *If* $L \in P^\infty(\mu)_*$, $a \in L^2(\mu)$, *and* $b \in L^2(\mu|A_r)$ *such that* $\|L - a \otimes b\|_* < \delta^4$, *then there exist functions* x *in* $P^2(\mu)$ *and* y *in* $L^2(\mu|A_r)$ *such that:*

(a) $\|x\| < 3\delta$;
(b) $\|b + y\| < \delta^2 + (1/(1 - 2\delta))\|b\|$;
(c) $\|L - (a + x) \otimes (b + y)\|_* < \varepsilon$;

(d) $\|x \otimes h_j \chi_{\mathbb{D}}\|_* < \gamma_j$ for $1 \leq j \leq m$;
(e) $|a + x| \geq (1 - 2\delta)|a|$ a.e. $[\mu]$ on $\partial \mathbb{D}$.

PROOF. By Lemma 5.6 there are functions u in $P^2(\mu)$ and v in $L^2(\mu|A_r)$ such that the following properties hold:

(i) $\|u\| < \delta^2$ and $\|v\| < \delta^2$;
(ii) $\|L - a \otimes b - u \otimes v\| < \varepsilon/6$;
(iii) $\|u \otimes b \chi_{\mathbb{D}}\|_* < \varepsilon/6$;
(iv) $\|a \otimes v \chi_{\mathbb{D}}\|_* < \varepsilon/6$;
(v) $\|u \otimes h_j \chi_{\mathbb{D}}\|_* < \gamma_j/6$ for $1 \leq j \leq m$.

Define g to be the outer function in H^∞ such that for ζ in $\partial \mathbb{D}$, $|g(\zeta)| = 2/\delta$ if $|a(\zeta)| \leq |u(\zeta)|/\delta$ and $|g(\zeta)| = 1$ otherwise. Let p be a sufficiently large positive integer such that if $f = z^p g$, then

$$(5.8) \quad \|uf \chi_{\mathbb{D}}\|_2 < \frac{1}{6} \min \left\{ \frac{\varepsilon}{\|b\| + 1}, \frac{\gamma_1}{\|h_1\| + 1}, \ldots, \frac{\gamma_m}{\|h_m\| + 1} \right\}.$$

Put $x = (1 + f)u$; since $f \in P^\infty(\mu)$, $x \in P^2(\mu)$. Also, $\|x\|_2 \leq \|1 + f\|_\infty \|u\|_2 \leq (1 + 2/\delta)\delta^2 = (\delta + 2)\delta < 3\delta$, proving (a).

On the subset of $\partial \mathbb{D}$ where $|a| \leq (1/\delta)|u|$ we have that $|a + u| \leq (1 + 1/\delta)|u|$. But $|f| = 2/\delta$ and so $|uf| = (2/\delta)|u|$. Hence $|a + x| = |a + u + uf| \geq |u|(2/\delta - 1 - 1/\delta) = (1/\delta - 1)|u| \geq (3 - 1)|u| \geq |u|$ since $\delta < \frac{1}{3}$. Thus $|a + x| \geq |u|$. Also $|a + x| \geq |u|(1/\delta - 1) + |u|((1 - \delta)/\delta)$. Hence $|(a + x)/a| = |(a + x)/u| |u/a| \geq ((1 - \delta)/\delta)\delta = 1 - \delta$.

Now suppose we are on the subset of $\partial \mathbb{D}$ where $|a| > 1/\delta|u|$. Here $|f| = 1$ and so $|x| \leq 2|u|$. Hence $|a + x| \geq |u|(1/\delta - 2) \geq |u|$ since $\delta < \frac{1}{3}$. In addition, $|a| \leq |a + x| + |x| \leq |a + x| + 2|u| \leq |a + x| + 2\delta|a|$ and so $(1 - 2\delta)|a| \leq |a + x|$.

Combining the last two paragraphs we get that

$$(5.9) \quad |a + x| \geq |u| \quad \text{a.e. } [\mu] \text{ on } \partial \mathbb{D}$$

and

$$|a + x| \geq (1 - 2\delta)|a| \quad \text{a.e. } [\mu] \text{ on } \partial \mathbb{D}.$$

This last inequality is part (e) of the lemma.

Now define $y = v$ on $\mathbb{D}$ and $y = w$ on $\partial \mathbb{D}$, where

$$w = \frac{\bar{u}v - \bar{x}b}{\bar{a} + \bar{x}}$$
$$= \frac{\bar{u}}{\bar{a} + \bar{x}}(v - (1 + \bar{f})b).$$

By (5.9) $w \in L^2(\mu|\partial \mathbb{D})$ so that $y \in L^2(\mu|A_r)$. Also

$$\begin{aligned}
\|y + b\|^2 &= \int_{\mathbb{D}} |v + b|^2 \, d\mu + \int_{\partial \mathbb{D}} \left| \frac{\bar{u}}{\bar{a} + \bar{x}} v + \frac{\bar{a}}{\bar{a} + \bar{x}} b \right|^2 d\mu \\
&= \int_{\mathbb{D}} |v|^2 \, d\mu + 2 \operatorname{Re} \int_{\mathbb{D}} v \bar{b} \, d\mu + \int_{\mathbb{D}} |b|^2 \, d\mu \\
&\quad + \int_{\partial \mathbb{D}} |v|^2 \left| \frac{u}{a + x} \right|^2 d\mu + 2 \operatorname{Re} \int_{\partial \mathbb{D}} v \bar{b} \frac{\bar{u}}{\bar{a} + \bar{x}} \frac{a}{a + x} d\mu \\
&\quad + \int_{\partial \mathbb{D}} \left| \frac{a}{a + x} \right|^2 |b|^2 \, d\mu \\
&\leq \int_{\operatorname{cl} \mathbb{D}} |v|^2 \, d\mu + 2 \int_{\mathbb{D}} |v| |b| \, d\mu + \int_{\mathbb{D}} |b|^2 \, d\mu \\
&\quad + \frac{2}{1 - 2\delta} \int_{\partial \mathbb{D}} |v| |b| \, d\mu + \frac{1}{(1 - 2\delta)^2} \int_{\partial \mathbb{D}} |b|^2 \, d\mu \\
&\leq \int |v|^2 \, d\mu + \frac{2}{1 - 2\delta} \int |v| |b| \, d\mu + \frac{1}{(1 - 2\delta)^2} \int |b|^2 \, d\mu \\
&\leq \delta^4 + \frac{2}{1 - 2\delta} \delta^2 \|b\| + \frac{1}{(1 - 2\delta)^2} \|b\|^2 \\
&= \left(\delta^2 + \frac{1}{1 - 2\delta} \|b\| \right)^2,
\end{aligned}$$

proving (b).

Using (v) and (5.8) we get that for $1 \leq j \leq m$, $\|x \otimes h_j \chi_{\mathbb{D}}\|_* \leq \|u \otimes h_j \chi_{\mathbb{D}}\|_* + \|uf \otimes h_j \chi_{\mathbb{D}}\|_* \leq \gamma_j/6 + \|uf\chi_{\mathbb{D}}\| \|h_j\| < \gamma_j$, establishing part (d) of the lemma. Similarly, we also have $\|x \otimes b \chi_{\mathbb{D}}\|_* < \varepsilon$.

Finally, write

$$L - (a + x) \otimes (b + y) = L - a \otimes b - x \otimes y - a \otimes y - x \otimes b$$

and let's consider each of these summands separately. First

$$\begin{aligned}
x \otimes y &= u \otimes y + uf \otimes y \\
&= u \otimes (v\chi_{\mathbb{D}} + w\chi_{\partial \mathbb{D}}) + uf \otimes (v\chi_{\mathbb{D}} + w\chi_{\partial \mathbb{D}}) \\
&= u \otimes v + uf \otimes v\chi_{\mathbb{D}} + Z
\end{aligned}$$

where $Z = u \otimes (-v\chi_{\partial \mathbb{D}} + w\chi_{\partial \mathbb{D}}) + uf \otimes w\chi_{\partial \mathbb{D}}$. Using $a \otimes y = a \otimes v\chi_{\mathbb{D}} + a \otimes w\chi_{\partial \mathbb{D}}$ and $x \otimes b = x \otimes b\chi_{\mathbb{D}} + x \otimes b\chi_{\partial \mathbb{D}}$ we get that

$$\begin{aligned}
L - (a + x) \otimes (b + y) &= [L - a \otimes b - u \otimes v] - uf \otimes v\chi_{\mathbb{D}} - a \otimes v\chi_{\mathbb{D}} \\
&\quad - x \otimes b\chi_{\mathbb{D}} - Z - a \otimes w\chi_{\partial \mathbb{D}} - x \otimes b\chi_{\partial \mathbb{D}}.
\end{aligned}$$

But on $\partial \mathbb{D}$, $Z + a \otimes w + x \otimes b = -u \otimes v + u \otimes w + uf \otimes w + a \otimes w + u \otimes b + uf \otimes b = 1 \otimes [-\bar{u}v + \bar{u}w + \bar{u}fw + \bar{a}w + \bar{u}b + \bar{u}fb] = 1 \otimes [(\bar{a} + \bar{x})w - \bar{u}v + \bar{u}b + \bar{u}fb] =$

§5 A FACTORIZATION THEOREM

$1 \otimes 0 = 0$. Therefore,

$$\|L - (a+x) \otimes (b+y)\|_* \leq \|L - a \otimes b - u \otimes v\|_* + \|uf \otimes v\chi_D\|_*$$
$$+ \|u(1+f) \otimes b\chi_D\|_* + \|a \otimes v\chi_D\|_*$$
$$= s_1 + s_2 + s_3 + s_4.$$

Now $s_1 < \varepsilon/6$ by (ii). Using (5.8) we get that $s_2 \leq \|uf\chi_D\|_2 \|v\|_2 < (\varepsilon/6)\delta^2 < \varepsilon/6$. Also $\|u \otimes b\chi_D\|_* < \varepsilon/6$ by (iii) and $\|uf \otimes b\chi_D\|_* \leq \|uf\chi_D\|_2 \|b\|_2 < \varepsilon/6$ by (5.8). Hence $s_3 < \varepsilon/3$. Finally $s_4 < \varepsilon/6$ by (iv) so that we have that $\|L - (a+x) \otimes (b+y)\|_* < \varepsilon$ as desired. □

The next lemma will be combined with the preceding one to prove Theorem 5.1.

5.10 LEMMA. *Suppose S is any subnormal operator on $\mathcal{H}$ with minimal normal extension N on $\mathcal{K}$. If $L \in P^\infty(S)_*$, $\{x_n\}$ is a sequence in $\mathcal{H}$, and $\{y_n\}$ is a sequence in $\mathcal{K}$ such that $\|x_n - x\| \to 0$, $y_n \to y$ weakly, and $\langle Tx_n, y_n \rangle \to L(T)$ for every T in $P^\infty(S)$, then $L(T) = x \otimes y$.*

PROOF. For T in $P^\infty(S)$ and $n \geq 1$, $|L(T) - x \otimes y(T)| \leq |L(T) - \langle Tx_n, y_n \rangle| + \|T\| \|x_n - x\| \|y_n\| + |\langle Tx, (y_n - y) \rangle|$. Since $\{y_n\}$ must be uniformly bounded, we see that the right-hand side of this inequality converges to 0 and the lemma follows. □

PROOF OF THEOREM 5.1.

CASE 1. Assume (5.3).

First suppose that $\|L\| < (\frac{1}{3})^4$. Choose a sequence of positive constants ε_n that decrease to 0 with $\varepsilon_1 < \frac{1}{3}$ and such that the infinite series $\sum_n \varepsilon_n$ and the infinite product $\prod_n (1 - 2\varepsilon_n)^{-1}$ both converge. Let M be the constant

$$M = \max\left\{ 3\sum_n \varepsilon_n, \left(\sum_n \varepsilon_n^2\right)\left(\prod_n \left(\frac{1}{1-2\varepsilon_n}\right)\right) \right\}.$$

Apply Lemma 5.7 with $a = b = 0$ and $\delta = \varepsilon_1 < \frac{1}{3}$ to get x_1 in $P^2(\mu)$ and y_1 in $L^2(\mu)$ such that $\|x_1\| < 3\varepsilon_1$, $\|y_1\| < \varepsilon_1^2$, and $\|L - x_1 \otimes y_1\|_* < \varepsilon_2^4$. Now apply Lemma 5.7 with $a = x_1$, $b = y_1$, and $\delta = \varepsilon_2$ to get a function x_2 in $P^2(\mu)$ and a y_2 in $L^2(\mu)$ such that $\|x_2 - x_1\| < 3\varepsilon_2$, $\|y_2\| < \varepsilon_2^2 + (1 - 2\varepsilon_2)^{-1}\|y_1\|$, and $\|L - x_2 \otimes y_2\|_* < \varepsilon_3^4$. Continuing, we get a sequence $\{x_n\}$ in $P^2(\mu)$ and a sequence $\{y_n\}$ in $L^2(\mu)$ such that for every n, $\|x_n - x_{n-1}\| < 3\varepsilon_n$, $\|y_n\| < \varepsilon_n^2 + (1 - 2\varepsilon_n)^{-1}\|y_{n-1}\|$, and $\|L - x_n \otimes y_n\|_* < \varepsilon_{n+1}^4$.

Since the series $\sum_n \varepsilon_n$ converges, $\{x_n\}$ is a Cauchy sequence and so there is an x in $P^2(\mu)$ such that $x_n \to x$. Manipulation with the above inequality for $\|y_n\|$ shows that

$$\|y_n\| < \sum_{k=1}^n \varepsilon_k^2 \left[\prod_{m=k+1}^n \frac{1}{1-2\varepsilon_m}\right]$$

for all n. (If in the preceding product $k = n$, then the product should be interpreted as 1.) Thus $\|y_n\| \leq M$ for all $n \geq 1$ and so there is a subsequence $\{y_{n_k}\}$ and a function y in $L^2(\mu)$ such that $y_{n_k} \to y$ weakly. Since $x_{n_k} \to x$ in norm and $\|L - x_{n_k} \otimes y_{n_k}\|_* \to 0$, Lemma 5.10 implies that $L = x \otimes y$. Note that $\|x\|$ and $\|y\| \leq M$.

Now let L be an arbitrary element of $P^\infty(\mu)_*$ with no restriction on its norm. The preceding argument applies to $(1/82\|L\|)L$ and so there is a function x_1 in $P^2(\mu)$ and a function y_1 in $L^2(\mu)$ such that $\|x_1\|, \|y_1\| \leq M$ and $(1/82\|L\|)L = x_1 \otimes y_1$. If $x = \sqrt{82\|L\|} x_1$ and $y = \sqrt{82\|L\|} y_1$, we have that $L = x \otimes y$ and $\|x\|, \|y\| \leq C\sqrt{\|L\|}$, where $C^2 = 82M^2$. Replacing y by its projection onto $P^2(\mu)$ proves Case 1.

CASE 2. Only assume (5.3)(ii). So S is any subnormal operator acting on a Hilbert space $\mathcal{H}$ with minimal normal extension N acting on $\mathcal{K}$ and scalar-valued spectral measure μ, but we assume that $P^\infty(\mu) = H^\infty(\mathbb{D})$. Proposition V.17.14 implies that there is a vector u in $\mathcal{H}$ such that u is a separating vector for N. Let $\mathcal{H}_u$ be the closed linear span of $\{S^n u : n \geq 0\}$. So $S|\mathcal{H}_u$ is a cyclic subnormal operator and there is a measure ν such that if $Wp = p(S)u$ for every polynomial p, then W extends to an isomorphism $W: P^2(\nu) \to \mathcal{H}_u$ satisfying $WS_\nu W^{-1} = S|\mathcal{H}_u$. Since u is a separating vector for N, $[\nu] = [\mu]$ and so $P^\infty(\nu) = P^\infty(\mu) = H^\infty(\mathbb{D})$. For the same reasons, the restriction map $T \to T|\mathcal{H}_u$ defines an isomorphism from $P^\infty(S)$ onto $P^\infty(S|\mathcal{H}_u)$. Thus L induces a weak-star continuous linear functional $\tilde{L}$ on $P^\infty(S|\mathcal{H}_u)$ by means of the formula $\tilde{L}(T|\mathcal{H}_u) = L(T)$. Thus Case 1, applied to $\tilde{L}$, implies there are vectors x and y in $\mathcal{H}_u$ with $\|x\|$ and $\|y\| \leq C\sqrt{\|\tilde{L}\|}$ and $\tilde{L} = x \otimes y$. But $\|\tilde{L}\| = \|L\|$ and is easy to see that $L = x \otimes y$.

CASE 3. Let S be arbitrary with any scalar-valued spectral measure μ, but assume that $P^\infty(\mu)$ is antisymmetric. Thus $P^\infty(\mu) = H^\infty(G)$ for a simply connected region G such that if $\tau: G \to \mathbb{D}$ is the Riemann map, τ is a weak* sequential generator of $P^\infty(\mu)$. Proposition 1.7 implies that $\nu = \mu \circ \tau^{-1}$ is a scalar-valued spectral measure for $\tau(S)$, $P^\infty(\nu) = H^\infty(\mathbb{D})$, and $\phi \to \phi \circ \tau$ defines a dual algebra isomorphism of $P^\infty(\nu)$ onto $P^\infty(\mu)$. Define $L_1: P^\infty(\nu) \to \mathbb{C}$ by $L_1(\phi) = L(\phi \circ \tau)$. So $L_1 \in P^\infty(\nu)_*$ and the preceding case implies there are vectors x and y in $\mathcal{H}$ such that $L_1(\phi) = \langle \phi(\tau(S))x, y \rangle$. But if $\psi \in P^\infty(\mu)$, then $\psi \circ \tau^{-1} \in P^\infty(\nu)$ and $L(\psi) = L_1(\psi \circ \tau^{-1}) = \langle \psi \circ \tau^{-1}(\tau(S))x, y \rangle = \langle \psi(S)x, y \rangle$. That is, $L = x \otimes y$. The estimates on $\|x\|$ and $\|y\|$ readily follow with the same constant C.

CASE 4. The general case. Here we can write $P^\infty(\mu) = L^\infty(\mu_0) \oplus \bigoplus_i P^\infty(\mu_i)$, where $\mu = \sum_i \mu_i$, $\mu_i \perp \mu_j$ for $i \neq j$, and each $P^\infty(\mu_i)$ is antisymmetric for $i \geq 1$. If $L \in P^\infty(\mu)_*$, then $L = \bigoplus_i L_i$, where $L_i \in P^\infty(\mu_i)_*$ for $i \geq 1$ and $L_0 \in L^\infty(\mu_0)_*$. Note that $\|L\| = \sum_i \|L_i\|$. From Corollary 1.6 we have a decomposition of the Hilbert space $\mathcal{H} = \mathcal{H}_0 \oplus \mathcal{H}_1 \oplus \mathcal{H}_2 \oplus \cdots$ and the subnormal operator $S = N_0 \oplus S_1 \oplus S_2 \oplus \cdots$. The previous case applies to each S_i. Thus for each $i \geq 1$ there are vectors x_i and y_i in $\mathcal{H}_i$ such that $\|x_i\|^2$, $\|y_i\|^2 \leq C\|L_i\|$ and $L_i(\phi(S_i)) = \langle \phi(S_i)x_i, y_i\rangle$ for each ϕ in $P^\infty(\mu_i)$. Theorem 5.1 can easily be proved for normal operators (see Exercise 1), so there are vectors x_0 and y_0 in $\mathcal{H}_0$ with $\|x_0\|^2$ and $\|y_0\|^2 \leq C\|L_0\|$ and such that $L_0(\phi) = \langle \phi(N_0)x_0, y_0\rangle$ for all ϕ in $L^\infty(\mu_0)$. If $x = \oplus_i x_i$ and $y = \oplus_i y_i$, it is easy to check that $\|x\|^2$, $\|y\|^2 \leq C\|L\|$ and $L = x \otimes y$. □

This section concludes with what is essentially a restatement of Lemma 5.7 in the form it will be used in the next section. First, if μ is a compactly supported measure on the plane and $\mu \in L^2(\mu)$ then say that u is an *ample function* if the identity map on polynomials extends to an isometric isomorphism of $P^\infty(|u|^2\mu)$ onto $P^\infty(\mu)$. Put $\nu = |u|^2\mu$; so $\nu \ll \mu$ and since the definition of the L^∞ space of a measure depends only on the measure class and not on the specific measure, we can discuss the restriction map from $L^\infty(\mu)$ onto $L^\infty(\nu)$. Note that the restriction map is a contraction and it always takes $P^\infty(\mu)$ into $P^\infty(\nu)$. The requirement that u is an ample function is equivalent to the requirement that the restriction takes $P^\infty(\mu)$ isometrically onto $P^\infty(\nu)$.

Now the only way that $L^\infty(\mu)$ and $L^\infty(\nu)$ can be naturally isomorphic is if μ and ν are mutually absolutely continuous; that is, that $u \neq 0$ a.e. $[\mu]$. However u can be an ample function even if it vanishes on a set of positive measure. For example, if μ is area measure on the open unit disk and u is the characteristic function of the annulus $\{z : r < |z| < 1\}$, then u is an ample function.

Maintaining the notation of the previous paragraph, note that the fact that $P^\infty(\mu) = P^\infty(\nu)$ implies that any weak* continuous linear functional L on $P^\infty(\mu)$ defines in a natural way a weak* continuous functional on $P^\infty(\nu)$.

5.11 LEMMA. *Assume* (5.3). *Let* $\varepsilon, \gamma_1, \ldots, \gamma_m > 0$, $0 < r < 1$, $0 < \delta < \frac{1}{3}$, *and* $h_1, \ldots, h_m \in L^2(\mu)$. *If* u *is an ample function in* $L^2(\mu)$ *and* $L \in P^\infty(\mu)_*$, $a \in L^2(\mu)$, *and* $b \in L^2(\mu|A_r)$ *such that* $\|L - a \otimes b\|_* < \delta^4$, *then there exist functions* x *in* $\mathcal{H}_u$ *and* y *in* $L^2(\mu|A_r)$ *such that:*

(a) $\|x\| < 3\delta$;
(b) $\|b + y\| < \delta^2 + (1 - 2\delta)^{-1}\|b\|$;
(c) $\|L - (a+x) \otimes (b+y)\|_* < \varepsilon$;
(d) $\|x \otimes h_j \chi_\mathbb{D}\|_* < \gamma_j$ for $1 \leq j \leq m$;
(e) $|a+x| \geq (1-2\delta)|a|$ a.e. $[|u|^2\mu]$ on $\partial\mathbb{D}$.

PROOF. Let $\nu = |u|^2\mu$ and put $\mathcal{H}_u =$ the closed linear span of $\{\phi u : \phi \in L^\infty(\mu)\}$. The map $f \to fu$ is an isomorphism of $L^2(\nu)$ onto $\mathcal{H}_u$ that takes $P^2(\nu)$ onto $\mathcal{H}_u$. Define the function a_1 to be a/u where $u \neq 0$ and $a_1 = 0$ otherwise; similarly define $b_1 = b/u$. So $a_1 \in L^2(\nu)$ and $b_1 \in L^2(\nu|A_r)$. Let L_1 be the functional on $P^\infty(\nu)$ corresponding to L. There is a subtlety here in that $a_1 \otimes b_1$ as a functional on $P^\infty(\nu) \cong P^\infty(S_\nu)$ is defined using the inner product of $L^2(\nu)$. Thus for ϕ in $P^\infty(\nu)$, $\langle a_1 \otimes b_1, \phi \rangle = \int \phi a_1 \bar{b}_1 d\nu = \int \phi a_1 \bar{b}_1 d\mu = L(\phi)$. Thus $L_1 = a_1 \otimes b_1$. Now apply Lemma 5.7 to obtain functions x_1 in $P^2(\nu)$ and y_1 in $L^2(\nu|A_r)$ and let $x = ux_1$ and $y = uy_1$. The execution of the details is left to the reader. □

Exercises

1. Prove Theorem 5.1 for normal operators with $C = 1$.
2. Give an easy proof of the version of Theorem 5.1 that requires only that x and y lie in $\mathcal{H}$.
3. Give an example to show that the constant C in Theorem 5.1 cannot be taken to be 1.
4. Show that if $L \in P^\infty(m)_*$, there is a function h in $L^1(m)$ with $\|h\|_1 = \|L\|_*$ and $\|h - L\|_* = 0$.
5. Give an example of a compactly supported measure μ on $\mathbb{C}$ such that there is a weak* continuous linear functional L on $P^\infty(\mu)$ with $\|L\|_* < \|h\|_1$ for every function h in $L^1(\mu)$ for which $\|h - L\|_* = 0$.
6. In Lemma 5.6, show that if it is also assumed that S is pure, then we can get $\|x \otimes h_j\|_* < \eta_j$ for $1 \le j \le n$. (*Hint*: If $h \in L^2(\mu)$ and $\eta > 0$ such that $\|x_n \otimes h\|_* \ge \eta$ for all $n \ge 1$, let $\phi_n \in$ ball $P^\infty(\mu)$ such that $|x_n \otimes h(\phi_n)| \ge \eta$, and let w be a weak cluster point of $\{\phi_n x_n\}$ in $P^2(\mu)$. Show that $w = 0$ a.e. $[\mu]$ off $\partial \mathbb{D}$ but $w \ne 0$. Now get a contradiction.) This assumption of purity can be used to simplify the proof of all the succeeding lemmas and theorems in this section and the next.

§6 An infinite factorization theorem. Here we will prove a theorem on factorization of an infinite sequence of functionals in $P^\infty(\mu)$. This will then be applied in the following sections to derive various properties of subnormal operators.

6.1 THEOREM. *If S is a subnormal operator acting on a Hilbert space $\mathcal{H}$ and $\{L_n\}_1^\infty$ is a sequence of weak* continuous linear functionals on $P^\infty(S)$, then there is a vector x in $\mathcal{H}$ and a sequence of vectors $\{y_n\}$ in $\mathcal{H}$ such that for each $n \ge 1$, $L_n = x \otimes y_n$.*

This result is similar to Theorem 5.1 though there are distinct differences. Of course the obvious difference is the fact that this theorem factors a sequence of weak* continuous linear functionals rather than a single one. But also no control of the norms of the vectors is imposed.

The proof begins with a lemma whose conclusion is stronger than required for the immediate purpose.

6.2 LEMMA. *Assume (5.3), let $0 < \delta < \frac{1}{3}$, let $a \in L^2(\mu)$, and let u be an ample function. If for some positive integer n, $0 < r_k < 1$, $b_k \in L^2(\mu|A_{r_k})$, and $L_k \in P^\infty(\mu)_*$ for $1 \le k \le n$ such that $\|L_k - a \otimes b_k\|_* < \delta^4$, then there is*

§6 AN INFINITE FACTORIZATION THEOREM

a function x in $\mathcal{H}_u$ and functions y_k in $L^2(\mu|A_{r_k})$ such that for $1 \leq k \leq n$ the following conditions are satisfied.
  (a) $\|x\| < 3n\delta$.
  (b) $\|y_k\| < \delta^2(1-2\delta)^{-n+k} + (1-2\delta)^{-n}\|b_k\|$.
  (c) $\|L_k - (a+x) \otimes y_k\|_* < \varepsilon$.

PROOF. Let $\mu_k = \mu|A_{r_k}$ and apply Lemma 5.11 with $L = L_1$ and $b = b_1$ to obtain a function x_1 in $\mathcal{H}_u$ and a function y_1 in $L^2(\mu_1)$ such that:
$$\|x_1\| < 3\delta;$$
$$\|y_1\| < \delta^2 + (1-2\delta)^{-1}\|b_1\|;$$
$$\|L_1 - (a+x_1) \otimes y_1\|_* < \varepsilon/n;$$
$$\|x_1 \otimes b_k \chi_{\mathbb{D}}\|_* < \delta^4 - \|L_k - a \otimes b_k\|_* \text{ for } 1 \leq k \leq n;$$
$$|a + x_1| > (1-2\delta)|a| \text{ a.e. } [|u|^2\mu] \text{ on } \partial\mathbb{D}.$$

Define $w_1 = (a/a+x_1)^*$ on $\partial\mathbb{D}$. (The $*$ is used here to denote complex conjugation; also interpret $0/0$ to be 0 in this proof.) If $\zeta \in \partial\mathbb{D}$ and $u(\zeta) = 0$, then $x_1(\zeta) = 0$ and so $|w_1(\zeta)| \leq 1$. If $u(\zeta) \neq 0$ and $(a+x_1)(\zeta) \neq 0$, then $|w_1(\zeta)| < (1-2\delta)^{-1}$. Thus $|w_1| \leq (1-2\delta)^{-1}$ a.e. $[\mu]$ on $\partial\mathbb{D}$.

Note that
$$\delta^4 > \|L_k - a \otimes b_k\|_* + \|x_1 \otimes b_k \chi_{\mathbb{D}}\|_*$$
$$\geq \|L_k - a \otimes b_k - x_1 \otimes b_k \chi_{\mathbb{D}}\|_*$$
$$= \|L_k - (a+x_1) \otimes b_k[\chi_{\mathbb{D}} + w_1 \chi_{\partial\mathbb{D}}]\|_*.$$

Since $b_k \in L^2(\mu_k)$ and $w_1 \chi_{\partial\mathbb{D}}$ is bounded, $b_k[\chi_{\mathbb{D}} + w_1 \chi_{\partial\mathbb{D}}] \in L^2(\mu_k)$ and $\|b_k[\chi_{\mathbb{D}} + w_1 \chi_{\partial\mathbb{D}}]\| \leq (1-2\delta)^{-1}\|b_k\|$.

Now apply Lemma 5.11 with $L = L_2$, a replaced by $a + x_1$, and $b = b_2(\chi_{\mathbb{D}} + w_1 \chi_{\partial\mathbb{D}})$ to get a function x_2 in $\mathcal{H}_u$ and a function y_2 in $L^2(\mu_2)$ such that:
$$\|x_2\| < 3\delta;$$
$$\|y_2\| < \delta^2 + (1-2\delta)^{-2}\|b_2\|;$$
$$\|L_2 - (a+x) \otimes y_2\|_* < \varepsilon/n;$$
$$\|x_2 \otimes y_1 \chi_{\mathbb{D}}\|_* < \varepsilon/n;$$
$$\|x_2 \otimes b_k \chi_{\mathbb{D}}\|_* < \delta^4 - \|L_k - (a+x_1) \otimes b_k[\chi_{\mathbb{D}} + w_1 \chi_{\partial\mathbb{D}}]\|_* \text{ for } 2 \leq k \leq n;$$
$$|a + x_1 + x_2| \geq |a + x_1| \text{ a.e. } [|u|^2\mu] \text{ on } \partial\mathbb{D}.$$

Define $w_2 = [(a+x_1+x_2)/(a+x_1)]^*$ on $\partial\mathbb{D}$. It follows that $|w_2| \leq (1-2\delta)^{-1}$ a.e. $[\mu]$ on $\partial\mathbb{D}$.

Note that for $2 \leq k \leq n$
$$\delta^4 > \|L_k - (a+x_1) \otimes b_k[\chi_{\mathbb{D}} + w_1 \chi_{\partial\mathbb{D}}] - x_2 \otimes b_k \chi_{\mathbb{D}}\|_*$$
$$= \|L_k - (a+x_1+x_2) \otimes b_k[\chi_{\mathbb{D}} + w_1 w_2 \chi_{\partial\mathbb{D}}]\|_*.$$

Again $b_k[\chi_{\mathbb{D}} + w_1 w_2 \chi_{\partial\mathbb{D}}] \in L^2(\mu_2)$ and $\|b_k[\chi_{\mathbb{D}} + w_1 w_2 \chi_{\partial\mathbb{D}}]\| \leq (1-2\delta)^{-2}\|b_k\|$.

Continuing this line of reasoning, for $1 \leq k \leq n-1$ we get functions x_k in $\mathscr{H}_u$ and functions y_k in $L^2(\mu_k)$ such that:

$\|x_k\| < 3\delta$;
$\|y_k\| < \delta^2 + (1-2\delta)^{-k}\|b_k\|$;
$\|L_k - (a + \sum_{j=1}^{k} x_j) \otimes y_k\|_* < \varepsilon/n$;
$\|x_k \otimes y_j \chi_{\mathbb{D}}\|_* < \varepsilon/n$ for $1 \leq j < k$;
$\|x_k \otimes b_i \chi_{\mathbb{D}}\|_* < \delta^4 - \|L_i - (a + \sum_{j=1}^{k} x_j) \otimes b_i(\chi_{\mathbb{D}} + w_1 \cdots w_{k-1})\chi_{\partial\mathbb{D}}\|_*$
for $k < i \leq n$;
$|a + \sum_{j=1}^{k} x_j| \geq (1-2\delta)|a + \sum_{j=1}^{k-1} x_j|$ a.e. $[|u|^2\mu]$ on $\partial\mathbb{D}$.

Define $w_k = [(a + \sum_{j=1}^{k-1} x_j)/(a + \sum_{j=1}^{k} x_j)]^*$ on $\partial\mathbb{D}$ so that $|w_k| \leq (1-2\delta)^{-1}$ a.e. $[\mu]$ on $\partial\mathbb{D}$.

To choose x_n and y_n apply Lemma 5.11 with $L = L_n$, a replaced by $a + \sum_{k=1}^{n-1} x_k$, and $b = b_n(\chi_{\mathbb{D}} + w_1 \cdots w_{n-1})\chi_{\partial\mathbb{D}}$ to get x_n in $\mathscr{H}_u$ and y_n in $L^2(\mu_n)$ such that:

$\|x_n\| < 3\delta$;
$\|y_n\| < \delta^2 + (1-2\delta)^{-n}\|b_n\|$;
$\|L_n - (a + \sum_{k=1}^{n} x_k) \otimes y_n\|_* < \varepsilon/n$;
$\|x_n \otimes y_j \chi_{\mathbb{D}}\|_* < \varepsilon/n$ for $1 \leq j < n$;
$|\sum_{k=1}^{n} x_k| \geq (1-2\delta)|a + \sum_{k=1}^{n-1} x_k|$ a.e. $[|u|^2\mu]$ on $\partial\mathbb{D}$.

Define $w_n = [(a + \sum_{k=1}^{n-1} x_k)/(\sum_{k=1}^{n} x_k)]^*$ on $\partial\mathbb{D}$ so that $|w_n| \leq (1-2\delta)^{-1}$ a.e. $[\mu]$ on $\partial\mathbb{D}$.

Put $x = \sum_{k=1}^{n} x_k$. So $x \in \mathscr{H}_u$ and $\|x\| < 3n\delta$. Put $h_n = y_n$ and for $1 \leq k < n$, let

$$h_k = y_k[\chi_{\mathbb{D}} + w_{k+1} \cdots w_n \chi_{\partial\mathbb{D}}].$$

Thus for $1 \leq k < n$

$$\|h_k\| < \left(\frac{1}{1-2\delta}\right)^{n-k} \|y_k\|$$
$$< \left(\frac{1}{1-2\delta}\right)^{n-k} \left(\delta^2 + \left(\frac{1}{1-2\delta}\right)^k \|b_k\|\right)$$
$$= \delta^2 \left(\frac{1}{1-2\delta}\right)^{n-k} + \left(\frac{1}{1-2\delta}\right)^n \|b_k\|.$$

This establishes (a) and (b).

For $1 \leq k < n$

$$\|L_k - (a+x) \otimes h_k\|_*$$
$$= \left\|L_k - \left(a + \sum_{j=1}^{n} x_j\right) \otimes y_k \left[\chi_{\mathbb{D}} + \left(\frac{a + \sum_{j=1}^{k} x_j}{a + \sum_{j=1}^{n} x_j}\right)^* \chi_{\partial\mathbb{D}}\right]\right\|_*$$

$$= \left\| L_k - \left(a + \sum_{j=1}^n x_j\right) \otimes y_k \chi_{\mathbb{D}} \right.$$
$$- \left(a + \sum_{j=1}^n x_j\right) \otimes y_k \left(\frac{a + \sum_{j=1}^k x_j}{a + \sum_{j=1}^n x_j}\right)^* \chi_{\partial \mathbb{D}} \Bigg\|_*$$
$$= \left\| L_k - \left(a + \sum_{j=1}^k x_j\right) \otimes y_k \chi_{\mathbb{D}} - \left(\sum_{j=k+1}^n x_j\right) \otimes y_k \chi_{\mathbb{D}} \right.$$
$$- \left(a + \sum_{j=1}^k x_j\right) \otimes y_k \chi_{\partial \mathbb{D}} \Bigg\|_*$$
$$\leq \left\| L_k - \left(a + \sum_{j=1}^k x_j\right) \otimes y_k \right\|_* + \sum_{j=k+1}^n \| x_j \otimes y_k \chi_{\mathbb{D}} \|_*$$
$$< \frac{\varepsilon}{n} + (n-k)\frac{\varepsilon}{n}$$
$$= \varepsilon$$

by the choices that were made. This establishes (c). □

This next lemma takes care of the crucial special case of the theorem with a little extra for future use.

6.3 LEMMA. *Assume* (5.3). *If* $\{L_n\}_1^\infty$ *is a sequence in* $P^\infty(\mu)_*$, $a \in L^2(\mu)$, $\{r_n\}$ *is a sequence of numbers in the open unit interval, and* u *is an ample function, then for every* $\varepsilon > 0$ *there is a function* x *in* $\mathscr{H}_u$ *with* $\|x\| < \varepsilon$ *and there are functions* h_n *in* $L^2(\mu | A_{r_n})$ *such that for all* $n \geq 1$,
$$L_n = (a + x) \otimes h_n.$$

PROOF. Let $\{\varepsilon_n\}_0^\infty$ be a decreasing sequence of positive numbers all of which satisfy $\varepsilon_n < \frac{1}{3}$ and such that we also have that
$$\sum_{n=1}^\infty n\varepsilon_n \quad \text{and} \quad \prod_{n=0}^\infty \left(\frac{1}{1-2\varepsilon_n}\right)^{n+1}$$
converge. Also assume that $\varepsilon > 3\varepsilon_n + \sum_n n\varepsilon_n$. Without loss of generality (since there is no requirement on the norms of the functions h_n) we may assume that $\|L_n\| < \varepsilon_{n-1}^4$ for all $n \geq 1$.

We will prove the following induction statement. For each $n \geq 0$ there is a function g_n in $\mathscr{H}_u$ and functions h_{nk} in $L^2(\mu|A_{r_k})$ such that $g_0 = h_{0k} = 0$ for all $k \geq 0$ and for $1 \leq k \leq n$,

$(6.4)_n$ $$\|g_n - g_{n-1}\| < 3n\varepsilon_{n-1};$$

$(6.5)_n$ $$\|h_{nk}\| < \left[\sum_{p=k-1}^{n-1} \varepsilon_p^2\right] \prod_{j=k}^{n-1}\left(\frac{1}{1-2\varepsilon_j}\right)^{j+1};$$

(if $k = n$, then this product is to be interpreted as being equal to 1.)

$(6.6)_n$ $\qquad \|L_k - (a + x_n) \otimes h_{nk}\|_* < \varepsilon_n^4;$

For $n = 1$, apply Lemma 5.11 with $L = L_1$, $a = a$, $b = 0$, and $\delta = \varepsilon_0$ to get a function g_1 in $\mathscr{H}_u$ and a function h_{11} in $L^2(\mu|A_{r_1})$ such that $\|g_1\| < 3\varepsilon_0$, $\|h_{11}\| < \varepsilon_0^2$, and $\|L_1 - (a + g_1) \otimes h_{11}\|_* < \varepsilon_1^4$. With these choices, the induction statement holds for $n = 1$.

Now assume that $(6.4)_k$, $(6.5)_k$, and $(6.6)_k$ hold for $1 \leq k \leq n-1$. Apply Lemma 6.2 with a replaced by $a + g_{n-1}$, $\delta = \varepsilon_{n-1}$, and, for $1 \leq k \leq n-1$, $b_k = h_{n-1,k}$ to get a function x_n in $\mathscr{H}_u$ and functions y_k in $L^2(\mu_k)$ such that:

$\|x_n\| < 3\varepsilon_{n-1};$
$\|y_k\| < \varepsilon_{n-1}^2 (1 - 2\varepsilon_{n-1})^{-n+k} + (1 - 2\varepsilon_{n-1})^{-n} \|h_{n-1,k}\|$ for $1 \leq k \leq n-1;$
$\|y_n\| < \varepsilon_{n-1}^2;$
$\|L_k - (a + g_{n-1} + x_n) \otimes y_k\|_* < \varepsilon_{n+1}^4.$

Let $g_n = g_{n-1} + x_n$ and $h_{n,k} = y_k$. Clearly $(6.4)_n$ and $(6.6)_n$ hold. To establish $(6.5)_n$, fix k, $1 \leq k \leq n-1$, and using the estimate on $\|y_k\|$ as well as $(6.5)_{n-1}$ we get

$$\|h_{nk}\| \leq \varepsilon_{n-1}^2 \left(\frac{1}{1 - 2\varepsilon_{n-1}}\right)^{n-k} + \left(\frac{1}{1 - 2\varepsilon_{n-1}}\right)^n \|h_{n-1,k}\|$$

$$\leq \varepsilon_{n-1}^2 \left(\frac{1}{1 - 2\varepsilon_{n-1}}\right)^{n-k} + \left(\frac{1}{1 - 2\varepsilon_{n-1}}\right)^n \left[\sum_{p=k-1}^{n-2} \varepsilon_p^2\right] \prod_{j=k}^{n-2} \left(\frac{1}{1 - 2\varepsilon_j}\right)^{j+1}$$

$$= \varepsilon_{n-1}^2 \left(\frac{1}{1 - 2\varepsilon_{n-1}}\right)^{n-k} + \left[\sum_{p=k-1}^{n-2} \varepsilon_p^2\right] \prod_{j=k}^{n-1} \left(\frac{1}{1 - 2\varepsilon_j}\right)^{j+1}$$

$$\leq \varepsilon_{n-1}^2 \left(\frac{1}{1 - 2\varepsilon_{n-1}}\right)^n + \left[\sum_{p=k-1}^{n-2} \varepsilon_p^2\right] \prod_{j=k}^{n-1} \left(\frac{1}{1 - 2\varepsilon_j}\right)^{j+1}$$

$$\leq \varepsilon_{n-1}^2 \prod_{j=k}^{n-1} \left(\frac{1}{1 - 2\varepsilon_j}\right)^{j+1} + \left[\sum_{p=k-1}^{n-2} \varepsilon_p^2\right] \prod_{j=k}^{n-1} \left(\frac{1}{1 - 2\varepsilon_j}\right)^{j+1}$$

$$= \left[\sum_{p=k-1}^{n-1} \varepsilon_p^2\right] \prod_{j=k}^{n-1} \left(\frac{1}{1 - 2\varepsilon_j}\right)^{j+1}.$$

Also $\|h_{nn}\| = \|y_n\| < \varepsilon_{n-1}^2$ so that $(6.5)_n$ holds. This completes the induction argument.

From the conditions on the sequence $\{g_n\}$ and the constants ε_n we have that $\{g_n\}$ is a Cauchy sequence in $\mathscr{H}_u$ and so there is a function x in $\mathscr{H}_u$

such that $g_n \to x$. Moreover, $\|x\| \le \|g_1\| + \sum \|g_{n+1} - g_n\| < \varepsilon$. Also we have that for k a fixed position integer, $\{h_{nk}\}_n$ is a bounded sequence in $L^2(\mu_k)$. Hence there is a subsequence $\{h_{n_ik}\}$ that converges weakly to a function h_k in $L^2(\mu_k)$. But $(6.6)_n$ implies that $\|L_k - (a + g_{n_i}) \otimes h_{n_ik}\|_* \to 0$ as $n_i \to \infty$. By Lemma 5.10, $L_k = (a + x) \otimes h_k$. □

PROOF OF THEOREM 6.1.

CASE 1. Assume (5.3). Here $\mathscr{H} = P^2(\mu)$ and so Lemma 6.3 with $u = 1$ and $a = 0$ gives a function x in $P^2(\mu)$ and functions h_n in $L^2(\mu_n)$ such that $L_n = x \otimes h_n$. Let $y_n = Ph_n$, where P is the orthogonal projection of $L^2(\mu)$ onto $P^2(\mu)$. It is easy to check that $L = x \otimes y_n$.

We have the following cases.

CASE 2. S is an arbitrary subnormal operator but $P^\infty(\mu) = H^\infty(\mathbb{D})$.

CASE 3. S is an arbitrary subnormal operator with any scalar-valued spectral measure but assume that $P^\infty(\mu)$ is antisymmetric.

CASE 4. The general case.

The proof of the theorem for each of these cases follows from the preceding case by using arguments like those used to prove the corresponding cases in Theorem 5.1. The reader should supply the details. □

§7 Full analytic subspaces. Recall the definition and properties of bounded point evaluations (see §II.7). We want to examine this concept for cyclic invariant subspaces of a subnormal operator, though for what is planned it is probably best phrased for normal operators. So let N be a normal operator on $\mathscr{H}$ with scalar-valued spectral measure μ. If $x \in \mathscr{H}$, let $\mathscr{H}_x \equiv \bigvee\{N^n x : n \ge 0\}$; so $\mathscr{H}_x$ is the smallest invariant subspace for N that contains x and $S_x = N|\mathscr{H}_x$ is a cyclic subnormal operator. As such S_x is unitarily equivalent to multiplication by z on $P^2(\mu_x)$ for some measure μ_x. Clearly $\mu_x \ll \mu$, but what else can be said? The first proposition details this representation. The proof is left to the reader.

7.1 PROPOSITION. *Let N be a normal operator on a Hilbert space $\mathscr{H}$ and let $N = \int z\,dE$ be the spectral representation of N. If $x \in \mathscr{H}$ and $\mu_x(\Delta) = \langle E(\Delta)x, x \rangle$ for all Borel sets Δ, then the map $Wp(N)x = p$ extends to an isomorphism $W: \mathscr{H}_x \to P^2(\mu_x)$ such that $WS_x = S_{\mu_x}W$ and $Wx = 1$.*

By virtue of Proposition 7.1 we can speak of analytic bounded point evaluations for $\mathscr{H}_x$ for any x in $\mathscr{H}$; viz., let abpe$(\mathscr{H}_x)$ = abpe(μ_x). If $\lambda \in$ abpe$(\mathscr{H}_x)$, then there is a unique vector k_λ in $\mathscr{H}_x$, called the reproducing kernel, such that $p(\lambda) = \langle p(N)x, k_\lambda \rangle$ for every polynomial p. Maintaining the notation of the preceding proposition and identifying k_λ with its corresponding function in $P^2(\mu_x)$, this means, as in §II.7, that $p(\lambda) = \int p\bar{k}_\lambda\,d\mu_x$ for polynomials p.

Now assume that there is a bounded point evaluation λ for $P^2(\mu_x)$. Since $P^2(\mu_x) \subseteq L^1(\mu_x)$, we get that $\phi \to \int \phi \bar{k}_\lambda\,d\mu_x$ is a weak* continuous linear functional on $P^\infty(\mu_x)$. It is not hard to check that this functional is, in fact, multiplicative. Therefore λ must belong to the interior of the Sarason hull of μ_x, $\Sigma(\mu_x)$ (VI.7.14). Since $\mu_x \ll \mu$, $\lambda \in \text{int}\,\Sigma(\mu)$. We have just seen that for every x in $\mathscr{H}$, $\text{abpe}(\mathscr{H}_x) \subseteq \text{int}\,\Sigma(\mu)$.

7.2 DEFINITION. If N is a normal operator on $\mathscr{H}$ with scalar-valued spectral measure μ and $x \in \mathscr{H}$, then the subspace $\mathscr{H}_x$ is called a *full analytic subspace* if $\text{abpe}(\mathscr{H}_x) = \text{int}\,\Sigma(\mu)$. If it is desirable to emphasize the roll of the scalar-valued spectral measure μ, we will say that $\mathscr{H}_x$ is a μ-full analytic subspace. If $\mathscr{H}_x = P^2(\mu)$ and $N = N_\mu$, we say that $P^2(\mu)$ is a full analytic subspace if $\text{abpe}(\mu) = \text{int}\,\Sigma(\mu)$.

Note that by Theorem II.7.7, $\mathscr{H}_x$ is a full analytic subspace if and only if $\sigma(S_x) \setminus \sigma_{ap}(S_x) = \text{int}\,\Sigma(\mu)$.

So full analytic subspaces have as many analytic bounded point evaluations as is possible. If you wish, full analytic subspaces are those with as much analytic structure as possible.

It is a result due to Olin and Thomson [**1980b**] that full analytic subspaces exist in abundance. Before stating and proving this theorem, let's examine a few examples and properties of full analytic subspaces.

7.3 EXAMPLE. There is a measure μ supported by $\text{cl}\,\mathbb{D}$ such that $\mu|\mathbb{D}$ is area measure on $\mathbb{D}$ and $[\mu|\partial\mathbb{D}] = [m]$ (as usual, m denotes normalized arc length measure on $\partial\mathbb{D}$) and such that $P^2(\mu) = L_a^2(\mathbb{D}) \oplus L^2(\mu|\partial\mathbb{D})$. For the details of this construction (due to A. L. Volberg), see Havin, Hruscev, and Nikolskii [**1984**], page 442. Notice that $P^\infty(\mu) = H^\infty(\mathbb{D})$ so that $\text{int}\,\Sigma(\mu) = \mathbb{D}$ and $P^2(\mu)$ is itself a full analytic subspace even though $S = S_\mu$ is not a pure subnormal operator. This example also illustrates the fact that the reproducing kernels k_λ need not span $P^2(\mu)$. The reader interested in the splitting phenomenon portrayed in this example should also consult Kriete [**1979**] and Kriete and Trent [**1977**].

The preceding example is, to an extent, typical of full analytic subspaces for which the evaluation kernels do not span. We'll see this in a moment, but observe that there is no loss of generality in studying the problem of full analytic subspaces to assume that $P^\infty(\mu)$ has no L^∞-summand. Indeed, suppose $P^\infty(\mu) = L^\infty(\mu_0) \oplus P^\infty(\mu_1)$ with $P^\infty(\mu_1) = H^\infty(\text{int}\,\Sigma(\mu))$ and let $\mathscr{H} = \mathscr{H}_0 \oplus \mathscr{H}_1$ be the corresponding decomposition of $\mathscr{H}$. If $\mathscr{H}_x$ is a full analytic subspace and $x = x_0 \oplus x_1$, then $\mathscr{H}_x = \mathscr{H}_{x_0} \oplus \mathscr{H}_{x_1}$, $\mathscr{H}_{x_1}$ is a full analytic subspace, and $S|\mathscr{H}_{x_0}$ is a reductive normal operator.

Recall that if $\lambda \in \text{bpe}(\mu)$ with reproducing kernel k_λ and $f \in P^2(\mu)$, then $\hat{f}(\lambda) = \langle f, k_\lambda \rangle$ equals f a.e. $[\mu]$ on $\text{bpe}(\mu)$ (II.7.3). If $\phi \in L^\infty(\nu) \cap P^2(\mu)$, then $(\widehat{\phi f})(\lambda) = \hat{\phi}(\lambda)\hat{f}(\lambda)$ (II.7.3).

7.4 PROPOSITION. *Assume that ν is a measure with compact support such that $P^\infty(\nu)$ has no L^∞-summand, $P^2(\nu)$ is a full analytic subspace, and for each λ in $\text{int}\,\Sigma(\nu)$, k_λ denotes the reproducing kernel. If $\mathscr{H}_1 =$*

$\bigvee\{k_\lambda : \lambda \in \operatorname{int}\Sigma(\nu)\}$ and $\mathscr{H}_0 = \mathscr{H}_1^\perp$, then $\mathscr{H}_0$ and $\mathscr{H}_1$ reduce $S = S_\nu$, $\mathscr{H}_1$ is a full analytic subspace, and $S|\mathscr{H}_0$ is a cyclic normal operator with scalar-valued spectral measure ν_0 absolutely continuous with respect to harmonic measure for $\Sigma(\nu)$. Moreover, if $\nu_1 = \nu - \nu_0$, then $\nu_0 \perp \nu_1$, $\mathscr{H}_1 = P^2(\nu_1)$, and $\mathscr{H}_0 = L^2(\nu_0)$.

PROOF. Since $S^* k_\lambda = \bar\lambda k_\lambda$ for all λ in $G \equiv \operatorname{int}\Sigma(\nu)$, $\mathscr{H}_1$ is invariant for S^* and so $\mathscr{H}_0 \in \operatorname{Lat} S$. Let $f \in \mathscr{H}_0$ and let $\{p_n\}$ be a sequence of polynomials such that $\int |p_n - f|^2 d\nu \to 0$. Since $f \in \mathscr{H}_0$, $\hat{f}(\lambda) = 0$ for all λ in G and hence $f = 0$ a.e. $[\nu]$ on G. Also, $p_n(\lambda) \to 0$ uniformly on compact subsets of G. Fix a λ in G and let q_n be the polynomial such that $p_n - p_n(\lambda) = (z - \lambda)q_n$. Let $r > 0$ such that $\overline{B} = \overline{B}(\lambda; r) \subseteq G$. If $|z - \lambda| = r$, then

$$|q_n(z)| = \left|\frac{p_n(z) - p_n(\lambda)}{z - \lambda}\right| \le \frac{1}{r}[\|p_n\|_{\overline{B}} + |p_n(\lambda)|].$$

By the Maximum Modulus Theorem, this inequality holds throughout $\overline{B}$. Thus $q_n \to 0$ uniformly on $\overline{B}$. On the other hand,

$$\limsup_n \int_{G\setminus \overline{B}} |q_n|^2 d\nu \le \frac{1}{r^2} \limsup_n \int_G |p_n - p_n(\lambda)|^2 d\nu$$
$$= \int_G |f|^2 d\nu$$
$$= 0.$$

In addition, since $\lambda \notin \partial G$, $q_n \to (z-\lambda)^{-1} f$ in $L^2(\nu|\partial G)$. Combining these facts, we get that $q_n \to (z-\lambda)^{-1} f$ in $L^2(\nu)$. Hence $g \equiv (z-\lambda)^{-1} f \in P^2(\nu)$.

If $\zeta \in G$, $\zeta \ne \lambda$, then $0 = \hat{f}(\zeta) = (\zeta - \lambda)\hat{g}(\zeta)$, so that $\hat{g} = 0$ on $G\setminus\{\lambda\}$. But $\hat{g}$ is analytic so that $\hat{g} = 0$ on G; that is, $g \perp k_\zeta$ for all ζ in G so that $g \in \mathscr{H}_0$. This shows that $\sigma(S|\mathscr{H}_0) \subseteq \partial G = \partial \Sigma(\nu)$. But $R(\Sigma(\nu))$ is a Dirichlet algebra and so $R(\partial G) = C(\partial G)$ (V.14.8). By Theorem II.9.9, $S|\mathscr{H}_0$ is normal and so $\mathscr{H}_0$ reduces S. It follows that $\mathscr{H}_1$ reduces S.

Since S has a cyclic minimal normal extension, $N_0 \equiv S|\mathscr{H}_0$ is a cyclic normal operator. Since $\sigma(N_0) \subseteq \partial G$ and $P^\infty(\nu)$ has no L^∞-summand, ν_0 is absolutely continuous with respect to harmonic measure for G. It is easy to check that $\mathscr{H}_1 = P^2(\nu_1)$. If $S_1 = S|\mathscr{H}_1$, then $S = S_1 \oplus N_0$. □

Note that if $\mathscr{H}_x$ is a μ-full analytic subspace and S_x is pure, then $P^\infty(\mu_x) \cong H^\infty(\operatorname{int}\Sigma(\mu)) \cong P^\infty(\mu)$ though there is nothing to imply that μ is a scalar-valued spectral measure for $N|\mathscr{H}_x$. The positive part of this comment is summarized below.

7.5 PROPOSITION. *If N is a normal operator on $\mathscr{H}$ with scalar-valued spectral measure μ, $x \in \mathscr{H}$ such that $\mathscr{H}_x$ is a full analytic subspace, and μ_x is the scalar-valued spectral measure for $N|\mathscr{H}_x$, then $\Sigma(\mu_x) = \Sigma(\mu)$.*

PROOF. Since $\mu_x \ll \mu$ it is easy to see that $\Sigma(\mu_x) \subseteq \Sigma(\mu)$. On the other hand, since $\mathscr{H}_x$ is a full analytic subspace, $\operatorname{int}\Sigma(\mu) = \operatorname{abpe}(\mu_x) \subseteq \Sigma(\mu_x)$. □

We now prove the existence of full analytic subspaces. First we will attack the case of a cyclic operator. This result will be parlayed to the general case but it also has applications later.

7.6 THEOREM (Olin and Thomson [1980b]). *Assume that μ is a measure such that $P^\infty(\mu)$ has no L^∞-summand. If $a \in L^2(\mu)$ and $\varepsilon > 0$, then there is an x in $P^2(\mu)$ with $\|x\| < \varepsilon$ such that $\mathscr{H}_{a+x} \equiv \bigvee\{z^n(a+x) : n \geq 0\}$ is a full analytic subspace.*

PROOF. We consider three cases.

CASE 1. Assume that $P^\infty(\mu) = H^\infty(\mathbb{D})$. Let $e_0 \colon P^\infty(\mu) \to \mathbb{C}$ be defined by $e_0(\phi) = \phi(0)$ for all ϕ in $P^\infty(\mu)$ and apply Lemma 6.2 with $u = 1$, $L_n = e_0$ for all n, and with the r_n chosen such that they increase to 1. Thus there is a function x in $P^2(\mu)$ with $\|x\| < \varepsilon$ and there are functions h_n in $L^2(\mu|A_{r_n})$ such that for all $n \geq 1$ and $e_0 = (a+x) \otimes h_n$.

Put $\nu = |a+x|^2\mu$ and $\mathscr{H} = \mathscr{H}_{a+x}$; by Proposition 7.1, $N|\mathscr{H} \cong S_\nu$. It must be shown that $\mathrm{abpe}(\nu) = \mathbb{D}$. If $\nu_n = \nu|A_{r_n}$, then for any polynomial p,

$$|p(0)| = \left|\int_{A_{r_n}} p(a+x)\bar{h}_n\, d\mu\right|$$

$$\leq \|h_n\| \left[\int |p|^2\, d\nu_n\right]^{\frac{1}{2}}.$$

Thus $0 \in \mathrm{bpe}(\nu_n)$ and so $0 \in \sigma(S_{\nu_n})$. But $\varnothing = B(0; r_n) \cap \mathrm{supp}\, \nu_n = B(0; r_n) \cap \sigma(N_{\nu_n})$ so that by Theorem II.2.11, $B(0; r_n) \subseteq \sigma(S_{\nu_n})$. But, for the same reason, $B(0; r_n) \cap \sigma_{\mathrm{ap}}(S_{\nu_n}) = \varnothing$. Hence Theorem II.7.7 implies that $B(0; r_n) \subseteq \mathrm{abpe}(\nu_n)$.

Now if K is any compact subset of $B(0; r_n)$, there is a constant $C > 0$ such that $|p(\lambda)| \leq C[\int |p|^2\, d\nu_n]^{\frac{1}{2}}$ for all λ in K and all polynomials p. Since $\nu_n \leq \nu$, Proposition II.7.6 implies that $K \subseteq \mathrm{abpe}(\nu)$. Thus $B(0; r_n) \subseteq \mathrm{abpe}(\nu)$. Since r_n can be chosen arbitrarily close to 1, we get that $\mathbb{D} \subseteq \mathrm{abpe}(\nu)$.

CASE 2. Assume that $P^\infty(\mu)$ is antisymmetric. Adopt the notation of Proposition 1.7. As in the proof of part (e) of that result, if $W \colon L^2(\nu) \to L^2(\mu)$ is defined by $Wf = f \circ \tau$, then W is an isomorphism and $WP^2(\nu) = P^2(\mu)$. Let $b \in L^2(\nu)$ such that $a = Wb = b \circ \tau$.

From Case 1 we know there is a function y in $P^2(\nu)$ with $\|y\| < \varepsilon$ such that $\mathscr{L}_{b+y} \equiv \bigvee\{z^n(b+y) : n \geq 0\} = \bigvee\{(b+y)\phi : \phi \in H^\infty(\mathbb{D})\}$ is a full analytic subspace of $L^2(\nu)$.

Put $x = Wy = y \circ \tau$. So $\|x\| < \varepsilon$. It is routine to check that $W\mathscr{L}_{b+y} = \mathscr{H}_{a+x}$ and so $\mathscr{H}_{a+x}$ is a full analytic subspace.

CASE 3. The general case. As in Corollary 1.6, let $\mu = \sum_n \mu_n$ so that $P^\infty(\mu) = \oplus_n P^\infty(\mu_n)$ and each $P^\infty(\mu_n)$ is antisymmetric and let $\{G_n\}$ be the corresponding components of $G = \operatorname{int} \Sigma(\mu)$. It follows that $P^2(\mu) = \bigoplus_n P^2(\mu_n)$. Let a_n be the projection of a into $L^2(\mu_n)$. By Case 2, for each $n \geq 1$ there is a vector x_n with $\|x_n\| < \varepsilon/2^n$ such that $\mathscr{H}_n \equiv \bigvee \{z^n(a_n + x_n) : n \geq 1\}$ is a full analytic subspace. If $x = \bigoplus_n x_n$, the reader can verify that the conditions of the theorem are satisfied. □

7.7 THEOREM (Olin and Thomson [**1980b**]). *Assume S is a subnormal operator on $\mathscr{H}$ with scalar-valued spectral measure μ and that $P^\infty(\mu)$ has no L^∞-summand. If $h \in \mathscr{H}$ and $\varepsilon > 0$, then there is a vector x in $\mathscr{H}$ with $\|h - x\| < \varepsilon$ such that $\mathscr{H}_x$ is a full analytic subspace.*

PROOF. By Proposition VI.16.4, there is a vector e in $\mathscr{H}$ with $\|e - x\| < \varepsilon/2$ that is separating for the minimal normal extension N of S. Thus there is a scalar-valued spectral measure μ for N such that if $\mathscr{L} = \bigvee \{N^{*n}N^m e : n, m \geq 0\}$, then there is an isomorphism $W : \mathscr{L} \to L^2(\mu)$ with $We = 1$ and $WN_\mu = NW$. It follows that W maps $\mathscr{H}_e$ onto $P^2(\mu)$. According to the preceding theorem there is a function x_1 in $P^2(\mu)$ such that $\|1 - x_1\|_2 < \varepsilon/2$ and $\bigvee\{z^n x_1 : n \geq 0\}$ is a full analytic subspace. Let $x = W^{-1} x_1$. Therefore $x \in \mathscr{H}_e \subseteq \mathscr{H}$, $\|x - h\| < \varepsilon$, and $W\mathscr{H}_x = \bigvee\{z^n x_1 : n \geq 0\}$, so that $\mathscr{H}_x$ is a full analytic subspace. □

Exercises

1. If $P^\infty(\mu)$ is antisymmetric, $\mathscr{H}_x$ is a μ-full analytic subspace of $\mathscr{H}$ such that S_x is pure, and y is a nonzero element of $\mathscr{H}_x$, show that $\mathscr{H}_y$ is a μ-full analytic subspace.
2. If $P^2(\mu)$ is a full analytic subspace and S_μ is pure, show that $P^2(\mu) \cap L^\infty(\mu) = P^\infty(\mu)$.

§8 Reflexivity for subnormal operators. Recall that for any operator A on a Hilbert space $\mathscr{H}$, Lat A is the lattice of all closed subspaces of $\mathscr{H}$ that are invariant for A. For any collection $\mathscr{L}$ of closed subspaces of $\mathscr{H}$, Alg $\mathscr{L}$ denotes the set of all bounded operators T on $\mathscr{H}$ such that $T\mathscr{M} \subseteq \mathscr{M}$ for every $\mathscr{M}$ in $\mathscr{L}$. It is not difficult to see that Alg $\mathscr{L}$ is a subalgebra of $\mathscr{B}(\mathscr{H})$ that is closed in the weak operator topology.

Observe that for any operator A and any polynomial p, $p(A) \in \operatorname{Alg Lat} A$. Thus $W(A)$, the weak operator topology closed algebra generated by A and the identity, is contained in Alg Lat A.

8.1 DEFINITION. An operator A on a Hilbert space $\mathscr{H}$ is *reflexive* if Alg Lat $A = W(A)$.

So reflexive operators have an enormous supply of invariant subspaces.

In light of Theorem 5.2, a subnormal operator S is reflexive if and only if Alg Lat $S = P^\infty(S)$. It is known that every normal operator is reflexive (Sarason [**1966a**]; see also [**ACFA**], page 298). In this section it will be shown that subnormal operators are reflexive. The approach will be to prove this result first for pure subnormal operators and then show that it is true for arbitrary subnormal operators.

8.2 LEMMA. *If $S = S_\mu$ on $P^2(\mu)$, $P^2(\mu)$ is a full analytic subspace, and S is pure, then S is reflexive.*

PROOF. Let $G = \operatorname{int}\Sigma(\mu)$ and for each λ in G let k_λ be the reproducing kernel. If $T \in \operatorname{Alg Lat} S$, then T^* leaves $\ker(S-\lambda)^* = \mathbb{C}k_\lambda$ invariant. Thus for each λ in G, there is a complex number $\phi(\lambda)$ such that $T^*k_\lambda = \overline{\phi(\lambda)}k_\lambda$. Clearly $|\phi(\lambda)| \leq \|T\|$ for all λ in G. If $f \in P^2(\mu)$ and $\lambda \in G$, then $\widehat{(Tf)}(\lambda) = \langle Tf, k_\lambda \rangle = \langle f, T^*k_\lambda \rangle = \langle f, \overline{\phi(\lambda)}k_\lambda \rangle = \phi(\lambda)\hat{f}(\lambda)$. In particular, $\widehat{(T1)}(\lambda) = \phi(\lambda)$. Thus ϕ is a bounded analytic function on G and hence $\phi \in P^\infty(\mu)$. Now $\phi(S) = M_\phi$ is a bounded operator on $P^2(\mu)$ and the preceding calculation shows that for every f in $P^2(\mu)$, $(Tf - \phi f) \perp k_\lambda$ for all λ in G. Since S is pure, Proposition 7.4 implies that $Tf = \phi f$. □

8.3 LEMMA. *If S is a pure cyclic subnormal operator and $T \in \operatorname{Alg Lat} S$, then $TS = ST$.*

PROOF. Let $S = S_\mu$ on $P^2(\mu)$.

CASE 1. Assume that $P^\infty(\mu)$ is antisymmetric. By Theorem 7.6 there is a function x in $P^2(\mu)$ such that $\mathscr{H}_x$ is a μ-full analytic subspace. Once again apply Theorem 7.6 to obtain a function y in $\mathscr{H}_x$ such that $\mathscr{H}_{1+y}$ is a μ_x-full analytic subspace. Note that since $\mathscr{H}_x$ is a μ-full analytic subspace and $\mathscr{H}_{1+y}$ is a μ_x-full analytic subspace, then $\mathscr{H}_{1+y}$ is a μ-full analytic subspace. Also the fact that $\mathscr{H}_x$ is full analytic and $y \in \mathscr{H}_x$ implies that $\mathscr{H}_y$ is also a full analytic subspace (Exercise 7.1).

Clearly $T|\mathscr{H}_{1+y} \in \operatorname{Alg Lat} S_{1+y}$. Thus Lemma 8.2 implies $\ker(ST - TS) \supseteq \mathscr{H}_{1+y}$. Similarly $\ker(ST - TS) \supseteq \mathscr{H}_y$. Therefore for any polynomial p, $(ST - TS)p = (ST - TS)(1+y)p - (ST - TS)yp = 0$. Hence $ST - TS = 0$.

CASE 2. S is any pure cyclic subnormal operator. Decompose S as $S = \bigoplus_1^\infty S_n$, where $P^\infty(S_n)$ is antisymmetric for each n. Since $T \in \operatorname{Alg Lat} S$, $T = \bigoplus_1^\infty T_n$, where $T_n \in \operatorname{Alg Lat} S_n$ for all $n \geq 1$. This case now easily follows from Case 1. □

8.4 LEMMA. *A pure cyclic subnormal operator is reflexive.*

PROOF. Assume that $S = S_\mu$ acting on $P^2(\mu)$ and $G = \operatorname{int}\Sigma(\mu)$; so $P^\infty(\mu) = H^\infty(G)$. According to (IV.5.6) there are mutually singular measures $\{\mu_n\}$ such that $\mu = \sum_n \mu_n$, $P^2(\mu) = \bigoplus_1^\infty P^2(\mu_n)$, and $S_n = S_{\mu_n}$ is irreducible for every $n \geq 1$. Let $G_n = \operatorname{int}\Sigma(\mu_n)$; so $P^\infty(\mu_n) = H^\infty(G_n)$. Since S is pure so is each S_n and hence $\dim P^2(\mu_n) = \infty$. Also $G_n \subseteq G$ by (VI.7.14).

Let $T \in \operatorname{Alg Lat} S$. By Lemma 8.3, $ST = TS$. Let $T_n \equiv T|P^2(\mu_n)$; so $S_n T_n = T_n S_n$ for all $n \geq 1$. By Yoshino's Theorem for each n there is a function ϕ_n in $P^2(\mu_n) \cap L^\infty(\mu_n)$ such that $T_n = \phi_n(S_n)$; let $\phi = \bigoplus_1^\infty \phi_n$. So $\phi \in P^2(\mu) \cap L^\infty(\mu)$ and $T = \phi(S)$.

By Theorem 7.6 there is a function x in $P^2(\mu)$ such that $\mathcal{H}_x$ is a full analytic subspace; let x_n be the component of x in $P^2(\mu_n)$. If $x_n = 0$, then again apply Theorem 7.6 to get a function y_n in $P^2(\mu_n)$ such that $\mathcal{H}_{y_n}$ is a μ_n-full analytic subspace and $\|y_n\| < 1/n$. So $\text{abpe}(\mathcal{H}_{y_n}) = G_n \subseteq G$. If $x_n \neq 0$ let $y_n = x_n$. thus $y = \bigoplus_1^\infty y_n \in P^2(\mu)$.

Since $\mathcal{H}_x$ is a μ-full analytic subspace, for a compact subset K of G there is a constant C such that $|p(\lambda)|^2 \leq C^2 \int |p|^2 |x|^2 \, d\mu$ for all polynomials p and all λ in K. But $|x| \leq |y|$ a.e. $[\mu]$ so that $|p(\lambda)|^2 \leq C^2 \int |p|^2 |y|^2 \, d\mu$ for all polynomials p and all λ in K. Thus $\text{abpe}\,\mathcal{H}_y = G$ and so $\mathcal{H}_y$ is a μ-full analytic subspace with $y_n \neq 0$ for all n.

According to Lemma 8.2 there is a bounded analytic function ψ on G such that $Tf = \psi(S)f$ for all f in $\mathcal{H}_y$. Thus $\ker(\phi(S) - \psi(S)) \supseteq \mathcal{H}_y$. But $\phi(S) - \psi(S) \in \{S\}'$ so that $\ker(\phi(S) - \psi(S))$ reduces S (II.5.6). By Corollary IV.5.7, for each $n \geq 1$, either $P^2(\mu_n) \subseteq \ker(\phi(S) - \psi(S))$ or $P^2(\mu_n) \perp \ker(\phi(S) - \psi(S))$. But $y \in \ker(\phi(S) - \psi(S))$ and $y_n \neq 0$ so the latter of these alternatives cannot happen. That is, $P^2(\mu_n) \subseteq \ker(\phi(S) - \psi(S))$ for all n; hence $T = \phi(S) = \psi(S) \in P^\infty(S)$ and so S is reflexive. □

8.5 THEOREM (Olin and Thomson [1980b]). *Every subnormal operator is reflexive.*

PROOF.

CASE 1. Assume S is a pure subnormal operator. Let S act on $\mathcal{H}$ and let $T \in \text{Alg Lat}\, S$. Let x and y be separating vectors for N that lie in $\mathcal{H}$. According to Corollary V.17.15, if we replace x by a suitable multiple of x, we may assume that $x + y$ is also a separating vector for N. By the preceding lemma, there are functions ϕ, ψ, and θ in $P^\infty(\mu)$ such that $T|\mathcal{H}_x = \phi(N)|\mathcal{H}_x$, $T|\mathcal{H}_y = \psi(N)|\mathcal{H}_y$, and $T|\mathcal{H}_{x+y} = \theta(N)|\mathcal{H}_{x+y}$. Hence $\theta(N)(x+y) = Tx + Ty = \phi(N)x + \psi(N)y$. Transposing we get the equation

(8.6) $\qquad [\theta(N) - \phi(N)]x = [\psi(N) - \theta(N)]y.$

Applying T to both sides of (8.6) we get that $\phi(N)[\theta(N) - \phi(N)]x = \psi(N)[\psi(N) - \theta(N)]y$. Using (8.6) once again, this implies that

(8.7) $\qquad \phi(N)[\theta(N) - \phi(N)]x = \psi(N)[\theta(N) - \phi(N)]x.$

A similar algebraic manipulation shows that

(8.8) $\qquad \phi(N)[\psi(N) - \theta(N)]y = \psi(N)[\psi(N) - \theta(N)]y.$

But x and y are separating vectors for N so that

$$\phi(\theta - \phi) = \psi(\theta - \phi) \text{ a.e. } [\mu]$$

and

$$\phi(\psi - \theta) = \psi(\psi - \theta) \text{ a.e. } [\mu].$$

Removing parentheses, bringing everything to one side of the equal sign, and then adding the two resulting equations together gives that

$$0 = \phi^2 + \psi^2 - 2\phi\psi$$
$$= (\phi - \psi)^2.$$

But ϕ and ψ belong to $P^\infty(\mu)$ which is isomorphic to an algebra of analytic functions. Therefore we get that $\phi = \psi$ a.e. $[\mu]$.

What we have shown is that there is a unique function ϕ in $P^\infty(\mu)$ such that for any vector x in $\mathscr{H}$ that is separating for N, $Tx = \phi(S)x$. But if $h \in \mathscr{H}$ then for any $\varepsilon > 0$ there is a vector x in $\mathscr{H}$ that is a separating vector for N with $\|h - x\| < \varepsilon$ (V.17.14). Thus $\|Th - \phi(S)h\| \leq \|Th - Tx\| + \|\phi(S)x - \phi(S)h\| \leq \varepsilon[\|T\| + \|\phi(S)\|]$. Hence $T = \phi(S)$.

CASE 2. S is arbitrary. Again let S act on $\mathscr{H}$ and let $S = N_0 \oplus S_1$, where N_0 is a normal operator and S_1 is a pure subnormal operator. If $T \in \text{Alg Lat } S$, then $T = T_0 \oplus T_1$, where $T_0 \in \text{Alg Lat } N_0$ and $T_1 \in \text{Alg Lat } S_1$. Since N_0 and S_1 are reflexive, $T_0 \in P^\infty(N_0)$ and $T_1 \in P^\infty(S_1)$. Thus $T \in P^\infty(N_0) \oplus P^\infty(S_1)$, a weak* closed algebra that contains $P^\infty(S)$ as a weak* closed subalgebra. To show that $T \in P^\infty(S)$, suppose L is a weak* continuous linear functional on $P^\infty(N_0) \oplus P^\infty(S_1)$ such that $L(A) = 0$ for all A in $P^\infty(S)$. It is easy to see that $L = L_0 \oplus L_1$, where L_0 is weak* continuous on $P^\infty(N_0)$ and L_1 is weak* continuous on $P^\infty(S_1)$. According to Theorem 5.1, for $j = 0, 1$ there are vectors x_j and y_j in $\mathscr{H}_j$ such that $L_j = x_j \otimes y_j$. If $x = x_0 \oplus x_1$ and $y = y_0 \oplus y_1$, it follows that $L = x \otimes y$. But then $\mathscr{M} = \bigvee\{p(S)x : p \text{ is a polynomial}\} \in \text{Lat } S$ and $y \perp \mathscr{M}$. Therefore $Tx \in \mathscr{M}$ and so $0 = \langle Tx, y \rangle = L(T)$. By the Hahn-Banach Theorem, $T \in P^\infty(S)$. □

Remarks. There is a considerable literature on reflexivity. Conway and Dudziak [**1990**] generalized Theorem 8.5 to the class of von Neumann operators. This result utilizes a theorem from Brown and Chevreau [**1988**] that certain contractions are reflexive. It is worth the time of the serious reader to look at Wogen [**1986**].

Exercises

1. If A and B are reflexive operators, show that $W(A \oplus B) = W(A) \oplus W(B)$ if and only if $\text{Lat}(A \oplus B) = \text{Lat } A \oplus \text{Lat } B$.
2. Let μ_1 be a measure such that $P^\infty(\mu_1) = H^\infty(\mathbb{D})$, $P^2(\mu_1)$ is a full analytic subspace, and $S_1 = S_{\mu_1}$ is pure. Let Γ be a Borel subset of $\partial\mathbb{D}$ with $m(\Gamma) > 0$ and $\mu_1(\Gamma) = 0$ and let $N_0 =$ multiplication by z on $L^2(m|\Gamma)$. If $S = S_1 \oplus N_0$, then there is an invariant subspace $\mathscr{N}$ for S such that $\mathscr{N} \cap ((0) \oplus L^2(m|\Gamma)) = (0)$ and $\mathscr{N}$ contains an element $f \oplus g$ such that $g \neq 0$.

§9 Filling in the holes of the spectrum of a normal operator. We know that if S is a subnormal operator with minimal normal extension N and U is a bounded component of the complement of $\sigma(N)$, then either $U \subseteq \sigma(S)$ or $U \cap \sigma(S) = \varnothing$ (II.2.11). For any compact subset K of $\mathbb{C}$, a bounded component of $\mathbb{C}\backslash K$ is called a *hole* of K. The question arises as to which holes of $\sigma(N)$ can be filled by a subnormal operator for which N is the minimal normal extension? Precisely, if N is a given normal operator and $\mathscr{S}(N)$ denotes the collection of all subnormal operators S such that $N = \operatorname{mne} S$, which holes of $\sigma(N)$ are contained in the spectrum of some S in $\mathscr{S}(N)$? Equivalently, what is the set

$$\bigcup \{\sigma(S) : S \in \mathscr{S}(N)\}.$$

Let's inspect an example. Let $C = \{z : |z - 3| = 1\}$, put $\nu = \sum_n 2^{-n} \delta_{a_n}$, where $\{a_n\}$ is a dense sequence in C, and let $\mu = \nu + m$, where m is normalized Lebesgue measure on $\partial \mathbb{D}$. If $N = N_\mu$, then $\sigma(N) = C \cup \partial \mathbb{D}$ so that $\sigma(N)$ has two holes. However, $P^\infty(\mu) = L^\infty(\nu) \oplus H^\infty(\mathbb{D})$ and if $S \in \mathscr{S}(N)$, then $S = N_\nu \oplus S_1$ where S_1 is a unilateral shift of multiplicity 1. Thus the inside of the circle C is never filled by any subnormal operator in $\mathscr{S}(N)$. In fact, some reflection reveals that no hole of the spectrum of a normal operator that arises from the reductive part of N can ever be filled. This is, however, the extent of the prohibition.

9.1 THEOREM. *If N is a normal operator with scalar-valued spectral measure μ, then there is a subnormal operator S in $\mathscr{S}(N)$ such that $\sigma(S) = \Sigma(\mu) \cup \sigma(N)$.*

PROOF. According to Proposition 11.6 a scalar-valued spectral measure μ can be chosen for N such that $P^2(\mu) \cap L^\infty(\mu) = P^\infty(\mu)$. Also there is a normal operator M whose scalar-valued spectral measure is absolutely continuous with respect to μ and such that $N \cong N_\mu \oplus M$. Let $S \cong S_\mu \oplus M$. Clearly $S \in \mathscr{S}(N)$ and $\sigma(S) \subseteq \Sigma(\mu) \cup \sigma(N)$.

If $\alpha \notin \sigma(S)$, then $\alpha \notin \sigma(S_\mu)$ and $\alpha \notin \sigma(M)$. In particular, $(S - \alpha)^{-1} \cong (S_\mu - \alpha)^{-1} \oplus (M - \alpha)^{-1}$. In particular, $(S_\mu - \alpha)^{-1} \in \{S_\mu - \alpha\}'$ and so, by Yoshino's Theorem, $(z - \alpha)^{-1} \in P^2(\mu) \cap L^\infty(\mu) = P^\infty(\mu)$. Thus $\alpha \notin \operatorname{int} \Sigma(\mu)$ and so $\alpha \notin \Sigma(\mu) \cup \sigma(N)$. $\square$

In Conway and Olin [1977] it was shown that $\Sigma(N) \backslash \sigma(N) = \bigcup \{\sigma(S) \backslash \sigma(N) : S \in \mathscr{S}(N)\}$. The fact that there is one subnormal operator S in $\mathscr{S}(N)$ satisfying this relation was pointed out to me by John McCarthy.

9.2 COROLLARY. *If N is a normal operator with scalar-valued spectral measure μ and $P^\infty(\mu)$ has no L^∞-summand, then there is a subnormal operator S in $\mathscr{S}(N)$ such that $\sigma(S) = \Sigma(N)$.*

The proof of the next result is left to the reader.

9.3 Proposition (Olin and Thomson [**1977b**]). *Let N be a normal operator with scalar-valued spectral measure μ and let $U_1, U_2, \ldots$ be the components of $\Sigma(\mu) \setminus \sigma(N)$. If J is a nonempty subset of $\mathbb{N}$, then the following statements are equivalent.*

(a) *There is a subnormal operator S in $\mathscr{S}(N)$ such that $\sigma(S) = K_J \equiv \sigma(N) \cup \bigcup \{U_n : n \in J\}$.*

(b) *If $n \notin J$, then for some α in U_n, $(z - \alpha)^{-1} \notin R^\infty(K_J, \mu)$.*

(c) *If $n \notin J$, then some α in U_n belongs to the envelope $E(K_J, \mu)$.*

A recent result from McCarthy [**1990**] implies that the measure μ can be chosen such that $\text{abpe}\{R^2(K_j, \mu)\} = \Sigma(K_J, \mu)$ (see VI.2.3 for the definition). Given this, the three conditions in the preceding corollary can be shown to be equivalent to

(d) *If $n \notin J$, then U_n is contained in*
$$\Sigma(K_J, \mu).$$

Another problem concerning $\mathscr{S}(N)$ is the following.

9.4 Open Problem. What is a necessary and sufficient condition on a normal operator N that $\mathscr{S}(N)$ contain a pure operator?

If μ is a scalar-valued spectral measure for N, then in order for $\mathscr{S}(N)$ to contain a pure operator, it is clearly necessary that $P^\infty(\mu)$ has no L^∞-summand. In Exercise 2 the reader is asked to show that for a cyclic normal operator this condition is also sufficient. The proof of this fact is like the proof of Proposition 9.6 below.

Olin and Thomson [**1980a**] studied this problem and made significant progress. Indeed, they show that if N is a normal operator with scalar-valued spectral measure μ, if $G = $ the interior of the Sarason hull of μ, and if $P^\infty(\mu) = H^\infty(G) = P^\infty(\mu|G)$, then N is the minimal normal extension of a pure subnormal operator. Here is a pertinent example. (See Olin and Thomson [**1980a**] for another.)

9.5 Example. There is a noncyclic normal operator N with scalar-valued spectral measure μ such that $P^\infty(\mu)$ has no L^∞ summand and N is not the minimal normal extension of any pure subnormal operator.

Let m be normalized Lebesgue measure on $\partial \mathbb{D}$ and let $U = N_m$ on $L^2(\partial \mathbb{D})$. Put $N = U \oplus 0$ on $\mathscr{K} = L^2(\partial \mathbb{D}) \oplus \mathbb{C}^2$. The scalar-valued spectral measure of N is $\mu = m + \delta_0$ and N is not cyclic. Suppose $S = N|\mathscr{H} \in \mathscr{S}(N)$ and S is pure. We will work towards a contradiction.

Note that if $0 \oplus \mathbb{C} \subseteq \mathscr{H}$, then S cannot be pure since S would have a nontrivial kernel. So $(0 \oplus \mathbb{C}) \cap \mathscr{H} = (0)$. Let $\mathscr{E}_1$ be the collection of first coordinates of elements of $\mathscr{H}$. It follows that for every f in $\mathscr{E}_1$ there is a unique scalar α such that $f \oplus \alpha \in \mathscr{H}$. That is, $\mathscr{H}$ is the graph of a linear map $\xi: \mathscr{E}_1 \to \mathbb{C}$. But linear functionals with closed graphs are bounded (Exercise 3). Also the fact that $\mathscr{H} = \text{graph } \xi$ is invariant for N implies that $N_\mu \mathscr{E}_1 \subseteq \mathscr{E}_1$ and $\xi(N_\mu f) = 0$ for all f in $\mathscr{E}_1$. It follows that $\mathscr{E}_1$ is closed (Why?).

§9 FILLING IN THE HOLES OF THE SPECTRUM OF A NORMAL OPERATOR

Since $N = \text{mne } S$, $N_\mu = \text{mne}(N_\mu | \mathscr{E}_1)$. Writing $L^2(\mu) = L^2(m) \oplus L^2(\delta_0) = L^2(\partial \mathbb{D}) \oplus \mathbb{C}$ and reasoning as above, it follows that $\mathscr{E}_1 = \{g \oplus \zeta(g) : g \in \mathscr{E}_2\} \subseteq L^2(\partial \mathbb{D})\}$, where $\mathscr{E}_2$ is closed in $L^2(\partial \mathbb{D})$ and $\zeta \in \mathscr{E}_2^*$.

Combining these results, we get that for some θ in $\mathscr{E}_2^*$

$$\mathscr{H} = \{g \oplus \zeta(g) \oplus \theta(g) : g \in \mathscr{E}_2\} \subseteq L^2(\partial \mathbb{D}) \oplus \mathbb{C} \oplus \mathbb{C}.$$

Moreover, for g in $\mathscr{E}_2$, $Ug \in \mathscr{E}_2$ and $\zeta(Ug) = \theta(Ug) = 0$. Define $R: \mathscr{E}_2 \to \mathscr{H}$ by $Rg = g \oplus \zeta(g) \oplus \theta(g)$. It is easy to see that R is invertible and $R(U|\mathscr{E}_2) = SR$; that is, $S \approx U|\mathscr{E}_2$. But $U|\mathscr{E}_2$ must be cyclic (I.3.14) so that S is cyclic. But $N = \text{mne } S$ and N is not cyclic, the sought for contradiction.

If μ is a scalar-valued spectral measure of N, then $\mathscr{S}(N) \subseteq \mathscr{S}(\widehat{\sigma(N)}, \mu)$, where $\mathscr{S}(K, \mu)$ is defined in §2. By expanding our attention to encompass $\mathscr{S}(K, \mu)$ rather than the subnormal operators with a fixed minimal normal extension, the analogue of the preceding question becomes easier to answer.

9.6 PROPOSITION. *There is a pure subnormal operator in $\mathscr{S}(K, \mu)$ if and only if $R^\infty(K, \mu)$ has no L^∞ summand.*

PROOF. Suppose $R^\infty(K, \mu) = L^\infty(\mu_0) \oplus R^\infty(K, \mu_1)$ with $\mu_0 \neq 0$. If N is any normal operator with μ as its scalar-valued spectral measure, then any subspace $\mathscr{H}$ invariant for $f(N)$ for every f in $R(K)$, must be invariant for $\phi(N)$ for every ϕ in $L^\infty(\mu_0)$. Thus $N|\mathscr{H}$ cannot be pure for any invariant subspace. That is, $\mathscr{S}(K, \mu)$ has no pure elements.

Conversely, assume that $R^\infty(K, \mu)$ has no L^∞ summand. By Proposition VI.1.11, there is a measure η such that $\eta \perp R(K)$ and $[|\eta|] = [\mu]$. Let $d\eta/d\mu = f_1 \bar{f}_2$, where f_1 and $f_2 \in L^2(\mu)$. Let $N = N_\mu$, put $\mathscr{H} = \bigvee\{gf_1 : g \in R(K)\}$, and let $S = N|\mathscr{H}$. Note that $f_2 \perp \mathscr{H}$ so that $\mathscr{H} \neq L^2(\mu)$. Also $\mu(f_1^{-1}(0)) = 0 = \mu(f_2^{-1}(0))$. Therefore the Stone-Weierstrass Theorem implies $L^2(\mu) = \bigvee\{\bar{z}^k gf_1 : k \geq 0 \text{ and } g \in R(K)\}$; that is, $N = \text{mne } S$ so that $S \in \mathscr{S}(K, \mu)$. To see that S is pure, assume that $\mathscr{L}$ is a reducing subspace for N that is contained in $\mathscr{H}$. Since N is a cyclic normal operator, there is a Borel set Δ such that $\mathscr{L} = \chi_\Delta L^2(\mu)$. But $f_2 \perp \mathscr{H}$, so that $f_2 \perp \chi_\Delta L^2(\mu)$, implying that $f_2 = 0$ a.e. $[\mu]$ on Δ. As observed before, this implies that $\mu(\Delta) = 0$. □

Exercises

1. Show that if N is a nonreductive normal operator, then the cardinality of the nonreducing invariant subspaces for N is the cardinality of $\mathbb{R}$.
2. Let N be a cyclic normal operator with scalar-valued spectral measure μ. Show that N is the minimal normal extension of a pure subnormal operator if and only if $P^\infty(\mu)$ has no L^∞ summand.
3. Show that a linear functional on a dense manifold of a Banach space whose graph is closed must be bounded.

4. (Olin and Thomson [**1977b**]) Let L be the open line segment from i to $-i$ and let ν be a purely atomic measure carried by L and whose support is the closure of L. Let $\mu = \nu + m$, where m is Lebesgue measure on $\partial \mathbb{D}$. So $\Sigma(\mu) = \operatorname{cl} \mathbb{D}$; let U_1 and U_2 be the two half disks that are the components of $\Sigma(\mu) \setminus \operatorname{supp} \mu$. If $N = N_\mu$, show there is no subnormal operator S in $\mathscr{S}(N)$ with $U_1 \subseteq \sigma(S)$ and $U_2 \cap \sigma(S) = \varnothing$.

5. Prove Proposition 9.3.

§10 Quasisimilarity revisited. In §II.13 a number of results on similarity and quasisimilarity were presented. In this section that study is rejoined, but now with the force of the functional calculus to propel it further.

10.1 THEOREM (Conway [**1980**]). *If S_1 and S_2 are subnormal operators such that for $i, j = 1, 2$ there is a quasi-invertible operator X_{ij} with $X_{ij} S_j = S_i X_{ij}$, then there is a dual algebra isomorphism $\rho: R^\infty(S_1) \to R^\infty(S_2)$ that has the following additional properties.*

(a) $\rho(S_1) = S_2$.
(b) $X_{21} A = \rho(A) X_{21}$ and $X_{12} \rho(A) = X_{12}$ for every A in $R^\infty(S_1)$.

A lemma is required for this proof.

10.2 LEMMA. *Let $\mathscr{A}_1$ and $\mathscr{A}_2$ be algebras of subnormal operators and let $\mathscr{B}_1$ and $\mathscr{B}_2$ be their weak* sequential closures. If $\rho: \mathscr{A}_1 \to \mathscr{A}_2$ is a contractive monomorphism and X is a quasi-invertible operator such that $XA = \rho(A)X$ for all operators A in $\mathscr{A}_1$, then ρ extends to a contractive monomorphism $\tilde{\rho}: \mathscr{B}_1 \to \mathscr{B}_2$ such that $XB = \tilde{\rho}(B) X$ for all operators B in $\mathscr{B}_1$.*

PROOF. Let $B \in \mathscr{B}_1$ and pick any sequence $\{A_n\}$ in $\mathscr{A}_1$ such that $A_n \to B$ weak*. Let $M \geq \|A_n\|$ for all n. Thus $\|\rho(A_n)\| \leq M$ for all $n \geq 1$ and so there is a subsequence $\{A_{n_k}\}$ and an operator C such that $\rho(A_{n_k}) \to C$ weak*. It follows that $C \in \mathscr{B}_2$ and, since $XA_n = \rho(A_n)X$ for all n, $XB = CX$. But B and C are subnormal and X is quasi-invertible so that Theorem II.13.4 implies that $\sigma(C) \subseteq \sigma(B)$. Hence $\|C\| \leq \|B\|$. If we put $\tilde{\rho}(B) = C$ we can check that $\tilde{\rho}$ is well-defined and all the required properties are possessed. The details are left to the reader. □

PROOF OF THEOREM 10.1. Put $K = \sigma(S_1) = \sigma(S_2)$ (II.13.5). Let $R_0(S_j) = \{f(S_j) : f \in \operatorname{Rat}(K)\}$ and let $R_1(S_j)$ be the weak* sequential closure of $R_0(S_j)$ in $\mathscr{B}(\mathscr{H}_j)$. We inductively define a space of operators $R_\alpha(S_j)$ for each ordinal number α as follows. If α has an immediate predecessor, $\alpha - 1$, let $R_\alpha(S_j)$ be the weak* sequential closure of $R_{\alpha-1}(S_j)$; if α is a limit ordinal and $R_\beta(S_j)$ is defined whenever $\beta < \alpha$, let $R_\alpha(S_j) = \bigcup \{R_\beta(S_j) : \beta < \alpha\}$.

Now algebraic manipulation shows that $X_{ij} f(S_j) = f(S_i) X_{ij}$ for every f in $\operatorname{Rat}(K)$. Define $\rho_0: R_0(S_1) \to R_0(S_2)$ and $\eta_0: R_0(S_2) \to R_0(S_1)$ by $\rho_0(f(S_1)) = f(S_2)$ and $\eta_0(f(S_2)) = f(S_1)$ for each f in $\operatorname{Rat}(K)$. It follows that both ρ_0 and η_0 are contractive monomorphisms and $\rho_0 = \eta_0^{-1}$.

By induction and Lemma 10.2, for each ordinal α there are contractive monomorphisms $\rho_\alpha \colon R_\alpha(S_1) \to R_\alpha(S_2)$ and $\eta_\alpha \colon R_\alpha(S_2) \to R_\alpha(S_1)$ such that $X_{21}A = \rho_\alpha(A)X_{21}$ for A in $R_\alpha(S_1)$, $X_{12}B = \eta_\alpha(B)X_{12}$ for B in $R_\alpha(S_2)$, and $\rho_\alpha = \eta_\alpha^{-1}$.

Now standard Banach space theory (See page 166 in [**ACFA**]) implies there is a countable ordinal α such that $R_\alpha(S_j) = R^\infty(S_j)$, $j = 1, 2$. Let $\rho = \rho_\alpha$ and $\eta = \eta_\alpha$. Since both ρ and η are contractive and $\rho = \eta^{-1}$, ρ is an isometry. From the definitions, ρ is weak* sequentially continuous. By Proposition I.2.10, ρ is a dual algebra isomorphism. □

A particular subalgebra of $R^\infty(S)$ is $P^\infty(S)$. Restricting the map ρ in Theorem 10.1 gives the following.

10.3 COROLLARY. *With the same hypothesis as Theorem 10.1, there is a dual algebra isomorphism $\rho \colon P^\infty(S_1) \to P^\infty(S_2)$ that has the following additional properties.*

(a) $\rho(S_1) = S_2$.
(b) $X_{21}A = \rho(A)X_{21}$ and $X_{12}\rho(A) = AX_{12}$ for every A in $P^\infty(S_1)$.

Let μ_j be the scalar-valued spectral measure for S_j. Since $P^\infty(S_j) \cong P^\infty(\mu_j)$ and Sarason's Theorem gives the structure of $P^\infty(\mu_j)$, we can interpret this corollary to some advantage. The details are left to the reader.

10.4 COROLLARY. *Let S_1, S_2 be subnormal operators with scalar-valued spectral measures μ_1, μ_2; let $(\Sigma(\mu_j), \tilde{\mu}_j)$ be the Sarason pair of μ_j. If $S_1 \sim S_2$, then the following conditions are satisfied.*

(a) $[\mu_1 - \tilde{\mu}_1] = [\mu_2 - \tilde{\mu}_2]$.
(b) $\Sigma(\mu_1) = \Sigma(\mu_2)$.
(c) *The identity map on polynomials extends to a dual algebra isomorphism of $P^\infty(\mu_1)$ onto $P^\infty(\mu_2)$.*

The last corollary can be interpreted as another result of the type that says that quasisimilarity between subnormal operators preserves various parts of the spectrum. (For example, see II.13.12.)

We now turn our attention to the study of subnormal operators that are quasisimilar to the unilateral shift. Once again, m denotes normalized Lebesgue measure on $\partial \mathbb{D}$ and we let S_m be the unilateral shift of multiplicity 1. This notation will remain fixed for the remainder of the section.

10.5 LEMMA. *If μ is a measure on $\mathrm{cl}\,\mathbb{D}$, $\eta = \mu|\mathbb{D}$, and $\nu \equiv \mu|\partial \mathbb{D} = \nu_a + \nu_s$, where $\nu_a \ll m$ and $\nu_s \perp m$, then $P^2(\mu) = P^2(\eta + \nu_a) \oplus L^2(\nu_s)$.*

PROOF. Apply Proposition VI.4.9. □

10.6 LEMMA. *With the notation of Lemma 10.5, if $Y \colon P^2(\mu) \to H^2$ is a nonzero operator such that $YS_\mu = S_m Y$, then $\ker Y = L^2(\nu_s)$ and the* *logarithm of the Radon-Nikodym derivative of ν_a with respect to m belongs to $L^1(m)$. Moreover, if $\psi = Y(1)$, then $Yf = \psi f$ for all f in $P^2(\mu)$.*

PROOF. Put $Y_s = Y|L^2(\nu_s)$. So $Y_s: L^2(\nu_s) \to H^2$ and $Y_s N_{\nu_s} = S_m Y_s$. Since N_{ν_s} is a reductive unitary operator, it is easy to show that $Y_s = 0$. (This also follows from Exercise 1. In any case, it is left to the reader.) Thus $L^2(\nu_s) \subseteq \ker Y$. Therefore it suffices to assume that $\nu_s = 0$ and to show that Y must be injective.

Since $Y \neq 0$, we may assume that $\|Y\| = 1$. If $\psi = Y(1)$, then $Y(p) = \psi p$ for every polynomial p. If $n \geq 1$, then $\int |p|^2 |\psi|^2 dm = \int |p|^2 |\psi|^2 |z|^{2n} dm = \|Y(z^n p)\|^2 \leq \int |z|^{2n} |p|^2 d\mu$. Letting $n \to \infty$ shows that $\int |p|^2 |\psi|^2 dm \leq \int |p|^2 d\nu$. If p is a polynomial with no constant term, this implies that $\int |p-1|^2 |\psi|^2 dm \leq \int |p-1|^2 d\nu$. Since $\psi \in H^2$, Szegö's Theorem implies that $-\infty < 2\int \log|\psi| dm \leq \int \log(d\nu/dm) dm$; hence $\log(d\nu/dm) \in L^1(m)$. In particular, $[\nu] = [m]$.

Now let $f \in P^2(\mu)$ and select a sequence of polynomials $\{p_n\}$ such that $\int |p_n - f|^2 d\mu \to 0$. Thus $Y(p_n) = \psi p_n \to Y(f) = g$ in H^2. By passing to a subsequence we may also assume that $p_n \to f$ a.e. $[\mu]$ and $\psi p_n \to g$ a.e. $[m]$. But $[\nu] = [m]$ and so $\psi p_n \to \psi f$ a.e. $[m]$. Hence $g = \psi f$ a.e. $[m]$. That is, $Y(f) = \psi f$ for all f in $P^2(\mu)$. Now the assumption that $Y \neq 0$ implies that $\psi \neq 0$. So if $f \in P^2(\mu)$ and $0 = Y(f) = \psi f$, $f = 0$ a.e. $[m]$ on $\partial \mathbb{D}$. On the other hand, $(Yp_n)(z) = \psi(z)p_n(z) \to 0$ uniformly on compact subsets of $\mathbb{D}$. By selecting circles in $\mathbb{D}$ on which ψ does not vanish, we see that $p_n(z) \to 0$ uniformly on circles of radius arbitrarily close to 1. By the Maximum Modulus Theorem, $p_n(z) \to 0$ in $\mathbb{D}$. Thus $f(z) = 0$ a.e. $[\mu]$ in $\mathbb{D}$ and hence $f = 0$ and Y is injective. □

10.7 LEMMA. *With the notation of Lemma 10.5, if $\nu \ll m$ and $\log(d\nu/dm) \in L^1(m)$, then there is a quasi-invertible operator $Y: P^2(\mu) \to H^2$ such that $YS_\mu = S_m Y$.*

PROOF. Let ψ be the outer function in H^2 such that $|\psi|^2 = (d\nu/dm)$. If p is a polynomial, $\int |p|^2 |\psi|^2 dm = \int |p|^2 d\nu \leq \int |p|^2 d\mu$. Hence $Y(p) = \psi p$ extends to a contraction $Y: P^2(\mu) \to H^2$. Clearly, $YS_\mu = S_m Y$. Since $Y(1) = \psi$, Lemma 10.6 implies that $Y(f) = \psi f$ for all f in $P^2(\mu)$ and Y is injective. Since ψ is an outer function, ran Y is dense. □

We can now produce one of the main theorems of this section.

10.8 THEOREM (Clary [1973]). *If S is a subnormal operator on $\mathcal{H}$, the following statements are equivalent.*

  (a) $S \sim S_m$.
  (b) *S is a pure cyclic contraction and there is a nonzero operator $Y: \mathcal{H} \to H^2$ such that $YS = S_m Y$.*

(c) *S is cyclic and for every measure μ such that $S \cong S_\mu$ we have that* $\operatorname{supp}\mu \subseteq \operatorname{cl}\mathbb{D}$, $\nu \equiv \mu|\partial\mathbb{D} \ll m$, *and* $\log(d\nu/dm) \in L^1(m)$.
(d) *There is a measure μ on $\operatorname{cl}\mathbb{D}$ such that $S \cong S_\mu$, $\nu \equiv \mu|\partial\mathbb{D} \ll m$, and* $\log(d\nu/dm) \in L^1(m)$.

PROOF. (b) *implies* (c). Because S is cyclic, there is a measure μ such that $S \cong S_\mu$. Since S is a contraction, $\operatorname{supp}\mu \subseteq \operatorname{cl}\mathbb{D}$. Thus (b) implies there is a nonzero operator $Y: P^2(\mu) \to H^2$ such that $YS_\mu = S_m Y$. Put $\nu = \mu|\partial\mathbb{D}$. Since S is pure, $\nu \ll m$ by Lemma 10.5. By Lemma 10.6, $\log(d\nu/dm) \in L^1(m)$, establishing (c).

(d) *implies* (a). Put $\eta = \mu|\mathbb{D}$ and let $S = S_\mu$. By Lemma 10.7 it suffices to find a quasi-invertible operator $X: H^2 \to P^2(\mu)$ such that $XS_m = SX$. If $\hat{\mu}$ is the sweep of μ to $\partial\mathbb{D}$, then $\hat{\mu} = \nu + \hat{\eta}$. Hence $\int \log(d\hat{\mu}/dm)\,dm \geq \int \log(d\nu/dm)\,dm > -\infty$. Thus $\log(d\hat{\mu}/dm) \in L^1(m)$. Let ϕ be an outer function in H^2 such that $|\phi|^2 = \log(d\hat{\mu}/dm)$ and define $U: P^2(\hat{\mu}) \to H^2$ by $Up = \phi p$. It is easy to check that U is an isomorphism and $US_{\hat{\mu}} = S_m U$.

If p is any analytic polynomial, a routine computation shows that $\bar{\partial}\partial|p|^2 = |\partial p|^2$. By (V.9.2), $|p|^2$ is subharmonic. Let u be a continuous function on $\operatorname{cl}\mathbb{D}$ such that u is harmonic on $\mathbb{D}$ and $u = |p|^2$ on $\partial\mathbb{D}$. Hence, $\int |p|^2 d\mu \leq \int u\,d\mu = \int |p|^2 d\hat{\mu}$. So if $X_1(p) = p$, X_1 extends to a contraction $X_1: P^2(\hat{\mu}) \to P^2(\mu)$ satisfying $X_1 S_{\hat{\mu}} = SX_1$. Let $X = X_1 U^{-1}: H^2 \to P^2(\mu)$. From the properties of U and X_1, $XS_m = SX$. Also $\operatorname{ran} X = \operatorname{ran} X_1$, which contains the polynomials and is therefore dense. By (II.13.9), X is injective and therefore quasi-invertible.

Since it is clear that (a) implies (b) and (c) implies (d), this completes the proof. □

A measure η is a *Carleson measure* if η is carried by $\mathbb{D}$ and there is a positive constant C such that for every polynomial p, $\int |p|^2 d\eta \leq C \int |p|^2 dm$. The usual definition of a Carleson measure η involves the amount of measure η assigns to certain curvilinear rectangles in $\mathbb{D}$. However, it is a standard result about Carleson measures that the above definition is equivalent to the usual one. See Duren [**1970**], page 157.

10.9 THEOREM (Clary [**1973**]). *If S is a subnormal operator, the following statements are equivalent.*

(a) $S \approx S_m$.
(b) *There is a Carleson measure η such that if $\mu = \eta + m$, then $S \cong S_\mu$.*
(c) *S is cyclic and if μ is any measure such that $S \cong S_\mu$, then μ is supported by $\operatorname{cl}\mathbb{D}$, $\nu \equiv \mu|\partial\mathbb{D} \ll m$, $\log(d\nu/dm) \in L^1(m)$, and if ψ is the outer function in H^2 such that $|\psi|^2 = (d\nu/dm)$, then $\psi^{-1} \in P^2(\mu)$ and $|\psi|^{-2}\mu|\mathbb{D}$ is a Carleson measure.*

PROOF. (a) *implies* (c). Let $S = S_\mu$. By (a) there is an invertible operator $R: P^2(\mu) \to H^2$ such that $RS = S_m R$. By Theorem 10.8, $\nu \ll m$ and $\log(d\nu/dm) \in L^1(m)$. Let ψ be the outer function as described in (c) and put $\tau = R(1)$ in H^2 and $\sigma = R^{-1}(1)$ in $P^2(\mu)$. From Lemma 10.6 and some similar arguments applied to R^{-1}, we get that $R(f) = \tau f$ for each f in $P^2(\mu)$ and $R^{-1}(g) = \sigma g$ for each g in H^2. Now if $c_1 = \|R^{-1}\|^{-2}$ and $c_2 = \|R\|^2$, then for every polynomial p, $c_1 \int |p|^2 d\mu \le \int |p|^2 |\tau|^2 dm \le c_2 \int |p|^2 d\mu$. Substituting $z^n p$ for p in these inequalities and letting $n \to \infty$, we see that $c_1 \int |p|^2 |\psi|^2 dm \le \int |p|^2 |\tau|^2 dm \le c_2 \int |p|^2 |\psi|^2 dm$ for every polynomial p. Therefore if we set $X(\psi p) = \tau p$ for all polynomials p, we have a densely defined linear transformation that extends to a bounded operator $X: H^2 \to H^2$. But X clearly commutes with S_m and so there is a bounded analytic function ϕ such that $X = T_\phi$. Thus $\tau p = X(\psi p) = \phi \psi p$ for all polynomials p. Therefore $\tau/\psi = \phi \in H^\infty$. Now $\tau \sigma = RR^{-1}(1) = 1$ and so $\sigma = \tau^{-1}$ a.e. $[\mu]$. Thus $\psi^{-1} = \phi \tau^{-1} = \phi \sigma \in P^2(\mu)$. In particular $|\psi|^{-2} \mu$ is a finite measure. In addition, if p is a polynomial, then $\int_\mathbb{D} |p|^2 |\psi|^{-2} d\mu = \int_\mathbb{D} |p|^2 |\phi|^2 |\sigma|^2 d\mu \le \|\phi\|_\infty^2 \int |\sigma p|^2 d\mu = \|\phi\|_\infty^2 \|R^{-1}(p)\|^2 \le \|\phi\|_\infty^2 \|R^{-1}\|^2 \int |p|^2 dm$. Therefore $|\psi|^{-2} \mu|\mathbb{D}$ is a Carelson measure.

(c) *implies* (b). Let μ, ν, and ψ be as described in (c) and define the measure λ on cl$\mathbb{D}$ by $\lambda = m + |\psi|^{-2} \mu|\mathbb{D}$. Thus $\lambda|\partial \mathbb{D} = m$ and $\lambda|\mathbb{D}$ is a Carleson measure. It remains to show that $S_\mu \cong S_\lambda$. If f is any function in $L^2(\mu)$, then $\int |f\psi|^2 d\lambda = \int_\mathbb{D} |f\psi|^2 |\psi|^{-2} d\mu + \int |f|^2 (d\nu/dm) dm = \int |f|^2 d\mu$. Therefore $U: L^2(\mu) \to L^2(\lambda)$ define by $Uf = \psi f$ is an isomorphism. (Verify surjectivity.) We want to show that U maps $P^2(\mu)$ onto $P^2(\lambda)$.

Let $\{p_n\}$ be a sequence of polynomials such that $\int |p_n - \psi|^2 dm \to 0$. Since $|\psi|^{-2} \mu|\mathbb{D}$ is a Carleson measure, it is easy to check that $\{p_n\}$ is a Cauchy sequence in $L^2(\lambda)$ and so there is a function h in $P^2(\lambda)$ such that $\int |p_n - h|^2 d\lambda \to 0$. By passing to a subsequence, we may assume that $p_n \to h$ a.e. $[\lambda]$. But $p_n \to \psi$ in H^2 and so $p_n(z) \to \psi(z)$ for $|z| < 1$. Thus $\psi = h \in P^2(\lambda)$. Therefore if p is any polynomial, $Up = \psi p \in P^2(\lambda)$ and so $UP^2(\mu) \subseteq P^2(\lambda)$. On the other hand, since $U^{-1} g = \psi^{-1} g$ for all g in $L^2(\lambda)$ and $\psi^{-1} \in P^2(\mu)$, a similar argument shows that $U^{-1} P^2(\lambda) \subseteq P^2(\mu)$. Therefore $U: P^2(\mu) \to P^2(\lambda)$ is an isomorphism and $US = S_\lambda U$ so that $S \cong S_\lambda$.

(b) *implies* (a). Let μ and η be as in part (b) and let C be the constant such that $\int |p|^2 d\eta \le C \int |p|^2 dm$ for all polynomials p. Thus $\int |p|^2 d\mu = \int |p|^2 d\eta + \int |p|^2 dm \le (C+1) \int |p|^2 dm$ and $\int |p|^2 dm \le \int |p|^2 d\mu$ for all polynomials p. Therefore the identity map on polynomials extends to a bounded invertible operator $R: H^2 \to P^2(\mu)$. Clearly $RS_m = S_\mu R$ so that $S_\mu \cong S_m$. □

Remarks. 1. Clary [**1973**] also shows that any subnormal operator that is quasisimilar to the unilateral shift has its invariant subspace lattice isomorphic to $\operatorname{Lat} S_m$.

2. Olin and Thomson [**1984**] and [**1986**] study subnormal operators S having the property that if $\mathcal{M}$ and $\mathcal{N}$ are nontrivial invariant subspaces for S, then $\mathcal{M} \cap \mathcal{N} \neq (0)$. Such operators are called *cellular indecomposable*. The unilateral shift has this property and if ϕ is a weak* generator of H^∞, $T_\phi = \phi(S_m)$ has this property. If S is a subnormal operator that is quasisimilar to such a Toeplitz operator T_ϕ, then S is also cellular indecomposable. (See Exercise 4.) This raises the following question.

10.10 OPEN QUESTION. If S is a cellular indecomposable subnormal operator, must S be quasisimilar to T_ϕ for some weak* generator of H^∞?

Olin and Thomson [**1984 and 1986**] have answered this question affirmatively in many special cases and have derived a number of necessary conditions for a subnormal operator to be cellular indecomposable. They conjectured that this question has an affirmative answer.

3. Some additional references on quasisimilarity are Conway [**1982**], Fong [**1977**], McCarthy [**preprint**], Raphael [**1982, 1985a, 1985b, and 1986**], Yan [**1988**], Yang [**1990**].

Exercises

1. If N is a normal operator on $\mathcal{K}$ and $Y : \mathcal{K} \to H^2$ is a bounded operator such that $YN = S_m Y$, then $Y = 0$.
2. Show that if μ is a measure on $\operatorname{cl}\mathbb{D}$ and $\nu \equiv \mu|\partial\mathbb{D} \ll m$ with $\log(d\nu/dm)$ in $L^1(m)$, then $\operatorname{abpe}(\mu) = \mathbb{D}$.
3. If μ is a measure on $\partial\mathbb{D}$ such that there is a positive constant C with $\int |p|^2 d\mu \leq C \int |p|^2 dm$ for every polynomial p, what can be said about μ?
4. If S is quasisimilar to T_ϕ, where ϕ is a weak* generator of H^∞, show that S is cellular indecomposable.

CHAPTER VIII

Bounded Point Evaluations

In this chapter the recent solution by James E. Thomson (Thomson [**preprint**]) of the long-standing problem on the existence of bounded point evaluations is presented. In fact, Thomson's work reveals a structure theory for the spaces $P^2(\mu)$ that answers some questions about these spaces that had previously been open and promises to enable researchers to answer many more. Some of these will be discussed in §6.

The problem about the existence of bounded point evaluations seems to have been raised first in Mergeljan [**1955**], though only for the restriction of area measure to a bounded set. It was raised in full generality in Brennan [**1971a**].

The earliest work on bounded point evaluations concentrated on the case of measures that are absolutely continuous with respect to area measure. Indeed, prior to Thomson's Theorem even this special case was not settled. The results for measures of the form $\mu = \omega\mathscr{A}$ always assumed additional conditions on the Radon-Nikodym derivative ω. The best results along these lines were obtained by Brennan and independently by Hruscev; see Brennan [**1979b**]. The earlier papers cannot be dismissed because they contain additional information. See Brennan [**1971a, 1971b, 1973, 1979a, and 1979b**].

§1 **A coloring scheme.** Form a grid in the plane with horizontal and vertical lines at a distance 2^{-n} apart and passing through the points on the axes that are integer multiples of 2^{-n}. Let $\mathscr{S}(n) = \{S_{np}\}_{p=1}^{\infty}$ be the resulting collection of open squares and refer to this collection of squares as the nth *generation*. We want to consider all the squares from all the different generations and devise a method of coloring some of them. The way we color the squares will depend on some point a in the plane, a positive integer m which selects the starting generation, and a fixed non-negative Lebesgue integrable function Φ with compact support.

Say that two squares are *adjacent squares* if the side of one is contained in the side of the other. A *path of squares* is a finite sequence $\{T_j\}_{j=1}^{m}$ such that T_j and T_{j-1} are adjacent for $2 \le j \le m$. When this happens, say that T_1 and T_m are *joined by a path of squares*.

Let Φ be a non-negative function in $L^1(\mathscr{A})$ with compact support. This function will remain fixed in this section. In the applications Φ will be of the form $\sum_k |\hat{\nu}_k|$, where $\nu_1, \ldots, \nu_n$ are measures with compact support that annihilate the polynomials. But at this time we will maintain the generality because the particular form has nothing to do with the coloring scheme to be defined. In fact the only role of Φ is to divide the squares in $\mathscr{S}(n)$ into two classes. Say that a square S in $\mathscr{S}(n)$ is *light* if $\int_S \Phi d\mathscr{A} \leq \mathscr{A}(S)^2$; otherwise say that S is *heavy*. Note that squares that are disjoint from the support of Φ are light.

Suppose we are given a point a in the plane, a non-negative Lebesgue integrable function Φ with compact support, and choose a positive integer m which signals the generation of squares in which the coloring commences. Choose any square S in $\mathscr{S}(m)$ such that its closure contains the point a. Color this square yellow and let $\Gamma_m = \partial S$. This completes the coloring we will do of squares in the mth generation and we proceed to the $(m+1)$-st generation.

Let $n = m + 1$; the role of m is suppressed here as the general induction step of coloring the nth generation squares once the $(n-1)$-st squares are colored is the same. We start by coloring some squares green. If S is a light square in $\mathscr{S}(n)$ such that S is outside Γ_{n-1} and has a side lying on Γ_{n-1}, color S green. Refer to this as step 1 in the green coloring. For step 2, color green any light square S in $\mathscr{S}(n)$ that lies outside Γ_{n-1} and can be joined to a green square from step 1 by a path of light squares in $\mathscr{S}(n)$ each of which is outside Γ_{n-1}. These are all the green squares in the nth generation; denote the collection of such squares by $\mathscr{G}(n)$. So, loosely speaking, $\mathscr{G}(n)$ consists of all the light squares outside of Γ_{n-1} that can be joined to this polygon by a path of light squares from the nth generation.

It may well be that $\mathscr{G}(n)$ is empty; if so proceed to the next stage of coloring described below. It may also be that $\mathscr{G}(n)$ is infinite. Indeed, once we get a green square that lies in the unbounded component of the complement of the support of Φ, every square in that unbounded component will be colored green. If this does indeed occur, stop the whole process; don't color anymore, the coloring is complete. If $\mathscr{G}(n)$ is infinite, then we say that the (Φ, m, a) coloring scheme produces *a light route to* ∞.

Suppose there were a finite number of green squares in the nth generation. From the construction process, the union of Γ_{n-1} together with the closures of the squares in $\mathscr{G}(n)$ is connected and hence its polynomially convex hull is also connected. Let γ_n be the boundary of this polynomially convex hull; from the connectedness, γ_n is a polygon. (If it is the case that $\mathscr{G}(n) = \varnothing$, then $\gamma_n = \Gamma_{n-1}$.) If $S \in \mathscr{S}(n)$, S is outside γ_n, and S has a side lying on γ_n, color S red. Let $\mathscr{R}(n)$ be the collection of all the red squares in the nth generation. Notice that each red square is heavy. In fact, by the construction there can be no light squares bordering on γ_n that are outside of γ_n since they would be joined to a green square. Refer to this collection of red squares as a *heavy barrier* or *red barrier* about a. It is when these barriers exist in abundance (one for each generation) that we will be able to show that the point a is an analytic bounded point evaluation.

To get an idea of what will take place, it is helpful to make an analogy with Runge's Theorem. If K is a compact subset of $\mathbb{C}$, then uniform limits of polynomials will have an analytic extension to a neighborhood of the point a exactly when a is in the interior of the polynomially convex hull of K; that is, exactly when K forms a barrier between a and ∞. If there is a path from a to ∞, then a is outside the polynomially convex hull of K and limits of polynomials do not have natural analytic continuations to a. Here we will not be taking uniform limits but $L^2(\mu)$ limits of polynomials and the effect of constructing a profusion of red barriers is to construct an $L^2(\mu)$ barrier around a that will enable functions in $P^2(\mu)$ to have analytic continuations to a neighborhood of a. The role of μ in this coloring scheme will be in the determination of the function Φ.

Now we construct a buffer zone of yellow squares in $\mathscr{S}(n)$ that insulates the red barrier from the next generation of green and red squares. If $S \in \mathscr{S}(n)$, color S yellow if it satisfies the following three properties:

(i) S is outside γ_n;
(ii) S has no side lying in γ_n;
(iii) the distance from S to some red square in $\mathscr{R}(n)$ is less than or equal to $n^2 2^{-n}$.

Let $\mathscr{Y}(n)$ be the collection of all yellow squares in the nth generation. The reason for the conditions (i) and (ii) is to keep the yellow squares distinct from the previously colored squares and to push further out away from a. The purpose of condition (iii) is to guarantee that we do not push too far away. The whole idea of the yellow squares is that they will insulate what we have colored from the squares that will be colored in the next generation.

This concludes the coloring of squares in the nth generation. So the picture of the squares that are colored in the nth generation is a core of green squares connected to the polygon Γ_{n-1}, then a collar of red squares surrounding the green squares followed by a buffer zone of yellow squares. In any generation there may be no green squares but there are always red and yellow squares.

Now to proceed to color squares from the $(n + 1)$-st generation. First observe that the union of the closures of all the red and yellow squares in the nth generation must be connected. Indeed, first note that the union of the closures of the squares in $\mathscr{R}(n)$ is connected since this union contains γ_n and each square in $\mathscr{R}(n)$ has a side in γ_n. Moreover, each square in $\mathscr{Y}(n)$ can be connected to a square in $\mathscr{R}(n)$ by a line segment in the union. To see this fix Y in $\mathscr{Y}(n)$ and let $\delta =$ the distance from Y to the union of all the red squares in $\mathscr{R}(n)$. Pick R in $\mathscr{R}(n)$ such that $\delta = \text{dist}(Y, R)$. If $\delta = 0$ there is nothing to prove; so assume that $\delta > 0$. Let L denote the straight line segment of length δ connecting $\text{cl}\, Y$ to $\text{cl}\, R$. Now the set $\{z : \text{dist}(z, Y) < \delta\}$ does not meet any red square so that L does not meet any red square. Thus L never meets γ_n. Therefore by condition (iii) every square in $\mathscr{S}(n)$ that meets the segment L must be a yellow square. This shows that the union of the closures of all the red and yellow squares must be connected. So if Γ_n is the boundary of the polynomially convex hull of the union of the closures of all squares in $\mathscr{R}(n)$ and $\mathscr{Y}(n)$, then Γ_n is a polygon. Note that the segments that make up Γ_n consist of sides of squares that are yellow.

Now the process of coloring squares in the $(n+1)$-st generation can begin by coloring green all the light squares in $\mathscr{S}(n+1)$ that lie outside Γ_n and have a side in Γ_n, as well as those light squares in $\mathscr{S}(n+1)$ that be joined to one of these by a path of light squares from $\mathscr{S}(n+1)$. Continue as before.

Remember that this coloring scheme depends on the triple consisting of the point a, the non-negative function Φ in $L^1(\mathscr{A})$ with compact support, and the integer m which signals the generation with which the coloring scheme begins. We will refer to this coloring scheme as the (Φ, m, a) coloring scheme.

The next proposition is obvious, but it expresses the fundamental dichotomy that is the essence of future proofs.

1.1 PROPOSITION. *If Φ is a non-negative function in $L^1(\mathscr{A})$ with compact support and $a \in \mathbb{C}$, then precisely one of the following statements is true.*

(a) *For every positive integer m, the (Φ, m, a) coloring scheme produces a light route to ∞.*
(b) *There is a positive integer m such that the (Φ, m, a) coloring scheme produces a sequence of heavy barriers about a.*

Now for some properties of this technicolored plane. If a straight line segment is contained in the segment joining the centers of two adjacent green squares, call it a *green segment*. The *green part of a polygonal path* is the union of all the green segments in it. The *orange part of a polygonal path* is the part of the path that is not green. In particular, if part of the polygonal path is a segment that is neither horizontal nor vertical, then this segment is orange.

In the results that follow we are still assuming that a, m, and Φ are fixed.

1.2. LEMMA. *If $R \in \mathscr{R}(n)$, then $\text{dist}(R, \Gamma_n) > n^2 2^{-n}$ and so $\text{dist}(\Gamma_n, \Gamma_{n-1}) > n^2 2^{-n}$.*

PROOF. The first inequality follows by the construction of the buffer zone of yellow squares in $\mathscr{Y}(n)$; R lies inside this buffer zone and Γ_n lies outside. Now any line segment from Γ_{n-1} to Γ_n must cut through γ_n and hence must meet the closure of a red square. So there is a square R in $\mathscr{R}(n)$ such that $\text{dist}(\Gamma_n, \Gamma_{n-1}) \geq \text{dist}(R, \Gamma_n) > n^2 2^{-n}$. □

1.3 LEMMA. *There is a universal constant $C_1 > 0$ such that if S is any colored square in $\mathscr{S}(n)$, then there is a polygonal path from the center of S*

to the center of a yellow square in $\mathscr{Y}(n-1)$ such that the orange part of this path can be covered by a finite number of open disks with radii $\{r_i\}$ satisfying

$$\sum_i r_i^{1/2} < C_1 n 2^{-n/2}.$$

PROOF. First assume that $S = G \in \mathscr{G}(n)$. Here there is a green path from the center of G to the center of a square G_1 in $\mathscr{G}(n)$ which has a side on Γ_{n-1}. But then there is a square Y in $\mathscr{Y}(n-1)$ that is adjacent to G_1 and lies on the other side of Γ_{n-1}. The orange part of the resulting path only contains a line segment of length less than 2^{-n+1}. This won't cause us any problem as long as $C_1 > 2$.

Now suppose $S = R \in \mathscr{R}(n)$. Either R is adjacent to a square Y in $\mathscr{Y}(n-1)$ (this happens if $\mathscr{G}(n) = \varnothing$), or there is a square G in $\mathscr{G}(n)$ that is adjacent to R lying on the inside of γ_n. If the former happens, just join the center of R and Y and, as in the last paragraph, no problem will arise as long as $C_1 > 2$. If the latter is the case, join the centers of R and G and then join the center of G to a square Y in $\mathscr{Y}(n-1)$ as in the preceding paragraph. So the orange part of this path has only two line segments each of length at most 2^{-n+1}. Be sure that $C_1 > 4$.

Finally, assume that $S \in \mathscr{Y}(n)$. By definition, there is a square R in $\mathscr{R}(n)$ such that $\mathrm{dist}(S, R) < n^2 2^{-n}$. Join the center of S to the center of R by a straight line segment; this segment must have length smaller than $(n^2 + 2) 2^{-n}$ and so is contained in a disk of that radius. From the preceding paragraph we know that the center of R can be joined to the center of a square Y in $\mathscr{Y}(n-1)$ which can be covered by disks of the correct radius. It is left to the reader to make the final assembly of these pieces and complete the proof. □

1.4 LEMMA. *There is a universal constant $C_2 > 0$ such that if S is any colored square in $\mathscr{S}(n)$, then there is a polygonal path from the center of S to the center of the solitary yellow square in the mth generation such that the orange part of this path can be covered by a finite number of open disks of radii $\{r_i\}$ satisfying $\sum_i r_i^{1/2} < C_2 m 2^{-m/2}$.*

PROOF. Apply Lemma 1.3 $(n-m)$ times and let P be the union of all the resulting polygonal paths. If $\{r_i\}$ are the radii of all the disks involved, then

$$\sum_i r_i^{1/2} < \sum_{k=m+1}^{n} C_1 \frac{k}{2^{k/2}}.$$

Note that such a series can be estimated using

$$\sum_{k=m+1}^{\infty} kx^k = x^{m+1} \sum_{k=1}^{\infty}(k+m)x^{k-1}$$
$$= x^{m+1}\left[\frac{d}{dx}\left(\frac{1}{1-x}\right) + m\left(\frac{1}{1-x}\right)\right]$$
$$= x^{m+1}\left[\frac{1}{(1-x)^2} + \frac{m}{1-x}\right].$$

Thus

$$\sum_i r_i^{1/2} \leq C_1 \sum_{k=m+1}^{\infty} \frac{k}{2^{k/2}}$$
$$= C_1 2^{-m/2}\left(\frac{\sqrt{2}}{\sqrt{2}-1} + m\right)\frac{1}{\sqrt{2}-1}$$
$$\leq 5m2^{-m/2}C_1.$$

The choice for C_2 is clear. □

We return to analytic capacity γ (V.12.2) and a result not needed until now. The proof of this proposition is rather involved and will not be given here.

1.5 PROPOSITION (Davie [1972b]). *There are constants ε_0 and δ_0 such that if K is any compact connected subset of $\mathbb{C}$ with $d = \operatorname{diam} K$ and $\{\Delta_i\}$ is a sequence of open disks with radii $\{r_i\}$ such that $\sum_i r_i^{1/2} < \varepsilon_0 d^{1/2}$, then $\gamma(K \setminus \bigcup_i \Delta_i) \geq \delta_0 d$.*

1.6 LEMMA. *There is a universal constant C_3 with the property that if $\delta > 0$ and a is any point in $\mathbb{C}$ such that for every positive integer m the (Φ, m, a) coloring scheme produces a light route to ∞, then for all sufficiently large $m \geq 1$ there is a polygonal path contained in $\overline{B}(a; 3\delta/4)$ whose green part has analytic capacity greater than $C_3 \delta$.*

PROOF. Without loss of generality we may assume that $a = 0$. Let ε_0 and δ_0 be as in the preceding proposition and let C_2 be as in Lemma 1.4. Choose m such that $2^{-m} < \delta/4$ and $C_2 m 2^{-m/2} < \varepsilon_0(\frac{1}{2}\delta)^{1/2}$. Fix this integer m and consider the $(\Phi, m, 0)$ coloring scheme. By hypothesis there is an integer $n \geq m$ such that $\mathscr{G}(n)$ is infinite. Let G be a square in $\mathscr{G}(n)$ such that $G \cap \overline{B}(0; \delta) = \varnothing$. According to Lemma 1.4 there is a polygonal path P from the center of G to the center of the yellow square Y in $\mathscr{Y}(m)$ such that the orange part of P is covered by open disks $\{\Delta_i\}$ of radii $\{r_i\}$ with $\sum_i r_i^{1/2} \leq C_2 m 2^{-m/2}$. Parametrize P as $z: [0, 1] \to \mathbb{C}$ so that $z(0)$ is the center of G. Let $r = \max\{t: |z(t)| = 3\delta/4\}$ and let $s = \min\{t: |z(t)| = \delta/4\}$. Let Q be the polygonal path $\{z(t): r \leq t \leq s\}$ and note that $\operatorname{diam} Q \geq \delta/2$

§1 A COLORING SCHEME

and so $\sum_i r_i^{1/2} < \varepsilon_0 (\operatorname{diam} Q)^{1/2}$. By the preceding proposition, $\gamma(Q \setminus \bigcup_i \Delta_i) \geq (\operatorname{diam} Q)\delta_0 \geq \delta_0(\delta/2)$. But the disks $\{\Delta_i\}$ cover the orange part of Q so that $Q \setminus \bigcup_i \Delta_i$ consists entirely of green part. □

Let's point out that the preceding lemma proves the existence of an infinite sequence of polygonal paths with the stated property, one for each of the sufficiently large integers m. Also the paths depend on a and Φ as well, since the definition of its green part depends on the coloring scheme.

We now perform a construction when the (Φ, m, a) coloring scheme produces a light route to ∞. Assume that the (Φ, m, a) coloring scheme ends with the ith generation. That is, $\mathscr{G}(i)$ has an infinite number of squares. As before $\mathscr{S}(k) = \{S_{kp}\}_{p=1}^{\infty}$; let z_{kp} be the center of S_{kp}. We define another collection of squares (which is multigenerational) by enlarging some of the squares from $\bigcup\{\mathscr{S}(k) : m \leq k \leq i\}$ as follows. If $m < k < i$ and S_{kp} lies between Γ_{k-1} and Γ_k, let Q_{kp} be the square with the same center as S_{kp}, z_{kp}, but with each side enlarged by a factor of $5/4$. If $k = m$, enlarge the one yellow square and call it Q_{mp}. If $k = i$, enlarge each S_{ip} that lies outside the polygon Γ_{i-1} by a factor of $5/4$ calling these enlargements Q_{ip}.

Note that the collection of all the squares Q_{kp} that have been thus defined form an open cover of the plane. Let's relabel these enlarged squares and call them $\{Q_r\}$; let z_r be the center of Q_r and let δ_r be the length of one of its sides. If Q_r is the enlargement of a square from the nth generation, say that Q_r is an nth generation square.

Here are some elementary facts about these squares that the reader can easily verify by geometrical considerations.

1.7 LEMMA.

(a) For each k, $Q_{kp} \subseteq B(z_{kp}; 2^{-k})$.
(b) If Q_{kt} and Q_{kp} are disjoint, then $\operatorname{dist}(Q_{kt}, Q_{kp}) \geq (3/4)2^{-k}$.
(c) Each point in the plane lies in at most 4 of the squares Q_r.
(d) For any k, one of the squares Q_{kp} meets at most 10 of the squares Q_r from the kth generation.

1.8 PROPOSITION. *If the (Φ, m, a) coloring scheme ends after the ith generation and $\{Q_r\}$ is the sequence of squares described preceding the proposition, then the following statements hold.*

(a) *There is a C^1 partition of unity $\{\phi_r\}$ subordinate to the cover $\{Q_r\}$ such that $\|\overline{\partial}\phi_r\|_\infty \leq 100/\delta_r$ for every $r \geq 1$.*
(b) *There is a universal constant C_4 such that if z lies inside Γ_{n-2} or outside Γ_{n+1} and Q_r is an nth generation square, then $\operatorname{dist}(z, Q_r) \geq C_4 n^2 2^{-n}$.*
(c) *There is a universal constant C_5 such that for every z in $\mathbb{C}$*

$$\sum_r \min\left\{1, \frac{\delta_r^3}{|z - z_r|^3}\right\} \leq C_5.$$

(d) *If Q_r and Q_s are enlargements of adjacent squares in $\mathscr{G}(n)$ and L is the segment from z_r to z_s, then for t distinct from r and s, $\operatorname{dist}(L, Q_t) \geq \delta_r/4$.*

PROOF. The proof of (a) is in Davie [1972b], page 417, or the reader can use Lemma V.22.10.

(b) Let Q_r be an enlargement of S_{np}; by definition S_{np} lies between Γ_{n-1} and Γ_n. If z is inside Γ_{n-2}, then Lemma 1.2 implies that $\operatorname{dist}(z, S_{np})$ $\geq \operatorname{dist}(z, \Gamma_{n-1}) \geq \operatorname{dist}(\Gamma_{n-2}, \Gamma_{n-1}) \geq (n-1)^2 2^{-(n-1)}$. If z lies outside Γ_{n+1}, then again Lemma 1.2 implies that $\operatorname{dist}(z, S_{np}) \geq \operatorname{dist}(z, \Gamma_n) \geq \operatorname{dist}(\Gamma_{n+1}, \Gamma_n) \geq (n+1)^2 2^{-(n+1)}$. Since Q_r is an enlargement of S_{np} by the factor $5/4$, part (b) easily follows.

(c) First it will be shown that there is a constant C_6 such that for all z in $\mathbb{C}$ and for $m \leq k \leq i$,

$$(1.9) \qquad \sum_p \min\left(1, \frac{2^{-3k}}{|z - z_{kp}|^3}\right) \leq C_6 \min\left(1, \frac{2^{-k}}{\operatorname{dist}(z, \bigcup_p Q_{kp})}\right),$$

where the sum on the left of this inequality is over all p for which Q_{kp} is defined. (This same proscription will apply for the remainder of the proof of this part.) To see this claim, fix z and let $\delta = \operatorname{dist}(z, \bigcup_p Q_{kp})$ and assume that $\delta > 2^{-k}$ so that the minimum on the right-hand side of (1.9) is $2^{-k}/\delta < 1$. Let $P_n \equiv \{p : n\delta < |z - z_{kp}| \leq (n+1)\delta\}$. Note that $|z - z_{kp}| > \operatorname{dist}(z, Q_{kp}) \geq \delta$ so that every integer p belongs to a set P_n for a unique value of n. Let $N(n)$ be the number of integers in P_n. If $p \in P_n$, then $Q_{kp} \subseteq B(z_{kp}; 2^{-k}) \subseteq B(z_{kp}; \delta) \subseteq \{w : (n-1)\delta \leq |z - w| \leq (n+2)\delta\} \equiv A$. Since each point in the plane is contained in at most 4 of the squares Q_{kp}, this implies that $N(n) \cdot \operatorname{Area}(Q_{kp}) \leq 4 \operatorname{Area}(A)$. Performing the necessary calculations we get that for $n \geq 1$,

$$N(n)\left(\frac{5}{4}\right)^2 (2^{-k})^2 \leq 4\{\pi(n+2)^2 \delta^2 - \pi(n-1)^2 \delta^2\}$$
$$= 4\pi\delta^2(6n + 3)$$
$$\leq 144\delta^2 n.$$

Thus for all $n \geq 1$,

$$N(n) \leq \frac{100\delta^2}{(2^{-k})^2} n.$$

Also if $p \in P_n$, then $2^{-k}/|z - z_{kp}| \leq 2^{-k}/n\delta$. Therefore in this case

$$\sum_p \min\left(1, \frac{2^{-3k}}{|z - z_{kp}|^3}\right) \leq \sum_{n=1}^{\infty} \sum_{p \in P_n} \frac{2^{-3k}}{|z - z_{kp}|^3}$$

$$\leq \sum_{n=1}^{\infty} N(n) \left(\frac{2^{-k}}{n\delta}\right)^3$$

$$\leq \frac{100\delta^2 (2^{-k})^3}{\delta^3 (2^{-k})^2} \sum_{n=1}^{\infty} \frac{n}{n^3}$$

$$= \left(100 \sum_{n=1}^{\infty} \frac{1}{n^2}\right) \frac{2^{-k}}{\delta}.$$

So we want to have $C_6 \geq 100 \sum_n n^{-2}$.

For the other case where $\delta \leq 2^{-k}$ and the minimum of the right-hand side of (1.9) is 1, let $P_1 = \{p : \min(1, 2^{-3k}|z - z_{kp}|^{-3}) = 1\}$ and let $P_2 = \{p : \min(1, 2^{-2k}|z - z_{kp}|^{-3}) < 1\}$. If $p \in P_1$, then $|z - z_{kp}| \leq 2^{-k}$ and so $\text{dist}(z, Q_{kp}) \leq 2^{-k} - (5/8)2^{-k} = (3/8)2^{-k}$. Thus (1.7b) implies that if z belongs to any square Q_{kt}, Q_{kp} and Q_{kt} must intersect. Since z belongs to at most 4 such squares Q_{kt}, (1.7d) implies that P_1 has at most 40 elements.

If $p \in P_2$, then $|z - z_{kp}| > 2^{-k}$. So if for each $n \geq 1$, $P_{2,n} = \{p : n2^{-k} < |z - z_{kp}| \leq 2^{-k}(n+1)\}$, then $P_2 = \bigcup_n P_{2,n}$. Let $N(n)$ be the number of integers p in $P_{2,n}$. Arguing as above it can be shown that $N(n) \leq 100n$ from which it follows that in this case

$$\sum_p \min\left(1, \frac{2^{-3k}}{|z - z_{kp}|^3}\right) \leq 40 + 100 \sum_{n=1}^{\infty} \frac{1}{n^2}.$$

So if we let $C_6 = 40 + 100 \sum_n n^{-2}$, we have (1.9).

Now suppose z lies between Γ_{n-1} and Γ_n. For $k \leq n-2$, z is outside Γ_{k+1} and so $\text{dist}(z, \bigcup_p Q_{kp}) \geq C_4 k^2 2^{-k}$ by part (b). Likewise, if $k \geq n+2$, then z is inside Γ_{k-2} and so $\text{dist}(z, \bigcup_p Q_{kp}) \geq C_4 k^2 2^{-k}$.

Thus for z between Γ_{n-1} and Γ_n,

$$\sum_r \min\left\{1, \frac{\delta_r^3}{|z-z_r|^3}\right\} = \sum_k \sum_p \min\left(1, \frac{(\frac{5}{4})^3 2^{-3k}}{|z-z_{kp}|^3}\right)$$

$$= \sum_{k=n-1}^{n+1} \min\left(1, \frac{(\frac{5}{4})^3 2^{-3k}}{|z-z_{kp}|^3}\right)$$

$$+ \sum_{|k-n|\geq 2} \min\left(1, \frac{(\frac{5}{4})^3 2^{-3k}}{|z-z_{kp}|^3}\right)$$

$$\leq 3\left(\frac{5}{4}\right)^3 C_6$$

$$+ \left(\frac{5}{4}\right)^3 \sum_{|k-n|\geq 2} C_6 \min\left(1, \frac{2^{-k}}{\operatorname{dist}(z, \bigcup_p Q_{kp})}\right)$$

$$\leq 3\left(\frac{5}{4}\right)^3 C_6 + \left(\frac{5}{4}\right)^3 \frac{C_6}{C_4} \sum_{|k-n|\geq 2} \frac{2^{-k}}{k^2 2^{-k}}$$

$$\leq 3\left(\frac{5}{4}\right)^3 C_6 + \left(\frac{5}{4}\right)^3 \frac{C_6}{C_4} \sum_{k=1}^{\infty} \frac{1}{k^2}$$

$$= C_5.$$

(d) It suffices to restrict our attention to a consideration of the case that Q_t meets either Q_r or Q_s. It follows from the construction that Q_t must be in either the $(n-1)$-st or the nth generation. Thus its sides are at most of length $(5/4)2^{-(n-1)}$. Since the squares Q_r, Q_s, and Q_t are enlargements of disjoint squares, the desired estimate follows. □

It will be necessary to renew our acquaintance with the Vitushkin localization operators (§V.5). Indeed, this is the reason for part (a) of the preceding proposition. This will come to the fore in Section 3.

§2 A sufficient condition for the existence of bounded point evaluations. In this section the following sufficient condition for the existence of bounded point evaluations due to Thomson [**preprint**] will be proved. Recall that if a measure with compact support annihilates the polynomials, then its Cauchy transform vanishes off the polynomially convex hull of its support and thus the Cauchy transform has compact support.

2.1 THEOREM. *If $\nu_1, \ldots, \nu_s$ are compactly supported measures on $\mathbb{C}$ that annihilate the polynomials, $\Phi = \max\{|\hat{\nu}_j| : 1 \leq j \leq s\}$, $\mu = |\nu_1| + \cdots + |\nu_s|$, and a is a complex number such that for some integer $m \geq 1$ the (Φ, m, a) coloring scheme produces a sequence of heavy barriers, then there is a constant $C > 0$ and a neighborhood V of a such that $|p(z)| \leq C \int |p| d\mu$ for every polynomial p and every point z in V.*

The perspicacious reader has noted that this theorem is not actually a sufficient condition for the existence of analytic bounded point evaluations. In fact there are two difficulties with any such claim.

First this only guarantees that the point a is an analytic bounded point evaluation for the Banach space $P^1(\mu)$, but as we shall see this is not a serious impediment. More serious is that the type of measure μ seems restricted, but an application of Proposition VI.1.11 can be used to counteract this objection. This will be explored later.

The second difficulty with this theorem is the fact that we have no idea when heavy barriers exist or whether they exist at all; this difficulty is short lived. Proposition 1.1 tells us that for such a function Φ every point a in the plane satisfies one of two incompatible conditions; either a satisfies the hypothesis of this theorem or for every integer m the (Φ, m, a) coloring scheme results in a light route to ∞. In the next section (Theorem 3.1) we will see that the set of points a satisfying the hypothesis of Theorem 2.1 is sufficiently large that the closure of this set contains the support of μ.

The method of proof of Theorem 2.1 is an employment of a variation of the "Scott Brown process" (S. W. Brown [1978]) to improve a sufficient condition for the existence of analytic bounded point evaluations due to Brennan [1971a] (see Exercise 1).

If Γ is a rectifiable Jordan curve in $\mathbb{C}$ and K is the union of Γ with its inside, then $R(K)$ is a Dirichlet algebra and so we can consider $H^\infty(\Gamma) = H^\infty(\partial K)$ as in §VI.4. Recall that $H^\infty(\Gamma)$ is identified with $H^\infty(\operatorname{ins}\Gamma)$ so that each bounded analytic function on $\operatorname{ins}\Gamma$ is identified with its boundary values on Γ. Let ω be harmonic measure for the inside of Γ evaluated at some point inside Γ. (This point won't be specified here.) So $H^\infty(\Gamma) \cong P^\infty(\omega)$ (VI.4.9). In particular, we can talk about the Banach space predual of $H^\infty(\Gamma)$, $H^\infty(\Gamma)_*$, and its norm, $\|\cdot\|_*$. So for L in $H^\infty(\Gamma)_*$, $\|L\|_* \equiv \sup\{|L(\phi)|: \phi \in \operatorname{ball} H^\infty(\Gamma)\}$. Also recall from general Banach space theory that $H^\infty(\Gamma)_*$ can be identified with the quotient space $L^1(\omega)/H^\infty(\Gamma)^\perp$.

For a final bit of notation before beginning the Scott Brown process, if w is any point inside Γ, let e_w be the element of $H^\infty(\Gamma)_*$ defined by $e_w(\phi) = \phi(w)$ for all ϕ in $H^\infty(\Gamma)$. Note that $\|e_w\|_* = 1$.

The first lemma is a rephrasing of Schwarz's Lemma.

2.2 LEMMA. *If $\phi \in H^\infty(\partial \mathbb{D})$ and $z \in \mathbb{D}$, $|\phi(z) - \phi(0)| \leq 2\|\phi\||z|$.*

To return to earlier considerations, let $\nu_1, \ldots, \nu_s$ be measures with compact support that annihilate the polynomials and put $\Phi = \max\{|\hat{\nu}_j|: 1 \leq j \leq s\}$. Suppose a is a complex number for which a sequence of heavy barriers can be obtained. Specifically, assume $a \in \mathbb{C}$ and there is an integer m such that the (Φ, m, a) coloring scheme results in a sequence of heavy barriers. This produces the sequences of polygons $\{\gamma_n\}$ and $\{\Gamma_n\}$ as in Section 1. Fix this notation for the present as well as the remaining notation from Section 1. Also for $n \geq m$, let

$$\|\cdot\|_{*(n)} = \text{the norm in } H^\infty(\Gamma_n)_*.$$

2.3 LEMMA. *If $n > m$, $w \in \gamma_n$, and $R \in \mathscr{R}(n)$ such that $w \in \operatorname{cl} R$, then for every z in R,*

$$\|e_z - e_w\|_{*(n)} \leq \frac{3}{n^2}.$$

PROOF. From the construction of the yellow buffer zone, $\operatorname{dist}(R, \Gamma_n) \geq n^2 2^{-n}$ and so $\operatorname{dist}(w, \Gamma_n) \geq n^2 2^{-n}$; also $|w - z| \leq (3/2) 2^{-n}$. So if $|\zeta| < 1$, $w + n^2 2^{-n} \zeta \in \operatorname{ins} \Gamma_n$ and therefore whenever $\phi \in H^\infty(\Gamma_n)$, $\psi(\zeta) = \phi(w + n^2 2^{-n} \zeta) \in H^\infty(\partial \mathbb{D})$. By Lemma 2.2, if $\phi \in \operatorname{ball} H^\infty(\Gamma_n)$, $|\psi(\zeta) - \psi(0)| \leq 2|\zeta|$ for $|\zeta| < 1$. This implies that for every z in R, $|\phi(z) - \phi(w)| \leq 2|z - w| n^{-2} 2^n \leq 3n^{-2}$. Since ϕ was an arbitrary function in ball $H^\infty(\Gamma_n)$, this proves the lemma. □

Let's keep track of the crucial properties of red squares that are being used. In the lemma just concluded, the only property of a red square that is used is that its distance from Γ_n is at least $n^2 2^{-n}$. The fact that red squares are heavy did not enter into the argument.

As in the preceding lemma, let $n > m$ and let w be a point in γ_n. From the definition of γ_n, there is a square R in $\mathscr{R}(n)$ such that $w \in \operatorname{cl} R$. Since R is red, R is heavy and so $\mathscr{A}(R)^2 < \int_R \Phi \, d\mathscr{A}$; define a constant c by $c^{-1} = \int_R \Phi \, d\mathscr{A}$ and let τ_w be the probability measure $c\Phi \mathscr{A}|R$. (Note R, and hence τ_w, may not be unique. This won't cause any difficulty because the measures τ_w will only have transitory interest.) Since $\operatorname{cl} R \subseteq \operatorname{ins} \Gamma_n$, the measure τ_w defines (via integration) a weak* continuous linear functional on $H^\infty(\Gamma_n)$ (Exercise 2). Also the fact that R is heavy implies that $c < \mathscr{A}(R)^{-2} = 2^{4n}$.

2.4 LEMMA. *If $n > m$, $\mathscr{R}(n) \neq \varnothing$, and $w \in \gamma_n$, then $\|e_w - \tau_w\|_{*(n)} \leq 3n^{-2}$.*

PROOF. If ϕ is an arbitrary function in ball $H^\infty(\Gamma_n)$, then

$$\left|\phi(w) - \int \phi \, d\tau_w\right| = \left|\int [\phi(w) - \phi] \, d\tau_w\right|$$

$$\text{(by (2.3))} \quad \leq \frac{3}{n^2} \|\tau_w\|$$

$$= \frac{3}{n^2}. \quad \square$$

Note that if $m < k < n$ and $L \in H^\infty(\Gamma_k)_*$, then L defines a weak* continuous linear functional on $H^\infty(\Gamma_n)$ by "restriction." That is, if $\phi \in H^\infty(\Gamma_n)$, then the restriction of ϕ to Γ_k defines a function in $H^\infty(\Gamma_k)$ and so $\phi \to L(\phi|\Gamma_k)$ is a functional in $H^\infty(\Gamma_n)_*$. So whenever $L \in H^\infty(\Gamma_k)_*$ we will also think of L as defining a functional in $H^\infty(\Gamma_n)_*$ for all $n \geq k$.

2.5 LEMMA. *If $n > m$, $\mathscr{R}(n) \neq \varnothing$, and $L \in H^\infty(\Gamma_{n-1})_*$ then there are points $w_1, \ldots, w_q$ in γ_n and scalars $\alpha_1, \ldots, \alpha_q$ with $\sum_j |\alpha_j| \leq \|L\|_{*(n-1)}$*

such that

$$\left\| L - \sum_{j=1}^{q} \alpha_j \tau_{w_j} \right\|_{*(n)} < \frac{4}{n^2} \|L\|_{*(n-1)}.$$

PROOF. It suffices to assume that $\|L\|_{*(n-1)} = 1$. Note that if $\phi \in H^\infty(\Gamma_n)$, then $|L(\phi)| \leq \|\phi\|_{\Gamma_{n-1}} \leq \|\phi\|_{\gamma_n}$. So if C is the closed convex hull in $H^\infty(\Gamma_n)_*$ of $\{\alpha e_w : \alpha \in \mathbb{C} \text{ with } |\alpha| \leq 1 \text{ and } w \in \gamma_n\}$, then $|L(\phi)| \leq \sup\{|F(\phi)| : F \in C\}$. By the Hahn-Banach Theorem, $L \in C$. Thus there are points $w_1, \ldots, w_q$ in γ_n and scalars $\alpha_1, \ldots, \alpha_q$ with $\sum_j |\alpha_j| \leq 1$ such that $\|L - \sum_j \alpha_j e_{w_j}\|_{*(n)} < n^{-2}$. The present lemma now follows from the preceding one. □

The next lemma just combines the preceding one with the definition of the measures τ_w and the observation that for the constant c that occurs in the definition of τ_w, $c < \mathscr{A}(R)^{-2} = 2^{4n}$. Specifically, if $R_1, \ldots, R_q$ are the red squares corresponding to the points $w_1, \ldots, w_q$ obtained in the preceding lemma and $c_1, \ldots, c_q$ are the corresponding constants, let $h = \sum_j \alpha_j c_j \chi_{R_j}$ and we get the following lemma. Note that this is the first place that we will use the fact that red squares are heavy.

2.6 LEMMA. *If $n > m$, $\mathscr{R}(n) \neq \varnothing$, and $L \in H^\infty(\Gamma_{n-1})_*$, then there is a function h in $L^\infty(\mathscr{A})$ such that $\|h\|_\infty \leq 2^{4n}\|L\|_{*(n-1)}$, $h = 0$ off the union of the squares in $\mathscr{R}(n)$, and*

$$\|L - h\Phi\mathscr{A}\|_{*(n)} < \frac{4}{n^2} \|L\|_{*(n-1)}.$$

Now we must assume that the coloring scheme continues without end.

2.7 LEMMA. *If the (Φ, m, a) coloring scheme results in a sequence of heavy barriers, then for each integer $n > m$ there is a constant $M(n)$ such that for any L in $H^\infty(\Gamma_{n-1})_*$ there is a function h in $L^\infty(\mathscr{A})$ with $\|h\|_\infty \leq M(n)\|L\|_{*(n-1)}$ and*

$$L(p) = \int p h \Phi \, d\mathscr{A}$$

for every polynomial p.

PROOF. It suffices to assume that $\|L\|_{*(n-1)} = 1$. According to (2.6) there is a function h_n in $L^\infty(\mathscr{A})$ with $\|h_n\|_\infty \leq 2^{4n}$, $h = 0$ off the union of the red squares in $\mathscr{R}(n)$, and $\|L - h_n \Phi \mathscr{A}\|_{*(n)} < 4n^{-2}$. Let $L_n = L - h_n \Phi \mathscr{A}$; so $L_n \in H^\infty(\Gamma_n)_*$. Once again Lemma 2.6 implies that there is a function h_{n+1} in $L^\infty(\mathscr{A})$ with $\|h_{n+1}\|_\infty \leq (4/n^2)2^{4(n+1)}$ and $\|L_n - h_{n+1}\Phi\mathscr{A}\|_{*(n+1)} \leq 4(n+1)^{-2}[4n^{-2}]$. Continue. So for every $k \geq 0$ there is a function h_{n+k} in $L^\infty(\mathscr{A})$ with

$$\|h_{n+k}\|_\infty \leq \frac{4^k 2^{4(n+k)}}{[n(n+1)\ldots(n+k-1)]^2},$$

$$\|L - (h_n + \ldots + h_{n+k})\Phi\mathscr{A}\|_{*(n+k)} < \frac{4^{k+1}}{[n(n+1)\ldots(n+k)]^2}.$$

It is now routine to show that $h = \sum_{k=0}^{\infty} h_{n+k}$ is a function that works and

$$\|h\|_\infty \leq M(n) \equiv \sum_{k=0}^{\infty} \frac{4^k 2^{4(n+k)}}{[n(n+1)\ldots(n+k-1)]^2}.$$

Note that $M(n)$ depends only on n and not on the functional L. □

PROOF OF THEOREM 2.1. According to the hypothesis, there is a positive integer m such that the (Φ, m, a) coloring scheme does not end. Note that $a \in V \equiv \text{ins}\,\Gamma_{m+1}$. Lemma 2.7 implies there is a constant $M(= M(m+2))$ such that if b is any other point in V, then there exists a function h in $L^\infty(\mathscr{A})$ with $\|h\| \leq M$ and $p(b) = \int ph\Phi\,d\mathscr{A}$ for every polynomial p. Fix such a point b inside Γ_{m+1}.

Let E be the support of Φ, let $E_1, \ldots, E_s$ be a Borel partition of E, and let $h_1, \ldots, h_s$ be Borel functions such that $h\Phi = h_j \hat{\nu}_j$ on E_j, and $h_j = 0$ off E_j. If w is any complex number and the Cauchy transform $\hat{\nu}_j$ exists at w, then Proposition V.8.9 implies that

$$p(w)\hat{\nu}_j(w) = \int \frac{p(z)}{z-w} d\nu_j(z).$$

Therefore

$$|p(b)| = \left| \sum_{j=1}^{r} \int_{E_j} p(w) h_j(w) \hat{\nu}_j(w) \, d\mathscr{A}(w) \right|$$

$$= \left| \sum_{j=1}^{r} \int_{E_j} \left[\int \frac{p(z)}{z-w} d\nu_j(z) \right] h_j(w) \, d\mathscr{A}(w) \right|$$

$$= \left| \sum_{j=1}^{r} \int p(z) \left[\int_{E_j} \frac{h_j(w)}{z-w} d\mathscr{A}(w) \right] d\nu_j(z) \right|$$

(by Exercise V.2.6) $\leq \sum_{j=1}^{r} 2\sqrt{\pi} \|h_j\|_\infty \sqrt{\mathscr{A}(E_j)} \int |p|\,d|\nu_j|$

$$\leq 2\sqrt{\pi} M \sqrt{\mathscr{A}(E)} \int |p|\,d\mu$$

$$= C \int |p|\,d\mu. \quad \square$$

Exercises

1. (Brennan [1971a]) Suppose that ν is a measure with compact support and A is an annulus such that $|\hat{\nu}|$ is bounded away from 0 on A. Prove that for $\mu = |\nu|$ every point inside the inner radius of A is in $\text{abpe}(\mu)$.

2. If Γ is a rectifiable Jordan curve and τ is a measure whose support is contained inside Γ, show that $L(\phi) = \int \phi\,d\tau$ defines a weak* continuous linear functional on $H^\infty(\Gamma)$.

§3 Heavy barriers exist.

In the preceding section we saw that when a coloring scheme for a certain type of function Φ produces heavy barriers, this implies the existence of analytic bounded point evaluations for a certain $P^2(\mu)$ space. The purpose of this section is to prove the following theorem which states that the set of points a such that the (Φ, m, a) coloring scheme, for some integer $m \geq 1$, produces a sequence of heavy barriers is quite large. In the process additional results will be obtained about light routes and heavy barriers.

3.1 THEOREM. *Let $\nu_1, \ldots, \nu_s$ be measures in the plane with compact support that annihilate the polynomials. If $\Phi(z) = \max\{|\hat{\nu}_j(z)|: 1 \leq j \leq s\}$ and*

$$U = \{a \in \mathbb{C}: \text{there is an integer } m \geq 1 \text{ such that the } (\Phi, m, a) \text{ coloring}$$
$$\text{scheme produces a sequence of heavy barriers}\},$$

then $\operatorname{cl} U$ contains the supports of each of the measures ν_j, $1 \leq j \leq s$.

The proof needs some additional preliminary work as well as the aid of the Vitushkin localization operators (§V.5).

If f is a function that is analytic in a neighborhood of ∞ and $z_0 \in \mathbb{C}$, there is a Laurent expansion

$$f(z) = f(\infty) + \frac{a_1}{z - z_0} + \frac{a_2}{(z - z_0)^2} + \cdots$$

valid near ∞. We have $f'(\infty) = a_1$ and we will designate the coefficient a_2 by

$$\beta(f, z_0) = a_2.$$

Note that the first coefficient is independent of the choice of the point z_0 while the second is not. The first lemma is the result of a straightforward application of the Cauchy estimates.

3.2 LEMMA. *Let $\delta > 0$, $z_0 \in \mathbb{C}$, and put $D = \overline{B}(z_0; \delta)$. If f is a bounded analytic function on $\mathbb{C}_\infty \setminus D$, then $|f'(\infty)| \leq \delta \|f\|$ and $|\beta(f, z_0)| \leq \delta^2 \|f\|$.*

In the following lemmas the reference to terms associated with the coloring scheme introduced in the last section will always come from some triple (Φ, m, a). Either the members of this triple will be clear from the statement of the results or they will be some general choice of these parameters.

3.3 LEMMA. *Let $\nu_1, \ldots, \nu_s$ be measures with compact support that annihilate the polynomials and put $\Phi = \max\{|\hat{\nu}_j|: 1 \leq j \leq s\}$. If $d > 0$, S is any light square with side of length d and center z_0, and α and β are complex numbers with $|\alpha|, |\beta| \leq 1$, then there is a continuous function f on $\mathbb{C}_\infty$ having the following properties:*

(a) $\|f\| \leq 6$;
(b) f is analytic off $\operatorname{cl} S$;
(c) $f(\infty) = 0$;

(d) $f'(\infty) = \alpha d$;
(e) $\beta(f, z_0) = \beta d^2$;
(f) $|\int f \, d\nu_j| \le 7d^3$ for $1 \le j \le s$.

Proof. Without loss of generality we may assume that $z_0 = 0$. Put $b = \frac{1}{2}d$, let $D = B(0; b)$, and let τ be the measure $(1/\pi)\mathscr{A}|D$. We will compute the Cauchy transform $\widehat{\tau}$. First note that

$$\int_0^{2\pi} (re^{it} - w)^{-1} dt = \int_{|z|=r} \frac{1}{z-w} \frac{1}{iz} dz.$$
$$= \begin{cases} -\frac{2\pi}{w} & \text{if } r < |w| \\ 0 & \text{if } r > |w| \end{cases}.$$

Therefore for $|w| > b$,

$$\widehat{\tau}(w) = \frac{1}{\pi} \int_0^b \int_0^{2\pi} (re^{it} - w)^{-1} dt \, r dr$$
$$= \frac{1}{\pi} \int_0^b \left(\frac{-2\pi}{w}\right) r dr$$
$$= -\frac{b^2}{w}.$$

On the other hand, for $|w| < b$

$$\widehat{\tau}(w) = \frac{1}{\pi} \int_0^{|w|} \left(\frac{-2\pi}{w}\right) r dr$$
$$= -\overline{w}.$$

But the Cauchy transform of τ is a continuous function and so we get that $\widehat{\tau}(w) = -\overline{w} = -b^2/w$ for $|w| = b$.

Notice that $D \subseteq S$ and so, since S is a light square,

$$\int \Phi \, d\tau = \frac{1}{\pi} \int_D \Phi \, d\mathscr{A}$$
$$\le \frac{1}{\pi} \int_S \Phi \, d\mathscr{A}$$
$$\le \frac{1}{\pi} \mathscr{A}(S)^2$$
$$= \frac{d^4}{\pi}.$$

Thus

$$\left|\int \hat{\tau} d\nu_j\right| = \left|\int \left[\int \frac{1}{z-w} d\tau(z)\right] d\nu_j(w)\right|$$

$$= \left|\int \hat{\nu}_j d\tau\right|$$

$$\leq \int \Phi d\tau$$

$$\leq \frac{d^4}{\pi}.$$

Let $g(w) = -\hat{\tau}(w)/b$. So g is a continuous function on $\mathbb{C}_\infty$ and it has the following properties, the verification of which is straightforward.

(3.4)
$$\begin{cases} \text{(i)} & \|g\|_\infty \leq 1. \\ \text{(ii)} & g \text{ is analytic off cl } D. \\ \text{(iii)} & g(\infty) = 0. \\ \text{(iv)} & g'(\infty) = b. \\ \text{(v)} & \beta(g, 0) = 0. \\ \text{(vi)} & |\int g \, d\nu_j| < \frac{2d^3}{\pi} \text{ for } 1 \leq j \leq s. \end{cases}$$

Now let η = the measure $\hat{\tau}\tau = (1/\pi)\hat{\tau}\mathscr{A}|D$. From (V.3.3) we know that in the sense of distributions $\bar\partial(\hat\eta) = -\pi\eta = -\pi\hat\tau\tau$. Also $\bar\partial(\tfrac12\hat\tau^2) = \hat\tau\bar\partial(\hat\tau) = -\pi\hat\tau\tau$. Thus $\hat\eta = \tfrac12\hat\tau^2$. Therefore for $1 \leq j \leq s$,

$$\left|\int \hat\eta d\nu_j\right| = \left|\int \hat\nu_j d\eta\right|$$

$$= \left|\int_D \hat\nu_j \hat\tau d\tau\right|$$

$$\leq b \int_D \Phi d\tau$$

$$\leq \frac{bd^4}{\pi}.$$

Let $h = 2\hat\eta/b^2$. For w outside D, $h(w) = (2/b^2)\tfrac12(\hat\tau(w))^2 = b^2/w^2$; while for w in cl D, $h(w) = \overline{w}^2/b^2$. Therefore $h \in C(\mathbb{C}_\infty)$. In addition h has the properties

(3.5)
$$\begin{cases} \text{(i)} & \|h\|_\infty = 1. \\ \text{(ii)} & h \text{ is analytic off cl } D. \\ \text{(iii)} & h(\infty) = 0. \\ \text{(iv)} & h'(\infty) = 0. \\ \text{(v)} & \beta(h, 0) = b^2. \\ \text{(vi)} & |\int h \, d\nu_j| < \frac{4d^3}{\pi} \text{ for } 1 \leq j \leq s. \end{cases}$$

Once again these properties are easy to verify.

Let $f = 2\alpha g + 4\beta h$. It is claimed that this function works. Indeed, to verify this we need only examine (3.4) and (3.5) and use the fact that $|\alpha|$ and $|\beta| \leq 1$. □

The next lemma is the linchpin in the proof of Theorem 3.1.

3.6 LEMMA. *There is a universal constant C_7 such that if $\nu_1, \ldots, \nu_s$ are compactly supported measures on $\mathbb{C}$ that annihilate polynomials, $\Phi = \max\{|\hat{\nu}_j| : 1 \leq j \leq s\}$, δ and ε are positive numbers, and α and β are complex numbers with $0 \leq |\alpha|$, $|\beta| \leq 1$, $a \in \mathbb{C}$ and $D = B(a; \delta)$, and if for every integer m the (Φ, m, a) coloring scheme produces a sequence of light routes to ∞, then there is a function f in $C(\mathbb{C}_\infty)$ having the following properties:*

(a) $\|f\|_\infty \leq C_7$;
(b) f *is analytic on* $\mathbb{C}_\infty \setminus \operatorname{cl} D$;
(c) $f(\infty) = 0$;
(d) $f'(\infty) = \alpha\delta$;
(e) $\beta(f, a) = \beta\delta^2$;
(f) $|\int f \, d\nu_j| \leq \varepsilon$ *for* $1 \leq j \leq s$.

PROOF. Choose m_0 sufficiently large so that for $m \geq m_0$, $2^{-m} < \delta/8$ and the (Φ, m, a) coloring scheme produces a polygonal path P in $B(a; 3\delta/4)$ whose green part has analytic capacity greater than $C_3\delta$ (Lemma 1.6). The dependence of P on m is being suppressed here. Later we will specify m by choosing it so that in addition to the above properties it is also sufficiently large that $2^{-m} C_{10} \mathscr{A}(D) < \varepsilon$, where C_{10} is a universal constant that will be defined later. So assume that m is so chosen, fix it, and fix the polygonal path P. Let P_g denote the green part of P.

From the definition of analytic capacity there is a function g analytic on $\mathbb{C}_\infty \setminus P_g$ and satisfying

$$\|g\|_\infty \leq 1, \ g(\infty) = 0, \ \text{and } g'(\infty) \geq C_3\delta.$$

Also Lemma 3.2 implies that

$$\beta(g, a) \leq \frac{9}{16}\delta^2.$$

Suppose the ith generation is the last in the (Φ, m, a) coloring scheme; that is, i is the first integer after m for which $\mathscr{G}(i)$ is infinite. Put $U = \{z : \operatorname{dist}(z, P_g) < \frac{1}{4} 2^{-i}\}$. Redefine g on U so that g is continuous on $\mathbb{C}_\infty$ and bounded by 1. Since g is left unchanged on $\mathbb{C}_\infty \setminus U$, it remains analytic on $\mathbb{C}_\infty \setminus \operatorname{cl} U$ and it retains the properties listed above.

Define the squares $\{Q_r\}$ as was done just before Lemma 1.8. Note that if Q_r is not the enlargement of a green square and $Q_r \cap \operatorname{cl} U \neq \varnothing$, then $\operatorname{dist}(Q_r, P_g) \leq \frac{1}{4} 2^{-i}$. Hence there is line segment L that is part of P, is contained in the segment joining the centers of two adjacent squares in $\mathscr{G}(n)$ for some $n \leq i$, and is such that $\operatorname{dist}(Q_r, L) \leq \frac{1}{4} 2^{-i}$. But Proposition 1.8(d) implies that $\operatorname{dist}(Q_r, L) \geq (5/4) \frac{1}{4} 2^{-i} > \frac{1}{4} 2^{-i}$, a contradiction. Hence $Q_r \cap \operatorname{cl} U = \varnothing$ whenever Q_r is not the enlargement of a green square and so the function g is analytic inside Q_r.

§3 HEAVY BARRIERS EXIST

Consider the function g^2. Expanding g in a power series in $(z-a)^{-1}$ in a neighborhood of ∞ and squaring it, we find that

$$\|g^2\|_\infty \leq 1, \quad g^2(\infty) = (g^2)'(\infty) = 0, \quad \beta(g^2, a) \geq C_3^2 \delta^2.$$

Let λ and η be the scalars

$$\lambda = \frac{\alpha \delta}{g'(\infty)}, \quad \eta = \frac{1}{\beta(g^2, a)}[\beta \delta^2 - \lambda \beta(g, a)].$$

Note that

$$|\lambda| \leq \frac{1}{C_3}, \quad |\eta| \leq \frac{1}{C_3^2}\left[1 + \frac{9}{16}\frac{1}{C_3}\right].$$

Put $h = \lambda g + \eta g^2$. It follows that h is continuous on $\mathbb{C}_\infty$ and there is a universal constant C_8 such that h has the following properties.

(3.7) $\begin{cases} \text{(i)} & \|h\|_\infty \leq C_8. \\ \text{(ii)} & h(\infty) = 0. \\ \text{(iii)} & h'(\infty) = \alpha \delta. \\ \text{(iv)} & \beta(h, a) = \beta \delta^2. \\ \text{(v)} & h \text{ is analytic on every square that is not the enlargement} \\ & \text{of a green square.} \end{cases}$

According to Proposition 1.8(a) there is a C^1 partition of unity $\{\phi_r\}$ subordinate to the cover $\{Q_r\}$ such that $\|\bar{\partial}\phi_r\|_\infty \leq (100/\delta_r)$ for all r. Let $h_r = T_{\phi_r} h$, where T_{ϕ_r} is the Vitushkin localization operator (V.5.1).

CLAIM. $h_r \equiv 0$ whenever h is analytic on Q_r.

In fact, (V.5.3) implies that h_r is analytic wherever h is so that h_r is analytic on Q_r. Also, h_r is analytic off the suport of ϕ_r so that h_r is an entire function. But $h_r(\infty) = 0$ so that $h_r \equiv 0$.

For each value of r we will define a function f_r. Since the cover $\{Q_r\}$ is a locally finite one, at most a finite number of these squares meet $\operatorname{cl} U$. Since h is analytic off $\operatorname{cl} U$, the claim implies that $h_r \equiv 0$ for all but finitely many r. Whenever r is such that $h_r \equiv 0$, let f_r be the identically 0 function.

Now assume that h_r is not identically 0 and let's find estimates of its vital parameters. According to the claim and (3.7)(v), Q_r is the enlargement of a green square S_{kp} in $\mathscr{G}(k)$. Using the estimate from Proposition V.5.6,

$$\|h_r\|_\infty \leq \frac{2}{\sqrt{\pi}} \operatorname{osc}(h, \operatorname{supp} \phi_r) \|\bar{\partial}\phi_r\|_\infty \sqrt{\operatorname{Area}(\operatorname{supp} \phi_r)}$$
$$\leq \frac{4}{\sqrt{\pi}} C_8 \frac{100}{\delta_r} \sqrt{\delta_r^2}$$
$$\equiv C_9.$$

Hence Lemma 3.2 implies that $|h_r'(\infty)| \leq C_9 \delta_r$ and $|\beta(h_r, z_r)| \leq C_9 \delta_r^2$. If $d =$ the length of a side of $S_{kp} = 2^{-k}$, then $\delta_r = (5/4)d$ so that

$$|h_r'(\infty)| \leq \frac{5}{4} d C_9 \quad \text{and} \quad |\beta(h_r, z_r)| \leq \frac{25}{16} d^2 C_9.$$

Thus if $\alpha = h'_r(\infty)/2dC_9$ and $\beta = \beta(h_r, z_r)/2d^2C_9$, then $|\alpha|$ and $|\beta|$ are easily seen to be less than 1. By Lemma 3.3 there is a function f_0 in $C(\mathbb{C}_\infty)$ that is analytic off cl S_{kp} and satisfies: $\|f_0\|_\infty \leq 6$; $f_0(\infty) = 0$; $f'_0(\infty) = \alpha d$; $\beta(f_0, z_r) = \beta d^2$; $|\int f_0 d\nu_j| \leq 7d^3$ for $1 \leq j \leq s$. Define $f_r = 2C_9 f_0$. It is easy to check that $f'_r(\infty) = h'_r(\infty)$ and $\beta(f_r, z_r) = \beta(h_r, z_r)$. Thus, recalling that $d = 2^{-k} \leq 2^{-m}$, we have that

(3.8) $\begin{cases} \text{(i)} & \|f_r\|_\infty \leq 12C_9. \\ \text{(ii)} & f_r \text{ is analytic off cl } Q_r. \\ \text{(iii)} & f_r - h_r \text{ has a triple zero at } \infty. \\ \text{(iv)} & |\int f_r d\nu_j| \leq 14 C_9 2^{-3k} \leq C_{10} 2^{-m} \mathscr{A}(S_{kp}) \text{ for } 1 \leq j \leq s, \end{cases}$

where $C_{10} = 14 C_9$.

Let $f = \sum_r f_r$; we must now verify that this function f is the one promised in the lemma. Since at most a finite number of the functions f_r are not identically 0, this sum defines a function f in $C(\mathbb{C}_\infty)$ with $f(\infty) = 0$. We also have that

(3.9) $$h = \sum_r h_r.$$

Indeed, let $\rho > 0$ such that cl $U \subseteq B = B(0; \rho)$ and let $R = \{r : Q_r \cap B \neq \emptyset\}$; note that R is a finite set. If $r \notin R$, then h is analytic on Q_r and so $h_r \equiv 0$. Put $H = h - \sum_r h_r$. Since h is analytic off cl U so is each h_r; thus H is analytic on $\mathbb{C}_\infty \setminus \text{cl } U$ and $H(\infty) = 0$. On B, Proposition V.5.3 implies that in the sense of distributions $\bar{\partial} H = \bar{\partial} h - \sum_{r \in R} \bar{\partial}(T_{\phi_r} h) = \bar{\partial} h - \sum_{r \in R} \phi_r \bar{\partial} h = 0$ by virtue of the fact that $\sum_{r \in R} \phi_r = 1$ on B. By Weyl's Lemma (V.2.2), H is analytic on B and hence an entire function. Thus $H \equiv 0$ and (3.9) is established.

Therefore $f'(\infty) = \sum_r f'_r(\infty) = \sum_r h'_r(\infty) = h'(\infty) = \alpha\delta$ and, similarly, $\beta(f, a) = \beta(h, a) = \beta\delta^2$. This establishes (c), (d), and (e) in the lemma.

If $U \cap Q_r \neq \emptyset$, then $\text{dist}(Q_r, P_g) \leq (5/4)2^{-m} + (1/4)2^{-i} \leq (3/2)2^{-m}$. Hence $\text{dist}(a, Q_r) \leq \text{dist}(a, P_g) + \text{dist}(P_g, Q_r) \leq (3/4)\delta + (3/2)2^{-m} < \delta$ by the choice of m. Thus $Q_r \subseteq D$ and so f_r is analytic off cl D. On the other hand, if $U \cap Q_r = \emptyset$, then $h_r \equiv 0$ and so the same holds for f_r. Therefore f is analytic on $\mathbb{C}_\infty \setminus \text{cl } D$ establishing part (b).

Again if $f_r \neq 0$ and Q_r is an enlargement of S_{kp}, then $S_{kp} \subseteq Q_r \subseteq D$. Hence by (3.8)(iv)

$$\left|\int f d\nu_j\right| \leq \sum_r \left|\int f_r d\nu_j\right| \leq \sum C_{10} 2^{-m} \mathscr{A}(S_{kp}) \leq 2^{-m} C_{10} \mathscr{A}(D),$$

where the last sum is over all S_{kp} with $f_r \neq 0$. Thus the choice of m implies we have (f).

Finally to obtain (a) note that $f_r - h_r$ is analytic off $\{z : |z - z_r| < \delta_r\}$ with a triple zero at ∞ and $\|f_r - h_r\|_\infty \leq 13 C_9$. By the Maximum Modulus

Theorem $|(z - z_r)^3(f_r - h_r)| \leq 13C_9\delta_r^3$ if $|z - z_r| \geq \delta_r$. Therefore for all complex numbers z

$$|f_r(z) - h_r(z)| \leq \min\left\{13C_9, \frac{13C_9\delta_r^3}{|z - z_r|^3}\right\}.$$

By Proposition 1.8 this implies that $\sum_r |f_r - h_r| \leq 13C_9C_5$. Since $f = h + \sum(f_r - h_r)$, this gives that $\|f\|_\infty \leq \|h\|_\infty + 13C_9C_5 \leq C_8 + 13C_9C_5 \equiv C_7$. □

Now we introduce a covering of the plane by open squares for each integer $k \geq 1$. Let $\mathscr{S}(k) = \{S_{kp}\}_p$ be the kth generation of squares and for each p let T_{kp} be the square with the same center as S_{kp} and with sides of length $(5/4)2^{-k}$. So these squares are obtained in a fashion similar to the way that the squares $\{Q_r\}$ were obtained, except that the squares $\{Q_r\}$ were a multigenerational covering of the plane while the squares being considered at present are all from a fixed generation. Indeed since the generation k is fixed, denote these squares by $\{T_p\}$ and let z_p be the center of T_p. A small variation in the proof of (1.8) will prove the following result. The details are left to the reader.

3.10 PROPOSITION. *If $\{T_p\}$ is the covering of $\mathbb{C}$ obtained above, then it has the following properties.*

(a) *No point of $\mathbb{C}$ belongs to more than 4 of these squares.*
(b) *There is a C^1 partition of unity $\{\phi_p\}$ on $\mathbb{C}$ that is subordinate to the cover $\{T_p\}$ and satisfies $\|\bar{\partial}\phi_p\|_\infty \leq 80 \cdot 2^k$.*
(c) *There is a universal constant C_{11} such that for every z in $\mathbb{C}$ and every finite set P of positive integers,*

$$\sum_{p \in P} \min\left\{1, \frac{2^{-3k}}{|z - z_p|^3}\right\} \leq C_{11} \min\left\{1, \frac{2^{-k}}{\text{dist}(z, \bigcup_{p \in P} T_p)}\right\}.$$

We can now prove the main result of the section.

PROOF OF THEOREM 3.1. Let V be any open set disjoint from the set U appearing in the statement of the theorem. It must be shown that $|\nu_j|(V) = 0$ for $1 \leq j \leq s$. To do this fix an arbitrary continuous function h with compact support contained in V. It suffices to show that $\int h \, d\nu_j = 0$ for each j.

Let $\varepsilon > 0$ and choose a positive integer k sufficiently large that $2^{-k} < \frac{1}{2} \text{dist}(U, \text{supp } h)$. For the kth generation of squares form the cover $\{T_p\}$ as before the statement of the preceding proposition. Note that by the choice of k no square T_p meets both U and $\text{supp } h$. Let $\{\phi_p\}$ be a partition of unity on the plane subordinate to $\{T_p\}$ as in (3.10). Put $h_p = T_{\phi_p}(h)$. If $\rho = \sup\{|h(w) - h(z)| : |z - w| < 2^{-k+1}\}$, then the fact that the diameter of

T_p is less than 2^{-k+1} implies, with the aid of (V.5.6), that

$$\|h_p\|_\infty \leq \frac{2}{\sqrt{\pi}} \operatorname{osc}(h; \operatorname{supp} \phi_p) \|\overline{\partial} \phi_p\|_\infty \sqrt{\mathscr{A}(\operatorname{supp} \phi_p)}$$
$$\leq \frac{200}{\sqrt{\pi}} \rho$$
$$= C_{12} \rho.$$

Later we will want to choose k sufficiently large that ρ is very small. So even though the dependence of the functions $\{h_p\}$ on k is being suppressed, later in the proof this suppression will have to be suppressed and the dependence resurrected.

Since h has compact support and is thus analytic on $\mathbb{C}_\infty \setminus \operatorname{supp}(h)$, it follows, as in the proof of Lemma 3.6, that $h_p \equiv 0$ for all but a finite number of p. Let N be the number of functions h_p that are not identically 0. We also get that $h = \sum_p h_p$.

We will now approximate each h_p by a function that comes arbitrarily close to annihilating each ν_j. Elementary geometric considerations show that each h_p is analytic off the disk $B(z_p; 2^{-k})$. Thus Lemma 3.2 implies that $|h_p'(\infty)| \leq \|h_p\|_\infty 2^{-k}$ and $|\beta(h_p, z_p)| \leq \|h_p\|_\infty 2^{-2k}$. If $h_p \equiv 0$, let $f_p = 0$. If h_p is not identically 0, then it must be that T_p meets the support of h (Why?). By the choice of k this implies that $U \cap T_p = \varnothing$. In particular $z_p \notin U$ and so there is a sequence of light routes from z_p to ∞. If we apply Lemma 3.6 with $\delta = 2^{-k-1}$, then we get a function f_0 in $C(\mathbb{C}_\infty)$ and satisfying:

(3.11)
$$\begin{cases} \text{(i)} & \|f_0\|_\infty \leq C_7; \\ \text{(ii)} & f_0 \text{ is analytic off } \overline{B}(z_p; 2^{-k-1}); \\ \text{(iii)} & f_0(\infty) = 0; \\ \text{(iv)} & f_0'(\infty) = \frac{1}{4} \frac{h_p'(\infty)}{\|h_p\|_\infty}; \\ \text{(v)} & \beta(f_0, z_p) = \frac{1}{4} \frac{\beta(h_p, z_p)}{\|h_p\|_\infty}; \\ \text{(vi)} & |\int f_0 \, d\nu_j| \leq \frac{\varepsilon}{4N\|h_p\|_\infty} \text{ for } 1 \leq j \leq s. \end{cases}$$

Let $f_p = 4\|h_p\|_\infty f_0$. It is straightforward to check that there is a universal constant C_{13} such that

(3.12)
$$\begin{cases} \text{(i)} & \|f_p\|_\infty \leq \rho C_{13}; \\ \text{(ii)} & f_p \text{ is analytic off } T_p; \\ \text{(iii)} & f_p(\infty) = 0; \\ \text{(iv)} & f_p'(\infty) = h_p'(\infty); \\ \text{(v)} & \beta(f_p, z_p) = \beta(h_p, z_p); \\ \text{(vi)} & |\int f_p \, d\nu_j| \leq \varepsilon/N \text{ for } 1 \leq j \leq s. \end{cases}$$

Thus $f_p - h_p$ has a triple zero at ∞ and so, as in the proof of Lemma 3.6, there is a universal constant C_{14} such that

$$|f_p - h_p| \leq \min\left\{\rho C_{14}, \rho C_{14} \frac{2^{-3k}}{|z - z_p|^3}\right\}$$

everywhere on $\mathbb{C}$. Now using Proposition 3.10 we get a universal constant C_{15} such that

$$\sum_p |f_p - h_p| \leq \rho C_{15}.$$

Let $f = \sum_p f_p$. Thus $\|f - h\|_\infty \leq \rho C_{15}$ and for $1 \leq j \leq s$,

$$\left|\int f\,d\nu_j\right| \leq \sum_p \left|\int f_p\,d\nu_j\right| \leq \varepsilon.$$

Therefore

$$\left|\int h\,d\nu_j\right| \leq \varepsilon + \left|\int (f-h)\,d\nu_j\right|$$
$$\leq \varepsilon + \rho C_{15}\|\nu_j\|.$$

Now resurrect the dependence of ρ on k so as to choose k sufficiently large that this last sum is less than 2ε for $1 \leq j \leq s$. Since ε was arbitrary, this shows that $\int h\,d\nu_j = 0$ for $1 \leq j \leq s$, completing the proof. □

§4 Bounded point evaluations exist. In this section the previous results of this chapter will be used to prove that bounded point evaluations exist. It will also be shown that every bounded point evaluation is an analytic bounded point evaluation.

The first result here is an easy consequence of the work we have done.

4.1 LEMMA. *If ν is a compactly supported measure on $\mathbb{C}$ that annihilates polynomials and $U = \{a :$ there is a neighborhood V of a and a constant $C > 0$ such that for every polynomial p and every point z in V, $|p(z)| \leq C \int |p|d|\nu|\}$, then the support of ν is contained in the closure of U.*

PROOF. Apply Theorem 2.1 for the case where $s = 1$; so $\mu = |\nu|$. Accordingly, the set U of this lemma contains the set of points for which the $(|\hat{\nu}|, a, m)$ coloring scheme results in a sequence of heavy barriers. Now apply Theorem 3.1. □

Now for the main result of this section.

4.2 THEOREM. *If μ is a positive measure such that the subnormal operator S_μ is pure, then the closure of the analytic bounded point evaluations for μ contains the support of μ.*

PROOF. Since S_μ is pure, $P^\infty(\mu)$ has no L^∞-summand and so Proposition VI.1.11 implies there is a measure ν that annihilates the polynomials such that $[\nu] = [\mu]$ and g, the Radon-Nikodym derivative of $|\nu|$ with respect to μ, belongs to $L^2(\mu)$. Let $U = \{a:$ there is a neighborhood V of

a and a constant $C > 0$ such that for every polynomial p and every point z in V, $|p(z)| \leq C \int |p| d|\nu|\}$. By Lemma 4.1, supp μ = supp $|\nu| \subseteq \operatorname{cl} U$. If $a \in U$ (with V and C as in the definition of U), then for every polynomial p and every z in V,

$$|p(z)| \leq C \int |p| g \, d\mu$$
$$\leq C \|g\|_2 \|p\|_2.$$

Thus a is an analytic bounded point evaluation for μ (II.7.6). □

If μ is a compactly supported measure in $\mathbb{C}$ such that $P^2(\mu) \neq L^2(\mu)$, then there is a nonzero function g in $L^2(\mu)$ such that $g \perp P^2(\mu)$. So $\nu = \overline{g}\mu$ annihilates polynomials. An application of Lemma 4.1 also proves the following. The details are left to the reader.

4.3 THEOREM. *If μ is a compactly supported measure on $\mathbb{C}$ such that $P^2(\mu) \neq L^2(\mu)$, then* abpe$(\mu) \neq \varnothing$.

This section concludes with a result that has theoretical significance and psychological merit.

4.4 THEOREM. *If μ is a compactly supported measure on $\mathbb{C}$ such that S_μ is pure, then every bounded point evaluation for μ is an analytic bounded point evaluation.*

PROOF. Without loss of generality assume that $0 \in \operatorname{bpe}(\mu)$. Put $\tau = \mu + \delta_0$, where δ_0 is the unit point mass at 0, and let $R_1 : L^2(\tau) \to L^2(\mu)$ be the restriction map. Clearly R_1 takes polynomials into polynomials. Thus $R = R_1 | P^2(\tau)$ defines a bounded linear map $R : P^2(\tau) \to P^2(\mu)$. Clearly R is bounded below so that R is injective and has closed range; since ran R contains the polynomials, R is bijective. It is therefore clear that 0 is an analytic bounded point evaluation for μ if and only if 0 is an analytic bounded point evaluation for τ.

Also note that $RS_\tau = S_\mu R$ so that S_μ and S_τ are similar subnormal operators. By (II.13.7) S_τ is also a pure subnormal operator. It will be shown that 0 is an analytic bounded point evaluation for τ.

Suppose $0 \notin \operatorname{abpe}(\tau)$. By (VI.1.11) there is a function g in $P^2(\tau)^\perp$ such that $|g| > 0$ a.e. $[\tau]$. Note that $\overline{g}\tau$ and $z\overline{g}\tau$ are annihilating measures for the polynomials. Put $\Phi(z) = \max\{|\widehat{\overline{g}\tau}(z)|, |\widehat{z\overline{g}\tau}(z)|\}$ and consider the $(\Phi, m, 0)$ coloring scheme for $m \geq 1$. The fact that 0 is not an analytic bounded point evaluation for the measure τ implies, by the Cauchy-Schwarz inequality, that 0 is not an analytic bounded point evaluation for the space $P^1((1 + |z|)|g|\tau)$. Hence Theorem 2.1 implies that for every integer $m \geq 1$ the $(\Phi, m, 0)$ coloring scheme produces a light route to ∞.

Therefore Lemma 3.6 implies that for every $n \geq 1$ there is a function f_n in $C(\mathbb{C}_\infty)$ having the following properties.

(i) $\|f_n\|_\infty \leq C_7$;

(ii) f_n is analytic on $\mathbb{C}_\infty \setminus \{z : |z| \leq 2^{-n}\}$;

(iii) $f_n(\infty) = 0$;
(iv) $f'_n(\infty) = 2^{-n}/C_7$;
(v) $\beta(f_n, 0) = 0$;
(vi) $|\int f_n \bar{g} \, d\tau|, |\int f_n z \bar{g} \, d\tau| < 2^{-2n}$.

Let $h_n(z) = 2^n z f_n(z)$. It is easy to verify that $h_n(\infty) = 2^n f'_n(\infty) = 1/C_7 < 1$ and if $|z| \leq 2^{-n}$, then $|h_n(z)| \leq 1$. By the Maximum Modulus Theorem $\|h_n\|_\infty \leq 1$. We also have that $h_n(0) = 0$ and

(4.5) $$\left|\int h_n \bar{g} \, d\tau\right| < 2^{-n}.$$

But $\{h_n\}$ is a normal family on $\mathbb{C}_\infty \setminus \{0\}$ and so by passing to a subsequence we may assume that there is an analytic function h on $\mathbb{C}_\infty \setminus \{0\}$ such that $h_n(z) \to h(z)$ uniformly on compact subsets of $\mathbb{C}_\infty \setminus \{0\}$. But the function h is bounded by 1 and so 0 is a removable singularity. By Liouville's Theorem h must be a constant function. Indeed, $h \equiv c = 1/C_7$.

Therefore, $c - h_n \to c\chi_{\{0\}}$ pointwise boundedly on $\mathbb{C}_\infty$ and so $\int (c - h_n) \bar{g} \, d\tau \to c\overline{g(0)}\tau(\{0\}) \neq 0$. But (4.5) implies that

$$\left|\int (c - h_n) \bar{g} \, d\tau\right| = \left|\int h_n \bar{g} \, d\tau\right| < 2^{-n}$$

which does converge to 0, yielding a contradiction. □

§5 Thomson's Theorem. Here the necessary components from the preceding sections will be assembled and combined with additional arguments to produce the main result of the chapter.

5.1 THOMSON'S THEOREM. *If μ is any compactly supported measure on $\mathbb{C}$, then there is a Borel partition $\{\Delta_0, \Delta_1, \ldots\}$ of the support of μ such that if $\mu_n = \mu|\Delta_n$, then the following statements are true.*

(a) $P^2(\mu) = L^2(\mu_0) \oplus P^2(\mu_1) \oplus \cdots$.
(b) *If $n \geq 1$, then S_{μ_n} is irreducible. Equivalently, $P^2(\mu_n)$ contains no nontrivial characteristic functions.*
(c) *If $n \geq 1$ and $G_n = \mathrm{abpe}(\mu_n)$, then G_n is a simply connected region with $\mathrm{supp}(\mu_n) \subseteq \mathrm{cl}\, G_n$.*
(d) *If S_μ is pure (that is, the set Δ_0 above is empty) and $f \in P^2(\mu)$ such that f vanishes a.e. $[\mu]$ on the set of analytic bounded point evaluations for μ, then $f = 0$. Equivalently, the reproducing kernels for the analytic bounded point evaluations have dense span in $P^2(\mu)$.*
(e) *If S_μ is pure and G is the set of analytic bounded point evaluations for μ, then the map $\phi \to \hat{\phi}$ is a dual algebra isomorphism of $P^2(\mu) \cap L^\infty(\mu)$ onto $H^\infty(G)$.*

Parts (a) and (b) of this theorem are not new and can be found in Corollary IV.5.3. The fact that each G_n is simply connected was also shown in

Proposition II.7.12. The fact that G_n is large enough to include supp(μ_n) was essentially done in the preceding section. So the really new parts of this theorem are parts (d) and (e).

It should be pointed out for the interested reader, that Thomson has actually proved this theorem (with the appropriate changes) for the spaces $P^p(\mu)$, $1 \leq p < \infty$. The case where $p = \infty$ is, of course Sarason's Theorem.

Two additional lemmas are needed concerning $P^2(\mu)$ in the pure case. To fix some notation, for any measure μ and a in abpe(μ), let k_a be the reproducing kernel. Recall that $\hat{f}(a) = \langle f, k_a \rangle$ for each f in $P^2(\mu)$ and a in abpe(μ). (Again the reader must be vigilant and tolerant since tradition dictates that the possibility of confusion must exist in that the same notation is used for both the Cauchy transform and the analytic extension of a function in $P^2(\mu)$ to its analytic bounded point evaluations. As usual the context will provide the distinction.)

5.2 LEMMA. *Assume that S_μ is pure, let U be a component of the set of analytic bounded point evaluations for μ, and let $f \in P^2(\mu)$. If $\hat{f} = 0$ and $f = 0$ a.e. [μ] off ∂U, then f is the zero function.*

PROOF. Let $G = $ abpe(μ) and let $\{p_n\}$ be a sequence of polynomials such that $\int |p_n - f|^2 \, d\mu \to 0$. Using Proposition II.7.6 it is not hard to see that $p_n(z) \to \hat{f}(z)$ uniformly on compact subsets of G. Since $\hat{f} = 0$ on G, an elementary argument, left to the reader, shows that for every w in G,

$$\frac{p_n(z) - p_n(w)}{z - w} \to \frac{\hat{f}(z)}{z - w} = 0$$

uniformly on compact subsets of G. Since $f = 0$ a.e. [μ] off ∂U and $\hat{f} = 0$ on G,

$$\int \left| \frac{p_n(z) - p_n(w)}{z - w} - \frac{f}{z - w} \right|^2 d\mu = \int_{\mathbb{C} \setminus \partial U} \left| \frac{p_n(z) - p_n(w)}{z - w} \right|^2 d\mu$$
$$+ \int_{\partial U} \left| \frac{p_n(z) - p_n(w)}{z - w} - \frac{f(z)}{z - w} \right|^2 d\mu$$
$$\leq \int_{\mathbb{C} \setminus \partial U} \left| \frac{p_n(z) - p_n(w)}{z - w} \right|^2 d\mu$$
$$+ \frac{1}{d} \int_{\partial U} |p_n(z) - p_n(w) - f(z)|^2 d\mu,$$

where $d = $ dist($w, \partial U$). Since $p_n(w) \to \hat{f}(w) = 0$, we get that $(z-w)^{-1} f \in P^2(\mu)$ for all w in G.

Fix w in G and put $f_1 = (z - w)^{-1} f$; so $0 = \hat{f} = (z - w)\hat{f}_1$. But $\hat{f}$ and $\hat{f}_1$ are analytic on G and so $\hat{f}_1 = 0$. Applying the argument of the preceding paragraph we get that $(z - w)^2 f \in P^2(\mu)$. Continuing we obtain

that
(5.3) $$(z-w)^{-k}f \in P^2(\mu) \quad \text{for all } k \geq 1$$
and all w in G.

In fact, we will now see that (5.3) is valid whenever $w \notin \partial U$. To do this, first assume that in addition $w \in \operatorname{supp}\mu$. In this case, Theorem 4.2 implies there is a sequence $\{w_n\}$ in G such that $w_n \to w$. From (5.3) we know that $(z-w_n)^{-k}f \in P^2(\mu)$. Moreover,

$$\int \left| \frac{f}{(z-w_n)^k} - \frac{f}{(z-w)^k} \right|^2 d\mu = \int_{\partial U} |f|^2 \left| \frac{1}{(z-w_n)^k} - \frac{1}{(z-w)_k} \right|^2 d\mu$$

$$\leq M \left\| \frac{1}{(z-w_n)^k} - \frac{1}{(z-w)_k} \right\|^2_{\partial U}$$

which converges to 0. Thus (5.3) holds for every w in $\operatorname{supp}\mu$ that is off the boundary of U.

Now assume that $w \notin \operatorname{supp}\mu$. If $g \perp P^2(\mu)$, let $\nu = \bar{g}\mu$. So ν is a measure that annihilates polynomials. Since $w \notin \operatorname{supp}\mu$, $\hat{\nu}(w)$, the Cauchy transform of ν at w, is defined. If $\hat{\nu}(w) \neq 0$, then it is not difficult to show that w is a bounded poinit evaluation for μ. (Start by using Proposition V.8.9 and make an estimate.) By Theorem 4.4 this implies that $w \in G$, a case already handled in (5.3). So assume $w \notin G$ and consequently $\widehat{\bar{g}\mu}(w) = 0$ whenever $g \perp P^2(\mu)$. Thus $g \perp (z-w)^{-1}$ for every g in $P^2(\mu)^\perp$ and so $(z-w)^{-1} \in P^2(\mu)$. But the fact that $w \notin \operatorname{supp}\mu$ implies that $(z-w)^{-1}$ is bounded and hence belongs to $P^2(\mu) \cap L^\infty(\mu)$. Therefore $(z-w)^{-k} \in P^2(\mu) \cap L^\infty(\mu)$ for all $k \geq 1$ and so (5.3) holds for all w not in ∂U.

But this implies that $uf \in P^2(\mu)$ for all u in $\operatorname{Rat}(\partial U)$. Now every point on ∂U belongs to the boundary of a component of the complement of ∂U (viz, U) and so every point of ∂U is a peak point of $R(\partial U)$ (V.13.3). According to (V.11.10) $R(\partial U) = C(\partial U)$. Therefore $\bigvee\{uf : u \in R(\partial U)\}$ is a reducing subspace for S_μ and the restriction of S_μ to this space is normal. Since it is hypothesized that S_μ is pure this means that $f = 0$. □

5.4 LEMMA. *Assume that S_μ is pure. If U is a component of $\operatorname{abpe}(\mu)$ and $\psi \in H^\infty(U)$, then there is a function ϕ in $P^2(\mu) \cap L^\infty(\mu)$ such that $\hat{\phi} = \psi \chi_U$ and $\phi = 0$ a.e. $[\mu]$ off $\operatorname{cl} U$.*

PROOF. First we'll show that it suffices to prove this lemma under the added assumption that the measure behaves a certain way inside $G = \operatorname{abpe}(\mu)$. Let $\{D_n\}$ be a sequence of nontrivial compact disks, each of which is contained in G and such that each component of G contains one of these disks. For each $n \geq 1$ there is a constant $M_n > 0$ such that $|\hat{f}(z)| \leq M_n \|f\|_2$ for all f in $P^2(\mu)$ and all z in D_n. Let $\nu = \mu + \sum_n M_n^{-2} \mathscr{A}|D_n$. Since $\mu \leq \nu$ this shows that the restriction map $R: P^2(\nu) \to P^2(\mu)$ is a contraction. Now if D is the union of all the disks D_n and $f \in P^2(\nu)$, then

$\int |f|^2 d\nu \leq \int |f|^2 d\mu + \int_D \|f\|_2^2 d\mathscr{A} \leq [1 + \mathscr{A}(D)]\|f\|_2^2$. Thus R is bounded below and hence must be a bijection. This implies that $\text{abpe}(\mu) = \text{abpe}(\nu)$ and so it suffices to prove the lemma for the measure ν. (Verify!)

So we now assume, in addition to the hypothesis stated in the lemma, that each of the disks D_n is contained in the support of μ.

Let $k \geq 1$ and put $V_k = \{z : \text{dist}(z, \partial U) < \frac{1}{3}2^{-k}\}$. There is no loss in generality in assuming that $\|\psi\|_\infty = 1$. Let u_k be a continuous function on $\mathbb{C}$ such that $|u_k| \leq 1$, $u_k(z) = 0$ on $\mathbb{C} \setminus \text{cl}(U \cup V_k)$, and $u_k(z) = \psi(z)$ for z in $U \setminus \text{cl} V_k$. The idea of the proof is to produce a function ϕ_k in $L^\infty(\mu)$ that approximates u_k on $\mathbb{C} \setminus V_k$. This sequence will be uniformly bounded and it will turn out that a weak* cluster point will be the function ϕ desired in the conclusion of the lemma. But for now hold k fixed.

Let $\{T_p\}$ be the open covering of the plane by squares with centers $\{z_p\}$ and sides of length $(5/4)2^{-k}$ as described just prior to Proposition 3.10. (For notational convenience we will suppress the dependence of this open cover on k.) Let $\{g_p\}$ be a C^1 partition of unity subordinate to this cover with $\|\bar{\partial} g_p\|_\infty \leq 80 \cdot 2^k$. Put $P = \{p \geq 1 : T_p \cap V_k \neq \emptyset\}$ and let $N =$ the number of elements in P. Let $y_p = T_{g_p}(u_k)$. Recall that $\sum_p y_p = u_k$.

Now $y_p(\infty) = 0$ and each y_p is analytic on $\mathbb{C}_\infty \setminus \text{cl} V_k$ as well as off the support of g_p, which is contained in T_p. Thus $y_p \equiv 0$ for p not in P. There is a universal constant C_{15} such that if $p \in P$, $\|y_p\|_\infty \leq C_{15}$, $y_p(\infty) = 0$, $|y'_p(\infty)| \leq C_{15} 2^{-k}$, and $|\beta(y_p, z_p)| \leq C_{15} 2^{-2k}$.

Temporarily fix a p in P. So $T_p \cap V_k \neq \emptyset$ and hence there is a point a in ∂U with $\text{dist}(a, T_p) < \frac{1}{3}2^{-k}$. Thus $T_p \subseteq B(a; 3 \cdot 2^{-k})$ and y_p is analytic off $B(a; 3 \cdot 2^{-k})$. Since $\|y_p\|_\infty \leq C_{15}$, Lemma 3.2 implies that $|\beta(y_p, a)| \leq 9 C_{15} 2^{-2k}$.

Let $\{h_i\}$ be a countable dense subset of $P^2(\mu)^\perp$ and put $\Phi_k = \max\{|\widehat{\bar{h}_i \mu}| : 1 \leq i \leq k\}$. Now no point of ∂U is an analytic bounded point evaluation for μ so, in particular, $a \notin G$. By Theorem 2.1 (and a routine argument) for every integer $q \geq 1$ there is a sequence of light routes to ∞ for the (Φ_k, q, a) coloring scheme. Thus Lemma 3.6 implies there is a function f_p in $C(\mathbb{C}_\infty)$ satisfying the following:

(i) $\|f_p\|_\infty \leq 9 C_7 C_{15}$;
(ii) f_p is analytic off $\overline{B}(a; 2^{-k})$;
(iii) $f_p - y_p$ has a triple zero at ∞;
(iv) $|\int f_p \bar{h}_i d\mu| \leq \frac{1}{kN}$ for $1 \leq i \leq k$.

We also have that

(5.5) $\qquad \|f_p - y_p\|_\infty \leq 9 C_7 C_{15} + C_{15} = C_{16}.$

Since $f_p - y_p$ is analytic off $\overline{B}(z_p; 3 \cdot 2^{-k})$ with a triple zero at ∞, property

(5.5) and the Maximum Modulus Theorem imply that

$$|(z-z_p)^3(f_p - y_p)| \leq 3^3 2^{-3k} C_{16}$$

for $|z - z_p| \geq 3 \cdot 2^{-k}$. Thus for every z in $\mathbb{C}$

$$|f_p(z) - y_p(z)| \leq \min\left\{C_{16}, C_{17}\frac{2^{-3k}}{|z - z_p|^3}\right\},$$

where $C_{17} = 27 C_{16}$. By Proposition 3.10(c) there is a constant C_{18} such that for all z in $\mathbb{C}$

(5.6) $$\sum_{p \in P} |f_p(z) - y_p(z)| \leq C_{18} \min\left\{1, \frac{2^{-k}}{\text{dist}(z, \bigcup_{p \in P} T_p)}\right\}.$$

Define ϕ_k (we now resume the dependence on k) by

$$\phi_k = \sum_{p \in P} f_p = u_k + \sum_{p \in P}(f_p - y_p).$$

By (5.6) the sequence $\{\phi_k\}$ is uniformly bounded by $\|u_k\|_\infty + C_{18} \leq 1 + C_{18}$. Also fix z in $\mathbb{C}\setminus \partial U$ and a positive constant M (arbitrary for the moment, but large). Choose k large enough that $\text{dist}(z, \partial U) > M 2^{-k}$. Note that for such a choice of k, $u_k(z) = \psi(z)\chi_U(z)$. From the definition of V_k and the fact that $\text{diam } T_p < 2 \cdot 2^{-k}$ we get that if $p \in P$, then for all w in T_p, $|w - z| > (M - 3)2^{-k}$. Hence the distance from z to $\bigcup\{T_p : p \in P\}$ is larger than $(M - 3)2^{-k}$. Thus (5.6) implies that

$$\sum_{p \in P} |f_p(z) - y_p(z)| \leq C_{18} \frac{1}{M - 3}.$$

Since $f_p = 0$ for p not in P, this shows that for any $\varepsilon > 0$ and any z in $\mathbb{C}\setminus \partial U$, $|\phi_k(z) - \psi(z)\chi_U(z)| = |\phi_k(z) - u_k(z)| < \varepsilon$ for all sufficiently large k. Therefore

(5.7) $$\phi_k(z) \to \psi(z)\chi_U(z) \quad \text{for all } z \text{ in } \mathbb{C}\setminus \partial U.$$

We also have that

(5.8) $$\left|\int \phi_k \overline{h}_i \, d\mu\right| < \frac{1}{k} \quad \text{for } 1 \leq i \leq k.$$

Now $\{\phi_k\}$ is a uniformly bounded sequence in $L^\infty(\mu)$, which is the dual of a separable Banach space. Thus by passing to a subsequence we may assume that $\phi_k \to \phi$ weak* in $L^\infty(\mu)$ and (5.7) and (5.8) remain valid. Replacing $\{\phi_k\}$ by a sequence of convex combinations of itself, we may also assume that $\phi_k \to \phi$ a.e. $[\mu]$ (V.22.3). Because of (5.8) we have that $\phi \perp h_i$ for all i and so $\phi \in P^2(\mu) \cap L^\infty(\mu)$. Condition (5.7) implies that $\phi = \psi\chi_U$ a.e. $[\mu]$ off ∂U, so that, in particular, $\phi = 0$ a.e. $[\mu]$ off $\text{cl } U$. Now since $\hat{\phi} = \phi$ a.e. $[\mu]$ on G and both $\hat{\phi}$ and $\psi\chi_U$ are analytic on G, the fact that we have assumed that each component of G contains a disk that belongs to the support of μ implies that $\hat{\phi} = \psi\chi_U$ on all of G. □

PROOF OF THEOREM 5.1. From operator theory basics, it suffices to only prove the theorem in the case that S_μ is pure. Let $\{G_n\}$ be the components of $G = \mathrm{abpe}(\mu)$. By the preceding lemma, for each $n \geq 1$ there is a function χ_n in $P^2(\mu) \cap L^\infty(\mu)$ with $\hat{\chi}_n = \chi_{G_n}$ and $\chi_n = 0$ a.e. $[\mu]$ off $\mathrm{cl}\, G_n$. Now $\chi_n(1-\chi_n) \in P^2(\mu) \cap L^\infty(\mu)$, $\chi_n(1-\chi_n) = 0$ a.e. $[\mu]$ off ∂G_n, and $\widehat{\chi_n(1-\chi_n)} = 0$ on G_n. Thus Lemma 5.2 implies that $\chi_n(1-\chi_n) = 0$. Hence each χ_n is a characteristic function; let $\chi_n = \chi_{\Delta_n}$ for some Borel set Δ_n. Another application of Lemma 5.2 shows that $\chi_n\chi_m = 0$ for $n \neq m$. Since $\chi_{\Delta_n \cap \Delta_m} = \chi_n\chi_m$, we may change the sets Δ_n by sets of μ measure 0 so that they are pairwise disjoint.

Put $\chi = \sum_n \chi_n = \chi_\Delta$, where Δ is the union of the sets $\{\Delta_n\}$. We want to show that $\chi = 1$ and so $\{\Delta_n\}$ is a Borel partition of the support of μ. But this infinite sum converges to χ weak* in $L^\infty(\mu)$ and so $\hat{\chi} = 1$. Let $g \in P^2(\mu)^\perp$ such that $|g| > 0$ a.e. $[\mu]$ (V.17.10) and define the measure $\nu = (1-\chi)\bar{g}\mu$. Note that ν annihilates the polynomials and is not the zero measure if and only if $\chi \neq 1$. Assume that $\nu \neq 0$. According to Lemma 4.1 there is an open set V in $\mathbb{C}$ and a constant $C > 0$ such that for every w in V there is a function k_w in $L^\infty(|\nu|)$ with $\|k_w\|_\infty \leq C$ and such that $p(w) = \int p k_w \, d\nu$ for all polynomials. Thus $|p(w)| = |\int p k_w(1-\chi)\bar{g}\, d\mu| \leq C'\|p\|_2$. That is, $V \subseteq G$. Fix a point w in V and let $\{p_j\}$ be a sequence of polynomials such that $\int |p_j - \chi|^2 \, d\mu \to 0$. Since $\hat{\chi}(w) = 1$, these polynomials may be chosen so that $p_j(w) = 1$ for all $j \geq 1$. But $\chi(1-\chi) = 0$ since χ is a characteristic function and so

$$0 = \int \chi k_w (1-\chi)\bar{g}\, d\mu$$
$$= \lim_j \int p_j k_w (1-\chi)\bar{g}\, d\mu$$
$$= 1,$$

the most basic of all contradictions. Thus $\nu = 0$ and so $\chi = 1$ and $\{\Delta_n\}$ is a Borel partition of $\mathrm{supp}\,\mu$.

If $f \in P^2(\mu)$ then it is easy to see that $\chi_n f \in P^2(\mu_n)$ and $f = \sum_n \chi_n f$, where the convergence is in the $P^2(\mu)$ norm. This proves (a).

To establish (c) we first show that $G_n = \mathrm{abpe}(\mu_n)$. Note that $\mathrm{abpe}(\mu_n) \subseteq \mathrm{abpe}\,\mu$. So if $w \in \mathrm{abpe}(\mu_n)$ there is a component G_m such that $w \in G_m$ and hence $\hat{\chi}_m(w) = 1$. But if $m \neq n$, then $0 = \chi_n\chi_m$ and so $0 = \widehat{\chi_n\chi_m}(w) = \hat{\chi}_n(w)\hat{\chi}_m(w) = 1$. Thus $m = n$ and so $\mathrm{abpe}(\mu_n) \subseteq G_n$. Conversely, if $w \in G_n$, then there is a function k_w in $P^2(\mu)$ such that $\hat{f}(w) = \int f k_w \, d\mu$ for all f in $P^2(\mu)$. If $f \in P^2(\mu_n)$, then $\hat{f}(w) = \hat{f}(w)\hat{\chi}_n(w) = \int f \chi_n \bar{k}_w \, d\mu = \int f \bar{\chi}_n \bar{k}_w \, d\mu$. Since $\chi_n k_w \in P^2(\mu_n)$, this implies that $w \in \mathrm{abpe}(\mu_n)$ and so

$G_n = \mathrm{abpe}(\mu_n)$. The fact that G_n is simply connected follows from (II.7.12). Theorem 4.2 implies that $\mathrm{cl}\, G_n$ contains the support of μ_n, completing the proof of (c).

For part (d), assume that $f \in P^2(\mu)$ and $\widehat{f} = 0$. For each $n \geq 1$, $\widehat{f\chi_n} = 0$ and $f\chi_n = 0$ a.e. $[\mu]$ off ∂G_n. By Lemma 5.2, $f\chi_n = 0$. Hence $f = 0$.

To prove (b), suppose g is a characteristic function in $P^2(\mu_n)$. Thus $\widehat{g} = \widehat{g}^2$ and so $\widehat{g}$ is a characteristic function that is analytic on G_n. Thus either $\widehat{g} = \widehat{\chi}_n$ or $\widehat{g} = 0$. By part (d), this proves (b).

For part (e), let $\rho: P^2(\mu) \cap L^\infty(\mu) \to H^\infty(G)$ be the map defined by $\rho(\phi) = \widehat{\phi}$. It is routine to verify that ρ is a homomorphism that is weak* continuous. Thus for each w in G, $\phi \to \widehat{\phi}(w)$ is a homomorphism on the Banach algebra $P^2(\mu) \cap L^\infty(\mu)$ and has norm 1. Thus ρ is a contraction. From part (d), ρ is injective.

Fix n for the moment and let χ_n be the characteristic function in $P^2(\mu)$ such that $\widehat{\chi}_n = \chi_{G_n}$. So $\chi_n \in P^2(\mu) \cap L^\infty(\mu)$. Let $\psi \in H^\infty(G_n)$; according to Lemma 5.4 there is a function ϕ in $P^2(\mu) \cap L^\infty(\mu)$ with $\widehat{\phi} = \psi\chi_{G_n}$. Thus $\widehat{\phi\chi_n} = \widehat{\phi}$ and so $\phi = \phi\chi_n$ by (d); hence $\phi \in P^2(\mu_n)$. This shows that the restriction of ρ to $P^2(\mu_n) \cap L^\infty(\mu)$ defines a bijection $\rho_n: P^2(\mu_n) \cap L^\infty(\mu) \to H^\infty(G_n)$. Let $c > 0$ such that $c\|\phi\| \leq \|\rho_n(\phi)\| \leq \|\phi\|$ for all ϕ in $P^2(\mu_n) \cap L^\infty(\mu)$. If there is a function ϕ in $P^2(\mu_n) \cap L^\infty(\mu)$ with $\|\rho(\phi)\| < \|\phi\| = 1$, then $c = c\|\phi^k\| \leq \|\rho_n(\phi^k)\| = \|\rho(\phi)\|^k \to 0$, a contradiction. Thus ρ_n must be an isometry.

It now easily follows that ρ itself must be a surjective isometry. The fact that ρ is a dual algebra isomorphism follows by an application of Proposition I.2.10. □

The next result gives a converse to Theorem 5.1. The method of proof is related to that of Proposition II.10.8.

5.9 PROPOSITION. *If G is a bounded simply connected region in the plane, there is a measure μ such that $G = \mathrm{abpe}\,\mu$.*

PROOF. Let $\tau: \mathbb{D} \to G$ be the Riemann map and let $\{D_n\}$ be an increasing sequence of closed disks in $\mathbb{D}$ such that $\mathscr{A}(G\setminus\tau(D_n)) < 2^{-n}$. Let $\{a_j\}$ be a sequence consisting of a dense subset of ∂G with each element repeated infinitely often. By the Jordan Curve Theorem, $\tau(D_n)$ is polynomially convex and so there is a polynomial p with

$$\|p_n - (z - a_n)^{-1}\|_{\tau(D_n)} < \frac{1}{n}.$$

Let c_n be a constant such that $0 < c_n < 1$ and

$$\int_{G\setminus\tau(D_n)} |p_n - (z - a_n)^{-1}|^2 \, d\mathscr{A} < \frac{1}{nc_n}.$$

Now define a sequence of measures $\{\mu_n\}$ as follows. Let $\mu_1 = \mathscr{A}|G$. If $\mu_1, \ldots, \mu_n$ have been defined let $\mu_{n+1} = \mu_n|\tau(D_n) + c_n \mu_n|(G\setminus\tau(D_n))$. Note that $\mu_{n+1} \leq \mu_n$ for all $n \geq 1$ and $\|\mu_n - \mu_{n+1}\| \leq (1-c_n)\mu_n(G\setminus\tau(D_n)) \leq (1-c_n)\mathscr{A}(G\setminus\tau(D_n)) \leq 2^{-n}$. Hence there is a positive measure μ such that $\|\mu_n - \mu\| \to 0$ and $\mu \leq \mu_n$ for all $n \geq 1$. Also

$$\int |p_n - (z-a_n)^{-1}|^2 \, d\mu \leq \int |p_n - (z-a_n)^{-1}|^2 \, d\mu_{n+1}$$
$$\leq \frac{1}{n^2}\mu_n(\tau(D_n)) + \frac{1}{n}.$$

Since each a_n is repeated an infinite number of times, this shows that $(z-a_n)^{-1} \in P^2(\mu)$ for all $n \geq 1$.

It is not difficult to see that for each $n \geq 1$, $\mu|\tau(D_n)$ and $\mathscr{A}|\tau(D_n)$ are boundedly mutually absolutely continuous. Thus $G \subseteq \text{abpe}\,\mu$.

We now show that no point a_n is an analytic bounded point evaluation for μ. Indeed if $a_n \in \text{abpe}\,\mu$, then the fact that $(z-a_n)^{-1} \in P^2(\mu)$ implies that

$$1 = [(z-a_n)(z-a_n)^{-1}]^\wedge(a_n) = [z-a_n]^\wedge(a_n)[(z-a_n)^{-1}]^\wedge(a_n) = 0.$$

Thus $a_n \notin \text{abpe}\,\mu$ for all $n \geq 1$. Since the set of analytic bounded point evaluations is open, $\partial G \cap (\text{abpe}\,\mu) = \varnothing$. But $\text{cl}\,G = \text{supp}\,\mu \subseteq \text{cl}[\text{abpe}\,\mu]$. Hence $G = \text{abpe}\,\mu$. $\square$

The results of this section can be used to almost recapture Sarason's Theorem (VI.7.1). There is, however, an essential ingredient of that theorem that cannot be obtained as a corollary to Theorem 5.1. In fact if μ is any compactly supported measure on $\mathbb{C}$, then Proposition II.11.6 implies there is a measure ν with $[\nu] = [\mu]$ and $P^2(\nu) \cap L^\infty(\nu) = P^\infty(\nu) = P^\infty(\mu)$. If G is the set of analytic bounded point evaluations for ν, Theorem 5.1(e) implies that $P^\infty(\mu) \cong H^\infty(G)$. Now this is not the full power of Sarason's Theorem because we know that $R(\text{cl}\,G)$ is a Dirichlet algebra and, according to the last proposition, any simply connected open set can arise as the set of analytic bounded point evaluations of some measure.

As a final exclamation point to this discussion, let G be the slit disk. By Proposition 5.9 there is a measure μ with $[\mu] = [\mathscr{A}|G]$ and $G = \text{abpe}\,\mu$, while $\text{abpe}(\mathscr{A}|G) = $ the entire disk.

Finally the question arises as to whether the above methods can be carried over to $R^2(K, \mu)$. Obviously the answer is no since there is an example of a Swiss cheese K and a measure μ on K such that $R^2(K, \mu) \neq L^2(\mu)$ but $R^2(K, \mu)$ has no bounded point evaluations (see Brennan [**1971a**] and Fernstrom [**1976**]). However, since the arguments used in this chapter are all local, they can be used to show that $R^2(K, \mu)$ has analytic bounded point evaluations whenever multiplication by z on $R^2(k, \mu)$ is pure and $\text{supp}\,\mu$ has nonempty interior.

§6 Some applications of Thomson's Theorem.

In addition to solving an old problem, Thomson's Theorem holds the promise of many applications in the theory of subnormal operators. In this section we will present some of these applications.

One of the first consequences of the existence of bounded point evaluations is the existence of nontrivial invariant subspaces for subnormal operators. Indeed, to prove this we need only consider the case of a cyclic subnormal operator S_μ which we can assume to be pure. If λ is a bounded point evaluation for μ and k_λ the reproducing kernel, then we have already seen that $S_\mu^* k_\lambda = \bar{\lambda} k_\lambda$. Hence $\mathscr{M} = [k_\lambda]^\perp$ is a nontrivial invariant subspace for S_μ.

We now prove a result from Raphael [1982]. This has recently been established for all subnormal operators by Liming Yang [1990] (see Theorem II.13.12).

6.1 PROPOSITION. *Two quasisimilar cyclic subnormal operators have the same essential spectrum.*

PROOF. Let S_1 and S_2 be the subnormal operators and assume that $S_1 \sim S_2$. For $j = 1, 2$ let μ_j be the measure such that $S_j \cong S_{\mu_j}$ and let $G_j = \mathrm{bpe}(\mu_j)$. An application of Proposition II.13.10 shows that it suffices to assume that these operators are pure. From Thomson's Theorem we know that $\sigma(S_j) = \mathrm{cl}\, G_j$. Thus (II.13.5) $\mathrm{cl}\, G_1 = \mathrm{cl}\, G_2$. But $\sigma_p(S_j^*) = G_j^*$ and since $S_1^* \sim S_2^*$, this implies that $G_1 = G_2$. But then $\sigma_e(S_j) = \partial G_j$ and so S_1 and S_2 must have the same essential spectrum. □

The next result was proved by Raphael [1986] for the rationally cyclic case. (Aso see McCarthy [**preprint**].)

6.2 PROPOSITION (Raphael [1986]). *Two quasisimilar cyclic subnormal operators have isomorphic commutants.*

PROOF. Adopt the notation from the preceding proof. Again (II.13.10) implies we may assume the operators are pure. According to Thomson's Theorem, in this case we have $\{S_j\}' \cong P^2(\mu_j) \cap L^\infty(\mu_j) \cong H^\infty(G_j)$. But the hypothesis of quasisimilarity implies that $G_1 = G_2$ so that the proposition follows. □

Recall the notation from (II.2.8). If S is a subnormal operator acting on $\mathscr{H}$ with minimal normal extension N acting on $\mathscr{K}$, then N can be represented as an operator matrix

(6.3) $$N = \begin{bmatrix} S & X \\ 0 & T^* \end{bmatrix},$$

on $\mathscr{K} = \mathscr{H} \oplus \mathscr{H}^\perp$. The operator T is also subnormal and, provided S is pure, N^* is the minimal normal extension of T (II.2.10). The operator T is called the *dual* of S. (See the remarks following Proposition II.2.10.) The proof of the next result is left to the reader.

6.4 PROPOSITION (Conway **[1981b]**). *If S is a pure subnormal operator and T is its dual, then $\sigma(S) = \sigma(T)^*$.*

A subnormal operator S is said to be *self-dual* if S is unitarily equivalent to its dual. The unilateral shift is an example of a self-dual subnormal operator and there are others (see Conway **[1981b]**). The next result answers a question raised in Conway **[1981a]**.

6.5 PROPOSITION. *If S is a self-dual irreducible cyclic subnormal operator, then $R(\sigma_n(S)) = C(\sigma_n(S))$.*

PROOF. Let $S = S_\mu$ and let $G = \text{bpe}(\mu)$. Using the notation established in (6.3) and the fact that N is normal it is easy to check that $XX^* = S^*S - SS^*$. Since S is cyclic the Berger-Shaw Theorem implies that XX^* must be trace class and so X is a compact operator. This implies that $\sigma_e(N) = \sigma_e(S) \cup \sigma_e(T^*) = \sigma_e(S) \cup \sigma_e(S)^*$ since $S \cong T$. Now Proposition 6.4 implies that $\sigma_e(S)$ is symmetric with respect to the real axis since S is self-dual. Thus $z \to \bar{z}$ is a homeomorphism of $\text{cl}\,G$ onto itself. Thus we also have that $\sigma_e(S) = \partial G$ is symmetric with respect to the real axis. Therefore $\sigma_e(N) = \sigma_e(S) = \partial G$. But the spectrum of N differs from $\sigma_e(N)$ by a collection of isolated eigenvalues of finite multiplicity (**[ACFA]**, page 359). Thus $\sigma(N) = \partial G \cup \{a_n\}$, where $\{a_n\}$ is a sequence of isolated points in G whose only limit points are in ∂G. Hence every point of $\sigma(N)$ is a boundary point of a component of the complement of $\sigma(N)$ (viz, $G\setminus\{a_n\}$). Thus every point of $\sigma(N)$ is a peak point (V.13.3) and so $R(\sigma(N)) = C(\sigma(N))$ (V.11.10). □

Epilogue

There are several topics connected with subnormal operator theory that have not been treated here. In this section some of these will be pointed out and references will be given so that the interested reader can educate himself or herself.

There is a growing body of recent work on unitary invariants and model theory for subnormal operators. The recent book by Martin and Putinar [**1989**] contains the results connected with hyponormal operators. Also D. Xia [**1987a** and **1987b**] deals exclusively with subnormal operators. It would be interesting to see if the model for hyponormal operators developed in Martin and Putinar [**1987**] could be used to capture the Xia model. Can the subnormal operators be characterized amongst the hyponormal operators by means of the Martin and Putinar invariant?

The theory of unbounded hyponormal operators and subnormal operators is growing. See Janas [**1983c** and **1989**], Jin [**1989**], McDonald and Sundberg [**1986**], Stochel and Szafraniec [**1985, 1989, preprint a, preprint b, preprint c, and preprint d**]. Part of the difficulty here is that all the conditions that are equivalent to a bounded operator being subnormal are not equivalent for unbounded operators. Indeed, many do not make sense. There are a large number of examples, however, and so the area looks like a fruitful one for the future.

The investigation of subnormal tuples of operators is off and running. A d-tuple of subnormal operators $(S_1, \ldots, S_d)$ on a Hilbert space $\mathscr{H}$ is *subnormal* if there is a Hilbert space $\mathscr{K}$ containing $\mathscr{H}$ and a commuting d-tuple of normal operators $(N_1, \ldots, N_d)$ on $\mathscr{K}$ such that for $1 \leq k \leq d$, $N_k \mathscr{H} \subseteq \mathscr{H}$ and $S_k = N_k|\mathscr{H}$. The investigation of this topic opens the door to a conjuction of several complex variables and subnormal operator theory. See Athavale [**1987, 1990, preprint a,** and **preprint b**], Athavale and Pedersen [**preprint**], Conway [**preprint a** and **preprint b**], Curto [**1981**], Curto and Muhly [**1985**], Curto and Salinas [**1985**], Janas [**1983a**], Putinar [**1984**]. There are some difficulties in formulating the definition of a hyponormal d-tuple, though it is generally accepted that the definition introduced in Athavale [**1988**] is the most appropriate. A general survey of the status of joint hyponormality is in Curto [**preprint**]. Additional papers on the subject are Conway and Szymanski [**1988**], Curto, Muhly, and Xia [**1988**], Janas [**1984**], McCullough and Paulsen [**1989**], and Slocinski [**1975**].

Finally there is the theory of bundle shifts. These subnormal operators were introduced in Abrahamse and Douglas [**1976**] and haven't received the attention they deserve. They constitute the only collection of pure subnormal operators for which it is possible to give a verifiable set of unitary invariants. I was very inclined to devote a chapter of this book to this subject but I decided that too much background information was needed. See Abrahamse and Bastian [**1978**], Rudol [**1982, 1988, and 1989**]. (*Note*: Rudol has confirmed to me that there are some mistakes in Rudol [**1988**]. He will rectify these in a future publication.)

Bibliography

Two references are prerequisites and are referred to in the text as:

[**ACFA**] J. B. Conway, *A course in functional analysis*, Springer-Verlag, New York (1985).

[**FOCV**] J. B. Conway, *Functions of one complex variable*, Springer-Verlag, New York (1986).

Following each of the remaining references are the page numbers on which the reference is cited.

M. B. Abrahamse [1974], *Toeplitz operators in multiply connected regions*, Amer. J. Math. **96**, 261–297. (210)

―――[1975], *Analytic Toeplitz operators with automorphic symbol*, Proc. Amer. Math. Soc. **52**, 297–302. (84)

―――[1976], *Subnormal Toeplitz operators and functions of bounded type*, Duke Math. J. **43**, 597–604. (84)

―――[1978], *Commuting subnormal operators*, Illinois J. Math. **22**, 171–176. (84)

―――[preprint], *Some examples on lifting the commutant of a subnormal operator.* (79)

M. B. Abrahamse and J. A. Ball [1976], *Analytic Toeplitz operators with automorphic symbol*, II, Proc. Amer. Math. Soc. **59**, 323–328. (84)

M. B. Abrahamse and J. J. Bastian [1978], *Bundle shifts and Ahlfors functions*, Proc. Amer. Math. Soc. **72**, 95–96. (210, 410)

M. B. Abrahamse and R. G. Douglas [1976], *A class of subnormal operators related to multiply connected domains*, Adv. Math. **19**, 106–148. (209, 210, 410)

M. B. Abrahamse and T. L. Kriete [1973], *The spectral multiplicity of a multiplication operator*, Indiana Math. J. **22**, 845–857. (317)

J. Agler [1979], *An invariant subspace theorem*, Bull. Amer. Math. Soc. **1**, 425–427. (78)

―――[1980], *An invariant subspace theorem*, J. Funct. Anal. **38**, 315–323. (78)

___ [1985a], *Hypercontractions and subnormality*, J. Operator Theory **13**, 203–217. (35)

___ [1985b], *Rational dilation on an annulus*, Ann. Math. **121**, 537–563. (78)

___ [1988], "An abstract approach to model theory," *Surveys of some recent results in operator theory*, Vol. 2, J. B. Conway and B. B. Morrel, Editors, Research Notes in Mathematics, Longman, London, pp. 1–23. (35)

P. R. Ahern and D. Sarason [1967a], *The H_p spaces of a class of function algebras*, Acta Math. **117**, 123–163. (210)

___ [1967b], *On some hypo-Dirichlet algebras of analytic functions*, Amer. J. Math. **89**, 932–941. (210)

L. V. Ahlfors and A. Beurling [1950], *Conformal invariants and function theoretic null sets*, Acta Math. **83**, 101–129. (167, 217, 222)

J. Akeroyd [1987], *Polynomial approximation in the mean with respect to harmonic measure on crescents*, Trans. Amer. Math. Soc. **303**, 193–199. (146, 210)

___ [1989], *Point evaluations and polynomial approximation in the mean with respect to harmonic measure*, Proc. Amer. Math. Soc. **105**, 575–581. (146, 210)

___ [preprint], *Polynomial approximation in the mean with respect to harmonic measure on crescents*, II. (146, 210)

J. Akeroyd, D. Khavinson, and H. S. Shapiro [preprint], *Remarks concerning cyclic vectors in Hardy and Bergman spaces*. (73, 156, 210)

H. Alexander [1973], *Projections of polynomial hulls*, J. Funct. Anal. **3**, 13–19. (176)

T. Andô [1963a], *Matrices of normal extensions of subnormal operators*, Acta Sci. Math. (Szeged) **24**, 91–96. (40)

___ [1963b], *On hyponormal operators*, Proc. Amer. Math. Soc. **14**, 290–291. (49)

___ [1963c], *Note on invariant subspaces of a compact normal operator*, Ark. Mat. **14**, 337–340. (315)

C. Apostol, H. Bercovici, C. Foiaş, and C. Pearcy [1985], *Invariant subspaces, dilation theory, and the structure of the predual of a dual algebra*, I, J. Funct. Anal. **63**, 369–404. (74)

W. B. Arveson [1969], *Subalgebras of C^*-algebras*, Acta Math. **123**, 141–224. (78)

___ [1976], *An invitation to C^*-algebras*, Springer-Verlag, New York. (91, 159)

A. Athavale [1987], *Holomorphic kernels and commuting operators*, Trans. Amer. Math. Soc. **304**, 101–110. (409)

___ [1988], *On joint hyponormality of operators*, Proc. Amer. Math. Soc. **103**, 417–423. (409)

___ [1990], *Subnormal tuples quasisimilar to the Szegö tuple*. (409)

___ [preprint a], *On the duals of subnormal tuples*. (409)

____[preprint b], *Model theory on the unit ball in* $\mathbb{C}^m$. (409)

A. Athavale and S. Pedersen [preprint], *Moment problems and subnormality*. (35, 409)

S. Axler [1982], *Multiplication operators on Bergman spaces*, J. Reine Angew. Math. **336**, 26–44. (91, 335)

____[1986], *Harmonic functions from a complex analysis viewpoint*, Amer. Math. Monthly **93**, 246–258. (231)

S. Axler, J. B. Conway, and G. McDonald [1982], *Toeplitz operators on Bergman spaces*, Canad. J. Math. **34**, 466–483. (68, 69, 91)

S. Axler and J. Shapiro [1985], *Putnam's theorem, Alexander's spectral area estimate, and VMO*, Math. Ann. **271**, 161–183. (149, 176, 177)

J. A. Ball, R. F. Olin, and J. E. Thomson [1978], *Weakly closed algebras of subnormal operators*, Illinois J. Math. **22**, 315–326. (87)

R. Bañuelos and T. Wolff [1985], *Note on H^2 of planar domains*, Proc. Amer. Math. Soc. **95**, 217–218. (210)

J. J. Bastian [1976], *A decomposition of weighted translation operators*, Trans. Amer. Math. Soc. **224**, 217–230. (35)

J. J. Bastian and K. J. Harrison [1974], *Subnormal weighted shifts and asymptotic properties of normal operators*, Proc. Amer. Math. Soc. **42**, 475–479. (60)

H. Behncke [1968], *Structure of certain non-normal operators*, J. Math. Mech. **18**, 103–107. (159)

____[1970], *Generators of W^*-algebras*, Tôhoku Math. J. **22**, 541–546. (92, 161)

S. K. Berberian [1959], *Note on a theorem of Fuglede and Putnam*, Proc. Amer. Math. Soc. **10**, 175–182.

____[1961], *Introduction to Hilbert space*, Oxford Univ. Press, New York. (46)

____[1962], *A note on hyponormal operators*, Pacific J. Math. **12**, 1171–1175. (49)

____[1966], *Notes on spectral theory*, Van Nostrand, Princeton. (30)

____[1969], *An extension of Weyl's theorem to a class of not necessarily normal operators*, Michigan Math. J. **16**, 273–279. (49)

____[1978], *Extensions of a theorem of Fuglede and Putnam*, Proc. Amer. Math. Soc. **71**, 113–114. (50)

H. Bercovici [1984], *On dominating sequences in the unit disc*, Math. Z. **188**, 33–43. (306)

____[1988], *Factorization theorems and the structure of operators on Hilbert space*, Ann. Math. **128**, 399–413. (341)

H. Bercovici and J. B. Conway [1988], *A note on the algebra generated by a subnormal operator*, Oper. Theory: Adv. Appl. **32**, 53–56. (341)

H. Bercovici, C. Foiaş, and C. Pearcy [1985], *Dual algebras with applications to invariant subspaces and dilation theory*, CBMS Regional Conf. Ser. in Math., no. 56, Amer. Math. Soc., Providence, R.I. (12, 74, 184)

C. A. Berger [1978], *Sufficiently high powers of hyponormal operators have rationally invariant subspaces*, Integral Equations Operator Theory **1**, 444–447. (49)

C. A. Berger and B. I. Shaw [1973a], *Selfcommutators of multicyclic hyponormal operators are trace class*, Bull. Amer. Math. Soc. **79**, 1193–1199. (155)

―― [1973b], "Intertwining, analytic structure, and the trace norm estimate," Proc. Conf. Operator Theory, Lecture Notes in Math., vol. 345, Springer-Verlag, Berlin and New York, pp. 1–6. (155)

―― [preprint], *Hyponormality: its analytic consequences.* (155, 156, 158)

S. Bergman [1947], *Sur les fonctions orthogonales de plusiers variables complexes avec les applications à la théorie des fonctions analytiques*, Gauthier-Villars, Paris. (72)

―― [1950], *The kernel function and conformal mapping*, Math. Surveys, no. V, Amer. Math. Soc., Providence, R.I. (72)

A. Beurling [1949], *On two problems concerning lienar transformations in Hilbert space*, Acta Math. **81**, 239–255. (23)

E. Bishop [1957], *Spectral theory for operators on a Banach space*, Trans. Amer. Math. Soc. **86**, 414–445. (36)

N. Bourbaki [1959], *Éléments de mathématique, Livre VI, Intégration, chap. 6, Intégration vectorielle*, Hermann et Cie, Paris (1959). (320)

J. Bram [1955], *Subnormal operators*, Duke Math. J. **22**, 75–94. (30, 31, 35, 41, 51, 80, 81, 82, 232)

J. E. Brennan [1971a], *Invariant subspaces and rational approximation*, J. Funct. Anal. **7**, 285–310. (65, 66, 74, 182, 375, 385, 388, 406)

―― [1971b], *Point evaluations and invariant subspaces*, Indiana Univ. Math. J. **20**, 879–881. (66, 74, 375)

―― [1973], *Invariant subspaces and weighted polynomial approximation*, Ark. Mat. **11**, 167–189. (66, 74, 375)

―― [1977], *Approximation in the mean by polynomials on non-Carathéodory domains*, Ark. Mat. **15**, 117–168. (72, 73)

―― [1979a], *Point evaluations, invariant subspaces and approximation in the mean by polynomials*, J. Funct. Anal. **34**, 407–420. (66, 73, 375)

―― [1979b], *Invariant subspaces and subnormal operators*, Proc. Symp. Pure Math., vol. 35, Part 1, Amer. Math. Soc., Providence, R.I, pp. 303–309. (66, 375)

A. Browder [1969], *Introduction to function algebras*, W. A. Benjamin, New York. (327)

A. Brown [1953], *On a class of operators*, Proc. Amer. Math. Soc. **4**, 723–728. (29, 43, 44, 45)

L. Brown, A. L. Shields, and K. Zellar [1960], *On absolutely convergent sums*, Trans. Amer. Math. Soc. **96**, 162–183. (306)

S. W. Brown [1978], *Some invariant subspaces for subnormal operators*, Integral Equations Operator Theory **1**, 310–333. (183, 341, 344, 385)

S. W. Brown [1987], *Hyponormal operators with thick spectra have invariant subspaces*, Ann. Math. **125**, 93–103. (49, 149, 184)

_____ [1988], *Full analytic subspaces for contractions with rich spectrum*, Pacific J. Math. **132**, 1–10.

S. Brown and B. Chevreau [1988], *Toute contraction à calcul fonctionnel isométrique est réflexive*, C. R. Acad. Sci. Paris **307**, 185–188. (364)

J. W. Bunce [1970], *Characters on singly generated C^*-algebras*, Proc. Amer. Math. Soc. **25**, 297–303. (88, 89)

_____ [1978], *A universal diagram property of minimal normal extensions*, Proc. Amer. Math. Soc. **69**, 103–108. (40, 90)

J. W. Bunce and J. A. Deddens [1977], *On the normal spectrum of a subnormal operator*, Proc. Amer. Math. Soc. **63**, 107–110. (30)

S. L. Campbell [1975], *Subnormal operators with nontrivial quasinormal extensions*, Acta Sci. Math. (Szeged) **37**, 191–193. (43)

R. W. Carey and J. D. Pincus [1975], *Commutators, symbols, and determining functions*, J. Funct. Anal. **19**, 50–80. (49, 149)

_____ [1977], *Mosaics, principal functions, and mean motion in von Neumann algebras*, Acta Math. **138**, 153–218. (49, 149)

_____ [1979], *Principal functions, index theory, geometric theory and function algebras*, Integral Equations Operator Theory **2**, 441–483. (49, 149)

_____ [1981], *An integrality theorem for subnormal operators*, Integral Equations Operator Theory **4**, 10–44. (49)

L. Carleson [1962], *Interpolation by bounded analytic functions and the corona problem*, Ann. Math. (2) **76**, 547–559. (139)

_____ [1967], *Selected problems on exceptional sets*, Van Nostrand, Princeton, N.J. (216)

J. Chaumat [1974], *Adhérence faible étoile d'algèbres de fractions rationelles*, Ann. Inst. Fourier **24**, 93–120. (246, 287, 301)

_____ [1975], *Adhérence faible étoile d'algèbres de fractions rationelles*, Publ. Math. Orsay **147**. (328, 329)

G. Choquet [1955], *Theory of capacities*, Ann. Inst. Fourier **5**, 131–295. (216)

J. Cima and A. Matheson [1985], *Approximation in the mean by polynomials*, Rocky Mountain J. Math. **15**, 729–738. (72)

K. F. Clancey [1979], Seminormal operators, Lecture Notes in Math., vol. 742, Springer-Verlag, New York. (49, 149, 161)

K. F. Clancey, J. B. Conway, and M. Raphael [1983], *On a conjecture of Carey and Pincus*, Integral Equations Operator Theory **6**, 158–159. (84)

K. F. Clancey and B. B. Morrell [1974], *The essential spectrum of some Toeplitz operators*, Proc. Amer. Math. Soc. **44**, 129–134. (84)

K. F. Clancey and C. R. Putnam [1972], *The local spectral behavior of completely subnormal operators*, Trans. Amer. Math. Soc. **163**, 239–244. (180)

W. S. Clary [1973], *Quasi-similarity and subnormal operators*, Ph.D. thesis, University of Michigan. (98, 370, 371, 373)

W. S. Clary [1975], *Equality of spectra of quasi-similar hyponormal operators*, Proc. Amer. Math. Soc. **53**, 88–90. (74)

L. A. Coburn [1966], *Weyl's theorem for non-normal operators*, Michigan Math. J. **13**, 285–288. (49)

B. J. Cole and T. W. Gamelin [1982], *Tight uniform algebras and algebras of analytic functions*, J. Funct. Anal. **46**, 158–220. (190, 250, 276)

____[1985], *Weak-star continuous homomorphisms and a decomposition of orthogonal measures*, Ann. Inst. Fourier **35**, 149–189. (190, 250, 276)

J. B. Conway [1973], *A complete Boolean algebra of subspaces which is not reflexive*, Bull. Amer. Math. Soc. **79**, 720–722. (140)

____[1977], *The direct sum of normal operators*, Indiana Univ. Math. J. **26**, 277–289. (308)

____[1980], *Quasisimilarity for subnormal operators*, Illinois J. Math. **24**, 689–702. (95, 96, 97, 368)

____[1981a], *Subnormal operators*, Pitman, London. (408)

____[1981b], *The dual of a subnormal operator*, J. Operator Theory **5**, 195–211. (40, 408)

____[1982], *Quasisimilarity for subnormal operators*, II, Canadian Math. Bull. **25**, 37–40. (98, 373)

____[1985a], *Arranging the disposition of the spectrum*, Proc. Royal Irish Acad. **85A**, 139–142. (35)

____[1985b], "A survey of some results on subnormal operators," in Operators and function theory, S. C. Power, D. Reidel, Editors, Dordrecht, Holland, pp. 19–37. (35)

____[1987], *Spectral properties of certain operators on Hardy spaces of planar regions*, Integral Equations Operator Theory **10**, 659–706. (209, 211)

____[1989], *The minimal normal extension of a function of a subnormal operator*, Analysis at Urbana II, (Proc. Special Year in Modern Analysis at University of Illinois, 1986–1987), Cambridge Univ. Press, Cambridge, pp. 128–140. (317)

____[1990], *Towards a functional calculus for subnormal tuples: the minimal normal extension and approximation in several complex variables*, Proc. Symp. Pure Math., vol. 51, part 1, Amer. Math. Soc., Providence, R.I., pp. 105–112. (409)

____[preprint b], *Towards a functional calculus for subnormal tuples: The minimal normal extension*, Trans. Amer. Math. Soc. (to appear). (409)

J. B. Conway and J. J. Dudziak [1990], *Von Neumann operators are reflexive*, J. Reine Angew. Math. **408**, 34–56. (78, 262, 341, 364)

J. B. Conway, J. J. Dudziak, and E. Straube [1987], *Isometrically removable sets for functions in the Hardy space are polar*, Michigan Math. J. **34**, 267–273. (210) J. B. Conway and D. W. Hadwin [1983], *Strong limits of normal operators*, Glasgow Math. J. **24**, 93–96. (36)

J. B. Conway and P. McGuire [1984], *Operators with C^*-algebra generated by a shift*, Trans. Amer. Math. Soc. **284**, 153–161. (92)

J. B. Conway and R. F. Olin [1976], *A functional calculus for subnormal operators*, Bull. Amer. Math. Soc. **82**, 259–261.

____[1977], *A functional calculus for subnormal operators*, II, Mem. Amer. Math. Soc., no. 184. (43, 86, 279, 316, 317, 326, 365)

J. B. Conway and C. R. Putnam [1985], *An irreducible subnormal operator with infinite multiplicities*, J. Operator Theory **13**, 291–297. (161)

J. B. Conway and W. Szymanski [1988], *Linear combinations of hyponormal operators*, Rocky Mount. Math. J. **18**, 695–705. (410)

J. B. Conway and P. Y. Wu [1977], *The splitting of $\mathscr{A}(T_1 \oplus T_2)$ and related questions*, Indiana Univ. Math. J. **26**, 41–56. (25, 26)

____[1982], *The structure of quasinormal operators and the double commutant property*, Trans. Amer. Math. Soc. **270**, 641–657. (45, 46)

C. C. Cowen [1978], *The commutant of an analytic Toeplitz operator*, Trans. Amer. Math. Soc. **239**, 1–31. (84)

____[1980a], *The commutant of an analytic Toeplitz operator*, II, Indiana Univ. Math. J. **29**, 1–12. (84)

____[1980b], *An analytic Toeplitz operator that commutes with a compact operator and a related class of Toeplitz operators*, J. Funct. Anal. **36**, 169–184. (84)

____[1984], *Subnormality of the Cesàro operator and a semigroup of composition operators*, Indiana Univ. Math. J. **33**, 305–318. (35)

M. J. Cowen and R. G. Douglas [1978], *Complex geometry and operator theory*, Acta Math. **141**, 187–261. (64)

R. E. Curto [1981], *Spectral inclusion for doubly commuting subnormal n-tuples*, Proc. Amer. Math. Soc. **83**, 730–734. (409)

____[1990], *Joint hyponormality: A bridge between hyponormality and subnormality*, Proc. Symp. Pure Math., vol. 51, part 2, Amer. Math. Soc., Providence, R.I., pp. 69–91. (410)

R. E. Curto and P. S. Muhly [1985], *C^*-algebras of multiplication operators on Bergman spaces*, J. Funct. Anal. **64**, 315–329. (409)

R. E. Curto, P. S. Muhly, and J. Xia [1988], *Hyponormal pairs of commuting operators*, Oper. Theory: Adv. Appl. **35**, 1–22. (410)

R. E. Curto and N. Salinas [1985], *Spectral properties of cyclic subnormal m-tuples*, Amer. J. Math. **107**, 113–138. (409)

A. M. Davie [1972a], *Bounded limits of analytic functions*, Proc. Amer. Math. Soc. **32**, 127–133. (267)

____[1972b], *Analytic capacity and approximation problems*, Trans. Amer. Math. Soc. **171**, 409–444. (380, 382)

J. A. Deddens [1971], *Intertwining analytic Toeplitz operators*, Mich. Math. J. **18**, 243–246. (79, 84)

____[1972], *Analytic and Toeplitz composition operators*, Canadian J. Math. **24**, 859–865.

J. A. Deddens and J. G. Stampfli [1973], *On a question of Douglas and Fillmore*, Bull. Amer. Math. Soc. **79**, 327–330. (60)

J. A. Deddens and T. K. Wong [1973], *The commutatnt of analytic Toeplitz operators*, Trans. Amer. Math. Soc. **186**, 261–273. (84)

R. G. Douglas [1966], *On majorization, factorization, and range inclusion of operators on Hilbert space*, Proc. Amer. Math. Soc. **17**, 413–415. (50)

____ [1972], *Banach algebra techniques in operator theory*, Academic Press, New York. (89, 91, 145, 146)

R. G. Douglas and V. I. Paulsen [1986], *Completely bounded maps and hypo-Dirichlet algebras*, Acta. Sci. Math. (Szeged) **50**, 143–157. (78)

J. J. Dudziak [1981], *Spectral mapping theorems for subnormal operators*, Ph.D. thesis, Indiana University. (163, 276)

____ [1984], *Spectral mapping theorems for subnormal operators*, J. Funct. Anal. **56**, 360–387. (326, 330, 333, 334)

____ [1986], *The minimal normal extension problem for subnormal operators*, J. Funct. Anal. **65**, 314–338. (315, 317, 324)

____ [1989], *A weak-star rational approximation problem connected with subnormal operators*, Proc. Amer. Math. Soc. **107**, 679–686. (316, 317)

N. Dunford and J. Schwartz [1963], *Linear operators* II, Interscience, New York. (2)

P. L. Duren [1970], H^p-*spaces*, Academic Press, New York. (99, 117, 371)

J. Dyer, E. Pedersen, and P. Porcelli [1972], *An equivalent formulation of the invariant subspace conjecture*, Bull. Amer. Math. Soc. **78**, 1020–1023. (314)

N. Elias [1988], *Toeplitz operators on weighted Bergman spaces*, Integral Equations Operator Theory **11**, 310–331. (73, 91, 158)

M. R. Embry [1973], *A generalization of the Halmos-Bram criterion for subnormality*, Acta. Sci. Math. (Szeged) **31**, 61–64. (30, 31)

M. R. Embry-Wardrop [1981], *Quasinormal extensions of subnormal operators*, Houston J. Math. **7**, 191–204. (45, 84)

M. R. Embry and A. Lambert [1977], *Subnormal weighted translation semigroups*, J. Funct. Anal. **24**, 268–275. (35)

O. J. Farrell [1934], *On approximation to an analytic function by polynomials*, Bull. Amer. Math. Soc. **40**, 908–914. (71)

C. Fernstrom [1976], *Bounded point evaluations and approximation in L^p by analytic functions*, Spaces of Analytic Functions, Lecture Notes in Math., vol. 512, Springer-Verlag, Berlin and New York, pp. 65–68. (65, 406)

P. A. Fillmore [1970], *Notes on operator theory*, Van Nostrand-Reinhold, New York. (30)

S. D. Fisher [1983], *Function theory on planar domains*, Wiley, New York. (210)

C. Foiaş [1959], *Some applications of spectral sets* I: *harmonic spectral measure*, Acad. R. P. Roumaine Stud. Cerc. Math. **10**, 365–401; English transl., Amer. Math. Soc. Transl. (2) **61** (1967), 25–62. (78)

C. K. Fong [1977], *Quasi-affine transforms of subnormal operators*, Pacific J. Math. **70**, 361–368. (373)

R. Frankfurt [1975], *Subnormal weighted shifts and related function spaces*, J. Math. Anal. Appl. **52**, 471–489. (59)

____[1976], *Subnormal weighted shifts and related function spaces, II*, J. Math. Anal. Appl. **55**, 2–17. (59)

____[1977a], *Function spaces associated with radially symmetric measures*, J. Math. Anal. Appl. **60**, 502–541. (59)

____[1977b], *Quasicyclic subnormal semigroups*, Canad. J. Math. **29**, 1230–1232. (35)

T. W. Gamelin [1969], *Uniform algebras*, Prentice Hall, Englewood Cliffs, N.J. (163, 212, 222, 248, 253, 296)

____[1973], "Uniform algebras on plane sets," *Approximation theory*, Academic Press, New York, pp. 100–149. (190, 250, 276)

____[1985], *On an estimate of Axler and Shapiro*, Math. Ann. **272**, 189–196. (177)

____[notes], *Rational approximation theory*, unpublished lecture notes. (163, 276)

T. W. Gamelin and J. Garnett [1969], *Constructive techniques in rational approximation*, Trans. Amer. Math. Soc. **143**, 187–200.

____[1970], *Distinguished homomorphisms and fiber algebras*, Amer. J. Math. **92**, 455–474. (335)

____[1971], *Pointwise bounded approximation and Dirichlet algebras*, J. Funct. Anal. **8**, 360–404. (297, 298, 332)

T. W. Gamelin and D. Khavinson [1989], *The isoperimetric inequality and rational approximation*, Amer. Math. Monthly **96**, 18–30. (167)

T. W. Gamelin, P. Russo, and J. E. Thomson [1989], *A Stone-Weierstrass theorem for weak-star approximation by rational functions*, J. Funct. Anal. **87**, 170–176. (316) J. Garnett [1972], *Analytic capacity and measure*, Lecture Notes in Math., vol. 297, Springer-Verlag, Berlin and New York. (222)

____[1981], *Bounded analytic functions*, Academic Press, New York. (140)

R. Gellar [1977], *Circularly symmetric subnormal and normal operators*, J. Analyse Math. **32**, 93–117. (59)

R. Gellar and L. J. Wallen [1970], *Subnormal weighted shifts and the Halmos-Bram criterion*, Proc. Japan Acad. **46**, 375–378. (58)

P. Ghatage [1976], *Subnormal shifts with operator-valued weights*, Proc. Amer. Math. Soc. **57**, 107–108. (60)

B. K. Ghosh, B. E. Rhoades, and D. Trutt [1977], *Subnormal generalized Hausdorff operators*, Proc. Amer. Math. Soc. **66**, 261–265. (35)

F. Gilfeather [1971], *On the Suzuki structure of non self-adjoint operators on Hilbert space*, Acta. Sci. Math. (Szeged) **32**, 239–249. (159)

I. Glicksberg [1962], *Measures orthogonal to algebras and sets of antisymmetry*, Trans. Amer. Math. Soc. **105**, 415–435. (216)

____[1967], *The abstract F. and M. Riesz Theorem*, J. Funct. Anal. **1**, 109–122. (250)

D. Hadwin and E. Nordgren [1988], *Extensions of the Berger-Shaw theorem*, Proc. Amer. Math. Soc. **102**, 517–525. (153, 155, 158)

P. R. Halmos [1950], *Normal dilations and extensions of operators*, Summa Bras. Math. **2**, 125–134. (29, 30, 46)

____[1952], *Spectra and spectral manifolds*, Ann. Polon. Math. **25**, 43–49. (29, 41)

____[1961], *Shifts on Hilbert space*, J. Reine Angew. Math. **208**, 102–112. (17, 26)

____[1982], *A Hilbert space problem book*, Springer-Verlag, New York. (13, 17, 24, 58, 62)

P. R. Halmos, G. Lumer, and J. J. Schäffer [1953], *Square roots of operators*, Proc. Amer. Math. Soc. **4**, 142–149. (73)

W. W. Hastings [1975], *A Carleson measure theorem for Bergman spaces*, Proc. Amer. Math. Soc. **52**, 237–241. (74, 98)

____[1978], *Commuting subnormal operators simultaneously quasisimilar to unilateral shifts*, Illinois J. Math. **2**, 506–519. (98)

____[1979a], *Subnormal operators quasisimilar to an isometry*, Trans. Amer. Math. Soc. **256**, 145–161. (96, 98)

____[1979b], *A construction in Hilbert spaces of analytic functions*, Proc. Amer. Math. Soc. **74**, 295–298. (82, 83, 84, 85)

____[1981], *The approximate point spectrum of a subnormal operator*, J. Operator Theory **5**, 119–126. (43)

M. Hasumi [1983], *Hardy classes on infinitely connected Riemann surfaces*, Lecture Notes in Math., vol. 1027, Springer-Verlag, Berlin and New York. (207, 210)

V. P. Havin, S. V. Hruscev, and N. K. Nikolskii [1984], *Linear and complex analysis problem book*, Lecture Notes in Math., vol. 1043, Springer-Verlag, Berlin and New York. (358)

L. I. Hedberg [1972], *Approximation in the mean by analytic functions*, Trans. Amer. Math. Soc. **163**, 157–171. (73)

D. A. Hedjal [1973], *Classification theory for Hardy classes of analytic functions*, Ann. Acad. Sci. Fenn. Ser. A I Math., **566**, 1–28. (204, 210)

M. Heins [1969], *Hardy classes on Riemann surfaces*, Lecture Notes in Math., vol. 98, Springer-Verlag, Berlin and New York. (210)

H. Helson [1964], *Invariant subspaces*, Academic Press, New York. (26)

____[1983], *Boundedness from measure theory*, Operator Theory: Adv. Appl. **11**. (248)

J. W. Helton and R. Howe [1973], *Integral operators, commutator traces, index and homology*, Proc. Conf. Operator Theory, Springer Lecture Notes, **345** (Heidelberg), pp. 141–209. (49)

D. A. Herrero [1972], *Eigenvectors and cyclic vectors for bilateral weighted shifts*, Rev. Un. Mat. Argentina **26**, 24–41. (58, 59)

____ [1978], *On multicyclic operators*, Integral Equations Operator Theory **1**, 57-102. (60, 66)

____ [preprint], *Spectral pictures of hyponormal bilateral operator weighted shifts*, Proc. Amer. Math. Soc. (to appear). (49, 60)

K. Hoffman [1962], *Banach spaces of analytic functions*, Prentice-Hall, Englewood Cliffs, N.J. (99, 117)

T. B. Hoover [1972], *Quasi-similarity of operators*, Illinois J. Math. **16**, 678-686. (98)

C. Horowitz [1974], *Zeros of functions in Bergman spaces*, Duke Math. J. **41**, 693-710. (74)

R. Howe [1974], *A functional calculus for hyponormal operators*, Indiana Univ. Math. J. **23**, 631-644. (49)

T. Ito [1958], *On the commutative family of subnormal operators*, J. Fac. Sci. Hokkaido Univ. **14**, 1-15. (35, 84)

T. Ito and T. K. Wong [1972], *Subnormality and quasinormality of Toeplitz operators*, Proc. Amer. Math. Soc. **34**, 157-164. (50)

N. Ivanovski [1973], *Subnormality of operator valued shifts*, Ph.D. thesis, Indiana University. (60)

J. Janas [1983a], *Spectral inclusion theorems for commuting subnormal pair*, Ann. Math. Polon. **23**, 219-229. (409)

____ [1983b], *A note on invariant subspaces under multiplication by z in Bergman space*, Proc. Royal Irish. Acad. **83A**, 157-164. (74)

____ [1983c], *Some applications of functions of several complex variables to Toeplitz and subnormal operators*, Ann. Polon. Math. **40**, 185-192. (409)

____ [1984], *Spectral properties of doubly commuting hyponormal operators*, Ann. Pol. Math. **44**, 185-195. (410)

____ [1989], *On unbounded hyponormal operators*, Ark. Mat. **27**, 273-281. (409)

K. H. Jin [1989], *On unbounded Bergman operators*, Ph.D. thesis, Indiana University. (73, 158, 409)

R. V. Kadison and I. M. Singer [1957], *Three test problems in operator theory*, Pacific J. Math. **7**, 1101-1106. (95)

E. Kay, H. Soul, and D. Trutt [1976], *Some subnormal operators and hypergeometric kernel functions*, J. Math. Anal. Appl. **53**, 237-242. (35)

G. E. Keough [1981], *Subnormal operators, Toeplitz operators, and spectral inclusion*, Trans. Amer. Math. Soc. **263**, 125-135. (92, 93)

S. Kobayashi [1978], *On a classification of plane domains for Hardy spaces*, Proc. Amer. Math. Soc. **68**, 79-82. (210)

H. Koenig and G. L. Seever [1969], *The abstract F. and M. Riesz theorem*, Duke Math. J. **36**, 791-797. (250)

P. Koosis [1980], *Introduction to H_p spaces*, London Math. Soc. Lecture Note Ser., vol. 40, Cambridge Univ. Press, Cambridge. (99, 117, 199)

T. L. Kriete [1979], *On the structure of certain $H^2(\mu)$ spaces*, Indiana Univ. Math. J. **28**, 757-773. (66, 358)

T. L. Kriete and T. T. Trent [1977], *Growth near the boundary in $H^2(\mu)$ spaces*, Proc. Amer. Math. Soc. **62**, 83–88. (66, 358)

T. L. Kriete and D. Trutt [1971], *The Cesàro operator in l^2 is subnormal*, Amer. J. Math. **93**, 215–225. (35)

―― [1974], *On the Cesàro operator*, Indiana Univ. Math. J. **24**, 197–214. (35)

R. Kulkarni [1970], *Subnormal operators and weighted shifts*, Ph.D. thesis, Indiana University. (60)

A. Lambert [1976], *Subnormality and weighted shifts*, J. London Math. Soc. **14**, 476–480. (60)

R. G. Lautzenheiser [1973], *Spectral sets, reducing subspaces, and function algebras*, Ph.D. thesis, Indiana University. (78, 161, 260)

―― [1979], *Spectral theory for subnormal operators*, Trans. Amer. Math. Soc. **255**, 301–314. (161)

A. Lebow [1963], *On von Neumann's theory of spectral sets*, J. Math. Anal. Appl. **7**, 64–90. (78)

A. Lubin [1977], *Weighted shifts and products of subnormal operators*, Indiana Univ. Math. J. **26**, 839–845. (35, 60, 80, 84)

―― [1978], *A subnormal semigroup without normal extension*, Proc. Amer. Math. Soc. **68**, 176–178. (35, 84)

―― [1979], *Weighted shifts and commuting normal extensions*, Proc. Amer. Math. Soc. **27**, 17–26. (84)

―― [1980], *Lifting subnormal double commutants*, Studia Math. **67**, 315–319. (84)

N. G. Makarov [1987], *Perturbations of normal operators and stability of the continuous spectrum*, English transl. Math. USSR-Izv. **29**, 535–558. (27)

A. I. Markusevic [1934], *Conformal mapping of regions with variable boundary and applications to the approximation of analytic functions by polynomials*, Dissertation, Moskow. (71)

M. Martin and M. Putinar [1987], *A unitary invariant for hyponormal operators*, J. Funct. Anal. **73**, 297–323. (49, 149, 409)

―― [1989], *Lectures on hyponormal operators*, Birkhäuser, Basel-Boston, MA. (49, 149, 409)

P. Masani [1978], *Dilations as propagators of Hilbertian varieties*, SIAM J. Math. Anal. **9**, 414–456. (35)

J. E. McCarthy, [1990], *Analytic structures for subnormal operators*, Integral Equations Operator Theory **13**, 251–270. (53, 366)

―― [preprint], *Quasisimilarity of rationally cyclic subnormal operators*, J. Operator Theory (to appear). (98, 373, 407)

S. McCullough and V. Paulsen [1989], *A note on joint hyponormality*, Proc. Amer. Math. Soc. **107**, 187–195. (75, 410)

G. McDonald and C. Sundberg [1986], *On the spectra of unbounded subnormal operators*, Canad. J. Math. **38**, 1135–1148. (409)

P. McGuire [1988a], C^*-*algebras generated by a subnormal operator*, J. Funct. Anal. **79**, 423–445. (92)

____[1988b], *On the spectral picture of an irreducible subnormal operator*, Proc. Amer. Math. Soc. **104**, 801–808. (43, 161)

S. N. Mergeljan [1953], *On the completeness of systems of analytic functions*, Uspeki Math. Nauk **8**, 3–63; English transl., Amer. Math. Soc. Transl. **19** (1962), 109–166. (72, 73)

____[1955], Общий метрический критерий полноты сичтемы полиномов, Dokl. Akad. Nauk CCCP **105**, 901–904. (375)

S. Merril [1968], *Maximality of certain algebras $H^\infty(dm)$*, Math. Z. **106**, 261–266. (296)

T. L. Miller, R. F. Olin, and J. E. Thomson [1986], *Subnormal operators and representations of bounded analytic functions and other uniform algebras*, Mem. Amer. Math. Soc. **63**, no. 354.

W. Mlak [1967], *Hyponormal contractions*, Colloq. Math. **18**, 137–142. (49)

____[1971], *Commutants of subnormal operators*, Bull. Acad. Polon. Sci. **19**, 837–842. (84)

____[1972a], *Intertwining operators*, Studia Math. **43**, 219–233. (84)

____[1972b], *Partitions of spectral sets*, Ann. Polon. Math. **25**, 273–280. (78, 161, 260)

B. B. Morrel [1973], *A decomposition for some operators*, Indiana Univ. Math. J. **23**, 497–511. (40, 161, 162)

G. J. Murphy [1982], *Self-dual subnormal operators*, Comment. Math. Univ. Carolin. **23**, 467–473. (40)

B. Sz-Nagy [1960], *Extensions of linear transformations in Hilbert space which extend beyond this space*, Appendix to Functional analysis by F. Riesz and B. Sz-Nagy, Fredrick Ungar Publ., New York. (78)

B. Sz-Nagy and C. Foiaş [1970], *Harmonic analysis of operators on Hilbert space*, North Holland, Amsterdam. (26, 78)

____[1975], *An application of dilation theory to hyponormal operators*, Acta Sci. Math. (Szeged) **37**, 155–159. (49)

N. K. Nikolskii [1969], *On perturbations of the spectrum of unitary operators*, Mat. Zametki **5**, 341–349; English transl., Math. Notes **5** (1969), pp. 207–211. (27)

E. Nordgren [1967], *Reducing subspaces of analytic Toeplitz operators*, Duke Math. J. **34**, 175–181. (84)

A. E. Nussbaum [1976], *Semigroups of subnormal operators*, J. London Math. Soc. (2) **14**, 340–344. (35)

M. Ohtsuka [1970], *Dirichlet problem, extremal length, and prime ends*, Van Nostrand Reinhold, New York. (202)

R. F. Olin [1976], *Functional relationships between a subnormal operator and its minimal normal extension*, Pacific J. Math. **63**, 221–229. (42, 324)

____[1977], *A class of pure subnormal operators*, Mich. Math. J. **24**, 115-118. (65, 314)

R. F. Olin and J. E. Thomson [1977a], *Limaçons, normal operators, and polar factorizations*, J. Reine Angew. Math. **291**, 133-144. (307)

____[1977b], *The spectrum of a normal operator and the problem of filling in holes*, Indiana Univ. Math. J. **26**, 541-544. (366, 368)

____[1979], *Lifting the commutant of a subnormal operator*, Canad. J. Math. **31**, 148-156. (84)

____[1980a], *Some index theorems for subnormal operators*, J. Operator Theory **3**, 115-142. (366)

____[1980b], *Algebras of subnormal operators*, J. Funct. Anal. **37**, 271-301. (341, 344, 346, 358, 360, 361, 363)

____[1980c], *Irreducible operators whose spectra are spectral sets*, Pacific J. Math. **91**, 431-434. (161)

____[1982], *Algebras generated by a subnormal operator*, Trans. Amer. Math. Soc. **271**, 299-311. (90, 91)

____[1984], *Cellular-indecomposable subnormal operators*, Integral Equations Operator Theory **7**, 392-430. (373)

____[1986], *Cellular-indecomposable subnormal operators*, II, Integral Equations Operator Theory **9**, 600-609. (373)

R. F. Olin, J. E. Thomson, and T. T. Trent [preprint], *Subnormal operators with finite rank self-commutator*, Trans. Amer. Math. Soc. (to appear). (161)

M. Parreau [1951], *Sur les moyennes des fonctions harmoniques et la classification des surfaces de Riemann*, Ann. Inst. Fourier **3**, 103-197. (210)

V. I. Paulsen [1988], *Toward a theory of K-spectral sets*, Surveys of some recent results in operator theory, vol. 1, J. B. Conway and B. B. Morrel, Editors, Research Notes in Mathematics, Longman, London, pp. 221-240. (78)

C. R. Putnam [1963], *On the structure of semi-normal operators*, Bull. Amer. Math. Soc. **69**, 818-819. (150)

____[1967], *Commutation properties of Hilbert space operators and related topics*, Springer-Verlag, Ergeb. Math., vol. 36, New York. (49, 152, 156)

____[1970], *An inequality for the area of hyponormal spectra*, Math. Z. **116**, 323-330. (156)

____[1971a], *The spectra of completely hyponormal operators*, Amer. J. Math. **93**, 699-708. (156)

____[1971b], *The spectra of subnormal operators*, Proc. Amer. Math. Soc. **28**, 473-477. (156)

____[1972], *Trace norm inequalities for the measure of hyponormal spectra*, Indiana Univ. Math. J. **21**, 775-779. (156)

____[1974a], *Spectra of polar factors of hyponormal operators*, Trans. Amer. Math. Soc. **188**, 419-428. (49, 161)

____[1974b], *Invariant subspaces of certain subnormal operators*, Indiana Univ. Math. J. **17**, 262–272. (161)

____[1974c], *The role of zero sets in the spectra of hyponormal operators*, Proc. Amer. Math. Soc. **43**, 137–140. (49)

____[1976a], *Generalized projections and reducible subnormal operators*, Duke Math. J. **43**, 101–108. (161)

____[1976b], *Almost isolated spectral parts and invariant subspaces*, Trans. Amer. Math. Soc. **216**, 267–277. (161)

____[1977a], *Rational approximation and Swiss cheeses*, Mich. Math. J. **24**, 193–196. (161)

____[1977b], *Peak sets and subnormal operators*, Illinois J. Math. **21**, 388–394. (314)

M. Putinar [1984], *Spectral inclusion for subnormal tuples*, Proc. Amer. Math. Soc. **90**, 405–406. (409)

M. Radjabalipour [1975], *On subnormal operators*, Trans. Amer. Math. Soc. **211**, 377–389. (65, 313)

____[1977], *Some decomposable subnormal operators*, Rev. Roumaine Math. Pures Appl. **22**, 341–345.

H. Radjavi and P. Rosenthal [1971], *On roots of normal operators*, J. Math. Anal. Appl. **34**, 653–664. (82, 84)

M. Raphael [1982], *Quasisimilarity and essential spectra for subnormal operators*, Indiana Univ. Math. J. **31**, 243–246. (98, 373, 407)

____[1985a], *Quasisimilar operators in the commutant of a subnormal operator*, Proc. Amer. Math. Soc. **94**, 265–268. (373)

____[1985b], *The uniform algebra associated with a cyclic subnormal operator*, Integral Equations Operator Theory **8**, 557–572. (98, 373)

____[1986], *Commutants of quasisimilar subnormal operators*, Pacific J. Math. **122**, 449–454. (98, 373, 407)

W. C. Ridge [1970], *Approximate point spectrum of a weighted shift*, Trans. Amer. Math. Soc. **147**, 349–356. (57)

J. R. Ringrose [1971], *Compact non-self-adjoint operators*, Van Nostrand Reinhold, New York. (2)

J. R. Robertson [1965], *On wandering subspaces for unitary operators*, Proc. Amer. Math. Soc. **16**, 233–236. (17)

P. Rosenthal [1968], *Completely reducible operators*, Proc. Amer. Math. Soc. **19**, 826–830. (315)

L. A. Rubel and A. L. Shields [1964], *Bounded approximation by polynomials*, Acta Math. **112**, 145–162. (72)

____[1966], *The space of bounded analytic functions*, Ann. Inst. Fourier (Grenoble) **16**, 235–277. (140)

W. Rudin [1955], *Analytic functions of class H_p*, Trans. Amer. Math. Soc. **78**, 46–66. (210)

K. Rudol [1982], *The functional model for a class of subnormal operators*, Bull. Acad. Polon. Sci. Math. **30**, 71–77. (210, 410)

___[1988], *The generalized Wold decomposition for subnormal operators*, Integral Equations Operator Theory **11**, 420–436. (210, 410)

___[1989], *On bundle shifts and cluster sets*, Integral Equations Operator Theory **12**, 444–448. (410)

___[preprint], "A model for analytic Toeplitz operators". (84)

T. Saito [1972], *Hyponormal operators and related topics*, Lectures on operator algebras, Lecture Notes in Math., vol. 247, Springer-Verlag, Berlin and New York, pp. 534–664. (49)

N. Salinas [1975], *Subnormal limits of nilpotent operators*, Acta Sci. Math. (Szeged) **37**, 117–124.

D. Sarason [1965a], *The H^p space of an annulus*, Mem. Amer. Math. Soc. No. 56, Providence. (210)

___[1965b], *On spectral sets having connected complements*, Acta Sci. Math. (Szeged) **26**, 289–299. (78, 260)

___[1965c], *A remark on the Volterra operator*, J. Math. Anal. Appl. **12**, 244–246. (139)

___[1966a], *Invariant subspaces and unstarred operator algebras*, Pacific J. Math. **17**, 511–517. (361)

___[1966b], *Weak-star generators of H^∞*, Pacific J. Math. **17**, 519–528. (307)

___[1968], *A remark on the weak-star topology of l^∞*, Studia Math. **30**, 355–359. (307)

___[1972], *Weak-star density of polynomials*, J. Reine Angew. Math. **252**, 1–15. (87, 297, 301, 306)

R. Schatten [1960], *Norm ideals of completely continuous operators*, Springer-Verlag, Berlin. (2)

J. F. Scroggs [1959], *Invariant subspaces of a normal operator*, Duke Math. J. **26**, 95–111. (314)

K. Seddighi [1983], *Essential spectra of operators in the class $B_n(\Omega)$*, Proc. Amer. Math. Soc. **87**, 453–458. (68)

___[1984], *Weak-star closed algebras and generalized Bergman kernels*, Proc. Amer. Math. Soc. **90**, 233–239.

G. L. Seever [preprint], *Operator representations of uniform algebras.* (260)

B. J. Shelburne [1982], *The operator M_z on Hilbert spaces of analytic functions*, Indiana Univ. Math. J. **31**, 191–207. (74)

A. L. Shields [1974], *Weighted shift operators and analytic function theory*, Math. Surveys, no. 13, Amer. Math. Soc., Providence, R.I., pp. 49–128. (54, 57)

A. L. Shields and L. J. Wallen [1971], *The commutants of certain Hilbert space operators*, Indiana Univ. Math. J. **20**, 777–788. (68, 73)

M. Slocinski [1975], *Normal extensions of commutative subnormal operators*, Studia Math. **54**, 259–266. (84, 410)

B. M. Solomyak [1986], *On the multiplicity of the spectrum of analytic Toeplitz operators*, Dokl. Akad. Nauk CCCP **286**; English transl., Soviet Math. Dokl. **33**, 286–290. (84)

B. M. Solomyak and A. L. Volberg [1989], *Multiplicity of analytic Toeplitz operators*, Oper. Theory: Adv. Appl. **42**, 87–192. (146)

J. Spraker [1990], *The minimal normal extension for M_z on the Hardy space of a planar region*, Trans. Amer. Math. Soc. **318**, 57–67. (209)

____ [preprint], *Multiplicity and uniformization maps*. (209)

J. G. Stampfli [1962], *Hyponormal operators*, Pacific J. Math. **12**, 1453–1458. (47, 48, 49)

____ [1965], *Hyponormal operators and spectral density*, Trans. Amer. Math. Soc. **117**, 469–476. (49)

____ [1966], *Which weighted shifts are subnormal?*, Pacific J. Math. **17**, 367–379. (40, 58, 60)

____ [1969], *On hyponormal and Toeplitz operators*, Math. Ann. **183**, 328–336. (49)

____ [1980], *An extension of Scott Brown's invariant subspace theorem: K-spectral sets*, J. Operator Theory **3**, 3–21.

J. G. Stampfli and B. L. Wadhwa [1976], *An asymmetric Putnam-Fuglede theorem for dominant operators*, Indiana Univ. Math. J. **25**, 359–365. (82)

____ [1977], *On dominant operators*, Monatsh. Math. **84**, 143–153. (50)

J. Stochel and F. H. Szafraniec [1984], *A characterization of subnormal operators*, Operator Theory: Adv. Appl. **14**, pp. 261–263. (35)

____ [1985], *On normal extensions of unbounded operators*, I, J. Operator Theory **14**, 31–55. (409)

____ [1989], *Unbounded weighted shifts and subnormality*, Integral Equations Operator Theory **12**, 146–153. (409) ____ [preprint a], *The normal part of an unbounded operator*, Indag. Math. (to appear). (409)

____ [preprint b], *A few assorted questions about unbounded subnormal operators*, Univ. Iagel. Acta Math. (to appear). (409)

____ [preprint c], *On normal extensions of unbounded operators*, II. (409)

____ [preprint d], *On normal extensions of unbounded operators*, III: *spectral properties*. (409)

E. L. Stout [1971], *The theory of uniform algebras*, Bogden and Quigley, Tarrytown-on-Hudson. (163, 296)

M. A. Subin [1967], *Factorization of parameter-dependent matrix functions in normal rings and certain related questions in the theory of Noetherian operators*, Mat. Sb. **73** (**113**), 610–629; English transl., Math. USSR-Sb. **2**, 543-560. (64)

N. Suzuki [1966], *The algebraic structure of non selfadjoint operators*, Acta Sci. Math. (Szeged) **27**, 173–184. (159)

F. H. Szafraniec [1982], *Subnormals in C^*-algebras*, Proc. Amer. Math. Soc. **84**, 533–534. (35)

W. Szymanski [1987], *Dilations and subnormality*, Proc. Amer. Math. Soc. **101**, 251–259. (35)

____ [preprint], *The boundedness condition of dilation characterizes subnormals and contractions*, Rocky Mountain J. Math. (to appear). (30)

J. E. Thomson [1975], *Intersections of commutants of analytic Toeplitz operators*, Proc. Amer. Math. Soc. **52**, 305–310. (84)

____ [1976a], *The commutants of certain analytic Toeplitz operators*, Proc. Amer. Math. Soc. **54**, 165–169. (84)

____ [1976b], *The commutant of a class of analytic Toeplitz operators*, II, Indiana Univ. Math. J. **25**, 793–800. (84)

____ [1977], *The commutant of a class of analytic Toeplitz operators*, Amer. J. Math. **99**, 522–529. (84)

____ [1986], *Invariant subspaces for algebras of subnormal operators*, Proc. Amer. Math. Soc. **96**, 462–464. (183, 184)

____ [1988a], "Invariant subspace for subnormal operators," Surveys of some recent results in operator theory, vol. 1, J. B. Conway and B. B. Morrel, Editors, Research Notes in Mathematics, Longman, London, pp. 241–259. (341)

____ [1988b], *Factorization over algebras of subnormal operators*, Indiana Univ. Math. J. **37**, 191–199. (341, 344)

____ [preprint], *Approximation in the mean by polynomials*, Ann. Math. (to appear). (53, 161, 375, 384)

T. T. Trent [1979a], $H^2(\mu)$ *spaces and bounded point evaluations*, Pacific J. Math. **80**, 279–292. (64, 65, 66)

____ [1979b], *Extension of a theorem of Szegö*, Mich. Math. J. **26**, 373–377. (66)

____ [1981], *New conditions for subnormality*, Pacific J. Math. **93**, 459–464. (35)

____ [1984], *Carleson measure inequalities and kernel functions in* $H^2(\mu)$, J. Operator Theory **11**, 157–169. (66)

____ [1985], *A characterization of* $P^2(\mu) \neq L^2(\mu)$, J. Funct. Anal. **64**, 163–177. (66)

____ [1987], *Invariant subspaces for operators in subalgebras of* $L^\infty(\mu)$, Proc. Amer. Math. Soc. **99**, 268–272. (184)

T. T. Trent and J. L. Wang [1984], $P^2(\mu)$ *and bounded point evaluations*, Proc. Amer. Math. Soc. **91**, 421–425. (66)

M. Tsuji [1975], *Potential theory in modern function theory*, Chelsea, New York. (198, 202, 205, 215)

D. Voiculescu [1980], *A note on quasitriangularity and trace-class self-commutators*, Acta Math. Sci. (Szeged) **42**, 195–199. (153, 155)

J. von Neumann [1951], *Eine Spektraltheorie für allgemeine Operatoren eines unitären Raumes*, Math. Nachr. **4**, 258–281. (75, 77)

B. L. Wadhwa [1973], *A hyponormal operator whose spectrum is not a spectral set*, Proc. Amer. Math. Soc. **38**, 83–85. (75)

J. Wermer [1952], *On invariant subspaces of normal operators*, Proc. Amer. Math. Soc. **3**, 276–277. (314, 315)

———[1955], "Report on subnormal operators," Report on an international conference on operator theory and group representations, National Acad. Sci.-National Research Council, Harriman, New York, pp. 1–3. (51, 180)

———[1960], *Dirichlet algebras*, Duke Math. J. **27**, 373–382. (240)

———[1964], *Seminar über funktionen algebren*, Lecture Notes in Math., vol. 1, Springer-Verlag, Berlin and New York. (296)

R. Whitley [1978], *Normal and quasinormal composition operators*, Proc. Amer. Math. Soc. **70**, 114–118. (45)

J. P. Williams [1967], *Minimal spectral sets of compact operators*, Acta Sci. Math. (Szeged) **28**, 93–106. (78)

L. R. Williams [1980a], *Equality of essential spectra of certain quasisimilar seminormal operators*, Proc. Amer. Math. Soc. **78**, 203–209. (98)

———[1980b], *Equality of spectra of quasisimilar quasinormal operators*, J. Operator Theory **3**, 57–69. (98)

W. R. Wogen [1971], *On special generators for properly infinite von Neumann algebras*, Proc. Amer. Math. Soc. **28**, 107–113. (92, 161)

———[1978], *On some operators with cyclic vectors*, Indiana Univ. Math. J. **27**, 163–171. (234)

———[1979], *Quasinormal operators are reflexive*, Bull. London. Math. Soc. **11**, 19–22. (45)

———[1985], *Subnormal roots of subnormal operators*, Integral Equations Operator Theory **8**, 432–436. (85)

———[1986], *Counterexamples in the theory of nonselfadjoint operator algebras*, Bull. Amer. Math. Soc. **15**, 225–227. (364)

D. Xia [1987a], *The analytic model of a subnormal operator*, Integral Equations Operator Theory **10**, 258–289. (161, 409) ———[1987b], *Analytic theory of subnormal operators*, Integral Equations Operator Theory **10**, 880–903. (161, 409)

S. Yamashita [1968], *On some families of analytic functions on Riemann surfaces*, Nagoya Math. J. **31**, 57–68. (210)

K. Yan [1985], *U-selfadjoint operators and self-dual subnormal operators*, J. Fudan Univ. Natur. Sci. **24**, 459–463 (Chinese). (40)

———[1988], *On the quasisimilarity of subnormal operators*, Acta Math. Sinica **4**, 76–82. (373)

———[preprint], *Invariant subspaces for joint subnormal systems*. (184)

L. Yang [1990], *Equality of essential spectra of quasisimilar subnormal operators*, Integral Equatins Operator Theory **13**, 433–441. (98, 373, 407)

T. Yoshino [1969], *Subnormal operators with a cyclic vector*, Tôhoku Math. J. **21**, 47–55. (52)

———[1973], *On commuting extensions of nearly normal operators*, Tôhoku Math. J. **25**, 263–272. (45, 84)

———[1976], *A note on a result of J. Bram*, Duke Math. J. **43**, 875. (81)

Index

$A(K)$, 163, 253, 254
abpe, 63
absolutely continuous, 150
Abstract F. and M. Riesz Theorem, 250
adjacent squares, 375
Ahlfors function, 218
almost open, 326
ample function, 351
analytic bounded point evaluation, 63, 397, 398, 399
analytic capacity, 217
analytic disk, 239
analytic Toeplitz operator, 22, 145
antisymmetric, 277
approximate point spectrum, 17
Aren's Theorem, 190

band generated by $\mathscr{S}$, 245, 251
band of measures, 244
Berger-Shaw Theorem, 66, 73, 152
Bergman operator, 28, 156
Bergman space, 28, 66
Beurling's Theorem, 23, 135
bilateral shift, 15, 19
bilateral weighted shift, 53
Bishop's Localization Theorem, 178
Blaschke product, 121
Blaschke sequence, 121
boundary, 191
bounded characteristic, 120
bounded point evaluation, 61, 181, 182, 397, 398, 399

boundedly mutually absolutely continuous, 200
bpe, 61
bundle shifts, 210, 410

canonical factorization (of an inner function), 136
capacity, 217
Carathéodory region, 69, 71
Carleson measure, 371
carrier of a Gleason part, 239
Cauchy transform, 174
cellular indecomposable, 373
Cesàro Means, 21, 99
Cesàro operator, 35
Chaumat map, 289
Chaumat's Theorem, 288
Chaumat transform, 289
cluster set, 334
cohyponormal, 47
commutant, 25
commutator, 152
complementary band, 245
completely subnormal, 29
complex representing measure, 194
contains no L^∞ summand, 281
continuous analytic capacity, 217
convolution, 103, 104
cornucopia, 70
crescent, 297
Curtis's Criterion, 222
curve generating system, 230
cyclic operator, 51
cyclic vector, 51

Davie's Theorem, 267
Dirichlet algebra, 226, 297, 299
Dirichlet region, 196
Dirichlet set, 196
disintegration, 318
disk algebra, 132
distribution, 170
dominant operator, 50
dominating sequences, 306
double commutant of S, 25
dual, 40, 407
dual algebra, 12
dual algebra homomorphism, 12
dual algebra isomorphism, 12

envelope, 282
essential boundary point, 68
essential spectrum, 17
essentially Dirichlet, 326
essentially normal, 158
expectation operator, 318

F. and M. Riesz Theorem, 130, 250
Fatou's Theorem, 111
finitely connected, 224, 230, 254
finitely multicyclic, 152
Fourier transform, 18
Fuglede-Putnam Theorem, 50
full analytic subspace, 358
function algebra, 163

generating vectors, 152
Gleason part, 235
Gleason-Whitney Theorem, 145
Gonchar's Criterion, 223
greatest common divisor, 136
green part, 378
green segment, 378
Green's Theorem, 169

Hardy operator, 208
Hardy spaces, 19, 117, 204
harmonic majorant, 204
harmonic measure, 199, 201
Hartogs-Rosenthal Theorem, 175
heavy barrier, 376
heavy square, 376
Herglotz's Theorem, 108
Hilbert-Schmidt operator, 8

hole, 365
hyponormal, 46, 149

inner function, 125
interpolating sequence, 335
irreducible, 91, 159

Jensen's Inequality, 115
joined by a path of squares, 375

Lautzenheiser-Mlak Decomposition Theorem, 260
Lavrentiev's Theorem, 232
least common multiple, 137
least harmonic majorant, 205
Lebesgue decomposition, 245
left essential spectrum, 17
left spectrum, 17
lift, 79
light route to ∞, 376
light square, 376
log integrable, 125

m-multicyclic, 152
Mergelyan's Theorem, 253
minimal normal extension, 38, 315
minimal unitary extension, 16
Minimax Theorem, 248
modulus of triangularity, 155
mollification, 171
mollifier, 171
multiplicity, 13, 15
mutually absolutely continuous, 200

n-connected, 230
nth generation, 375
natural projection, 266
Nevanlinna class, 120, 123, 127
nontangentially, 110
nontrivial, 235
nontrivial part, 235
normal spectrum, 40
norming point, 207

orange part, 378
oscillation, 186
outer function, 126

$P(K)$, 163, 232, 253
Painlevé null set, 220
partition of unity, 172
path of squares, 375
peak interpolating set, 212
peak point, 211
peak set, 216
Poisson kernel, 105
polar decomposition, 3
polynomially convex hull, 41
POM, 30
positive operator valued measure, 30
positive upper area density, 224
pseudosymmetric, 277
pure, 15, 38
Putnam's Inequality, 156, 177

quasi-invertible, 94
quasiaffinity, 94
quasinormal, 29, 44, 45
quasisimilar, 94, 407
quasisimilarity, 368

$R(K)$, 76, 163, 253, 254
rationally cyclic operator, 51
red barrier, 376
reducing band, 250
reductive, 310
reflexive operator, 361
regularization, 171
regularizer, 171
remainder operator, 269
removable boundary point, 68
representing measure, 193, 194
reproducing kernel, 62
restriction algebra, 85
Riemann map, 240
right essential spectrum, 17

Sarason hull, 306
Sarason pair, 306
Sarason process, 302
Sarason's Theorem, 301, 303
satisfies the Maximum Principle, 196
scalar-valued spectral measure, 85
Schatten p-class, 2
self commutator, 152

self-dual, 408
seminormal, 47
separating vector, 249
Shilov boundary, 192
similar, 94
simply connected, 296
singular band, 251
singular inner function, 125
smooth exhaustion, 204
solvable function, 195
SOT, 10
spectral mapping theorem, 326, 333
spectral set, 75
Stoltz angle, 110
string of beads, 239
strong operator topology, 10
subharmonic, 115
subnormal, 27, 34, 409
subordinate to a cover, 173
superharmonic, 115
support, 168
sweep of μ, 201
Swiss Cheese, 165
symbol, 89
Szegö's Theorem, 143

T-invariant algebra, 187, 281
test function, 170
the bilateral shift, 15
Toeplitz operator, 84, 89
trace, 4
trace class, 1, 7, 8
trace norm, 1
trigonometric polynomial, 19
trivial, 235
trivial band, 244
trivial part, 235

unilateral shift, 13, 17, 22, 23, 26, 136, 370, 371
unilateral weighted shift, 53
upper area density, 224

Vitushkin Cover, 270
Vitushkin localization operator, 184
Vitushkin partition of unity, 270
von Neumann operator, 75

von Neumann-Wold decomposition, 14

weak operator topology, 10
weak-star generator, 306
weak-star topology, 11
weighted shift, 53
Wermer Embedding Theorem, 240
Weyl spectrum, 49
Weyl's lemma, 172
Weyl's Theorem, 49
Wilkin's Theorem, 251
WOT, 10

List of Symbols

$\mathscr{B}(\mathscr{H})$, 1
$\mathscr{B}_1(\mathscr{H}), \mathscr{B}_1$, 1
$\|A\|_1$, 2
$\mathscr{B}_2(\mathscr{H}), \mathscr{B}_2$, 2
$\|A\|_2$, 2
$\mathscr{B}_p, \mathscr{B}_p(\mathscr{H})$, 2
$\mathscr{B}_o, \mathscr{B}_o(\mathscr{H})$, 3
tr A, 4
$\mathscr{B}_{oo}$, 4
$e \otimes f$, 4
ball $\mathscr{B}(\mathscr{H})$, 10
$\sigma_{ap}(T)$, 17
$\sigma_l(T), \sigma_{le}(T)$, 17
$\sigma_{re}(T)$, 17
$\sigma_e(T)$, 17
$\sigma(T)$, 17
$\hat{f}$, 18
$\mathscr{SF}$, 19
H^p, 19
k_λ, 20
T_ϕ, 22
$P^\infty(S)$, 22
Lat S, 23
$T^{(n)}, \mathscr{H}^{(n)}$, 24
$\mathscr{S}', \mathscr{S}''$, 25
$L_a^2(G)$, 28
$P^2(\mu)$, 28
S_μ, 28
Rat(K), 28
$R^2(K,\mu)$, 28
$E^\perp$, 29
$C^*(S)$, 30
mne S, 39
$\sigma_n(S)$, 40

$\widehat{K}$, 41
$U \otimes A$, 45
$\sigma_w(A)$, 49
bpe(K,μ), 61
bpe(μ), 61
k_λ, 61
abpe(K,μ), 63
abpe(μ), 63
$L^p(G)$, 66
$L_a^p(G)$, 66
$\mathscr{A}$, 66
$P^2(G)$, 69
$R^2(G)$, 69
$\|\cdot\|_E$, 75
$R(K)$, 76
$\mathscr{R}(N,\mathscr{H})$, 85
$\phi(S)$, 85
$\mathscr{M}^\infty(\mu)$, 87
$A_1 \approx A_2$, 94
$A_1 \sim A_2$, 94
$\mu * \nu$, 103
usc, 115
$M_p(r,f)$, 117
H^p, 117
N, 120
$\mathbf{I}$, 136
$\mathscr{M}_1 \vee \mathscr{M}_2$, 136
$\mathscr{M}_1 \wedge \mathscr{M}_2$, 136
gcd(ϕ_1,ϕ_2), 136
lcm(ϕ_1,ϕ_2), 137
A_0, 141
$P_0^2(\mu)$, 141
$P(K), A(K)$, 163
$\mathscr{M}_A$, 164

LIST OF SYMBOLS

$C^n(G), C^n$, 168
$C_c^n(G), C_c^n$, 168
$\partial u, \overline{\partial} u$, 169
$L_{loc}^1(G)$, 170
$\tilde{\mu}$, 173
$\hat{\mu}$, 174
L_c^∞, 178
$R(K, E)$, 179
$R(E)$, 179
$T_g f$, 184
osc$(f; X)$, 186
$A(K, U)$, 189
∂_A, 191
M_ρ, 194
$\widehat{\mathscr{P}}(u, G), \widecheck{\mathscr{P}}(u, G)$, 197
$\hat{u}, \check{u}$, 197
$\hat{\mu}$, 201
$L^\infty(\partial K), L^1(\partial K)$, 202
$H^p(G)$, 204
$\gamma(F), \alpha(F)$, 217
$\rho(\alpha, \beta)$, 234
$\mathscr{B}_Q$, 251
$L^\infty(\mathscr{B})$, 254
μF, 256
$A^\infty(\mu), A^\infty(\mathscr{B})$, 257

$F(T)$, 262
R_g, 269
$E(K, \mu)$, 282
E_μ, 282
$\Sigma(K, \mu)$, 283
$R^\infty(E, \mu)$, 287
$\check{\phi}$, 289
$H^\infty(\partial K)$, 292
$\|f\|_\mu$, 292
$\hat{f}$, 292
$\Sigma(\mu)$, 306
$\mathscr{S}(K, \mu)$, 315
$\mathscr{A}^\infty(\phi, \mu)$, 316
λ_ζ, 318
clu$(\phi; \sigma_e(S))$, 334
$x \otimes y$, 342
e_λ, 342
$\mathscr{H}_x$, 357
Alg $\mathscr{L}$, 361
$\mathscr{S}(N)$, 365
$\mathscr{S}(n)$, 375
$\mathscr{G}(n)$, 376
$\mathscr{R}(n)$, 376
$\mathscr{Y}(n)$, 377